The Biochemistry of Fruits and their Products

Volume 2

FOOD SCIENCE AND TECHNOLOGY

A SERIES OF MONOGRAPHS

Maynard A. Amerine, Rose Marie Pangborn, and Edward B. Roessler, PRINCIPLES OF SENSORY EVALUATION OF FOOD. 1965.

C. R. Stumbo, THERMOBACTERIOLOGY IN FOOD PROCESSING. 1966.

Gerald Reed, ENZYMES IN FOOD PROCESSING. 1966.

S. M. Herschdoerfer, QUALITY CONTROL IN THE FOOD INDUSTRY. Volume I—1967. Volume II—1968. Volume III—in preparation.

Hans Riemann, FOOD-BORNE INFECTIONS AND INTOXICATIONS. 1969.

Irvin E. Liener, TOXIC CONSTITUENTS OF PLANT FOODSTUFFS. 1969.

Martin Glicksman, GUM TECHNOLOGY IN THE FOOD INDUSTRY. 1970.

L. A. Goldblatt, AFLATOXIN. 1970.

Maynard A. Joslyn, METHODS IN FOOD ANALYSIS, second edition. 1970.

A. C. Hulme, THE BIOCHEMISTRY OF FRUITS AND THEIR PRODUCTS. Volume 1—1970. Volume 2—1971.

G. Ohloff and A. F. Thomas, GUSTATION AND OLFACTION. 1971.

The Biochemistry of Fruits and their Products

Edited by
A. C. HULME
A.R.C. Food Research Institute, Norwich, England

VOLUME 2
1971

ACADEMIC PRESS London and New York

ACADEMIC PRESS INC. (LONDON) LTD
24–28 Oval Road,
London NW1

U.S. Edition published by

ACADEMIC PRESS INC.
111 Fifth Avenue,
New York, New York 10003

Library of Congress Catalog Card Number: 77–117145
SBN: 12–361202–0

Printed in Great Britain by
John Wright & Sons Ltd., at the Stonebridge Press, Bristol

Contributors to Volume 2

J. B. ADAMS, Fruit and Vegetable Preservation Research Association, Chipping Campden, England.

J. A. ATTAWAY, State of Florida, Department of Citrus, Lake Alfred, Florida, U.S.A.

J. B. BIALE, University of California, Los Angeles and Riverside, California, U.S.A.

H. A. W. BLUNDSTONE, Fruit and Vegetable Preservation Research Association, Chipping Campden, England.

C. O. CHICHESTER,* Department of Food Science and Technology, University of California, Davis, California, U.S.A.

I. D. CLARKE, Joint FAO/IAEA Division of Atomic Energy in Food and Agriculture, Vienna, Austria.

J. N. DAVIES, A.R.C. Glasshouse Crops Research Institute, Littlehampton, Sussex, England.

G. G. DULL, U.S.D.A., Southeastern Marketing and Nutrition Research Division, Athens, Georgia, U.S.A.

M. J. F. FERNANDEZ DIEZ, Instituto de la Grasa y sus Derividos, Sevilla, Spain.

B. J. FINKLE, Western Regional Research Laboratory, U.S.D.A., Albany, California, U.S.A.

A. GREEN, Beecham Products (U.K.) Ltd., Coleford, England.

G. E. HOBSON, A.R.C. Glasshouse Crops Research Institute, Littlehampton, Sussex, England.

A. C. HULME, A.R.C. Food Research Institute, Norwich, England.

S. ITO, Horticultural Research Station, Ministry of Agriculture and Forestry, Okitsu, Shimizu, Shikuoka-Ken, Japan.

W. G. JENNINGS, University of California, Davis, California, U.S.A.

M. A. JOSLYN, Department of Nutritional Sciences, University of California, Berkeley, California, U.S.A.

D. MCG. MCBEAN, Division of Food Research, C.S.I.R.O., New South Wales, Australia.

* Present address: Department of Food and Resource Chemistry, University of Rhode Island, Kingston, Rhode Island 02881, U.S.A.

R. McFeeters,* Department of Food Science and Technology, University of California, Davis, California, U.S.A.

F. S. Nury, Department of Food Science, Fresno State College, Fresno, California, U.S.A.

M. Olliver, Consultant, Histon, Cambridge, England.

J. K. Palmer, Department of Nutrition and Food Science, Massachusetts Institute of Technology, Cambridge, Massachusetts, U.S.A.

E. Peynaud, Station Agronomique et Oenologique de Bordeaux, France.

A. Pollard, Department of Agriculture and Horticulture, University of Bristol, England.

H. K. Pratt, Department of Vegetable Crops, University of California, Davis, California, U.S.A.

M. J. C. Rhodes, A.R.C. Food Research Institute, Norwich, England.

P. Ribérau-Gayon, Faculté des Sciences de Bordeaux, France.

R. J. Romani, University of California, Davis, California, U.S.A.

R. M. Silverstein, State University College of Forestry, Syracuse, New York, U.S.A.

C. F. Timberlake, Department of Agriculture and Horticulture, University of Bristol, England.

S. V. Ting, State of Florida, Department of Citrus, Lake Alfred, Florida, U.S.A.

J. S. Woodman, Fruit and Vegetable Preservation Research Association, Chipping Campden, England.

R. E. Young, University of California, Los Angeles and Riverside, California, U.S.A.

* Present address: Department of Food Science, Michigan State University, East Lansing, Michigan 48823, U.S.A.

Preface

Volume 1 has dealt with the chemistry of the main constituents of fruits, with growth and pre-harvest factors and with physiological disorders. In the present volume, Part I concerns the biochemistry and physiology of some commercially important fruits, written by research workers who have themselves contributed substantially to the knowledge of the fruits with which they deal. In Part II, an attempt is made to explore some of the biochemical changes associated with processing procedures, although here there are wide gaps in our knowledge. Indeed, a considerable proportion of this part of the book is concerned with observed effects rather than with causal mechanisms. Undoubtedly, processed fruit products could be greatly improved if more was known of the sequence of chemical changes taking place during heating and freezing, particularly freezing of fruit tissues under a variety of environmental conditions.

Some fruits, for example the avocado pear and the apple, have been the subject of research for many years and many of the biochemical changes which they undergo during development and ripening are well documented. Others, for example the mango, are only just beginning to receive serious attention using modern techniques. There is little detailed biochemical knowledge concerning the olive, a fruit much too astringent for consumption as harvested. With this fruit, attention has tended to be focused on gross changes occurring during mild processing to remove this astringency, and on oil production, to the exclusion of systematic studies of all aspects of the development of the fruit.

It is inevitable, therefore, that the chapters will be uneven in the depth of the biochemical coverage from fruit to fruit. This is not the fault of the authors; it is a reflection of the paucity of experimentation and of data on many fruits. One of the objects of including fruits about which little strictly biochemical data are available has been to draw attention to the wide gaps in our knowledge of such fruits, in the hope that they will receive the attention lavished on such fruits as the apple and the avocado. This will undoubtedly be to the advantage of both the biochemistry of plant growth and senescence generally and the producer and consumer of edible fruits.

A. C. HULME

NORWICH
JULY, 1971

Acknowledgements

The editor is greatly indebted to Miss Mamie Olliver for assistance in the preparation of this volume, especially Part II where Miss Olliver's specialized knowledge has proved invaluable.

The editor is also grateful to Professor Claude Lance for the translation from the French of Chapter 4.

Dr. Wenzel, owing to other commitments, was unable to contribute the chapter on Canned Citrus Fruits and the editor is greatly indebted to Messrs. Adams, Blundstone and Woodman for stepping into the breach at the eleventh hour.

Once again, the editor wishes to thank Academic Press for their excellent presentation of this second volume.

Contents

Contributors to Volume 2 v
Preface vii
Acknowledgements viii
Abbreviations xv
Contents of Volume 1 xvii

Part I Biochemistry and Physiology of Commercially Important Fruits

1. The Avocado Pear

JACOB B. BIALE AND ROY E. YOUNG

I Botany and Horticulture 2
II Composition 6
III Compositional Changes Associated with Development and Storage 16
IV Fat Synthesis 24
V Enzymes 27
VI The Gas Exchange during Respiration 28
VII Regulation of Respiratory Rates 32
VIII Production and Role of Ethylene 36
IX Metabolic Pathways 40
X Metabolic Events in Fruit Ripening 50
XI Concluding Remarks 58
References 60

2. The Banana

JAMES K. PALMER

I Brief History and World Production 65
II The Fruit 66
III Post-harvest Physiology and Biochemistry 73
IV Biochemical Mechanisms of Ripening.. 94
References 101

3. Citrus Fruits

S. V. TING AND J. A. ATTAWAY

I Introduction 107
II The Fruit 109
III Changes during Fruit Development and Maturation 113
IV Post-harvest Physiology 149
V Metabolism of Citrus Fruits 156
References 161

4. The Grape

E. PEYNAUD AND P. RIBÉREAU-GAYON

I Introduction .. 172
II Constitution and Development of Grapes .. 174
III Carbohydrates .. 178
IV Organic Acids .. 181
V Nitrogen Compounds .. 186
VI Phenolic Compounds .. 188
VII Other Compounds .. 192
VIII Respiration during the Development of the Grape .. 195
IX Factors Affecting the Maturation and Quality of Grapes .. 199
X Grape Technology .. 201
References .. 204

5. Melons

HARLAN K. PRATT

I Introduction .. 207
II Origins .. 209
III Botanical Aspects .. 209
IV Growth and Development .. 210
V Growth Regulation .. 212
VI Maturity and Quality .. 213
VII Composition .. 213
VIII Enzymes .. 221
IX Respiration, Ethylene Production and the Initiation of Ripening .. 222
X Tissue Investigations .. 228
XI Summary .. 228
References .. 229

6. The Mango

A. C. HULME

I Introduction .. 233
II Botanical Aspects .. 234
III Growth and Development .. 235
IV Biochemical Indices of Maturity .. 239
V Chemical Composition of Developing and Maturing Fruits .. 244
VI Biochemical Changes in Storage .. 250
References .. 253

7. The Olive

M. J. FERNANDEZ DIEZ

I Introduction .. 255
II Fresh Olives .. 260
III Processed Olives .. 267
IV Olives for Oil .. 274
V Conclusion .. 276
References .. 277

8. The Persimmon

SABURO ITO

I Introduction 281
II Principal Varieties of Commercial Significance 283
III Constituents of the Fruit 288
IV Post-harvest Treatments 298
V Conclusion 300
References 301

9A. The Pineapple: General

GERALD G. DULL

I Introduction 303
II Composition 305
III Developmental Biochemistry 309
IV Post-harvest Physiology 316
References 323

9B. The Pineapple: Flavour

R. M. SILVERSTEIN 325

10. Pome Fruits

A. C. HULME AND M. J. C. RHODES

I Introduction 333
II History and Distribution 334
III Morphology of the Fruit 335
IV The Biochemistry of Pome Fruits 336
V Conclusion 368
References 369

11. Soft Fruits

AUDREY GREEN

I Introduction 375
II Characteristics of Soft Fruits and their Changes with Maturation .. 375
III Post-harvest Changes 405
References 407

12. Stone Fruits

R. J. ROMANI AND W. G. JENNINGS

I Introduction 411
II Botanical Aspects 412
III Origins 412
IV Growth and Development 413
V Maturation, Ripening and Senescence 416
VI Maturity Indices 421

VII The Respiratory Climacteric 424
VIII Flavour and Quality 425
IX Enzymes and Organelles 429
X Summary 430
References 431

13. The Tomato

G. E. Hobson and J. N. Davies

I Introduction and General Survey 437
II Fruit Composition and Chemical Changes during Growth and Maturation 440
III Aspects of Metabolism during Maturation and Senescence .. 462
IV Factors Affecting Tomato Fruit Ripening 468
V The Control of Tomato Fruit Ripening—A Summary 473
References 475

Part II The Biochemistry of Fruit Processing

14. General Introduction

Mamie Olliver

I Introduction 485
II The Preservation of Foods 486
III Fruit for Processing 488
IV Processes for Fruit Utilization and Preservation 496
V Developments in Biochemical Studies of Fruits 500
References 505

15. Canned Fruits other than Citrus

J. B. Adams and H. A. W. Blundstone

I Introduction 507
II Organoleptic Changes during the Canning Processes 508
III Vitamin Changes during the Canning Processes 526
IV Interaction between the Fruits and their Containers 530
V Selection of Varieties of Fruit for Canning 535
References 537

16. Canned Citrus Products

H. A. W. Blundstone, J. S. Woodman and J. B. Adams

I Introduction 543
II Colour 544
III Flavour 548
IV Texture 555
V Vitamin Changes during Processing and Storage 560
VI Fruit Can Interactions 567
References 569

17. Fruit Juices

A. POLLARD AND C. F. TIMBERLAKE

I Introduction 573
II Fruit and Juice Composition 575
III Processing Operations 579
IV Chemical Changes in Stored Juices 595
V Conclusions 611
References 612

18. Dehydrated Fruit

D. MCG. MCBEAN, M. A. JOSLYN AND F. S. NURY

I Introduction 623
II Maturity and Variety 625
III Colour Changes 626
IV Taste and Flavour 629
V Preparation for Drying 630
VI Drying 637
VII Storage 642
VIII Rehydration 647
IX Conclusion 648
References 648

19. Freezing Preservation

BERNARD J. FINKLE

I Introduction 653
II Physical and Chemical Aspects of Freezing Preservation 655
III Enzymological Considerations in Freezing 667
IV Factors in Freezing Preservation of Fruit 671
V Fruit Stability during Frozen Storage 677
VI Conclusions 680
References 682

20. Effects of Radiation Treatments

I. D. CLARKE

I Introduction 687
II Extension of Storage Life of Fresh Fruits 690
III Microbiological Aspects 698
IV The Radurization of Fruit Juices 699
V Public Health Clearance 701
VI Prospect for Commercial Application 701
References 702

21. Pigment Degeneration During Processing and Storage

C. O. CHICHESTER AND ROGER MCFEETERS

I Introduction 707
II Carotenoids 708
III Flavonoids 711
IV Chlorophylls 714
References 717

22. Quality

GERALD G. DULL AND A. C. HULME 721

Author Index 727
Subject Index 763

Abbreviations

Å	Angstrom unit
ACP	acyl carrier protein
ADH	alcohol dehydrogenase
AIS	alcohol insoluble solids
AMP, ADP, ATP	adenosine-5′-mono, di-, and tri-phosphate
B9 (Alar)	N-dimethylaminosuccinamic acid
CA storage	controlled atmosphere storage ("gas" storage)
CCC	(2-chloroethyl)-trimethylammonium chloride
CH	cycloheximide
CMP, CDP, CTP	cytidine-5′-mono, di-, and tri-phosphate
CoA	co-enzyme A
2:4-D	2:4-dichlorophenoxyacetic acid
DEAE cellulose	diethylaminoethyl cellulose
DHA	dehydro-L-ascorbic acid
DIECA	diethyldithiocarbamate
DNA	deoxyribonucleic acid
DNP	2:4-dinitrophenol
Dr. wt.	dry weight
EDTA	ethylenediaminetetraacetic acid
EMP	Embden–Meyerhof–Parnas pathway
Ethrel	2 chloroethylphosphonic acid
FAD	flavin-adenine dinucleotide
FMN	flavin mononucleotide
Fr. wt.	fresh weight
GA_1–GA_9	gibberellic acids
GDP	guanosine-5′-diphosphate
GLC	gas-liquid chromatography
G-6-P	glucose-6-phosphate
HMP	hexose monophosphate pathway
IAA	indolyl-3-acetic acid
krad	kilorad
MAK	methylated albumen on Kieselguhr
MD	malate dehydrogenase
MDHA	mono-dehydroascorbic acid
ME	malic enzyme (NADP-malate dehydrogenase)
Mrad	megarad
NAD, $NADH_2$	oxidized and reduced nicotinamide adenine dinucleotide
NADP, $NADPH_2$	oxidized and reduced nicotinamide adenine dinucleotide phosphate

OAA	oxalacetic acid
PD	pyruvate decarboxylase
PE, PME	pectin esterase
PEP	phosphoenolpyruvate
PG	polygalacturonase
PGA	phosphoglyceric acid
PMG	polymethylgalacturonase
P/O ratio	ratio of micromoles inorganic phosphate esterified/microatoms oxygen absorbed
p.p.m.	parts per million
PPP	pentose phosphate pathway
PVP	polyvinylpyrrolidone
RNA	ribonucleic acid
m-RNA	messenger-RNA
t-RNA	transfer-RNA
RQ	respiratory quotient (CO_2 production/O_2 absorption)
2,4,5-T	2,4,5-trichlorophenoxyacetic acid
TCA cycle	Krebs tricarboxylic acid cycle
TLC	thin-layer chromatography
TMS	trimethylsilyl
TPP	thiamine pyrophosphate
UMP, UDP, UTP	urindine-5′-mono-, di- and tri-phosphate
UDPG	uridine diphosphoglucose

Contents of Volume I

Part I CONSTITUENTS OF FRUITS

1. Sugars, by G. C. WHITING
2. Hexosans, Pentosans and Gums, by F. A. ISHERWOOD
3. Pectic Substances and other Uronides, by W. PILNIK AND A. G. J. VORAGEN
4. Organic Acids, by R. ULRICH
5. Amino Acids, by L. F. BURROUGHS
6. Proteins, by ELMER HANSEN
7. Protein Patterns in Fruits, by R. L. CLEMENTS
8. Enzymes, by DAVID R. DILLEY
9. Lipids, by P. MAZLIAK
10. Volatile Compounds: The Aroma of Fruits, by H. E. NURSTEN
11. Fruit Phenolics, by J. VAN BUREN
12. Carotenoids and Triterpenoids, by T. W. GOODWIN AND L. J. GOAD
13. Vitamins in Fruits, by L. W. MAPSON

Part II GROWTH AND PRE-HARVEST FACTORS

14. The Physiology and Nutrition of Developing Fruits, by E. G. BOLLARD
15. Hormonal Factors in Growth and Development, by J. P. NITSCH

Part IIIa BIOCHEMISTRY OF MATURATION AND RIPENING

16. The Ethylene Factor, by W. B. McGLASSON
17. The Climacteric and Ripening of Fruits, by M. J. C. RHODES

Part IIIb PHYSIOLOGICAL DISORDERS OF FRUITS

18. Physiological Disorders of Fruit After Harvesting, by B. G. WILKINSON
19. Apple Scald, by D. F. MEIGH

To R. G. Tomkins in gratitude for many years
of encouragement and support

Part I

Biochemistry and Physiology of Commercially Important Fruits

Chapter 1

The Avocado Pear

JACOB B. BIALE AND ROY E. YOUNG

University of California, Los Angeles and Riverside, U.S.A.

I Botany and Horticulture 2
A. Avocado Races and Varieties 2
B. Fruit Anatomy 4
C. Fruit Growth 5
II Composition 6
A. Major Components 6
B. Lipids 8
C. Carbohydrate 11
D. Free Amino Acids 12
E. Vitamins 13
F. Minerals 14
G. Tannins 15
H. Components with Pharmacological Activity 15
III Compositional Changes Associated with Development and Storage .. 16
A. Changes in Lipids and in Carbohydrates 16
B. Changes in Protein 20
C. Changes in Pectins 23
IV Fat Synthesis 24
V Enzymes 27
VI The Gas Exchange During Respiration 28
A. The Climacteric in the Intact Fruit 28
B. The Internal Atmosphere and Gas Diffusion 30
C. The Climacteric in Slices of Tissue 31
VII Regulation of Respiratory Rates 32
A. Temperature Effects 32
B. Oxygen Tension 34
C. Effects of Changes in CO_2 Concentration 35
VIII Production and Role of Ethylene 36
A. Identification of Ethylene 36
B. Ethylene as Ripening Hormone 37
C. Ethylene Incorporation 38
D. Irradiation Effects 40
IX Metabolic Pathways 40
A. The Electron Transport Chain 40
B. The Tricarboxylic Acid Cycle 42

X Metabolic Events in Fruit Ripening 50
A. Oxidative Activities in Mitochondria 50
B. Phosphate Metabolism in Avocado Slices 55
C. Protein Metabolism in the Avocado 56
D. Nucleic Acid Metabolism in the Avocado 58
XI Concluding Remarks 58
References 60

I. BOTANY AND HORTICULTURE

The avocado tree (*Persea americana* Mill.) belongs to the family Lauraceae and is one of the few commercially significant members of the genus *Persea*. Progenitors of the modern avocado date back to the Middle Eocene and Pliocene, some ten to sixty million years ago. Collections of fossil avocados were made in several widely spread areas of California as well as in other parts of America. As man sought gold by mining he released prehistoric specimens which were identified as species of the genus *Persea*. The identifications are based on morphological variations in leaf form, patterns of venation, relation of veins to midrib, etc.; the fragile flowers and fruits were not preserved. In some sites in which detailed exploration was pursued the avocado fossils accounted for most of the specimens discovered. Botanical relatives of the prehistoric fossils can be found in Sonoran Mexico. Palaeobotanical studies testify to the change in climate which must have caused the elimination of the wild avocado in California. In modern times selection and hybridization practices coupled with irrigation made it possible to establish an avocado industry in a state with a climate of dry summers and of rather wide seasonal and diurnal variations in temperature. From California several varieties were introduced to other sub-tropical and tropical regions of the world.

A. Avocado Races and Varieties

The horticulturist distinguishes between three general ecological groups or races of the avocado: Mexican, Guatemalan and West Indian. The Mexican race, which originated in the mountains of Mexico and Central America, is characterized by an anise odour of fruits, leaves and young growth. The fruits are relatively small, ranging from 75 to 300 g, and have a thin, smooth skin. The avocados of the Guatemalan race are native to the highlands of Central America and are not as resistant to low temperatures as those of the Mexican race. The fruit is large, averaging 500–600 g per unit, and the skin is thick and brittle. The West Indian race is native to the lowlands of Central America and northern South America. The fruit size is intermediate between the other races and the peel is smooth, leathery and sometimes glossy. In addition to these races there are hybrids of considerable commercial importance such as the Fuerte variety, which is considered to be a cross between the Mexican and Guatemalan races.

The differences between the several races and varieties of the avocado which are of special concern to the fruit physiologist relate to maturity and oil content. The West Indian and Mexican races blossom and mature in a much shorter season than the Guatemalan race. At maturity the oil content of avocados of the Mexican race is much higher than of fruit of the other two races and especially of the West Indian.

The two most prominent varieties in the California avocado industry are the Fuerte, a Guatemalan/Mexican hybrid, and the Hass, which originated from a Guatemalan seedling. The maturation season for the Fuerte is October to March and for the Hass April to August, depending on the district. Under suitable conditions fruit of both varieties can be stored on the tree beyond the months indicated since no softening takes place in attached avocados so long as the stem remains healthy. Ripening in the Hass is associated with change of the skin colour from green to black while the Fuerte retains its green colour even when soft.

In Florida, several varieties of avocados are grown both of the West Indian race and hybrids which originated in mixed plantings of Guatemalan and West Indian trees. Some of the commercially important Florida varieties are Lula, Booth 8, Waldin, Booth 7 and Pollock. In all cases the oil content at maturity is decidedly lower than that of the leading California varieties. Susceptibility to temperatures slightly below the freezing point of water is a major problem in avocado growing. It might also be of major concern to the investigator in the selection of fruit for physiological and biochemical studies. The research worker must also take cognizance of the fact that avocados, unlike other fruits, can be stored on the tree after maturation. The implication is that fruit may vary considerably in age, especially since the period of flowering may extend over weeks or months. Frequently "off-bloom" fruits are encountered in an advanced stage of maturity while fruit of the regular crop may be rather immature. An experienced picker will recognize the differences and a seasoned investigator resorts to objective physiological responses (to be described below) which tell him the past history of the material.

The length of the period from blossoming to maturation is affected by unique flower behaviour, by the nutritional competence of the tree and by external conditions during the growing period. While the flower of the avocado is perfect in form, its function is not. The general tendency is for the pistil to mature before the pollen is shed and for the flower to have normally two periods of opening—a condition referred to as dichogamy. At the first opening the pistil becomes receptive, and usually a day or so later the stamens shed their pollen at the second opening. This condition is not conducive to self-pollination. Overlapping flower sets and the opportunity to become self-fruitful are brought about at times by weather conditions such as a sudden drop in temperature.

Once the pollen is transferred to the stigma, germination is rapid and so is the growth of the pollen tube through the stylar tissue of the pistil. Under favourable conditions fecundation occurs in the span of one day or slightly more. As a result of zygote formation cell division is stimulated in the ovary wall and the growth of the fruit commences. The condition of so-called alternating dichogamy coupled with sensitivity to adverse climatic conditions accounts for the low percentage of blossoms developing into fruit. Winter blossoming habits, such as encountered in the Fuerte under coast conditions, has been quoted as the major reason for low yields. Temperature is a decisive factor in both fruit setting and fruit development.

B. Fruit Anatomy

The anatomy, morphology and cellular development of several varieties of the avocado were studied by Cummings and Schroeder (1942), Schroeder (1953) and by Valmayor (1967). Botanists refer to the fruit as a berry consisting of a single carpel and single seed. The pericarp, which is the fruit tissue proper excluding the seed, comprises the rind known as the exocarp, the flesh edible portion or mesocarp, and a thin layer next to the seed coat, the endocarp.

The outer layer of the exocarp is a thin wax-like cuticle covering the surface of the fruit. In addition to this film the exocarp consists of 1 layer of epidermis, 1–3 layers of brick-shaped hypodermal cells, several layers of parenchyma cells and a layer of sclerenchyma or stone cells limiting the inner surface of the peel. The parenchyma cells are about 40 μ in diameter and contain chloroplasts, tannin and some oil. Varietal differences in the thickness of the rind have been ascribed to the amount and density of stone cells. Varieties of the thick-skinned Guatemalan race are characterized by densely packed sclerenchyma, while in thin-skinned Mexican varieties the stone cells are scattered and loosely packed. The epidermal cell wall was found by Schroeder (1950) to be impregnated with cutin, while Valmayor (1967) reported cellulose in the epidermal walls of the Florida varieties he studied.

In the peel of young fruit, stomata consisting of a pore about 10 μ in length and surrounded by two guard cells can be readily detected. Schroeder's (1950) measurements indicate that a typical stoma cell is 21 μ long and 14 μ wide. In older fruits many of the stomata are forced upward by lenticels and are not visible. In smooth-skinned varieties the lenticels are slightly raised patches of tissue which usually do not break through the epidermis.

In rough-skinned varieties the lenticels have extensive cork and they rupture the epidermis.

The fleshy mesocarp is composed mostly of uniform isodiametric parenchyma cells of about 60 μ in diameter in mature fruit. Throughout the tissue there are specialized oil cells although small droplets of oil can also be detected in the parenchyma cells. The oil cells, or idioblasts, are distinguished by their large size and lignified walls. Scott *et al.* (1963) observed that the suberized idioblasts were frequently impermeable to stains which react with regular parenchyma tissue. They also reported, with the aid of the electron microscope, the development of microfibrillar pit patterns in the walls, lamellae in chloroplasts and the path of plasmodesmata which form cytoplasmic connections between cells.

The mesocarp is completely permeated with conductive tissue, the vascular system. Starting with a cylinder at the stem end the vascular tissue separates at some distance below into six major strands which divide and anastomose until the entire mesocarp is permeated by conducting vessels (Cummings and Schroeder, 1942). The vascular system of the avocado is asymmetrical as is the fruit as a whole. Since the seed develops off centre, a "thick" and "thin" side are normally formed. On the thicker side, well developed strands of vascular tissue enter the seed coat and from there branch out into fine bundles penetrating all parts of the seed. On the thinner side, bundles branch out to the various regions of the pericarp. When dark-coloured fibres are observed in over-ripe, improperly ripened or chilled avocados this phenomenon is due to discoloration of tracheal elements in the vascular tissue.

The endocarp, the inner layer of the pericarp, consists of a few rows of parenchyma cells which are smaller than those of the mesocarp and which adhere to the seed coat. Care has to be taken to exclude from homogenates this adhering layer as well as seed tissue if biochemically active mitochondria are desired.

The large seed of the avocado consists of two fleshy cotyledons, plumule, hypocotyl, radicle and two thin seed coats adhering to each other. The endosperm disappears in the course of development. The cotyledons consist of parenchyma tissue interspersed with idioblasts and contain starch as the main storage material.

C. Fruit Growth

The growth of the avocado from a fertilized ovum to a mature fruit is similar in some respects but differs in others from the development of other fruits. As in other species, cell division is very rapid in the early stages, but, unlike other species, the avocado is unique in that mitotic figures can be observed in mature fruit (Schroeder, 1953). In the avocado, therefore, the ultimate size of the fruit is determined not only by cell division in the early stage and final cell size but also by cell multiplication throughout the entire growth period. In some cases cell enlargement stops when the fruit reaches 50% of its size at full maturity, while cell division accounts for continued growth. The growth pattern is S-shaped but quantitative differences have been observed between different varieties. In early maturing varieties the growth curve is steep and the fruit increases in size when mature, while in late varieties the growth increments are smaller and decrease considerably before harvest time (Valmayor, 1967). In general, avocados tend to continue growing while attached to the tree. The differences in size between avocado varieties is determined more by cell division than cell enlargement, since the average size of 40–60 μ in diameter of the parenchyma cells in the mesocarp is characteristic of the species. Because of essential uniformity in growth patterns, in tissue types and in cell size one might expect similarities in physiological and biochemical responses of fruit from different races or varieties. This contention will be examined first with respect to the respiratory behaviour of the detached mature fruit.

II. COMPOSITION

A. Major Components

The high nutritive value and unusual composition of the avocado fruit were proclaimed early by enthusiastic supporters of the fruit. In a review by Wardlaw (1937) the statement was made that the avocado has "calorific value three times that of banana and one and one-half times that of beefsteak . . . with abundant amounts of vitamins A, B and E . . . and mineral matter higher than other fruit". Most other fruits are low in fats, protein and minerals and are high in sugars. Early studies by Church (1921–22) indicated the unusual composition of the avocado fruits.

In Table I the data for Fuerte and Dickenson are taken from Jaffe and Gross (1923), who made a critical survey of the composition of sixty-eight California varieties from several locations. The information on the Hass avocado is from Hall *et al.* (1955) and represents the average of 5–14 determinations taken during one season only. The results for three Florida varieties are those of Wolfe *et al.* (1934) where each sample represents the average of 10–100 mature fruits.

The outstanding compositional feature of avocado fruit is the high fat content. There is a great deal of variation in the lipid level, ranging in mature California varieties from 8 to 31·6%. Of the 68 California varieties analysed by Jaffe and Gross (1923) 15 showed more than 25% fat and 14 others in excess of 20%. Lipid level in Florida avocados is, in general, somewhat lower, from 4·7 to 18·8% considering all varieties reported by Wolfe *et al.* (1934). The lower fat level is presumably due to a more rapid development of the fruit.

TABLE I. Composition of several varieties of avocado fruit

Variety	Location	Fruit wt. (g)	% fr. wt. Edible portion	Moisture	Protein	Fat	Carbo-hydrates	Ash
Fuerte	Altadena, California	256	71·3	65·7	1.51	26.6	4.62	1.60
Fuerte	Yorba Linda, California	566	73·5	68·3	1·36	24·2	4·82	1·27
Hass	California	200	75·0	68·4	1·80	20·0	7·80	1·20
Dickenson	California	254	70·0	72·0	1·56	20·4	4·69	1·35
Lula	Florida	496	63·3	73·9	1·21	13·6	1·78	0·92
Trapp		422	72·2	83·5	0·90	6·3	1·56	0·64
Taylor		298	64·8	76·9	1·40	13·0	1·52	0·87

Data from Jaffe and Gross (1923), Hall *et al.* (1955) and Wolfe *et al.* (1934).

The three races of avocado discussed in the first section of this chapter differ markedly in fat content. West Indian varieties are lowest with 4–7%; Guatemalan fruit vary from 10 to 13%, while the Mexican race yields 10–15% in Mexico and 15–25% in California. There are exceptions to this generalization as demonstrated by varieties showing appreciably different lipid content and varying in other constituents as well, depending on habitat.

Protein levels are high for avocados. While most fruit contain less than 1% protein, Jaffe and Gross (1923) reported extreme values for the 68 varieties tested from 0·86 to 4·39% with the average at 2·1. The three most important California varieties listed in Table I show somewhat lower figures than the average, but nevertheless are high for fruit. Florida varieties have a slightly lower protein content.

The level of sugars is extremely variable. In the data of Table I, values recorded as "total carbohydrate" are noted to include crude fibre with an average content of about 1·5%. The authors did not indicate the method used for the carbohydrate determination, but their values are higher than those reported by other authors. Bean (1958), for example, cites "total sugars" as 1·75% in early season and 0·68% in mid-season for the Fuerte variety. The level of sugar decreases rapidly during storage and ripening and thus analyses may vary considerably depending on growth conditions, the exact picking time and the length of storage before the analysis is carried out. Jaffe and Gross (1923) reported carbohydrate values as high as 10% in the Chappelow variety.

Jaffe and Gross (1923) compared the composition of the several varieties grown in two different locations. Data for the Fuerte variety are shown in Table I. Even though the fruit grown in Yorba Linda was over twice as large as those picked in Altadena, the composition was very similar.

B. Lipids

The fatty acid composition of avocado lipid has been studied extensively and was reviewed by Hilditch and Williams (1964). Their comparisons of fatty acid composition of lipids from various habitats is probably not valid as different varieties and different analytical methods were used. These factors would lead to differences greater than those attributed to habitat.

Mazliak (1965a) analysed the fatty acids derived from hydrolysis of the lipids from various parts of the fruit. Acids were separated by gas chromatography of the methyl esters using two different liquid phases. The fatty acid composition as per cent of total fatty acids of the lipids of the tissue fraction is shown in Table II.

The edible portion of the fruit is rich in oleic, palmitic, linoleic and palmitoleic acids, while stearic acid is present only in trace amounts. Little difference was noted in the fatty acid complement between mesocarp and endocarp, and even the exocarp. The fatty acids of the seed are markedly different from those of the pericarp, being higher in linoleic and linolenic acids, as is typical for seeds. It should be noted that the seed is low in fat, having only about 1% on a fresh weight basis.

TABLE II. Fatty acid composition of lipids of Fuerte avocado fruit (% of total fatty acids)

Fraction	C 14:0	C 16:0	C 16:1	C 18	C 18:1	C 18:2	C 18:3	C 20:0
Exocarp	t	12–22	2·5–5·5	t	59–70	12–15	1·2–2·3	t–0·3
Mesocarp	t	13–17	3·0–5·1	t	67–72	10–12	t–1·5	t
Endocarp	t	13–20	5·0–7·3	t	62–70	10–12	t–1·2	t
Seed	0·8	22	3·2	0·6	25	42	5·1	t

t = trace; from Mazliak (1965a).

TABLE III. Percentages of the various classes of lipids in mesocarp of mature Fuerte avocados

Class	% fr. wt.
Free fatty acids	0·10
Triglycerides	19·96
Diglycerides	1·29
Monoglycerides	0·78
Phospholipids	0·39
Others[a]	0·28
Total	22·80

[a] Other substances extracted in chloroform : methanol (2 : 1). Data from Kikuta (1968).

Kikuta (1968) made an exhaustive study of the lipids of avocado fruit of the Hass and Fuerte varieties grown in southern California. He separated the various classes of lipids by silicic acid chromatography, then hydrolysed each purified lipid fraction and determined the fatty acid composition by gas chromatography. Mature Fuerte avocados were used and the results are given in Table III. The free fatty acid level was found to be very low while 86% of the lipid fraction consisted of triglycerides in mature Fuerte fruit harvested in February. An earlier sample gave an identical proportion of triglycerides. The diglyceride level, which was composed of three fractions, was just over 1% of the fresh weight or 5% of the lipid fraction, while monoglycerides made up 3·4% and phospholipids 1·7% of the lipid fraction. The phospholipid fraction was separated into five phospholipids on thin-layer chromatography.

Kikuta (1968) hydrolysed each lipid fraction and determined the fatty acid composition by gas–liquid chromatography as shown in Table IV. It is interesting that the free fatty acid composition closely resembles the fatty acid composition of the triglycerides which make up the major part of the lipid. All of the lipid fractions are low in stearic acid, diglyceride I, and the monoglyceride fractions are more unsaturated than the triglycerides, while diglyceride II is similar to the triglyceride fraction. Glycolipids I and II as well as the phospholipid fraction are very different from all of the other fractions and are made up of a higher proportion of unsaturated acids. The

TABLE IV. Fatty acid composition as per cent of each fraction

	Fatty acid							
	16:0	16:1	18:0	18:1	18:2	18:3	20:0	UK[a]
Free fatty acid	20·3	9·7	0·4	43·7	22·5	3·0		0·4
Triglyceride	25·4	7·0	0·5	54·3	12·3		0·5	
Diglyceride I	15·0	9·5		45·0	28·0	3·0		
Diglyceride II	18·4	3·9	0·7	64·8	12·2			
Glycolipid I	6·7	2·5	1·6	13·1	76·1			
Monoglyceride	17·1	7·2	2·7	43·2	24·3	1·0	0·9	3·6
Glycolipid II	3·8	2·2	1·2	12·8	74·1	6·0		
Phospholipid	16·9	4·4	3·3	20·5	36·1	9·8		9·0

[a] 16:0, 16:1, 18:0, 18:1, 18:2, 18:3, 20:0 and UK represent palmitic, palmitoleic, stearic, oleic, linoleic, linolenic, arachidic and unknown fatty acids, respectively. Data from Kikuta (1968).

TABLE V. Composition of the cuticular wax of Fuerte avocado exocarp

Carbon numbers	Acids	%Total fraction	Alcohols		Paraffins	
13			Tridecanol	1·6		
14	Myristic	0·2	Tetradecanol	16·3		
15			Pentadecanol	3·3		
16	Palmitic	13·3	Hexadecanol	22·5		
17			Heptadecanol	1·6		
18	Oleic	41·5	Octadecanol	39·5		
19	Nonadecanoic	2·9				
20	Arachidic	0·9	Eicosanol	2·5		
21	Heneicosanoic	0·4	Henicosanol	1·6	Henicosane	0·2
22	Behenic	3·0	Docosanol	1·6	Docosane	0·6
23	Tricosanoic	0·5			Tricosane	2·3
24	Lingnoceric	15·5	Tetracosanol	2·4	Tetracosane	1·5
25	Pentacosanoic	0·7			Pentacosane	4·8
26	Cerotic	20·8			Hexocosane	2·5
27					Heptacosane	17·5
28					Octacosane	1·5
29					Nonacosane	43·0
30					Triacontane	1·4
31					Hentriacontane	24·5

Data from Mazliak (1965b).

phospholipid fraction has more stearic acid than any other fraction and 9% of the hydrolysed fatty acids is made up of one or more unidentified acids, possibly the polar acid noted by Mazliak (1965a).

Mazliak (1965b) examined the composition of the cuticular wax of the Fuerte variety by gas–liquid chromatography. Many acids, alcohols and paraffins were found in trace amounts. Principal components are shown in Table V. The presence of C-24 and C-26 acids in substantial amounts is notable along with small amounts of acids of odd carbon number. The majority of the paraffins are of odd carbon number.

Eckey (1954) has collected data on the physical and chemical properties of avocado oil from all available sources. Some of the properties listed by Eckey are shown in Table VI. This table indicates a surprising large range in the ratio of unsaturated to saturated fatty acids while the solidification point, specific gravity and refractive index show remarkably small differences for all samples tested.

TABLE VI. Properties of avocado oil

Property	Minimum	Maximum
Acid value[a]	1	7
Saponification number[b]	177	198
Iodine value[c]	71	95
Thiocyanogen value[d]	(Iodine value of 87·2)	71·8
Hydroxyl value[e]	8	10
R-M value[f]	1·7	15·9
Polenske value[g]	0·2	8·0
Unsaponifiable (%)	0·8	1·6
Refractive index nD, 40°C	1·461	1·465
Specific gravity 20–25°C	0·910	0·916
Solidification point (0°C)	7	9

Data from Eckey (1954). Range of values from several varieties in different countries.

[a] 2% oleic acid.

[b] mg KOH to saponify 1 g lipid.

[c] g iodine absorbed/100 g sample.

[d] g thiocyanogen absorbed/100 g sample.

[e] mg KOH necessary to combine with acetic acid liberated by saponification of 1 g of acetylated fat.

[f] Reichert–Meissel value. ml 0·1N NaOH required to neutralize the volatile fatty acids obtained from 5 g of mixed triglycerides.

[g] ml 0·1N KOH required to neutralize the non-volatile fatty acids obtained from 5 g of a saponified fat or oil.

C. Carbohydrate

The avocado fruit is a fertile source of unusual carbohydrates. The 7 carbon alcohol, perseitol (see Fig. 1 for structure I) was described by Maquenne in 1890. La Forge and Hudson (1917) identified the 7 carbon sugar, D-mannoheptulose (II). Glucose, fructose and sucrose are also present at all stages of ripening (Davenport and Ellis, 1959). Charlson and Richtmyer (1960) found an octulose, D-glycero-D-manno-octulose (V) and the corresponding alcohol, D-erythro-D-galacto-ocitol. The presence of D-talo-heptulose (III) was also confirmed. Sephton and Richtmyer (1963a, b, 1966) isolated, purified and determined the structure of D-glycero-D-galacto-heptose (IV), D-glycero-L-galacto-octulose (VI), D-erythro-L-gluco-nonulose (VII) and D-erythro-L-galacto-nonulose (VIII). These were the first octuloses and nonuloses to be found in nature. Racker and Schroeder (1957) demonstrated an octulose-8-phosphate which was formed when D-ribose-5-phosphate and D-fructose-6-phosphate were incubated in the presence of transaldolase.

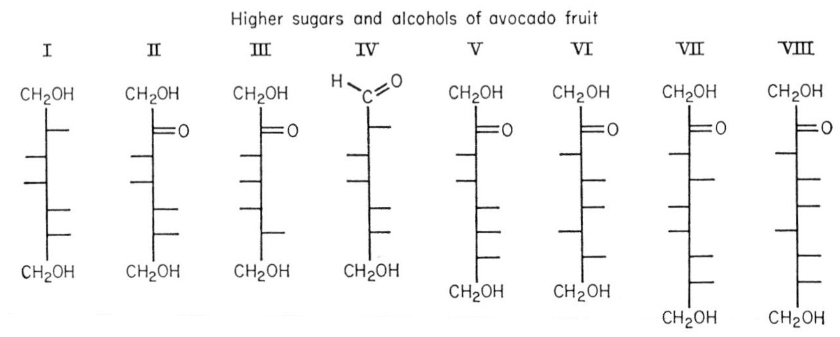

FIG. 1. Structure of sugars and alcohols discovered in avocado mesocarp.

The mechanism of biosynthesis of the unusual sugars is still not clear. Sephton and Richtmyer (1966) suggested that two systems may be involved. Jones and Sephton (1960) have shown that synthesis of those sugars of the D series with the hydroxyl on C-3 on the left when written in the Fischer projections can be catalysed by aldolase from dihydroxyacetone phosphate and the suitable aldehyde. Since only three of the ketoses have hydroxyl groups of C-3 on the left and of C-4 on the right, as must result from the aldolase reaction, another mechanism must exist for synthesis of the other sugars. The biological functions of these sugars are unknown.

D. Free Amino Acids

The free amino acid content of Fuerte avocados was found by Joslyn and Stepka (1949) to be higher than in any other fruit tested (apricots, apples, prunes and pears). The principal amino acids were asparagine, aspartic acid, glutamine and glutamic acid. Serine, threonine, alanine, valine and cystine were detected. Cysteine and lysine were tentatively identified. γ-Amino butyric acid was shown also to be present.

Spitzer (1931) has shown an o-dihydroxyphenyl derivative in apples and other fruit which he believed to be dihydroxyphenylalanine. He proposed that it is involved in enzymatic browning in fruit. Joslyn (1941) stated that amino acids might be responsible for non-enzymatic browning. Joslyn and Stepka (1949) could not detect either tyrosine or dihydroxyphenylalanine in avocado tissue and concluded that browning in this fruit was not due to a specific dopa oxidase.

E. Vitamins

Avocado fruit, being a high fat fruit, might be expected to be rich in vitamins, especially the lipid soluble ones. In Table VII vitamin data were collected from several sources. In comparison with other fruits, avocado is rather poor in vitamins A, C, K and folic acid, but is rich in vitamin B. The pyridoxine level of avocado is highest among fruit, except for banana, which has 0·59 mg/100 g (Polansky and Murphy, 1966). The pantothenic acid level is highest among fruit with the exception of passion fruit (*Passiflora laurifolia*), which was reported by Asenjo and Muniz (1955) to have 1·55 mg/100 g. The vitamin K level of 8 μg/100 g is low compared to 12 μg/100 g for tomato fruit and 176 μg for spinach leaves. The folic acid level of 11 μg/100 g is low compared to peas and beans at 50 μg/100 g.

Hall *et al.* (1955) compared the vitamin B of three commercially important varieties of California, Fuerte, Hass and Anaheim. The first samples were taken at commercial maturity and at intervals during a three month picking season. The Anaheim variety had four- to five-fold more folic acid than did Fuerte or Hass fruit, but other vitamin levels were not strikingly different.

TABLE VII. Vitamin content of avocados

	Vitamin	mg/100 g fr. wt.	Reference
A	Carotene	0·13–0·51	French and Abbot (1948)
B	Thiamine	0·08–0·12	Hall *et al.* (1955)
	Riboflavin	0·21–0·23	Hall *et al.* (1955)
	Pyridoxine[a]	0·45	Polansky and Murphy (1966)
	Niacin	1·45–2·16	Hall *et al.* (1955)
	Pantothenic acid	0·90–1·14	Hall *et al.* (1955)
	Folic acid	0·018–0·040	Hall *et al.* (1955)
	Biotin	0·003–0·006	Hall *et al.* (1955)
C	Ascorbic acid	13·0–37·0	French and Abbot (1948)
D	Calciferol	0·01	Schwob (1951)
E	α-Tocopherol	3·0	Schwob (1951)
K	2-Methyl-1,4-naphthoquinone	0·008	Schwob (1951)

[a] Includes pyridoxal and pyridoxamine.

French and Abbot (1948) compared two Florida varieties in vitamins A and C and their results are also shown in Table VIII. Pollock variety was much higher in both vitamins than Lula variety.

TABLE VIII. Vitamin content of various varieties of avocado

Vitamin	California varieties		
	Fuerte	Hass	Anaheim
Thiamine	0·12	0·09	0·08
Riboflavin	0·22	0·23	0·21
Niacin	1·45	2·16	1·56
Pantothenic acid	0·90	1·14	1·11
Pyridoxine[a]	0·61	0·62	0·39
Folic acid	0·03	0·04	0·018
Biotin	0·005	0·006	0·0034
	Florida varieties		
	Lula		Pollock
Carotene	0·130		0·510
Ascorbic acid	13·0		37·0

[a] Includes pyridoxal and pyridoxamine.
All values in mg/100 g fr. wt.
Data for California varieties from Hall *et al.* (1955) and for Florida varieties from French and Abbot (1948).

Hall *et al.* (1955) followed the levels of thiamine and riboflavin in the Fuerte variety throughout the normal picking season of January to April. There was no significant change in the levels of these two vitamins throughout the picking season.

F. Minerals

According to Jaffe and Gross (1923), the mineral content of avocado fruit is higher than that recorded for any other fresh fruit. The minimum ash in

TABLE IX. Mineral composition of avocado mesocarp (% of total ash)

K_2O	26·2
Na_2O	18·6
CaO	4·7
MgO	5·3
Fe_2O_3	1·51
Al_2O_3	2·58
Mn	Trace
P_2O_5	17·40
SO_4	11·24
SiO_2	0·50
Cl	14·36

Data from Jaffe and Gross (1923).

California avocados (Rhoad variety) was 0·54% which is about the average for most fresh fruits. Ash analysis without reference to variety is presented in Table IX. The high phosphate and iron content is of interest nutritionally. Weatherby (1935) found that rats which had become anaemic on an iron-free diet formed haemoglobin when their diet was supplemented by 1–5 g of avocado tissue. The copper content was found to be 1·79 p.p.m. by Giral and Castillo (1953), which is about the same level as in banana pulp and in beans grown in Mexico.

G. Tannins

Joslyn and Smit (1954) were able to separate tannins by column chromatography into four groups with distinct spectra. Avocado fruit tissue was shown to have two such groups, a catechin (270–280 nm absorption) and a flavone (250–260 nm absorption).

Avocado seeds are a rich source of a complex mixture of polyphenolic compounds ranging from simple (+) catechin and (−) epicatechin to highly polymeric substances. Geissman and Dittmar (1965) used single directional paper chromatograms which revealed a streak of proanthocyanins (leuco-anthocyanins) from the origin to the R_f of catechin. A dimer with the following structure was isolated and purified from avocado seed:

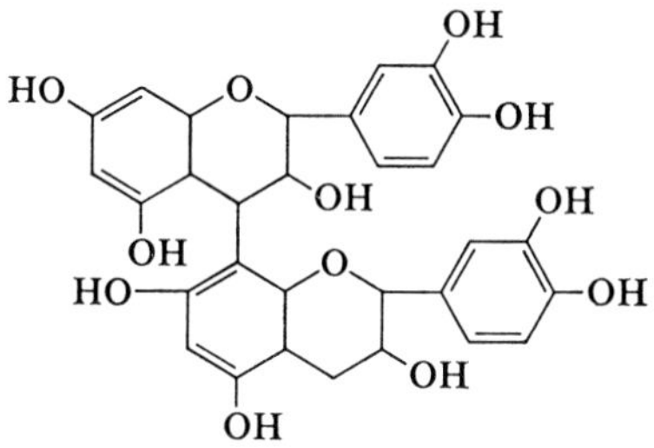

H. Components with Pharmacological Activity

Antibacterial activity of extracts of avocado fruit and seed has been investigated in several studies. Spoehr *et al.* (1949) showed some antibacterial activity against *Staphylococcus aureus*, and Gallo and Valeri (1954) obtained an extract with some antibacterial activity against *Micrococcus pyrogenes*, but none against *Escherichia coli* or *Bacillus subtilis*. On the other hand, Valeri and Gimeno (1954) reported some activity of certain avocado extracts against *Escherichia coli*.

Young and Biale (1967) have noted the ability of slices of Fuerte and Hass avocado fruit to resist invasion of bacteria and fungi so long as they are in the pre-climacteric state. Immediately after the peak of the climacteric, the slices became an excellent substrate for micro-organisms. It is possible that in the pre-climacteric phase, tannins or some other component contribute to the resistance against micro-organisms. Variable results of the antimicrobial experiments may be due to the failure to use pre-climacteric fruit consistently.

Seegers *et al.* (1955) investigated polysaccharides from a number of plant sources, including avocado, as blood plasma extenders. Avocado gum was effective as an extender, but like dextran and some other natural compounds, inhibited prothrombin activation somewhat.

Grant (1960) substituted one-half of the dietary fat of male adult hospital patients with avocado fruit. Serum cholesterol declined in the eight patients on avocado diet compared to patients receiving animal fat. It was not clear whether the decrease was due to the unsaturation of avocado fat, the sitosterol present in the unsaponifiable fraction or to some other factor. The unsaponifiable fraction of avocado mesocarp may be as high as 2% of the lipid.

III. COMPOSITIONAL CHANGES ASSOCIATED WITH DEVELOPMENT AND STORAGE

A. Changes in Lipids and in Carbohydrates

1. *Seasonal changes in air*

Changes in fat, sugar and protein associated with development and storage have been studied by Church (1921–22), Appleman and Noda (1941), Davenport and Ellis (1959), Mazliak (1965b), Dolendo *et al.* (1966), Appleman (unpublished)† and by Kikuta (1968).

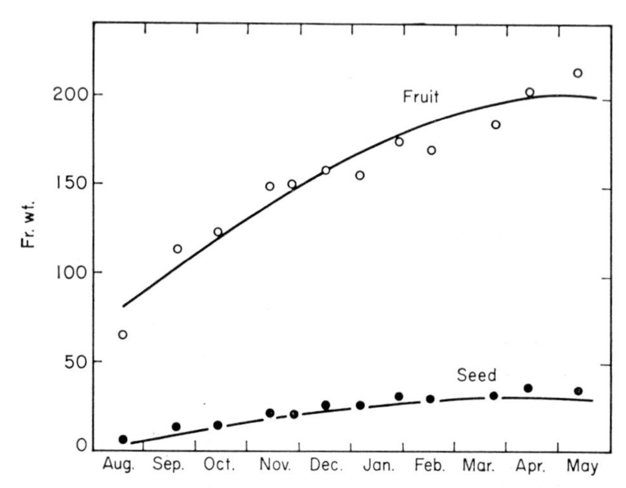

FIG. 2. Seasonal changes in fruit and seed weight (in grams) of the Fuerte avocado (by courtesy of Professor D. Appleman).

† The authors are greatly indebted to Professor David Appleman who offered us unpublished data for inclusion in this chapter.

Appleman followed changes in the size of the fruit and the seed, the amount of fats and sugars as well as the changes in the properties of the fats. His first measurements were in August when the fruit weighed 60 g and the seed weighed about 5 g (Fig. 2). The weight of the fruit and seed both increased rapidly through December when the fruit became "horticulturally mature", that is attained the capacity to ripen normally and, from the legal standpoint in California, attained 8% fat. Fruit picked before December 15th shrivelled badly, did not soften normally and had less than 8% fat.

From December until May the fruit continued to increase in size but at a slower rate than during the earlier period. Fat increased slowly until late November and then more rapidly until March, when the rate of accumulation tended to level off. Moisture content decreased more rapidly during this same period, as shown in Fig. 3.

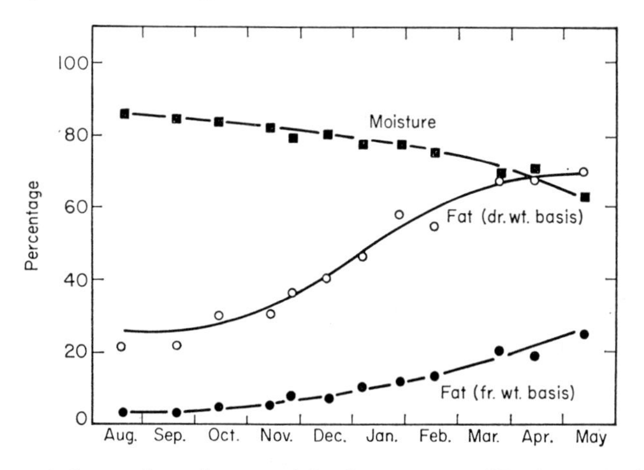

FIG. 3. Seasonal changes in moisture and fat (by courtesy of Professor D. Appleman).

Total and reducing sugars as measured by Appleman (Fig. 4) were high and constant until about December, when the very rapid growth phase was replaced by a slow one. From December until May free sugars decreased at a linear rate. Davenport and Ellis (1959) separated the free carbohydrates on paper chromatograms. They found sucrose, manoheptulose, fructose, glucose and an unknown disaccharide. The 7-carbon alcohol, perseitol, was also present. These authors did not make precise quantitative determinations of the carbohydrates but estimated the changes in sugars on the basis of intensity of spots of the chromatograms. Throughout the picking season of April, May and June, there was no marked change except for perseitol, which was highest in May. During storage after picking, perseitol and manoheptulose almost disappeared, sucrose remained unchanged and fructose and the unknown disaccharide increased markedly.

The properties of the lipids changed dramatically as the fruit developed (Fig. 5). The iodine number, which is an index of degree of unsaturation, of the 60 g fruit in October was about 190 and decreased rapidly until mid-December when the value of 107 was recorded. The iodine number changed much more slowly in the mature fruit from January until May. Saponification

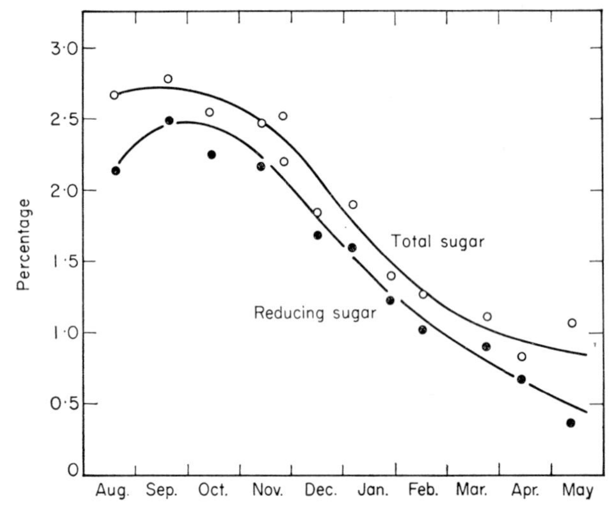

FIG. 4. Seasonal changes in sugars (by courtesy of Professor D. Appleman).

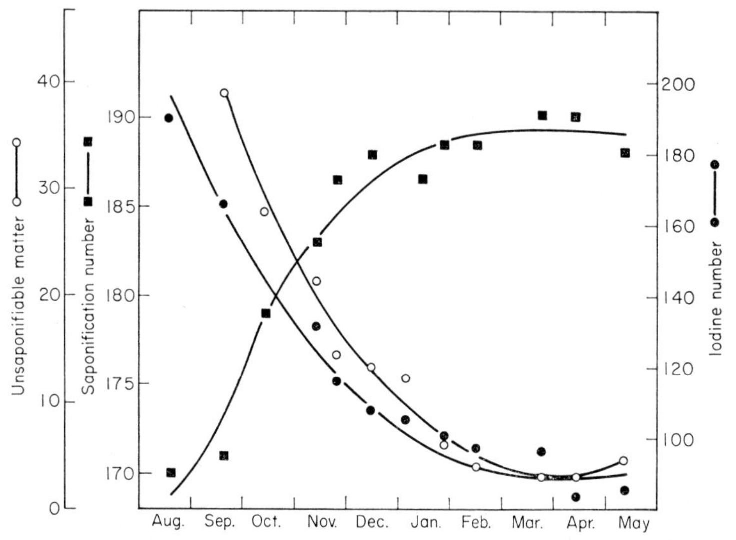

FIG. 5. Seasonal changes in degree of unsaturation (iodine no.), saponification and unsaponifiable matter of lipids (by courtesy of Professor D. Appleman).

number increased rapidly from a value of 170 in October to 188 in December. There was no further change throughout the season.

The chemical properties of the oil changed at the time of maturity. Considerable effort has been expended to find an easily measurable property

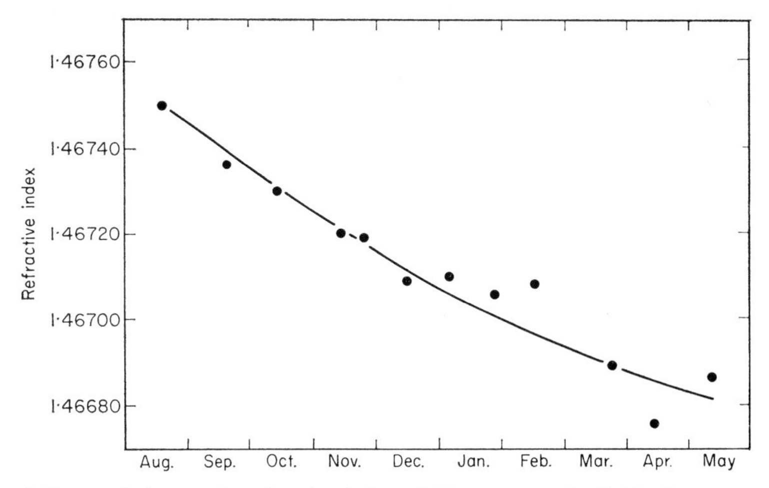

FIG. 6. Seasonal changes in refractive index of Fuerte avocado lipids (by courtesy of Professor D. Appleman).

which would provide a positive criterion of maturity, that is the ability of the fruit to ripen after picking to a fruit of constant qualities of texture, flavour and appearance. Unfortunately, changes in the chemical properties of the extracted oil are too complicated for a field maturity test. Appleman, in an effort to find a simple maturity test, measured the specific gravity of the pericarp tissue and the refractive index of the pressed oil. The specific gravity of the pericarp was about 1·000 throughout development and ripening. The refractive index did change in a regular manner (Fig. 6) from 1·4750 in August when the immature fruit weighed 60 g and the fat content was 3% to 1·4687 in May when the fruit had been mature for several months. Refractive index of a tissue homogenate mixed with Halowax has been used as a measure of maturity. This change in refractive index is closely related to the fat content. Fuerte avocados picked in California before the fat content reaches 8% tend to shrivel and fail to soften in a manner characteristic of more mature fruit.

Kikuta (1968) has measured the change in the classes of lipids associated with development of Fuerte avocado fruit. A histogram of the changes is shown in Fig. 7. In September, when the fruit weighed about 100 g or less, the phosopholipid, free fatty acid and hydrocarbon levels remained unchanged on a per cent fresh weight basis. The monoglyceride fraction decreased slightly while the diglyceride fraction increased slightly. Virtually all of the change in fatty acid composition was in the triglyceride fraction. Kikuta also measured the changes in fatty acid distribution of the lipids associated with development as shown in Fig. 8. Linolenic acid remained unchanged throughout the development period. Palmitic, palmitoleic and linoleic increased slightly while the major change was a large increase in oleic acid. Dolendo *et al.* (1966) measured the fatty acid composition throughout a ripening storage period (11 days at 15°C). There was no significant change in the distribution of fatty acids of the hydrolysed lipids during ripening.

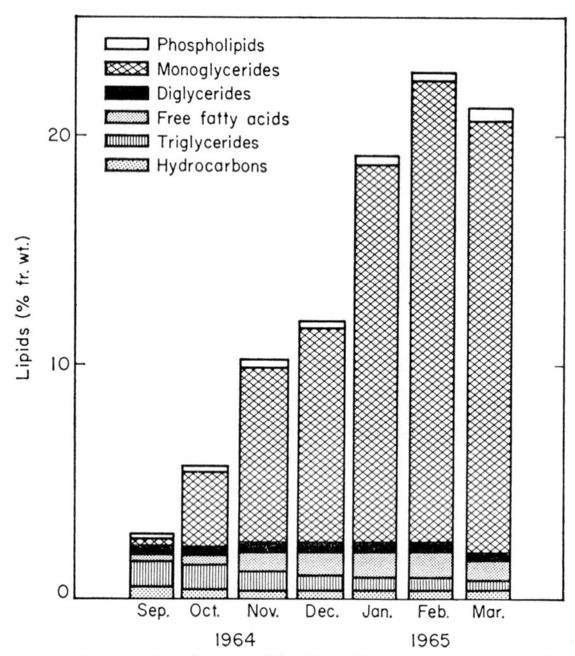

FIG. 7. Seasonal changes in classes of lipids of Fuerte avocado (Kikuta, 1968).

2. *Changes in lipids in relation to atmospheric composition*

Mazliak (1965c) studied changes in the fatty acids of the hydrolysed lipids in response to storage atmosphere enriched in carbon dioxide and depleted in oxygen. An atmosphere high in carbon dioxide and low in oxygen tended to cause an increase in the amount of palmitic and palmitoleic acid and a decrease in the percentage of oleic acid. Decreasing the oxygen to 1·5% had a greater effect than increasing the carbon dioxide to 5–7%.

B. Changes in Protein

Several claims have been made for substantial increases in protein associated with the climacteric rise in respiration and ripening. Using the extraction procedure of Hulme (1936), Rowan *et al.* (1958) reported a change in the protein nitrogen to total nitrogen ratio (PN/TN) of 0·5–0·75 associated with the ripening process. These authors followed the respiration of individual fruit and extracted whole fruit at particular points in the ripening cycle based on the respiratory rate. Their data indicated a trend towards an increase in the PN/TN, but the scatter of points was such that no definite conclusion

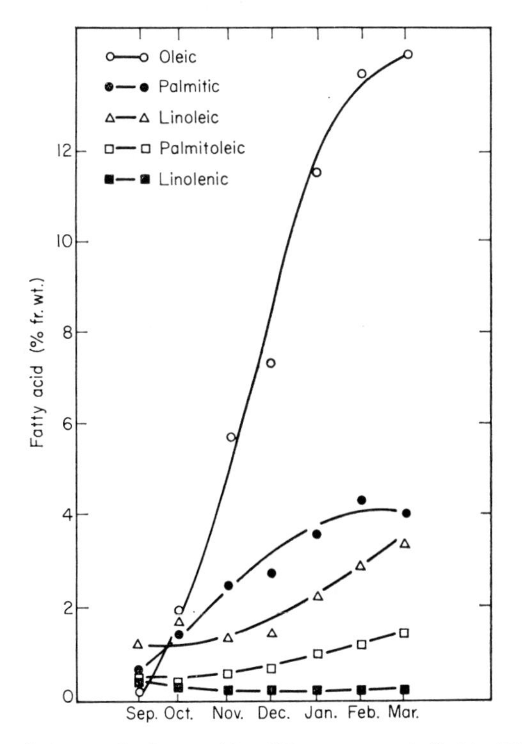

FIG. 8. Seasonal changes in fatty acids of Fuerte avocado lipids (Kikuta, 1968).

could be drawn. They oven dried the tissue, extracted with 70% ethanol and assumed the nitrogen left in the tissue as representing protein nitrogen. Davenport and Ellis (1959) also reported small increases in protein using a cold 80% ethanol–petroleum ether extraction of the fresh tissue and assumed the total nitrogen of the extracted residue to be protein.

Young, Popper and Robertson (unpublished) restudied extraction procedures and protein determinations in detail. In order to be able to choose fruit at a particular stage of ripening, respiration of individual fruit was followed on the oxygen analyser (Young and Biale, 1962). It was found with the Fuerte variety, as Rowan *et al.* (1958) reported with the Anaheim variety in Australia, that the variability of nitrogen within individual fruit was so great that no conclusion could be drawn as to the change in protein with ripening.

The problem of the considerable differences in the total and protein nitrogen of individual fruit was solved by taking samples throughout the climacteric from a single fruit. This was made possible by the finding that freshly picked fruit could be peeled and cut into 16 longitudinal sections with respiration and ripening occurring just as in the intact fruit except that the climacteric was induced soon after cutting. Fruit were cut under sterile conditions and placed in sterile respirometers. There was, however, no tendency for micro-organisms to invade the tissue until the post-climacteric phase.

Even using single fruit, nitrogen distribution around the fruit was not symmetrical (see Section IB). Nitrogen content of sections taken around the fruit varied in a regular way. Uniform samples were obtained by cutting the fruit in 16 longitudinal sections and taking every fourth section for each of 4 samples at (1) pre-climacteric; (2) early climacteric rise; (3) late climacteric rise, and (4) climacteric peak. This technique provided four samples with identical total nitrogen content.

Variable results in protein nitrogen were still obtained depending on the drying and fat extraction procedures used. When the slices were lyophilized, soxhlet extracted with diethyl ether and petroleum ether followed by regrinding and re-extraction with ether and then soxhlet extraction with 80% ethyl alcohol, reproducible non-extractable nitrogen values were obtained. This procedure gave a PN/TN = 0·74 for pre-climacteric and 0·78 for climacteric peak tissue. These results indicate that there is no significant increase in net protein content during ripening.

A completely different method was adopted to check the ethanol extraction procedure. Bevenue and Williams (1959) reported that 92–95% of the nitrogen could be extracted from bean and pea seeds by an alkaline solution of the detergent Dupanol, sodium dodecylsulphate. The lyophilized fat-free tissue was extracted three times with Dupanol and the residue filtered off and washed. The protein in the extract was precipitated with trichloroacetic acid and nitrogen determined by the semi-micro Kjeldahl method. Nitrogen was also determined in the residue and in the TCA filtrate. The results of these determinations for these three fractions from pre-climacteric and climacteric peak fruit are shown in Table X. The amount of nitrogen which could not be extracted from the tissue was greater in the pre-climacteric. If one assumes that this residue nitrogen is not protein, then the PN/TN ratios are 0·65 for pre-climacteric and 0·66 for climacteric peak showing no net increase in protein synthesis. If on the other hand one assumes that all of the nitrogen left in the residue is protein, then there is an actual decrease in the ratio of protein to total nitrogen in the course of the climacteric as shown in Table X.

We conclude that no increase in the total protein is associated with ripening of avocado fruit. There may be, of course, changes in the kind of proteins present. The capacity of the tissue for protein synthesis is demonstrated through studies of amino acid incorporation as discussed in Section X of this chapter.

Clements (1965) separated the proteins of avocado fruit on gel electrophoresis. He used extracts of an acetone powder for his source of protein. Unripe gave a clearer pattern than did ripe, but resolution on the gel of even the unripe was very diffuse. Clements indicated in diagrams that four major bands and six minor bands were distinguishable on the pH 8·9 gels of extract from unripe fruit. None of the proteins was identified.

TABLE X. Nitrogen fractionation in relation to the climacteric[a]

	Pre-climacteric	Climacteric peak
Total N in sample	32·3	36·7
Dupanol extract		
TCA filtrate	8·8	10·3
TCA precipitate	20·9	24·2
Residue of Dupanol extraction	3·2	1·5
PN/TN (assuming residue N is not protein)	0·65	0·66
PN/TN (assuming residue N is protein)	0·75	0·70

[a] mg N/g fat-free lyophilized mesocarp.
PN = Protein nitrogen; TN = Total nitrogen.

C. Changes in Pectins†

The most obvious change in avocado fruit associated with ripening is the softening process. It was assumed early that changes in the pectic components were mainly responsible for the softening. This led to investigation of the pectic enzymes.

McCready *et al.* (1955) separated polygalacturonase, (E.C. 3.2.1.15), (poly-α-1,4-galacturonide glycanohydrolase), by salt extraction followed by pH precipitation and ammonium sulphate fractionation. The enzyme appeared to have the same specificity as preparations isolated from fungi and tomato fruit. Reymond and Phaff (1965) studied the properties of the enzyme in more detail. They purified their preparation with ion exchange chromatography and calcium phosphate gel and obtained a thirty-five-fold purification.

† See also p. 30.

The purified enzyme had a pH optimum of 5·5 with very narrow pH-activity range. Only 50% of the activity was observed at pH 5·0 or 6·0. Addition of $NaHSO_3$, Na_2S, EDTA, NaCN or Triton X-100 did not increase the activity and in some cases decreased it. NH_4^+, K^+ and $H_2PO_4^-$ inhibited the enzyme. Oligogalacturonides were hydrolysed at random positions; at 50% hydrolysis, mono-, di- and trigalacturonides were present. Reymond and Phaff believe that avocados, particularly immature fruit, contain an inhibitor of the enzyme because at successive purification steps more than 100% of the original activity could be recovered.

Hobson (1962) measured polygalacturonase activity in various ripe fruits including the avocado. He found 0·24 units per 100 g fresh weight in avocado while tomato gave 3·68 units and pear 0·06 units. Units of activity referred to reducing power in grams of galacturonic acid per 100 g fresh weight of ripe fruit. While the fruit were reported to be "ripe" it was not clear as to whether or not they were post-climacteric.

Pectin changes in MacArthur avocados were followed by Dolendo *et al.* (1966). Softening was accompanied by a rapid decrease in protopectin and an increase in water-soluble pectin. There was also a decrease in the degree of esterification of the pectin which contributes to the softening process.

TABLE XI. The change in polygalacturonase activity associated with ripening in Fuerte avocado[a]

Days from picking	Polygalacturonase activity μmoles reducing groups/min/mg protein	
	Blossom end	Stem end
1	1·7	2·5
2	5·4	2·5
3	4·0	3·1
4	9·1	8·8

[a] Fruit kept at 23–25°C.
Data from Raymond and Phaff (1965).

Reymond and Phaff (1965) cut plugs of tissue from individual Fuerte avocado fruit and analysed for polygalacturonase activity. They removed plugs from the same fruit at intervals until the fruit were soft and ripe, which generally required 4 days. Their results are shown in Table XI. The activity of the enzyme increased as the fruit softened and occurred more rapidly in the blossom end than in the stem end.

IV. FAT SYNTHESIS

The avocado fruit possesses a remarkable ability to synthesize fats of unusual character and to accumulate levels as high as 30% on a fresh weight basis. Stumpf and co-workers at the University of California, Davis, have studied the fat synthesizing system of avocado fruit extensively, and Kikuta and Erickson (1969a, b) have completed a study on fat synthesis in avocado slices.

Slices of avocado tissue were shown to incorporate acetate-^{14}C into lipids (Stumpf, 1961). Stumpf and Barber (1957) observed that particulate preparations possessing the properties of mitochondria could also incorporate C^{14}-acetate into palmitic and oleic acid if ATP, Mn^{2+} and CoA were supplied. Squires *et al.* (1958) found a soluble extract of mitochondria could form long-chain fatty acids from acetate provided CO_2, ATP, Mn^{2+} and CoA were added. Barron *et al.* (1961) reported that if malonyl-CoA and acetyl-CoA were supplied, CO_2, ATP and CoA were not required for synthesis. This soluble system however synthesized only palmitate and stearate.

TABLE XII. A comparison of fatty acid composition and acetate incorporation into fatty acids of lipids

Preparation	Fatty acid composition of lipids (%) Palmitic	Stearic	Oleic	Linoleic
Mesocarp	20	1	60	15
Mitochondria	15	1	70	15
	% Incorporation of ^{14}C-acetate			
Slice	48	10	31	3
Mitochondria	50	20	10–20	1
Soluble enzyme	25–40	60–75	0	0

Data from Stumpf (1961).

Stumpf (1961) examined the normal avocado mesocarp lipids as well as those of mitochondria and compared these values with the relative amounts of ^{14}C-acetate incorporated into the various fatty acids by (1) avocado slices; (2) mitochondria isolated from mesocarp, and (3) the soluble synthetase system (mitochondrial acetone powder extract). The data are shown in Table XII. Surprisingly, even incorporation into the slice was markedly different from the fatty acid composition of the mesocarp tissue. It is possible that the stage of ripening of the tissue had some effect as ripening was not controlled, but, in general, fruit was just softening and presumably was on the climacteric rise of respiration. However, in both mesocarp and mitochondria, stearic acid was present only to the extent of 1% while slices incorporated 10% of the ^{14}C-acetate into stearate. In the soluble system, 60–75% was incorporated into stearic acid. Obviously, the intact fruit is capable of making stearic acid, but is prevented from doing so. Likewise, oleic and linoleic acids show low ^{14}C-acetate incorporation while they are the major acids of the intact fruit.

Mudd and Stumpf (1961) studied factors of importance in determining the kind of fatty acid synthesized by the mitochondrial system. They showed that four factors could influence the ratio of saturated to unsaturated acid formed: (1) If the mitochondria were suspended at pH 7·4, much oleic acid was formed; at pH 9 only saturated acids were synthesized; (2) At 25°C, 72% of the total acid synthesized from ^{14}C-acetate was palmitic and 27% stearic; at 46°C the distribution was 38% palmitic, 24% stearic and 38% oleic; (3) Oxygen markedly affected the acids synthesized as is shown in Table XIII; (4) A dilute suspension of mitochondria favoured oleate formation while dense suspensions produced mainly saturated acids. Both high temperature and dilute suspensions probably act by permitting a better oxygen supply in the suspension. These results suggest that oleate may be derived from the saturated acids by an aerobic conversion.

TABLE XIII. Distribution of labelled acetate in the lipids of mitochondria as influenced by oxygen

Gas phase	Acetate incorporated (m μmoles)	Distribution of labelled acetate (%) Palmitate	Stearate	Oleate
Air	11·9	53	1	46
Helium	11·8	61	39	0

Data from Mudd and Stumpf (1961).

Overath and Stumpf (1964) explored the properties of thc soluble fatty acid synthetase. They found that the synthetase could be separated easily into heat-labile and heat-stable fractions which when combined would mediate long-chain saturated fatty acid synthesis from malonyl CoA. Yang and Stumpf (1965a) showed that neither acetate nor acetyl-CoA would serve as substrate. The product was 78% triglyceride, 6% free fatty acid, 6% phospholipid and 9% acyl-CoA. Hydrolysis of the neutral lipid fraction showed the fatty acid composition to be 62% stearate and 38% palmitate. No unsaturated acid was formed.

Yang and Stumpf (1965b) demonstrated a chain elongation system in sonicated avocado mitochondria which converts 12-hydroxy-9-octadecenoyl-CoA+acetyl-CoA to 14-hydroxy-11-eicosenoyl-CoA+CoA. The physiological role of this reaction is not clear since neither the substrate nor the product is found in avocados and since stearic, palmitic or myristic acids do not serve as substrates. It is possible that this system may be involved in the insertion of double bonds in some still unknown way. Simoni *et al.* (1967) suggested that it is probable that aerobic desaturation of stearyl-ACP gives rise to oleic acid. Hawke and Stumpf (1965) have shown that the anaerobic pathway for synthesis of double bonds does not exist in higher plant tissue.

V. ENZYMES

Chase (1921–22) provided the first reports of enzymatic activities in avocado fruit. He demonstrated catalase (E.C. 1.11.1.6 hydrogen peroxide: hydrogen peroxide oxidoreductase), peroxidase (E.C. 1.11.1.7 donor: hydrogen-peroxide oxidoreductase), and emulsin (E.C. 3.2.1.21 β-D-glucoside glucohydrolase). Lipase (E.C. 3.1.1.3 glycerol-ester hydrolase) could not be demonstrated by Chase. D. Appleman (personal communication) has searched for evidence of a fat degrading system and found none.

Polyphenol oxidase (E.C. 1.10.3.1 o-diphenol:oxygen oxidoreductase) has been purified from avocado fruit by Knapp (1965) by acetone precipitation and ammonium sulphate fractionation. The pH optimum was 4·8 with 50% of the activity being apparent at pH 4·0 or 5·3. The enzyme was most active with catechol. Substrate preference is shown in Table XIV. Resorcinol acted as a competitive inhibitor of catechol oxidation while hydroquinone inhibited non-competitively. Copper binding agents inhibited the enzyme strongly, 1 mm diethyldithiocarbamate inhibiting the oxidation of catechol by 95%.

TABLE XIV. Substrate specificity of polyphenol oxidase

Substrate	Relative reaction rate
Catechol	100
Chlorogenic acid	33
Caffeic acid	33
Catechin	81
Dihydroxyphenylalanine	12
Quercetin	6
Hydroquinone	0
Resorcinol	0

Data from Knapp (1965).

Makower and Schwimmer (1957) studied the darkening of avocado slices, presumably due to a polyphenol oxidase. They observed that addition of dihydroxyphenylalanine had no effect on either the rate of formation or the final colour developed on the surface of the slice. This is in harmony with Knapp's (1965) finding that dihydroxyphenylalanine was a poor substrate for the enzyme. Makower and Schwimmer observed that addition of solutions of ascorbic acid or ATP delayed the initial rate of darkening of the slice but had little effect on the final colour after some hours. The mechanism of action of ATP is not clear, but Makower and Schwimmer (1954) suggest it is not a direct inhibition of competition, but related to mitochondrial activity. Addition of the uncoupling agent, 2,4-dinitrophenol prevented the action of

ATP while AMP, ADP or PP were not effective in delaying the rate of browning when potato was used as the test tissue.

Anderson *et al.* (1952) demonstrated glucose-6-phosphate dehydrogenase (E.C. 1.1.1.49 D-glucose-6-phosphate:NADP oxidoreductase) and malic enzyme (E.C. 1.1.1.40 L-malate:NADP oxidoreductase (decarboxylating)), linked to the reduction of NADP. They further demonstrated an enzyme which would catalyse the reduction of oxidized glutathione linked to NADPH.

Fat-synthesizing enzymes are very active in avocado mesocarp tissue. Two have been purified. ACP was isolated and purified by Simoni *et al.* (1967) from both Fuerte and Hass avocado fruit mesocarp. The protein had a molecular weight of 11,400 which is larger than the ACPs of spinach leaves or *E. coli* (9800 and 9900 respectively). The amino acid composition was determined and shown to be different from both the spinach and *E. coli* proteins. The avocado and spinach ACPs were similar in their reaction with fatty acid synthetase of spinach in making 20% palmitic and 80% stearic acid.

Galliard and Stumpf (1968) isolated a thiokinase from Fuerte avocado tissue which is useful in the preparation of long-chain S-acyl co-enzyme A.

VI. THE GAS EXCHANGE DURING RESPIRATION

A. The Climacteric in the Intact Fruit

Physiologists dealing with tropical and subtropical fruits of high metabolic activity are particularly cognizant of the contributions made by the early investigators working with temperate zone fruits. It was the apple with its relatively low rate of respiration in which Kidd and West (1925) observed the upsurge in CO_2 evolution and coined the term "climacteric" rise. They realized its importance as a transition stage between development and senescence. For the avocado, credit should be given to Wardlaw and Leonard (1935) who reported a rise in CO_2 production from 40 mg/kg/hour at the pre-climacteric stage to 170 mg/kg/hour during the peak at 21°C. They observed that the fruit softening followed the maximum respiratory activity and concluded, therefore, that ripening is not related to the climacteric. Biale (1941) viewed this matter differently. He noted that softening was associated with the climacteric because any treatment which suppressed this respiratory pattern resulted in suppression of softening. More detailed studies on single fruits have shown noticeable changes in the texture of the mesocarp during the respiratory rise though the most desirable edible condition is reached only some 1–4 days after the peak in air at 15°C. At 20°C the coincidence between softening and climacteric peak is closer than at 15°C.

The temperature of 15°C has been used in most of the studies on intact fruit conducted at this laboratory (Biale, 1960a, b; Biale and Young, 1962). At this temperature the rate of respiration is sufficiently slow to observe the characteristic decline to a pre-climacteric minimum following harvesting. At 15°C the metabolic activity is high enough to exhibit a 100–200% change between the minimum and the peak. The pre-climacteric and the climacteric peak values appear to be characteristic for a given species under a given set of conditions. On a comparative basis the avocado is more active than all the temperate zone fruits and most tropical fruits for which information is available. The rate in ml O_2-uptake or CO_2-output was calculated (Biale, 1960b) to be 0·91/100 mg protein N and 41/100 mg phosphorus. Since the avocado is relatively high in these constituents the differences with other fruits are not as striking as when expressed per unit fresh or dry weight. For example, the rate of oxygen uptake or CO_2 evolution is two and one-half times higher in avocado than in the mango on a wet weight basis, while it is nearly the same on a protein basis. Differences between varieties or races might also be a function of method of calculation. For studies in which the objective is to observe the response of a given sample, the universally used kg/hour as the denominator is satisfactory.

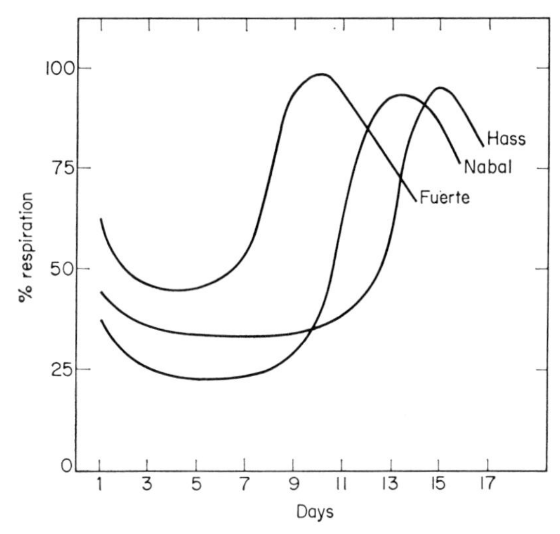

FIG. 9. The climacteric pattern in three varieties of avocado.

The climacteric pattern of respiration for three varieties of avocado will be seen from Fig. 9. The significant feature common to all varieties is the decline following harvesting and the rapid rise to peak values of about the same magnitude. While the lag period, that is the time between harvest and the onset of the rise, varies from approximately 7 days for the Fuerte to 13 days for the Hass, this is not a specific varietal characteristic but rather a reflection of the state of maturity. Since avocados can be stored on the tree, considerable variations in the state of maturity at time of harvest may be encountered. Early season fruit tends to have a long lag period. Uniformity in a composite sample is another matter deserving of close scrutiny. The lag period has been observed to vary from 25 to 16 days in an experiment with individual fruits of the Nabal variety. In a composite sample it is likely that fruits of an irregular bloom ("off bloom") as well as those of a regular bloom will be included, leading to respiration values which deviate from those characteristic of the species.

Changes in pectins are among the most ubiquitous transformations in ripening fruit. A close correlation between disappearance of protopectin, formation of soluble pectins and the rise in respiration was demonstrated by Dolendo *et al.* (1966) for the MacArthur avocado. Their results are reproduced in Fig. 10.

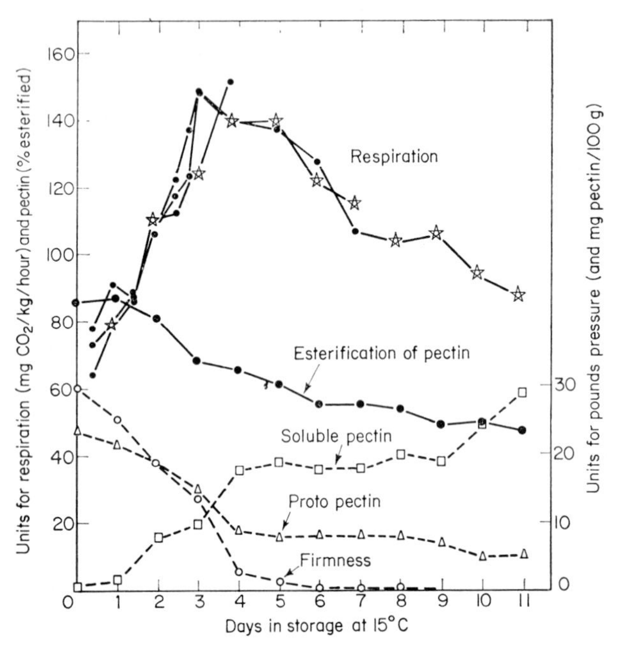

FIG. 10. Pectic changes in the avocado in relation to the climacteric pattern (Dolendo *et al.*, 1966).

B. The Internal Atmosphere and Gas Diffusion

Attempts were made to understand the physiological behaviour of the avocado by following the composition of the internal atmosphere in relation to the respiratory rise. Ben-Yehoshua *et al.* (1963) observed that in freshly picked Hass fruit placed at 20°C, the oxygen content inside the fruit was 15–19% and the CO_2 concentration 1–3%. A drop in oxygen and increase in CO_2 levels occurred with the onset of the climacteric. At the peak of the climacteric oxygen dropped to 5–10% while CO_2 rose to 5–10%. In the post-climacteric stage the O_2 concentration fell to 2% and the CO_2 level showed a plateau at about 13%. In general, the trends were similar at 15°C, though the sharp drop in oxygen and rise in CO_2 took place some 2 to 4 days after the climacteric peak. The change in relationship between respiratory rate and the intercellular composition suggested that in the course of softening changes occur in the resistance of the tissue to gaseous diffusion.

The resistance of fruit to diffusion calculated according to the formulae of Trout *et al.* (1942) is related directly to the volumetric concentration of the gas and inversely to the respiration rate. For the avocado the trends for resistance to gaseous flow were the same for O_2 and CO_2. Prior to the climacteric peak these values were low and fairly constant, but with softening a three- to ten-fold increase in resistance ensued. Similar behaviour was noticed in peeled fruit except for the marked reduction in the lag period. Peeling and placing the fruit at 15°C hastened the onset of the climacteric rise by some 10 to 12 days. The trend in composition of the internal atmosphere, which was not materially different in peeled than unpeeled fruit, was ascribed by Ben-Yehoshua (1964) to the formation of a new periderm after peeling. The general conclusion from these studies was that improvement in gaseous exchange accelerated but did not induce the ripening process.

Burg and Burg (1965b) working with the Choquette avocado, a Florida variety, also observed changes in resistance to CO_2 flow during the climacteric. They noted that such trends do not invalidate Fick's law of diffusion but indicate rather that, with ageing, changes occur either in the diffusion coefficient, the effective thickness of the barrier to diffusion or in that fraction of the surface area through which gas exchange takes place. Burg and Burg (1965b) also confirmed the finding of Ben Yehoshua *et al.* (1963) that a substantial CO_2 gradient existed across the peel of the avocado, that the peel is an important barrier to gas diffusion, and that the pulp also restricts the movement of gases. The avocado was found to differ from the apple with its relatively porous pulp, and from the tomato and pepper in which the opening around the pedicel contributes substantially to gas flow.

C. The Climacteric in Slices of Tissue

The most extensive studies on the climacteric phenomenon in the avocado has been carried out with detached intact fruit but some experiments have been done on fruit parts and tissue discs. Earlier work (Millerd *et al.*, 1953) with slices of Fuerte avocado taken from fruit at four different stages of ripeness (immediately after harvesting, at the climacteric minimum, on the rise, and at the peak) have shown no significant trends in oxygen uptake. The measurements were made in Warburg vessels over a period of 1 or 2 hours in phosphate buffer, pH 5. It has been shown since by Ben-Yehoshua (1964) that the tissue discs have to be maintained in moist air and not in a solution in order to demonstrate a respiratory rise. Under such conditions the rate of oxidation of tissue slices from Hass avocado increased from about 100 to 190 ml O_2/kg/hour. These values were about twice as high as corresponding rates for intact fruit, but conformed rather closely to the results obtained with fruit components such as peel and pulp. Softening of the slices at the peak of respiration was cited as evidence for the occurrence of the climacteric in this material. Additional criteria must be applied to determine the relationship between gas exchange and ripening in tissue slices of the avocado, as Palmer and McGlasson (1968) have done for banana slices.

VII. REGULATION OF RESPIRATORY RATES

A. Temperature Effects

The climacteric rise of respiration is a suitable pattern for studying the response of the fruit to changes in atmospheric conditions such as temperature, gaseous composition, ionizing radiations, etc. The relatively constant climacteric maxima supply guidelines for comparisons.

TABLE XV. Comparative effects of temperature on Fuerte avocado respiration

5°	7·5°	10°	15°	20°	25°	30°C
2	11	21	41	63	100	64

The rate of respiration at 25°C was 180 ml O_2/kg/hour and was equated to 100. All values were for the climacteric peak except for 5°C which was an average for several readings. No climacteric rise was observed at 5°C (Pratt and Biale, 1944).

It has been shown by Biale and Young (1962) that relatively small changes in temperature have a marked effect on the slope of the climacteric rise, on the magnitude of the peak and on the lag phase, the period between picking and the onset of the rise. The occurrence of the peak is accelerated by 1 or 2 days at 25°C compared to 20°C and delayed at 7·5°C. The respiratory rates at different temperatures are given in Table XV. At 30°C the climacteric cycle was found to deviate considerably from the expected pattern. The initial and the pre-climacteric rates were higher than at 25°C but the maximum was lower by about 40%. The minimum to peak increase was 30% at 30°C as compared to 250% for 25°C. This deviation in the climacteric was reflected also in the ripening process as seen in Fig. 11. The characteristic feature of high temperatures (25 and 20°C) is rapid and uniform softening. The non-uniformity in a composite sample is counteracted by interactions such as the ethylene produced by ripe avocados accelerating the ripening of the less advanced fruit. For Florida varieties, Hatton *et al.* (1965) reported 15·5°C as the most desirable ripening temperature from the standpoint of appearance and quality.

A suppression of the climacteric pattern was observed also at 5°C. No detectable rise could be discerned on account of the low rates of about 10 ml O_2/kg/hour. However, some physiological changes which are associated with the climacteric do occur even at 5°C. Of special significance is the production of ethylene as observed by the triple response of pea seedlings (Pratt and Biale, 1944). This was noted at 5°C some 6 to 8 weeks after picking. Though the fruits tend to soften at this temperature the ripening process is not as desirable as at the higher temperatures. The skin shows dark discoloration and the colour and texture of the flesh are abnormal. These symptoms

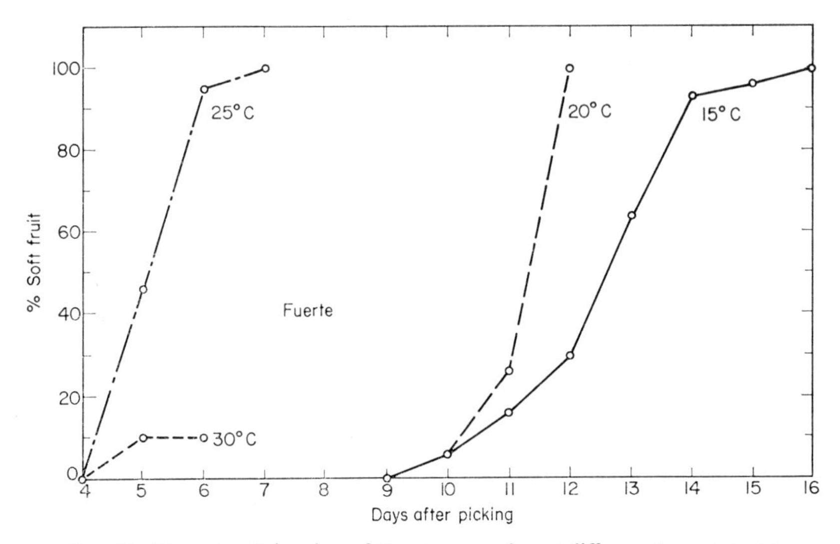

FIG. 11. The rate of ripening of Fuerte avocados at different temperatures.

indicative of chilling injury show up in the avocado at a considerably higher temperature than in many temperate zone fruits, but lower than in some fruits of the tropics. A physiological criterion may be applied to the choice of proper storage conditions of the avocado. It consists of determining the pattern of respiration at fairly high temperatures such as 15°C after exposure to low temperatures which suppress the climacteric. If upon transfer to 15°C a rise followed by a decline is observed it may be concluded that the fruits were physiologically pre-climacteric at the time of transfer. This was found to be the case with Fuerte avocados placed at 15°C (Biale, 1941) after 3, 4 and 5 weeks at 4–5°C. However, after 2 months at this temperature a declining rate of respiration rather than typical climacteric pattern was recorded and the fruit was definitely inferior in quality.

Hatton *et al.* (1965) described the chilling injuries occurring in some Florida varieties stored for a prolonged period at 40°F and in others at 50°F. The most common symptom is a greyish brown discoloration of the mesocarp which frequently shows up first in the vascular bundles. In severe cases the entire flesh discolours, undesirable flavour develops, the skin darkens and normal ripening is arrested. At times the avocados may have a satisfactory appearance while at the low temperature, but the chilling symptoms appear upon transfer to a higher temperature. For the avocado, as for many other fruits, temperature limitations have directed the attention of investigators to alterations in the gaseous atmosphere as a possible solution to the storage problem.

B. Oxygen Tension

The avocado occupies a unique place among fruits so far as the role of oxygen is concerned. The intact fruit is highly sensitive to anaerobic conditions unlike other fruits which are able to switch to a fermentative metabolism when deprived of oxygen. Studies on the Fuerte avocado by Biale (1946) have shown that upon transfer from air to nitrogen the fruit exhibited a temporary rise for half a day in CO_2 evolution followed by a sharp decline. At 15°C the rate levelled off to a value of 5 mg CO_2/kg/hour, which was about 3% of the maximal rate in air. Both the climacteric and ripening were totally suppressed. Upon transfer from anaerobic conditions to air a threefold increase in CO_2 production ensued, but the fruit discoloured rapidly, was attacked by fungi and never softened.

When a range of oxygen tensions from 0·5 to 100% was investigated, the onset of the climacteric in Fuerte avocados was delayed and the rate of respiration diminished, especially in oxygen atmospheres below air. It appears that the O_2 content of air was sufficiently high for maximal metabolic activity. Raising the oxygen levels to 50 and 100% resulted in slight or no delay in the lag period or in a change of the slope during the climacteric rise. The responses of the Nabal variety to oxygen are illustrated in Fig. 12. Here, too, we must note a departure in the avocado from other fruits. In citrus and in temperate zone fruits a threshold oxygen level was found above and below which the rate of CO_2 output increases. In the avocado no such critical oxygen concentration was detected, suggesting that the nature of the metabolism is unchanged by oxygen depletion.

The effect of oxygen tension was followed in tissue slices in which diffusion is not as much of a limiting factor as in the intact fruit. Slices 1 cm in diameter and 1 mm thick were taken from fruit at different stages of the climacteric and oxygen uptake determined manometrically. The slices were placed in 3 ml phosphate buffer at pH 5·3. The respiration measured under these conditions for a period of 1 to 2 hours was linear and was a function of

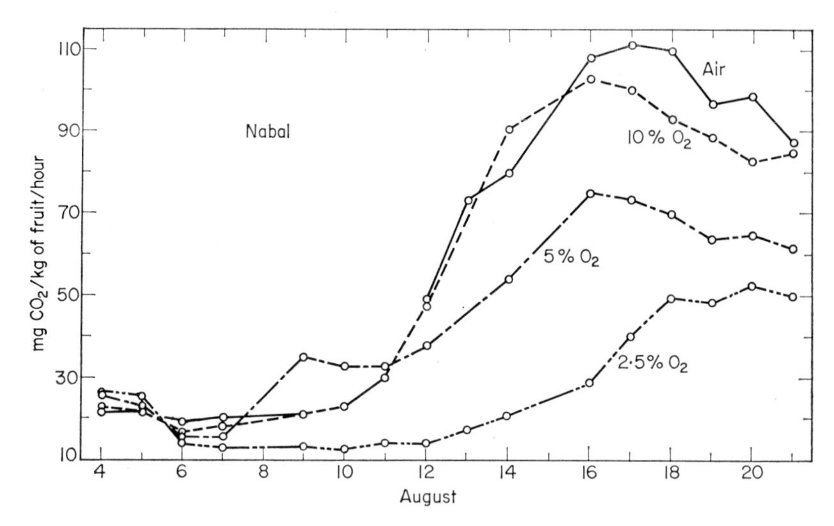

FIG. 12. The effects of oxygen concentration on CO_2 evolution of Nabal avocados.

oxygen tension at oxygen levels below 10%. A higher level of oxygen was required for maximal activity in slices of ripe fruit than in those of the other two stages as seen in Table XVI. In general the pattern of response to aerobiosis was similar in the slice as in the intact, detached fruit.

TABLE XVI. Relative respiration of tissue slices of Fuerte avocado in relation to oxygen tension

				% Oxygen			
	Air	9·3	7·4	5·4	2·4	1·6	0·7
At harvest	100	100	103	80	54		16
Pre-climacteric minimum	100	101	92	74	48	37	16
Climacteric peak	100	74	65	52	37		

The respiration rates in μlitre O_2 per g fr. wt. in air were 154, 147 and 231 for slices taken from fruit at harvest, at minimum and at climacteric maximum, respectively.

C. Effects of Changes in CO_2 Concentration

The response of avocados to alterations in CO_2 levels appears to be generally similar to the response of other fruit. In one study (Young *et al.*, 1962) the effects of 5 and 10% CO_2 in air and in 10% O_2 were followed. When the CO_2 concentration was 5% but the oxygen level was that of air (21%), the pre-climacteric rate was affected only slightly and the onset of the rise was delayed by 3 days only. However, the peak rate was 40% below that of air and was attained after 21 days compared to 8 days for the control. When the gas mixture consisted of 10% CO_2 and 21% O_2 the rate of oxygen uptake was low and constant for more than 3 weeks at 15°C. Upon transfer to air there was an increase in respiration, indicating that no climacteric had taken place under the 10% CO_2 treatment. The response of avocados to CO_2 in 10% O_2 was qualitatively the same as at the higher oxygen level but the magnitude of the respiratory rates was reduced. In all cases normal ripening ensued upon transfer to air. The longest storage period of 45 days at 15°C was attained by an atmosphere of 5% oxygen and 10% CO_2. Raising the CO_2 content to 15% resulted in some injury.

The experimental work on modified air storage (also referred to as "gas storage" or controlled atmosphere storage) has not been as extensive with the avocado as with some of the temperate zone fruits, notably the apple. However, certain conclusions can be drawn from the limited studies. To prolong storage it is essential to lower the oxygen to 5%. This level allows for a margin of error in case the content drops to 3%. A very marked retardation of the ripening process will result from this treatment. Additional benefit will accrue from a rise in CO_2 level to 10% but not higher. Lowering the partial pressure of ethylene by ventilation or, preferably, removal by adsorption is essential for best results.

VIII. PRODUCTION AND ROLE OF ETHYLENE

The subject of ethylene as related to the physiology of the avocado has been discussed within the past decade in the reviews of Biale (1960b, 1964), Biale and Young (1962), Burg (1962), Burg and Burg (1965a) and Pratt and Goeschl (1969).

A. Identification of Ethylene

Denny and Miller (1935) supplied the first evidence of a volatile product of avocado, which, like ethylene, caused epinasty in potato plants. Pratt and Biale (1944) observed the triple response (stem thickening, reduced growth in length, diageotropism) in pea seedlings subjected to air passed first over avocados. They noted that at 15 and 25°C the triple response was most pronounced during the climacteric rise. At 5°C no rise in CO_2 evolution was evident, but the effect on the seedlings was seen some 6 weeks later than in the case of fruit kept at 15°C. These observations suggest that, normally, ethylene evolution in harvested fruits is associated with the climacteric, but that the respiratory rise is not a requirement for its manifestation.

The identification of the epinasty producing emanation as ethylene was accomplished by Pratt *et al.* (1948). A quantitative manometric assay capable of detecting the production of 0·2 μlitre of ethylene per hour was developed by Young *et al.* (1952) and was used for a study of the relationship of ethylene evolution to respiration by Biale *et al.* (1954). By the use of this procedure it was shown that the peak of ethylene evolution coincided with the peak of respiration and that rates of production in excess of 0·2 μlitre per hour could not be detected before the initiation of the climacteric. With a much more sensitive technique Young (1965) demonstrated that the peak of ethylene production preceded the respiratory peak by 24 hours. He has also shown that the rate of 0·2 μlitre per hour is exceeded for the first time at or a few hours after the induction of climacteric rise (Fig. 14(a)).

B. Ethylene as Ripening Hormone

The role of ethylene as the ripening hormone of fruit has been the subject of considerable controversy. Biale *et al.* (1954) advanced the hypothesis that "native ethylene is a product of the ripening process rather than a causal agent". They found by the manometric technique of Young *et al.* (1952) that in some fruits the first measurable ethylene followed the rise in respiration by 3 days. In the avocado the increase in CO_2 evolution coincided with ethylene production within the limits of the determination.

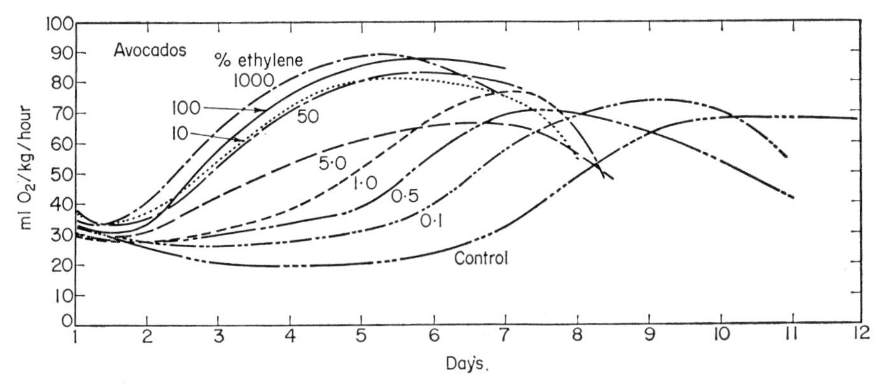

FIG. 13. Ethylene concentration effects on the climacteric pattern.

Ethylene production of Choquette avocados was studied by Burg and Burg (1962b, 1965a). They measured ethylene in the internal atmosphere and found that a significant amount of ethylene, of the order of 0·1 p.p.m., existed in pre-climacteric fruit and concluded that this was sufficient to initiate the rise. This concentration was found some 2 hours after harvest and increased to 0·3–0·6 p.p.m. in 10 hours. At that time a steady state of ethylene production prevailed and existed for 30–48 hours at which time both ethylene and CO_2 production increased simultaneously. The maximum ethylene production of 300–700 p.p.m. coincided with the climacteric peak. The assumption of Burg and Burg (1962b) that 0·1 p.p.m. of ethylene was sufficient to induce the climacteric was not borne out for the Fuerte avocado as deduced from the curves presented by Biale (1960b) and as seen in Fig. 13.

When 0·1 p.p.m. of ethylene was applied in the external atmosphere, respiration was accelerated slightly within 2 days of application. It took some 4–5 days for 0·1 p.p.m., and at least 2 days for 1 p.p.m., to initiate the climacteric rise in Fuerte avocados. Treatment with 10 p.p.m. brought about an immediate rise in respiration. Apparently a relatively high concentration of ethylene is required for induction without a lag period.

Additional evidence for the importance of ethylene in the internal atmosphere as a factor in inducing the climacteric was cited by Burg and Burg (1966) from low pressure experiments. They reduced the pressure around Pollock avocado fruit to one-fifth of an atmosphere of pure oxygen, making the partial pressure of oxygen the same as that of air at one atmosphere. The ripening time was doubled, suggesting that something other than oxygen depletion is involved is the delay of ripening. This factor may be the decrease in ethylene level due to the low pressure.

Some fruits are attached to the tree for a long time in an unripe condition but ripen in a few days when harvested. Fuerte avocados, for example, mature as early as November, but may be left on the tree until the following June or July. During this period, they are held in the pre-climacteric state with the only apparent change being some increase in size and fat content. Biale (1960b) noted that fruit picked early in the season required higher ethylene levels to induce a rapid respiratory rise than did late season fruit, and that the time from picking to the climacteric decreased as a log function of the applied ethylene concentration. Burg and Burg (1962a, 1964) have shown by girdling experiments that both avocado and mango fruit will not ripen on the tree if attached to a branch having functional leaves. When the fruits were detached, mangoes, like avocados, reached the climacteric sooner as more ethylene was applied up to a value between 4 and 40 p.p.m. The response was again a log function of the ethylene concentration. These data suggest that the leaves of the tree supply a hormone to the fruit which prevents ripening and that the effect of this hormone can be counteracted by application of ethylene or by allowing sufficient time to pass for the hormone level to be depleted.

C. Ethylene Incorporation

Studies endeavouring to trace the fate of ethylene applied to avocado fruit have been carried out by Buhler *et al.* (1957). They treated ripe avocados with 0·2 mc ^{14}C-ethylene at a concentration of 1000 p.p.m. for several days and found that 0·05% of the amount supplied was absorbed and that one-half of the ^{14}C fixed was in the organic acid fraction. Fumaric and succinic acids were separated, purified and degraded by a procedure which distinguishes between the carboxyl and the methylene groups. Three-fold more activity was detected in the carboxyl group, suggesting that ethylene was incorporated

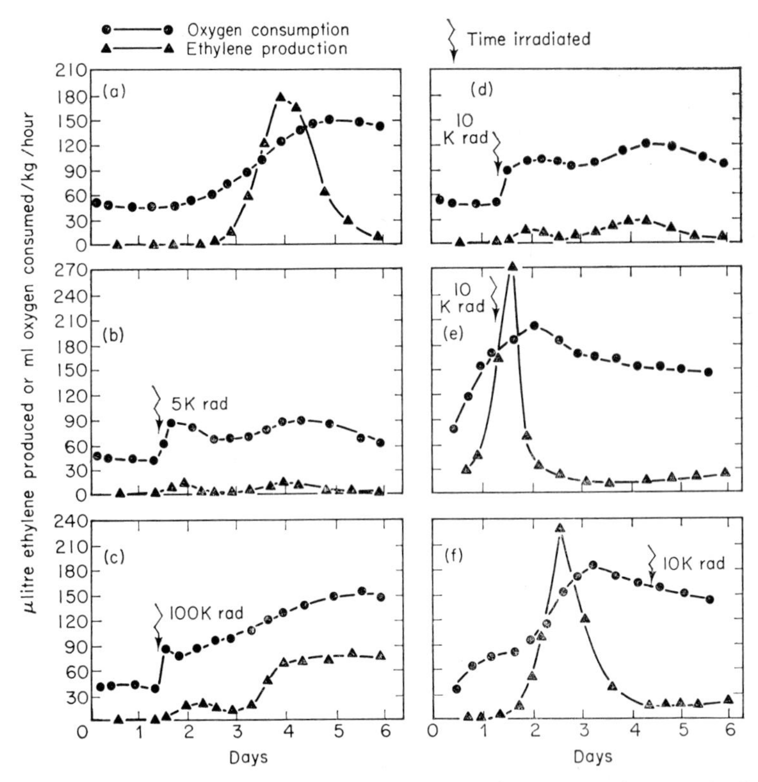

FIG. 14. Respiration as measured by oxygen consumption and ethylene production of individual Fuerte avocado fruit: (a) not irradiated; (b) irradiated in the pre-climacteric phase with 5 krads; (c) irradiated in the pre-climacteric phase with 100 krads; (d) irradiated in the pre-climacteric phase with 10 krads; (e) irradiated on the climacteric rise with 10 krads; (f) irradiated in the post-climacteric phase with 10 krads (from Young, 1965).

into an unsymmetrical intermediate in such a way that randomization had occurred preferentially in one of the carbon atoms of the ethylene skeleton. In their experiments no ethylene was converted to carbon dioxide.

Jansen (1964) treated avocados with 20 mc of ^{14}C-ethylene for 4 hours. He found some activity in CO_2, some in organic acids and 4% of the absorbed activity in benzene and toluene. Experiments with tritium-labelled ethylene (Jansen 1963) showed tritium incorporation in the methyl group of toluene. The mechanism of incorporation into benzene and toluene is unknown. The differences in the results between the two groups of workers may be ascribed to the application of labelled ethylene at different stages of the ethylene production cycle or to the differences in period of treatment.

D. Irradiation Effects

Young (1965) studied the effect of gamma radiation on respiration, ethylene production and ripening of Fuerte avocados. Ethylene evolution and respiration of untreated fruit are presented in Fig. 14(a). The effect of irradiation in the pre-climacteric phase at 5 and 10 kilorads respectively is shown in Fig. 14(b, d). This low dose caused an immediate doubling of respiration and a small ethylene production. The climacteric occurred between days 4 and 5 just as in the control fruit, but with only slight ethylene production. The high dose of 100 kilorads shown in Fig. 14(c) also caused a doubling of respiration and a small ethylene production, but severe injury resulted and the fruit did not ripen. Irradiation after the climacteric was initiated (Fig. 14(e)) or in the post-climacteric phase (Fig. 14(f)) was without any effect on ethylene production or respiration, nor was there any effect on the appearance or quality of the fruit. Figures 14(b) and 14(d) indicate that a substantial amount of ethylene can exist in the fruit without an immediate induction of the climacteric. In both cases the production rate is in excess of 10 μlitre/kg/hour which, according to the conversion factor of Burg and Burg (1962a) of 2·2, would be equivalent to a concentration of at least 22 p.p.m. of ethylene in the internal atmosphere. At the time of the "pseudo" climacteric, the fruit produced only 10–15% of the normal amount of ethylene. Thus radiation has somehow inhibited the ability of the fruit to produce ethylene, but allowed the fruit to ripen in the normal way, so long as the radiation dose was low.

IX. METABOLIC PATHWAYS

A. The Electron Transport Chain

The sequence of enzymes responsible for the transfer of electrons from substrate to oxygen has been established for animal systems but is not fully elucidated for plant materials. The diversity of oxidizable substrates, the unconventional response to inhibitors and the presence of phenolases have contributed to the lack of definitive knowledge in plant systems. The problem is compounded with fruits because of these considerations, because of the relatively slow rates of oxidation, and because of the difficulties encountered in extracting active enzymes from tissue with low pH. The avocado is a notable exception due to the rather high pH of the flesh and due to the ease of obtaining measurable and reproducible metabolic activities.

1. *Spectrophotometric evidence*

The elucidation of metabolic pathways requires evidence for the existence of the appropriate enzymes as well as a demonstration of their activities commensurate with physiological processes. In the case of the avocado, Chow and Biale (1957) have shown that particles obtained by procedure A (see Table XVIII) were able to reduce oxidized cytochrome c upon addition of NADH. A particulate enzyme also caused a pronounced decrease in optical density at 550 mμ when reduced cytochrome c was supplied. The absorption spectrum of particles incubated in nitrogen was identical with that of reduced cytochrome c. The existence of the oxidase was demonstrated by an assay in which ascorbic acid served as electron donor, cytochrome c as carrier, particles as oxidase and DIECA as inhibitor of phenolases. It was also shown that NAD can act as intermediate carrier in the avocado particulate system.

2. *Electron chain inhibitors of tissue oxidation*

The question as to the physiological operation of the electron transport chain was studied by Lips and Biale (1966) in tissue discs of the Hass and Fuerte varieties of the avocado. When amytal (amobarbital) cyanide and azide, respiratory chain inhibitors, were added individually, the oxygen uptake was actually stimulated in slices of pre-climacteric and, to a lesser extent, of climacteric peak fruit. A stimulation of 80% was noted with 5 mM azide. The tissue was able to esterify ^{32}P in the presence of either azide or amytal. The stimulation was reversed and ^{32}P esterification markedly lowered when the two inhibitors were applied together. Amytal is known to block the transfer of electrons from substrate to cytochrome b via NAD, while CN^- and N_3^- prevent the acceptance of electrons by oxygen because of the inhibition of cytochrome oxidase (cytochromes a–a_3). On the basis of these results the hypothesis was advanced that the electron transfer system of the avocado fruit consists of a phosphorylative cytochrome chain linked to a non-phosphorylative pathway. Electrons may be transferred from one of the pathways to the other. The phosphorylating ability of the tissue disappears when all key enzymes of the cytochrome chain are blocked by two inhibitors. The stimulation of the endogenous oxidation by individual inhibitors was explained as the result of the increased supply of limiting phosphorylative co-factor to the unblocked sites. The presence of branches in the chain permits the bypass of the inhibited site. The fact that inhibitors are required to force electrons into the alternate pathway can be explained by differences in enzyme affinities at the branching point. While these ideas explain the unique observations on the avocado they require substantiation by the reconstruction of events in isolated systems.

3. *Terminal oxidase inhibition of mitochondrial oxidations*

The results of recent studies (Biale, 1969) on the effects of inhibitors on the oxidative activities of mitochondrial preparations of Fuerte avocados appear to be in contradiction with the tissue slice work of Lips and Biale (1966) as revealed by Table XVII. The terminal oxidase was blocked by cyanide in both states 3 and 4 and in both stages of ripeness, though the effect on climacteric peak particles was not so pronounced as on those of unripe fruit. Azide at the level of 1 mM in the assay medium also caused greater inhibition of state 3 than of state 4 oxidation. In no instance was a stimulation of oxygen uptake recorded as in the case of the tissue slice. The discrepancies may be reconciled when more is known about the endogenous substrate oxidized in the tissue slice, and when the relative contributions of the soluble and the particulate systems to the overall oxygen uptake in the intact cell have been determined.

TABLE XVII. Effect of 1 mM KCN on oxygen uptake by Fuerte avocado mitochondria[a]

	Preclimacteric				Climacteric			
	Control		KCN		Control		KCN	
Substrate	S3	S4	S3	S4	S3	S4	S3	S4
Succinate	1200	560	100	110	1600	570	560	270
Malate	570	340	170	125	1040	420	330	260

[a] μlitres O_2 per mg protein N per hour.
S3 and S4 refer to state 3 and state 4 oxidations in accordance with the definitions of Chance and Williams (1955).
The data are based on studies by Biale (1969).

B. The Tricarboxylic Acid Cycle

1. *Preparation of particulate fractions from cytoplasm*

The early attempts to test for the operation of the TCA in tissue slices and in homogenates of the avocado fruit were beset with difficulties due to high endogenous rates of oxidation in the former and low activities in the latter. However, with the increased knowledge in the isolation of cellular organelles, it has been possible to obtain metabolically active mitochondria. The avocado was among the first plant materials to yield a particulate fraction with the capacity to oxidize organic acids and to generate ATP through oxidative phosphorylation.

During the period 1953 to 1968 the methodology has evolved so that it is now possible to obtain particles showing high degrees of purity, high rates of oxidation and reasonably high respiratory control. Nearly all the available information is based on one of the two preparative procedures outlined in Table XVIII or on slight modifications of these procedures.

The two procedures have in common the requirement for rapid isolation at low temperatures of close to 0°C. They differ, however, in a number of important features. The sucrose concentration was raised in the currently

TABLE XVIII. Comparison of two procedures used in the isolation of avocado mitochondria

Used in studies reported	Procedure A 1953–62	Procedure B 1964–present
Grinding tools	Blender, mortar	mortar
Grinding medium		
Amount	300 ml	240 ml
Sucrose	0·25 M or 0·5 M	0·4 M
Phosphate	0·05 M	
BSA		0·75 mg/ml
Cysteine		4 mM
$MgCl_2$		1 mM
KCl		10 mM
Tris		50 mM
EDTA		10 mM
pH	7·2	8·1
Dilution medium		
Amount		160 ml
Sucrose		0·4 M
KCl		10 mM
BSA		0·75 mg/ml
Resuspension medium		
Amount	200 ml	320 ml
Sucrose	0·5 M	0·4 M
KCl		10 mM
BSA		0·75 mg/ml
Tris-phosphate, pH 7·2		10 mM
Centrifugation		
First	500 **g** 10 min	1500 **g** 15 min
Second	17,000 **g** 15 min	12,000 **g** 20 min
Third	17,000 **g** 15 min	2500 **g** 10 min
Fourth		10,000 **g** 15 min

used method (procedure B) though no clear evidence for increased tonicity was demonstrated. The addition of bovine serum albumin, cysteine and EDTA was found necessary when respiratory control became a criterion for activity in addition to oxidative and phosphorylative capacities. Omission of phosphate in procedure B was adopted to minimize tendency for uncoupling. In the earlier procedure the grinding was done in a Waring blender at low speed for 1 minute. A change in speed of blending caused by raising the voltage from 46 to 60 V lowered the succino-oxidase activity markedly and suppressed completely the oxidation of α-ketoglutarate. In procedure B the tissue is first passed through a stainless steel grater and dropped directly into a mortar containing the grinding medium. The use of sand or any other abrasive was found to be detrimental. While grinding with a pestle for 5 min the pH was maintained at 7·6. The centrifugation scheme introduced in procedure B accomplished the removal of mitochondrial aggregates through spinning at 2500 **g** following the washing of the pellet obtained by 12,000 **g** with the resuspension medium. In both procedures the final pellet was suspended with a small volume of either the preparative medium (A) or the washing medium (B). The objective was to have a mitochondrial suspension with a relatively high protein content of 1–2 mg nitrogen/ml.

The method of measuring rates of oxidation has also changed. Warburg manometry of the early period has been replaced by the Clark oxygen electrode which gives the opportunity of determining initial rates immediately following the addition of substrate, co-factors and inhibitors. By means of the electrode it has been more feasible to determine oxidative activity not only in the absence of phosphate acceptor but also upon its depletion. The credit for supplying evidence for the operation of key metabolic pathways is shared by both procedures.

Baker *et al.* (1968) observed that mitochondria prepared by the regular procedure of differential centrifugation contained a complex mixture of chloroplast fragments, oil droplets and other cellular debris. Much of this contamination was removed by continuous or discontinuous density gradient preparations as can be seen in the electron micrograph presented in Fig. 15.

Mitochondria from pre-climacteric avocados accumulated in the 1·2 M sucrose layer of a discontinuous gradient of 0·4–1·6 M, while continuous gradients of 0·4–1·6 M proved to be more satisfactory for the general separation of particles throughout all stages of ripeness. The purified mitochondria appeared quite similar to those of the regular preparations when isolated in 0·4 M sucrose. They contained a somewhat dense continuous matrix, triple layered inner and outer membranes and dilated cristae. Respiratory control was obtained with succinate as substrate but there was no improvement over regular preparations. No shifts in ultrastructural form were observed with change in metabolic state as has been reported by Hackenbrock (1966) for mammalian mitochondria.

2. *Mitochondrial oxidations*

The first indispensable evidence for the operation of the Krebs cycle is the ability of the system to oxidize pyruvate and the di- and tricarboxylic acids of the cycle. As can be seen from Table XIX, the particulate fractions obtained by both procedures are capable of carrying on oxidations of the TCA cycle acids at high rates. The comparison is made for particles obtained from avocados immediately after harvesting while higher activities for some of the acids are observed in mitochondria from fruit at the climacteric peak. The striking differences between the two procedures might be ascribed at least partially to the fact that the recorded traces of the oxygen electrode

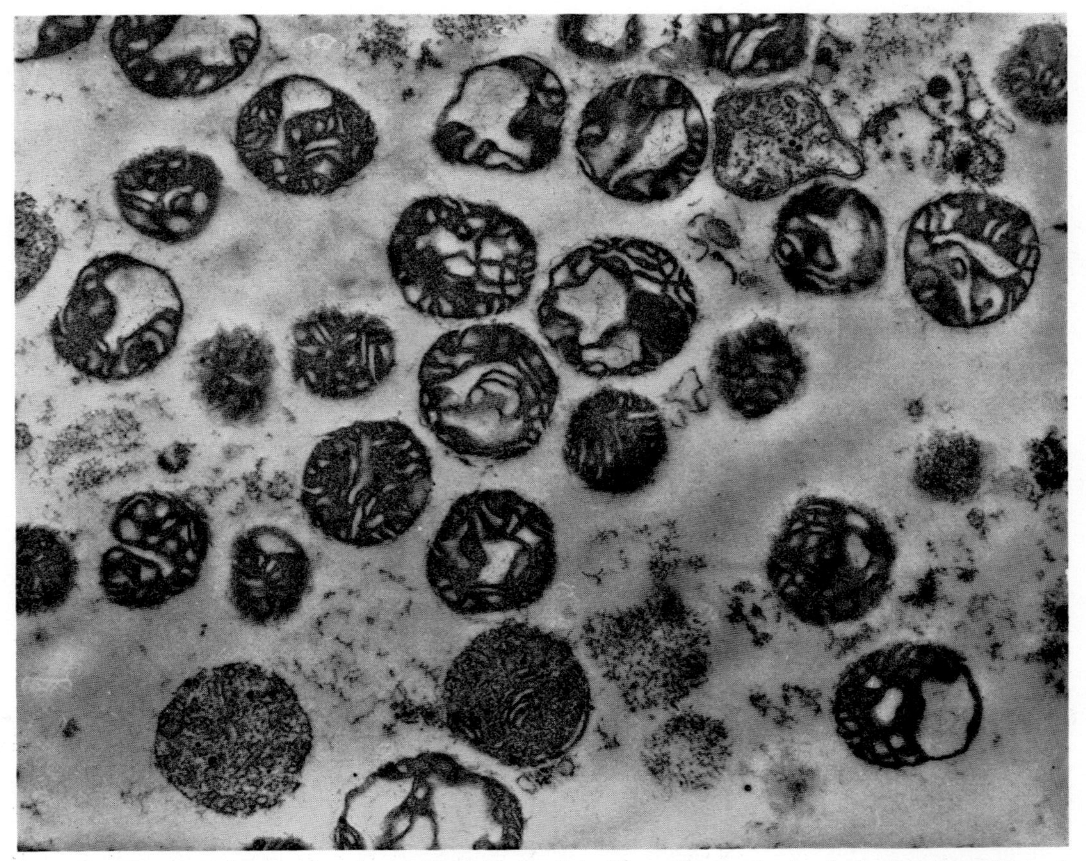

FIG. 15. Electron micrographs of avocado mitochondria from climacteric rise fruit. Fraction obtained by density gradient centrifugation. For explanations of procedure and structure see Baker *et al.* (1968).

TABLE XIX. Oxidation of Krebs cycle acids by avocado mitochondria[a]

Substrate	Procedure A[b]	Procedure B[c]
Pyruvate	334	720
Citrate	263	
Cis-aconitate	224	
α-oxoglutarate	247	660
Succinate	525	1440
Fumarate	130	
Malate	259	620

[a] μ litre O_2/mg N/hour.
[b] Based on Biale *et al.* (1957) and Avron and Biale (1957a).
[c] Based on Hobson *et al.* (1966).

(used in procedure B) are more nearly representative of initial rates than the readings of the manometric deflections of procedure A.

3. *Co-factor requirements for oxidation of Krebs cycle acids*

For maximal activity, all mitochondrial preparations of avocado mesocarp require a phosphate acceptor and magnesium ions, although the optimal concentration of Mg^{2+} may vary between the pre-climacteric and climacteric stages. This generalization holds in different degrees depending on the washing procedure in preparing the particles and on the substrate studied.

If one washing only is used (that is, a single high-speed centrifugation) there may be sufficient adenylate adhering to the particles to satisfy the need for the oxidation of succinate which appears to have the lowest requirement for phosphate acceptor. With highly washed preparations a striking difference in the co-factor requirements is observed. With α-oxoglutarate oxidation one factor omission is most pronounced if Mg^{2+} is left out. Omission of NAD alone had little effect. Cumulative omissions of co-factors during α-oxoglutarate oxidation show greater changes in rate than with citrate and succinate substrate. The response to thiaminepyrophosphate (TPP) will be discussed in greater detail in Section XA. A detailed study of pyruvate oxidation by single factor omission showed that CoA was not required, but that the elimination of a 4-carbon "sparker", such as malate, reduced the activity to less than 10% of the control. However, when the cumulative omission procedure was followed, the deletion of CoA yielded only one-half of the rate of the complete system. Here again the absolute requirement for "sparker" acid supplied substantial evidence for the condensation reaction between pyruvate or a product of pyruvate, presumably acetyl-CoA, and the 4-carbon dicarboxylic acid. The need for malate rather than oxalacetate for the formation of citrate will be explained below by a discussion of the unique reactions of the oxo acid.

4. *Detection of reaction products*

The role of the "sparker" acid becomes evident when products of the oxidation processes are identified. This was done (Avron and Biale, 1957a) by deproteinizing the reaction mixtures, applying an aliquot of the supernatant on chromatography paper, developing the chromatograms with pentanol–water–formic acid and identifying the spots. It was observed by this procedure that citrate was the product of the oxidation of pyruvate and di- or tricarboxylic acids; citrate was converted into malate and α-oxoglutarate; anaerobically, cis-aconitate became citrate; aerobically α-oxoglutarate was changed into malate and fumarate; malonate caused accumulation of succinate from α-oxoglutarate by blocking the oxidation of succinate; arsenite inhibited oxo acid oxidation. These results are in complete agreement with the operation of the tricarboxylic acid cycle.

5. *Effects of oxalacetate on the oxidation of pyruvate and succinate*

The avocado fruit was not only the first ovarian tissue to yield active mitochondria but it disclosed also certain unique features in the effects of one acid of the Krebs cycle on the oxidation of other acids.

Avron and Biale (1957b), in the course of testing the oxidative capability of the acids of the tricarboxylic acid cycle, noted that when oxalacetate (OAA) was used in substrate concentrations no oxygen uptake was observed manometrically for a period of an hour or more. However, when the reaction mixture was chromatographed there was no doubt that citrate and malate appeared as products of OAA oxidation. It was necessary, therefore, to account for a pathway of oxidation in which OAA was converted to citrate and malate without oxygen uptake. The following reactions were postulated:

$$\text{Oxalacetate} \longrightarrow \text{pyruvate} + CO_2 \qquad (1)$$

$$\text{Oxalacetate} + \text{pyruvate} \xrightarrow{-2H} \text{citrate} + CO_2 \qquad (2)$$

$$\text{Oxalacetate} \xrightarrow{+2H} \text{malate} \qquad (3)$$

In support of this scheme it was observed that the lag in oxygen uptake with pyruvate as substrate was a function of OAA concentration. The lag period for 0·005 M OAA was four times shorter than for 0·018 M. When arsenite was added to a reaction mixture containing 30 micromoles OAA and other factors required for oxidation, the rate of CO_2 evolution was about one-half of that of the control. This can be explained by the fact that arsenite does not affect reaction (1) of the proposed scheme but is known to interfere with reaction (2), the citrate condensation reaction. With arsenite neither citrate nor malate appeared on the chromatogram. The coupled dismutation reaction was blocked by the inhibitor. The action of OAA is of particular interest since it took place aerobically, unlike the situation in animal preparations in which the dismutation has been observed only in the absence of air. In avocado mitochondria the competition between the electron transport chain and the dismutation appears to be in favour of the latter. The coupled oxidation–reduction reaction proposed for the avocado must hold also for other plant tissue as judged from chromatographic results and from the fact that malate is the best "sparker" acid in pyruvate oxidation. The system is also present in apples (Hulme *et al.*, 1967).

The effect of OAA on succinate oxidation has been shown (Avron and Biale, 1957b) to resemble and to differ from the effect on pyruvate oxidation. In this case, too, a lag period in oxygen uptake was observed to be proportional to the OAA concentration. While the manometric picture was similar to the one with pyruvate the chromatographic picture clearly indicated the presence of a large succinate spot, suggesting therefore a true inhibition rather than a coupled oxidation–reduction. The reversal of succinate inhibition with time was attributed to the breakdown of the labile inhibitor. The inhibition was also reversed by magnesium and ATP. The addition of calcium ions to a succinoxidase system inhibited by OAA markedly accentuated the inhibition. In this case, too, Mg and ATP reversed the inhibition. In the control neither Ca nor ATP showed any effect while some enhancement was observed with Mg. The effects of the several co-factors may be explained either by the formation of a complex which would be more or less effective as inhibitor or by accelerating or decelerating the breakdown of oxalacetate. The available evidence does not support the idea that the inhibition is due to malonate formed by decarboxylation of the oxo acid. A similar situation occurs in the apple (Hulme *et al.*, 1967).

6. *Oxidative phosphorylation*

The capacity of avocado mesocarp for phosphorylation was evident from the early studies in which it was shown that oxidations of Krebs cycle acids by mitochondrial suspensions depended on adenylates as phosphate acceptor. Direct evidence for phosphate esterification was obtained by following the disappearance of inorganic phosphate in a reaction mixture containing particles, AMP or ADP, glucose, hexokinase, Mg and phosphate. Under these conditions the P/O ratios (microatoms P esterified per microatom oxygen taken up) were highest for the single step oxidation of α-oxoglutarate to succinate and lowest for succinate. In all cases the values were one-half or less of those expected theoretically. The situation changed materially with the advent of the polarographic technique and the introduction of the concept of respiratory control or acceptor control ratio.

Respiratory control was defined by Chance and Williams (1955) as the ratio of the rate of oxidation immediately following the addition of ADP to a system containing particles and substrate (state 3 oxidation) to the rate recorded upon depletion of this phosphate acceptor (state 4 oxidation). The polarographic traces also afford the opportunity of calculating the ADP/O ratio on the assumption that all the ADP is used up between the onset of state 3 and state 4. The acceptor control and ADP/O ratios for avocado mitochondria are given in Table XX.

TABLE XX. Oxidative phosphorylation by avocado mitochondria

	Malate	Succinate	α-oxoglutarate	
			−Malonate	+Malonate
Acceptor control ratio	4·3	2·7	3·0	4·8
ADP/O	1·9	1·5	2·3	2·7

From Lance *et al.* (1965).
The source of mitochondria was mesocarp from avocados at climacteric peak. For changes in course of ripening see next section.

The higher acceptor control ratios for α-oxoglutarate were ascribed (Wiskich *et al.*, 1964) to tighter coupling to oxidation for substrate level phosphorylation. A consistent difference between the several substrates was the fact that α-oxoglutarate was not oxidized at all prior to addition of ADP, while malate and succinate were oxidized to some extent in the absence of ADP. This is another indication for the tight coupling between oxidation and phosphorylation at the substrate level site of α-oxoglutarate oxidation.

State 4 oxidation was stimulated by ADP or an uncoupling agent such as DNP (2,4-dinitrophenol) or CCP (carbonyl cyanide m-chlorophenylhydrazone). DNP concentration of 30 μM was optimal for succinate oxidation, while CCP was equally effective at 0·5 μM. The contention that the DNP effect might be attributed to a stimulation of adenosine triphosphatase causing a recycling of ADP was disproved by the use of oligomycin, an antibiotic known to inhibit phosphorylation and ATP-ase action. Oligomycin brought about state 4, ADP had no effect, while DNP reversed the oligomycin induced inhibition with both succinate and malate. This pointed clearly to the uncoupling mechanism of DNP with these substrates.

When α-oxoglutarate was oxidized oligomycin did not lower the ADP stimulated oxidation to that of the state 4 rate. The oxidation inhibited by oligomycin could be recovered by addition of DNP or CCP. In the presence of oligomycin DNP did not stimulate the state 4 respiration. These responses are quite different from those observed in the oxidation of succinate and malate. They could be explained by the assumption that substrate level phosphorylation which is unaffected by oligomycin and DNP is the rate-limiting step in α-oxoglutarate oxidation. The role of these two substances may be attributed to the stimulation of ATPase by DNP and its inhibition by oligomycin.

X. METABOLIC EVENTS IN FRUIT RIPENING

A. Oxidative Activities in Mitochondria

For the past fifteen years a great deal of attention has been focused on mitochondrial processes in the avocado. It was hoped that an understanding of oxidative and phosphorylative reactions in the particles which are the sites of respiratory activities might help in elucidating the course of the climacteric rise. It took nearly a decade to develop procedures which would yield preparations of high activity and which would show respiratory control. At first this was accomplished in climacteric peak fruit (Wiskich *et al.*, 1964) and later at all stages of ripening (Hobson *et al.*, 1966). In view of the changes

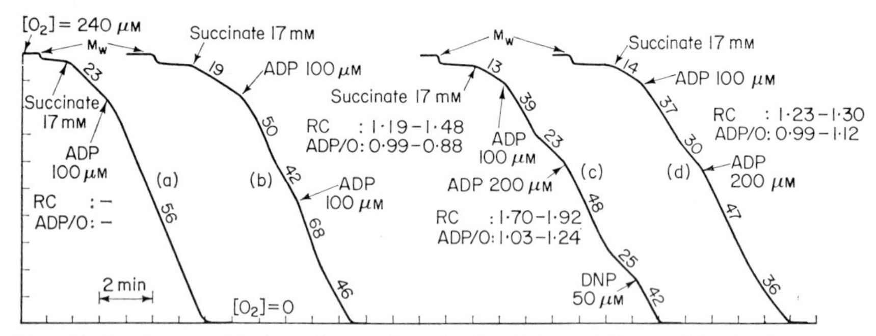

FIG. 16. Succinate oxidation of avocado mitochondria as influenced by bovine serum albumin (BSA) (Lance *et al.*, 1965). (a) No BSA; (b) BSA in assay medium only; (c) BSA throughout; (d) BSA in extraction medium only.

in texture of the fruit with ripening, suitable techniques of grinding and an appropriate preparative medium had to be developed. The critical break-through occurred with the introduction of low fatty acid bovine serum albumin (BSA). The role played by this substance is illustrated in Fig. 16. Respiratory control could not be demonstrated in the absence of BSA, was partially restored on addition to the reaction mixture, and the best results were obtained when used in the course of homogenization, washing and during the assay. In the case of the avocado BSA gave better protection than polyvinylpyrrolidone, ficoll, dextran or thiogel. It is assumed that BSA protects the mitochondria against the uncoupling action of free fatty acids (Lehninger, 1964).

1. *Succinate oxidation*

With the introduction of the concept of respiratory control and with the improvement in methodology, it has been possible to give a fresh appraisal of the role of mitochondria in the respiratory upsurge associated with ripening. On the basis of the differences observed in mitochondrial behaviour, the ripening process has been divided into four phases: early pre-climacteric, late pre-climacteric, climacteric rise to climacteric peak and post-climacteric. An analysis of these four phases was undertaken for the oxidation of several substrates of the Krebs cycle.

When succinate was oxidized, the "basic" rate (that is the rate in the presence of substrate but prior to addition of ADP) was reasonably constant throughout the climacteric. So was the state 3 oxidation, the rate in the presence of ADP. $Q_{O_2(N)}$ expressed in μlitres oxygen consumed/hour/mg mitochondrial nitrogen varied from 1270 to 1590 without any trend related to ripeness. There was an indication that state 4 oxidation resulting from depletion of ADP was higher in particles from early pre-climacteric fruit than from later stages. In consequence there was a rise in respiratory control with a maximum of 2·8 at the peak of the climacteric. The ADP/O ratio also improved from stage to stage giving the highest value in particles from fully ripe avocados. The observation was made with succinate, as well as with other acids, that subsequent additions of ADP gave respiratory control ratios which were higher than the one following the first addition. Apparently the initial introduction of ADP contributes to mitochondrial membrane configuration which results in tighter coupling.

2. *Malate oxidation*

The oxidation of malate in relation to the climacteric pattern presents a more complicated picture than that of succinate. The basic respiration increased with ripening but the chief difference between the several stages was due to co-factors. The response to ADP in mitochondria from early and late pre-climacteric fruit was slight. Addition of TPP was essential to bring about a gradual acceleration in oxidative activity ending in respiratory control (Fig. 17). During the rise and in the post-climacteric period, rates of $Q_{O_2(N)}$ in excess of 1000 were obtained upon addition of ADP without TPP. However, the high activity was not constant but fell with time, bringing about in succession an inhibited state 3, reduced state 4 and finally normal state 4. Addition of either ADP or DNP during state 4 restored the oxidation to the high rate. The length of the plateau during the inhibitions appeared to be proportional to the amount of ADP added. TPP, which was not required for maximal oxidation in mitochondria from ripe fruit, counteracted the inhibitions resulting in a linear state 3. With malate, as with succinate, the respiratory control and the ADP/O ratio rose with ripening.

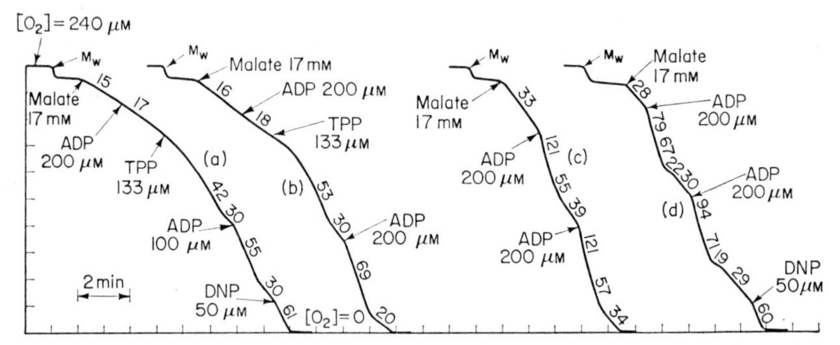

FIG. 17. Malate oxidation by avocado mitochondria. (a) early pre-climacteric particles; (b) late pre-climacteric; (c) near beginning of rise; (d) near climacteric peak (Lance *et al.*, 1965).

3. *The oxidation of α-oxoglutarate*

The α-oxoglutarate oxidation resembled that of malate in some respects but it also exhibited unique features. Here, too, TPP enhanced the ADP stimulated oxidation in early pre-climacteric mitochondria, but was not necessary in later stages (Fig. 18). This oxidation was studied in the absence

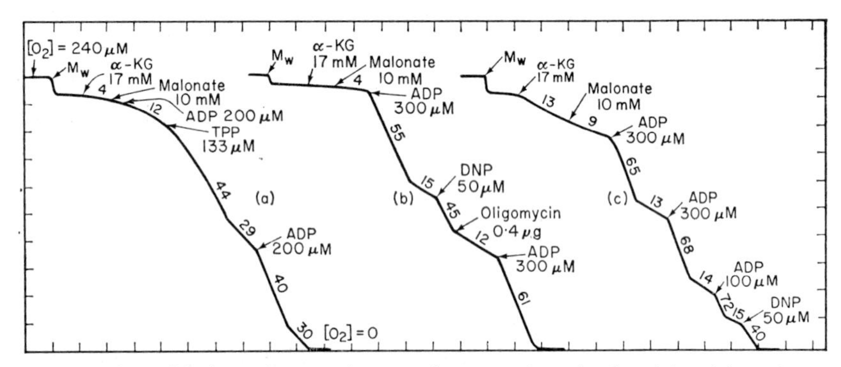

FIG. 18. The oxidation of α-oxoglutarate by avocado mitochondria. (a) early pre-climacteric; (b) late pre-climacteric; (c) climacteric peak particles (Lance *et al.*, 1965).

and in the presence of malonate which inhibits succinic dehydrogenase and limits the oxidation to a single step. The behaviour was similar in both cases but with malonate there was a marked decrease in the state 4 rates with concomitant improvement in respiratory control and ADP/O ratios. These ratios increased to 5 and 3, respectively, as ripening progressed.

The α-oxoglutarate oxidation appeared to be more tightly coupled than the oxidations of succinate and malate. This was evidenced not only by the levelling of the state 4 but also by the very low rates of the basic oxidation. The tight coupling was attributed here, as in the study of Wiskich *et al.*

(1964), to the predominance of substrate level phosphorylation. Evidence for this suggestion was obtained from the inhibition by oligomycin of the DNP stimulated oxidation and subsequent reversal by ADP. In this respect particles from all stages of ripening behaved similarly.

4. *Pyruvate and malate oxidation*

The pattern of malate oxidation suggested a probable inhibiting role of a product of the reaction. This suggestion was based on several facts: (1) the initial state 4 rate is significantly lower than the basic rate; (2) the lag period between the inhibited and the true state 4 is a direct function of ADP added and hence of the amount of malate oxidized, and (3) upon addition of TPP the rate became linear. OAA was suspected to be the inhibitor since it is a direct product of malate oxidation and since it was reported (Kun, 1963) to inhibit malate dehydrogenase. It was previously demonstrated (Avron and Biale, 1957b) that a lag in pyruvate oxidation was a function of OAA concentration and it was shown also (Wiskich *et al.*, 1964) that the OAA inhibition may be relieved through a transamination with glutamate. A study was therefore undertaken on the relationship between malate and pyruvate oxidation (Lance *et al.*, 1967) in order to localize the regulatory action of TPP in the course of the climacteric. Table XXI is selected to demonstrate some of the features of this relationship.

TABLE XXI. Oxidation of pyruvate by pre-climacteric and climacteric avocado mitochondria

	μlitre O_2 hour^{-1} mg^{-1} N			
	Pre-climacteric		Climacteric	
Addition Malate (mM)	1·2	17	1·2	17
None	27	185	162	745
Pyruvate (17 mM)	68	190	264	740
TPP (133 μM)	95	500	262	1105
Pyruvate (17 mM) + TPP (133 μM)	222	720	620	1140

Results are the averages of four different experiments in both cases. No significant oxygen uptake in the absence of malate was observed. Results are given for a "sparker" (1·2 mM) and for a substrate (17 mM) concentration of malate.

From Lance *et al.* (1967).

The most pronounced difference was obtained at substrate concentration 17 mM of malate in which TPP increased the rate of oxidation by nearly 200% for pre-climacteric mitochondria while the stimulation for climacteric particles was less than 50%. TPP caused a striking increase of pyruvate oxidation at a "sparker" concentration (1·2 mM) of malate for mitochondria from both stages. In this case also the activity was higher in particles from ripe than from unripe fruit. Several tests were employed to see whether

OAA is the intermediate between malate and pyruvate metabolism. It was pointed out earlier and proven again in these experiments that when TPP was present in the medium the state 3 rate was linear and the transient state 4 inhibition was relieved. Upon addition of arsenite the effect of TPP was entirely abolished and glutamate restored the oxidation, presumably by transamination since glutamic dehydrogenase was found to be absent in avocado mitochondria (Wiskich *et al.*, 1964). The data suggested strongly that either TPP or glutamate served to accelerate the removal of oxaloacetate. However, a direct test for the pathway of malate oxidation was necessary and this was obtained in experiments with labelled malate.

5 *Oxidation of* ^{14}*C-malate*

Distribution of radioactivity during the oxidation of uniformly labelled malate was conducted on both pre-climacteric and climacteric-peak particles. Samples were taken immediately after addition of mitochondria to the reaction mixture and after successive periods of oxidation in the presence of ADP, TPP, arsenite and glutamate.

The data indicated clearly that label was transferred to both OAA and pyruvate. In pre-climacteric mitochondria the activity was high in pyruvate when ADP alone was added and the reaction proceeded at a slow rate. The pyruvate formed after addition of TPP did not accumulate but was converted into citrate which showed high activity. Production of citrate stopped when arsenite was supplied but labelling in pyruvate increased. Addition of glutamate caused the activity in pyruvate and citrate to remain at a steady level, but new label showed up in aspartate. In climacteric mitochondria also supplied with ^{14}C–malate, the activity in OAA built up during state 3 to reach a maximum during the inhibited states 3 and 4. When the normal state 4 was reached about half of the OAA was depleted. Pyruvate accumulated during the rapid state 3 and was present at much higher levels than OAA. These results can be explained by the following scheme from Lance *et al.* (1967):

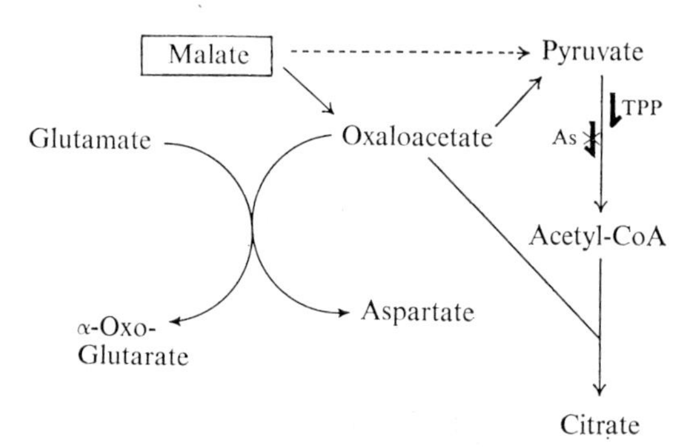

It is not known whether OAA is converted to pyruvate in avocado mitochondria by an enzyme reaction such as OAA carboxylase or non-enzymatically by Mg^{2+} or other ions. The conversion of OAA to malate and citrate was demonstrated (Avron and Biale, 1957b) and discussed previously.

The mode of action of OAA can be deduced from the work on the mechanism of malic dehydrogenase reaction. It is possible that when present in excess, even very slightly, it binds with the NAD site of the enzyme and causes inhibition. Apparently OAA does not accumulate in mitochondria but undergoes decarboxylation, condensation or transamination as soon as it leaves the catalytic site. Consequently the rate of malate oxidation depends greatly on the rate of OAA removal.

The scheme for malate conversion explains also the differences in the oxidation pattern between the several stages of ripeness. In the pre-climacteric mitochondria the rate limiting step in the oxidation of malate is the production of acetyl-CoA from pyruvate. Addition of TPP accelerates the decarboxylation rate of pyruvate. In mitochondria extracted from climacteric avocados, TPP is not required for high rates of malate oxidation but for the removal of OAA. Under these conditions the pyruvate oxidase complex might be working at full capacity and is still limited by TPP. It is assumed, and some preliminary evidence is available, that the concentration of TPP in mitochondria changes with the climacteric.

B. Phosphate Metabolism in Avocado Slices

It has been shown by Millerd *et el.*, (1953) that 2,4-dinitrophenol (DNP) of the order of 10 to 100 μM brought about an increase in oxygen uptake in tissue slices from pre-climacteric but not from climacteric avocados. On the basis of this finding and on the response of cytoplasmic particles isolated from different stages, the idea was advanced that an endogenous uncoupler was formed at the induction of the climacteric and caused a release of oxygen uptake from phosphorylation. According to the uncoupling theory one would expect a decrease in P/O ratios with ripening. Romani and Biale (1957) showed, however, that these ratios actually increased as the climacteric increased. More recently Wiskich *et al.* (1964) and Lance *et al.* (1965) found good respiratory control and high ADP/O ratios in mitochondria from fruit close to and at the peak of ripening. Because of the key role played by the adenylate system in mitochondrial reactions, it was thought relevant to follow the ADP and ATP status in intact cells. This was done by Young and Biale (1967) by incubating tissue discs of pre-climacteric and climacteric peak Fuerte avocado in ^{32}P solution for 10 min. Extraction, chromatography and assay of radioactivity were followed essentially according to the procedure of Bieleski and Young (1963). The uptake of phosphate by discs from both pre-climacteric and climacteric peak fruit was linear for an hour or longer, but the incorporation was three times higher in the ripe material. When the slices were washed in ice water over 90% of label was lost from climacteric peak tissue, and only 30% from pre-climacteric. The substances lost included sugar phosphates and nucleotide phosphates in addition to inorganic phosphate. Climacteric peak tissue lost about the same amount of esterified phosphate as P_i, while in pre-climacteric slices only one-third of the nucleotides and half of the sugar phosphates were lost. The failure to maintain permeability barriers is a striking feature of the ripening avocado as is the case in the banana (Sacher, 1966).

The study of ^{32}P-uptake by avocado slices has also shown that 10 times more phosphate was incorporated into climacteric peak than into pre-climacteric material. Of the amount incorporated, 1·4% was esterified in pre-climacteric and 8% in climacteric peak slices. DNP did not lower significantly the amount of P_i accumulated but it reduced the amount esterified by about 35% in both stages. The rate of esterification was strikingly different. About 50 times more labelled phosphate esters were found in climacteric peak than in pre-climacteric. This trend was obvious in the sugars and in the individual nucleotides. Of the three adenylates examined ADP had the highest activity. The ADP/ATP ratio decreased from 2·30 to 0·88 with the transition from pre-climacteric to climacteric. This finding is in agreement with the results of Rowan *et al.* (1961) for the cantaloupe and sheds doubt on the notion that deficiency of ADP is responsible for the low rates of respiration in the pre-climacteric phase. Similarly, the enhancement of phosphorylating capacity in climacteric slices and the sensitivity to DNP point to the operation of coupled oxidations and argue against control by an endogenous uncoupler.

C. Protein Metabolism in the Avocado

The mitochondrial studies clearly indicate that the machinery for energy generation remains fully active throughout the climacteric cycle. There was actually an improvement in the phosphorylative capacity of the particles with ripening. The question is raised whether and to what extent the generated ATP is used for synthetic reactions. Rowan *et al.* (1958) reported an increase in protein nitrogen and in high energy phosphate in the avocado during the climacteric rise. The work on protein synthesis is beset with difficulties because the methodology of extraction may not be equally applicable to all stages of ripeness. Richmond and Biale (1966) approached this problem by following amino acid incorporation into tissue discs.

Fuerte avocado discs, 5 mm in diameter and 1 mm thick, were taken from fruit of predetermined respiratory history and incubated in 5×10^{-4} M phosphate buffer, pH 6·5, containing uniformly labelled ^{14}C L-leucine or ^{14}C L-valine. Measurements were made of total uptake, active uptake (label retained inside the permeability barrier) and incorporation into a protein fraction which was precipitated by 10% trichloracetic acid. Proteolytic cleavage and acid hydrolysis confirmed that the label was incorporated into a polypeptide. Additional evidence that the incorporated label was bound to a newly synthesized protein was obtained by the use of puromycin, the inhibitor of protein synthesis. Precautions were taken against interference by micro-organisms and assurance that the incubation medium was sufficiently free of contamination was derived from the fact that soft tissue which is susceptible to microbial attack showed a very low rate of incorporation.

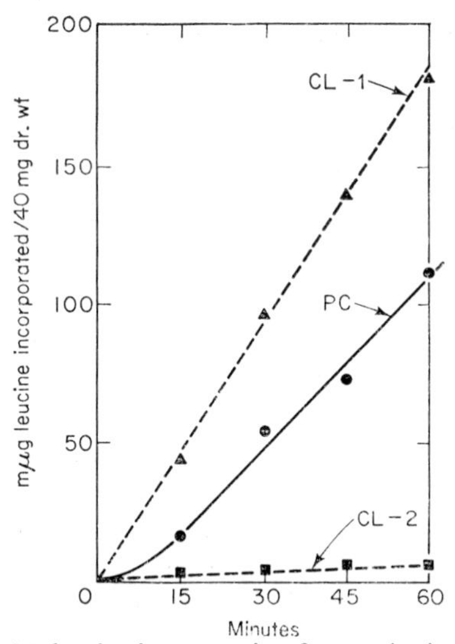

FIG. 19. Incorporation of L-leucine into protein of avocado tissue slices. Leucine conc. 100 μM, 2×10^4 cpm/mμmole. PC, pre-climacteric; CL–1, early climacteric rise; CL–2, near climacteric peak (Richmond and Biale, 1966).

From the standpoint of amino acid incorporating activity three stages were distinguished: pre-climacteric (PC) extending from harvesting to the minimum preceding the rise, early climacteric (Cl-1) up to the middle of the rise, and late climacteric (Cl-2) close to the peak of O_2 uptake. Total uptake of the two amino acids was generally a function of the concentration in the incubation medium. At low levels a steady state uptake was maintained in pre- and in climacteric material. At higher concentrations the rapid initial rate of uptake gradually declined especially in stage Cl-1. Active uptake to within the permeability barrier behaved similarly, showing highest values in stage Cl-1. When incorporation of label into proteins was related to active uptake (I/U ratio) there was a marked increase from PC to Cl-1 with a drastic reduction at Cl-2. In some cases no incorporation could be detected in slices near the peak (Fig. 19). Endogenous dilution was ruled out since the I/U ratio remained constant for a wide range of concentrations at any given stage of ripeness. The rising I/U ratios throughout most of the climacteric period indicates that the Cl-1 stage has a higher incorporating potential than the other stages. Conceivably, with the onset of the rise, there is an induction of the formation of enzymes which catalyse the climacteric process and cause ultimate breakdown. Puromycin had no effect on O_2-uptake by the slices, while it interfered with amino acid incorporation, suggesting that the respiratory rise is not coupled energetically to protein synthesis.

D. Nucleic Acid Metabolism in the Avocado

The changes in amino acid incorporation into protein during the course of the climacteric led Richmond and Biale (1967) to study the changes in RNA over the same period. They incubated Fuerte avocado discs in sucrose-citrate solution containing labelled phosphate, pH 6·0, at 20°C. At the end of the incubation period the samples were washed with cold phosphate buffer, RNA extracted in cold phenol and fractionated on a MAK (methylated albumin-Kieselguhr) column. Absorbance and radioactivity were determined on 3 ml fractions.

The experiments demonstrated marked differences in the synthesis of RNA with ripening. In pre-climacteric and early climacteric discs, ^{32}P was incorporated chiefly into a heavy ribosomal fraction which was thought to be messenger RNA (mRNA) on comparative considerations. Later in the climacteric rise, when the oxygen uptake was close to 80% of the peak rate, a very definite decrease in ^{32}P uptake by the ribosomal and mRNA fractions was noted. At the same time a rise in incorporation into soluble RNA region occurred. At the peak of the climacteric the incorporation of ^{32}P took place into soluble RNA; incorporation into ribosomal RNA was reduced sharply and none was found in mRNA. On the basis of these results the hypothesis was advanced that degeneracy of the avocado cell is caused by failure to synthesize ribosomal and messenger RNA.

XI. CONCLUDING REMARKS

It is hoped that this review of the physiology and biochemistry of the avocado has stressed both the conventional and the unique aspects of the studies with a somewhat unusual type of material. Though typically a high fat fruit the avocado also contains rare sugars of high carbon number and is relatively rich in certain vitamins, minerals and nitrogenous substances. The biosynthetic capacity persists throughout maturation as evidenced in the numerous studies of Stumpf and colleagues on fat metabolism in mesocarp slices, mitochondrial and soluble systems. Thanks to these contributions the knowledge of lipid metabolism is perhaps better understood in the avocado than in any other mesocarp tissue.

Despite the high fat content of the mature avocado, the evidence does not support the idea of lipid utilization as respiratory substrates during the course of the climacteric. Sugars disappear with ripening and so do insoluble pectins, but whether they account for all the fuel required in respiration remains to be investigated. Information on the endogenous substrates is particularly important for an understanding of the nature of the respiratory process.

The trends in overall oxygen uptake and CO_2 evolution by the intact, detached organ are qualitatively similar to the pattern in other climacteric fruits. The response to temperature is essentially the same as in other fruits with a temperature span of 5–25°C as the physiological range for most varieties. As far as fruit quality is concerned it is desirable not to store at temperatures below 5°C and not to ripen at temperatures above 25°C. Since chilling injury occurs in the region of 5°C, extended storage life may be obtained under modified gaseous conditions of lowered O_2 and increased CO_2. With the Fuerte variety, best results have been obtained with 5% O_2 and 10% CO_2 at 15°C.

The biochemical changes associated with ripening have been studied extensively at the cellular and subcellular levels. It has been shown by the use of inhibitors at the dehydrogenase and at the terminal oxidase sites that oxygen uptake by tissue slices can be stimulated if not all sites are blocked. This is especially the case during the pre-climacteric stage. It appears that independent but interlinked phosphorylative and non-phosphorylative pathways are operative at the cellular level. When one or two sites are blocked, limiting phosphorylation, co-factors are shunted to the unblocked site or sites resulting in increased respiration. Before any of these ideas are accepted, however, it is necessary to reconcile the tissue slice work with recent studies on mitochondrial systems showing sensitivity to the inhibitors of the conventional electron transport chain.

Convincing evidence has accumulated on the role of one co-factor, thiamine pyrophosphate (TPP), in the course of the climacteric. In the case of malate and α-oxoglutarate the oxidation rate is increased strikingly by TPP when added to the reaction mixture with pre-climacteric mitochondria. It is not required for maximal oxidation rates in climacteric particles, but it prevents the occurrence of inhibited state 3 and state 4 oxidations. The inhibition has been traced to oxalacetate and its mechanism studied in detail. For pyruvate oxidation, TPP is required in both stages of ripeness.

With the appropriate additions to the preparative and assay medium, oxidative phosphorylation progresses actively and improves with the development of the climacteric. Similarly, studies of the phosphate metabolism in tissue slices have provided convincing evidence of the ability of the material to generate the energy required for cellular metabolism. Structure and function of the ATP forming machinery remain fully intact and operative throughout the climacteric. Changes, possibly leading to breakdown and senescence, have been observed in the protein synthesizing capacity and in the incorporation of labelled phosphate into certain nucleic acid fractions during the early stages of the climacteric.

ACKNOWLEDGEMENTS

The studies conducted in this laboratory and reported in this chapter were supported in part by Public Health Research Grant GM–08224, by NSF GB–7408, by the American Cancer Society Grant E–293, by the Cancer Research Committee of the University of California and by the Committee on Research, UCLA.

The technical assistance of Mrs. Catherine Popper, Mrs. Wynona Appleman and Mr. Donald Barcus is gratefully acknowledged.

REFERENCES

Anderson, D. G., Stafford, H. A., Conn, E. and Vennesland, B. (1952). *Pl. Physiol., Lancaster* **27**, 675.

Appleman, D. and Noda, L. (1941). *Calif. Avocado Soc. Year Book* 60.

Asenjo, C. F. and Muniz, A. (1955). *Fd Res.* **20**, 47.

Avron, M. and Biale, J. B. (1957a). *Pl. Physiol., Lancaster* **32**, 100.

Avron, M. and Biale, J. B. (1957b). *J. biol. Chem.* **225**, 699.

Baker, J. E., Elfvin, L. G., Biale, J. B. and Honda, S. I. (1968). *Pl. Physiol., Lancaster* **43**, 2001.

Barron, E. J., Squires, C. and Stumpf, P. K. (1961). *J. biol. Chem.* **236**, 2610.

Bean, R. C. (1958). *Calif. Avocado Soc. Year Book*, **42**, 90.

Ben-Yehoshua, S. (1964). *Physiologia. Pl.* **17**, 71.

Ben-Yehoshua, S., Robertson, R. N. and Biale, J. B. (1963). *Pl. Physiol., Lancaster* **38**, 194.

Bevenue, A. and Williams, K. T. (1959). *J. Ass. off. agric. Chem.* **42**, 441.

Biale, J. B. (1941). *Proc. Am. Soc. hort. Sci.* **39**, 137.

Biale, J. B. (1946). *Amer. J. Bot.* **33**, 363.

Biale, J. B. (1960a). *In* "Handbuch der Pflanzenphysiologie" (W. Ruhland, ed.), Vol. 12, Part II, pp. 536–592. Springer Verlag, Berlin.

Biale, J. B. (1960b). *Adv. Fd Res.* **10**, 293.

Biale, J. B. (1964). *Science, N.Y.* **146**, 880.

Biale, J. B. (1969). *Qualitas Pl. Mater. Veg.* **19**, 141.

Biale, J. B. and Young, R. E. (1962). *Endeavour* **21**, 164.

Biale, J. B., Young, R. E. and Olmstead, A. J. (1954). *Pl. Physiol., Lancaster* **29**, 168.

Biale, J. B., Young, R. E., Popper, C. S. and Appleman, W. E. (1957). *Physiologia. Pl.* **10**, 48.

Bieleski, R. L. and Young, R. E. (1963). *Analyt. Biochem.* **6**, 54.

Buhler, D. R., Hansen, E. and Wang, C. H. (1957). *Nature, Lond.* **179**, 48.

Burg, S. P. (1962). *A. Rev. Pl. Physiol.*, **13**, 265.

Burg, S. P. and Burg, E. A. (1962a). *Pl. Physiol., Lancaster* **37**, 179.

Burg, S. P. and Burg, E. A. (1962b). *Nature, Lond.* **194**, 398.

Burg, S. P. and Burg, E. A. (1964). *Pl. Physiol., Lancaster* Suppl. **39**, x.

Burg, S. P. and Burg, E. A. (1965a). *Science, N.Y.* **148,** 1190.
Burg, S. P. and Burg, E. A. (1965b). *Physiologia. Pl.* **18,** 870.
Burg, S. P. and Burg, E. A. (1966). *Science, N.Y.* **153,** 314.
Chance, B. and Williams, G. R. (1955). *Nature, Lond.* **176,** 250.
Charlson, A. J. and Richtmyer, N. (1960). *J. Am. chem. Soc.* **82,** 3428.
Chase, E. M. (1921–22). *Calif. Avocado Assoc. Ann. Rept.*, 52.
Chow, C. T. and Biale, J. B. (1957). *Physiologia. Pl.* **10,** 64.
Church, C. G. (1921–22). *Calif. Avocado Assoc. Ann. Rept.*, 40.
Clements, R. L. (1965). *Analyt. Biochem.* **13,** 390.
Cummings, K. and Schroeder, C. A. (1942). *Calif. Avocado Soc. Year Book*, 56.
Davenport, J. B. and Ellis, S. C. (1959). *Aust. J. biol. Sci.* **12,** 445.
Denny, F. E. and Miller, L. P. (1935). *Contrib. Boyce Thompson Inst.* **7,** 97.
Dolendo, A. L., Luh, B. S. and Pratt, H. K. (1966). *J. Fd Sci.* **31,** 332.
Eckey, E. W. (1954). "Vegetable Fats and Oils," p. 836. Reinhold Publishing Corp., New York.
French, R. B. and Abbot, O. D. (1948). *Florida Agr. Expt. Sta. Tech. Bull.* **444,** 21 pp.
Galliard, T. and Stumpf, P. K. (1968). *Biochem. Prep.* **12,** 66.
Gallo, P. and Valeri, H. (1954). *Rev. med. vet. y parasitol.* (*Caracas*) **13,** 63.
Geissman, T. A. and Dittmar, H. F. K. (1965). *Phytochemistry* **4,** 359.
Giral, J. and Castillo, M. T. (1953). *Men. congr. cient. mex., IV Centenario Univ. Mex.* **2,** 232.
Grant, W. C. (1960). *Proc. Soc. Exptl Biol. Med.* **104,** 45.
Hackenbrock, C. R. (1966). *J. cell Biol.* **30,** 269.
Hall, A. P., Moore, J. G. and Morgan, A. F. (1955). *J. Agr. Food Chem.* **3,** 250.
Hatton, T. T. Jr., Reeder, W. S. and Campbell, C. W. (1965). *USDA, A.R.S. Marketing Research Rept.* 697, pp. 1–13.
Hawke, J. C. and Stumpf, P. K. (1965). *J. biol. Chem.* **240,** 4746.
Hilditch, T. P. and Williams, P. N. (1964). "The Chemical Constitution of Natural Fats." John Wiley & Sons, New York
Hobson, G. E. (1962). *Nature, Lond.* **195,** 804.
Hobson, G. E., Lance, C., Young, R. E. and Biale, J. B. (1966). *Nature, Lond.* **209,** 1242.
Hulme, A. C. (1936). *Biochem. J.* **30,** 258.
Hulme, A. C. (1954). *J. exp. Bot.* **5,** 159.
Hulme, A. C., Rhodes, M. J. C. and Wooltorton, L. S. C. (1967). *J. exp. Bot.* **18,** 277.
Jaffe, M. E. and Gross, H. (1923). *Univ. Calif. Agr. Exp. Sta. Bull.* **365,** 630.
Jansen, E. F. (1963). *J. biol. Chem.* **238,** 1552.
Jansen, E. F. (1964). *J. biol. Chem.* **239,** 1664.
Jones, J. K. N. and Sephton, H. H. (1960). *Can. J. Chem.* **38,** 753.
Joslyn, M. A. (1941). *Ind. Engng Chem.* **33,** 308.
Joslyn, M. A. and Smit, C. J. B. (1954). *Mitt. Klosternenburg, Ser. B, Obst. n. Garten* **4,** 141.
Joslyn, M. A. and Stepka, W. (1949). *Fd Res.* **14,** 459.
Kidd, F. and West, C. (1925). *Gt. Brit. Dept. Sci. Ind. Research Food Invest. Board Rept.* 27.
Kikuta, Y. (1968). *J. Fac. Agr., Hokkaido Univ.* **55,** 469.
Kikuta, Y. and Erickson, L. C. (1969a). *Plant Cell Physiol.* **10,** 563.
Kikuta, Y. and Erickson, L. C. (1969b). *Plant Cell Physiol.* **10,** 759.
Knapp, F. W. (1965). *J. Fd Sci.* **30,** 930.

Kun, E. (1963). *In* "The Enzymes" (Boyer, Lardy and Myrback, eds.), Vol. 7, 149. Academic Press, London and New York.
La Forge, F. B. and Hudson, C. G. (1917), *J. biol. Chem.* **28,** 511.
Lance, C., Hobson, G. E., Young, R. E. and Biale, J. B. (1965). *Pl. Physiol., Lancaster* **40,** 1116.
Lance, C., Hobson, G. E., Young, R. E. and Biale, J. B. (1967). *Pl. Physiol., Lancaster* **42,** 471.
Lehninger, A. A. (1964). *In* "The Mitochondrion". W. A. Benjamin, Inc., New York.
Lips, S. H. and Biale, J. B. (1966). *Pl. Physiol., Lancaster* **41,** 797.
Makower, R. U. and Schwimmer, S. (1954). *Biochim. biophys. Acta* **14,** 156.
Makower, R. U. and Schwimmer, S. (1957). *J. Agr. Food Chem.* **5,** 768.
Maquenne, M. (1890). *Ann. de Chimie et de Physique* **19** (ser. 6), 5.
Mazliak, P. (1965a). *Fruits, Paris* **20,** 49.
Mazliak, P. (1965b). *Fruits, Paris* **20,** 117.
Mazliak, P. (1965c). *Fruits, Paris* **20,** 120.
McCready, R. M., McComb, E. A. and Jansen, E. F. (1955). *J. Fd Res.* **20,** 186.
Millerd, A., Bonner, J. and Biale, J. B. (1953). *Pl. Physiol., Lancaster* **28,** 521.
Mudd, J. B. and Stumpf, P. K. (1961). *J. biol. Chem.* **236,** 2602.
Overath, P. and Stumpf, P. K. (1964). *J. biol. Chem.* **239,** 4103.
Palmer, J. K. and McGlasson, W. B. (1968). *Aust. J. biol. Sci.* **22,** 87.
Polansky, M. H. and Murphy, E. (1966). *J. Am. Dietet. Assoc.* **48,** 109.
Pratt, H. K. and Biale, J. B. (1944). *Pl. Physiol., Lancaster* **19,** 519.
Pratt, H. K. and Goeschl, J. D. (1969). *Ann. Rev. Pl. Physiol.* **20,** 541.
Pratt, H. K., Young, R. E. and Biale, J. B. (1948). *Pl. Physiol., Lancaster* **23,** 526.
Racker, E. and Schroeder, E. (1957). *Archs biochem. Biophys.* **66,** 241.
Raymond, D. and Phaff, H. J. (1965). *J. Fd Sci.* **30,** 266.
Richmond, A. and Biale, J. B. (1966). *Pl. Physiol., Lancaster* **41,** 1247.
Richmond, A. and Biale, J. B. (1967). *Biochim. biophys. Acta* **138,** 625.
Romani, R. J. and Biale, J. B. (1957). *Pl. Physiol., Lancaster* **32,** 692.
Rowan, K. S., Pratt, H. K. and Robertson, R. N. (1958). *Aust. J. biol. Sci.* **11,** 329.
Rowan, K. S., McGlasson, W. B., and Pratt, H. K. (1961). *Fedn Am. Socs. exp. Biol.* **20,** 374.
Sacher, J. A. (1966). *Pl. Physiol., Lancaster* **41,** 701.
Schroeder, C. A. (1950). *Calif. Avocado Soc. Year Book* **35,** 169.
Schroeder, C. A. (1953). *Proc. Am. Soc. hort. Sci.* **61,** 103.
Schwob, R. (1951). *Fruits* **6,** 177.
Scott, F. M., Bystrom, B. G. and Bowler, E. (1963). *Bot. Gaz.* **124,** 423.
Seegers, W. H., Levine, W. G. and Johnson, S. A. (1955). *J. appl. Physiol.* **7,** 617.
Sephton, H. H. and Richtmyer, N. K. (1963a). *J. org. Chem.* **28,** 1691.
Sephton, H. H. and Richtmyer, N. K. (1963b). *J. org. Chem.* **28,** 2388.
Sephton, H. H. and Richtmyer, N. K. (1966). *Carbohydrate Res.* **2,** 289.
Simoni, R. D., Criddle, R. G. and Stumpf, P. K. (1967). *J. biol. Chem.* **242,** 573.
Spitzer, K. (1931). *Biochem. Z.* **231,** 309.
Spoehr, H. H., Smith, J. H. C., Strain, H. H., Milner, H. W. and Hardin, G. J. (1949). *Carnegie Inst. Wash. Pub. No.* **586,** 67 pp.
Squires, C. L., Stumpf, P. K. and Schmidt, C. (1958). *Pl. Physiol., Lancaster* **33,** 365.
Stumpf, P. K. (1961). *Proc. 5th Intern. Congr. Biochem., Moscow*, 1961, **2,** 63.
Stumpf, P. K. and Barber, G. A. (1957). *J. biol. Chem.* **227,** 407.

Trout, S. A., Hall, E. G., Robertson, R. N., Hackney, F. M. V. and Sykes, S. M. (1942). *Aust. J. exp. Biol. med. Sci.* **20,** 219.
Valeri, H. and Gimeno, F. (1954). *Rev. med. vet. y parasitol.* (*Caracas*) **13,** 37.
Valmayor, R. V. (1967). *The Philippine Agr.* **L**, 907.
Wardlaw, C. W. (1937). *Trop. Agric., Trin.* **14,** 34.
Wardlaw, C. W. and Leonard, E. R. (1935). *Trinidad Low Temp. Res. Sta. Imp. Coll. Trop. Agr.*, Memoir No. 1, 12.
Weatherby, L. G. (1935). *Calif. Avocado Assoc. Year Book*, 53.
Wiskich, J. T., Young, R. E. and Biale, J. B. (1964). *Pl. Physiol., Lancaster* **39**, 312.
Wolfe, H. S., Toy, L. R. and Stahl, A. L. (1934). *Univ. of Florida Agr. Exp. Sta. Bull.* **272,** 46.
Yang, S. F. and Stumpf, P. K. (1965a). *Biochim. biophys. Acta* **98,** 19.
Yang, S. F. and Stumpf, P. K. (1965b). *Biochim. biophys. Acta* **98,** 27.
Young, R. E. (1965). *Nature, Lond.* **205,** 1113.
Young, R. E. and Biale, J. B. (1962). *Pl. Physiol., Lancaster* **37,** 409.
Young, R. E. and Biale, J. B. (1967). *Pl. Physiol., Lancaster* **42,** 1357.
Young, R. E., Pratt, H. K. and Biale, J. B. (1952). *Analyt. Chem.* **24,** 551.
Young, R. E., Romani, R. J. and Biale, J. B. (1962). *Pl. Physiol., Lancaster* **37,** 416.

Chapter 2

The Banana

JAMES K. PALMER

Department of Nutrition and Food Science,
Massachusetts Institute of Technology,
Cambridge, Massachusetts, U.S.A.

I Brief History and World Production 65
II The Fruit 66
A. Principal Varieties of Commercial Significance 66
B. Morphology 67
C. Compositional Changes during Growth and Development 69
D. Effect of Cultural Practices, Pests and Diseases 71
III Post-harvest Physiology and Biochemistry 73
A. Introduction 73
B. Initiation of Ripening 73
C. Respiratory and Compositional Changes during Ripening 76
D. Diseases and Physiological Disorders 92
IV Biochemical Mechanisms of Ripening 94
A. Introduction 94
B. Permeability Changes during Ripening 94
C. Protein Synthesis in Ripening Bananas 96
D. Mitochondria from Bananas 98
E. Conclusions 100
References 101

I. BRIEF HISTORY AND WORLD PRODUCTION

Geographic, climatic and genetic studies suggest that wild bananas have been used as a food in South-east Asia for as long as man has existed in this area. The steps leading to domestication and to selection of the edible, seedless bananas remain obscure. The banana was first referred to in writings from India dated about 500 B.C. Bananas were subsequently introduced into Africa (ca. A.D. 500), Polynesia (ca. A.D. 1000) and finally to the Canary Islands and the Americas in the fifteenth and sixteenth centuries. Significant banana export trades developed only in the late nineteenth century.

With the exception of New South Wales (Australia), Taiwan and Israel, all the significant banana growing areas lie within the north and south 30° lines of latitude. Within these broad limits exist a wide variety of climates, ranging from wet tropical to dry sub-tropical. Bananas are also cultivated on soils of varied origins, physical structure and composition. Good drainage is the common and crucial feature of these soils.

Bananas are an important world food crop, the total banana production of the world being estimated at 18 million metric tons per year and second only to the grape. Africa produces about 50% of the total, Asia (including Pacific Islands and Australia) 20% and the Americas 30%. Most bananas are consumed locally, only about 20% being transported or exported to more or less distant markets. Bananas represent 40% by weight of all world trade in fruits, fresh or dried. World exports in 1963 were about 4 million metric tons, with a total value estimated at 1·1 billion U.S. dollars, ranking bananas seventh in tonnage and ninth in value of all world agricultural and fish export crops, exclusive of forest crops. Export tonnage increased to 5 million metric tons in 1965 and is expected to reach 7 million tons by 1970.

In 1965, 80% of the export bananas came from the Americas, 10% from Africa and the rest from Australasia. Fifty per cent of these bananas were imported by the United Kingdom and Western Europe, 40% by the United States and Canada. The remainder went to a few countries in South America, Africa and Australasia.

About one-half of all bananas produced are eaten in the cooked state; one-half as a raw, fresh fruit. Bananas eaten in the cooked state are often termed "plantains", but this nomenclature is ambiguous (Simmonds, 1966, p. 57). Cooked bananas are important locally in Africa as a staple foodstuff and are sometimes eaten in large quantities (5–35 bananas per person per day). The consumption of raw fruit averages 1–2 bananas per person per week in most countries. Virtually all export bananas are eaten as fresh fruits.

An almost insignificant proportion of bananas is preserved by canning, freezing, drying and acid fermentation, although these products may be important locally.

Loesecke (1950), May and Plaza (1958), Simmonds (1966) and Freiberg (1969) provide detailed information and references on the history, world production and use of bananas.

II. THE FRUIT

A. Principal Varieties of Commercial Significance

Bananas are classified in the family Musaceae of the order Zingiberales. The Musaceae family has two genera: *Musa* and *Ensete*. All edible cultivars (varieties) are classified into *Musa*. The genus *Musa* contains four sections: *Eumusa*, *Rhodochlamys*, *Australimusa* and *Callimusa*. *Callimusa* and *Rhodochlamys* species are of ornamental interest only. *Australimusa* species are utilized across a large area of the Pacific as a cooked vegetable. However, the section

Eumusa is the largest and most widespread geographically and contains all the major edible species of bananas. Most edible bananas are derived from two members of the section *Eumusa: Musa acuminata* and *Musa balbisiana.*

Simmonds (1966) provides details of the cultivars which are grown around the world to produce edible bananas. The number is bewildering, and the wide geographic distribution has often led to a single cultivar having ten or fifteen synonyms. However, export trade is limited primarily to Gros Michel and to a few Cavendish mutants. Cavendish varieties have recently replaced Gros Michel as the principal export type, largely because Cavendish cultivars are resistant to the devastating *Fusarium* wilt disease known as Panama Disease.

B. Morphology

Wardlaw (1961) and Simmonds (1966) have reviewed the vegetative morphology and general course of development of the banana plant. Barker and Steward (1962a, b) present a comprehensive view of the vegetative growth and development of the banana plant, including the transformation from the vegetative to the floral shoot. Fahn *et al.* (1963) have studied changes in the shoot apex as it becomes reproductive. The origin of the inflorescence and the development of flowers has been studied by Ram *et al.* (1962).

Banana varieties which produce fruit of commercial use are parthenocarpic. They are propagated from a rhizome. The leaves originate from a meristematic region located at the apex of the rhizome, at about the level of the soil surface. The leaves are built by cell division of marginal meristems and emerge in sequence. After the first leaf expands, subsequent leaves emerge through the centre of the previous leaf sheath. The leaf emerges tightly rolled and gradually unrolls as it grows upward out of the encircling leaf sheath. The overlapping and tightly packed leaf sheaths form the visible "trunk" or "pseudostem" of the banana plant, which may be up to 20 feet tall.

Some 9 months after planting and after about 45 leaves have been produced, there is a switch from the vegetative to the reproductive phase. The relatively quiescent central cell zone of the apical dome springs into activity and rapidly produces a succession of bracts and flower primordia. Then, as the true stem elongates, the floral apex is forced up the inside of the pseudostem and eventually emerges ("shoots") out the top of the pseudostem. The inflorescence consists of many groups of flowers, each subtended and covered by a bract.

Commercial varieties produce from 6 to 15 floral bracts containing female flowers, each bract normally containing 15 to 20 individual flowers. The bracts drop off in a few days, leaving the female flowers to develop into the horticulturally mature fruit in the next 90 to 150 days. Immediately below the pendant fruit structure are a few groups of hermaphroditic flowers, which may develop into small bananas. Still further down are numerous groups of male flowers. These mature sequentially as the terminal portion of the flower stalk continues to elongate throughout fruit development.

Ram *et al.* (1962) have made a comprehensive study of the morphology of developing banana fruits, both seeded and parthenocarpic varieties. Figure 1 shows the general course of development of the parthenocarpic fruit.

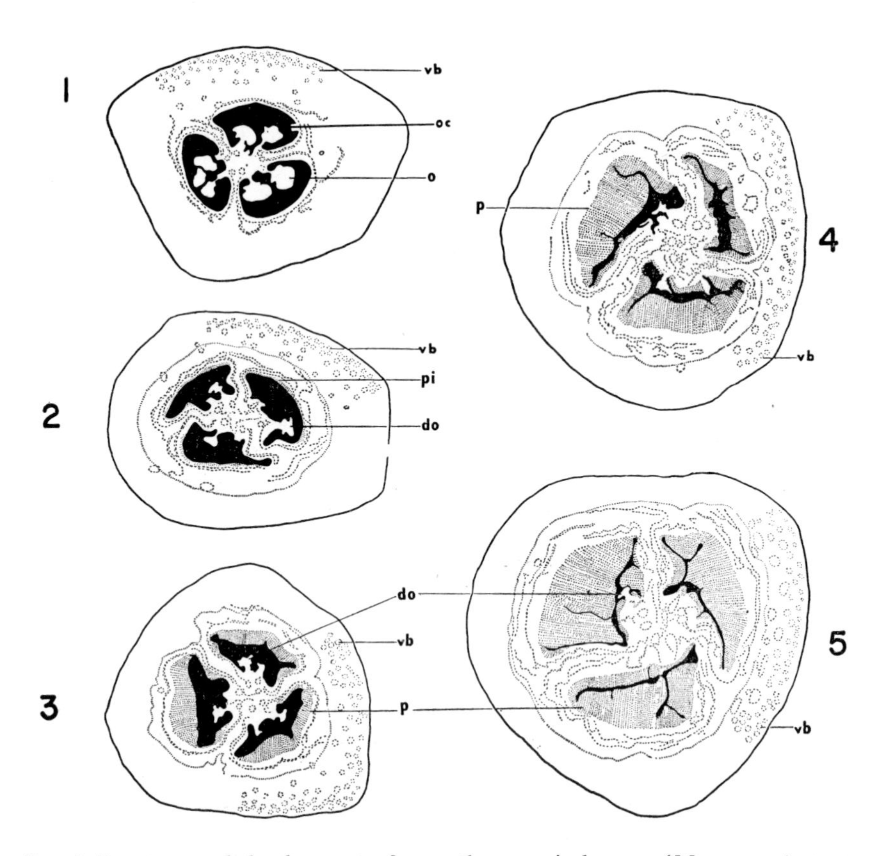

Fig. 1. Structure and development of a parthenocarpic banana (*Musa acuminata* cv Pisang lilin). 1–5. Interpretive diagrams of transverse sections of the fruit at 0, 2, 4, 8 and 12 weeks after emergence. Abbreviations used: do, degenerating ovules; o, ovule; oc, ovarian cavity; p, pulp; pi, pulp-initiating cells; vb, vascular bundle. From Ram *et al.* (1962).

Details of the peel structure are readily seen under the light microscope. It consists of an outer cuticle and epidermis, several layers of hypodermal parenchyma (containing chloroplasts and scattered raphides) and a broad region of parenchyma cells interspersed with latex vessels, vascular bundles and air spaces. The hypodermal cells and the innermost pulp-initiating cells tend to be smaller and more tightly packed than the rest of the cells. Scattered starch grains are visible.

Comparable details of pulp structure are not available. Casual observation reveals that the cells are virtually obscured by the large numbers of starch grains in mature, pre-climacteric tissue. During ripening, the pulp cells become progressively depleted of starch and more details of the individual cells are revealed. The only other change visible during ripening is the coagulation and browning of the contents of the latex vessels (Barnell and Barnell, 1945).

One characteristic feature of banana fruits is the relatively large proportion of peel (exocarp) tissue which makes up about 80, 40 and 33% of the fresh weight of juvenile, mature and fully ripe fruits, respectively. The peel undoubtedly makes a significant contribution to the overall metabolism of the banana fruit.

C. Compositional Changes during Growth and Development

Since bananas for export are invariably picked green and ripened off the plant, there has been little incentive to study chemical changes in the fruit during development on the plant. Barnell (1940), however, has described certain changes in the pulp and peel of Gros Michel bananas sampled up to 130 days after emergence, which is well past the normal harvest time of 80–100 days. The dry matter increased steadily, reaching about 25% at normal harvest date. Starch accumulated for about 100 days and then hydrolysed extensively from 110 to 130 days. Sugar content was below 1% fr. wt. for 120 days and then increased to 4% at 130 days. Sucrose was the primary sugar in the early stages of development, with glucose and fructose predominating later. [Other evidence suggests that sucrose is the predominant sugar arising from starch hydrolysis in bananas (Poland *et al.*, 1938); Loesecke (1950) suggested that the high value for fructose and glucose resulted from failure to inactivate invertase prior to analysis.] Acidity was high in the young fruits, fell steadily until 90–100 days and then increased slightly. Although the bananas did not ripen "properly" (as compared with detached fruits—see p. 76), the changes occurring during the period 110–130 days clearly indicated that ripening processes were initiated on the plant.

The alcohol-soluble nitrogen of banana pulp changes markedly during development (Fig. 2). The period of nitrogen depletion corresponds to the period of most rapid expansion of the fruit, and Steward *et al.* (1960b) suggested that the rapidly growing fruit utilizes nitrogen reserves for protein synthesis faster than they can be replenished. As the fruit approaches maturity and growth slows, soluble nitrogen reserves again accumulate. No data on protein content were provided. Perhaps the most striking feature of this study was the marked change in the composition of the soluble nitrogen during the development sequence. In the young inflorescence, 80% of the nitrogen was in alanine, arginine, proline, γ-aminobutyric, aspartic and glutamic acids. These were depleted during growth and at harvest maturity (93–115 days after emergence) about 70% of the soluble nitrogen was in asparagine, glutamine and histidine. There were significant variations in the pattern of nitrogen distribution between fruits grown in the hotter, summer months and those grown in the cooler, winter months, reflecting the slower growth of the latter.

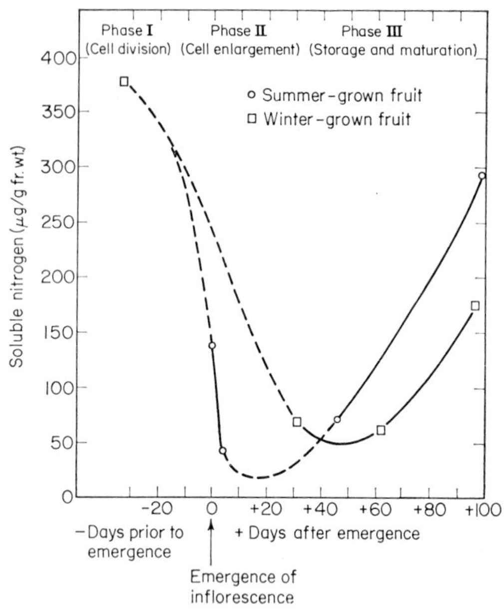

FIG. 2. Change in total alcohol-soluble nitrogen of Gros Michel pulp during development. From Steward *et al.* (1960b).

Total soluble nitrogen in the peel followed the same general trends as in the pulp, although the nitrogen content was much lower (ca. 5 μg/g fr. wt. at the minimum and 40 μg/g fr. wt. at harvest). γ-Aminobutyric acid was the predominant amino acid in the immature peel, making up 65–68% of the soluble nitrogen in the period 60–90 days after emergence. From this period on, asparagine and glutamine accumulated rapidly and these amides plus γ-aminobutyric acid made up 60% of the soluble nitrogen at harvest.

The ethylene content of bananas is essentially constant at about 0·2 p.p.m. throughout growth and development, until ripening commences (Burg and Burg, 1965b).

Buckley (1965) studied the synthesis and accumulation of dopamine (the primary substrate for browning reactions) in the peel of developing banana fruits. Each green peel contained about 70 mg of dopamine (1·0–1·2 mg/

g fr. wt.) at harvest. Some 10–15% of the dopamine accumulated prior to emergence, 85–90% accumulated during the first month after emergence and there was little or no change thereafter until ripening was initiated. The plateau in dopamine content during the last two months of development apparently resulted from a balance between synthesis and utilization, since ^{14}C tracer studies showed that dopamine was actively synthesized in the peel throughout fruit development.

D. Effect of Cultural Practices, Pests and Diseases

1. *Cultural practices*

The effect of cultural practices on banana yields has been extensively studied. Since few of these studies touch directly on the biochemistry or physiology of the fruit, only a brief summary is given here, drawn largely from Wardlaw (1961) and Simmonds (1966). Freiberg (1966) should also be consulted for a comprehensive review of fertilization practices.

For maximum yields, bananas require liberal and properly spaced applications of nutrients, particularly nitrogen. The highest recorded yields are just under 30 tons/acre; average yields are substantially lower, perhaps 3–7 tons/acre. Simmonds (1966) points out that banana yields are roughly comparable with those obtained with potatoes.

Irrigation is necessary in drier areas and is desirable in some areas to supplement rainfall.

Since the banana plant is a stooling perennial, plantations are potentially everlasting. However, commercial plantings tend to be renewed after 5 to 20 years of continuous cultivation; in a few areas fields are replanted after each crop has been harvested.

There is little evidence concerning the effect of cultural practices on ripening behaviour, nutritive value, flavour or shelf-life. It is generally recognized that fruits grown under sub-optimal conditions take longer to reach harvest grades and are, therefore, more prone to premature ripening. Preliminary studies of the effect of nitrogen fertilization on flavour suggest that off-flavours tend to occur more often when the supply of nitrogen is restricted (J. K. Palmer, unpublished); for example, when fertilized with 2·5 lb nitrogen per plant, the flavour of Valery† bananas was uniformly good to excellent whilst at 0·5 lb per plant "earthy" or "musty" off-flavours developed. A high content of glutamine and asparagine in shipped fruit was correlated with stronger banana aroma in the ripe fruit, as measured by organoleptic tests and by gas chromatography (E. H. Buckley, unpublished).

† Valery® is the registered trademark of United Fruit Co. for a Cavendish variety which predominates in this Company's plantations.

2. *Pests*

Some 209 insect pests, including 7 mites and 17 eelworms, have been reported to attack the banana plant. Only 5 of these have been intensively studied and, broadly speaking, insects and other pests have not been as troublesome to banana growers as diseases. Nevertheless, they are a substantial problem in many areas and constant vigilance is required to keep them in check.

3. *Diseases*

There are four major diseases which attack the banana plant and affect the fruit. Wardlaw (1961) and Simmonds (1966) have reviewed the distribution, biology, host range and control of these diseases.

(a) *Panama Disease* (Banana Wilt) is caused by a soil-borne fungus and normally results in the death of the entire plant. Simmonds (1966) compares Panama Disease with wheat rust or potato blight, in terms of the damage done. The only effective control is by cultivation of resistant clones. In tropical America, where the disease has been responsible for the loss of thousands of acres of bananas, the non-resistant Gros Michel is rapidly being replaced by resistant Cavendish varieties. The disease is not a problem in most other growing areas, since Cavendish varieties have been grown from the start.

(b) *Sigatoka Disease* (Leaf Spot) is caused by an air-borne fungus, which attacks the leaves and thus reduces the area available for photosynthesis. The disease has little effect on vegetative growth, but it does retard fruit growth. The fruits from infected plants take longer to reach harvest grades and are often reduced in size. These "older" fruits are more prone to premature ripening. Sigatoka Disease can be controlled by repeated spray-applications of copper or zinc fungicides and/or selected oils. However, oil spraying can significantly reduce banana yields, even at 0·5–1 gallon/acre.

(c) *Bunchy-top* is a virus disease which devastated the Australian banana industry in the 1920s. It was subsequently controlled by "one of the most striking examples of disease control by purely phytosanitary methods in the whole history of plant pathology" (Simmonds, 1966). In affected plants, there is vegetative stunting and the fruit bunches develop abnormally, often splitting the pseudostem before they emerge. The fruits are small and generally unsaleable.

(d) *Bacterial Wilt* (Moko Disease) is a problem in tropical America. The external symptoms (wilting, yellowing, necrosis) are similar to Panama Disease. Infected plants generally die, but a few survive to produce fruit. The bunches from infected plants show premature ripening and rotting of isolated fingers, scattered randomly among green fruits. The disease has been brought under partial control by phytosanitary methods.

III. POST-HARVEST PHYSIOLOGY AND BIOCHEMISTRY

A. Introduction

Wardlaw (1961), Simmonds (1966) and Freiberg (1969) provide details of the harvest, transport and ripening practices used around the world. A brief description of the more common techniques will serve as introduction to the present consideration of post-harvest physiology and biochemistry.

All bananas for export are harvested in the green state. The maturity at harvest is governed largely by the time required to transport the bananas to market, although there are clonal and seasonal considerations. Growers try to harvest the fruit at the most advanced stage consistent with the fruit arriving green at the ripening rooms. This stage is determined by experience and judged largely by the visual appearance of the hanging "stem" or "bunch" and particularly by the angularity of the individual banana "fingers". The terminology and criteria vary, but the common practice is to designate bananas as Three-Quarters, (fruits at about one-half their possible maximum size, with clearly visible angles), Full Three-Quarters (fruits with less prominent angles) and Full (fruit from which the angles have virtually disappeared). Intermediate designations are often adopted.

The stems are usually harvested by cutting down the banana plant, while taking precautions to prevent damage to the fruits. The predominant practice at present is to cut the individual "hands" of fruit from the stalk, wash briefly to prevent staining by the exuded latex (which turns brown on exposure to air), treat the cut surfaces with fungicides and pack the hands in fibre-board cartons. The cartons may contain up to 42 lb of fruit and often contain polyethylene film to protect the bananas from abrasion en route.

Export bananas are generally transported on refrigerated ships at 11–13·5°C, precautions being taken to keep ethylene concentrations as low as possible (see Section IIIB). In some areas, transport is via ventilated rail car.

In tropical countries, fruit for local consumption is harvested at the Full stage and ripened by hanging the bunch in a shady spot for a few days. In commercial trade, the green bananas are placed in special ripening rooms and ripening is initiated by treatment with about 1000 p.p.m. of ethylene, at temperatures in the range of 14·5–21°C and at high relative humidity. These conditions may vary considerably to suit the variety, the trade conditions, etc. (United Fruit Co., 1964; Hall, 1967).

B. Initiation of Ripening

The questions of how and when ripening is initiated are of primary concern in the banana trade. The majority of post-harvest losses result from "ripes and turnings" on arrival of ships in port. Ripe fruits are virtually a total loss; "turning" fruits (fruits beginning to show yellow colour in the peel) are down-graded. Under carefully controlled conditions of culture, harvest and transport, 0·5–3% of the boxes in a cargo will contain ripes and turnings on the average; 5–10% is not uncommon (personal communication, J. Stier, United Fruit Co.) and, under less favourable conditions, 20–50% may be reached. A loss of 1% of the average cargo represents a loss of about 40,000 lb of bananas.

Studies on initiation of banana ripening have centred around the role of ethylene ever since Harvey (1928) showed that this gas could initiate ripening in bananas and Niederl *et al.* (1938) demonstrated that bananas produce ethylene naturally. Other key references are Gane (1937), Biale *et al.* (1954), Burg and Burg (1965b) and Mapson and Robinson (1966). A detailed analysis of the role of ethylene in fruit ripening is presented in Volume 1, Chapter 16. The subject has also been summarized by Burg and Burg (1965a). Discussion here is limited primarily to information which is pertinent in commercial transport and ripening of bananas.

Picking of green banana fruits hastens ripening, evidently by lowering the threshold of sensitivity to endogenous ethylene (Burg and Burg, 1965b). Thus, bananas harvested at the usual export grades (Three-Quarters or Full Three-Quarters) ripen within 1 to 2 weeks at 18·5–24°C, whereas comparable fruits left on the plant remain green for another 40–50 days. "Full" grades generally ripen immediately after picking but immature bananas often will not ripen without application of external ethylene. Summer-grown fruit reaches a particular grade sooner, is said to be "physiologically younger", than comparable winter-grown fruit, and tends to remain green longer after cutting. This makes it difficult to predict the ripening behaviour of harvested fruits in growing areas with relatively large seasonal variations in temperature, such as Australia. Peacock and Blake (1969) have measured the "green-life" (time elapsed between harvest and initiation of ripening) of Australian Cavendish bananas in relation to a range of cultural and storage conditions. They found "green-life" to be a useful measure of the physiological maturity of green fruits at harvest.

United Fruit Company scientists have measured the time for summer-grown, Central American bananas to ripen at 18·5°C after cutting, as affected by age (days from emergence to cutting) of the fruit. The time to ripen was proportional to age within the range 90–120 days (the grade presumably varying from Light-Full to Full). The 90-day Cocos (a Gros Michel sport) ripened in 17 days, 120-day Cocos in 9 days. Corresponding figures for Valery (a Cavendish variety) were 21 days and 14 days. The greater resistance of Valery to spontaneous ripening is reflected in fewer "ripes and turnings" in Valery shipments.

Burg and Burg (1965b) have studied the relation between ethylene production and ripening in bananas. Fruits at harvest contained 0·2 p.p.m. of ethylene. About 4 hours before initiation of ripening, the ethylene content abruptly increased, averaging about 0·5 p.p.m. at initiation.

TABLE I. Initiation of banana ripening by low concentrations of ethylene[a,b]

C_2H_4 p.p.m.	Days exposure to reach colour grade 2[c]	
	80-day fruits[d]	111-day fruits[d]
0	18	16
0·1	15	10
0·3	7	4
0·5	6	3·5
1·0	5	3
5·0	3	2

[a] United Fruit Co., unpublished work.

[b] Valery bananas exposed continuously to C_2H_4 concentrations noted and colour grade rated daily.

[c] Colour grade 2 is ripening stage when yellow colour is just perceptible (Loesecke, 1950), typically about 2 days after onset of the climacteric.

[d] Days from emergence to cutting.

Application of ethylene concentrations of 0·1 p.p.m. or higher to green bananas accelerates the onset of the climacteric (Biale *et al.*, 1954; Brady *et al.*, 1970b). Concentrations of about 1 p.p.m. or higher generally induce the climacteric within 12 hours; lower concentrations must be applied for longer periods. Table I shows data which illustrate this time–concentration relationship at low concentrations of ethylene. It was also shown that temperature markedly alters the effects of ethylene. At 13·5–15·5°C ripening was not initiated by 24 hours of exposure to 100 p.p.m. ethylene. Ripening was initiated by 16–20 hours exposure at 18·5–23°C. In practice it is only necessary to protect the fruit from a build-up of ethylene during the 2 to3 days from harvest until the fruit reaches the transport temperature of about 13°C.

Initiation of ripening can be delayed for weeks or months by holding the green fruit in an atmosphere of 1–10% oxygen, 5–10% CO_2 or a combination of low O_2 and high CO_2 (Young *et al.*, 1962; Mapson and Robinson, 1966; W. B. McGlasson, personal communication). Similar effects can be achieved by ventilating the green fruits with air at sub-atmospheric pressures (Burg and Burg, 1966). Increasing application of controlled atmosphere storage for the longer voyages, usually by sealing the fruit in polyethylene bags of selective permeability, with or without an ethylene absorbent (Scott and Roberts, 1966; Scott *et al.*, 1968; Badran, 1969), should virtually eliminate the problem of premature ripening.

Irradiation with doses of 25–35 krads are also reported to delay initiation of "natural" ripening without interfering with ethylene-induced ripening or affecting the quality of the fruit (Maxie and Sommer, 1968). Ferguson *et al.* (1966), however, report no effect on ripening from doses up to 200 krads. No respiration data were obtained in the latter study; probably ripening had already been initiated prior to irradiation.

Evidence for the role of growth-regulating substances in the initiation of ripening in bananas is somewhat conflicting. The ovary and immature fruit contain substances which promote growth of carrot root explants (Steward and Simmonds, 1954); indolyl-3-acetic acid (IAA) and two giberellin-like substances occur in developing banana fruits (Khalifah, 1966a, b). Results with whole bananas (Mitchell and Marth, 1944; Freiberg, 1955; Blake and Stevenson, 1959; Dedolph and Goto, 1960) indicate that 2,4-dichlorophenoxyacetic acid (2,4-D) or 2,4,5-trichlorophenoxyacetic acid (2,4,5-T) accelerate ripening. Recently Vendrell (1970) found IAA and 2,4-D to delay the ripening of ethylene-treated banana slices, despite the fact that respiration and ethylene production were accelerated. Wade and Brady (1970) demonstrated a delay in the onset of the climacteric and of other ripening changes in banana fruit slices treated with kinetin (10^{-4} M). Respiratory responses and pulp ripening were not affected when kinetin-treated slices were exposed to ethylene, but peel degreening was delayed.

C. Respiratory and Compositional Changes during Ripening

Despite obvious limitations in technology, workers such as Kidd and West, Gane, Leonard, Barnell, Wardlaw and also Poland's group at United Fruit Company laid a firm foundation in the period 1930 to 1945 for subsequent studies of the biochemistry of banana ripening. Loesecke (1950) has reviewed this earlier work in detail. More recent reviews (Biale, 1960a, b; Wick *et al.*, 1966) and the books by Wardlaw (1961) and Simmonds (1966) include information on banana physiology and biochemistry.

The purpose of this section will be to outline the chemical and respiratory changes which occur during banana ripening and to review the meagre information on the enzymes involved in these changes.

1. *Respiration*

The ripening banana fruit exhibits a climacteric pattern of respiration (Loesecke, 1950; Biale, 1960a; Palmer and McGlasson, 1969). After harvest, at 20°C, the respiration rate in 2–4 days rises from a steady value of about

20 mg CO_2/kg/hour in the hard, green fruit to about 125 at the climacteric peak and then falls to about 100 as ripening proceeds.

Considerable variation from this typical pattern has been reported. Some fruits pass through the classic pre-climacteric minimum (Tager and Biale, 1957; Sacher, 1967); in others the respiration rate rises continuously from harvest until the onset of the climacteric (Peacock and Blake, 1969). Pre-climacteric respiration rates at 16–24°C can vary from 8 to about 50 mg CO_2/kg/hour; climacteric rates from 60 to about 250 mg CO_2/kg/hour. The increase from pre-climacteric to climacteric varies from four- to ten-fold.

TABLE II. Respiratory Quotient of bananas induced to ripen with ethylene[a]

Time from application of ethylene (Hours)	Ethylene added to air stream (p.p.m.) 1		100	
	Respiration[b]	R.Q.[c]	Respiration[b]	R.Q.[c]
0	19·4	1·0 ±0·10	19·0	0·93±0·11
10·3	25·0	0·81±0·05	28·0	0·86±0·05
14·8	30·6	0·83±0·05	38·9	0·84±0·05
21·3	46·5	0·83±0·02	61·3	0·84±0·02
25·5	58·5	0·76±0·03	76·0	0·90±0·04
32·8	82·7	0·84±0·05	108·2	0·85±0·03
37·3	95·8	0·87±0·02	119·0	0·99±0·09
55·3	127·2	0·97±0·04	140·8	—
72	129·0	1·02±0·02	136·2	1·01±0·02

[a] From Brady *et al.* (1969b).
[b] mg CO_2 kg/hour. Means for 5 fruits.
[c] Mean R.Q. and range for 5 fruits.

Respiration rate increases with increasing temperature, Q_{10} (°C) being about 2·2 (Gane, 1936 as quoted by Simmonds, 1966).

There is a distinct lowering of the Respiratory Quotient (R.Q.) of whole banana fruits during the climacteric, followed by a return to near 1 in the post-climacteric period (Table II). The R.Q. of ripening slices remains close to 1 throughout the climacteric.

Little information on respiratory enzymes in bananas is available. Tager and Biale (1957) reported increases in carboxylase and aldolase activity during ripening, in agreement with earlier evidence (Tager, 1956) of a shift from the pentose shunt in pre-climacteric fruits to the glycolytic pathway in post-climacteric fruits. However, Young (1965) showed that the activity of these enzymes is constant during ripening if precautions are taken during extraction to prevent inactivation by tannins.

Despite the absence of certain key organic acids in banana pulp (Wyman and Palmer, 1964), the Krebs cycle is apparently operative, since functional mitochondria can be isolated throughout the climacteric (Tager, 1958; Haard, 1967; see Section IVD for further details on banana mitochondria). Cytochrome oxidase is assumed to act as terminal oxidase, although the banana fruit contains a highly active polyphenol oxidase (Palmer, 1963).

2. *Water relations*

Stomata occur on the surface of mature banana peel at average densities of 480/cm^2 (as compared with about 17,000/cm^2 on the surface of the banana leaf). These stomata are functional in excised green fruits and can be induced to open by exposure to low intensity light (10 f.c.) when the fruits are held at 90–100% relative humidity (Johnson and Brun, 1966).

Transpiration is relatively constant in the mature green fruit. Once ripening is initiated, a plot of transpiration rate versus time for individual bananas is remarkably similar to the climacteric curve for respiration. The peak transpiration rate is about twice that of the green fruit (Leonard, 1941, quoted by Simmonds, 1966).

Despite transpirational losses, the moisture content of banana pulp normally increases during ripening, from about 69% ($\pm$4%) to about 74% ($\pm$3%). The water derived from breakdown of carbohydrates, presumably during respiration, contributes to this net increase. Probably a more significant factor is the osmotic withdrawal of moisture from the peel. A marked difference in osmotic pressure between peel and pulp develops during ripening, largely because sugar content increases more rapidly in the pulp than in the peel (Stratton and Loesecke, 1931). The osmotic pressure of pre-climacteric peel and pulp is about 6 atmospheres. The peel pressure increases only slightly during the climacteric, then increases to about 11·5 atmospheres at the fully ripe stage. The pulp pressure increases to about 6·5 atmospheres during the climacteric and then rises sharply, reaching about 25–27 atmospheres when the fruit is fully ripe.

This osmotic transfer of moisture is reflected in changes in the weight ratio of pulp to peel which is about 1·2–1·6 in the green fruit and rises to 2·0–2·7 when the fruit is fully ripe. The pulp to peel ratio has been suggested as a "coefficient of ripeness" (Loesecke, 1950).

3. *Carbohydrates*†

The most striking chemical changes which occur during the post-harvest ripening of the banana are the hydrolysis of starch and the accumulation of sugars (Loesecke, 1950). About 20–25% of the pulp of the fresh green fruit is starch. In the week or so from initiation to completion of ripening, the

† See Volume 1, Chapters 1 and 2.

starch is almost completely hydrolysed, only 1–2% remaining in the fully ripe fruit. Sugars, normally 1–2% in the pulp of green fruits, increase to 15–20% in the ripe pulp. Total carbohydrate decreases 2–5% during ripening, presumably as sugars are utilized in respiration.

The green peel contains about 3% starch, localized mostly in cells adjacent to the pulp. This starch is also hydrolysed during ripening with concomitant accumulation of sugars.

Plantains differ from sweet bananas in having a higher starch content in the pulp throughout: 30% at harvest and 5–10% when ripe. The sugar content of plantains is similar to that of sweet bananas.

Poland *et al.* (1937) identified sucrose, glucose and fructose as the major sugars in banana pulp. The three sugars all increased during ripening, maintaining nearly constant proportions of 66% sucrose, 14% fructose and 20% glucose (Poland *et al.*, 1938). Maltose was the only other sugar found; it was present in trace amounts in ripe Gros Michel.

Bates *et al.* (1943) reported banana starch to contain 20·5% amylose; it has not been characterized otherwise.

Uridine diphosphoglucose (UDPG), an intermediate in the synthesis of sucrose by higher plants, is present in green Cavendish fruits (Rowan, 1959). Uridine diphosphogalactose was tentatively identified. Fructosyl-sucrose has also been tentatively identified in banana pulp (Henderson *et al.*, 1959).

Despite the massive synthesis and hydrolysis of starch in the developing and ripening banana, virtually nothing is known about the enzymes or the mechanisms involved. Yang and Ho (1958) proposed that starch breakdown was catalysed by phosphorylase, but presented little evidence to support their proposal.

4. *Pectic substances, cellulose and hemicelluloses*†

The interconversion of pectic substances is presumed to be involved in the characteristic softening which occurs during fruit ripening. In the pulp of bananas, insoluble protopectin decreases from about 0·5 to about 0·3% fr. wt. and soluble pectin shows a corresponding increase during ripening (Loesecke, 1950).

Banana pulp at harvest contains 2–3% cellulose and this decreases slightly during ripening.

The hemicelluloses make up 8–10% of the fresh banana pulp in the green fruit, decreasing to about 1% in the ripe fruit (Barnell, 1943). Barnell suggested that the hemicelluloses are labile reserve carbohydrates and a potential source of sugars, acids and other respiratory substances during ripening.

These substances were isolated by relatively crude differential extraction

† See Volume 1, Chapter 3.

procedures. Further work is needed to characterize these fractions and to explore their presumed role in softening or as reserve carbohydrates.

One of the enzymes involved in pectic conversions in fruits is pectin methyl esterase (PME). Three molecular forms of PME may be isolated from banana fruit pulp by successive extractions with water (Fraction I), 0·15 M NaCl (Fraction II) and 0·15 M NaCl at pH 7·5 (Fraction III) (Hultin and Levine, 1963; Hultin *et al.*, 1966). Conflicting results have been reported on the change of total PME activity in Gros Michel bananas during ripening since Hultin and Levine (1965) reported a four-fold increase, with the water-soluble fraction (I) showing the biggest change in activity (three-fold increase), while DeSwardt and Maxie (1967) found no change in total PME activity. The total PME activity of Australian Cavendish bananas also remains constant during ripening, but the water-soluble PME (corresponding to Hultin's Fraction I) increases three- to eight-fold (J. K. Palmer, unpublished). Precautions were taken in all of the PME studies to minimize inactivation by tannins, particularly in the pre-climacteric fruits. It is not clear why the results differ.

5. *Pigments*†

The change in colour of the peel from green to yellow is the most obvious change which occurs during ripening and serves as a rough guide to the stage of ripeness. Yellowing begins at or shortly after the climacteric peak and the fruit becomes fully yellow within about 3–7 days at normal ripening temperatures.

Loesecke (1929, 1950) presents a colour chart of the typical colour changes and summarizes information on the chemistry of the colour changes. The green banana peel contains about 50–100 μg/g fr. wt. chlorophyll, 5–7 μg/g fr. wt. xanthophyll and 1·5–3·5 μg/g fr. wt. carotene. During ripening all the chlorophyll is lost and total yellow pigment remains approximately constant. Brady *et al.* (1970b) found that peel tissue lost about 50% of its chlorophyll (originally 75 μg/g fr. wt.) in 2 days of exposure to 10 or 100 p.p.m. ethylene and the chlorophyll content was near zero at 6–7 days. In fruits ripened with 0·1 p.p.m. ethylene it required 5 days to deplete 50% of the chlorophyll and about 20% remained after 7 days. Looney and Patterson (1967) report that chlorophyllase activity in banana peels increases sharply at the onset of the climacteric, rises to a peak which coincides with the climacteric peak, and then falls to near zero in the post-climacteric period.

6. *Volatile constituents*

The banana fruit contains at least 200 individual volatile components (Wick *et al.*, 1969). Although identification of these substances has proceeded

† See Volume 1, Chapter 12.

quite rapidly (see Volume 1, Chapter 10), many remain unidentified and little is known about the contribution of individual volatiles to the characteristic flavour and aroma. Even less is known about the mechanism of biosynthesis of these volatiles. Wick *et al.* (1966) have reviewed the studies up to the end of 1965.

McCarthy and Palmer (1964) investigated the production of volatiles in ripening Gros Michel and Valery bananas. No significant quantities of volatiles appeared until about 24 hours after the climacteric. Valery bananas produced larger quantities and a more complex mixture of volatiles than Gros Michel and began to produce these earlier in the post-climacteric period (Fig. 3). This correlated well with the earlier development of flavour in Valery. McCarthy *et al.* (1963) extended these studies to confirm that the

TABLE III. Sensory impressions of major banana volatiles[a]

Banana-like	Fruity	Green, woody or musty
Isoamyl acetate	Butyl acetate	Methyl acetate
Amyl acetate	Butyl butyrate	Pentanone
Amyl propionate	Hexyl acetate	Butyl alcohol
Amyl butyrate	Amyl butyrate	Amyl alcohol
		Hexyl alcohol

[a] McCarthy *et al.* (1963).

increasing concentration and complexity of volatiles during ripening was correlated with increasing development of banana flavour, as measured by the flavour profile technique (Cairncross and Sjöström, 1950). They also concluded that the major volatile components could be classified according to three general sensory impressions (Table III). Valery bananas consistently had a fuller and more interesting flavour than Gros Michel, primarily because Valery contained higher concentrations of "fruity" constituents. The combined gas chromatography–flavour profile technique subsequently proved useful in the laboratories of United Fruit Co. to assess the flavour of additional varieties of bananas and to measure the effect of cultural and handling practices on banana flavour (A. I. McCarthy and J. K. Palmer, unpublished). Nevertheless, it is clear that these techniques are only rough approximations and that considerably more chemical and sensory information is required to evaluate fully and to define banana flavour.

There have been few studies concerning the biosynthesis of volatile constituents and only preliminary conclusions can be drawn from these. Hewitt *et al.* (1960) demonstrated that enzymes are required for flavour development. They reported enhancement of banana flavour when a crude banana enzyme was added to an aqueous extract of blanched bananas. Similarly Hultin and Proctor (1962) found that a crude enzyme preparation from ripe banana pulp

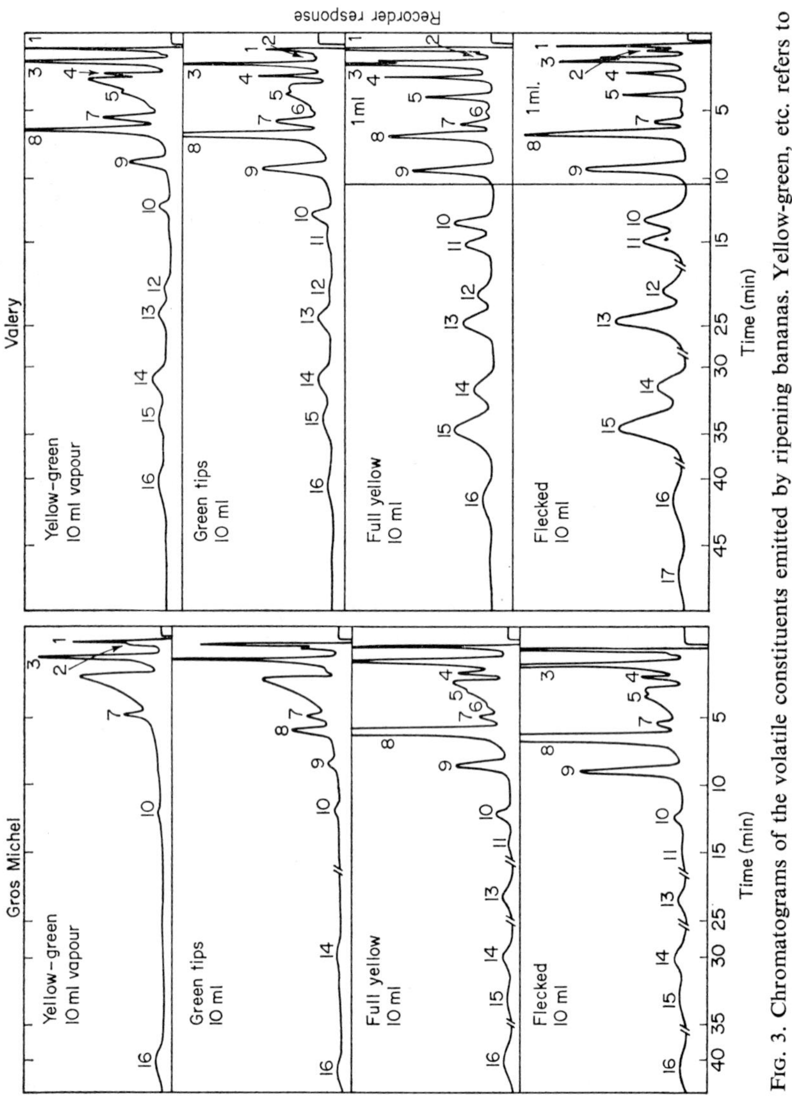

FIG. 3. Chromatograms of the volatile constituents emitted by ripening bananas. Yellow-green, etc. refers to the peel colour. Reproduced from McCarthy *et al.* (1963).

would stimulate development of banana aroma in heat-processed banana purée, particularly if supplemented with valine, oleic acid or pyruvic acid.

Leucine, isoleucine and valine have been tested as precursors of branched-chain alcohols, on the assumption that the biosynthetic pathways in bananas are similar to those involved in formation of fusel oil during fermentation (Guymon, 1966). Wyman *et al.* (1964), utilizing the technique of Buckley

TABLE IV. ^{14}C in banana volatiles after introducing ^{14}C-labelled amino acids—c.p.m. above background[a]

	Test precursors				
	L-Leucine	L-Isoleucine		L-Valine	
Compounds	U-^{14}C	U-^{14}C	1-^{14}C	U-^{14}C	1-^{14}C
Low boilers[b]	430±9[c]	630±10	8·8±4·0	996±13	3·2
i-Butyl acetate	31±4	36±5	2·1	1328±15	−0·9
i-Butyl alcohol	2·1	60±5	−2·9	161±6	2·5
n-Butyl acetate	19±4	228±7	2·9	60±5	2·0
n-Butyl alcohol + 2-pentyl acetate	53±5	32±5	−0·1	5·0±4·0	1·6
i-Amyl acetates	1638±16	473±9	15±4	80±5	9·4±4·0
i-Amyl alcohols[d]	160±6	50±5	−0·1	9·0±4·0	4·4±4·0
n-Amyl acetate	2·6	1·2	2·3	−0·3	3·0
n-Butyl butyrate	1·1	3·1	1·6	0·3	1·6
n-Amyl propionate	0·2	−0·1	−0·3	0·0	2·1
n-Hexyl acetate	−0·1	0·7	−0·1	0·2	0·5
i-Amyl butyrate	0·5	0·4	−1·8	0·1	0·1
n-Hexyl alcohol	0·2	0·7	1·9	−0·1	0·6
n-Amyl butyrate	0·7	0·2	1·2	−0·2	0·9
Blank	0·3	−0·1	0·8	0·3	−1·2
Blank	0·0	0·0	3·5	−0·2	−1·4

[a] 5–10 μc ^{14}C introduced into each slice. Only c.p.m. greater than 4 c.p.m. above background are significant at the 99·7% confidence limits.
[b] Includes methyl acetate, ethanol and 2-pentanone fractions.
[c] Three sigma errors of the mean count rate.
[d] 3-Methyl-n-butanol and 2-methyl-n-butanol.

(1962), incubated ripe banana slices with ^{14}C-labelled amino acids for 16 hours, collected a sample of the evolved volatiles and measured ^{14}C in appropriate fractions from a gas chromatograph. The labelling pattern (Table IV) suggested that valine was a precursor of isobutyl alcohol and isobutyl acetate. Similarly, the leucines appeared to be precursors of amyl alcohols and amyl acetates. Negligible labelling from C-1-labelled amino acids indicated decarboxylation of the amino acids during biosynthesis of the volatiles. Most of the ^{14}C in the oxo acid fraction was in α-oxoiso-caproic acid when ^{14}C leucine was fed. These results were all consistent with the fermentation scheme. However, these conclusions must be considered preliminary, as an appreciable proportion of the ^{14}C appeared in volatiles other than the predicted alcohols and esters, and later work on the banana volatiles indicated that the trapped fractions probably contained more than one component.

Myers *et al.* (1970) fed L-leucine-U-^{14}C to slices taken from bananas at various stages of ripeness. The volatiles were isolated by vacuum distillation and ether extraction of the distillate. Leucine uptake was rapid and about 1% of the leucine was converted to volatiles within 30 min. From 50 to 78% of the total radioactivity in the volatiles was in isoamyl alcohol after a 1 hour incubation, suggesting again that leucine is a precursor of this alcohol.

Drawert *et al.* (1966) found that 2-hexenal and hexanal were formed from linolenic and linoleic acids, respectively, when banana tissues were homogenised in the presence of air. The reactions were enzymatically catalysed.

7. *Phenolic substances and enzymic browning*†

The tannins and other phenolic substances of the banana fruit have been little studied, with the exception of dopamine (3,4-dihydroxy phenylethylamine), the primary substrate for browning reactions. In this section, dopamine is considered primarily as a phenolic substance and in relation to its role in browning. In sub-section 11, *Amines*, dopamine is considered as a catecholamine.

"Tannins" are chemically ill-defined phenolic substances, generally detected and estimated by various more or less specific colour reactions (Swain and Hillis, 1959; Goldstein and Swain, 1963). Certain tannins are believed to be responsible for the sensation of astringency, via cross-linking of the proteins in the mouth.

The pulp of green banana fruit is markedly astringent but this astringency is greatly reduced during ripening. Barnell and Barnell (1945) observed that banana tannins were mainly confined to the latex vessels of the pulp and skin and to small scattered cells of the outer and middle regions of the skin. In green, unripe bananas the contents of these cells (the "latex") was a viscous liquid which stained evenly with tannin reagents. During ripening the latex tended to dry out, finally becoming caked and brittle and withdrawn from the walls. The dried latex no longer reacted or reacted only faintly with the tannin reagents, although the tannins in the scattered cells of the peel still stained darkly. Some tannins appeared in cells adjacent to the latex vessels as ripening proceeded. "Active" tannins (tannins which precipitated diastase) decreased in the ripe pulp to about one-fifth of their value in the green, preclimacteric fruit. The peel contained 3 to 5 times more "active" tannins than the pulp; these were also sharply reduced during ripening. The decreases correlated well with the observed loss of astringency.

† See Volume 1, Chapter 11.

Goldstein and Swain (1963) have presented preliminary evidence that loss of astringency during ripening of bananas results from increased polymerization of the tannins.

The tannins in the green banana fruit interfere with the extraction of enzymes (Young, 1965). A number of tannin-precipitating agents have proved effective in reducing enzyme losses, including caffeine (Young, 1965), polyvinylpyrollidone (Badran and Jones, 1965), polyethylene glycols or derivatives (Palmer, 1963; Badran and Jones, 1965; Young, 1965) and proteins (Tager, 1958). Haard (1967) has compared the effectiveness of these substances in preventing complexing of mitochondrial proteins in the banana and has concluded that casein gave the most consistent results, although polyethylene glycol and caffeine were almost equally effective. It is clear that considerably more work is required on the banana tannins to elucidate their chemical structure, their role in astringency and their metabolic significance.

Robinson (1937) detected delphinidin in the hydrolysis products from banana pulp and Simmonds (1954) reported large amounts of leucoanthocyanidins in the peel and pulp of edible varieties, with delphinidin predominating over cyanidin. These components are apparently derived from the latex (D. E. Jones and W. G. C. Forsyth, unpublished).

Dopamine was reported to occur in high concentration (700 μg/g fr. wt.) in banana peel and to be present in pulp (8 μg/g fr. wt.) by Waalkes *et al.* (1958). Griffiths (1959) confirmed the presence of dopamine in bananas and demonstrated that it was the primary substrate in enzymatic browning. Griffiths (1961) later reported that dopamine was the only major phenolic constituent in the peel of 16 tested banana clones. The content of dopamine was determined by the genotype, being highest in *M. acuminata* cultivars, intermediate in hybrids and lowest in *M. balbisiana* itself.

These observations prompted a series of investigations on dopamine accumulation and synthesis (Buckley, 1965) and on banana polyphenoloxidase (Palmer, 1963, 1965; Palmer and Roberts, 1967), since peel discolouration is a problem in the banana industry. The peel contained 1·0–1·2 mg dopamine/g fr. wt. or about 70 mg/whole peel at harvest (see Section IIC for discussion of dopamine accumulation during development). Usually this increased sharply during ripening, reaching values 30–60% higher than at harvest. Some fruits accumulated more than this, but winter-grown Honduran fruit usually showed no accumulation of dopamine during ripening. As the peel developed obvious brown flecks, the dopamine content dropped sharply.

Buckley (1965) also studied the biosynthesis of dopamine by measuring incorporation of ^{14}C from various potential precursors. The results indicate that the normal pathway of biosynthesis is via tyramine (Fig. 4). The pathway via DOPA (3,4-dihydroxyphenylalanine) is the usual pathway in other organisms, but this pathway appears to have little or no significance in bananas. DOPA has not been detected in bananas.

The banana system also differed from animal systems in that phenylalanine was not converted to tyrosine or dopamine, but gave rise instead to a variety of other phenolic substances, three tentatively identified as ferulic, p-coumaric and caffeic acids.

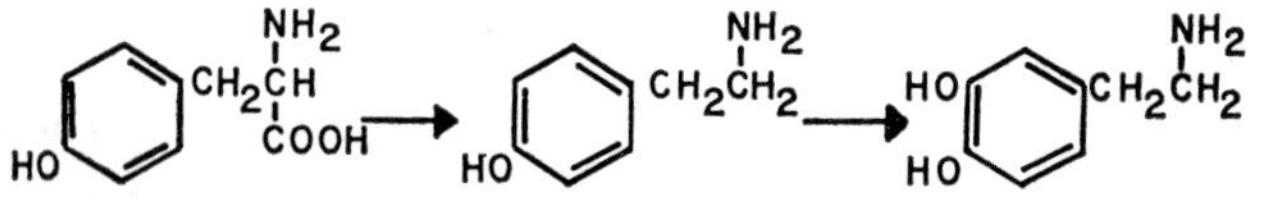

TYROSINE → TYRAMINE → DOPAMINE

FIG. 4. Biosynthetic pathway for dopamine in banana peel.

Palmer (1963) isolated a polyphenoloxidase (PPO) from banana pulp which catalysed the oxidation of a variety of diphenolic substances. Monophenols were not oxidized. Dopamine was the most reactive substrate for this enzyme. Typical Michaelis constants were $6{\cdot}3 \times 10^{-4}$ M for dopamine, $2{\cdot}6 \times 10^{-3}$ M for catechol, $3{\cdot}6 \times 10^{-3}$ M for L-arterenol and $6{\cdot}6 \times 10^{-2}$ M for L-dopa. The pH optimum of the enzyme is about 7·0. Spectrochemical studies indicate that dopamine is oxidized to brown pigments via 2,3-dihydroindole-5,6-quinone and indole-5,6-quinone; the reaction is inhibited by various reducing agents, chelating agents and structural analogues (Palmer, 1965). Sodium mercaptobenzothiazole is a particularly potent inhibitor of banana PPO (Palmer and Roberts, 1967). This compound delays the onset of substrate oxidation at 10^{-7} M and causes prolonged inhibition at 2×10^{-5} M or higher.

8. *Acidity: organic acids and oxo acids*†

The pH of banana pulp falls during ripening, from about 5·4 ($\pm$0·4) in pre-climacteric to about 4·5 ($\pm$0·3) in post-climacteric pulp. The titratable acidity of plantains (8 meq/100 g fr. wt.) is about twice that of sweet bananas (Loesecke, 1950).

Earlier studies (reviewed by Loesecke, 1950) indicated L-malic acid to be the principal non-volatile organic acid, but there were also reports of citric, oxalic and tartaric acids. More recently, several authors have reported citric and L-malic acids to predominate in banana pulp (Lulla, 1954; Wolf, 1958; Miller and Ross, 1963). Wolf (1958) also found oxalic and tartaric acids.

Steward *et al.* (1960a) reported malic and citric acids to predominate and identified shikimic, quinic, glycolic, glyceric, pyroglutamic, succinic and tartaric acids as minor constituents in ripe pulp of Gros Michel bananas. They also found 19 unidentified acidic substances. Colour tests indicated

† See Volume 1, Chapter 4.

that several of these were sugar acids; it would be interesting to know if these are products of the hydrolysis of hemicelluloses.

Wyman and Palmer (1964) found 14 organic acids, not including oxo acids, in Gros Michel pulp. Malic, citric and oxalic acids were major constituents; glutamic, aspartic, quinic, glyceric, glycolic, shikimic, succinic and glutaric acids were minor constituents (0·005–0·1 meq/100 g fr. wt.). Three acids were not identified and the absence of certain Krebs cycle acids, such as fumaric and isocitric acids, was noted.

Banana pulp appears to contain an exceptional variety of oxo acids. Steward *et al.* (1960a) found evidence for 22 oxo acids, 10 tentatively identified as succinic semialdehyde and α-oxoglutaric, pyruvic, β-hydroxypyruvic, α-oxoisovaleric, glyoxylic, oxalacetic, α-oxoisocaproic (and/or α-oxo, β-methyl valeric) and α-oxo, β-hydroxy pyruvic acids. Wyman and Palmer (1964) found 8 oxo acids and identified pyruvic, α-oxoglutaric, oxalacetic and glyoxylic acids. β-Hydroxypyruvic, α-oxoisocaproic and/or α-oxo-β-methylvaleric and 2 unknown acids were detected only as amino acid derivatives. In a later study (J. K. Palmer and A. Hynes, unpublished), the identification of pyruvic, α-oxoglutaric, oxalacetic and α-oxoisocaproic acids was confirmed and 2 additional acids, α-oxoisovaleric and phenylpyruvic, were identified.

Harris and Poland (1937) found malic acid to increase about 6·5-fold during ripening, from 0·8 to 5·3 meq/100 g fr. wt. Barker and Solomos (1962) reported a three to six-fold increase in malate, but presented no data. Wyman and Palmer (1964) reported that the total acidity (4·5 meq/100 g fr. wt.) of unripe Gros Michel pulp was 50% oxalic acid, 35% malic acid and 10% "citric peak" acidity (citric acid plus certain phosphates). During ripening (Table V), both malic acid and "citric peak" acidity increased three- to four-fold and oxalic acid dropped to 60% of its original value. The net result was a doubling in organic acidity during the climacteric, with malic acid becoming the major acid (65% of total acidity).

Oxalate generally occurs in plants as water-insoluble and metabolically-inert calcium oxalate. The oxalic acid in bananas is completely water-soluble (Wyman and Palmer, 1964) and the decrease during ripening suggested that the ripening banana was capable of metabolizing this acid. Feeding of oxalate-^{14}C to banana slices confirmed that it is metabolized (J. K. Palmer and H. Wyman, unpublished), but the products were not identified.

Steward *et al.* (1960b) reported substantial decreases in α-oxoglutaric and pyruvic acids during ripening of Gros Michel bananas while Barker and Solomos (1962) reported three- to six-fold increases in these same acids and in oxalacetic acid in a Cavendish variety.

Studies with ripening Valery bananas (J. K. Palmer and A. Hynes, unpublished) revealed yet a different pattern (Table VI). Pyruvic acid remained

TABLE V. CO_2 production, peel colour and organic acid content of bananas during ripening[a]

	Stages of ripening		
	Pre-climacteric	Climacteric	Post-climacteric
CO_2 production mg/100 g fr. wt. × hours	2·0–4·0 (steady)	10·0–17·5 (rising)	9·0–11·0 (erratic fluctuations)
Peel colours	Green	Yellow-green through yellow with green tips	Fully yellow through yellow flecked with brown
Organic acids, meq/ 100 g fr. wt.[b]			
Malic	1·36	5·37	6·20
"Citric peak"[c]	0·68	1·70	2·17
Oxalic	2·33	1·32	1·37
Other acids	0·19	0·16	0·17
Total organic acidity[d]	4·43	8·74	10·90

[a] From Wyman and Palmer (1964).
[b] Mean values for at least six samples from three independent experiments.
[c] Citric acid plus varying amounts of phosphates and trace of an unknown acid.
[d] Includes inorganic phosphate eluted immediately after "citric peak".

TABLE VI. Oxo acids in ripening Valery bananas[a]

	Peel Colour				
Acid	Green	Yellow-green	Green tips	Full yellow	Brown flecks
Pyruvic	+ + +[b]	+ + +	+ + +	+ + +	+ +
α-Oxoglutaric	+	+	+ +	+ + +	+ +
Oxalacetic	+	+	+ +	+ +	+ +
Phenylpyruvic	0	+ +	+	0	0
α-Oxoisovaleric	0	0	0	+ +	+ +
α-Oxoisocaproic	0	0	0	+ +	+ +

[a] Extracted from pulp tissue as phenylhydrazone derivatives by method of Isherwood and Niavis (1956). Separated on thin layer plates with benzene/acetone/acetic acid (85 : 10 : 5) or benzene/hexanol (1 : 1) equilibrated with 5% acetic acid.
[b] Intensity of colour following NaOH spray: 0 = not visible, + = faint, + + = moderate, + + + = strong.

essentially constant throughout ripening, while α-oxoglutaric and oxalacetic acids increased somewhat. Phenylpyruvic acid appeared and disappeared; the branched-chain acids, α-oxoisovaleric and α-oxoisocaproic, were detected only in fully-ripe fruits. The oxo acids of the banana fruit deserve further study.

9. *Lipids*†

Ether-extractable materials constitute between 0·2 and 0·5% of the fresh weight of bananas at all stages of ripeness (Loesecke, 1950). Grobois and

† See Volume 1, Chapter 9.

Mazliak (1964) identified palmitic, oleic and linolenic as the major fatty acids in both peel and pulp. They reported an increased proportion of fatty acids in the peel and a decreased proportion of unsaturated acids, especially palmitoleic, in the pulp during ripening. However, the magnitude of these changes was not reported.

Goldstein and Wick (1970) examined the lipids of the peel and pulp of pre-climacteric and fully-ripe Cavendish (var. Valery) bananas. Total lipids showed no significant change during ripening, averaging about 1% of the dry weight of the pulp and 6·5% of the peel. Table VII shows data on the

TABLE VII. Fatty acids of bananas mg/10 g dr. wt.[a]

Acid	Pulp		Peel	
	Unripe	Ripe	Unripe	Ripe
14:0	0	0	1·35	1·43
15:0	0·33	Trace	0	0
16:0	10·89	11·92	56·30	62·80
16:1	2·21	0·84	0	0
16:2	1·16	Trace	0	0
18:0	0·63	1·68	7·32	6·46
18:1	4·44	4·08	8·70	9·50
18:2	12·85	4·88	38·00	26·70
18:3	6·08	6·84	19·80	18·40
Total sat.	11·85	13·60	64·97	70·69
Total unsat.	26·47	16·64	66·50	54·60

[a] From Goldstein and Wick (1970).

fatty acid composition of the lipid fraction, determined by gas chromatography of the derived methyl esters. Palmitic (16:0), oleic (18:1), linoleic (18:2) and linolenic (18:3) were the major acids of the pulp. The tendency during ripening was towards loss of unsaturated acids (particularly linoleic acid) and towards slight increases in saturated acids, the net effect being a 20% loss of the total acids and a substantial increase in the degree of saturation of the fatty acids. In the peel, palmitic, linoleic and linolenic acids predominated. The peel acids changed only slightly during ripening, with the same tendency towards increased degree of saturation.

10. *Proteins and amino acids*†

The protein content of unripe banana pulp varies from 0·5 to 1·6% and there is no change during ripening (Stratton and Loesecke, 1930; Steward *et al.*, 1960b; Sacher, 1967; Brady *et al.*, 1970a) although Sacher (1967) reported a significant increase about 5 days before onset of the climacteric. Protein nitrogen makes up 60–65% of the total nitrogen (Steward *et al.*, 1960b; Brady *et al.*, 1970a). Total nitrogen remains constant in both peel and pulp

† See Volume 1, Chapters 5 and 6.

during ripening (N. W. Wade and C. J. Brady, 1970), indicating that the nitrogen data are not influenced by translocation. Reference should be made to Section IVC for information on synthesis of proteins during ripening.

Steward *et al.* (1960b), Buckley and Sullivan (1964) and Brady *et al.* (1970a) have determined the amino acids in the soluble-nitrogen fraction from several varieties of bananas. Table VIII shows some representative

TABLE VIII. Free amino acids of pre-climacteric banana fruit pulp (micromoles/g fr. wt. of pulp)

Amino acid	Steward *et al.* (1960b)[a]	Buckley and Sullivan (1964)[a]	Brady *et al.* (1970a)[b]
Aspartic acid	0·10	1·03	2·44
Glutamic acid	0·027	1·04	1·35
Serine	0·062	n.d.[c]	0·55
Glycine	0·030	0·24	0·55
Asparagine	0·45	n.d.	3·33
Threonine	0·024	n.d.	0·36
α-Alanine	0·11	0·38	0·56
Glutamine	0·39	n.d.	4·27
Histidine	0·14	2·10	6·09
Lysine	0·013	0·40	1·07
Arginine	0·014	1·42	1·25
Proline	0·017	0·18	0·18
Valine	0·008	0·37	0·11
Leucine } Isoleucine }	0·009	0·51	0·20
		0·17	0·12
Tyrosine	0·011	0·25	0·07
Phenylalanine	n.d.[c]	0·17	0·10
β-Alanine	Trace	n.d.	n.d.
γ-Aminobutyric acid	0·12	n.d.	0·55
Pipecolic acid	0·10	n.d.	n.d.

[a] Extracted with aqueous ethanol.
[b] Extracted with cold trichloroacetic acid.
[c] Not determined.

data for pre-climacteric pulp. The only outstanding features are the exceptionally high content of histidine and the presence of pipecolic acid. It is not clear why Steward *et al.* found so much lower concentrations of amino acids than Brady *et al.*, since the total soluble nitrogen in the samples analysed by the two groups was not appreciably different.

Manek and Fripp (1963) identified 18 ninhydrin-reacting compounds in acetone extracts of uncooked "Matoke" bananas (a variety of *M. paradisaca* eaten in large quantities as a boiled vegetable in Uganda). These included all the essential amino acids, plus β-alanine and γ-aminobutyric acid. Glutamine, asparagine, histidine, serine, arginine and the leucines were the predominant acids.

During the ripening of bananas, two consistent trends were noted by Steward *et al.* (1960b) and by Buckley and Sullivan (1964); histidine content increased at the expense of glutamic and aspartic acids and their amides, and there was a marked increase in the content of valine and leucine, presumed precursors for certain volatile constituents which contribute to banana flavour (see 6. *Volatile constituents*, above).

Steward *et al.* (1960b) also compared the amino acid content of morphological regions of the ripe pulp. The total soluble nitrogen in the inner or placental region of the pulp was almost four times that of the surrounding fleshy pericarp. The concentration of all of the amino acids was higher in the placental region, but the major difference was the virtual absence of amides in the pericarp.

11. *Amines*

Anderson *et al.* (1958) reported that ingestion of bananas resulted in increased urinary excretion of the serotonin (5-hydroxy tryptamine) metabolite, 5-hydroxyindoleacetic acid. Waalkes *et al.* (1958) found bananas to contain surprisingly high concentrations of the physiologically active amines serotonin, L-norepinephrine and dopamine. In the pulp, serotonin averaged 28 μg/g fr. wt., norepinephrine about 2 μg/g fr. wt. and dopamine about 8 μg/g fr. wt. The peel concentrations were much higher: 65 μg/g fr. wt. for serotonin, 122 for norepinephrine and 700 for dopamine. No 5-hydroxyindoleacetic acid could be detected, but evidence was obtained for the presence of at least two unidentified 5-hydroxyindole compounds, and at least one additional catecholamine. West (1958) corroborated the findings of Waalkes *et al.* (1958). Udenfriend *et al.* (1959) found the banana fruit to contain considerably more of these substances than other fruits and vegetables. The tyramine content was also high in bananas (7 μg/g fr. wt. in the pulp; 65 μg/g fr. wt. in the peel). The serotonin content of the peel increased sharply during ripening, from 74 to 161 μg/g fr. wt. in the outer peel and 13 to 170 μg/g fr. wt. in the inner peel. Pulp content increased about 50% during ripening, from 24 to 36 μg/g fr. wt.

Marshall (1959) reported the pulp of Matoke bananas to contain 3 μg/g fr. wt. norepinephrine and 16 μg/g fr. wt. serotonin. Epinephrine could not be detected.

Udenfriend *et al.* (1959) presented evidence that the serotonin or catecholamines ingested in bananas would have no physiological effects, although the possible contribution of these substances to urinary metabolites could cause erroneous diagnoses in some cases (Anderson *et al.*, 1958).

Smith and Kirshner (1960) isolated an enzyme (dopamino-β-oxidase) from bananas which converted dopamine to a racemic mixture of norepinephrines. The dopamine is derived from tyrosine, via tyramine (Buckley, 1965).

Udenfriend *et al.* (1959) could find no evidence for the expected tryptophan to 5-hydroxytryptophane to serotonin pathway in bananas.

The questions of why and how the banana fruit produces and accumulates such high concentrations of these physiologically active amines deserves further study.

12. *Vitamins*

Little has been done with vitamins, except to tabulate data on vitamin content (see McCance and Widdowson, 1960). Thornton (1938) reported that ascorbic acid is quickly destroyed when banana pulp is macerated in air. He also showed (Thornton, 1943) that ascorbic acid increased slightly in the pulp just after the climacteric, fell gradually to 10–12 mg/100 g fr. wt. at full ripeness and then held constant until the fruit began to spoil.

D. Diseases and Physiological Disorders

1. *Diseases*

There are several fungal diseases which cause serious post-harvest losses of bananas. The major causal organisms are *Thielaviopsis*, *Botryodiplodia*, *Gleosporium* and *Nigrospora* species, although many secondary pathogens, such as *Fusaria* and *Verticillia* species, may also be involved. These organisms are prevalent in banana-growing areas, and they enter the unripe fruit tissues through cut surfaces or other wounds.

The predominant system of transporting hands of bananas in cartons encourages infection, since it involves dissecting "hands" from the stalk and washing to remove the exuding latex. Once the organisms have entered the tissues, they are generally held in check (presumably by phenolic constituents) until the fruit is ripened. As soon as ripening is initiated, the organisms multiply rapidly, typically causing a progressive softening and blackening of peel and/or pulp tissues and rendering the fruit unsaleable.

Control measures include careful handling to reduce mechanical injury, reducing the potential inoculum by sanitation procedures at the packing stations (including treatment of the wash water), re-trimming of the cut surfaces just prior to packing, and treatment with various fungicides (see Wardlaw (1961) and Simmonds (1966) for details and references).

Recently the compound 2-(4′-thiazolyl)-benzimadazole (Merck "Thibenzole" or "Mertect") has been shown to be highly successful in protecting bananas from fungal rots (United Fruit Co., unpublished; Scott and Roberts, 1967; Palmer and McGlasson, 1969). It has been approved for use on bananas by the health authorities in the U.S.A. and Australia.

2. *Physiological disorders*

The only serious physiological disorder of banana fruits is chilling injury. Information on this disorder is scattered and often conflicting, but it is possible to make certain generalizations from published and unpublished information reviewed by Loesecke (1950) and Wardlaw (1961), and from the recent studies of Murata and his colleagues (1966a, b).

Chilling injury is a time by temperature effect. At temperatures between −1 and 7°C, exposure of green bananas for a few hours can cause sufficient injury to downgrade the fruit while after 12 hours or more the fruit will probably be unsaleable. At 10–11°C, the response is less predictable. Some bananas tolerate these temperatures for up to 2 weeks, others show significant injury in a few hours. Hence, bananas are normally transported overseas at temperatures between 11 and 13°C, which reduces chilling injury to a minimum. Even at 13°C, however, there may be some chilling injury on the longer (8 days or more) voyages. Where bananas must be transported overland by rail in temperate climates, some chilling injury is inevitable in winter.

Although most chilling injury occurs during transport, bananas can be severely chilled even in the plantation during winter in sub-tropical areas, where night temperatures are often just above freezing. In tropical plantations night temperatures of 10–13°C are not uncommon and can cause moderate to severe injury to growing fruits.

The extent of injury, and the time required for it to appear at a particular temperature varies considerably from clone to clone and even from banana to banana. "Full" fruits tend to be more susceptible than thinner grades and, in Australia, fruits maturing at higher temperatures (tropical or summer sub-tropical climates) appear to be more susceptible than fruits maturing at lower temperatures (sub-tropical winter) (W. B. McGlasson, personal communication).

In severe chilling, the green peel develops extensive sub-epidermal browning or blackening and the peel may become entirely black during ripening. When chilling is less severe, green fruits usually show no visible effect, but on ripening, the colour of the peel commonly varies from a dull yellow to greyish-yellow or grey. These symptoms arise from accumulation of oxidized phenolic substances in epidermal or sub-epidermal areas, accompanied by some retention of chlorophyll.

The effects of chilling on the ripening behaviour of bananas are variable. In general, where peel discoloration is slight to moderate, ripening processes in the pulp are little affected, although the fruits may be downgraded on their appearance. Conversely, if the peel symptoms on the ripening fruit are obvious (definite greying, large areas which blacken or fail to de-green), then pulp ripening will also be abnormal. Typically, the pulp softens unevenly, tends to be acid and astringent, and lacks flavour and sweetness.

The symptoms of severe chilling suggest a breakdown in the coordination of the various ripening processes, as if certain cells did not receive the "signal" which initiates ripening. Alternatively, as suggested by Murata and Ku (1966a), chilling may cause inhibition of certain key respiratory enzymes, resulting in the accumulation of ethanol and acetaldehyde, which further interfere with the normal metabolic pathways. It is clear that more research is required on the physiology and biochemistry of chilling injury.

Another physiological disorder can be induced by attempting to ripen bananas at relative humidities below the recommended 90–95%. It has long been recognized that high relative humidities in the ripening room favourably influence the ripening process by reducing shrinkage, enhancing flavour development, etc. Haard and Hultin (1969) have shown that the relative humidity must be at least 80% for normal ripening of Cavendish bananas. Below 80%, the bananas failed to undergo a climacteric or ripen normally, eventually exhibiting symptoms characteristic of severe chilling.

IV. BIOCHEMICAL MECHANISMS OF RIPENING

A. Introduction

Fruit ripening clearly involves the induction and/or acceleration of a variety of metabolic reactions, presumably most or all enzymatically catalysed. There are two general concepts of how this comes about. In the first, ripening results from a progressive increase in cell permeability, leading to increased contact between enzymes and substrates already present in the tissue (Blackman and Parija, 1928; Sacher, 1962, 1966, 1967; Bain and Mercer, 1964). In the second concept, ripening is a differentiation process, under genetic control and involving the programmed synthesis of specific enzymes required for ripening (Frenkel *et al.*, 1968, Hulme *et al.*, 1968; Brady *et al.*, 1970a).

The banana fruit has been utilized as a "model system" for testing these two concepts. The studies have been primarily concerned with estimating cell permeability and seeking evidence for increased rates of protein synthesis in ripening fruits. Recent studies on banana mitochondria are also reviewed because of the key role of these organelles in respiratory processes.

B. Permeability Changes during Ripening

Permeability has been estimated as the leakage or uptake of low molecular weight solutes by thin (1–3 mm) slices of pulp tissue, when bathed in an appropriate solution.

The permeability of banana pulp tissue increases markedly during the climacteric (Sacher, 1962; Baur and Workman, 1964). More detailed studies (Sacher, 1966, 1967) indicate that significant increases in permeability precede the climacteric by about 2 days, and that, by the climacteric peak, all the cells have become totally permeable to solutes by simple diffusion. Sacher (1967) concluded that the climacteric is a direct result of changes in permeability and protoplasmic compartmentalization. He also questioned the conclusion that ethylene initiates ripening in bananas (Burg and Burg, 1965b) and proposed that the observed enhancement of ethylene synthesis a few hours prior to the onset of the climacteric (Burg and Burg, 1965b) results from the earlier change in compartmentalization. The increasing concentration of ethylene would, according to Sacher, accelerate alterations in the membrane properties in other cells, resulting in the observed exponential rise in permeability during the climacteric.

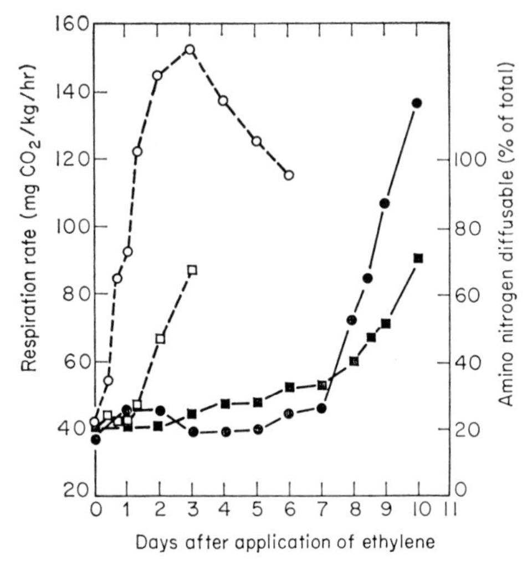

FIG. 5. Effect of ethylene on respiration and permeability of banana fruit slices. ——— control, no ethylene; - - - - - 100 p.p.m. ethylene; ○ ● respiration □ ■ amino acid diffusion. From Brady *et al.* (1970b).

Brady *et al.* (1970b) studied both permeability and total soluble sugars of banana pulp while varying the rate of ripening over wide limits by control of the ethylene concentration in the atmosphere. They first provided evidence that amino acid leakage arises from permeability changes (Sacher, 1967) and not simply from increases in endogenous sugar content, as proposed by Burg *et al.* (1964). They also found that permeability changes preceded the onset of the climacteric in fruits ripened in air. In fruits induced to ripen with 1–100 p.p.m. ethylene, the climacteric preceded the permeability changes by at least 24 hours. Results with 100 p.p.m. ethylene are shown in Fig. 5. Brady *et al.* (1970b) concluded that the development of the climacteric in bananas is not caused by changes in permeability.

Burg (1968) has challenged Sacher's concept that cells are totally permeable (100% free space) at the climacteric. He demonstrated that post-climacteric pulp has normal osmotic properties and estimated the osmotic volume to be at least 40%. Burg (1968) also concluded that there is little evidence to support the view that ethylene alters cellular permeability.

C. Protein Synthesis in Ripening Bananas

1. *Rates of protein synthesis*

The rate of protein synthesis in ripening banana pulp can be estimated by measurements of amino acid incorporation. However, it is difficult to assess the dilution of the introduced acids by the endogenous acids, a problem compounded by the increasing permeability of the ripening tissues. Also, it is difficult to isolate a "clean" and representative protein fraction from a tissue which contains relatively little protein, a high concentration of polysaccharides and considerable quantities of reactive phenolic substances.

TABLE IX. Effect of ethylene treatment on the uptake and incorporation of L-^{14}C-valine by banana pulp cells[a b]

	Time of exposure to ethylene, (hours)		
	0	13	25
Respiration, mg CO_2/kg × hours	16–22	36–45	50–58
Total uptake (%)	73·8	91·2	89·8
Incorporation			
% total uptake	22·4	40·9	29·0
D.p.m./mg protein	971	1607	1577
D.p.m. × 10^{-3}/μmole protein valine	1·95	4·03	4·68

[a] From Brady *et al.* (1970a).

[b] Banana slices were ventilated continuously with a humidified air stream containing 10–15 p.p.m. ethylene for the times indicated. Each slice was infiltrated with a solution containing 0·21 μc L-valine (and 50 μg/ml chloramphenicol to minimize microbial contamination). After incubation for 45 minutes, the central pulp zones were analysed for uptake and incorporation.

Sacher (1966) introduced ^{14}C amino acids into thin discs (1 × 3 mm) of pulp, cut from pre-climacteric and climacteric bananas. He made no attempt to estimate uptake of ^{14}C amino acid, but introduced increasing concentrations of amino acid, hoping thereby to reach a concentration too large to be significantly diluted by the endogenous pool. This was apparently achieved with 0·05 M phenylalanine and the incorporation data indicated a 50%

decline in the rate of protein synthesis during the climacteric. Sacher commented that protein degradation must show a corresponding decline, since it is well established that the protein content of banana pulp remains constant during ripening (Stratton and Loesecke, 1930; Steward *et al.*, 1960b; Sacher, 1966, 1967; Brady *et al.*, 1970a). A second possibility is that protein turnover is low in banana pulp.

Brady *et al.* (1970b) infiltrated ^{14}C-labelled amino acids into 6–8 mm thick transverse slices of banana fruits and estimated uptake from measurements of the proportion of total ^{14}C and of total amino nitrogen which diffused from thin slices of the infiltrated tissue. Incorporation was measured on a protein fraction rigorously separated from non-protein substances by phenol extraction and molecular sieve chromatography. The results showed increases in both uptake and incorporation of lysine and valine, following treatment of the slices with ethylene at concentrations (10–60 p.p.m.) which induced the slices to ripen (Palmer and McGlasson, 1969). Table IX shows typical results with valine, indicating substantial increases in the rate of protein synthesis during the early climacteric. Later in the climacteric (25–40 hours after application of ethylene), the rate of synthesis tended to decline.

2. *Effect of protein synthesis inhibitors*

Utilizing the techniques of Palmer and McGlasson (1969), Brady *et al.* (1970a) infiltrated inhibitors into banana slices and measured the effect on ethylene-induced ripening. Cycloheximide (1 μg/ml) and *p*-fluorophenylalanine (1000 μg/ml) inhibited de-greening and softening of the slices. L-Phenylalanine (1 mole/mole) reversed the inhibition by *p*-fluorophenylalanine. Puromycin (100 μg/ml) delayed ripening slightly; chloramphenicol (50–2000 μg/ml) consistently had no effect.

Cycloheximide and *p*-fluorophenylalanine also markedly reduced the respiratory climacteric and interfered with starch to sugar conversions. Figure 6 shows results obtained with cycloheximide at 2 μg/ml. The effect on the climacteric and associated ripening changes was progressively less at lower concentrations, but was still detectable at 0·25 μg/ml. Ripening was severly inhibited above 2 μg/ml but necrotic zones indicated cellular damage. Cycloheximide at 1–2 μg/ml had no effect on the respiration rate or appearance of pre-climacteric slices during the 2 days following infiltration. These results indicate that protein synthesis is required for ripening. They do not necessarily mean that inhibitors prevent synthesis of enzymes specifically required for ripening. The inhibition of ripening could result, for example, from the depletion of a critical enzyme or regulatory protein. The fact that unripe slices are unaffected for several days after treatment with cycloheximide suggests, however, that such a depletion is unlikely, at least during the critical 12–24 hours following application of ethylene.

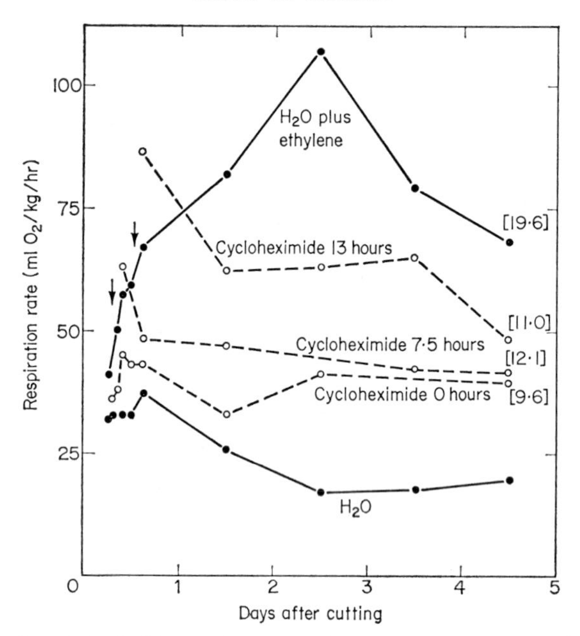

FIG. 6. Effect of cycloheximide on the respiration and sugar content of banana slices ventilated continuously with 10–15 p.p.m. ethylene. Slices were infiltrated with water or cycloheximide solution (2 μg/ml) immediately after cutting, or at the times (7·5 or 13 hours) indicated by the arrows on the figure. The bottom trace shows the respiration rate of slices not treated with ethylene. Numbers in brackets are the per cent soluble solids (approximately equal to total sugars) at the completion of the experiment. Reproduced with permission from Brady *et al.* (1970a).

3. *Changes in soluble proteins during ripening*

Brady *et al.* (1970a) prepared pulp extracts from fruits at six stages of ripeness (pre-climacteric to yellow ripe) and separated the proteins by gel electrophoresis. The gel patterns indicated a progressive change in the soluble proteins during the climacteric (Fig. 7). The proteins in extracts of post-climacteric fruits were poorly resolved, apparently because of interference from polysaccharides.

Changes in proteins during ripening are also indicated by the changing amino acid composition of the protein fraction (Steward *et al.*, 1960b).

D. Mitochondria from Bananas

Tager (1958) isolated mitochondria from banana pulp at all stages of ripening. He used dipotassium phosphate to control pH and egg albumin to reduce the precipitating effect of tannins. The mitochondria oxidized citrate, α-oxoglutarate, succinate and malate, but not pyruvate. The QO_2 (N) (microlitres O_2/mg mitochondrial nitrogen/hour) was low

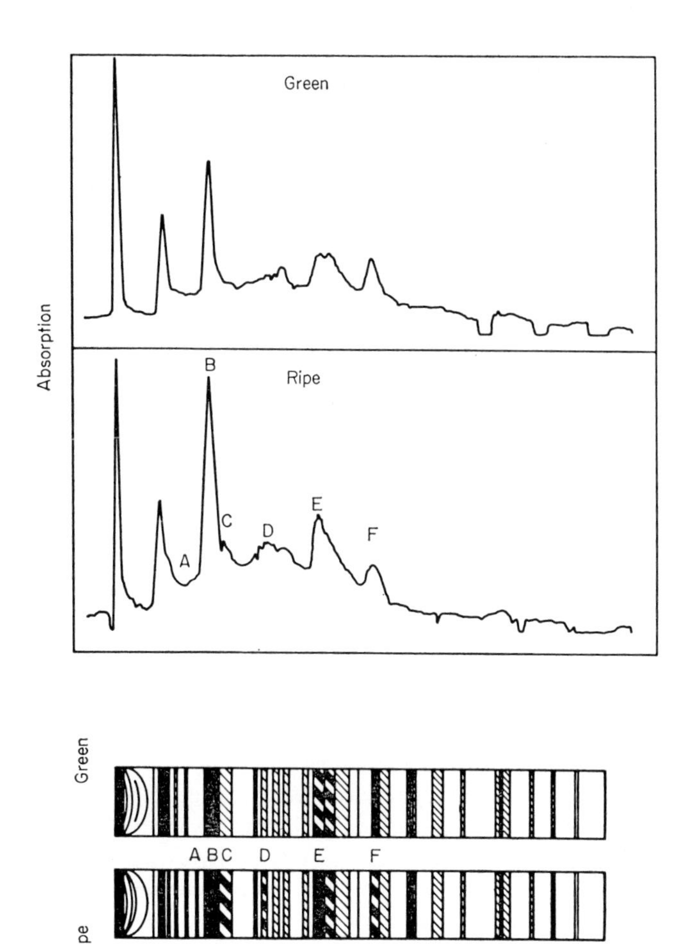

FIG. 7. Gel electrophoresis of the soluble proteins from pre-climacteric ("Green") and climacteric ("Ripe") banana pulp. Soluble proteins (170 μg/tube) were separated in 7·5% polyacrylamide gel at pH 8·3 and stained with Amido Schwarz. The lower schematic drawing indicates the separation achieved and the relative intensity of the protein bands. The upper curves are densitometer tracings of the gels. The increases at A, B, C, D and E developed progressively during the climacteric. Reproduced with permission from Brady *et al.* (1970a).

(17–235) and variable. No attempt was made to measure oxidative phosphorylation.

Haard (1967) has made an extensive study of the mitochondria from bananas. He developed an isolation technique (Haard and Hultin, 1968) which permitted the repeated isolation of mitochondria of high specific activity (QO_2 (N) generally 400–750 for Krebs cycle intermediates). The procedure involved freezing of the tissue in liquid nitrogen, grinding the frozen tissue to a fine powder and extraction of the powder with a special medium containing sucrose, $MgCl_2$, EDTA, casein (to react with tannins) and l-cysteine (to prevent browning). The mitochondria were isolated from the extracts by appropriate centrifugations.

Haard's mitochondria catalysed the oxidation of Krebs cycle acids, including pyruvate. The mitochondria exhibited respiratory control (R.C. = 3) and carried out oxidative phosphorylation, the P/O ratios approaching 2·0 (the theoretical maximum) with succinate. There was no significant change in the numbers, specific activity, respiratory control or oxidative phosphorylation during ripening. However, the activity of the mitochondria was greatly influenced by the composition of the test medium and Haard suggested that the climacteric in bananas involves activation of existing mitochondrial enzymes. He also pointed out the difficulty of predicting the *in vivo* behaviour of mitochondria when so little is known about the cellular environment.

Haard observed two significant differences between pre- and post-climacteric mitochondria, both involving calcium ions. First, the succinoxidase activity of mitochondria from pre-climacteric pulp was stimulated by Ca^{2+} (1 mM); post-climacteric mitochondria were not affected. Secondly, active uptake of Ca^{2+} by post-climacteric mitochondria was about three times that of pre-climacteric. Ca^{2+} accumulation was directly associated with oxidative phosphorylation, took place only when substrate (or ATP) and bovine serum albumin were added and was suppressed by ADP. Haard suggested that a study of the possible relationship between calcium ions, ion uptake, mitochondrial swelling and ethylene synthesis and action could shed light on the role of membrane permeability in banana ripening.

E. Conclusions

The evidence on balance supports the concept that banana ripening is primarily a differentiation process. Sequential changes in permeability and in compartmentation undoubtedly influence the course of ripening, particularly in the post-climacteric period, but the primary factor is the re-orientation of protein synthesis. However, confirmation of the differentiation concept will require unequivocal demonstration of increased synthesis of specific enzymes required for ripening.

It seems clear that changes in total cellular permeability do not cause banana ripening. It is also likely that the magnitude of the permeability changes during the climacteric has been considerably overestimated, probably as a result of damage to the tissues when cutting the thin slices utilized in measuring solute leakage. Nevertheless, the evidence indicates substantial increases in cell permeability during the climacteric in bananas. The resulting increased mixing of enzymes and substrates, or changes in the cellular environment of mitochondria, could easily be responsible for initiation or control of some of the characteristic ripening changes.

Finally, a few comments on experimental approaches are in order. In studying such things as the effect of inhibitors or metabolism of ^{14}C-labelled substrates during ripening, the usual practice has been to introduce the test substance into thin (0·5–1·0 mm) slices taken from fruits at appropriate stages of ripening. Palmer and McGlasson (1969) contend that such thin discs are not representative of the whole fruit, since cutting induces marked changes in the metabolism of the tissue. Ku *et al.* (1965) had earlier shown that the respiration rates of thin sections cut from banana fruits during the climacteric differ markedly from the intact fruit rate. Palmer and McGlasson (1969) have proposed the use of aseptically-prepared 6 mm thick transverse slices as a compromise between the thinner slices and the intact fruit. Slices from pre-climacteric fruits, after incubation for a few days to allow "induced" respiration to subside, appeared to be representative of the original fruit. The slices subsequently ripened "naturally" or ripening could be induced with ethylene. Test substances were readily vacuum-infiltrated into the slices, although infiltration did cause a significant increase in the respiration rate.

Tissue culture provides another possible approach to the study of fruit ripening. Ram and Steward (1964) have shown that sections taken from pre-climacteric bananas can be readily cultured. If the cultured tissues could be induced to ripen, they would be superb experimental material for studying the biochemical mechanism of fruit ripening.

ACKNOWLEDGEMENT

The author thanks United Fruit Company for financial assistance which enabled him to prepare this chapter and for permission to include unpublished information from their research files.

REFERENCES

Anderson, J. A., Ziegler, M. R. and Doeden, D. (1958). *Science, N.Y.* **127,** 236.
Badran, A. M. (1969). U.S. Patent No. 3,450,542.
Badran, A. M. and Jones, D. E. (1965). *Nature, Lond.* **206,** 622.
Bain, J. M. and Mercer, F. V. (1964). *Aust. J. biol. Sci.* **17,** 78.
Barker, J. and Solomos, T. (1962). *Nature, Lond.* **196,** 189.
Barker, W. G. and Steward, F. C. (1962a). *Ann. Bot. N.S.* **26,** 389.

Barker, W. G. and Steward, F. C. (1962b). *Ann. Bot. N.S.* **26,** 413.
Barnell, H. R. (1940). *Ann. Bot. N.S.* **4,** 39.
Barnell, H. R. (1943). *Ann. Bot. N.S.* **7,** 297.
Barnell, H. R. and Barnell, E. (1945). *Ann. Bot. N.S.* **9,** 77.
Bates, F. L., French, D. and Rundle, R. E. (1943). *J. Am. chem. Soc.* **65,** 142.
Baur, J. R. and Workman, M. (1964). *Pl. Physiol., Lancaster* **39,** 540.
Biale, J. B. (1960a). *Adv. Fd Res.* **10,** 293.
Biale, J. B. (1960b). *In* "Handbuch der Pflanzenphysiologie" (W. Ruhland, ed.), Vol. XII/2, pp. 536–586. Springer Verlag, Berlin.
Biale, J. B., Young, R. E., and Olmstead, A. J. (1954). *Pl. Physiol., Lancaster* **29,** 168.
Blackman, F. F. and Parija, P. (1928). *Proc. R. Soc.* **B103,** 412.
Blake, J. R. and Stevenson, C. D. (1959). *Queensland J. agric. Sci.* **16,** 87.
Brady, C. J., O'Connell, P. B. H., Smydzuk, J. and Wade, N. W. (1970). *Aust. J. biol. Sci.* In preparation.
Brady, C. J., Palmer, J. K., O'Connell, P. B. H. and Smillie, R. M. (1970a). *Phytochemistry* **9,** 1037.
Buckley, E. H. (1962). *Pl. Physiol., Lancaster* **37,** Suppl. xlvi.
Buckley, E. H. (1965). *In* "Phenolics in Normal and Diseased Fruits and Vegetables" (V. C. Runeckles, ed.), pp. 1–6. Plant Phenolics Group of North America, Montreal.
Buckley, E. H. and Sullivan, W. M. (1964). Report, Thayer Summer Science Programme, Thayer Academy, Braintree, Mass.
Burg, S. P. (1968). *Pl. Physiol., Lancaster* **43,** 1503.
Burg, S. P. and Burg, E. A. (1965a). *Science, N.Y.* **148,** 1190.
Burg, S. P. and Burg, E. A. (1965b). *Bot. Gaz.* **126,** 200.
Burg, S. P. and Burg, E. A. (1966). *Science, N.Y.* **153,** 314.
Burg, S. P., Burg, E. A. and Marks, R. (1964). *Pl. Physiol., Lancaster* **39,** 185.
Cairncross, S. E. and Sjöström, L. B. (1950). *Food Technol.* **4,** 308.
Dedolph, R. R. and Goto, S. (1960). *Hawaii Fm Sci.* **8,** 3.
DeSwardt, G. H. and Maxie, E. C. (1967). *S. Afr. J. agric. Sci.* **10,** 501.
Drawert, F., Heimann, W., Emberger, R. and Tressl, R. (1966). *Annln. Chemie,* **694,** 200.
Fahn, A., Stoler, S. and First, T. (1963). *Bot. Gaz.* **124,** 246.
Ferguson, W. E., Yates, A. R., MacQueen, K. F., and Robb, J. A. (1966). *Fd Technol.* **20,** 105.
Freiberg, S. R. (1955). *Bot. Gaz.* **117,** 113.
Freiberg, S. R. (1966). *In* "Nutrition of Fruit Crops" (N. F. Childers, ed.), pp. 77–100. Horticultural Publications, Rutgers University, New Brunswick, New Jersey.
Freiberg, S. R. (1969). *In* "Encyclopedia of Food and Food Science". John Wiley and Sons, Inc., New York.
Frenkel, C., Klein, I. and Dilley, D. R. (1968). *Pl. Physiol., Lancaster* **43,** 1146.
Gane, R. (1936). *New Phytol.* **35,** 382.
Gane, R. (1937). *New Phytol.* **36,** 170.
Goldstein, J. L. and Swain, T. S. (1963). *Phytochemistry* **2,** 371.
Goldstein, J. L. and Wick, E. L. (1970). *J. Fd Sci.* **34,** 482.
Griffiths, L. A. (1959). *Nature, Lond.* **184,** 58.
Griffiths, L. A. (1961). *Nature, Lond.* **192,** 84.
Grobois, M. and Mazliak, P. (1964). *Fruits* **19,** 55.
Guymon, J. F. (1966). *Developments in Indust. Microbiol.* **7,** 88.

Haard, N. F. (1967). The Isolation and Partial Characterization of the Mitochondria from the Pulp of the Ripening Banana. Ph.D. Thesis, University of Massachusetts.
Haard, N. F. and Hultin, H. O. (1968). *Anal. Biochem.* **24,** 299.
Haard, N. F. and Hultin, H. O. (1969). *Phytochemistry* **8**, 2149.
Hall, E. G. (1967). *Fd Preserv. Q.* **27,** 36.
Harris, P. L. and Poland, G. L. (1937). *Fd Res.* **2,** 135.
Harvey, R. B. (1928). *Minn. Agr. Expt. Sta. Bull.* 247.
Henderson, R. W., Morton, R. K. and Rawlinson, W. A. (1959). *Biochem. J.* **72,** 340.
Hewitt, E. J., Hasselstrom, T., MacKay, A. M., and Konigsbacher, K. S. (1960). U.S. Patent 2,924,521.
Hulme, A. C., Rhodes, M. J. C., Galliard, T. and Wooltorton, L. S. C. (1968). *Pl. Physiol., Lancaster* **43,** 1154.
Hultin, H. O. and Levine, A. S. (1963). *Archs Biochem. Biophys.* **101,** 396.
Hultin, H. O. and Levine, A. S. (1965). *J. Fd Sci.* **30,** 917.
Hultin, H. O. and Proctor, B. E. (1962). *Fd Technol.* **16,**(2) 108.
Hultin, H. O., Sun, B. and Bulger, J. (1966). *J. Fd Sci.* **31,** 320.
Isherwood, F. A. and Niavis, C. A. (1956). *Biochem. J.* **64,** 549.
Johnson, B. E. and Brun, W. A. (1966). *Pl. Physiol., Lancaster* **41,** 99.
Khalifah, R. A. (1966a). *Pl. Physiol., Lancaster* **41,** 771.
Khalifah, R. A. (1966b). *Nature, Lond.* **212,** 1471.
Ku, H., Murata, T. and Ogata, K. (1965). *J. Fd Sci. and Technol. (Japan)* **12,** 163.
Leonard, E. R. (1941). *Ann. Bot. N.S.* **5,** 89.
Loesecke, H. W. von (1929). *J. Am. chem. Soc.* **51,** 2439.
Loesecke, H. W. von (1950). "Bananas", 2nd ed. Interscience, New York.
Looney, N. E. and Patterson, M. E. (1967). *Nature, Lond.* **214,** 1245.
Lulla, B. S. (1954). *Curr. Sci., India* **23,** 362.
McCance, R. A. and Widdowson, E. M. (1960). "The Composition of Foods." H.M.S.O., London.
McCarthy, A. I. and Palmer, J. K. (1964). *In* "Proc. First Intl Cong. Food Sci. and Technol.", Vol. I (J. M. Leitch, ed.), pp. 483–488. Gordon and Breach, New York.
McCarthy, A. I., Palmer, J. K., Shaw, C. P. and Anderson, E. E. (1963). *J. Fd Sci.* **28,** 379.
Manek, P. V. and Fripp, P. J. (1963). *Biochem. J.* **89,** 79p.
Mapson, L. W. and Robinson, J. E. (1966). *J. Fd Technol.* **1,** 215.
Marshall, P. B. (1959). *J. Pharm. Pharmacol.* **11,** 639.
Maxie, E. C. and Sommer, N. F. (1968). *Chem. Engng. Prog.* **64,** 65.
May, S. and Plaza, G. (1958). "The United Fruit Company in Latin America." National Planning Association, Washington, D.C.
Miller, C. L. and Ross, E. (1963). *J. Fd Sci.* **28,** 193.
Mitchell, J. W. and Marth, P. C. (1944). *Bot. Gaz.* **106,** 199.
Murata, T. and Ku, H. (1966a). *J. Fd Sci. Technol. (Japan)* **13,** 466.
Murata, T. and Ogata, K. (1966b). *J. Fd Sci. Technol. (Japan)* **13,** 367.
Myers, M. J., Issenberg, P. and Wick, E. L. (1970). *Phytochemistry* **9,** 1693.
Niederl, J. B., Brenner, M. W. and Kelley, J. N. (1938). *Am. J. Bot.* **25,** 357.
Palmer, J. K. (1963). *Pl. Physiol., Lancaster* **38,** 508.
Palmer, J. K. (1965). *In* "Phenolics in Normal and Diseased Fruits and Vegetables" (V. C. Runeckles, ed.), pp. 7–12. Plant Phenolics Group of North America, Montreal.

Palmer, J. K. and McGlasson, W. B. (1969). *Aust. J. biol. Sci.* **22,** 87.
Palmer, J. K. and Roberts, J. B. (1967). *Science, N.Y.* **157,** 200.
Peacock, B. C. and Blake, J. R. (1969). *Qd J. Agric. Animal Sci.* In press.
Poland, G. L., von Loesecke, H. Brenner, M. W., Manion, J. T. and Harris, P. L. (1937). *Fd Res.* **2,** 403.
Poland, G. L., Manion, J. T., Brenner, M. W. and Harris, P. L. (1938). *Ind. Engng Chem.* **30,** 340.
Ram, H. Y. M. and Steward, F. C. (1964). *Can. J. Bot.* **42,** 1559.
Ram, H. Y. M., Ram, M. and Steward, F. C. (1962). *Ann. Bot. N.S.* **26,** 657.
Robinson, G. M. (1937). *J. chem. Soc.* **1937,** 1157.
Rowan, K. S. (1959). *Biochim. Biophys. Acta* **34,** 270.
Sacher, J. A. (1962). *Nature, Lond.* **195,** 577.
Sacher, J. A. (1966). *Pl. Physiol., Lancaster* **41,** 701.
Sacher, J. A. (1967). *In* "Aspects of the Biology of Ageing" (H. W. Woolhouse, ed.), pp. 269–303. Academic Press, New York and London.
Scott, K. J. and Roberts, E. A. (1966). *Aust. J. Exptl Agric. Anim. Husbandry* **6,** 197.
Scott, K. J. and Roberts, E. A. (1967). *Aust. J. Exptl Agric. Anim. Husbandry* **7,** 283.
Scott, K. J., McGlasson, W. B. and Roberts, E. A. (1968). *Agric. Gaz., N.S.W.* **79,** 52.
Simmonds, N. W. (1954). *Ann. Bot. N.S.* **18,** 471.
Simmonds, N. W. (1966). "Bananas", 2nd ed. Longmans, Green and Co., Ltd., London.
Smith, W. J. and Kirshner, N. (1960). *J. biol. Chem.* **235,** 3589.
Steward, F. C. and Simmonds, N. W. (1954). *Nature, Lond.* **173,** 1083.
Steward, F. C., Hulme, A. C., Freiberg, S. R., Hegarty, M. P., Pollard, J. K., Rabson, R. and Barr, R. A. (1960a). *Ann. Bot. N.S.* **24,** 83.
Steward, F. C., Freiberg, S. R., Hulme, A. C., Hegarty, M. P., Barr, R. A. and Rabson, R. (1960b). *Ann. Bot. N.S.* **24,** 117.
Stratton, F. C. and Loesecke, H. von (1930). Research Bulletin No. 32, United Fruit Co., Boston.
Stratton, F. C. and Loesecke, H. von (1931). *Pl. Physiol., Lancaster* **6,** 361.
Swain, T. S. and Hillis, W. E. (1959). *J. Sci. Fd Agric.* **10,** 63.
Tager, J. M. (1956). *S. Afr. J. Sci.* **53,** 167.
Tager, J. M. (1958). *S. Afr. J. Sci.* **54,** 324.
Tager, J. M. and Biale, J. B. (1957). *Physiologia. Pl.* **10,** 79.
Thornton, N. C. (1938). *Contrib. Boyce Thompson Inst.* **9,** 273.
Thornton, N. C. (1943). *Contrib. Boyce Thompson Inst.* **13,** 201.
Udenfriend, S., Lovenberg, W. and Sjoerdsma, A .(1959). *Archs Biochem. Biophys.* **85,** 487.
United Fruit Co. (1964). "Ripening Manual".
Vendrell, M. (1970). *Aust. J. biol. Sci.* **22,** In press.
Waalkes, T. P., Sjoerdsma, A., Creveling, C. R., Weissbach, H. and Udenfriend, S. (1958). *Science, N.Y.* **127,** 648.
Wade, N. L. and Brady, C. J. (1969). *Aust. J. biol. Sci.* In press.
Wardlaw, C. W. (1961). "Banana Diseases." John Wiley & Sons, Inc., New York.
West, G. B. (1958). *J. Pharm. Pharmacol.* **10,** 589.
Wick, E. L., McCarthy, A. I., Myers, M., Murray, E., Nursten, H. and Issenberg, P. (1966). *In* "Flavor Chemistry", Advances in Chemistry Series, No. 56 (R. F. Gould, ed.), pp. 241–260. American Chemical Society, Washington, D.C.

Wick, E. L., Yamanishi, T., Kobayashi, A., Valenzuela, S. and Issenberg, P. (1969). *J. Agric. Fd Chem.* **17,** 751.
Wolf, J. (1958). *Z. Lebensmitt. Untersuch.* **107,** 124.
Wyman, H., Buckley, E. H., McCarthy, A. I. and Palmer, J. K. (1964). Annual Research Report, United Fruit Co.
Wyman, H. and Palmer, J. K. (1964). *Pl. Physiol., Lancaster* **39,** 630.
Yang, S. and Ho, H. (1958). *J. Chin. chem. Soc.* **5,** 71.
Young, R. E. (1965). *Archs Biochem. Biophys.* **111,** 174.
Young, R. E., Romani, R. J. and Biale, J. B. (1962). *Pl. Physiol. Lancaster*, **37,** 416.

Chapter 3

Citrus Fruits

S. V. TING AND J. A. ATTAWAY

State of Florida, Department of Citrus, Lake Alfred, Florida, U.S.A.

I	Introduction	107
	A. Scope of Chapter	107
	B. Brief History of Citrus	108
	C. World Production and Importance	108
II	The Fruit	109
	A. Principal Varieties of Commercial Significance	109
	B. General Morphology and Composition	111
III	Changes during Fruit Development and Maturation	113
	A. General Morphological and Anatomical Changes	113
	B. Chemical Constituents and their Changes	114
	C. Effects of Environmental Factors and Cultural Practices on the Chemical Composition of the Fruit	143
	D. Growth Regulators and Abscission	146
IV	Post-harvest Physiology	149
	A. Changes in Chemical Constituents During Storage	149
	B. Respiratory Activity	151
	C. Ethylene	153
	D. Physiological Disorders in Storage	154
V	Metabolism of Citrus Fruits	157
	A. Citric Acid Synthesis	157
	B. The Citrus Mitochondria	158
	C. Other Citrus Enzymes	159
	References	161

I. INTRODUCTION

A. Scope of Chapter

The principal objects of this chapter are to present a general account of the chemical compositions of citrus fruits and to describe some of the physiological and biochemical changes that take place during the development, growth, maturation and senescence of the fruit. No attempt is made in this chapter to discuss in detail the chemistry of the chemical constituents since these have been covered in Volume 1.

The name "citrus" has been applied to several different fruits by earlier writers on the subject of citrus. Much confusion resulted over the classification of these fruits which had many similarities. Hume (1957) suggested that "citrus fruits" as commonly known in horticulture include the three genera in the family Rutaceae, namely, *Citrus*, *Fortunella* and *Poncirus*. The fruits of the two latter genera and some species of the genus *Citrus* are of no commerical value at present and will not be considered in this chapter. Of all citrus fruits cultivated today, oranges, grapefruit, tangerine, lemon and lime, and some of the newly-developed hybrids, make up almost all the world's production.

B. Brief History of Citrus

The original home of the citrus fruits included in the three distinct genera is the southern and eastern regions of Asia, China and Cochin China and the Malayan Archipelago. The introduction of the citron (*Citrus medica*) to Europe dates back to the third century B.C. when Alexander the Great conquered Western Asia. The orange and lemon were introduced to the Mediterranean countries when the Romans navigated directly from the Red Sea to India. From Europe, the citrus fruits spread to America, South Africa and finally to Australia (Tolkowsky, 1938).

Today, citrus fruits are grown in all regions of the world where the climate is not severe during winter and suitable soil conditions exist. Also, many desert regions have been opened for citrus culture by the development of irrigation facilities, and many thousands of acres of lowlands by drainage. Frost and cold protection techniques extend the growing regions further north where previously citrus could not be grown profitably.

C. World Production and Importance

Nearly 3·7 million acres of citrus were in production over the world in 1962 (Burke, 1967). This production has steadily increased in recent years and is greater than that of any other tree fruit (see Table I). The only figures which appear to be available for bitter (sour) oranges is that in 1952; 200,000 to 400,000 boxes were exported from Spain, one of the main producers of this orange which is used mainly for the production of marmalade.

The rapid growth of the citrus fruit industry in the past twenty-five years is largely due to population increase and improved economic conditions in the consuming nations of the world together with the rapid advance of agricultural sciences and technology of by-products. Also, because of the nutrition-conscious consuming public and the natural distinctive flavour of the citrus, the demand for citrus fruits and citrus products has increased and is likely to increase further. Citrus fruit is fast becoming a staple food product in the daily diet of many people, and large consumption of citrus fruit is also attributed to other types of food and beverage industries which require the flavour of citrus.

TABLE I. World citrus production 1940–66 (millions of boxes)[a]

Year	Oranges and mandarins	Grapefruit	Lemons	Limes
Average				
1940–49[b]	286·0	53·6	27·1	3·1
1950–59	329·9	48·1	34·1	3·5
1960–64	443·8	48·5	42·9	5·4
Annual				
1964	465·7	52·7	46·0	6·3
1965	536·6	60·2	47·9	7·4
1966	656·6	71·2	52·9	6·6
1967[c]	575·9	59·9	52·6	7·0

From Burke (1967); U.S. Dept. of Agric. (1967 and 1968).

[a] Estimated on the following weights: Orange, 70 lb boxes; grapefruit, 80 lb boxes; lemon and lime, 76 lb boxes.

[b] The 1940–59 figures did not include some countries in data recorded after 1960 as reported by U.S. Dep. of Agr. Foreign Agricultural Service.

[c] Preliminary.

II. THE FRUIT

A. Principal Varieties of Commercial Significance

The citrus fruits of commerce may generally be classified into four horticultural groups, all of which belong to the genus *Citrus*. These groups are: (i) the oranges and mandarins; (ii) the hybrid citrus fruits; (iii) the grapefruit and pummelo and (iv) the acid fruits which include the lemon, lime and citron.

1. *Oranges*

In this group are included the bitter (sour) orange (*C. aurantium*), the sweet orange (*C. sinensis*) and the mandarin orange (*C. reticulata*). The bitter orange, chiefly the Seville variety, is commercially grown in southern Europe mainly for citrus products such as marmalade. Its annual production is small in comparison with the other citrus crops. In the United States, bitter orange is extensively grown as a rootstock, and the seeds are widely sought at a good price. Bitter orange rootstock is especially adapted to the Hammock soils of Florida.

The most important of all citrus fruit is the sweet orange. It is widely grown in all regions of the world adapted to citrus, although each region usually has its own characteristic varieties. Over 100 varieties have been described and brought before the public as of varied commercial importance

(Hume, 1957). The distribution of varieties in citrus-growing countries is given by Burke (1967).

According to Hodgson (1967), the main varieties of sweet oranges grown around the world can be included in four groups, namely the common orange, the acidless orange, the pigmented orange and the navel orange. The acidless orange is of minor commercial importance and will not be discussed here. Among the common oranges are Valencia, Pineapple, Hamlin, Parson Brown, Jaffa and Shamouti which are generally grown in various parts of the world. In addition, Pera, Corriente and Bahianinha are some of this type especially grown in Latin America mostly from seedlings.

The pigmented oranges are widely grown in the Mediterranean area. Under favourable conditions, the fruit develops red pigmentation in the flesh and in the rind which is not due to carotenoids but rather to the anthocyanins. The fruit has a distinctive flavour and is preferred by people of the producing areas. Processing of the pigmented orange has not been successful due to the deterioration of the anthocyanins, but experimental preparations of frozen concentrate proved to be feasible. The principal cultivars of these fruits include the Doblefina, Eutrifina, Moro, Tarocco, Ovale, Sanguinello Commun, Ruby Blood and Maltese Blood.

Navel oranges are grown mostly for the fresh market because of their tendency to develop a bitter taste in processed products. The Washington Navel is probably the most important cultivar of this group. The Frost Washington and Dream are some of the promising new cultivars.

The mandarins are a group of loose-skinned oranges of primary importance to the Far East, but they are also becoming increasingly popular in the United States. The Satsumas are an important citrus crop in Japan; Dancy tangerine is widely grown in the United States; and Clementine is an important cultivar of the Mediterranean areas, especially Algeria. Other newer cultivars are being grown commercially. Among these are King, Robinson, Page, Ponkan and Murcott. The last-mentioned cultivar according to many horticulturists could be a tangor which is a hybrid of mandarin and orange. Temple, which is also believed to be a tangor, has excellent flavour and is being widely planted in the United States, especially in Florida. In 1967, Florida alone produced five million boxes of Temple (Fla. Dep. Agr.) and almost two million boxes of tangelo which is the hybrid of mandarins and grapefruit or pummelo. About a dozen of the tangelo cultivars have been named. The most important cultivars of the tangelo are Orlando and Mineola, both of which are the crosses between Dancy tangerine and Duncan grapefruit.

2. *Grapefruits* (*C. grandis; C. aurantium, var. grandis; C. decumana.*)

Grapefruit came from the West Indies as either a mutation or hybrid of Pummelo (Tolkowsky, 1938). It can be separated into the common grapefruit and the pigmented grapefruit. The coloration of the pigmented fruit is due to the carotenoid, lycopene. Grapefruit can also be grouped into seedy and seedless cultivars. Marsh Seedless grapefruit is probably the most predominant cultivar followed by Duncan, a seedy cultivar. The pigmented cultivars such as Thompson and Ruby are becoming very popular. Of the 71 million boxes of grapefruit produced in the world in 1966, 56 million of them were produced in the United States (U.S. Dep. Agr., 1967). Grapefruit is usually produced for the fresh fruit market except in Florida where 60% of the 1967 production was processed as single-strength juice, frozen concentrates or canned and chilled sections (Fla. Dep. Agr., 1967).

3. *Lemons* (*C. limonia*)

Lemons are an important crop of Italy and Israel. About one-third of the total citrus crop of Italy is in lemons and about one-tenth of that of Israel is reported to be in lemons. In the United States, lemons have been chiefly grown in California and Arizona, but in recent years appreciable numbers are also being grown in Florida.

The predominant cultivars in the Mediterranean countries is Femminello. Eureka and Lisbon are the main cultivars in California.

4. *Limes* (*C. acida; C. aurantifolia*)

Limes are best produced in warm and humid areas. India, Egypt, Africa, Mexico and the West Indies produce a large percentage of the world's lime crop. Southern Florida and Southern California also produce the Tahiti or Persian lime. The West Indian is the most important variety.

Sweet lemon and sweet lime are separate groups from the acid fruits. They are not of any commercial importance in world trade and are usually consumed locally. These fruits are of unusual interest to plant physiologists and biochemists in their study of citric acid synthesis in citrus fruits.

B. General Morphology and Composition

1. *Structure of the fruit*

Tissues of citrus fruit vary in the different component parts of a fruit which can be roughly divided into flavedo, albedo and carpel segments. The flavedo consists of several morphologically different tissues. The outermost layer of the fruit is the epidermis, a layer of isodiametric, polygonal cells covering the entire surface of the fruit except where the numerous stomates are located. Turell and Klotz (1940) estimated the number of stomates at the fruit surface to be $1{\cdot}386 \times 10^3/cm^2$. The outer cell walls of this layer are heavily cutinized and partly covered with a waxy substance. These cutinized walls and the wax prevent excess loss of water from the fruit. Directly under the epidermis are layers of collenchyma and parenchyma cells in which occur the chromatophores. At different depths of the parenchyma tissues are located numerous oblate, spherical-shaped oil glands from which the essential oils are obtained. These oil glands, ranging in size from 0·2 to 1·0 mm. in diameter, must be thoroughly broken and pressed before the oils are released.

Below the layers of the collenchyma and parenchyma cells is the white, spongy portion of the parenchyma cells called the albedo. These parenchyma cells do not contain any chromatophores. The albedo varies in thickness in different types of citrus. In mature tangerine or other loose-skinned oranges, this layer is relatively thin. In grapefruit or pummelo, it may be 1–3 cm in thickness.

The edible portions, sometimes called the pulp of the citrus fruit, are the carpels or the segments which are separated by carpel walls. Within the carpel are many juice vesicles developed from hair-like papillae on the segment membrane (Ford, 1942) and some seeds. These vesicles enlarge as the fruit develops and are attached to the membrane with thread-like stalks through which are translocated water and solutes from other parts of the plant. Within the juice vesicles are many thin-walled juice cells and small amounts of other cell substances (Davis, 1932). At maturity, the vacuoles of these juice cells are full of juice; when the membranes are broken, the

TABLE II. Changes in fruit weight and weight of component parts of the Hamlin orange and Marsh Seedless grapefruit

Month	Fr. wt./ fruit (g)[a]	Fresh weight of component parts (g)			
		Flavedo	Albedo	Membrane	Juice vesicles
			Hamlin orange		
May	24·1	4·4	13·3	2·1	4·3
June	56·3	7·5	26·2	4·6	18·0
July	78·6	10·6	28·1	4·7	27·8
August	96·6	11·5	32·5	5·0	47·7
September	129·0	14·1	34·1	6·0	74·8
October	158·2	15·8	42·1	6·5	93·6
November	174·9	17·4	42·2	6·9	107·5
December	173·8	22·5	42·4	6·6	102·6
			Marsh Seedless grapefruit		
May	73·1	10·5	50·1	4·5	8·0
June	127·5	19·6	70·0	6·4	31·5
July	185·2	30·2	77·5	7·2	70·4
August	254·5	32·1	96·9	12·5	113·1
September	305·0	35·8	100·9	13·5	156·9
October	335·9	41·4	105·5	13·3	175·7
November	349·8	44·5	104·0	13·2	195·5
December	379·8	54·2	110·5	13·0	195·5

From Ting and Vines (1966).
[a] Each figure represents the average of 25 fruit.

juice together with some of the thin cell walls and cell contents are released. In commercial processing the extracted juice may also contain broken juice vesicles and some tissues from the carpel membrane or the albedo. These large particles are removed by a finishing process through a screen. The residue is called the rag and pulp and is used in by-product manufacturing together with the peel. On a whole fruit basis, the distribution of component parts of citrus fruit differs both according to varieties and maturity (Table II). Since chemical analyses are often expressed as concentration of certain chemical constituents on the basis of the weight of a component part, a knowledge of the changes and characteristics of these components part is necessary to the study of biochemical changes of developing fruit as a whole.

2. *General composition*

In addition to the presence of normal cell constituents in each cell, each of the different component parts of the citrus fruits is richer in one or more particular constituents than the others. In the flavedo, the essential oils accumulate in the oil glands embedded in the parenchymatous tissue just below the outer layers of the flavedo. The yield of citrus oil varies from 1·8 to 9·7 lb per ton of peel residue from the cannery (Hendrickson and Kesterson, 1965). The flavedo also contains the carotenoid pigments and the steroids of citrus.

The albedo generally is rich in cellulose, hemicellulose, lignin, pectic substances and phenolic compounds. The segment membrane and the juice vesicle membrane have about the same chemical constituents as the albedo. Most of the sugars and nearly all the citric acid of the citrus fruit occur in the juice which also contains nitrogenous compounds, lipids, phenolic compounds, vitamins and inorganic substances.

III. CHANGES DURING FRUIT DEVELOPMENT AND MATURATION

A. General Morphological and Anatomical Changes

The morphological, anatomical and physiological changes in the developing Valencia oranges were studied by Bain (1958). According to this author, the growth from fruit-set to maturity can be divided into three definite stages.

Stage I, which lasts between 4–9 weeks after fruit-set, is distinguished as the cell division period. Increase in fruit size and weight during this stage is due mainly to growth of the peel by cell division and some cell enlargement. As peel increases in thickness after all cell division has ceased, pectic materials begin to associate with the newly-formed cells. Increases in the volume of the pulp and of the axis at this stage are also due to cell division in the septae between the segments and in the segment walls. Juice vesicle primodia continue to form by cell division up to the end of this state. Throughout Stage I, both fresh weight and dry weight increases.

Stage II is essentially a cell enlargement period. The fruit increases in size accompanied by cell enlargement, differentiation and expansion of the albedo spongy tissue. Juice vesicles become enlarged and juice content increases in the enlarging juice cells. The peel begins to change in colour when fruit approaches maturity.

This period in Valencia in Australia apparently corresponds to the rapid growth period of Valencia described by Marloth (1950) of South Africa. The rate of increase in diameter of fruit is parallel to that of increase in weight per fruit until the colour changes from green to yellow. After this period, the increase in fruit diameter slows down considerably while weight increase continues at its original rate for a period of 6 weeks but then decreases to a rate again parallel to that of diameter increase. During this period, a shortage of moisture in the soil would greatly affect the fruit size.

Stage III is the maturation period. During this period, there is a change of yellow peel colour to orange. The acid in the juice continues to decrease. The peel increases little in thickness, but its tangential growth keeps pace with the increase in volume of the pulp. Anatomically, the epidermal and hypodermal cells of the flavedo continue to divide and the juice vesicles increase in size. The rind volume increases, but the thickness is unchanged. Marloth (1950) found that rind thickness is directly proportional to fruit volume in the early stage.

Valencia oranges require a much longer period from blossom to maturity than most other varieties. In Florida, where all oranges bloom around March, the early season varieties begin to mature in October, the mid season ones in November and early December and the late season varieties around March of the succeeding year. Grapefruit begins to mature in October and may be harvested throughout the season. E. J. Deszyck and S. V. Ting (unpublished data) found that the pigmented grapefruit on rough lemon rootstock in Florida reached their highest soluble solids contents in August and decreased thereafter.

Although no data of a similar study, such as that made on Valencia by Bain (1958), are available for other varieties, fruit growth studies at other citrus-growing areas (Waynick, 1927; Reed, 1930; Furz and Taylor, 1939; Marloth, 1950; Goren and Monselise, 1965) showed that most citrus fruit have a period of rapid size increase. This period will vary in length according to the varieties and environmental factors.

B. Chemical Constituents and their Changes

1. *Colour of citrus fruit* (*carotenoids*)

When citrus fruit is immature, the dominant colour is green. During ripening, chlorophylls a and b break down, and the orange or yellow pigments in the peel begin to increase (Miller *et al.*, 1940b). A similar trend is found with the colour of orange juice (Miller *et al.*, 1941). Total carotenoids of both peel and juice increase as fruit ripens. In the peel, as chlorophyll disappears, the total carotenoid increases several fold. In mature green fruit, the methanol-soluble fraction (xanthophylls) is the predominant member of the group. Later, when the fruit has attained its maximum carotenoid content, the petroleum ether fraction (hydrocarbon, monol and esters) is higher (Table III). It has often been observed that in certain varieties, fruits on the

TABLE III. Distribution of carotenoids in the peel of oranges at beginning and end of the season

Cultivar	Date	Chlorophyll	Petroleum Ether fraction	Methanol fraction	Total carotenoid
			(mg/100 g fresh peel)		
Parson Brown	29.9.37	4·350	0·225	1·000	1·225
	1.2.38	0	4·150	1·060	5·210
Pineapple	Oct., 37	5·000	0·207	1·027	1·234
	22.2.38	0	5·720	2·815	8·535
Valencia	11.11.37	3·000	1·025	1·550	2·575
	23.2.38	0	4·370	1·225	5·595

From Miller *et al.* (1940b).

TABLE IV. Carotenoid content of peel of Pineapple oranges from different sides of the same tree

Position on tree	Petroleum ether fraction	Methanol fraction	Total carotenoid
	(mg/100 g fresh peel)		
Fruit from NE side	6·40	2·010	8·410
Fruit from SW side	3·42	1·400	4·820

From Miller *et al.* (1940b).

northeast side of the tree exhibited a deeper colour than those on the southwest side (Northern hemisphere). With fruits from the south or southwest side of the trees, the unexposed side of the fruit showed a deeper colour. Analysis of these fruits indicated that the higher colour was largely due to higher petroleum ether-soluble fraction (Table IV).

The colour of orange juice, except that of the "blood" orange see (p. 118), is due to carotenoids. Miller *et al.* (1941) reported that sweet orange contained 4–6 mg of total carotenoids/litre of juice and about 1 mg of carotene. Taylor and Witte (1938) found between 0·32 and 0·57 mg of carotene/litre of juice from Florida oranges and between 1·07 and 1·65 mg carotene/litre of juice from California oranges. During the maturation period, the Florida early and mid season oranges had a gradual increase in the carotenoid pigments in their juice until fully mature. The late season fruit showed an increase in juice pigment content until February or March followed by a decline (Miller *et al.*, 1941).

The carotenoid content of bitter oranges is much lower than that of sweet oranges. Sinclair (1961) quotes values for total carotenoids of the peel of sweet oranges ranging from 1·2 to 3·5 mg/100 g fr. wt. while the range given for bitter oranges is 0·2–0·6 mg/100 g. For the pulp of the fruit the values are: sweet, 0·13–0·34; bitter, 0·04–0·1 mg.

In a study of the carotenoids of orange peel and pulp, Curl and Bailey (1956) separated the saponified pigment extracts of Valencia orange peel and juice each into three major groups in a counter-current distribution between petroleum ether and 99% methanol. Fraction I was mainly hydrocarbons, Fraction II, compounds with one oxygen, and Fraction III, compounds with

TABLE V. Carotenoids of Valencia orange pulp and peel (assuming all constituents to have same specific extinction coefficient)

Constituent	% of total Carotenoids Orange pulp	Orange peel
Phytoene	4·0[a]	3·1[a]
Phytofluene	13·0	6·1
α-Carotene	0·5	0·1
β-Carotene	1·1	0·3
ζ-Carotene	5·4	3·5
OH-α-Carotene-like	1·5	0·3
Cryptoxanthin epoxide-like	—	0·4
Cryptoxanthin	5·3	1·2
Cryptoflavin-like	0·5	1·2
Cryptochrome-like	—	0·8
Lutein	2·9	1·2
Zeaxanthin	4·5	0·8
Capsanthin-like	—	0·3
Antheraxanthin	5·8	6·3
Mutatoxanthins	6·2	1·7
Violaxanthin	7·4	44·0
Luteoxanthins	17·0	16·0
Auroxanthin	12·0	2·3
Valenciaxanthin	2·8	2·2
Sinensiaxanthin	2·0	3·5
Trollixanthin-like	2·9	0·5
Valenciachrome	1·0	0·7
Sinensiachrome-like	—	0·2
Trollichrome-like	3·0	0·8

From Curl and Bailey (1956).
[a] Approximate values.

TABLE VII. Seasonal changes in the amino acids in the juice of Valencia orange

Date	Aspartic acid	Glutamic acid	Asparagine	γ Amino	Alanine	Serine	Proline	Arginine
			(mg/100 ml juice)					
2nd Jan.	42	12	60	6	6	18	80	72
19th Jan.	32	14	45	7	4	11	80	60
5th Feb.	29	12	43	7	6	10	90	70
19th Feb.	21	8	23	6	1	11	80	58
6th Mar.	27	7	45	11	tr.	11	95	62
22nd Mar.	25	5	35	12	2	8	88	55
7th Mar.	31	12	51	12	11	14	74	65
22nd Apr.	30	14	42	28	9	15	15	72

From S. V. Ting, unpublished data.
tr. = tracc.

of the proteins could be easily detected in the free form. These included leucine, isoleucine, ornithine, phenylalanine, threonine, tyrosine and valine.

Clements and Leland (1962) analysed the amino acids of the juice of various citrus fruits using ion-exchange chromatography. The amounts of each of the amino acids isolated for the different fruits are shown in Table VIII. Eleven unidentified compounds with various ninhydrin-positive reactions were also reported by these authors.

TABLE VIII. Amino acid content of citrus fruits

Amino acids	Navel orange	Valencia orange	Dancy tangerine	Eureka lemon	Marsh grapefruit
			(mg/100 ml juice)		
Alanine	12	13	7	9	9
γ-Aminobutyric acid	24	32	18	7	19
NH_3	+	+	+	++	+
Arginine	54	57	84	++	47
Asparagine	67	50	85	16	42
Aspartic acid	27	33	36	36	81
Glutamic acid	12	18	16	19	22
Glycine	+	+	+	+	+
Lysine	++	++	++	+	++
Phenylalanine, tyrosine	++	++	++	+	++
Proline	107	239	100	41	59
Serine	18	22	19	17	15
Valine	+	+	+	+	+

From Clements and Leland (1962).
+ less than 2 mg/100 ml.
++ less than 5 but more than 2 mg/100 ml.

TABLE IX. Phenolic amines in citrus juices

Fruit and variety	Octopamine	Synepherine	Tyramine	N-Methyl tyramine	Hordenine	Feruloyl-putresine
			(mg/litre)			
Oranges						
Hamlin	Nil	22	Nil	2	Nil	5
Navel	Nil	15	Nil	1	Nil	4
Pineapple	Nil	27	Nil	2	Nil	4
Valencia	Nil	19	Nil	1	Nil	4
Temple	tr.	27–43	1	Nil	Nil	Nil
Grapefruit						
Duncan	Nil	Nil	Nil	Nil	Nil	15–22
Marsh	Nil	Nil	Nil	Nil	Nil	41
Ruby	Nil	Nil	Nil	Nil	Nil	25
Tangerine						
Dancy	1	97–152	1	15	Nil	Nil
Murcott	tr.	50–52	tr.	Nil	Nil	Nil
Cleopatra	2	280	tr.	58	7	Nil
Lemon						
Meyer	4	2	25	Nil	Nil	Nil
Sweet	Nil	20	Nil	Nil	Nil	Nil
Rough	Nil	11	Nil	Nil	Nil	Nil

From Stewart and Wheaton (1964b); Wheaton and Stewart (1965b).
tr. = trace.

The presence and methods of determination of some sulphur-containing amino acids have been reported by Jansen and Jang (1952) and Miller and Rockland (1952). These sulphur-containing amino acids are not as readily determined as other amino acids. Rockland (1961) compiled data from many sources of twenty-six free amino acids which have been found in the juice of different citrus fruits. All these amino acids have been reported to be present in Valencia oranges except glycine, β-alanine, tyrosine and tryptophane which had only been reported in Japanese fruits. Clements and Leland (1962) reported the presence of glycine and possibly tyrosine in all citrus fruits.

The presence of several amines has been reported in citrus fruit. Swift and Veldhuis (1951) found ethanolamine in the lipid fractions of Valencia orange juice. Stewart *et al.* (1964) first isolated and identified synepherine from citrus leaves and Valencia orange juice. Octopamine was isolated from the leaves of Meyer lemon (Stewart and Wheaton, 1964a) and feruloylputrescine from the leaves and fruit of grapefruit and oranges, but not in tangerine and lemon varieties (Wheaton and Stewart, 1965a). Underfriend *et al.* (1959) reported the presence of tyramine in orange pulp. Wheaton and Stewart (1965b) were unable to find tyramine in orange juice, but found it in the tangerine, Temple and Meyer lemons. They also reported the occurrence of N-methyl tryamine and N-dimethyl tyramine in the juice of Cleopatra mandarin. The phenolic amine content of the juice of various citrus fruit is shown in Table IX.

3. *Carbohydrates in citrus fruit*

(a) *Sugars.* From 75 to 85% of the total soluble solids of orange juice are sugars (Bartholomew and Sinclair, 1943). The changes in sugar content of the juice of Valencia oranges on the tree were studied by Stahl and Camp (1936). The reducing, non-reducing and total sugars increased as fruit continued to ripen on the tree. Similar results were found with grapefruit and Temple orange by Harding and Fisher (1945), and Harding and Sunday (1953). Curl and Veldhuis (1948) concluded that the principal sugars in Florida Valencia orange juice were sucrose, glucose and fructose which occurred in an approximate ratio of 2 : 1 : 1. McCready *et al.* (1950), using paper chromatography, identified these three sugars to be the main ones in the juices of grapefruit, lemon, tangerine and orange. Alberala *et al.* (1967) identified α-and β-glucose, fructose, sucrose and a small amount of galactose in Valencia orange juice by gas chromatography using the TMS derivative technique. The presence of free galactose had not been reported previously.

As the early and mid season oranges and tangerines ripen on the tree, the total sugars in the juice increase rapidly due to an accumulation of sucrose. These fruit continue to mature as the mean season temperature decreases.

In Valencia, the fruit matures in the season when the daily mean temperature tends to increase, resulting in relatively small increments of sucrose (Table X). There appears to be no definite trend in sugar concentrations in the maturing grapefruit.

TABLE X. Seasonal changes in sugar composition of juice of citrus fruits

Date of harvest	Glucose	Fructose	Sucrose	Total sugars
		(g/100 ml juice)		
Hamlin orange				
22nd Sep.	1·17	2·26	4·16	7·59
12th Oct.	1·32	2·09	4·29	7·70
3rd Nov.	1·56	1·98	4·71	8·25
23rd Nov.	1·64	2·01	5·35	9·00
Pineapple orange				
9th Nov.	0·96	2·86	4·96	8·78
12th Dec.	1·34	2·78	5·61	9·73
5th Jan.	1·78	2·64	6·05	10·51
23rd Jan.	2·00	2·55	6·51	11·06
Valencia orange				
17th Feb.	1·74	2·44	4·31	8·49
29th Mar.	1·97	2·46	5·04	9·47
22nd Apr.	2·29	2·55	4·95	9·79
25th May	2·13	2·49	5·13	9·75
Dancy tangerine				
15th Sep.	1·19	1·55	2·19	4·93
26th Oct.	1·24	1·49	3·64	6·37
30th Nov.	1·02	1·58	4·97	7·57
29th Dec.	1·09	1·54	4·64	7·27
Marsh Seedless grapefruit				
23rd Sep.	1·55	1·71	2·62	5·88
19th Oct.	1·75	1·78	2·59	5·12
23rd Nov.	1·67	1·76	2·47	5·90

From Ting (1954).

The free sugars in citrus peel are predominantly sucrose, glucose and fructose although xylose is present in trace amounts (Ting and Deszyck, 1961). In a similar way to the sugars in juices, the sugars of the orange peel also accumulate with maturity; in general, the increases are due mainly to reducing sugars (Table XI). In grapefruit peel, sucrose increases in the early part of the ripening period but falls during the latter part. There is generally an increase in reducing sugars with maturity.

As the alcohol-soluble fractions of the peel increase, there is a corresponding decrease of the alcohol-insoluble solids which are mainly the cell wall and cytoplasmic constituents.

TABLE XI. Seasonal differences of sugar composition of orange and grapefruit peel

Date of harvest	Glucose	Fructose	Sucrose	Total sugars
		(g/100 g dr. wt.)		
Hamlin orange				
11th Oct.	7·6	9·6	13·9	31·1
15th Nov.	10·1	14·2	14·4	38·8
15th Dec.	10·3	16·8	11·8	38·8
15th Jan.	12·6	14·9	8·6	36·2
24th Feb.	14·0	17·5	10·2	41·7
Pineapple orange				
15th Oct.	6·3	10·4	14·3	31·0
15th Nov.	5·2	15·0	13·8	34·0
18th Dec.	8·3	15·8	10·4	34·5
17th Jan.	10·8	21·2	13·9	45·8
Valencia orange				
19th Dec.	7·8	12·1	5·1	25·1
15th Jan.	10·6	11·8	8·0	30·4
19th Feb.	12·6	11·5	6·0	30·0
15th Mar.	15·2	13·2	5·3	33·7
15th Apr.	10·9	10·9	16·5	37·4
Marsh Seedless grapefruit				
6th Aug.	11·7	10·4	4·0	26·2
12th Sept.	8·2	8·3	9·0	25·5
15th Oct.	7·2	6·8	16·2	30·3
15th Nov.	8·5	7·2	20·9	36·7
15th Dec.	11·6	12·8	14·3	38·7
15th Jan.	13·5	13·5	11·9	39·0

From Ting and Deszyck, 1961.

(b) *Polysaccharides.* The alcohol-insoluble solids contain all the polysaccharides and pectic substances together with the proteins. Ting (1958a) noted that from 19 to 25% of the alcohol-insoluble solids of the juice of Valencia oranges were proteinaceous, but only 5% of that of the peel. The hydrolysates of the alcohol insoluble solids (polysaccharides) of the peel of several cultivars of oranges and of the Marsh grapefruit were analysed by Ting and Deszyck (1961). Their results suggested that the main polysaccharides present in the tissue are polygalacturonic acid, glucosan, araban and galactan with smaller amounts of xylan. There were no marked differences between the various cultivars of orange, either qualitatively or quantitatively; the grapefruit contained a higher proportion of glucosan than the oranges. Hydrolysis of the hemicellulose fraction of both orange and grapefruit peel showed it to contain xylan, glucosan, galactan and araban in the ratio of

4 : 3 : 3 : 1, with trace amounts of uronic acids. The α-cellulose fraction after hydrolysis yielded between 80 and 90% glucose in the total monosaccharides. Arabinose, xylose and galacturonic acid composed the remainder of the sugars. Traces of mannose and galactose were also found in this fraction. There was no marked trend in the changes of concentrations of these fractions in the fruit throughout the maturation of oranges and grapefruit.

During the course of the development of orange fruits, starch is found in all components of the fruit including the juice vesicles (Webber and Batchelor, 1943). It is especially abundant in the albedo, but is also found in the flavedo when green. As the fruit grows older, the starch begins to disappear. When horticulturally mature, starch is usually absent from the fruit.

The pectic substances in citrus juice (see also Volume I, Chapter 3) are important to the processing industry because of its function as a "cloud" stabilizer in the juice.

The tissues of citrus fruits have high contents of pectic substances (Table XII). They are used as a source of commercial pectin.

TABLE XII. Sugar content of lemon and lime fruit

Fruit	Parts	Glucose	Fructose	Total reducing sugars	Sucrose
Lemon	Whole fruit	1·40	1·35	2·75	0·41
	Juice	0·52	0·92	1·44	0·18
Lime	Juice			3·48	0

From Widdowson and McCance (1935).

Gaddum (1934), determining pectin as calcium pectate, observed that as Valencia oranges ripened, the water-soluble pectin in both the albedo and the pulp increased to a peak and then decreased while the acid-extracted pectic material continued to decrease. The total pectin showed only slight decrease when the fruit was fully ripened. The total pectin in the juice showed a similar trend to that of the albedo and the pulp.

Sinclair and Joliffe (1961) made a study of the changes of pectic substances from the very immature to the near maturing stages of the fruit development of Valencia oranges. They found a rapid initial increase of both total and water-soluble pectic substance in the pulp and the peel during rapid growth of the fruit. The percentage of methylation of the carboxyl groups of the pectic compounds in the peel rose rapidly to approximately 80% and remained relatively constant during the rest of the period. In maturing oranges (Sinclair and Jolliffe, 1958), a decrease in total pectin and water-soluble pectic substances in the peel and pulp was observed.

Rouse and co-workers in a series of studies on the seasonal changes of pectic materials in different component parts of Pineapple oranges (Rouse *et al.*, 1962), Valencia oranges (Rouse *et al.*, 1964), Silver Cluster grapefruit (Rouse *et al.*, 1965), and lemons (Rouse and Knorr, 1969) separated the pectic materials into water-soluble, oxalate-soluble and sodium hydroxide-soluble fractions. There was no drastic change in any of these fractions in the different component parts as the fruit changed from the immature to mature stage, except that in all cases there was some increase of ammonium oxalate-soluble fractions. Table XIII shows the total sugar and pectin content of various parts of citrus fruits in relation to acidity (Money and Christian, 1950).

TABLE XIII. Total sugars, acidity and pectin content of some citrus fruits

Fruit	Total sugars (%)	Acidity ml 0·1 M per 100 g (as citric)	Pectin (%) as calcium pectate
Grapefruits			
Edible portion	6·74	1·30	—
Peel and pith	4·80	0·59	—
Juice	7·27	1.56	—
Lemons			
Edible portion	2·19	5·98	—
Peel and pith	4·14	0·49	—
Juice	1·92	7·20	—
Oranges (bitter)			
Edible portion	5·49	3·30	0·86
Peel and pith	5·86	0·46	0·89
Juice	5·74	3·77	—
Oranges (sweet)			
Edible portion	7·88	0·79	0·59
Peel and pith	6·81	0·27	—
Juice	8·47	1·17	0·13

From Money and Christian (1950).

4. *Organic acids*

The titratable acidity of the juice of most citrus fruit is due largely to citric acid. Citrus fruits also contain considerable amounts of cations, mainly potassium, calcium and magnesium. The free acid together with the salts forms a very effective buffer system (Sinclair, 1961). The pH of the citrus juices generally varies from about 2 for lemons and other acid fruit to about 5 in overmature tangerines or oranges. Sinclair (1961) lists the pH of the juice of Valencia and Washington Navel oranges as varying between 2·9 and 3·9.

In oranges and grapefruit, Sinclair and Ramsey (1944) found that the free acid per fruit increased in early growth and then became approximately constant. The decrease in *titratable* acidity was considered to be due to dilution as the fruit increased in size and in juice content. Decrease in the concentration of acid with the gradual increase in total sugars during development results in an increase in the ratio of total soluble solids to acidity, which is the basis for determining the legal maturity of the fruit as well as their palatability.

While citric acid is the principal acid of the endocarp of all citrus fruit except the sweet lemon and the acidless orange, the main acids of the citrus peel are oxalic, malic, malonic with some citric. Together they account for 30–50% of the anions present (Sinclair and Eny, 1947; Clements, 1964). Quinic acid was found in large quantities in the peel (Ting and Deszyck, 1959). Tartaric, benzoic and succinic acids have also been reported to be present (Braverman, 1949). In the juice of lemons, citric acid may account for 60–70% of the total soluble solids.

With the application of chromatographic and ion exchange techniques, the detection and determination of previously unknown acids became more feasible. As fruit developed, Rasmussen (1964) confirmed that the major acid of the pulp of Valencia oranges was citric acid. He found 55 meq/100 g dr. wt. in the pulp and 45 meq/100 g dr. wt. in the peel of young oranges (May). The total acidity in the pulp rose to a peak of 172 meq in September, while remaining fairly constant in the peel. He found malic to be the major acid in the peel with little citric acid. Clements (1964), using silicic acid column chromatography, found oxalate to be predominant in the peel of all citrus fruits with malonate and malate at considerably lower concentrations. In the juice, only citric and malic acids were determined, and no malonic acid was detected. In both the flavedo and the albedo of navel oranges, oxalate decreased and malonate increased with advancing maturity. In overmature oranges, the malonate concentration of the flavedo is nearly four times that of the albedo, while oxalate concentration decreases in both component parts.

Table XIV gives some of the results of Clements (1964) for the acids present in the juice and peel of various types and varieties of citrus fruits.

Sweet lemon is characterized by its lack of accumulation of citric acid. Erickson (1957) also reported 5·1% of reducing sugars in the sweet lemon and only 1·4% in the usual (bitter) lemon of commerce.

Quinic acid was isolated from the peel and pulp of different citrus fruit (Ting and Deszyck, 1959). The concentration of this acid in the young fruit is high, especially in the juice vesicles, approaching about one-fifth of that of citric acid and about one-half of that of malic acid in these vesicles (Ting and Vines, 1966). As citric acid increases in concentration, that of malic and quinic acid declines rapidly. Malic acid concentration rises again at the

TABLE XIV. Organic acids of juice and peel of oranges, tangerines, grapefruits, lemons and limes

Variety	Juice (g/100 ml) Malic	Citric	Peel (meq/g dr. wt.) Malic	Citric	Oxalic	Malonic
Orange						
Washington Navel I	0·06	0·56	0·02	0·01	0·11	0·02
Washington Navel III	0·20	0·93	0·02	0·01	0·10	0·03
Valencia	0·16	0·98	0·02	tr.	0·13	0·03
Tangerine						
Dancy I	0·18	1·22	0·06	0·02	0·15	0·01
Dancy II	0·21	0·86	0·09	0·02	0·20	0·02
Grapefruit						
Marsh (Calif.)	0·06	1·79	0·03	0·01	0·06	0·02
Arizona	0·04	2·10	0·10	0·03	0·12	0·02
Texas (pink)	0·06	1·19	0·08	0·01	0·08	0·02
Lemon						
Eureka I	0·17	4·00	0·04	0·04	0·15	0·03
Eureka II	0·26	4·38	0·02	0·03	0·12	0·04
Lime						
Palestine Sweet[a]	0·20	0·08	0·04	tr.	0·05	0·05

From Clements (1964).

[a] The pH of the juice of this variety was 5·7, considerably higher than for the other varieties.

tr. = trace.

end of the maturing period of the fruit. Quinic acid has been considered as a precursor in the synthesis of certain aromatic compounds in plants (Weinstein *et al.*, 1959). It is apparently high in very actively growing tissues of the young fruit and seems to follow a similar trend as the flavonoids in the fruit (Kesterson and Hendrickson, 1953; Hendrickson and Kesterson, 1954).

In fruit of the lime (*Citrus acida*) less than 1·5 cm in diameter, Varma and Ramakrishnan (1956) found succinic acid to be the predominant acid with two unknown acids. In fruit of larger size, citric acid became the main acid with some malic acid. In the juice of oranges damaged by freezing, succinic acid has been observed (S. V. Ting, unpublished). Succinic acid was reported to be present in apples injured by high concentrations of CO_2 (Hulme, 1956). Freeze-damaged oranges have a high respiration rate which, in turn, would increase the internal CO_2 concentration to such a level as to cause possible injuries to the tissues.

A major characteristic of the bitter orange is the high acidity. Titratable acids range from 2 to 6% of the fresh weight.

5. *Flavonoids and limonoids*

(a) *Flavonoids.* These compounds have been defined as a group of substances containing the C_6–C_3–C_6 carbon skeleton. The principal citrus flavonoids are the anthocyanins (see p. 118), flavones, flavonols or flavanones. The last-mentioned group contains hesperidin and naringin, the main flavonoids of oranges and grapefruit, respectively. Hesperidin is also found in mandarins, lemons, limes and hybrids. A list of flavonoid compounds in various citrus fruits is shown in Table XV.

TABLE XV. The major flavonoids of citrus fruit

Fruit	Flavonoids
Orange	Hesperidin
	Rutinoside of naringenin
	Isosakuranetin
	4-β-D glucoside of naringin
	4-β-D glucoside of naringenin rutinoside
	Neohesperidin
Grapefruit	Naringin
	Rutinoside of naringenin
	Neohesperidin
	Hesperidin
	Poncirin
	Isosakuranetin
	4-β-D glucoside of naringin
	4-β4D glucoside of naringenin rutinoside
Tangerine	Hesperidin
	Tangeritin
	Nobiletin
Lemon	Hesperidin
	Eriodicitrin
	Diosmin
	Limocitrin
	Isolimocitrin
	Isorhamnetin

From Gentili and Horowitz (1964, 1965); Horowitz (1961); Mizelle *et al.* (1965, 1967).

Hesperidin is the 7-β-rutinoside of hesperetin and can be readily isolated from the peel by the method of Hendrickson and Kesterson (1954). It has also been found in canned orange products. When oranges are damaged by freezing, hesperidin can be found in the segment membrane in the form of white spots.

The bitter principle of bitter (sour) oranges is naringin† which, in these

† See also Volume 1, Chapter 11.

fruits, takes the place of the hesperidin in sweet oranges. The cloudiness of marmalade made from sweet oranges is due to precipitation of hesperidin which is less soluble than naringin (Smith, 1953).

The main flavonoid compound of the grapefruit and shaddock is naringin, the bitter taste of which can be detected at a dilution of 1 : 50,000 (Zoller, 1918). Chemically, it is a rhamnoglucoside of the aglycone, naringenin; it can be easily recovered by extracting the peel in a similar manner as for hesperidin. The bitter taste of naringin was found to be due to the structure of the disaccharide moiety (Horowitz and Gentili, 1961, 1963). In the case of hesperidin, the rhamnose and glucose are in the form of rutinose; but in grapefruit, these two sugars link to form neohesperidose (2-0-α-L-rhamnopyranosyl-D-glucose), which is the sugar moiety of neohesperidin, an isomer of hesperidin (Zemplen and Tettamanti, 1938). Through the work of Horowitz and Gentili, it is now generally concluded that citrus flavanones containing neohesperidose are bitter; whereas, the corresponding flavanones containing the isomeric disaccharide, rutinose (6-0-α-L-rhamnopyranosyl-D-glucose), are "tasteless". Dunlap and Wender (1962) reported the presence in grapefruit of some minor flavonoids including neohesperidin, isosakurenetin-7-rhamnoglucoside, rhoifolin and kaempferol. Mizelle *et al.* (1965) isolated the 7-β-rutinoside of naringenin from grapefruit segments together with rutinosides of hesperetin and isosakuraetein and their neohesperioside isomers, namely neohesperidin and poncirin. Gentili and Horowitz (1965) characterized the rutinosides of naringenin and isosakuranetin isolated from peel of both navel and Valencia oranges. The presence of two triglycerides of flavanone in grapefruit segments was reported by Mizelle *et al.* (1967), namely the 4-β-D-glucoside of naringin and 4-β-D-glucoside of naringenin-7-β-rutinoside. They also indicated the presence of these two compounds in shaddock (*C. grandis*) and in sweet oranges (*C. sinensis*). Maier and Metzler (1967) reported the occurrence of dihydrokaempferol in enzyme-hydrolysed extracts of peel and endocarp of both mature and immature Marsh Seedless grapefruit.

The total flavonoid of all citrus fruit increases to a maximum during the early stage of fruit development and then remains constant. With increase in fruit size, the flavonoid concentration decreases. Hendrickson and Kesterson (1954) obtained maximum yield of hesperidin in the early season when the fruit was about 1–2 in. in diameter. The yield of flavonoid then decreased with increase in fruit size, possibly because of the large amount of fresh materials to be extracted. The flavonoid concentration of the component parts of citrus fruit decreased with maturity. Kesterson and Hendrickson (1953) reported that only 10% of the naringin of the whole fruit is in the juice of grapefruit; between 20 and 30% of the hesperidin of the whole fruit was found in the juice of the orange and mandarin (Hendrickson and Kesterson, 1954).

The juice vesicles of immature grapefruit are extremely bitter due to the high concentration of naringin. As the fruit ripens, the bitterness decreases in conjunction with a decrease of naringin. The determination of flavanone glycosides is based on its reaction with alkali to form, quantitatively, a yellow colour (Davis, 1947). Although the method is not specific, it is, nevertheless, used extensively in routine analysis due to its simplicity. The correlation of bitterness with the Davis test for naringin may be good in some years but not in others, or high during the early season and low in the late season. Ting (1958b) found an enzyme capable of hydrolysing naringin in an extract from a commercial pectic enzyme preparation, and showed that the grapefruit juice vesicles contained substances other than naringin which could form a yellow colour with the Davis reagent. The Davis method was used to determine naringin by the difference before and after hydrolysis. Thomas *et al.* (1958) separated naringinase from commercial pectic enzyme preparations. Enzymes capable of hydrolysing naringin were also found by Hall (1938). Shimokoriyama (1957) described the presence of a chalconase in the peel of grapefruit which catalyses the reaction between the flavanones and their chalcones.

Hagen *et al.* (1966) analysed grapefruit segments throughout the season for all six flavanone glycosides reported by Mizelle *et al.* (1965) and confirmed that the absolute amount of flavanone glycoside did not change. The decrease in concentration was due to an increase in fruit size.

For many years, the only flavonoid known to occur in lemon was hesperidin (Bartholomew and Sinclair, 1951). Horowitz (1956, 1957) and Horowitz and Gentili (1960a) found the precursors of diosmin (the flavone analogue of hesperidin) and eriodicitrin in the peel of lemon. Next to hesperidin, eriodicitrin and diosmin are probably the more abundant flavonoids in lemon peel. Eriodicitrin differs from hesperidin only in the B-ring where an orthodihydroxy group is present. The presence of these similar compounds in the same fruit points to the possibility of common precursors in their biosynthesis. Limocitrol, isolimocitrol and limocitrin were found in lemon flavonoid extracts (Gentili and Horowitz, 1964) and citronin, 2′-methoxy-5, 7-dihydroxyflavanone-7-rhamnoglucoside, was isolated and identified from the Pondorosa lemon (Horowitz and Gentili, 1960b).

Some methylated flavones and flavonols are also found in citrus fruit. Tangeritin ($C_{20}H_{20}O_7$) found in tangerine oil (Nelson, 1934) has been identified as a pentamethoxyflavone (Goldsworthy and Robinson, 1957). From the peel of a cultivar of *C. reticulata* (Tankan), Robinson and Tseng (1938) isolated a hexamethoxy flavone which they named "nobiletin". Several other pentamethoxy and hexamethoxy flavone isomers isolated and detected in various other citrus fruits have been recorded by Horowitz (1961). Swift (1964) isolated a pentamethoxy flavone (sinensetin) from pressed liquid from

Florida mid season oranges and identified it as similar to the compound reported by Born (1960). In the neutral fraction of a benzene extract of the orange peel juice the following substances were found: nobiletin, tangeretin, a heptamethoxyflavone, sinensetin and tetra-O-methyl scutellarein (Swift, 1965).

Auranetin and demethylnobiletin have been isolated from the bitter orange. Further details of flavonoids in citrus fruits are given by Goodall (1969).

(b) *Limonoids.*† Limonin has been recognized as a bitter principle in citrus fruit, and it is most commonly found in the juice of Washington Navel oranges. Higby (1938) isolated it from the edible portion of the navel orange and from both the pulp and the seeds of Valencia oranges. A number of limonoid bitter principles have been isolated from different component parts of citrus fruit. In addition to the work of Higby, Emerson (1948) found nomilin in the seeds of Valencia. Chandler and Kefford (1951, 1953) found limonin and limonexic acid in the peel, juice and seeds of Australian navel and Valencia oranges. Samish and Ganz (1950) detected limonin in the juice of Shamouti oranges. The occurrence of limonin in grapefruit concentrate in amounts up to 9·5 p.p.m. on a reconstituted basis was reported by Maier and Dreyer (1965). In the membrane and the central axis of grapefruit, as much as 140 p.p.m. was found.

It is generally known in the processing industry that juice of some varieties of oranges such as Washington Navel sometimes becomes unpalatably bitter a few hours after extraction. This bitterness is most intense with early fruit and becomes less marked in more matured fruit. The bitter taste also develops much faster when the temperature of the product is increased. The postulation that limonin exists as a water-soluble non-bitter precursor in the pulp and that this precursor is converted into limonin by either oxidation or enzymatic action is supported by the work of Higby (1938), Emerson (1949) and Samish and Ganz (1950). Kefford (1959) believed that this phenomenon of delayed bitterness was due to the physical process of diffusion of limonin from the suspended solids into the juice. Because of the low solubility of limonin, heating or prolonged standing increased its concentration. A concentration of 2 p.p.m. imparts a definite bitterness to the juice. Recently, Maier and Beverly (1968) reported the occurrence of a non-bitter precursor of limonin, a limonin mono-lactone. During juice manufacture, this compound is converted into the bitter compound. Heating the juice increases this rate of conversion.

The structure of limonin is known now to be a triterpenoid oxidation product containing one furan ring, one ketone group, one epoxide group and two lactone rings (Arigoni *et al.*, 1960). Methods of quantitative determination have been developed by many workers (Emerson, 1952; Dreyer, 1965; Chandler and Kefford, 1966; Wilson and Crutchfield, 1968; Maier

† See also Volume 1, Chapter 12.

and Beverly, 1968). The thin-layer chromatography method of Dreyer (1965) is reasonably accurate and is considered the most specific.

Bitter principles in citrus generally decrease with maturity of the fruit (Higby, 1938; Emerson, 1949; Samish and Ganz, 1950). In Valencia, the bitter principle completely disappeared when fully matured according to Kefford (1959). When fruit was stored at a high temperature (27–32°C), Rockland *et al.* (1957) found a rapid loss of the bitter principle. Navel oranges ripened in storage with ethylene produced a juice that did not turn bitter (Ball, 1949), but treating the entire tree with ethylene in a tent did not show any such effect (Emerson, 1949).

6. *Lipids*

In the lipid fraction of the rind, Matalack (1929) found oleic, linolenic, palmitic and stearic acids, and glycerol, phytoseterolin and ceryl alcohol. The pulp contained, in addition to those fatty acids and alcohols found in the peel, cerotic acid, phytosterol and pentacosane (Matalack, 1940). Huskins and Swift (1953) reported the composition of juice lipids of Florida Valencia oranges as shown in Table XVI. Swift (1952a) determined the various fatty acids from Valencia oranges and found that the main fatty acids were similar to those found by Matalack.

TABLE XVI. Composition of lipids from Florida Valencia orange juice (%)

Constituents	1947	1950
Unsaponifiable matters	14·81	16·53
Resin acids	12·41	10·31
Free fatty acids	21·00	34·30
Combined fatty acids	13·00	3·42
Sterol and sterol glycosides	1·48	4·93
Phospholipids	33·30	37·90

From Huskins and Swift (1953).

The dried seeds of oranges contain from 30 to 45% lipid, and those of grapefruit contain from 29 to 37%. A number of chemical and physical characteristics of different citrus seed oils are tabulated in Table XVII. There was a significant correlation between refractive indices and the degree of unsaturation as measured by the iodine values (Hendrickson and Kesterson, 1963). The citrus seed oil is a mixture of glycerides of various fatty acids. The distributions of these fatty acids, as analysed by Hendrickson and Kesterson (1965) using the gas chromatographic procedure, are tabulated in Table XVIII.

The unsaturated fatty acids occupy a large percentage of the total fatty acids of the citrus seed oils. This fact makes the citrus oil a desirable dietetic substitute for other unsaturated fats in food.

TABLE XVII. Chemical and physical characteristics of seed oils of various citrus fruits

Characteristics	Grapefruit	Lemon	Orange	Tangerine
Specific gravity	0·9197	0·914–0·917	0·916–0·920	0·9165
Refractive index N_D^{25c}	1·4698	1·471–1·472	1·468–1·470	1·4702
Iodine value	100·9	103–110	98–104	107·3
Saponification number	193	188–196	192–197	194
Unsaponifiable matter (%)	0·48	<1	0·4–1·0	0·5
Titre (softening point)	—	32–38° C	34–35° C	—
Acetyl value	2·4	13–33	2–11	8·2
Reichert–Meissl value	0·47	$<0·5$	$<0·5$	—
Polenske value	0·20	$<0·5$	$<0·5$	—

From Hendrickson and Kesterson (1965).

TABLE XVIII. Comparison of average fatty acid composition of seed oils from many species of citrus

Seed oil	Palmitic	Stearic	Oleic	Linoleic	Linolenic
Orange	31·0	4·2	26·0	36·0	2·8
Grapefruit	34·0	3·4	22·0	37·0	4·6
Tangerine	31·0	3·4	21·0	40·0	4·7
Key lime	31·0	4·3	20·0	35·0	9·7
Lemon	26·0	2·8	27·0	33·0	11·2
Calamondin	24·5	5·1	36·7	28·2	5·3

From Hendrickson and Kesterson (1965).

7. *Volatile compounds*

The most important volatile materials of citrus fruit are those associated with flavour and aroma. These include terpene hydrocarbons, carbonyl components, alcohols, esters and volatile organic acids; they are generally associated with the "peel oil" in the flavedo but are also found in oil sacs embedded in the juice vesicles (Davis, 1932). These compounds are dealt with in Volume 1, Chapter 10. One compound, namely the monoterpene hydrocarbon (+)-limonene, accounts for 80–96% by weight of all citrus oils. However, there is a significant variation in both the qualitative and quantitative composition of other volatiles from different citrus species. Other hydrocarbons include the monoterpenes: α-pinene, α-thujene, camphene, β-pinene, sabinene, myrcene, Δ-3-carene, α-phellandrene, α-terpinene, β-terpinene, p-cymene, terpinolene, p-isopropenyltoluene, 2,4-p-menthadiene and the sesquiterpenes: cubebene, copaene, elemene, caryophyllene, farnescene, α-humulene, valencene and Δ-cadinene (Hunter and Brogden, 1965a; Teranishi *et al.*, 1963). One of the more important sesquiterpenes is valencene (Hunter and Brogden, 1965b).

Esters, on a quantitative basis, account for only a small fraction of citrus oils. Among those identified are: ethyl formate, ethyl acetate, ethyl butyrate, ethyl isovalerate, ethyl caproate, ethyl caprylate, linalyl acetate, octyl acetate, nonyl acetate, decyl acetate, terpinyl acetate, geranyl acetate, ethyl 3-hydroxyhexanote, citronellyl butyrate, geranyl butyrate, methyl anthranilate and methyl N-methylanthranilate. In orange essence, the only ester found in appreciable concentration is ethyl butyrate (Wolford *et al.*, 1963; Attaway *et al.*, 1964).

A number of aldehydes and ketones make important contributions to citrus flavours (Stanley *et al.*, 1961). For example, the terpene aldehydes, neral and geranial, provide the characteristic flavour of the lemon; and the sesquiterpene ketone, nootkatone, is a major factor in the flavour of grapefruit juice (MacLeod and Buigues, 1964). The conversion of valencene to nootkatone is by oxidation of the sesquiterpene with tetrabutyl chromate to form the sesquiterpene ketone (Hunter and Brogden, 1965c). No one compound has been found that makes a dominant contribution to orange flavour. However, a number of aldehydes and ketones have been identified; the most important ones being 2-hexenal, n-octanal, n-decanal and geranial. Also found are traces of acetaldehyde, acetone, n-butyraldehyde, n-hexanal, methyl ethyl ketone, n-heptanal, n-nonanal, furfural, methyl heptenone, citronellal, n-undecanal, n-dodecanal, neral, carvone, perillaldehyde, piperitenone and β-sinensal.

Carbonyl compounds in trace amounts in lemon oil include n-hexanal, n-heptanal, n-octanal, methyl heptenone, n-nonanal, n-decanal, citronellal, n-undecanal, n-dodecanal and carvone (Ikeda *et al.*, 1962). In grapefruit oil, the major carbonyls in addition to nootkatone are n-octanal and n-decanal, with traces of citronellal, neral and geranial. The only carbonyls identified to date in tangerine oil are n-octanal, n-decanal, neral and geranial. Octanal is the most significant of these quantitatively.

There are large amounts of alcoholic components among the volatile flavour materials of citrus. However, their contribution to the overall flavour and aroma is probably less than that of the carbonyl components and terpenes. The predominant alcohol of orange oil and essence is linalool, but substantial amounts of octanol are also present. Significant amounts of terpinen-4-ol and α-terpineol have also been found along with traces of methanol, ethanol, n-propanol, isobutanol, n-butanol, isopentanol, n-pentanol, n-hexanol, 3-hexenol, n-heptanol, methyl heptenol, 2-nonanol, n-nonanol, n-deconol, citronellol, nerol, geraniol, carveol, undecanol and dodecanol. The major contributors to the identification of the alcohols of orange oil were Hunter and Moshonas (1965), while the alcohols of orange essences were identified by Attaway *et al.* (1962). Essentially the same alcohols, but in different proportions, are found in lemon, grapefruit and tangerine oils (Hunter and Moshonas, 1966).

The major volatile organic acids of orange juice essences are aectic, propionic, butyric, caproic and capric (Attaway *et al.*, 1964). Volatile acids present in trace amounts include isovaleric, valeric, isocaprioc and caprylic.

Attaway *et al.* (1967) studied the qualitative and quantitative changes in 14 of the major volatile flavour components of oranges, tangerines and grapefruit over the course of a season. The most significant changes were found to occur between the terpene hydrocarbons, particularly (+)-limonene, and the terpene alcohols, particularly linalool. In Hamlin oranges, the flavour oil contained 52% limonene and 27% linalool in May, but by October the limonene concentration had increased to 95% and the linalool concentration had decreased to 0·6%. In tangerine oil, the limonene increased from 20 to 87%, while the linalool dropped from 62 to 4%. In grapefruit oil, the limonene concentration was already 83% in May and increased to 93% as linalool decreased from 3% to 0·4%. Some of the other oxygenated terpenes also decreased in concentration during the season. For example, thymol, a constituent found only in tangerine oil, decreased from 3·4 to 0·2%. In orange oil, geranial decreased from 3·5 to 0·2%, geraniol from 0·5 to 0% citronellal from 0·55 to 0%, terpinen-4-ol from 5·5 to 0·6%, and α-terpineol from 4·3 to 0·2%. These same compounds also decreased in concentration in grapefruit oil. Only one other terpene hydrocarbon increased during the season. This was myrcene, which increased from 0·69 to 1·76% in orange oil, 0·46 to 1·4% in tangerine oil, and 1·2 to 1·9% in grapefruit oil.

TABLE XIX. Constituents of bitter orange oil

Constituent	Reference
Linalool	Komatsu *et al.*, 1930
Nonyl alcohol	Komatsu *et al.*, 1930
D-Terpineol	Komatsu *et al.*, 1930
Farnesol	Naves, 1946
D-Nerolidol	Naves, 1946
$C_{15}H_{23}OH$	Naves, 1946
Decyl aldehyde	Komatsu *et al.*, 1930
Nonyl aldehyde	Igolen and Sontag, 1941
N-Dodecenyl aldehyde	Igolen and Sontag, 1941
Limonene	Igolen and Sontag, 1941
α-Pinene	Ikeda and Fujita, 1930
Formic acid	Igolen and Sontag, 1941
Acetic acid	Igolen and Sontag, 1941
Pelargonic acid	Igolen and Sontag, 1941
Cinnamic acid	Igolen and Sontag, 1941
Decyl pelargonate	Igolen and Sontag, 1941
Neryl acetate	Igolen and Sontag, 1941
Geranyl acetate	Igolen and Sontag, 1941
Citronellyl acetate	Igolen and Sontag, 1941
Umbelliferone	Komatsu *et al.*, 1930
Ethylumbelliferone	Komatsu *et al.*, 1930
Auraptene	Komatsu *et al.*, 1930
Auraptin	Nomura, 1950

In more recent work using linalool labelled with ^{14}C, Attaway and Buslig (1968a, 1969) demonstrated that linalool could be converted to α-terpineol and other volatile materials by living citrus tissue. Various constituents isolated from bitter orange oil are shown in Table XIX.

8. *Miscellaneous lipid-solvent soluble compounds*

The compounds grouped in this section differ from one another in chemical properties, but are similar in the fact that they are all found in the residue after distillation of citrus oil, and that they are all soluble in lipid solvents. Among these compounds are waxes, coumarins, triterpenoids and steroids.

(a) *Waxes*. In the residue from the distillation of grapefruit oil, Markley *et al.* (1937) identified wax acids with mean molecular weight corresponding to $C_{32}H_{64}O$ and a group of other waxes and steroids. Waxes from peel of Valencia oranges were reported to contain C_{26} acids, stearic acids and other fatty acids (Warth, 1947). Hunter (1966) identified some paraffin wax fractions from residues after distillation of cold-pressed Valencia orange oil.

The epicuticular wax of Hamlin orange has been found to increase with advancing maturity of fruit (S. V. Ting, unpublished). The petroleum ether-soluble material in the epicuticular wax increases at a much faster rate than those insoluble in petroleum ether. Suberin, a complex substance consisting of high molecular weight, saturated fatty acids was found to deposit on the walls of the juice vesicles (King, 1947).

(b) *Steroids and triterpenoids.*† In the peel oil of grapefruit, Weizmann *et al* (1955) reported the occurrence of 22-dihydrostigmasterol (β-sitosterol) and a triterpenoid ketone, friedelin. Swift (1952b) first isolated β-sitosterol D-glucoside from Valencia orange juice. Mazur *et al.* (1958) identified citrostadienol from the peel oils of both orange and grapefruit. Williams *et al.* (1967) separated the sterol fraction of grapefruit peel into three groups, namely 4-4′-dimethyl sterol, 4-α-methyl sterols and desmethyl sterols. Each group of sterols could be further fractionated into several components. The 4-α-methyl sterol fraction of grapefruit peel was shown to contain several components, only one of which was the previously identified citrostadienol. The complex nature of these sterols in a single plant was considered to be significant in suggesting that some of them may be precursors in the biosynthesis of the major phytosterols such as β-sitosterol. Among the desmethyl sterols three components were found with GLC retention times similar to those of β-sitosterol, stigmasterol and campesterol.

(c) *Coumarins*. Coumarin compounds have been isolated from the oils of various citrus fruit. Stanley and Vannier (1957) reported the identification of five coumarin substances from lemon oil, including limettin and bergamotin. From grapefruit oil, Vannier and Stanley (1958) identified 7-geranoxycoumarin. They also proposed a procedure for the detection of grapefruit oil in lemon oil. Fisher and Nordby (1965) isolated 9 coumarins from grapefruit, 5 of which had been reported previously. Peyron (1963) separated and identified 4 coumarins from orange oil by the use of thin-layer chromatography and flourescence. These included auraptene, meranzine, bergaptol and isopentenyl psoralenes.

9. *Vitamins*

Vitamin C, or ascorbic acid, is by far the most abundant vitamin in citrus fruits, which are important sources of this vitamin. The peel is especially

† See also Volume 1, Chapter 12.

rich in ascorbic acid and Atkins *et al.* (1945) found that the concentration of ascorbic acid in the juice of oranges is only one-fifth of that of the flavedo and one-third of that of the albedo on a freshweight basis. In grapefruit, the juice had only one-seventh and one-fifth of the vitamin C concentration of that in flavedo and albedo, respectively. On a wholefruit basis, the juice contained about 25% of the total vitamin C of the orange and only 17% of that of grapefruit.

There is a considerable variation in the vitamin C content of juice of different citrus fruits. Oranges generally contain from 40 to 70 mg/100 ml, grapefruit, tangerine and lemon between 20 and 50 mg/100 ml. Ascorbic acid is usually high in immature oranges and grapefruit. As fruit ripen and increase in size, the concentration generally decreases (Harding *et al.*, 1940; Harding and Fisher, 1945). When calculated on a per fruit basis, the total ascorbic acid usually increases. Vitamin C was found to be higher in the stem half of the pulp of oranges and grapefruit than in the stylar halves, and higher around the central axis than on the outside nearest to the peel (Ting, 1968a).

A positive correlation between vitamin C and the soluble solids of the juice of Valencia oranges from the same tree was shown by Sites and Reitz (1951). These authors also reported that fruits at the top and on the outside of the tree had a higher vitamin C than those in the inside and at the lower level. When harvested at the same time, the green coloured oranges were found to be lower in vitamin C than the orange-coloured ones; fruits from the north side of the tree had a considerably lower concentration than those from the south side. The directional difference was not evident in fruits picked from inside the tree. Citrus fruit also contains many other vitamins. A list of vitamins as compiled from different published reports is presented in Table XX.

TABLE XX. Vitamins in citrus fruit juice

Vitamins	Units	Orange	Grapefruit	Tangerine	Lemon
Vitamin A (β-carotene)	I.U./100 ml	190–400	0–21	350–420	0–2
Vitamin C[a]	mg/100 g (pulp)	50	40	30	50
Thiamin	μg/100 ml	60–145	40–100	70–120	30–90
Niacin	μg/100 ml	200–300	200–220	200–220	100–130
Riboflavin	μg/100 ml	11–90	20–100	30	60
Pantothenic acid	μg/100 ml	130–210	290		
Biotin	μg/100 ml	0·1–2·0	0·4–3·0	0·5	
Folic acid	μg/100 ml	1·2–2·3	0·8–1·8	1·2	
Inositol	mg/100 ml	98–210	88–150	135	85
Tocopherol	mg/100 ml	88–121			

From U.S. Dep. Agr. (1957).
[a] Average values from Olliver (1967).

Inositol and tocopherol are present in fairly large amounts in the juice of citrus fruit, but the other vitamins also occur in amounts appreciable enough to be of dietetic importance (Rakieten *et al.* (1951). On a concentration basis, citrus juices are higher in vitamin A, thiamin and niacin than milk but lower in riboflavin (Plicher, 1947). Compared with other foods, citrus juice may supply larger amounts of several vitamins on a per calorie basis.

10. *The inorganic constituents*

The inorganic substances of the citrus fruit are all found in the ash which is a measure of the alkalinity of the tissue (Sinclair and Eny, 1947). Besides their importance in human nutrition, the minerals play vital roles in the biochemical reactions within the tree and the fruit. Mineral analysis of the leaves is used as a tool to determine the status of particular elements in tree nutrition. The inorganic constituents of the fruit are greatly influenced by fertilizer applications (Embleton *et al.*, 1956; Labanouskas *et al.*, 1963), and there are considerable variations in the mineral content of fruit from different positions of the same tree (Koo and Sites, 1956).

The juice of citrus fruit contains about 0·4% ash. Harding *et al.* (1940) found that the ash content of orange juice was generally the highest in immature fruit and gradually decreased as the fruit maturity progressed. Potassium is by far the most abundant element, and citrus juices are a good source of potassium in human nutrition.

The amounts and the constituents of ash of orange juices from different sources were analysed by Yufera and Iranzo (1967). The data are presented in Table XXI. While these values are typical, slightly higher values of potassium than the maximum recorded here have been reported, for instance in Florida. Roberts and Goddum (1937) reported the presence of iron, manganese, copper, zinc and boron. Strontium, barium, aluminium and chromium were found to occur consistently in the samples they analysed and titanium, lead, tin, nickel and silver were also reported. In oranges, anions found included phosphates, sulphates and chlorides. Bromine, iodine and flourine have also been found in the juice (Rakienten *et al.*, 1952; Stevens, 1954).

Since many of these elements are associated with enzyme systems in the fruit, they are extremely important in the metabolism of fruit. Deficiency symptoms of many of these elements have been described and reported elsewhere and are beyond the scope of this discussion (Chapman, 1968). Potassium, calcium and magnesium occur in fruit in combination with organic acids such as citric, malic and oxalic. Calcium is also usually associated with the pectic substances in the fruit. High potassium is related to high total acidity in the fruit (Sinclair and Eny, 1946). Copper is known to have a destructive effect on ascorbic acid (Lovett-Janison and Nelson, 1940).

TABLE XXI. The amounts and constituents of ash of orange juice from different geographic locations (mg/100 ml juice)

Source of fruit		Ash	Potassium	Phosphorus (mg/100 ml juice)	Calcium plus magnesium (as Ca)	Sodium
Spain	Maximum	410	190	18·3	37·0	1·6
	Minimum	310	122	9·8	26·4	0·4
	Mean	350	149	12·3	32·1	0·7
California	Maximum	550	232	18·8	40·0	2·8
	Minimum	380	150	11·0	26·5	1·6
	Mean	450	194	16·3	38·8	2·2
Florida	Maximum	620	248	11·8	42·5	2·8
	Minimum	420	181	5·0	25·0	0·8
	Mean	570	208	8·3	31·9	2·0

From Yufera and Iranzo (1967).

Details of the ash contents of the various types of citrus fruits are given by McCance and Widdowson (1960); see Table XXII.

TABLE XXII. Mineral composition of some citrus fruits

	Na	K	Ca	Mg	Fe	Ca	P	S	Cl
					mg/100 g				
Grapefruit (flesh only)	1·4	234	17·1	10·4	0·26	0·06	15·6	5·1	0·6
Lemon (whole—excluding seeds)	6·0	163	10·7	11·6	0·35	0·26	20·7	12·3	5·1
Lemon juice	1·5	142	8·4	6·6	0·14	0·13	10·3	2·0	2·6
Orange (flesh only)	2·9	197	41·3	12·9	0·33	0·07	23·7	9·0	3·2
Orange juice	1·7	179	11·5	11·5	0·30	0·05	21·7	4·6	1·2
Tangerine (flesh only)	2·2	155	41·5	11·2	0·27	0·09	16·7	10·3	2·4
Lime (whole—excluding seeds)[a]	2	102	33	—	0·6	—	18	—	—
Lime juice[a]	1	104	9	—	0·2	—	11	—	—

From McCance and Widdowson (1960).
[a] Watt and Merrill (1963).

C. Effects of Environmental Factors and Cultural Practices on the Chemical Composition of the Fruit

The chemical composition of a fruit is determined by varietal characteristics, influence of rootstock and external factors. The selection of variety and rootstock, once made, is difficult and uneconomical to change. To some extent, however, the external factors can be controlled or modified by the grower. The soil on which the trees are planted will determine the most suitable rootstock to be used, and the amounts and time of application of fertilizers. Selection of rootstock is also subject to restriction due to susceptibility to certain pests. The effects of these factors are undoubtedly interrelated. The mineral content of the fruit is influenced by the application of fertilizer to the soil (Jones and Parker, 1949), but the changes in mineral composition are only the direct and obvious effects of this application. The indirect effect on the organic composition and quality of the fruit is not well understood.

The prevailing temperature of a growing area not only influences the internal quality of the fruit but also its external appearance and texture. The development of external colour is greatly influenced by temperature. Stearn and Young (1942) showed that under conditions in Florida, early and mid season fruit changed from dark green to yellowish green only when the minimum temperature was below 13°C. Miller and Winston (1939) found that carotenoids began to increase in concentration rather early in the autumn. The rapid increase in colour seemed to coincide with the seasonal decline in mean temperature. Young and Erickson (1961) observed that soil, night-time air and daytime air temperatures all played a part in colour changes of Valencia fruit grafted on Cleopatra mandarin rootstock. Fruit produced the most colour with temperatures of 20°C for day-time, 7°C for night-time and 12°C for soil. Higher or lower temperatures in either group gave less colour. Colour change was due to a loss of chlorophyll and an increase of xanthophyll. The carotene content was not affected by the temperature variations. Regreening usually occurs in Valencia oranges and is related to temperature. Caprio (1956) noted that cooler than normal temperatures during the period immediately before full bloom and extending several months afterward, and a warmer than normal temperature during the remainder of the fruit-developing and maturing period, were the conditions favouring regreening. Valencia oranges on sour orange rootstock had a higher and deeper orange colour than those grown on other rootstocks (Harding *et al.*, 1940). The colour of navel oranges was deeper when phosphate was deficient (Chapman and Rayner, 1951). High nitrogen produces grapefruit with less coloration (Hilgeman, 1941; Jones *et al.*, 1945). The coloration of grapefruit appears when chlorophyll fades, since the yellow carotenoids are present from the early stages of fruit development; nitrogen appears to delay this disappearance of chlorophyll. The exact effect of different nutrients in the biosynthesis of carotenoids is not known. Colour development in orange is reported to be retarded by horticultural spray materials containing oil (Winston, 1942). Coggins *et al.* (1970) showed that lycopene accumulates in grapefruits and oranges treated with 2-(4-chlorophenylthio)-triethylamine HCl.

Environmental factors affecting those chemical constituents directly related to internal qualities of fruit, as opposed to the colour of the peel discussed above, are many; a few of these have been reported and will be discussed.

Citrus growers have long recognized that weather conditions are a prime factor influencing fruit quality. The soluble solid content of the juice differs in different years even in varieties which do not have a tendency for alternate bearing. Sites (1947) noted that during a 7 year period, fruit produced in 2 years had high solids, but he failed to find any one element of weather which could be considered as a cause of the difference. The high solids years had a lower rainfall and less cloudy days during the early fruit development period. Early bloom (approximately 2–3 weeks) characterized one of the other high solids years. The abundance of rainfall or excessive irrigation during the ripening period seems to have a dilution effect on the total soluble solids and acids of the fruit (Sites and Camp, 1955). The total soluble solids content of the juice is also influenced by rootstocks. Trees on rough lemon rootstock produce fruit with lower soluble solids and titratable acidity than those on bitter (sour) orange or on grapefruit stocks (Harding and Fisher, 1945).

Vitamin C content of the juice may also be affected by the rootstock (Harding *et al.*, 1940; 1940; Cohen, 1956).

Rootstocks also exert an effect on the nitrogen content of the juice. Marsh (1953) found a higher nitrogen content in Washington Navel on rough lemon rootstock than that on bitter orange. He also found the rootstock to affect the development of limonoid bitterness in Washington Navel orange juice. Juice from fruit grown on grapefruit rootstock developed no bitterness, whereas juice from fruit on rough lemon rootstock became intensely bitter very soon after extraction.

The rough lemon rootstock is extensively used in Florida because of its adaptation to the light sandy soil. Recent development in fruit tree nutrition gives the grower a tool to regulate his fruit production both in yield and quality.

Potassium is the main plant food affecting juice quality. High potash supply may result in increase in fruit size, coarse-textured rind, late maturity and low total soluble solids in the juice as compared with low potash fertilization (Reuther and Smith, 1952). Sites (1950) and Sites and Deszyck (1952) reported that grapefruit (Duncan) and orange (Hamlin) fruit from trees deficient in potassium were smaller and lower in soluble solids and total acids than normal fruit. A slight increase in the amount of potassium in the fertilizer increased the size of the fruit and its content of soluble solids. Further increases in this element reduced the total soluble solid content, but increased the titratable acidity of the juice.

Increases in nitrogen fertilization may cause a slight decrease in total soluble solids and vitamin C and an increase in titratable acidity of the juice (Jones and Parker, 1949; Reuther and Smith, 1952). Increase in phosphorus in the fertilizer depressed slightly the total soluble solids, titratable acidity and vitamin C content of the juice with no clear-cut effect either on the size or the yield of fruit (Jones and Parker, 1949; Smith *et al.*, 1952). Minor element nutrition of citrus fruit is especially important on light soil. Deficiency in magnesium, calcium, zinc, iron, copper, manganese, boron and molybdenum all contribute to various changes in fruit quality (Smith, 1966).

Another method which can be used to regulate fruit quality is in reducing the titratable acidity by spraying the tree with lead arsenate. The discovery that citrus fruit acidity can be affected by arsenicals applied to the tree was made nearly half a century ago (Gray and Ryan, 1921; Juritz, 1925), but the mechanism remains obscure. The spray can be applied post-bloom together with any orchard spray for pest control. It is usually most effective when applied about 2 months after fruit set. The effect is more pronounced on orange than on grapefruit. Deszyck and Ting (1958) reported that, in fruit from pigmented grapefruit trees receiving 1·6 lb of lead arsenate/100 gallons the acidity was reduced to less than half compared with those sprayed with

0·4 lb/100 gallons. The intermediate rate of 0·8 lb/100 gallons is recommended as the maximum for these fruit; 1 lb/100 gallons is the maximum on white grapefruit. The effect of arsenate was observed to be low in years of low rainfall during growing season. Whether this is due to a decrease of solubility of the arsenate and absorption by the tree is not known. In addition to affecting the acidity, the arsenic treatment also changes the sugar and flavonoid composition of grapefruit (Deszyck and Ting, 1960). However, the use of arsenate sprays is now prohibited in Florida except on grapefruit.

The ratio between the total soluble solids (sugars) and titratable acidity generally reflects the palatability of the fruit and is used as a regulatory criterion to prevent low quality fruit to enter the market. The lowering of the acidity greatly modifies this ratio.

Vines and Oberbacher (1965) showed that arsenate at 10^{-2} M completely uncoupled phosphorylation from oxidation in citrus mitochondria with citrate as a substrate without effecting a change in rates of oxidation. Metcalf and Vines (1965) also reported on inhibition of both oxygen uptake and phosphate esterification when arsenate is introduced into a reaction mixture of mitochondrial preparations. Increasing the phosphate content did not alter the inhibition of the oxygen uptake but markedly reduced the inhibition of phosphorylation. Arsenate had no effect on oxygen uptake in the absence of a phosphate acceptor (ADP). They concluded that arsenate may function as a competitive inhibitor with phosphate in the esterification reaction. Attaway and Buslig (1968b) made a study of the effect on citric acid levels of fruit by spraying the trees with several inhibitors under commercial conditions.

Recently, Bruemmer and Roe (1970) have studied the effect of various ratios of NAD and $NADH_2$ and NADP and $NADPH_2$ on the activity of malic and isocitric dehydrogenases present in orange fruits. They suggest that the decline in the acidity of citrus fruits during maturation may be the result of continued citric acid degradation while citric acid formation is reduced through depression of the activity of malic dehydrogenase.

D. Growth Regulators and Abscission

Auxins have long been known to affect abscission (Leopold, 1964), but it is only in recent years that our understanding of plant hormones has progressed to the point that we understand to some extent the balance of hormones necessary for prevention of abscission. It appears that auxins, gibberellin, kinins and abscissic acid make up this balance (van Overbeek, 1968), but recently ethylene has been proposed as a plant hormone (Burg and Burg, 1965).

1. *Growth regulators*

The premature dropping of citrus fruit in some varieties represents a serious economic loss to the growers. The successful use of growth regulators to reduce fruit drop in the apple industry has prompted its adoption to citrus. Stewart and Klotz (1947) reported that when 2,4-dichorophenoxy-acetic acid (2,4-D) was applied to Valencia orange trees even 2 weeks after severe drop had begun, fruit drop was reduced by as much as 78%. Stewart *et al.* (1947) reported the reduction in fruit drop of navel oranges from 27 to 96% when 5–25 p.p.m. of 2,4-D was applied as a water spray. Some epinastic effect was observed on the new growth. Sites (1954) successfully controlled the premature drop of Pineapple oranges with 20–60 p.p.m. of 2,4,5-trichlorophenoxypropionic acid (2,4,5-TPP) with no apparent leaf curling even at the highest concentration. The addition of soluble nitrogen such as urea seems to have a synergistic effect. Only 10 p.p.m. of 2,4,5-TPP was found to be sufficient to control fruit drop when 1 lb of urea/100 gallons of spray was added. The use of 2,4,5-TPP was specific for Pineapple orange; its application on Valencia orange was reported to be ineffective. Wilson and Hendershott (1967) found 2,4-D prevented abscission of citrus fruit at a concentration of 10^{-4} M. Higher concentrations increased senescence and lower concentrations delayed but did not prevent abscission.

The use of growth regulators to increase fruit-set of navel oranges and grapefruit was attempted by Pomeroy and Aldrich (1943) with no success. Stewart and Klotz (1947) were able to prevent the drop of blossoms in full bloom of Valencia orange for 8 to 10 weeks but fruit-set was not increased. Recently, gibberellins have been postulated as playing an important role in fruit-set (van Overbeek, 1968) as well as during early growth of many fruit. Gibberellin sprays of 25 p.p.m. effectively set fruit of Orlando tangelo (Krezdorn and Cohen, 1962), a weakly-parthenocarpic variety. Higher concentrations produced adverse effects, including heavy defoliation.

Potassium gibberellate sprayed onto orange trees increased the yield (Hield *et al.*, 1958) and delayed development of carotenoid colour (Coggins and Hield, 1958). Predominantly orange-coloured Valencia fruit treated with potassium gibberellate showed an increased chlorophyll content (Coggins and Lewis, 1962). In navel oranges treated with gibberellin A_3 sugar accumulation in the rind was reduced with a corresponding increase in respiration and a general delay in senescence (Lewis *et al.*, 1967).

Kinins or cytokinins are capable of altering many growth functions, including cell division, and are known to bring about an increase in the levels of protein, DNA and RNA in cells (Leopold, 1964). Crane (1964), however, in summarizing the effects of kinins on fruits, observed that proof of their existence in fruits has been by indirect methods. Although little information is available concerning the effects of kinins on citrus fruits *per se*, it would appear from the literature that auxin and kinetin play a large part in regulating abscission through their ability to control cellular senescence (Wilson, 1966a).

2. *The biochemistry of abscission*

It hastening the formation of the abscission layer of the fruit, the growth regulator involved must be able to promote senescence. Among these compounds are abscissic acid and ethylene and some non-hormone compounds (see Cooper *et al.*, 1968).

Abscissic acid or dormin is a senescence-producing hormone first extracted from cotton bolls (Addicott *et al.*, 1964). It is readily translocated, but evidence so far indicates that it is unable readily to penetrate the plant cuticle (Wilson, 1967a).

Ethylene is now considered a plant hormone principally as the result of studies by Burg and Burg (1965), and appears to be produced wherever abscission is stimulated by chemical means in explant studies (Rubinstein and Abeles, 1965). Wilson (1966b) found that ethylene was capable of producing fruit and leaf abscission more rapidly than any other chemical used in explant tests. Ethrel (2-chloroethane phosphonic acid) a known ethylene-producing material, leads to rapid leaf drop when sprayed on orange trees at concentrations above 200 p.p.m., but appears to have little effect on the fruit (Wilson and Coppock, 1968). Ethylene gas applied to whole trees under a canopy produced abscission in mature fruit but no leaf or young fruit drop when concentrations above 1 p.p.m. were applied for 6 hours (Wilson, 1967b). However, all leaves and fruit were removed when fumigation was in excess of 14 hours. Attempts to apply ethylene to whole trees through soil injection, foams and sprinkler irrigation systems have proved ineffective (Wilson, 1966b; 1967a).

Explant studies of citrus fruit (Wilson and Hendershott, 1967) have shown that sucrose and hexose sugars delay abscission; manitol, a sugar alcohol, increases abscission, while sorbitol and inositol have no effect.

The separation layers of mature orange fruits usually contain large quantities of starch (Wilson, 1966a; Wilson and Hendershott, 1968). Pectic materials were found to disappear gradually from the separation layer during the maturation of the fruit unless 2,4-D sprays had been applied. These sprays did not, however, prevent the development of the starch layer. The loss of pectic materials probably accounts for the ease of separation with increasing fruit maturity, as pectic substances have long been known to be the cementing materials between plant cell walls.

Iodoacetic acid has produced notable abscission under field conditions (Hendershott, 1964), but this material always produces some leaf fall and fruit scarring, and its effect is not consistent from year to year (Wilson and Coppock, 1968). Ethrel (2-chloroethane phosphonic acid) produced heavy leaf abscission, but had little effect on the fruit. Ethylene gas was impractical (Wilson, 1967a).

Attempts to loosen fruit by chemical treatments are achieving success in some cases. In Florida, ascorbic or erythorbic (isoascorbic) acids, alone or in combination with citric acid and cyclohexmide, produced consistent abscission of most varieties of oranges (Cooper and Henry, 1967; Wilson and Coppock, 1968; Buttram, 1970). Fruit scarring usually occurred, but leaf drop was non-existent or very light at low concentrations. It has been concluded that the primary mode of action in abscission is through fruit injury and ethylene formation (Wilson and Biggs, 1968). Cost prevents widespread use of ascorbic acid as an abscission agent, but cheaper acids are currently being tested for this purpose. It is apparent from the many acids tested, however, that the acid itself must have mild abscission-producing ability to achieve the desired properties.

Wilson and Coppock (1968) arbitrarily classified field abscission chemicals into two classes based on the concentrations necessary to cause abscission. "Hormone level" compounds were those which produced abscission activity at around 2000 p.p.m. and made up the majority of the compounds screened. "Mass action" or "weak acid" types, characterized by ascorbic acid, are compounds which required about 20,000 p.p.m. or more. The former class of compounds probably holds the most hope for development of an abscission chemical suitable for fresh fruit use in Florida, California and other areas. The "weak acid" types appear suitable for use with fruit for processing and may be an important adjunct to mechanical harvesting of these fruits. Cyclohexmide is used only at a concentration of 20 p.p.m. but is not considered as a "hormonal level" compound. Its mode of action is also in the production of ethylene through fruit injury. However, current studies indicate that a practical abscission chemical *per se* will not necessarily greatly improve the efficiency of all types of mechanical shakers (Wilson, 1967c; Wilson and Coppock, 1968).

IV. POST-HARVEST PHYSIOLOGY

A. Changes in Chemical Constituents During Storage

Storage of citrus fruit for an extended period is not practised to any great extent in citrus-producing areas. Unlike deciduous fruit, citrus fruit do not undergo rapid chemical or physical changes after removal from the tree. Unless there is a severe water stress, fruit can be kept as well on the tree as in storage. The picking period of late season fruit may extend into summer. If the temperature becomes exceedingly high and the fruit is losing excessive water during a water stress situation within the tree, early picking followed by storage is beneficial. Another instance is the curing of lemons which may be picked green and stored from 2 to 6 months to attain the right colour and texture. Grapefruit may be picked and allowed to colour naturally in transit-storage during shipment overseas (Oberbacher, 1962).

While on the tree, fruit continues to maintain a water balance with the tree. Once it is severed, it begins to lose water. Shrivelling through loss of water by stored fruit can be minimized by the application of wax, decreasing the storage temperature, increasing the relative humidity, or a combination of these practices.

Studies of changes in chemical composition of citrus fruit in storage usually deal with such constituents as affect the quality of fruit. Little work has been done on the changes in enzymes or cell organelles in storage. Loss of chlorophyll from the peel in storage is usually hastened by the use of ethylene gas. Apparently, carotenoid synthesis does not take place simultaneously with the disappearance of the chlorophyll (Miller and Winston, 1939). Over long storage periods, grapefruit has been observed to deepen in peel colour when stored at 15·5°C.

Harvey and Rygg (1936a, b) stored oranges and grapefruit at different temperatures and found that certain substances in the rind had definite relationships with storage temperatures. In navel oranges, sucrose showed a slight loss (8% of the original amount) at 0°C but a large loss (47%) at 11°C. Reducing sugars changed much less than sucrose, but in the same direction. There was a loss of soluble solids with corresponding increases in easily-hydrolysable polysaccharides. These authors calculated that there was a net synthesis of hesperidin by the oranges in storage and that the hesperidin synthesis was related to temperature and rate of respiration. In stored grapefruit, according to Harvey and Rygg (1936b), total sugars of the rind decreased, the decrease being the least at 0°C and greatest at 11°C. Reducing sugars of the rind increased upon storage at temperatures above 0°C. Sucrose in the stored samples, on the other hand, decreased strikingly except at 0°C. The naringin content showed an increase in some samples but decreased generally. The accumulation of reducing sugars and the decrease of non-reducing sugars at higher temperatures seem to be the result of rapid action of the invertase at these temperatures without the corresponding rapid decrease of monosaccharides due to respiration. At a temperature between 0 and 9°C, there seems to be a threshold point limiting the action of invertase.

Stahl and Camp (1936) stored fruit at various temperatures between 0 and 14·4°C and analysed the chemical composition during a 25 week storage period. With Valencia oranges there was a decrease in total weight, specific gravity, per cent juice and per cent acid but an increase in total sugars, total soluble solids and hydrolysable carbohydrates. The changes were less at the lower temperatures. In grapefruit there was a decrease in total weight of fruit, specific gravity and juice content, and an increase in soluble solids. The acidity, however, changed little.

At a storage of 4·4–5·6°C, Trout *et al.* (1938) reported an increase in the soluble solids of orange juice and a decrease in titratable acidity after 15 weeks in storage. In the rind there was a decrease in both reducing and total sugars and titratable acidity. The bitter principle of Washington Navel oranges decreased when stored at 26·7–32·2°C (Rockland *et al.*, 1957).

French and Abbot (1940) stored oranges and grapefruit for 5 months at 5·6°C and reported a slow loss of vitamin C. To study the changes of fruit quality during the marketing period for oranges, Harding (1954) stored oranges for 3–6 days at 10°C followed by 7 days at 21°C to simulate marketing conditions, and reported only a very slight loss of vitamin C. Eaks (1961) found that lemons stored at 12·8°C for 3 months lost little ascorbic acid: but at 24°C for the same period, significant loss of this vitamin occurred. The time of harvest seems to influence the effect of storage temperature on the loss of vitamin C (Khalifah and Kuykendall, 1965). Kefford (1966) concluded that the loss of vitamin C in fresh fruit is unlikely to exceed 10% under reasonable conditions of distribution and marketing.

Martin *et al.* (1939) compared the chemical changes of grapefruit in storage with those left on the tree. The percentage of total soluble solids tended to increase for a time in storage and then to decrease with prolonged storage, but it never decreased to the same level as it did on the tree. There was a slight increase in the percentage of citric acid in stored fruit at first followed subsequently by a decrease. Again, the citric acid of fruit in storage was always higher than that of fruit left on the tree and examined on the same date. They concluded that changes in acid and soluble solids taking place in grapefruit during storage are similar to those in fruit left on the tree, but that they proceed at a slower rate.

Tangerines stored at various temperatures lost ascorbic acid continually through the 8 week storage period (Bratley, 1939). The higher the storage temperature, the more rapid was the loss. There was also a loss of total acidity of tangerines with storage at different temperatures, greater loss being associated with higher temperatures.

Miller and Schomer (1939) found a decrease of total sugars within the peel and pulp of lemons in 15 weeks' storage. There was an increase in acidity and glucoside. Acetaldehyde tended to increase and was found to be higher in fruit stored at lower temperatures.

B. Respiratory Activity

It is generally assumed that the rate of respiration of a fruit is a measure of its metabolic activity. After the fruit is detached from the tree, the rate of respiration becomes an indication of its rate of loss of stored, respirable substrate.

With some fruits there is a dramatic rise in respiration—the respiration climacteric (see Volume 1, Chapter 17)—immediately preceding visible ripening. The attainment of a peak in respiration in "climacteric" type fruits is usually accompanied by chemical changes associated with ripening—changes in pectic substances, other polysaccharides and reserve materials (Biale, 1961); such chemical changes take place only gradually in citrus fruits.

Trout *et al.* (1938) observed a rise in respiration in Washington Navel oranges stored at 4·5°C. Biale and Young (1947) also found a climacteric type rise in mature lemons when placed in O_2-tensions of 34, 68 and 100%, but no such rise occurred in air (21% O_2). Later, Biale *et al.* (1954) and Bain (1958) could again find no respiratory rise in sound oranges stored in air and suggested that a rise sometimes observed (Biale, 1940; Biale and Shepherd, 1941) in citrus fruits was due to fungal attack. Fungi produce ethylene and exogenous ethylene will induce a rise in respiration in citrus fruits (see p. 153). However, Trout *et al.* (1960) again reported a rise in respiration of Washington Navel oranges held at 6·5–10°C. From a re-examination of their earlier results and extrapolation of respiration rates to higher temperatures, they suggested that at 21°C an ephemeral climacteric peak lasting only 1 day might occur and that this could easily have been missed by other investigators.

Aharoni (1968) investigated the pattern of respiration of various varieties of oranges and Marsh grapefruits picked at monthly intervals from May to the end of December; the size of the earliest picked fruit was 2·2 cm. He found typical climacteric rise patterns in the young, immature fruit. Simultaneous measurement of ethylene production by these young fruit gave curves which were parallel with the curves for respiration rate. The rise in ethylene production and respiration coincided with colour change and abscission of the *stem ends* of the fruit. Fruit harvested near horticultural maturity showed a gradual decline in rate of respiration and produced no ethylene. The production of ethylene by the young fruit could have been the cause of the increase in respiration and the colour change, since ethylene is known to stimulate the respiration and colour change in citrus fruits (Chace and Church, 1927; see also the section on Ethylene). More recently, Aharoni *et al.* (1969), in experiments in which ethylene was applied to oranges at various stages of development and then stored at 20°C, found that young fruits responded with a typical climacteric rise in respiration. At maturity, however, ethylene continued to stimulate respiration *after* a previously ethylene-stimulated peak in respiration had been passed and respiration was falling. In a true climacteric type fruit, ethylene will not bring about an increase in respiration once the climacteric has been passed (Kidd and West, 1935). Aharoni *et al.* concluded that oranges (and, presumably, other citrus fruits) differ materially from typical climacteric type fruits—and there the matter rests for the moment.

According to Hussein (1944) different tissues of citrus fruits have different rates of respiration, the rate being highest in the flavedo, lower in the albedo and juice sacs with the segment membrane having the lowest rate of all. He concluded that the cytochrome oxidase system was responsible for the major part of the O_2-uptake of the tissues of the fruit.

The temperature coefficient (Q_{10}) for the respiration of oranges has a maximum value between 0 and 10°C falling to a minimum of approximately 1·5 between 17 and 32°C and rising again above 32°C up to "lethal" temperatures (Haller *et al.*, 1945). The respiratory quotient is close to unity at 10–17°C rising somewhat at higher temperatures. Haller also found that the respiratory activity was higher in oranges grown on bitter orange stocks than those grown on rough lemon rootstocks.

Trout (1931) showed that acetaldehyde at concentrations between 1 and 1000 p.p.m. (by volume of air) markedly increased the respiration rate of oranges. The acetaldehyde probably enters into the normal respiratory pattern of the fruit.

C. Ethylene†

The subject of ethylene production by citrus fruits is still somewhat controversial. The tissue, undoubtedly, has the capacity to produce the gas since, as already stated, Biale and Young showed that both oranges and lemons produced ethylene in appreciable amounts when held in high O_2-tensions, and tissue slices of orange will produce ethylene a few hours after excission. Miller *et al.* (1940a) observed epinasty in tomato, potato and sunflower plants when subjected to the emanations of sound and decaying citrus fruits and concluded that this was due to ethylene present in the emanations. However, Elmer (1936) found that the emanations from the orange did not inhibit the sprouting of potatoes. Hall (1951), using absorption in permanganate as the method of determination, reported ethylene production by Valencia oranges. Recent analytical techniques employing GLC have made it possible to measure ethylene concentrations as low as 0·005 p.p.m. in air (Burg and Burg, 1962). Using these methods, Burg and Burg (1962, 1965) reported that both oranges and lemons produce very small amounts of ethylene, Valencia oranges giving 0·02–0·06 μlitre/kg/hour. Vines *et al.* (1968) found that oranges, tangerines and grapefruits suffering frost injury on the tree or suffering mild mechanical injury produced some ethylene and they concluded that ethylene production by citrus fruit is a stress symptom and not a normal metabolic process. It is clear that, unlike many fruits, citrus fruit do not give off ethylene at maturity, when in a sound

† See also Volume 1, Chapter 16.

condition, under normal conditions. Nevertheless, it remains a matter of interest and practical importance that the tissue of the fruit is capable of producing ethylene and, as we shall now see, of being influenced by relatively low concentrations of the gas.

Denny (1924) first showed that mature, green oranges turned yellow when treated with ethylene. In many citrus-producing areas where the night temperature is high during the maturing period for early varieties, the fruit becomes palatable when the flavedo is still green. It is a general practice in such cases to treat the green fruit with low concentrations of ethylene which "degreens" them efficiently without causing any major changes in the other ripening processes. Miller and Winston (1939) found that during the process of chlorophyll breakdown little change in the carotenoids occurred. Norman and Craft (1968) reported that ethylene treatment causes an increase in the production of volatiles by lemons.

Although the effects of ethylene on colour may be observed at a concentration as low as 1 p.p.m., the recommended conditions for degreening in commercial practice are 1 part of ethylene in 50,000 parts of air at 30°C and a relative humidity of 85–95% (Winston, 1955). Grierson and Newhall (1955) found that storage loss of citrus was significantly correlated with the concentration of ethylene used and the duration of the degreening process. Tangerines and Temple oranges are more susceptible to ethylene injury than Valencia oranges or Marsh grapefruits (Grierson and Newhall, 1956).

D. Physiological Disorders in Storage

Non-pathogenic disorders of citrus fruit in storage cause fruit to lose their economic value from unsightly appearance and loss of eating qualities. Damaged fruit are also susceptible to fungal decay both in storage and on the shelf. Being semi-tropical fruit, almost all citrus are susceptible to low temperature injury. Storage at high temperatures (15·5°C or room temperature), however, is undesirable because of high incidence of fungal attack and rapid deterioration of the fruit.

The most commonly occurring physiological breakdown in orange and grapefruit is pitting (Miller, 1946; Rose *et al.*, 1943). The fruit may have areas of blotches of sunken spots. In oranges, the early and mid season varieties are more affected than the late season ones. In grapefruit, the spots may first turn pink and then brown. Harvey and Rygg (1936b) injected solutions of narigenin into grapefruit and caused a pink coloration of the peel.

Another post-harvest disorder of citrus fruit is oleocellosis (oil spotting), which occurs in all citrus fruit and is especially severe in lemons and limes. In early oranges, the symptoms become most evident after degreening treatment.

Oranges may also suffer from brown stains or scald, which differs from pitting in appearance in that large areas are affected, and that the affected area is seldom as dark as in the case of pitting. This disorder usually develops if fruit is stored at 0°C according to Harvey and Rygg (1936a).

Watery breakdown of citrus fruits shows symptoms very similar to that of freeze damage. The affected tissue has a water-soaked effect. The occurrence of this disorder has been reported to be due to low temperature storage at between 0·6°C and 2°C (Miller, 1946). McCornack (1966) reported blossom-end "clearing" in certain susceptible varieties of grapefruit as being due to mechanical dropping of the fruit. Fruit susceptible to this type of injury were usually mature with thin rinds and seedless; water-soaked areas at the blossom end may develop within 24 hours at room temperature. Blossom end clearing may be related to watery breakdown which could be the manifestation of incipient bruises of the tissue after storage.

"Ageing" is a storage disorder where the rind around the stem becomes wilted and shrivelled, giving an unsightly appearance to the fruit. It is caused by loss of water from fruit, the collapse of the oil glands and subsequent death of the cells. The stem end and the stylar end rings of the Washington Navel and Valencia oranges behave differently in storage (Harvey and Rygg, 1936a). The stem end loses water and soluble solids more rapidly and has a higher acid content than the blossom end.

Lemons and limes are especially susceptible to cold storage breakdown. Some of the lemon disorders are membranos stain, albedo browning, red blotch and peteca. Membranos stain is the browning or darkening of the carpellary membrane between the segments, and albedo browning is the discoloration of the white spongy tissue. Red blotch is a scald-like reddish-brown area of the rind; and peteca resembles pitting except that the sunken areas usually have round edges. Eaks (1955) found that oleocellosis and surface pitting of limes are related to the methods of fruit handling during and after harvesting. Oleocellosis appeared in storage within a week and was definitely associated with bruises or pressures on the rind leading to breakage of the oil glands, and with the turgidity of the fruit. The more turgid the fruit, the more easily it was bruised. Careful picking by clipping of the fruit eliminated oleocellosis regardless of the fruit turgidity. Surface pitting was found to be associated with long storage periods at low temperatures. Fruit usually developed pitting when removed to room temperature. Fruits stored at 10°C showed no surface pitting after removal from storage to room temperature, but they began to turn yellow during storage.

The conditions under which the fruit is produced predetermines the development of physiological disorders in storage. Miller (1946) found that oranges and grapefruits produced in soils containing abundant moisture and high organic matter are more likely to breakdown in cold storage than those grown on sandy soils. The rainfall, mean temperature during the growing season, maturity of the fruit at harvest and manurial treatment all affect the storage behaviour of the fruit.

Oranges and tangerines can tolerate lower temperatures than limes, lemons and grapefruit. Suggested temperatures for oranges are 1–3·5°C (Miller, 1958); for grapefruit, 7–13°C in areas where stem-end decay is not serious, and, where this type of decay is a problem, 0–3·5°C although fruits removed from these lower temperatures and placed at 21°C are likely to show pit injury within a week. Lemons should be stored at 10–13°C, and limes, although more susceptible to low temperature injury, at 7–9°C. High temperatures hasten colour change in limes and this is undesirable in this fruit. Cold injury may be avoided by holding the fruit at a higher temperature for a period before placing it in low temperature storage. Hawkins and Barger (1926) found that curing grapefruit for 1 or 2 weeks at 21 or 24°C at a relative humidity of 68% could prolong the cold storage life of grapefruit. Five days at a temperature of 16°C was found sufficient as a curing period for grapefruit by Brooks and McColloch (1936). The duration of pre-conditioning the fruit at a higher temperature apparently varies according to the pre-harvest history of the fruit.

Khalifah and Kuykendall (1965) in Arizona, harvesting Valencia oranges in March and June and storing at 3·3°C and 6·6°C, found that the higher temperature storage gave less rind breakdown. Earlier harvested fruit stored best at the higher temperature but the June harvested fruit stored better at 3·3°C.

Although some workers have reported a reduction of rind breakdown by pre-treating the fruit with high concentrations of CO_2 (Stahl and Cain, 1937), most attempts to store citrus fruit in controlled atmosphere have given no special benefits and usually lead to rind injury (Harding *et al.*, 1966; Grierson *et al.*, 1966). Miller (1946) found increasing amounts of alcohol and acetaldehyde in citrus fruit in prolonged storage. Such accumulations were also noted in fruit subject to high CO_2 storage (Miller and Schomer, 1939) and may be the cause of physiological breakdown.

The effect of low temperatures on the physiological processes of oranges was studied by Eaks (1960). Fruits that had been stored at low temperatures and subsequently removed to 20°C had much higher rates of CO_2 evolution and O_2 uptake. There was an accumulation of time-temperature influence of chilling exposure on the degree of stimulation of CO_2 evolution. The stimulated respiratory activity of the chilled fruit was considered to be due to the accumulation of certain metabolic intermediates which became rapidly oxidized at the higher temperature. The accumulation of metabolic intermediates in cold stored apples was observed by Hulme *et al.* (1964). They found an accumulation of oxaloacetic acid in fruits subjected to cold storage breakdowns. A short interim period of storage at higher temperatures reduced both the accumulation of oxaloacetic acid and the intensity of breakdown.

V. METABOLISM OF CITRUS FRUITS

A. Citric Acid Synthesis

The accumulation of citric acid in the juice vesicles of citrus fruit is a subject of interest to the grower since it directly affects the quality of the fruit and palatability of the juice. In acid fruit such as lemon and lime, the amount of acid in the fruit determines to a large extent the value of the crop.

The site of citric acid synthesis in the fruit has been under question for some time. The acid was thought to be synthesized either in the leaves or in fruit peel and then translocated to the juice vesicles, or to be formed directly in the juice vesicles. At present, the latter assumption seems to be gaining support. While citric acid is dominant in the juice vesicles of most citrus fruits, the dominant acid in the peel of all citrus fruit is malic acid. The concentrations of organic acids in the peel are very low compared to that of the pulp. This led Sinclair and Eny (1947) to suggest that the location of citric acid synthesis is in the pulp. Erickson (1957) grafted sweet lemon fruit onto a sour lemon plant and vice versa. The results pointed to the fact that the leaves do not directly synthesize citric acid since the sweet lemon fruit grafted to the sour lemon plant remained low in citric acid content. On the other hand, the sour lemon fruit on the sweet lemon plant was high in acid.

The site of acid formation was also investigated by Huffaker and Wallace (1959) using the measurement of dark fixation of CO_2 as an indication of the possibility of acid synthesis capability. Homogenates from juice vesicles of Valencia oranges showed CO_2 fixation throughout the fruit growing and maturing period, but a much greater fixation was observed in fruit under 4 cm in diameter. Bean and Todd (1960) found that juice vesicles could assimilate more $^{14}CO_2$ in the dark than either the flavedo or albedo of orange. The $^{14}CO_2$ activity was found in malic acid, citric acid, serine and aspartic acid. In light, the activity was also found in sucrose in addition to the acids found in the dark. In intact fruit, the juice vesicles fixed CO_2 to the same extent regardless of light. Clark and Wallace (1963), however, found no correlation between dark CO_2 fixation rates and acid accumulation.

Huffaker and Wallace (1959) observed a slight decrease in CO_2 fixation in both the reaction of PEP carboxylase and the PEP carboxykinase systems in citrus fruits when treated with arsenate. However, this slight decrease in the dark CO_2 fixation could not account for the decrease of acidity since Clark and Wallace (1963) showed that no correlation existed between citric acid content and dark CO_2 fixation. Bogin and Wallace (1966a) demonstrated the existence of at least two enzymes in both sweet and sour lemon fruit that are responsible for CO_2 fixation to incorporate a 1-carbon fragment with a 3-carbon fragment to form a 4-carbon carboxylic acid.

Citric acid is apparently formed in citrus through the normal glycolytic scheme by which carbohydrates are first oxidized to pyruvic acid, which enters the Krebs cycle in the usual way (Ochoa and Stern, 1952). The final step, condensation of acetyl CoA with oxaloacetate, requires the presence of citrate synthase (citrate-condensing enzyme). Investigators at three different laboratories have demonstrated the presence of this enzyme in citrus fruit.

Srere and Senkin (1966) found citrate synthase activity in the leaves, peel and pulp of lemons and oranges. Leaves had the most activity and pulp the least in both oranges and lemons. Green orange peel and pulp were found to contain more of this enzyme than peel and pulp from ripe oranges, with green peel having considerably more than green pulp. In fact, the activity in green peel was of the same order of magnitude as that of leaves.

Bogin and Wallace (1966b) isolated citrate synthase from the mitochondrial fraction of sour lemon fruit and demonstrated that it was inhibited by ATP. It appeared that one effect of ATP was to increase the apparent K_m of citrate synthase for acetyl CoA. Mitochondrial preparations from sour lemon had a 30–50% lower oxygen uptake as compared with those of sweet lemon (Bogin and Wallace, 1966c). The sweet lemon preparations were found to be more active in oxidation, phosphorylation and pyruvate amination to alanine than those of sour lemon. It was also shown that sour lemon could form more citramalate from pyruvate than the sweet lemon, and citramalate was found to inhibit aconitase, thus causing a build-up of citrate in the sour fruit.

Vines (1968b) has recently made a thorough study of citrate synthase from citrus using mitochondria from mature grapefruit as the source of the enzyme. Activity was found only in the mitochondrial fractions, none being detected in the cytoplasm. The inhibitors N-ethylmaleimide (5×10^{-5} M) and fluorocitrate (8×10^{-7} M) were competitive with acetyl CoA and effectively prevented the formation of citric acid *in vitro*. Other inhibitors tested did not appear to interfere with the reaction.

B. The Citrus Mitochondria

Citrus mitochondria were first isolated almost concurrently by Vines (1964) from oranges, and Bogin and Erickson (1965) from lemons. The work of Vines was directed toward understanding the mechanism of citric acid accumulation in citrus. Mitochondrial isolation was accomplished by continuously adjusting the pH with KOH while grating the cells into an isotonic buffer containing mannitol and cysteine. The preferred pH range was 6·8–7·2. The P/O ratios of isolated mitochondria varied between 1·0 and 2·0 as the fruit matured, and finally exceeded 2·0 at maturity. The procedure of Bogin and Erickson also required neutralizing with KOH while grating the cold, peeled fruit into a 0·6 M sucrose solution containing 0·25 M Tris buffer. Their mitochondrial preparations oxidized citrate, α-ketoglutrate, succinate, malate and pyruvate. The oxidation was coupled to phosphorylation giving relatively high P/O ratios.

Vines and Oberbacher (1965) studied orange mitochondria in greater depth to determine the response of oxidation and phosphorylation in citrus mitochondria to arsenate, an inhibitor known to retard citric acid accumulation in citrus.

Metcalf and Vines (1965) and Vines and Metcalf (1967) isolated mitochondria from grapefruit for the first time and studied seasonal changes in oxidation and phosphorylation. They found that mitochondrial activity of preparations from immature fruit was low, but increased rapidly as the fruit matured, after which it generally declined during the latter stages of ripening. The rise and fall of activity coincided with the rise and fall of the total acidity of the juice, leading to the conclusion that the rise and fall of citric acid content may be correlated with the general phosphorylative ability of the grapefruit mitochondria. The citrate-condensing enzymes were demonstrated to be present in the mitochondrial fraction of both sour lemon (Bogin and Wallace, 1966b) and grapefruit (Vines, 1968b).

C. Other Citrus Enzymes

More work of citrus enzyme study has been made on the pectic enzymes than any other because of their effect on the quality of citrus juices (see also Volume 1, Chapter 3). The citrus juice is an opaque liquid with various insoluble particles suspended by the pectic substances in the medium. On standing, freshly extracted citrus juices soon lose this appearance when the solid particles settle to the bottom leaving a clear liquid portion on the top. This is termed "clarification" in citrus juice technology and can be readily inhibited by heat treatment, indicating that it is the action of enzymes (Wenzel *et al.*, 1951).

Pectin esterase (PE) is the main pectic enzyme of citrus. It de-esterifies the methylated carboxyl groups of polygalacturonic acid and causes precipitation of the calcium salts of pectinic acids. The removal of pectin results in the settling of the suspended particles and the clarification of the juice. In concentrated orange juice, low methoxyl pectin gel formation causes "gelation". Both "clarification" and "gelation" had resulted in heavy losses to the industry before proper heat treatment techniques were developed to inactivate the enzyme without undesirable heat damage to the product.

Orange pectin esterase was first isolated by MacDonnell *et al.* (1945). Rouse (1953) reported the distribution of PE in various component parts of citrus fruits from Florida. The highest activity was found in the juice sacs and the lowest in centrifuged juice with intermediate amounts in the flavedo, albedo and segment membrane. Clear juice had practically no pectin esterase activity which seems to be completely associated with insoluble solids. The

PE activity of the juice vesicles of citrus was found to increase with fruit maturity with oranges (Rouse *et al.*, 1962, 1964) but to decrease with grapefruit (Rouse *et al.*, 1965) and lemons (Rouse and Knorr, 1969). Pectin esterase activity is expressed in PEu or the milliequivalent of ester bond hydrolysed per minute per unit weight or volume of sample (Rouse and Atkins, 1955).

Free galacturonic acid has not been detected in freshly extracted orange juice although it is sometimes found in juices of other fruit (Almendiger *et al.*, 1954). Pratt and Powers (1953) reported the presence of depolymerizing enzymes in grapefruit juice which was more heat resistant than the PE. Earlier, MacDonnell *et al.* (1945) had failed to detect polygalacturonase in extracts of orange flavedo. Pectin depolymerase or polygalacturonase may partially or completely break the long chains of pectin.

Peroxidase, oxidase and catalase have been found in all citrus varieties and in all component parts (Ajon, 1926; Davis, 1942; Willimott and Wokes, 1926). Peroxidase was highest in the outer flavedo and lowest in the juice and pulp. Cytochrome oxidase was reported to be present in navel orange tissues (Hussein, 1944). The activity of this enzyme can be inhibited by cyanide and hydrogen sulphide.

Oberbacher and Vines (1963) found evidence for the occurrence of ascorbic acid oxidase in immature oranges. Among the different varieties, Marsh Seedless grapefruit had the highest ascorbic acid oxidase activity, while lemon and lime were very low in this enzyme (Vines and Oberbacher, 1962, 1963). The Marsh Seedless grapefruit also had the highest peroxidase activity. On fresh weight basis, ascorbic acid oxidase was highest in the flavedo and lowest in the juice. The activity in the juice and albedo decreased as the fruit increased in size. Fruit greater than 3·5 cm in diameter contained no ascorbic acid oxidase in the juice, but in the flavedo it increased with fruit development. Huelin and Stephens (1948) reported the occurrence of ascorbic acid oxidase in orange rind but found only negligible amounts in the juice.

Malic acid dehydrogenase activity, associated with grapefruit juice vesicle mitochondria and with cytoplasm was effective in catalysing the reaction:

$$\text{L-malate} + \text{NAD} \rightleftharpoons \text{oxaloacetate} + \text{NADH}.$$

The mitochondrial MDH differed from the cytoplasmic MDH in that the latter was in solution and was not affected by high concentrations of oxaoacetic acid. The enzymes of both sources were specific for L-malate and insensitive to D-malate (Vines, 1968a). Bogin and Wallace (1966a) showed that NADP-dependent malic dehydrogenase (malic enzyme) and PEP carboxylase are present in lemons.

Co-enzyme A is a necessary co-factor in the Krebs cycle reactions. Seifter (1954) reported its presence in higher plants including oranges. Citrate synthase catalyses the condensation of acetyl co-enzyme A (AcCoA) with oxaloacetic acid to form citric acid. Srere and Sekin (1966) and Bogin and Wallace (1966b) reported the occurrence of citrate synthase from orange and lemon. Vines (1968b) demonstrated the presence of citrate synthase in grapefruit juice vesicle mitochondria but not in cytoplasm.

The citrus fruit also contain a phosphatase (Axelrod, 1947) which is capable of hydrolysing phosphomonesters and various polyphosphates. The phosphatase activity is higher in the peel than in the juice of oranges, grapefruit and lemons; it is not necessarily related to the insoluble particles.

Acetyl esterase is a relatively non-specific esterase which can hydrolyse esters of acetic, propionic or butyric acids of various alcohols including the different acetyl esters. It occurs in high concentrations in the flavedo and decreases towards the centre of the fruit. It has been found in oranges, grapefruits and lemons (Jansen *et al.*, 1947). The enzyme in solution is reported to be inhibited by hexaethyltetrophosphate (HETP), an organic phosphatic insecticide, at a concentration of 0·2 p.p.m. Ball (1949) found HETP to be ineffective in inhibiting the acetyl esterase in orange fruit.

Manchester (1942) showed the possible presence of proteinase in oranges and lemons. Peptidase, a related proteolytice enzyme to proteinase, has also been found in citrus fruit (Jansen *et al.*, 1952). γ-Amino butyric acid is known to occur in orange juice (see p. 120) and increases with fruit maturity (Wedding and Horspool, 1955). It can be readily formed by decarboxylation of glutamic acid one of the more abundant free amino acids in citrus. Axelrod *et al.* (1955) isolated and characterized a glutamic acid decarboxylase from the flavedo of orange and lemon. The relative activity on a fresh weight basis was higher in the flavedo of Valencia orange than in that of navel orange. On a nitrogen basis, however, the activity of the two varieties was not significantly different.

ACKNOWLEDGEMENT

The authors wish to thank Dr. W. C. Wilson for his contribution to the section on Growth Regulators and Abscission.

REFERENCES

Addicott, F. T., Carns, H. R., Lyon, J. L., Smith, O. E. and McMeans, J. L. (1964). *In* "Regulateurs Naturels de la Croissance Végétale", Coll. Int. Centr. Nat. Rech. Sci. (Paris) **123,** 678.

Aharoni, Y. (1968). *Pl. Physiol., Lancaster* **43,** 99.

Aharoni, Y., Littar, F. S. and Monsclise, S. P. (1969). *Pl. Physiol., Lancaster* **44,** 1473.

Ajon, G. (1926). *Atti Congr. Naz. Chim. Pura Appl.* **2,** 1092. *Chem. Abstr.* **22,** 3682 (1928).

Alberala, J., Casas, A. and Primo, E. (1967). *Rev. Agroquim. Tecnol. Aliment.* **7,** 1.

Almendinger, V. V., Dillman, C. A. and Beisel, C. G. (1954). *Fd Technol.* **8,** 86.

Arigoni, D., Bartan, H. R., Corey, E. J. and Jeger, O. (1960). *Experientia* **16.** 41.

Atkins, C. D., Wiederhold, E. and Moore, E. L. (1945). *Fruit Prod. J.* **24,** 260.
Attaway, J. A. and Buslig, B. S. (1968a). *Biochim. biophys. Acta* **164,** 609.
Attaway, J. A. and Buslig, B. S. (1968b). *Proc. Fla St. hort. Soc.* **81,** 1.
Attaway, J. A. and Buslig, B. S. (1969). *Phytochemistry* **8,** 1671.
Attaway, J. A., Wolford, R. W., Alberding, G. E. and Edwards, G. J. (1962). *J. agric. Fd Chem.* **10,** 297.
Attaway, J. A., Wolford, R. W., Alberding, G. E. and Edwards, G. J. (1964). *J. agric. Fd Chem.* **12,** 118.
Attaway, J. A., Pieringer, A. P. and Barabas, L. J. (1967). *Phytochemistry* **6,** 25.
Axelrod, B. (1947). *J. biol. Chem.* **167,** 57.
Axelrod, B., Jang, R. and Lawrence, J. M. (1955). *J. agric. Fd Chem.* **17,** 531.
Bain, J. M. (1958). *Aust. J. Bot.* **6,** 1.
Ball, A. K. (1949). *Fd Technol.* **3,** 96.
Bartholomew, E. T. and Sinclair, W. B. (1943). *Pl. Physiol., Lancaster* **18,** 185.
Bartholomew, E. T. and Sinclair, W. B. (1951). "The Lemon Fruit, Its Composition, Physiology and Products." University of California Press, Berkeley.
Bean, R. C. and Todd, G. W. (1960). *Pl. Physiol., Lancaster* **35,** 425.
Biale, J. B. (1940). *Science, N.Y.* **91,** 458.
Biale, J. B. (1961). *In* "The Orange, Its Physiology and Biochemistry" (W. B. Sinclair, ed.), pp. 96–130. University of California Press, Berkeley.
Biale, J. B. and Shepherd, A. D. (1941). *Am. J. Bot.* **28,** 263.
Biale, J. B. and Young, R. E. (1947). *Am. J. Bot.* **34,** 301.
Biale, J. B., Young, R. E. and Olmstead, A. J. (1954). *Pl. Physiol., Lancaster* **29,** 168.
Bogin, E. and Erickson, L. C. (1965). *Pl. Physiol., Lancaster* **40,** 566.
Bogin, E. and Wallace, A. (1966a). *Proc. Am. Soc. hort. Sci.* **88,** 298.
Bogin, E. and Wallace, A. (1966b). *Biochim. biophys. acta* **128,** 190.
Bogin, E. and Wallace, A. (1966c). *Proc. Am. Soc. hort. Sci.* **89,** 182.
Born, R. (1960). *Chemy Ind.* 264.
Bratley, C. O. (1939). *Proc. Am. Soc. hort. Sci.* **37,** 526.
Braverman, J. B. S. (1949). "Citrus Products. Chemical Composition and Chemical Technology." Interscience, New York.
Brooks, C. and McColloch, L. P. (1936). *J. agric. Res.* **52,** 319.
Bruemmer, J. H. and Roe, B. (1970). *Phytochemistry.* In press.
Burg, S. P. and Burg, E. A. (1962). *Pl. Physiol., Lancaster* **37,** 179.
Burg, S. P. and Burg, E. A. (1965). *Science, N.Y.* **148,** 1190.
Burke, J. H. (1967). *In* "The Citrus Industry". Rev. Ed., Vol. 1, pp. 40–189 (W. Reuther, H. Webber and L. D. Batchelor, eds.). University of California Press, Berkeley.
Buttram, J. R. (1970). *Proc. Fla St. hort. Soc.* **83.** In press.
Cameron, S. H., Appleman, D. and Bialogowski, J. (1936). *Proc. Am. Soc. hort. Sci.* **33,** 87.
Caprio, J. M. (1956). *Proc. Am. Soc. hort. Sci.* **67,** 222.
Chace, E. M. and Church, C. G. (1927). *Ind. Engng Chem. ind. Edn* **19,** 1135.
Chandler, B. V. (1958). *Nature, Lond.* **182,** 933.
Chandler, B. V. and Kefford, J. F. (1951). *Aust. J. Sci.* **13,** 112; **14,** 24.
Chandler, B. V. and Kefford, J. F. (1953). *Aust. J. Sci.* **16,** 28.
Chandler, B. V. and Kefford, J. F. (1966). *J. Sci. Fd Agric.* **17,** 193.
Chapman, H. D. (1968). *In* "Citrus Industry", Vol. 2, pp. 127–289 (W. Reuther, H. J. Webber and L. D. Batchelor, eds.). University of California Press, Berkeley.

Chapman, H. D. and Rayner, D. S. (1951). *Hilgardia* **20,** 325.
Clark, R. B. and Wallace, A. (1963). *Proc. Am. Soc. hort. Sci.* **83,** 322.
Clements, R. L. (1964). *J. Fd Sci.* **29,** 276, 281.
Clements, R. L. and Leland, H. V. (1962). *Proc. Am. Soc. hort. Sci.* **80,** 300.
Coggins, C. W., Jr. and Hield, H. Z. (1958). *Calif. Agric.* **12,** 11.
Coggins, C. W., Jr. and Lewis, L. N. (1962). *Pl. Physiol., Lancaster* **37,** 625.
Coggins, C. W., Henning, G. L. and Yokoyama, H. (1970). *Science, N.Y.* **168,** 1589.
Cohen, A. (1956). *Res. Council (Israel) Sec. D. Bot. Bull.* **5** (2/3), 181.
Cooper, H. and Henry, W. H. (1967). *Proc. Fla. St. hort. Soc.* **80,** 7.
Cooper, W. C., Rasmussen, G. K., Rogers, B. J., Reece, P. C. and Henry, W. H. (1968). *Pl. Physiol., Lancaster* **43,** 1560.
Crane, J. C. (1964). *A. Rev. Pl. Physiol.* **15,** 303.
Curl, A. L. (1962). *J. Fd Sci.* **27,** 537.
Curl, A. L. (1965). *J. Fd Sci.* **30,** 13.
Curl, A. L. and Bailey, G. F. (1956). *J. agric. Fd Chem.* **4,** 150.
Curl, A. L. and Bailey, G. F. (1957a). *J. agric. Fd Chem.* **5,** 605.
Curl, A. L. and Bailey, G. F. (1957b). *Fd Res.* **22,** 63.
Curl, A. L. and Bailey, G. F. (1961). *J. Fd Sci.* **26,** 442.
Curl, A. L. and Veldhuis, M. K. (1948). *Fruit Prod. J.* **27,** 342.
Davis, W. B. (1932). *Am. J. Bot.* **19,** 101.
Davis, W. B. (1942). *Am. J. Bot.* **29,** 252.
Davis, W. B. (1967). *Analyt. Chem.* **19,** 476.
Denny, F. E. (1924). *J. agric. Res.* **27,** 757.
Deszyck, E. J. and Ting, S. V. (1958). *Proc. Am. Soc. hort. Sci.* **72,** 304.
Deszyck, E. J. and Ting, S. V. (1960). *Proc. Am. Soc. hort. Sci.* **75,** 266.
Deszyck, E. J. and Ting, S. V. (1961). *Fla. Ass. Sci. Teachers J.* **5,** 9.
Dreyer, D. L. (1965). *J. org. Chem.* **30,** 749.
Dunlap, W. J. and Wender, S. H. (1962). *Analyt. Biochem.* **4,** 110.
Eaks, I. L. (1955). *Proc. Am. Soc. hort. Sci.* **66,** 141.
Eaks, I. L. (1960). *Pl. Physiol., Lancaster* **35,** 632.
Eaks, I. L. (1961). *J. Fd Sci.* **26,** 593.
Eilati, S. K., Budowski, P. and Monselise, S. P. (1970). *XVIII Internat. hort. Congress*, paper 47.
Elmer, O. H. (1936). *J. agric. Res.* **52,** 609.
Embleton, T. W., Jones, W. W. and Kirkpatrick, J. D. (1956). *Proc. Am. Soc. hort. Sci.* **67,** 191.
Emerson, O. H. (1948). *J. Am. chem. Soc.* **70,** 545.
Emerson, O. H. (1949). *Fd Technol.* **3,** 248.
Emerson, O. H. (1952). *J. Am. chem. Soc.* **74,** 688.
Erickson, L. C. (1957). *Science, N.Y.* **125,** 994.
Fisher, J. F. and Nordby, H. E. (1965). *J. Fd Sci.* **30,** 869.
Florida Department of Agriculture (1967). *Fla Agr. Statist. Citrus Summary, Fla Dep. Agr., Tallahassee, Florida.*
Ford, E. S. (1942). *Bot. Gaz.* **104,** 288.
French, R. B. and Abbot, O. D. (1940). *J. Nutr.* **19,** 223.
Furz, J. R. and Taylor, C. A. (1939). *U.S. Dep. Agr. Tech. Bull.* **640,** 71 pp.
Gaddum, L. W. (1934). *Fla agric. exp. St. Bull.* **268,** 28 pp.
Gentili, B. and Horowitz, R. M. (1964). *Tetrahedron* **20,** 2313.
Gentili, B. and Horowitz, R. M. (1965). *Indian Inst. Sci. Bull.* **31,** 78.
Goldsworthy, L. J. and Robinson, R. (1957). *Chemy Ind.* 47.

Goodall, H. (1969). "The Composition of Fruits." B.F.M.I.R.A. (Leatherhead, Surrey), Scientific and Technical Surveys, No. 59.
Goren, R. and Monselise, S. P. (1965). *J. hort. Sci.* **40,** 83.
Gray, G. P. and Ryan, H. J. (1921). *Calif. Dep. Agr. Monthly Bull. Spec. Chem.* **10,** 11.
Grierson, W. and Newhall, W. F. (1955). *Proc. Am. Soc. hort. Sci.* **65,** 244.
Grierson, W. and Newhall, W. F. (1956). *Proc. Am. Soc. hort. Sci.* **67,** 236.
Grierson, W., Vines, H. M., Oberbacher, M. F., Ting, S. V. and Edwards, G. J. (1966). *Proc. Am. Soc. hort. Sci.* **88,** 311.
Hagen, R. E., Dunlap, W. J. and Wender, S. H. (1966). *J. Fd Sci.* **31,** 542.
Hall, D. H. (1938). *Chemy Ind.* **16,** 473.
Hall, W. C. (1951). *Bot. Gaz.* **113,** 55.
Haller, M. H., Rose, D. H., Lutz, J. M. and Harding, P. L. (1945). *J. agric. Res.* **71,** 327.
Harding, P. L. (1954). *Fd Technol.* **8,** 311.
Harding, P. L. and Fisher, D. F. (1945). *U.S. Dep. Agr. Tech. Bull.* 886, 100 pp.
Harding, P. L. and Sunday, M. B. (1953). *U.S. Dep. Agr. Tech. Bull.* 1072, 61 pp.
Harding, P. L., Winston, J. R. and Fisher, D. F. (1940). *U.S. Dep. Agr. Tech. Bull.* 753, 89 pp.
Harding, P. L., Chace, W. G., Smoot, J. J., Rygg, G. L. and Johnson, H. (1966). "Am. Soc. of Heating, Refrigeration and Air Conditioning Engineers Guide and Data Book", 1966, 656.
Hardy, F. and Warnerford, F. H. S. (1925). *Ind. Engng Chem. ind. Edn* **17,** 48.
Harvey, E. M. and Rygg, G. L. (1936a). *J. agric. Res.* **52,** 723.
Harvey, E. M. and Rygg, G. L. (1936b). *J. agric. Res.* **52,** 747.
Hawkins, L. A. and Barger, W. R. (1926). *U.S. Dep. Agr. Tech. Bull.* 1368.
Hendershott, C. H. (1964). *Proc. Am. Soc. hort. Sci.* **85,** 201.
Hendrickson, R. and Kesterson, J. W. (1954). *Fla agric. exp. St. Bull.* 545, 43 p.
Hendrickson, R. and Kesterson, J. W. (1963). *J. Am. Oil Chem. Soc.* **40,** 746.
Hendrickson, R. and Kesterson, J. W. (1965). *Fla agric. exp. St. Bull.* 698, 119 pp.
Hield, H. Z., Coggins, C. W., Jr. and Garber, M. J. (1958). *Calif. Agric.* **12,** 9.
Higby, R. H. (1938). *J. Am. chem. Soc.* **60,** 3013.
Hilgeman, R. H. (1941). *Proc. Am. Soc. hort. Sci.* **39,** 119.
Hodgson, R. W. (1967). *In* "The Citrus Industry". Rev. Ed., Vol. 1, pp. 431–592 (W. Reuther, H. J. Webber and L. D. Batchelor, eds.). University of California Press, Berkeley.
Horowitz, R. M. (1956). *J. org. Chem.* **21,** 1184.
Horowitz, R. M. (1957). *J. Am. chem. Soc.* **79,** 6561.
Horowitz, R. M. (1961). *In* "The Orange, Its Physiology and Biochemistry" (W. B. Sinclair, ed.), pp. 334–372. University of California Press, Berkeley.
Horowitz, R. M. and Gentili, B. (1960a). *J. Am. chem. Soc.* **82,** 2803.
Horowitz, R. M. and Gentili, B. (1960b). *Nature, Lond.* **185,** 319.
Horowitz, R. M. and Gentili, B. (1961). *Archs Biochem. Biophys.* **92,** 191.
Horowitz, R. M. and Gentili, B. (1963). *Tetrahedron* **19,** 773.
Huelin, F. E. and Stephens, I. M. (1948). *Aust. J. scient. Res.* (B–1), 58.
Huffaker, R. C. and Wallace, A. (1959). *Proc. Am. Soc. hort. Sci.* **74,** 348.
Hulme, A. C. (1956). *Nature, Lond.* **178,** 218.
Hulme, A. C., Smith, W. H. and Wooltorton, L. S. C. (1964). *J. Sci. Fd and Agr.* **15,** 303.
Hume, H. H. (1957). "Citrus Fruits", p. 13. MacMillan, New York.

Hunter, G. L. K. (1966). *Phytochemistry* **5,** 807.
Hunter, G. L. K. and Brogden, W. B., Jr. (1965a). *J. Fd Sci.* **30,** 1.
Hunter, G. L. K. and Brogden, W. B., Jr. (1965b). *J. Fd Sci.* **30,** 383.
Hunter, G. L. K. and Brogden, W. B., Jr. (1965c). *J. Fd Sci.* **30,** 876.
Hunter, G. L. K. and Moshonas, M. G. (1965). *Analyt. Chem.* **37,** 378.
Hunter, G. L. K. and Moshonas, M. G. (1966). *J. Fd Sci.* **31,** 167.
Huskins, C. W. and Swift, L. J. (1953). *Fd Res.* **18,** 305.
Hussein, A. A. (1944). *J. biol. Chem.* **155,** 201.
Igolen, G. and Sontag, D. (1941). *Chim. Ind.* (*Paris*) Suppl. No. 3, 157.
Ikeda, R. M., Tessakau, F. and Fujita, Y. (1930). *J. chem. Soc. Japan* **51,** 349.
Ikeda, R. M., Stanley, W. L., Rolle, L. A. and Vannier, S. H. (1962). *J. Fd Sci.* **27,** 593.
Jansen, E. F. and Jang, R. (1952). *Archs Biochem. Biophys.* **40,** 358.
Jansen, E. F., Jang, R. and Ball A. K. (1952). *Fedn Proc.* **11,** 236.
Jansen, E. F., Jang, R. and MacDonnell, L. R. (1947). *Archs Biochem.* **15.** 415.
Jones, W. W. and Parker, E. R. (1949). *Proc. Am. Soc. hort. Sci.* **53,** 91.
Jones, W. W., Van Horn, C. W. and Finch, A. H. (1945). *Arig. agric. exp. St. Bull.* **106,** 28 pp.
Juritz, C. J. (1925). *S. Afr., Dep. Agr. Sci. Bull.* 40, Div. Chem. Series 60.
Karrer, P. Jucker, E. (1944). *Helv. chim. Acta* **27,** 1695.
Karrer, P. and Jucker, E. (1947). *Helv. chim. Acta* **30,** 536.
Kefford, J. F. (1959). *Adv. Fd Res.* **9,** 351.
Kefford, J. F. (1966). *Wld Rev. Nutr. Dietet.* **6,** 197.
Kesterson, J. W. and Hendrickson, R. (1953). *Fla agric. exp. St. Bull.* 511, 29 pp.
Khalifah, R. A. and Kuykendall, R. (1965). *Proc. Am. Soc. hort. Sci.* **86,** 288.
Khan, M. U. D. and MacKinney, G. (1953). *Pl. Physiol., Lancaster* **28,** 550.
Kidd, F. and West, C. (1935). *Rep. Fd. Invest. Bd for* 1934, p. 119.
King, G. S. (1947). *Am. J. Bot.* **34,** 427.
Koo, R. C. J. and Sites, J. W. (1956). *Proc. Am. Soc. hort. Sci.* **68,** 245.
Komatsu, S., Tanaka, S., Ozawa, S., Kubo, R., Ono, Y. and Matsuda, Z. (1930). *J. chem. Soc. Japan* **51,** 478.
Krezdorn, A. H. and Cohen, M. (1962). *Proc. Fla Sta. hort. Soc.* **75,** 53.
Labanauskas, C. K., Jones, W. W. and Embleton, T. W. (1963). *Proc. Am. Soc. hort. Sci.* **82,** 142.
Leopold, A. C. (1964). "Plant Growth and Development." McGraw-Hill, New York.
Lewis, L. N., Coggins, C. W., Jr., Labanauskas, C. K. and Dugger, W. M., Jr. (1967). *Pl. Cell Physiol., Tokyo* **8,** 151.
Lime, B. J., Stephens, T. S. and Griffith, F. P. (1954). *Fd Technol.* **8,** 566.
Lovett-Janison, P. L. and Nelson, J. M. (1940). *J. Am. chem. Soc.* **62,** 1409.
McCance, R. A. and Widdowson, E. M. (1960). "The Composition of Foods." H.M.S.O., London.
McCornack, A. A. (1966). *Proc. Fla St. hort. Soc.* **79,** 258.
McCready, R. M., Walker, E. D. and Maclay, W. D. (1950). *Fd Technol.* **4,** 19.
MacDonnell, L. R., Jansen, E. F. and Lineweaver, H. (1945). *Archs Biochem.* **6,** 389.
MacLeod, W. D., Jr. and Buigues, N. M. (1964). *J. Fd Sci.* **29,** 565.
Maier, V. P. and Beverly, G. D. (1968). *J. Fd Sci.* **33,** 488.
Maier, V. P. and Dreyer, D. L. (1965). *J. Fd Sci.* **30,** 874.
Maier, V. P. and Metzler, D. M. (1967). *Phytochemistry* **6,** 763.
Manchester, T. C. (1942). *Fd Res.* **7,** 394.

Markley, K. S., Nelson, E. K. and Sherman, M. S. (1937). *J. biol. Chem.* **118,** 433.
Marloth, R. H. (1950). *J. hort. Sci.* **25,** 235.
Marsh, G. L. (1953). *Fd Technol.* **7,** 145.
Martin, W. E., Hilgeman, R. H. and Smith, J. G. (1939). *Proc. Am. Soc. hort. Sci.* **37,** 529.
Matalack, M. B. (1929). *J. Am. pharm. Ass.* **18,** 24.
Matalack, M. B. (1931). *Pl. Physiol., Lancaster* **6,** 729.
Matalack, M. B. (1940). *J. org. Chem.* **5,** 504.
Mazur, Y., Weizmann, A. and Sondheimer, F. (1958). *J. Am. chem. Soc.* **80,** 1007.
Metcalf, J. F. and Vines, H. M. (1965). *Proc. Ass. Sth. Agric. Wkrs* **62,** 242.
Miller, E. V. (1946). *Bot. Rev.* **12,** 393.
Miller, E. V. (1958). *Bot. Rev.* **24,** 43.
Miller, E. V. and Schomer, H. A. (1939). *J. agric. Res.* **59,** 601.
Miller, E. V. and Winston, J. R. (1939). *Proc. Fla Sta. hort. Soc.* **52,** 87.
Miller, E. V., Winston, J. R. and Fisher, D. F. (1940a). *J. agric. Res.* **60,** 269.
Miller, E. V., Winston, J. R. and Schomer, H. A. (1940b). *J. agric. Res.* **60,** 259.
Miller, E. V., Winston, J. R. and Fisher, D. F. (1941). *U.S. Dep. Agr. Tech. Bull.* **780,** 31 pp.
Miller, J. M. and Rockland, L. B. (1952). *Archs Biochem. Biophys.* **40,** 416.
Miller, R. L., Bassett, I. P. and Yother, W. W. (1933). *U.S. Dep. Agr. Tech. Bull.* **350,** 20 pp.
Mizelle, J. W., Dunlap, W. J., Hagen, R. E., Wender, S. H., Lime, B. J., Albach, R. F. and Griffith, F. P. (1965). *Analyt. Biochem.* **12,** 316.
Mizelle, J. W., Dunlap, W. J. and Wender, S. H. (1967). *Phytochemistry* **6,** 1305.
Money, R. W. and Christian, W. A. (1950). *J. Sci. Fd Agric.* **1,** 8.
Monselise, S. P. and Halevy, A. H. (1961). *Science, N.Y.* **133,** 1478.
Naves, Y. R. (1946). *Helv. chim. Acta* **29,** 1084.
Nelson, E. K. (1934). *J. Am. chem. Soc.* **56,** 1392.
Norman, S. and Craft, C. C. (1968). *Hort. Sci.* **3,** 66.
Nomura, D. (1950). *J. Jap. Chem.* **4,** 561.
Oberbacher, M. F. (1962). *Proc. Am. Soc. hort. Sci.* **80,** 308.
Oberbacher, M. F. and Vines, H. M. (1963). *Nature, Lond.* **197,** 1203.
Ochoa, S. and Stern, J. R. (1952). *A. Rev. Biochem.* **21,** 547.
Olliver, M. (1967). *In* "The Vitamins", Vol. 1 (Sebrell, W. H. and Harris, R. S., eds). Academic Press, London.
Peyron, L. (1963). *Soap, Perfums Cosm.* **36,** 673.
Plicher, R. W. (editor) (1947). "The Canned Food Manual", 3rd Ed., pp. 126–130. American Can Company, New York.
Pomeroy, C. S. and Aldrich, W. W. (1943). *Proc. Am. Soc. hort. Sci.* **42,** 146.
Porter, J. W. and Lincoln, R. E. (1950). *Archs Biochem.* **27,** 390.
Pratt, D. E. and Powers, J. J. (1953). *Wallestein Lab. Communi.* **10,** 133.
Rakieten, M. L., Newman, B., Falk, K. B. and Miller, I. (1951). *J. Am. diet. Ass.* **27,** 864.
Rakieten, M. L., Newman, B., Falk, K. B. and Miller, I. (1952). *J. Am. diet. Ass.* **28,** 1050.
Rasmussen, G. K. (1964). *Proc. Am. Soc. hort. Sci.* **84,** 181.
Reed, H. (1930). *Am. J. Bot.* **17,** 971.
Reuther, W. and Smith, P. F. (1952). *Proc. Am. Soc. hort. Sci.* **59,** 1.
Roberts, J. A. and Gaddum, L. W. (1937). *Ind. Engng Chem. ind. Edn* **29,** 574.
Robinson, R. and Tseng, Kwang-Fang (1938). *J. chem. Soc.* 1004.

Rockland, L. B. (1961). *In* "The Orange, Its Biochemistry and Physiology" (W. B. Sinclair, ed.). University of California Press, Berkeley.
Rockland, L. B., Beavens, E. A. and Underwood, J. C. (1957). "Removal of Bitter Principles from Citrus Fruits." U.S. Patent 2,816,835.
Rose, D. H., Brooks, C., Bratley, C. O. and Winston, J. R. (1943). *U.S. Dep. Agr. Misc. Bull.* **498**, 57 pp.
Rouse, A. H. (1953). *Fd Technol.* **7**, 360.
Rouse, A. H. and Atkins, C. D. (1955). *Fla agric. exp. St. Bull.* **570**, 19 pp.
Rouse, A. H. and Knorr, L. C. (1969). *Fd Technol.* **23**, 121.
Rouse, A. H., Atkins, C. D. and Moore, E. L. (1962). *J. Fd Sci.* **27**, 419.
Rouse, A. H., Atkins, C. D. and Moore, E. L. (1964). *J. Fd Sci.* **29**, 34.
Rouse, A. H., Atkins, C. D. and Moore, E. L. (1965). *Fd Technol.* **19**, 673.
Rubinstein, B. and Abeles, F. B. (1965). *Bot. Gaz.* **126**, 225.
Samish, Z. and Ganz, D. (1950). *Canner* **110** (23) 7; (24) 36; (25) 22, 24.
Seifter, E. (1954). *Pl. Physiol., Lancaster* **29**, 403.
Shimokoriyama, M. (1957). *J. Am. chem. Soc.* **79**, 4199.
Sinclair, W. B. (1961). *In* "The Orange." (W. B. Sinclair, ed.). University of California, Riverside.
Sinclair, W. B. and Eny, D. M. (1946). *Proc. Am. Soc. hort. Sci.* **47**, 119.
Sinclair, W. B. and Eny, D. M. (1947). *Bot. Gaz.* **108**, 398.
Sinclair, W. B. and Joliffe, V. H. (1958). *Bot. Gaz.* **119**, 217.
Sinclair, W. B. and Joliffe, V. H. (1961). *J. Fd Sci.* **26**, 125.
Sinclair, W. B. and Ramsey, R. C. (1944). *Bot. Gaz.* **106**, 140.
Sites, J. W. (1947). *Proc. Fla St. hort. Soc.* **60**, 55.
Sites, J. W. (1950). *Proc. Fla St. hort. Soc.* **63**, 60.
Sites, J. W. (1954). *Proc. Fla St. hort. Soc.* **67**, 56.
Sites, J. W. and Camp, A. F. (1955). *Fd Technol.* **9**, 361.
Sites, J. W. and Deszyck, E. J. (1952). *Proc. Fla St. hort. Soc.* **65**, 92.
Sites, J. W. and Reitz, H. J. (1951). *Proc. Am. Soc. hort. Sci.* **56**, 103.
Smith, A. H. (1925). *J. biol. Chem.* **63**, 71.
Smith, P. F. (1966). *In* "Nutrition of Fruit Crop" (N. F. Childers, ed.). 2nd Ed. pp. 174–207. Rutgers, The State University, New Brunswick, New Jersey.
Smith, P. F., Reuther, W. and Gardner, F. E. (1952). *Proc. Am. Soc. hort. Sci.* **53**, 85.
Smith, P. R. (1953). *B.F.M.I.R.A. Tech. Circ.* No. 44.
Srere, P. A. and Sekin, J. (1966). *Nature, Lond.* **212**, 506.
Stahl, A. L. and Cain, J. C. (1937). *Fla agric. exp. St. Bull.* 316, 44 pp.
Stahl, A. L. and Camp, A. F. (1936). *Fla agric. exp. St. Bull.* 303, 67 pp.
Stanley, W. L. and Vannier, S. H. (1957). *J. Am. chem. Soc.* **79**, 3488.
Stanley, W. L., Ikeda, R. M., Vannier, S. H. and Rolle, L. A. (1961). *J. Fd Sci.* **26**, 43.
Stearn, C. R., Jr. and Young, G. T. (1942). *Proc. Fla St. hort. Soc.* **55**, 59.
Stevens, J. W. (1954). *Fd Technol.* **8**, 88.
Stewart, I. and Wheaton, T. A. (1964a). *Science, N.Y.* **145**, 60.
Stewart, I. and Wheaton, T. A. (1964b). *Proc. Fla St. hort. Soc.* **77**, 318.
Stewart, I., Newhall, W. F. and Edwards, G. J. (1964). *J. biol. Chem.* **239**, 930.
Stewart, W. S. and Klotz, L. J. (1947). *Bot. Gaz.* **109**, 150.
Stewart, W. S., Klotz, L. J. and Hield, H. Z. (1947). *Calif. Citrog.* **33**, 49, 77.
Swift, L. J. (1952a). *Fd Res.* **17**, 8.
Swift, L. J. (1952b). *J. Am. chem. Soc.* **74**, 1099.
Swift, L. J. (1964). *J. Fd Sci.* **29**, 766.

Swift, L. J. (1965). *J. agric. Fd Chem.* **13,** 431.
Swift, L. J. and Veldhuis, M. (1951). *Fd Res.* **16,** 142.
Taylor, A. L. and Witte, P. J. (1938). *Ind. Engng Chem. ind. Edn* **30,** 110.
Teranishi, R., Schultz, T. H., McFadden, W. H., Ludin, R. E. and Black, D. R. (1963). *J. Fd Sci.* **28,** 541.
Thomas, D. W., Smythe, C. V. and Labee, M. D. (1958). *Fd Res.* **23,** 591.
Ting, S. V. (1954). *Fla Cit. exp. St. Mimeo Rpt* 55–5.
Ting, S. V. (1958a). *Fla agric. exp. St. Ann. Rep.* **1958,** 251.
Ting, S. V. (1958b). *J. agric. Fd Chem.* **6,** 546.
Ting, S. V. (1967). *Proc. Fla St. hort. Soc.* **80,** 257.
Ting, S. V. (1969). *J. Am. Soc. hort. Sci.* **94,** 515.
Ting, S. V. and Deszyck, E. J. (1958). *Proc. Am. Soc. hort. Sci.* **71,** 271.
Ting, S. V. and Deszyck, E. J. (1959). *Nature, Lond.* **183,** 1404.
Ting, S. V. and Deszyck, E. J. (1961). *J. Fd Sci.* **26,** 146.
Ting, S. V. and Vines, H. M. (1966). *Proc. Am. Soc. hort. Sci.* **88,** 291.
Tolkowsky, S. (1938). "Hesperides. A History of the Culture and Use of Citrus Fruits." 371 pp. John Bales, Sons and Curnow Ltd.
Townsley, P. M., Joslyn, M. A. and Smit, C. J. B. (1953). *Fd Res.* **18,** 522.
Trout, S. A. (1931). *Rep. Fd Invest. Bd* **1930,** 64.
Trout, S. A., Huelin, F. E. and Tindale, G. B. (1938). *Aust. Coun. Sci. Ind. Res. Pam.* **80,** 50 pp.
Trout, S. A., Huelin, F. E. and Tindale, G. B. (1960). *Aust. Commonwealth Sci. Ind. Res. Div. Food Preservation Transport Tech. Paper* **14,** 1.
Turrell, F. M. and Klotz, L. J. (1940). *Bot. Gaz.* **101,** 862.
Undenfriend, S., Lovenberg, W. and Sjoerdsma, A. (1959). *Archs Biochem. Biophys.* **85,** 487.
Underwood, J. C. and Rockland, L. B. (1953). *Fd Res.* **18,** 17.
U.S. Dep. Agr. (1957). Handbook No. 98. 99 pp. Agric. Research Service, U.S. Dep. Agr.
U.S. Dep. Agr. (1967; 1968). World Agricultural Production and Trade Statistical Report Foreign Agricultural Service. U.S. Dep. Agr.
Vannier, S. H. and Stanley, W. L. (1958). *J. Ass. off. agric. Chem.* **41,** 432.
van Overbeek, J. (1968). *Scient. Am.* **219,** 75.
Varma, T. N. S. and Ramakrishnan, C. V. (1956). *Nature, Lond.* **178,** 1358.
Vines, H. M. (1964). *Pl. Physiol. Suppl.* **39,** 59.
Vines, H. M. (1968a). *Proc. Am. Soc. hort. Sci.* **92,** 179.
Vines, H. M. (1968b). Unpublished. On file, Citrus Exp. Sta., Lake Alfred, Florida.
Vines, H. M. and Metcalf, J. F. (1967). *Proc. Am. Soc. hort. Sci.* **90,** 86.
Vines, H. M. and Oberbacher, M. F. (1962). *Proc. Fla St. hort. Soc.* **75,** 283.
Vines, H. M. and Oberbacher, M. F. (1963) *Pl. Physiol., Lancaster* **38,** 333.
Vines, H. M. and Oberbacher, M. F. (1965). *Nature, Lond.* **206,** 319.
Vines, H. M., Grierson, W. and Edwards, G. J. (1968). *Proc. Am. Soc. hort. Sci.* **92,** 227.
Warth, A. H. (1947). "The Chemistry and Technology of Waxes," p. 170. Reinhold Publishing Corp., New York.
Watt, B. K. and Merrill, A. L. (1963). "Composition of Foods", U.S. Dep. Agr. Handbook, No. 8. Washington.
Waynick, D. D. (1927). *Calif. Cartograph.* **12,** 150.
Webber, H. J. and Batchelor, L. D. (editors) (1943). "The Citrus Industry", Vol. 1, p. 696. University of California Press, Berkeley.

Wedding, R. T. and Horspool, R. P. (1955). *Citrus Leaves* **35,** 12.
Wedding, R. T. and Sinclair, W. B. (1954). *Bot. Gaz.* **116,** 183.
Weinstein, L. H., Porter, C. A. and Laurencot, H. J., Jr. (1959). *Contr. Boyce Thompson Inst. Pl. Res.* **20** (2), 1211.
Weizmann, A. and Mazur, V. (1958). *J. org. Chem.* **23,** 832.
Weizmann, A., Meisels, A. and Mazur, Y. (1955). *J. org. Chem.* **20,** 1173.
Wenzel, F. W., Moore, E. L., Rouse, A. H. and Atkins, C. D. (1951). *Fd Technol.* **5,** 454.
Wheaton, T. A. and Stewart, I. (1965a). *Nature, Lond.* **206,** 620.
Wheaton, T. A. and Stewart, I. (1965b). *Anal. Biochem.* **12,** 585.
Widdowson, E. M. and McCance, R. A. (1935). *Biochem. J.* **29,** 151.
Williams, B. L., Good, L. J. and Goodwin, T. W. (1967). *Phytochemistry* **6,** 1137.
Willimott, S. G. and Wokes, F. (1926). *Biochem. J.* **20,** 1008.
Wilson, K. W. and Crutchfield, C. A. (1968). *Agric. Fd Chem.* **16,** 118.
Wilson, W. C. (1966a). Doctoral Diss., University of Florida, Gainesville.
Wilson, W. C. (1966b). *Proc. Fla St. hort. Soc.* **79,** 301.
Wilson, W. C. (1967a). *Proc. Fla St. hort. Soc.* **80,** 227.
Wilson, W. C. (1967b). Unpublished data, Florida Citrus Comm.
Wilson, W. C. (1967c). *Am. Soc. Agr. Eng. Paper* No. 67, p. 145.
Wilson, W. C. and Biggs, R. H. (1968). *Abstracts Am. Soc. hort. Sci. 65th Ann. Meeting*, 1968.
Wilson, W. C. and Coppock, G. E. (1968). "Chemical Stimulation of Fruit Abscission." Proc. First Int. Citrus Symp., Vol. 3, p. 1125. University of California, Riverside.
Wilson, W. C. and Hendershott, C. H. (1967). *Proc. Am. Soc. hort. Sci.* **90,** 123.
Wilson, W. C. and Hendershott, C. H. (1968). *Proc. Am. Soc. hort. Sci.* **92,** 203.
Winston, J. R. (1942). *Proc. Fla St. hort. Soc.* **55,** 42.
Winston, J. R. (1955). U.S. Dep. Agr. Circ. 961. 13 pp.
Wolford, R. W., Attaway, J. A., Alberding, G. E. and Atkins, C. D. (1963). *J. Fd Sci.* **28,** 320.
Yokoyama, H. and Vandercook, C. E. (1967). *J. Fd Sci.* **32,** 42.
Yokoyama, H. and White, M. J. (1965a). *J. org. Chem.* **30,** 2481.
Yokoyama, H. and White, M. J. (1965b). *J. org. Chem.* **30,** 3994.
Yokoyama, H. and White, M. J. (1966a). *J. org. Chem.* **31,** 3452.
Yokoyama, H. and White, M. J. (1966b). *Phytochemistry* **5,** 1159.
Yokoyama, H. and White, M. J. (1967). *J. Agric. Fd Chem.* **15** (4), 693.
Yokoyama, H. and White, M. J. (1968). *Phytochemistry* **7,** 1031.
Yokoyama, H., White, M. J. and Vandercook, C. E. (1965). *J. org. Chem.* **30,** 2482.
Young, L. B. and Erickson, L. C. (1961). *Proc. Am. Soc. hort. Sci.* **78,** 197.
Yufera, E. P. and Iranzo, J. R. (1967). *Rev. Agroquim. Tecnol. Alimentos* **7,** 364.
Zechmeister, L. and Tuzson, P. (1936). *Ber. dt. chem. Ges.* **69,** 1878.
Zemplen, G. and Tettamanti, A. K. (1938). *Ber. dt. chem. Ges.* **71B,** 2511.
Zoller, H. F. (1918). *Ind. Engng Chem. ind. Edn* **10,** 364.

Chapter 4

The Grape†

E. PEYNAUD AND P. RIBÉREAU-GAYON

Station Agronomique et Oenologique de Bordeaux, France;
Faculté des Sciences de Bordeaux, France

I Introduction 172
II Constitution and Development of Grapes 174
A. Vegetative Cycle of the Vine 174
B. Anatomy of the Berry 176
C. General Composition of the Raceme 177
III Carbohydrates 178
A. General Metabolism 178
B. Carbohydrate Formation from Malic Acid 180
C. Carbohydrate Catabolism and the Tricarboxylic Acid Cycle .. 181
IV Organic Acids 181
A. General Composition 181
B. Malic Acid Metabolism 182
C. Tartaric Acid Metabolism 185
V Nitrogen Compounds 186
A. General Composition 186
B. Mechanism of Formation of Amino Acids 187
VI Phenolic Compounds 188
A. General Composition 188
B. Metabolism of Phenolic Compounds during Maturation 190
VII Other Compounds 192
A. Volatile Components 192
B. Pectins 193
C. Vitamins 194
D. Mineral Components 194
VIII Respiration during the Development of the Grape 195
IX Factors Affecting the Maturation and Quality of Grapes 199
X Grape Technology 201
A. Vinification 201
B. Grape Juice 203
C. Storage of Dessert Grapes 203
References 204

† Translated by C. Lance, Paris.

I. INTRODUCTION

The cultivation of the grape-vine dates from very ancient times. It can be traced back to the time of the Pharaohs in Egypt, many centuries before the Christian era. A juicy, sweet fruit, grapes can be eaten fresh, but, from antiquity, they have been used to make wine. If the juice of the grape is left to stand, it will spontaneously ferment into a smooth liquor which keeps well owing to the presence of alcohol. The cultivation of the grape-vine was mainly developed for wine-making. Wine is no ordinary drink; it is so strongly linked to the way of life of the Latin nations that one can speak of a grape-vine civilization.

The grape-vine originated in Asia Minor; more precisely from that part of Caucasia which lies south of the Black and Caspian Seas. From there the Phoenicians brought the cultivation of the grape-vine first to Greece, then to Rome and finally to the whole Mediterranean area and the south of France.

Today, about 25 million acres of land throughout the world are planted with the grape-vine. Annual production is as follows: 40 million tons of grapes are used for wine-making and yield some 75 thousand million gallons of wine (the production of grape juice amounts only to 26 million gallons); 60 million tons of grapes are used as dessert grapes and 1 million tons as raisins. The area used for the cultivation of the grape-vine increases very slowly, but production is steadily increasing owing to improvements in the yield per unit area.

The grape-vine can be readily grown between the 28th and the 48th Parallels in both hemispheres. There is, however, a little wine production in some tropical areas: Columbia, Peru, Brazil and India. In the Northern Hemisphere, the most northerly vineyards give the best quality grapes for wine-making. Farther south, the grapes mature more easily, the dessert grapes have more flavour and are sweeter, but the wine made from them is heavier and of a poorer quality.

The grape-vine grows from sea level to an elevation of 3000 feet. In tropical areas, elevation improves the quality of the grapes; in cold regions, it limits the area of cultivation. The grape-vine does not require a tillable soil and many vineyards are in areas where no other cultivated plant can be grown. Grown on a rich soil, the yield is better but the quality of the wine is not so good. Cultivation of the grape, nevertheless, requires considerable care, largely because the vine persists for several score years on the same soil and is subject to many vicissitudes.

The economical importance of grape-vine and wine explains the extent of the literature in this field. The grape-vine, grapes and wines present many interesting scientific problems, as testified by the fundamental researches of

Pasteur which made important contributions to the evolution of the biological sciences. For instance, the grape-vine is one of the few plants which contains tartaric acid as the major component acid. This fact raises many questions concerning the biosynthesis degradation of this acid during the development and maturation of the grape. The practical importance of tartaric acid lies in the fact that it gives wine a pH of about 3; of all fermented beverages, wine is both the most acidic and most alcoholic.

The grape-vine belongs to the genus *Vitis* (family Ampelidaceae) which includes several species. *Vitis vinifera* is the only species of European origin. It is subdivided into many varieties, bearing black or green grapes. Asian vines (*Vitis amurensis*) have no economical importance. American vines (*Vitis riparia*, *Vitis rupestris*, *Vitis labrusca*) generally bear black grapes.

TABLE I. The most common varieties of grapes grown in France for wine making

Time of maturation	Black varieties	White varieties
First period At the same time as the Chasselas variety[a]	Gamay Pinot Meunier Pinot noir	Aligoté Chardonnay Melon Traminer
Second period 2–15 days after Chasselas	Alicante-Bouschet Cabernet Franc Cabernet-Sauvignon Cinsaut Grolleau Jurançon noir Malbec Merlot Syrah	Altesse Chenin Gewurztraminer Jurançon blanc Mauzac Merlot blanc Meslier Muscadelle Muscat de Frontignac Riesling Roussanne Sauvignon Sémillon Sylvaner
Third period 24–30 days after Chasselas	Aramon Bouchalès Carignan Grenache Mérille Mourvèdre Petit-Verdot Tannat Terret Valdiguié	Aubun Clairette Colombard Folle Blanche Maccabeu Piquepoul Ugni-blanc

[a] See Table II.

Many are either not fructiferous or else produce grapes unsuitable for human consumption or wine-making. However, the crossing of American species with *Vitis vinifera* has resulted in the production of hybrids with good cultural qualities. Generally speaking, the costs of production of these hybrids are less than those with *Vitis vinifera*, but the quality of the grapes is not so good. The biochemical analysis of the anthocyanins present in the grapes makes

TABLE II. Varieties of dessert grapes

Black grape varieties	White grape varieties
Alphonse Lavallée	Admirable de Courtiller (early)
Cardinal (very early)	Chasselas doré
Gros Colman (late)	Dattier de Beyrouth (late)
Muscat de Hambourg	Emperor (late)
	Gros Vert (very late)
	Italia (late)
	Muscat Reine des Vignes (early)
	Muscat d'Alexandrie
	Panse précoce
	Servant (very late)
	Sultanine

it possible to differentiate between wines from *Vitis vinifera* and those from its hybrids. This characterization is used in the control of wine quality; great vintage wines originate from *Vitis vinifera* only, and not from hybrids.

American vines are also used as phylloxera-resistant stocks for the cultivation of *Vitis vinifera*.

The number of cultivated varieties is considerable as will be seen from Tables I and II. There are also many purely local varieties.

II. CONSTITUTION AND DEVELOPMENT OF GRAPES

A. Vegetative Cycle of the Vine

The vegetative cycle of the grape-vine is summarized in Table III. The bunches (racemes) are present in the buds in an embryonic state and they generally sprout as the third or fourth leaf develops. "Flowering" is when the flowers are in full bloom. At "nouaison" (fruit-set) the young grapes become clearly visible. The growth period continues until "veraison", when the colour of the grapes change and shoot growth stops. Then the grapes mature and at this point the whole vine begins to build up reserves; sugars are stored in the berries. The shoots synthesize starch, undergo lignification and take on a brown colour. The fall of leaves, which ends the vegetative cycle, takes place after harvest.

It is convenient to divide the life history of grapes into four phases.

(i) The herbaceous phase which lasts from nouaison to veraison. The bunches are green, contain chlorophyll and display a very active metabolism. As the berries increase in size, they remain green and firm and contain 20 g of sugar/kg and almost as much acid. During the whole of this period, the respiration rate is high.

TABLE III. Annual vegetative cycle of the grape vine in Europe

Bud sprouting →	April	
	May	Growth of shoots, leaves and grapes
Flowering →	June	
	July	
Veraison; growth stops →	August	
Grapes ripening →	September	Maturation and storage of reserves
	October	
Fall of leaves →	November	
	December, January, February, March	Winter dormancy

(ii) At veraison a rapid change in the appearance and constitution of the berries takes place. They swell, become elastic and translucid as a result of modifications in the pectocellulosic wall of the cell. The berries lose their chlorophyll; from green, white grapes turn yellow; black grapes turn light red and finally dark red. This colour change is very rapid and may take place in one day. Under normal conditions, all the grapes in the same vineyard colour in about two weeks. Carbohydrates increase and acids decrease markedly. During this period, the respiration rate decreases, but the activity of the enzymes of the fruit is still high.

(iii) The period of maturation lasts from veraison to the state of ripeness. During the 40–50 days of this period, grapes continue to swell, store carbohydrates and lose their acidity. Tannin synthesis increases as the skin becomes more coloured and the characteristic aroma develops. As with many fleshy fruits, the maturation of fruit and seeds does not always coincide. In varieties of grapes which are late, the seed matures long before grape-harvest and the stones may transmit some of their components to the surrounding pulp. In early varieties of dessert grapes, however, the pulp matures earlier than the seeds which are not viable when the grapes are harvested.

(iv) If the grapes are not harvested they become over-ripe. They use up their reserves, lose water and their juice becomes more concentrated whereas their respiration rate remains constant. In many areas, wines of quality can only be produced by allowing the grapes to become over-ripe on the vine-stocks. Over-ripening can also be artificially induced after harvesting ("passerillage"). Finally, there is the practice of "pourriture noble" when complex changes are completed by the action of *Botrytis cinerea*.

B. Anatomy of the Berry

The fruit of the grape-vine is a berry. Each berry consists of an epicarp (the skin), thin and elastic; a mesocarp, juicy and fleshy; the pulp, a fragile tissue which when ruptured yields the juice or must; and an endocarp, indistinguishable from the pulp, which surrounds the carpels containing the seeds (Fig. 1).

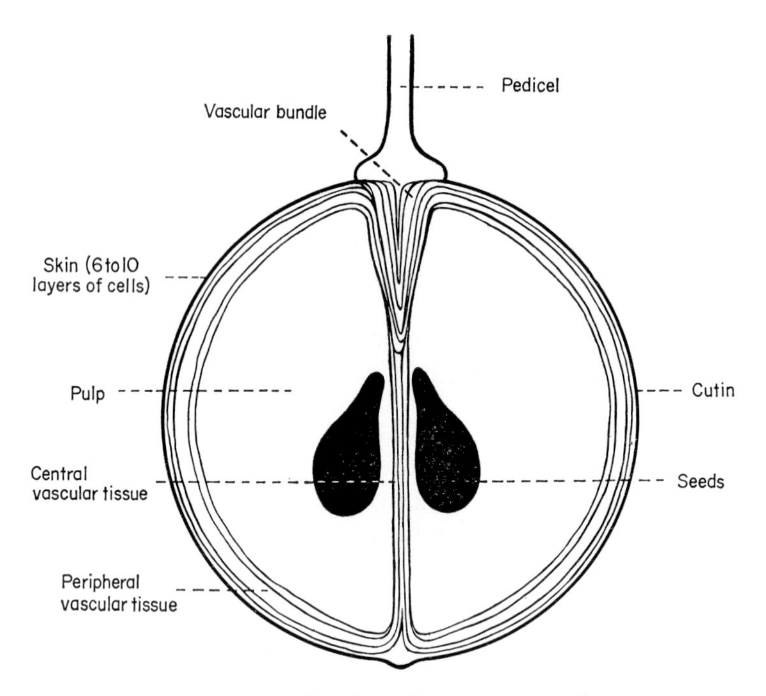

FIG. 1. Cross-section through a mature grape berry.

The skin consists of an epiderm and several underlying layers of cells. This is the ill-defined part which becomes detached when grapes are crushed. It is composed of 6–10 layers of small, thick-walled cells and the firmness of the berry depends on the coherence of this epicarp. The epidermis contains stomata and is covered with a cuticule on which lies a cuticular wax ("pruine") which forms microscopic scales. This cuticular wax consists mainly of oleanolic acid (79%); it also contains long chain alcohols (24–30 carbon atoms) and traces of esters, fatty acids, aldehydes and paraffins (Radler, 1965b, c).

The most characteristic components of the skin are red and yellow pigments, leucoanthocyanins and other compounds largely responsible for the aroma and flavour of the fruit.

The pulp consists of 25–30 layers of cells with large vacuoles containing the cell sap which is the major constituent of the juice or must obtained when the berries are crushed. Nutrients are supplied to the berry through vascular bundles which translocate the sap through the stem, peduncle and pedicel into the berry. Inside the berry, 10–12 vascular bundles divide into axial bundles which feed the seeds and the endocarp, and into peripherical bundles which branch out within the pulp.

TABLE IV. Relations between the number of seeds, the size and the composition of ripe Malbec grapes

Number of seeds per berry	Weight of berry (g)	Carbohydrates (g/litre of juice)	Acidity (meq/litre of juice)
1	1·91	188	134
2	2·52	160	142
3	2·96	153	154
4	3·25	145	160

From Ribéreau-Gayon and Peynaud (1960).

The grape berry normally contains four seeds, originating from the four ovules of the ovary. Often, however, the number of stones is less than four, because of the absence or abortion of one or several ovules. Sometimes the seeds are small but viable, while the absence of seeds in some varieties renders them particularly suitable for dessert use or for the production of raisins.

The development of the seeds influences the size and composition of the grapes (Table IV). The greater the number of stones, the heavier the weight of the berry, the lower the sugar content and the higher the acidity. Similarly, the tartaric acid content decreases and the malic acid content increases; there is three times as much citric acid in single-seed berries than in those with three seeds. The nitrogen content of the pulp decreases as the number of stones increases. These observations could be explained by the presence of growth factors in the stones. Iwahori *et al.* (1968) have demonstrated that during the whole of their development grapes of the variety Tokay have a higher gibberellin content than the berries of the same variety deprived of seeds.

C. General Composition of the Raceme

Tables V, VI and VII give examples of the constitution of the various components of the raceme for several varieties of grape.

TABLE V. Examples of the constitution of the racemes in two grape-vine varieties grown in the Bordeaux region

	Merlot	Sauvignon
Constitution of the bunch		
Stalk (%)	2·7	3·0
Berries	97·3	97·0
Constitution of the berries		
Average weight (g)	1·62	1·60
Pulp (%)	78·8	82·9
Skin (%)	16·4	14·2
Seeds (%)	4·8	2·9

TABLE VI. Chemical composition of the pulp and skin of grapes (variety Sauvignon) (1000 g fr. wt.)

	Pulp	Skin
Weight (g)	835	136
Reducing sugars (g)	188	14
pH	3·30	4·15
Free acids (meq)	106	131
Neutralized acids (meq)	38	207
Tartaric acid (meq)	77	138
Malic acid (meq)	68	175
Citric acid (meq)	2·1	13
Solubles phenols (g)	traces	3·6

TABLE VII. Distribution of organic acids between pulp, skin and stalk of the grape (variety Sauvignon) (Amounts in meq found in 100 grape berries)

Organic acid	Pulp	Skin	Stalk	Total
Free acids	13·9	2·1	1·0	17·0
Neutralized acids	4·3	3·4	1·2	8·9
Tartaric acid	10·0	2·2	1·1	13·3
Malic acid	8·8	3·0	0·9	12·7

III. CARBOHYDRATES

A. General Metabolism

The major carbohydrate compounds of the grapes are glucose and fructose. They amount in ripe grapes to about 150–250 g litre of juice. In unripe grapes, glucose accounts for 85% of the sugar content. At the period of veraison, the ratio of glucose to fructose is close to unity, while in ripe grapes there is generally a slight excess of fructose. The sucrose concentration in the ripe grape is less than 0·1% of the fresh weight (Guichard, 1953; Kliewer, 1966).

The pentoses, mainly arabinose and traces of xylose, are present in small amounts (0·3–1 g/litre of juice) in ripe grapes. The occurrence of stachyose and raffinose has been demonstrated in various parts of the vine by Carles (1962). There are also reports of the presence of melibiose.

We have seen that at veraison, a period of intense physiological activity in which pigment synthesis is visibly involved, a considerable accumulation of sugar takes place. Within 10 days or so, the sugar content increases by a factor of 6 or 7. It is difficult to explain this increase in terms of an increase in photosynthetic activity. It is possible that there is a sudden mobilization of the stem reserves stored during the vegetation cycle of the year before. This idea was put forward by Moreau and Vinet (1932) who showed that defoliated vines were able to produce mature grapes almost as readily as under normal conditions. However, more recent findings by Bouard (1960, 1966, 1967) do not support this conclusion. For instance, if the stem is ringed, a process which should impede translocation, the grapes will mature even earlier. According to this author, carbohydrate translocation from reserves within the vine will only take place under abnormal conditions such as those involved in defoliation.

Photosynthesis is the main mechanism for the production of the carbohydrates of the grape. The leaves are the main sites for the synthesis of sugars, which are then translocated to other organs, particularly to the berries. However, before veraison, the berries, which still contain chlorophyll, are able to synthesize carbohydrates although to a lesser extent than the leaves. After an exposure to $^{14}CO_2$ and light for 30 minutes, the leaves and immature grapes contain respectively 79 and 24% of the total radioactivity (Ribéreau-Gayon, 1966).

It is known that sucrose is the first sugar produced by photosynthesis; it may be translocated as such or it may be hydrolysed after leaving the leaves. Only traces of sucrose can be found in the berries, and Kliewer (1964) has shown that hydrolysis of sucrose is more rapid in immature than in ripe grapes; temperature increases the rate of hydrolysis. Hawker (1969) found that a large increase in the level of invertase in the grape coincided with the beginning of the accumulation of sugars. About the same time, an increase in the activities of sucrose synthetase, sucrose-phosphate synthetase and sucrose phosphatase was observed. Hawker suggested that carbohydrate translocation to, and accumulation in, the fruit involved the following sequence of reactions: phosphorylation of glucose and fructose, synthesis of sucrose-phosphate, hydrolysis of sucrose-phosphate into sucrose and followed by hydrolysis of sucrose. Table VIII illustrates the distribution of the radioactivity in various acids and sugars between leaves and berries after exposing the leaves to $^{14}CO_2$.

TABLE VIII. Translocation to the immature grape berries of compounds synthesized in the leaves[a]

Compound	Radioactivity (10^3 cpm/g)			
	Leaves 48 hours	Berries 6 hours	Berries 24 hours	Berries 48 hours
Total carbohydrates	265	10	50	130
Total organic acids	113	1	8	60
Total amino acids	12	0	1	17
Fructose	50	tr.	17	19
Glucose	154	tr.	14	61
Sucrose	53	tr.	17	38
Malic acid	104	0	6	41
Tartaric acid	1	0	1	14
Citric acid	2	0	tr.	1

[a] Leaves were fed $^{14}CO_2$ during a 48 hour period. Analyses were made of the leaves and adjacent grape berries of the same vine-stock after various times of exposure as shown.
tr. = trace.

Hardy (1968) fed uniformly labelled ^{14}C-sucrose to the cut surface of grapes and found that the radioactivity was unevenly distributed between glucose and fructose, the latter being more heavily labelled. This disparity he considered to be due to the more active metabolism of glucose.

B. Carbohydrate Formation from Malic Acid

Ribéreau-Gayon (1966) has shown that immature grapes can readily synthesize carbohydrate from malic acid. This phenomenon has also been observed in ripe grapes as can be seen from Table IX which shows the

TABLE IX. Transformations of labelled glucose and malic acid in mature grapes (figures represent percentages of the total radioactivity after 6 days)

	Radioactivity after the introduction of: Glucose	Radioactivity after the introduction of: Malic acid
Carbohydrates	97	84
Organic acids	1	11
Amino acids	2	5

From Ribéreau-Gayon (1966).

radioactivity in sugars, organic acids and amino acids 6 days after introducing ^{14}C-glucose and ^{14}C-malate into the berries. This appears to be the first proof of the conversion of malic acid to sugars in the grape and goes some way to explain the decrease in acidity and the increase in sugar content

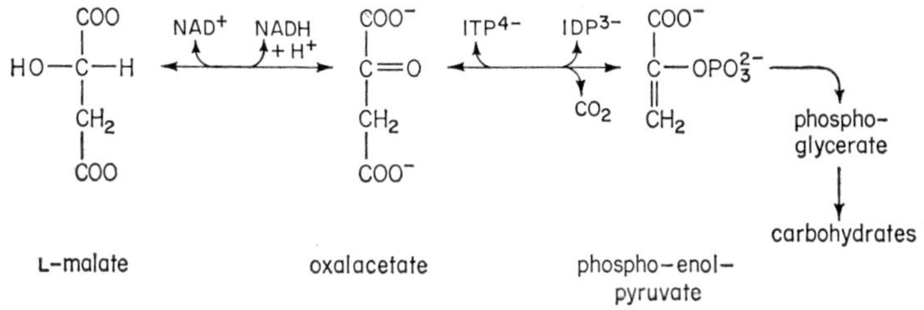

FIG. 2. Carbohydrate synthesis from L-malic acid (Karlson, 1964).

observed during ripening. The mechanism proposed for this conversion is shown in Fig. 2. A somewhat analogous mechanism was put forward by Peynaud and Guimberteau (1962) to explain the formation of ethanol from malic acid in ripe grapes stored in an oxygen-free atmosphere (intracellular fermentation).

C. Carbohydrate Catabolism and the Tricarboxylic Acid Cycle

As in plant tissues generally, carbohydrates are broken down through the pentose phosphate cycle or through glycolysis which transforms the glucose molecule into pyruvic acid. Pyruvic acid can give rise to malic acid through carboxylation or the operation of the tricarboxylic acid (TCA) cycle (see Volume 1, Chapter 4). Ribéreau-Gayon (1966) has shown that, in the grape, the cycle is particularly active in the region of the condensation of pyruvate (as acetyl CoA) with oxaloacetate to give α-oxoglutarate via citric acid. Introduction of ^{14}C-acetic acid results in the formation of highly labelled glutamic acid formed by amination of α-oxoglutarate. However, the next stages in the cycle from α-oxoglutarate back to oxaloacetate are much less active. There is no direct relation between the operation of the cycle as a whole and the accumulation of malic acid.

IV. ORGANIC ACIDS

A. General Composition

As in most fruits, malic acid is present in grapes. By comparison with other berry fruits, however, grapes are characterized by large amounts of tartaric acid and relatively small amounts of citric acid. In grapes, tartaric acid is present as the D-isomer and malic acid as the L-isomer. Other acids present in small amounts are: succinic, fumaric, pyruvic, α-oxoglutaric, glyceric, glycolic, dimethyl-succinic, shikimic and quinic acids. Colagrande (1959) reported the occurrence also of mandelic, *cis*- and *trans*-aconitic, maleic and isocitric acids.

TABLE X. Organic acid distribution in grapes during growth (Cabernet-Sauvignon variety in 1962) as percentage acidity expressed in meq

Acids	9th Jul.	5th Oct.
Malic	39	47
Tartaric	55	45
Citric	1	2
Succinic	1	2
Fumaric	1	1
Glyceric	·1	<1
Shikimic	0	0
Oxalic	0	0

From Ribéreau-Gayon (1966).

The acidity varies from 300 to 500 meq/litre of juice at veraison to 100 meq at maturity. Both tartaric acid and malic acid account for more than 90% of the total acidity (Table X). During maturation, the malic acid content decreases while the tartaric acid content of the berry remains more or less constant. In other respects, the content of tartaric and malic acids at half-veraison and at maturity varies from year to year for the same vineyard (Peynaud, 1946; Peynaud and Maurié, 1953, 1956) although the ratio of these acids fluctuates more than their sum. It would appear that the total content of organic acids is a physiological constant, implying that the conversion of one acid to another is relatively easy. However, this implication is not supported by experiments carried out with labelled coupounds on the biosynthesis of organic acids (Stafford and Loewus, 1958; Gyr, 1960; Hale, 1962; Kliewer, 1964; Ribéreau-Gayon, 1968).

B. Malic Acid Metabolism

For a long time, it has been thought that organic acids, and particularly malic acid, resulted from an incomplete oxidation of carbohydrates. This interpretation is not in agreement with the results of more recent experiments involving the use of labelled compounds which have led to the discovery of other mechanisms.

In immature grapes, Ribéreau-Gayon (1968) has demonstrated that the main biosynthetic pathway of malic acid is the carboxylation of pyruvic acid by atmospheric CO_2 according to the Wood–Werkman reaction. This synthesis follows the mechanism shown in Fig. 3; it is catalysed by phosphoenolpyruvate-carboxykinase. Although the reaction can take place in the dark, it is much more active in the light when radioactivity from $^{14}CO_2$ given to immature grape berries appears to a greater extent in malic acid than in carbohydrates. Aspartic acid, which like malic acid, is derived from oxaloacetic acid, is also labelled. The Wood–Werkman reaction is of general occurrence in plants, but it is remarkably active in grapes, particularly in immature grapes, where the metabolic rate is higher than in ripe grapes.

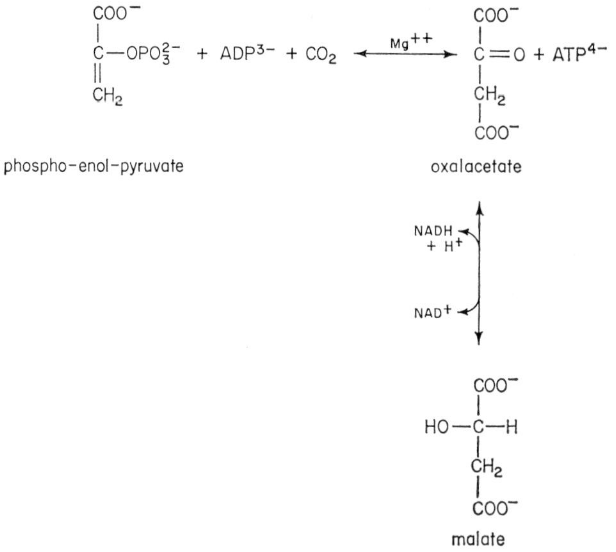

$$\underset{\text{phospho-enol-pyruvate}}{COO^- - C(-OPO_3^{2-}) = CH_2} + ADP^{3-} + CO_2 \xrightleftharpoons{Mg^{++}} \underset{\text{oxalacetate}}{COO^- - C(=O) - CH_2 - COO^-} + ATP^{4-}$$

$$\text{oxalacetate} \underset{NAD^+}{\overset{NADH + H^+}{\rightleftharpoons}} \underset{\text{malate}}{COO^- - CH(OH) - CH_2 - COO^-}$$

FIG. 3. Malic acid synthesis through carboxylation of pyruvic acid.

In the light, as in all higher plants, there is a competition for the utilization of atmospheric CO_2 between the carboxylation of ribulose-diphosphate involved in the synthesis of carbohydrates, and the carboxylation of phosphoenolpyruvate to form malic acid. In the leaves of the grape-vine, the first process prevails, while in the grape berry the synthesis of malic acid is more

TABLE XI. Carbohydrate and malic acid synthesis in leaves and immature grape berries, following a 10 sec exposure to $^{14}CO_2$, expressed as percentage radioactivity

	Leaves	Immature grapes
Phosphorylated sugars	46	21
Malic acid	6	30

important (Table XI). In the dark, photosynthesis is abolished and, if $^{14}CO_2$ is then supplied, 90% of the radioactivity is recovered as malic acid in both leaves and fruits.

The conversion of glucose to malate via phosphoglycerate and phosphoenolpyruvate can be demonstrated by supplying immature grapes with uniformly labelled glucose thus emphasizing the close interconnections between the metabolism of malic acid and of glucose. In immature grapes, carbohydrates are transformed into malic acid but the intensity of this reaction decreases as the berry ripens, and in ripe grapes, as has already been mentioned, the reverse process may take place.

It is known that the grape-vine can store citric acid in its roots (Alquier-Bouffard and Carles, 1963), probably as a result of the oxidation of carbohydrates via the TCA cycle. This accumulation is probably related to the regulation of the reaction: isocitrate + NAD + isocitrate dehydrogenase $\rightarrow$ α-oxoglutarate + CO_2NADH_2. In plants, isocitric acid is generally present in small amounts and its conversion to α-oxoglutaric acid could control the operation of the TCA cycle, and, more precisely, the accumulation of citric acid. This acid, stored in the roots, can migrate, during the next vegetative cycle, to the aerial parts of the plant where it becomes oxidized to malic acid. If, indeed, various types of labelled molecules are introduced into the roots of young grape-vine plants and the translocation of the synthesized products is followed, the ratio of radioactivity in malic acid to that in citric acid steadily increases from the roots to the leaves (Ribéreau-Gayon, 1966).

In both immature and ripe grapes, malic enzyme, malic dehydrogenase, phosphopyruvic and pyruvic carboxylases have been detected by several workers.

At the period of veraison, there is a rapid decrease in malic acid concentration and a rise in the respiratory quotient (see Section VIII). At the same time, the activities of malic enzyme, malic dehydrogenase and pyruvic carboxylase are at their highest levels in the berries according to Hawker (1969) who followed the changes in the activities of these enzymes during the whole development of the fruit.

Phosphopyruvic-carboxylase certainly plays an important role in CO_2-fixation and malic acid synthesis in immature grapes. Later on, as the berries develop, the rate of CO_2-fixation and the concentration of malic acid decrease together with a decrease in the activity of pyruvic carboxylase.

The metabolism of malic acid is probably dependent on its localization within the fruit. In immature grapes, malic acid will accumulate as a "storage" product probably inaccessible to enzymes. Malic enzyme might at this point take part in the dark-fixation of CO_2. As grapes ripen, the permeability of the cell membrane increases and malic acid might then again become available for decarboxylation by malic enzyme.

C. Tartaric Acid Metabolism

Until recently it was thought that tartaric acid is produced only in the leaves as a result of the oxidation of carbohydrates and thence translocated to the berries. It now appears probable, however, that grapes are able to synthesize their own tartaric acid during the early stages of development (Ribéreau-Gayon, 1966). At this period, tartaric acid synthesis is as active as malic acid synthesis but it decreases rapidly until at veraison it practically ceases. Generally speaking, the metabolism of tartaric acid is slow when compared with that of malic acid; once formed it does not undergo any rapid change. Experiments on the *in vivo* degradation of tartaric acid have failed to produce evidence of any intermediary products of its decomposition (Drawert and Stephan, 1965; Ribéreau-Gayon and Lefèbvre, 1967).

If labelled precursors are supplied to the tissue, radioactivity is recovered in both malic acid and tartaric acid, but its distribution between the different carbon atoms is not the same for both acids (Stafford and Loewus, 1958; Ribéreau-Gayon, 1966). It appears, therefore, that the mechanism of formation of these acids is different. According to Hardy (1968), the high percentages of ^{14}C found in organic acids only a few hours after supplying immature grapes with labelled sugars indicates that most of the malate and tartrate in the berries originates from glucose and fructose; translocation from the leaves is also possible. Hardy considered that malic acid and tartaric acids are not biochemically related. The small amounts of radioactivity found in tartaric acid after introducing labelled malic acid into grapes probably result from the prior conversion of malic acid into carbohydrates.

Several mechanisms have been put forward to explain the formation of tartaric acid in plants (Dupuy, 1960) but none of them has so far been convincing. In the case of the grape-vine, the most thorough study has been conducted by Ribéreau-Gayon (1966), who introduced $(^{14}C)_{-1}$-glucose and $(^{14}C)_{-6}$-glucose into young leaves and fruits of the grape. He found that the carbon atom 1 of glucose is more rapidly introduced into the tartaric acid molecule than carbon atom 6. Moreover, introduction of $(^{14}C)_{-1}$-glucose gave rise to tartaric acid mostly labelled in the carboxyl group. These results would indicate that the biosynthesis of tartaric acid requires a splitting of the glucose molecule between carbon atoms 4 and 5, following a prior oxidation of glucose to oxo-5-gluconic acid (Fig. 4).

Such a mechanism has been proposed by Vickery and Palmer (1954) for the tobacco leaf on the basis of the identity of the steric configuration of carbon atoms 2 and 3 in glucose and of the central carbon atoms in tartaric acid. In all probability, this mechanism is linked to the operation of the pentose-phosphate cycle which is very active in young organs; the first product of this pathway is gluconic acid, the most probable precursor of oxo-5-gluconic acid.

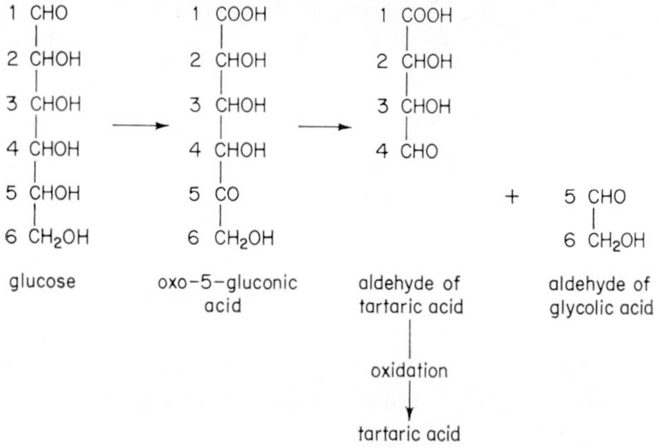

FIG. 4. Mechanism of formation of tartaric acid from glucose.

V. NITROGEN COMPOUNDS

A. General Composition

In grapes, nitrogen compounds are found as ammonium cations and as organic compounds—amino acids, peptides and proteins. Lafon-Lafourcade and Peynaud (1959), using microbiological techniques, have reported the presence of 17 amino acids in the following concentrations (mg/litre of juice): arginine 327, proline 266, threonine 258, glutamic acid 173, serine 69, glycine 22, leucine 20, lysine 16, histidine 11, isoleucine 7, valine 6, phenylalanine 5, aspartic acid 2, methionine 1, tryptophane 0·6; cysteine and tyrosine were absent. The amount of amino-nitrogen depends on the nature of the grape-vine variety and averages 50–100 mg of nitrogen/litre of juice. The varieties which have high levels of organic acids also have a high content of amino acids. Similarly, it is in the years when the grapes are most acid that they contain the highest amounts of amino-nitrogen. Changes in several amino acids during the growth of the berry have been determined by Lafon-Lafourcade (1962); see Fig. 5.

Peynaud (1939), Peynaud and Maurié (1953) have followed the evolution of the various forms of nitrogen in grapes. During the maturation phase, total nitrogen increases, mainly in the form of NH_4^+ ions, which are then progressively used for the synthesis of amino acids and proteins. Ammonium cations account for more than half of the total nitrogen in immature grapes but only for a quarter in mature grapes, even though the absolute amounts are almost the same.

At the end of the maturation period, the accumulation of nitrogen compounds in the berries stops. These compounds are then redistributed throughout the different regions of the berry. The mature seeds may even yield some of their constituents to the pulp, the nitrogen content of which increases in consequence.

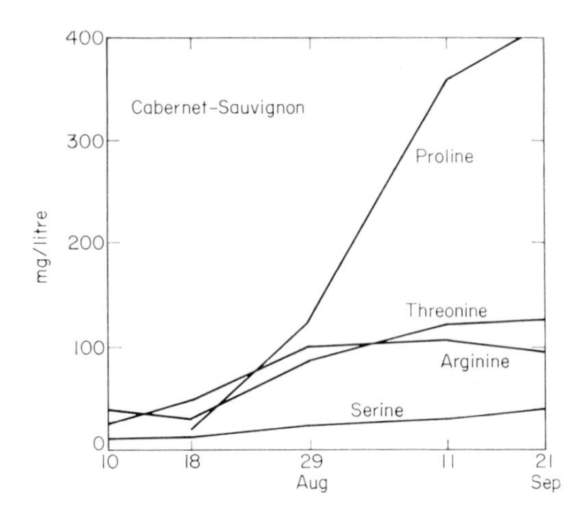

FIG. 5. Changes in various amino acids during maturation, as determined by microbiological techniques (Lafon-Lafourcade, 1962).

B. Mechanism of Formation of Amino Acids

Introduction of labelled compounds into immature grape berries generally results in a rapid synthesis of amino acids, labelling being maximal when the times of exposure are short (45 minutes to 3 hours). The amino acids lose their radioactivity rapidly since they are intermediary compounds in many metabolic reactions, particularly in the formation of organic acids and carbohydrates.

Serine, glycine, alanine and aspartic acid are highly labelled when immature grapes are exposed to $^{14}CO_2$. It may be assumed that the synthesis of the first three amino acids is related to the synthesis of carbohydrates (Fig. 6). The synthesis of aspartic acid results from transamination between glutamic and oxaloacetic acids. In the plant, aspartic acid might represent a storage form of oxaloacetic acid which, according to the circumstances, can be transformed into malic acid or into carbohydrates.

Proline is actively synthesized in grapes during the maturation phase; in all probability glutamic acid is its precursor (Davies *et al.*, 1964).

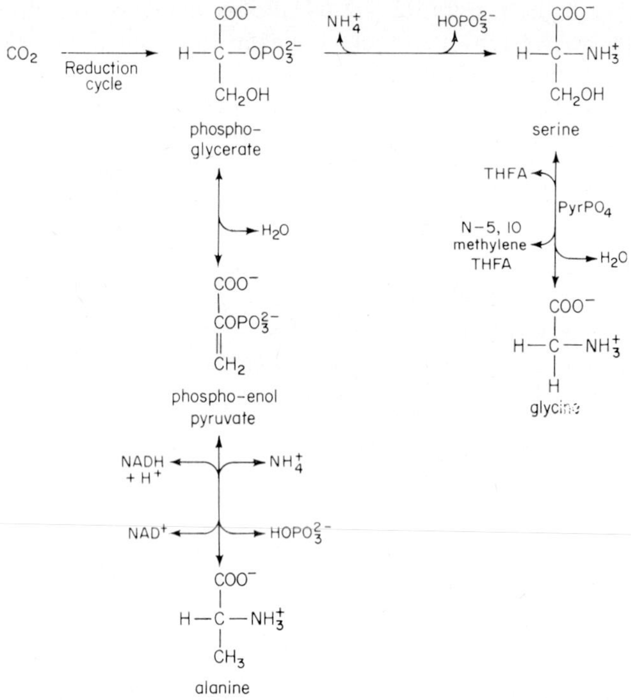

CO_2 → Reduction cycle → COO^-, $H—C—OPO_3^{2-}$, CH_2OH (phospho-glycerate)

NH_4^+ → $HOPO_3^{2-}$ → COO^-, $H—C—NH_3^+$, CH_2OH (serine)

serine → THFA, PyrPO$_4$, N-5, 10 methylene THFA, H_2O → COO^-, $H—C—NH_3^+$, H (glycine)

phospho-glycerate → H_2O → COO^-, $COPO_3^{2-}$, CH_2 (phospho-enol pyruvate)

phospho-enol pyruvate → NADH + H$^+$, NH_4^+, NAD$^+$, $HOPO_3^{2-}$ → COO^-, $H—C—NH_3^+$, CH_3 (alanine)

FIG. 6. Photosynthetic formation of serine, glycine and alanine. (THFA = tetrahydrofolate.)

VI. PHENOLIC COMPOUNDS

A. General Composition

Phenolic compounds have been particularly well studied in grapes, since they play an important role in the determination of the character of wines (Ribéreau-Gayon, 1964) to which they contribute the colour, taste and "body" during ageing. They include, particularly, the red pigments (the anthocyanins) and the tannins; in the main they account for the differences between red and white wines. White grapes also contain phenolic compounds, in lesser amounts than black grapes, but they contain no anthocyanins. In the preparation of white wines, the juice is rapidly separated from the remainder of the berry which contains most of the phenolic compounds. In this way, a minimum of phenolic compounds is included in the "must" for the production of white wines.

The most important phenolic compounds identified in grape skin are listed in Table XII.

Research has shown that the various species of the genus *Vitis* have different capacities for the glycosidation of anthocyanin pigments. *Vitis vinifera* can only synthesize mono-glucosides whereas other species, such as *Vitis riparia* and *Vitis rupestris*, synthesize both mono- and di-glucosides. The

TABLE XII. Phenolic compounds present in the grape

General formulae	Nature of specific compounds	Type of combination
Benzoic acids		
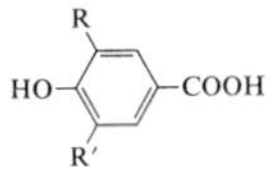 (structure: HO–C₆H₂(R)(R′)–COOH)	$R = R' = H$ gives p-hydroxy-benzoic acid $R = OH$, $R' = H$ gives proto-catechuic acid $R = OCH_3$, $R' = H$ gives vanillic acid $R = R' = OCH_3$ gives syringic acid	Combinations of an ester type, the nature of which is unknown
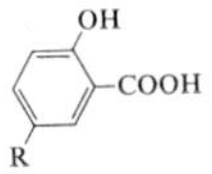 (structure: OH, COOH, R)	$R = H$ gives salicylic acid $R = OH$ gives gentisic acid	
Cinnamic acids		
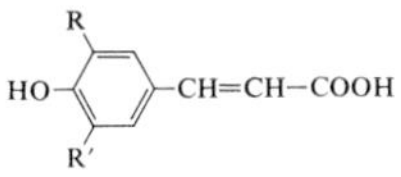 (structure: R, HO, R′, CH=CH–COOH)	$R = H$ gives p-coumaric acid $R = OH$ gives cafeic acid $R = OCH_3$ gives ferulic acid	Ester-type combinations with anthocyanins and tartaric acid
Flavonols		
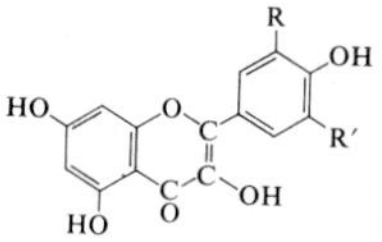 (structure: R, OH, HO, O, C, R′, C, C, OH, HO, O)	$R = R'$ gives kaempferol $R = OH$, $R' = H$ gives quercetin $R = R' = OH$ gives myricetin	Two or three glucosides are present and one glucuronoside
Anthocyanidins		
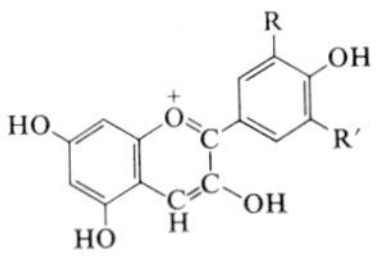 (structure: R, OH, HO, +, O, C, R′, C, C, OH, H, HO)	$R = OH$, $R' = H$ gives cyanidin $R = OCH_3$, $R' = H$ gives poenidin $R = R' = OH$ gives delphinidin $R = OCH_3$, $R' = OH$ gives petunidin $R = R' = OCH_3$ gives malvidin	Several types of glucosides and acyl glucosides depending on the species of *Vitis*
Tannin "precursors"		
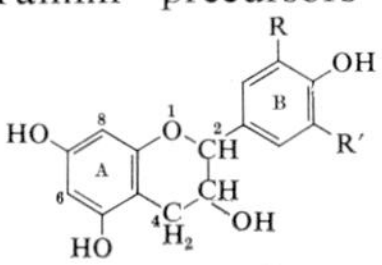 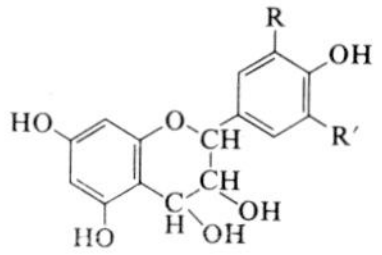 (structures: R, OH, B, HO, 8, 1, O, 2, CH, R′, A, CH, 6, 4, C, H₂, OH, HO; R, OH, HO, O, CH, R′, CH, C, OH, H, OH, HO)	$R = OH$, $R' = H$ gives catechin $R = R' = OH$ gives gallocatechin $R = OH$, $R' = H$ gives leucocyanidine $R = R' = OH$ gives leucodelphinidine	Tannins present are polymers of flavans, chiefly flavan-3,4-diols. These flavans are only present as monomers in small amounts

ability to synthesize di-glucosides is a biochemical character which is genetically transferred to the hybrids. It is a dominant character under the control of a single gene. This has led to the development of a method, now in general use, for differentiating between grapes and wines from *Vitis vinifera* and those from hybrids (Ribéreau-Gayon, 1953, 1963). Wines from *Vitis vinifera* never contain di-glucosides, whereas di-glucosides are frequently but not universally found in wines made from hybrids (Fig. 7).

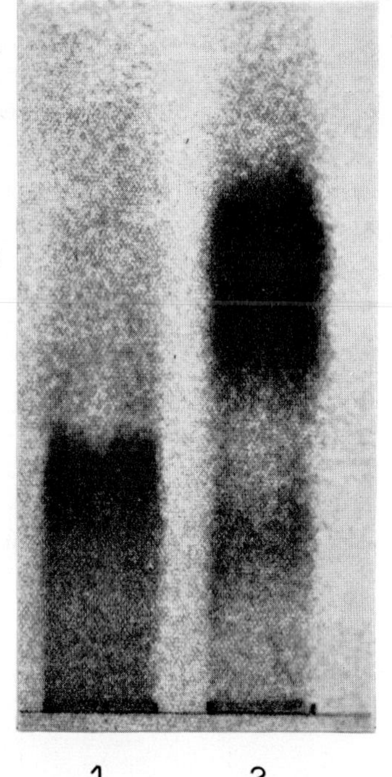

FIG. 7. Characterization of wines by their anthocyans. (1) Wines from *Vitis vinifera*; (2) Most of the wines from hybrids.

B. Metabolism of Phenolic Compounds during Maturation

In higher plants two mechanisms are responsible for the synthesis of the benzene ring, (1) the condensation of three molecules of acetyl co-enzyme A, and (2) the direct formation from sugar via shikimic acid (Ribéreau-Gayon, 1968). Although there is no direct proof that these reactions occur in the grape-vine, it is most likely that they are indeed responsible for the elaboration of the phenolic compounds present. Moreover, shikimic acid is a normal component of the various organs of the grape-vine. Experiments with $^{14}CO_2$

have shown that the radioactivity of this acid becomes much higher in the leaves than in the berries. In all probability, therefore, shikimic acid is synthesized in the leaves and then migrates to the grape berries, where the synthesis of phenolic compounds, particularly of anthocyanins, is very active.

The formation of anthocyanins is directly related to the metabolism of carbohydrates. The synthesis of these pigments begins at veraison, a period characterized by a considerable accumulation of sugars in the fruit. In grape varieties which can synthesize di-glucosides, the ratio of mono-glucosides to di-glucosides is markedly higher during veraison than during the maturation period which follows (Ribéreau-Gayon, 1964). During this period the synthesis of anthocyanins as mono-glucosides is probably more rapid than their transformation into di-glucosides.

TABLE XIII. Changes in anthocyanins in different grape varieties during ripening expressed as mg/100 g fr. wt.

Variety	6th Sep.	20th Sep.	4th Oct.	18th Oct.	2nd Nov.	14th Nov.
Baco	101	200	—	—	—	—
Cabernet B	1	16	51	89	—	—
Cabernet BB	1	6	32	49	60	57
Gamay Beaujolais	4	22	45	—	—	—
Gamay Chaudenay	22	85	161	—	—	—
Gamay Fréau	88	206	282	282	—	—
Grosiot	2	89	80	119	—	—

From Puissant and Léon (1967).

After veraison, the accumulation of anthocyanins steadily continues until past commercial maturity. Ripe black grapes may contain from 50 to 300 mg of anthocyanin per 100 g depending on the variety (Table XIII).

A more exact knowledge of the constitution of the tannins of the grape is urgently required. It is known that the condensed tannins in plant materials consist of a mixture of flavolans of molecular weight between 500 and 3000 resulting from the polymerization of elementary flavan units (Ribéreau-Gayon, 1968) (see Volume 1, Chapter 11).

The astringent properties of tannins (which give wine its harsh taste) depend on the total amount present, on the structure of the elementary units and on their degree of polymerization. It is believed that the degree of condensation of the tannins varies during maturation and ripening. The astringency of immature grapes could be due to tannins at an intermediate degree of condensation corresponding to maximum astringency. During ripening the decrease in astringency may be explained by an increase in the degree of polymerization of the tannin to a molecular weight in excess of 3000.

Peri (1967) has followed the development of these compounds during the ripening of several grape varieties from Italy. The analyses were made on extracts of whole bunches homogenized in 99% methanol. He found that the tannin content increased during veraison and then fell during naturation and ripening; an increase in the degree of polymerization was also observed.

VII. OTHER COMPOUNDS

A. Volatile Components

The odoriferous compounds which characterize each variety of grape and are responsible for the particular bouquet of the particular wine are located in both the pulp and skin of the fruit. Certain varieties of grapes, such as Muscat and Labrusca, especially when grown in a hot climate and harvested very ripe with a high sugar content, contain odoriferous compounds which pass unchanged into the wine. Other varieties contain specific compounds which contribute to the aroma of the wine after transformation during the fermentation of the wine. It is these varieties which produce the great vintage

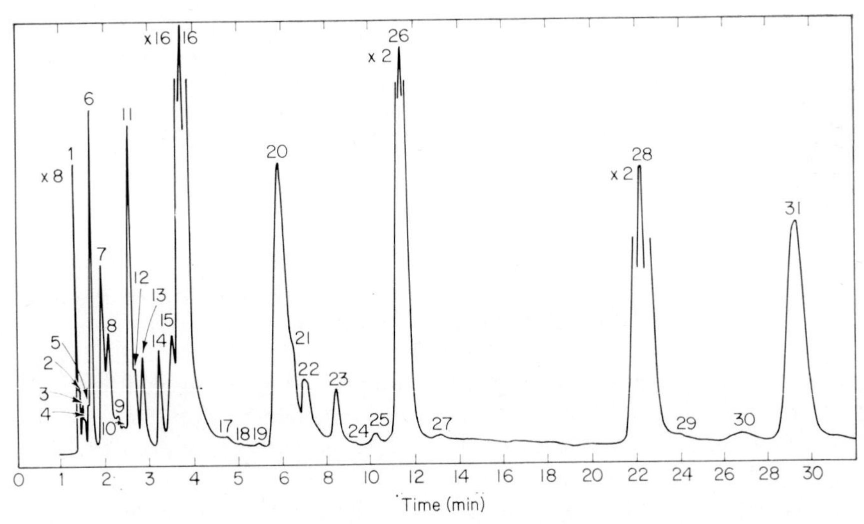

FIG. 8. Gas chromatogram of an extract of 250 ml of juice from Cabemet-Sauvignon grapes, using a column of Carbowax 400 at 100°C in a nitrogen stream (22 ml/min) and a flame-ionization detector (Bertrand *et al.*, 1967). Key to the peaks: 1–5, unknowns; 5, hexane (?) + ether; 7, ethanol; 8, methyl formate; 9, propanal; 10, isobutanal; 11, ethyl formate + methyl acetate; 12, acetone; 13, ethyl acetate; 14, 3-methylbutanal; 15, methanol; 16, ethanol; 17, i-propanol; 18 and 19, unknown; 20, hexenal; 21, 2-methyl-1-propanol; 22, 3-pentanol + water; 23, 2-pentanol; 24, 1-butanol; 25, unknown; 26, 2-hexenal + 2-methylbutan-1-ol + 3-methylbutan-1-ol; 27, unknown; 28, 1-hexanol; 29, unknown; 30, *cis*-hex-1-en-3-ol; 31, unknown.

wines (Carbernet in the Bordeaux region, Pinot in Burgundy). These compounds reach their highest levels and acquire a great "delicacy" in grapes ripening in the relatively cold regions at the limits of the grape-growing areas.

In recent years, the development of gas chromatography has enabled some progress to be made in separating and identifying the large number of volatile constituents of the grape, many of which are only present in extremely small amounts. As an example, Fig. 8 represents a gas-chromatogram of the volatile compounds present in Carbernet-Sauvignon grapes.

The typical aroma of Muscat grapes has been studied by Cordonnier (1956), using thin layer chromatography techniques, and by Webb *et al.* (1966), using gas chromatography techniques. Several compounds responsible for the aroma have been identified. The two main ones appear to be linalool and geraniol. With grape varieties which have the so-called "foxe" taste (Labrusca), it is probable that methyl anthranilate is responsible, although this has not been demonstrated unequivocally.

B. Pectins

The pectic substances found in the cell wall as an intercellular cement are a mixture of pectin and gums or pentosans (Table XIV). Soluble pectin is found in the juice while insoluble protopectin is strongly bound in the cell wall of the tissue.

TABLE XIV. Pectic compounds in musts of grape (g/litre)

	Merlot	Cabernet-Sauvignon
Total pectic compounds	3·10	4·43
Pectin	0·33	0·59
Gums	2·77	3·84

From Peynaud (1952).

The occurrence of pectolytic enzymes (pectin methyl-esterase, polygalacturonase) has been demonstrated in grapes. Protopectinase is responsible for the soluble pectin in the juice of crushed grapes. In the skin, the pectin methyl-esterase activity is generally twice as high as in the pulp, but it is much less in the seeds. The activity of the enzyme, measured by the liberation of methanol, as well as that of polygalacturonase, steadily increases during the ripening of the grape.

The gums, a group of sugar polymers, are mainly arabans and, less frequently, (see Volume 1, Chapter 2), galactans. From the juice of grapes, Büchi and Deuel (1954) have isolated a polysaccharide with a molecular weight of about 1000 containing galactose, mannose, arabinose, rhamnose and galacturonic acid units; it was present in amounts of 140 mg/litre of juice.

C. Vitamins

Vitamins of the B group have been measured in the juice of grapes by microbiological techniques (Peynaud and Lafourcade, 1958); results are shown in Table XV. Compared with other fruits, grapes have, together with blackcurrants, the highest content of thiamine and mesoinositol. They are relatively rich in nicotinamide and pantothenic acid. Most fruits have about the same content of pyridoxin and biotin, but grapes are relatively very poor in riboflavin. A litre of grape juice usually contains 1–2 μg of folic acid and 10–20 mg of p-aminobenzoic acid.

TABLE XV. Changes in the vitamins of group B in the grape (Merlot) during ripening

	26th Aug.	10th Sep.	20th Sep.	1st Oct.	11th Oct.	17th Oct.
Thiamine	228	268	264	350	450	316
Riboflavin[a]	—	2·6	6·3	14	12·2	8·1
Pantothenic acid[a]	490	550	960	640	660	470
Nicotinamide[a]	390	460	590	630	660	700
Pyridoxin[a]	120	190	220	240	240	190
Biotin[a]	5·0	4·2	1·5	1·3	2·5	2·0
Mesoinositol (mg)	118	184	212	238	238	297

[a] The figures express the amounts of vitamins in μg/1000 grape berries.

Generally, the vitamin content of grapes increases from veraison, except for biotin which decreases. Apart for nicotinamide and mesoinositol, which are still increasing at harvest, the vitamin content reaches its peak some days or weeks before the fruit is completely ripe.

Ascorbic acid increases during the growth of the grape berry, remains at a constant level during maturation and tends to decrease during ripening. Grapes are not particularly rich in ascorbic acid, averaging only 50 mg/litre of juice.

D. Mineral Components

The changes in mineral constituents of the grape during the maturation are shown in Table XVI. There is a constant uptake of mineral anions and cations from the soil and a distribution to the various parts of the plant including the berries; the berries are relatively poor in mineral elements compared with the other parts of the plant. During maturation, the cation content increases 2 or 3 times in the skin; 1·5–2·5 in the stalk, but only by 1·2–1·9 times in the pulp. During the same period, heavy metals increase by 50%. There is also an increase in anions, particularly phosphate, which is mostly concentrated in the seeds. The phosphate content also increases steadily in the peel and pulp as the fruit ripens.

TABLE XVI. Changes in the mineral components during the ripening of the Merlot grape[a]

	12th Aug.	30th Aug.	10th Sep.	20th Sep.	30th Sep.	7th Oct.
Total ash	1·7	2·1	3·7	2·8	3·9	5·0
Alkalinity of the ash	22·8	27·0	39·0	36·7	39·8	28·0
K^+	16·2	21·6	35·4	42·4	50·5	42·8
Na^+	0·5	0·9	0 8	1·0	0·9	1·7
$Ca^{++}+Mg^{++}$	7·0	12·6	11·5	8·8	9·5	13·0

From Bonastre (1959).

[a] The amounts of the cations and the alkalinity of the ash are expressed in meq in the juice from 100 grape berries. Total ash content is expressed in g/100 berries.

VIII. RESPIRATION DURING THE DEVELOPMENT OF THE GRAPE

Although gaseous exchange in grapes has been measured for many years there is a need for a more detailed examination of the phenomenon covering a range of varieties under carefully controlled conditions.

Direct measurements carried out on the whole berry do not reflect precisely the behaviour of the pulp cells, since they include the respiration of the seeds. A distinction has to be made between the respiration of the pericarp, which is relatively low with an RQ greater than 1·00, and the respiration of the seeds which is much greater and has an RQ averaging 0·75. Exact measurement of the respiration rate of the pericarp can only be performed on seedless varieties. Comparative measurements are significant only if performed on grape berries of about the same size and containing the same number of seeds. The skin and the cuticle with its waxy surface limit gaseous exchange, including transpiration. Whereas an untreated grape berry loses 1·7 mg of water/g/hour, a berry from which the cuticular wax has been removed by a solvent loses 8–9 mg of water while peeled grapes lose 14 mg/g/hour (Radler, 1965a). The permeability of grape skin to oxygen and carbon dioxide has not been studied.

Measurements of the respiration rate of immature grapes are complicated by the simultaneous operation of photosynthesis and dark fixation of CO_2. Even mature fruits may fix 25% of the CO_2 produced by respiration according to Lefèbvre (1968). Most determinations of respiration rate have been performed in closed vessels although it has been shown that respiration rate increases if the atmosphere surrounding the berry is continually renewed. The respiration rate of the fruit removed from the vine decreases within a few hours of picking while no such decrease is observed in berries remaining attached to the vine. Gerber (1897) was the first to measure, quantitatively, the respiration of the grape. His work was followed by that of Gore (1911), Luthra and Chima (1931) and, more recently, by Geisler and Radler (1963), Saulnier-Blache (1963) and Saulnier-Blache and Bruzeau (1967).

In immature grapes the respiration rate (O_2 uptake) at 30°C is 100–150 ml/kg/hour and this falls to 20–30 ml when the berries become ripe.

The respiratory rate of grapes increases 2·5 times for every 10°C rise in temperature. According to Gerber's results, the respiratory rate of grapes is maximal at 37°C.

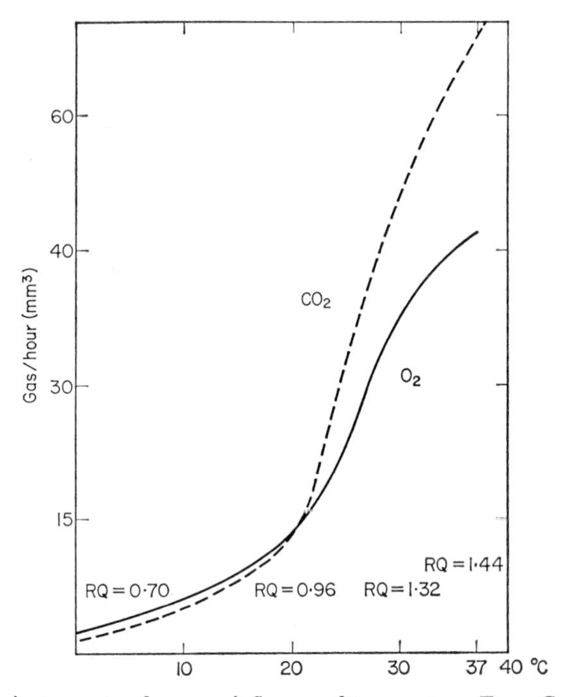

FIG. 9. Respiratory rate of grapes: influence of temperature. From Gerber (1897).

During the growth period of the grape berry, the respiratory rate, expressed on a fresh weight basis, decreases steadily. This could be explained by the dilution of the respiratory enzymes due to the large increase in the volume of the vacuoles.

If the respiratory rate is expressed on a per berry basis, the picture is somewhat different. According to Gerber (1897), the respiratory rate increases in the immature grape and then remains constant during the maturation phase as will be seen from Table XVII. According to Gatet (1939) and Geisler and Radler (1963) the respiratory rate of the berry continually decreases whether expressed on a fresh weight or per berry basis (Fig. 10). For this reason the grape is classified as a non-climateric fruit (Biale, 1960a, b). The results of Geisler and Radler (1963) shown in Fig. 11 illustrate the rates of respiration and of photosynthetic CO_2 fixation during the development of the grape.

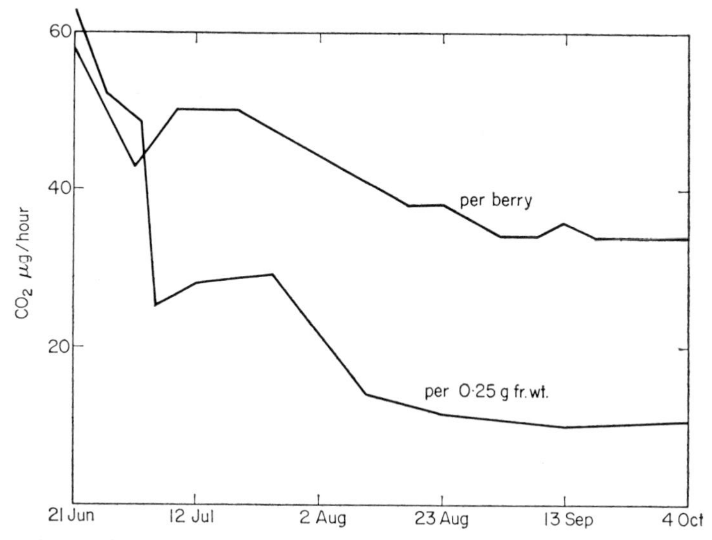

FIG. 10. Changes in the respiration rate of the grape during development. The rate is expressed in μg CO_2 evolved/hour/berry/0·25g fr. wt. From Geisler and Radler (1963).

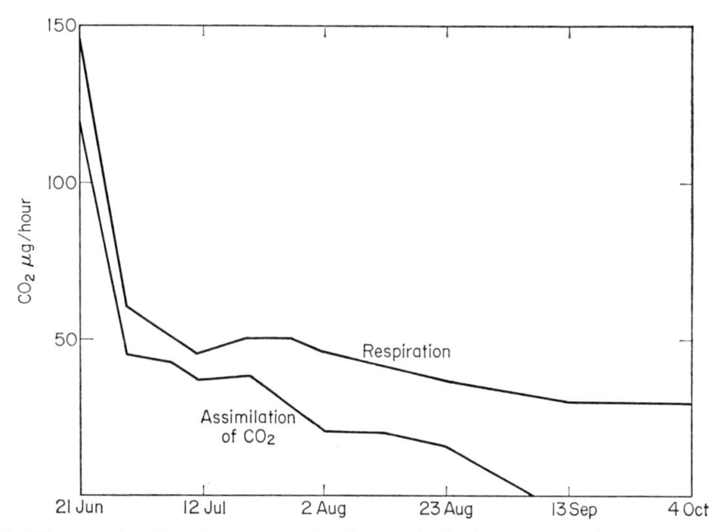

FIG. 11. Changes in the photosynthetic CO_2-assimilation and of the respiratory rate (CO_2-output) during development and maturation of grapes From Geisler and Radler (1963).

From accurate measurements carried out on carefully selected samples, Saulnier-Blache and Bruzeau (1967) concluded that the respiratory rate increases around the period of veraison (see Fig. 12). This increase, which is limited in duration and is followed by a period of gradual decrease, could be considered as a rudimentary climateric. Veraison (see Table III) is characterized by a sudden increase in the size of the grape berry, following a stationary phase (see Fig. 12, upper curves) and the increase in respiration at veraison is evidence of a rapid increase in metabolic activity. Respiratory quotients are generally lower in berries containing large seeds.

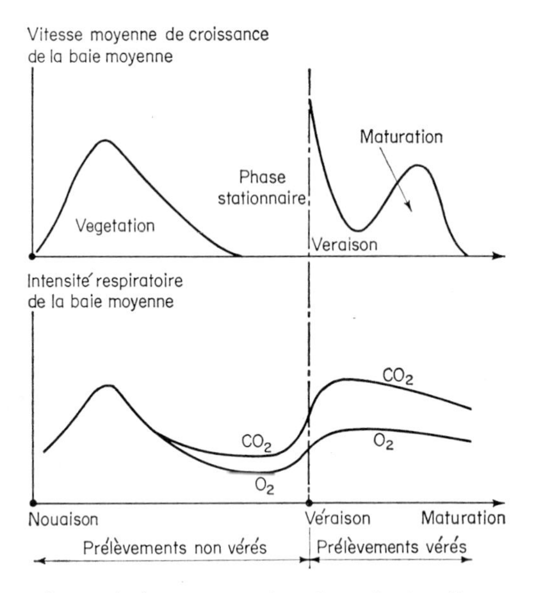

FIG. 12. Evolution of growth (upper curves) and respiration (lower curves) of grapes (variety Carignan) from nouaison through veraison to maturity. Schematic from Saulnier-Blanche and Bruzeau (1967).

As will be seen from Table XVII immature grapes have an RQ close to unity. At veraison the value rises to 1·3 or even higher. This suggests a change-over from oxidation of carbohydrates to the oxidation of malic acid. This hypothesis is in accordance with experimental data which show that the malic acid content tends to decrease at veraison (Table XVIII), and at this period the enzymes involved in malic acid metabolism are most active.

Gerber (1897) has shown that the RQ increased with temperature. He cites values of 0·70 at 0°C, 1·00 at 20°C, 1·33 at 30°C and 1·44 at 37°C. It would appear, therefore, that tartaric acid, the oxidation of which has an RQ of 1·60, is decarboxylated only at high temperatures.

TABLE XVII. Respiratory activities of Panse Muscat grapes at different phases of development (μl/hour/berry)

Date	Sugars (g/kg)	CO_2 evolution	O_2 uptake	R.Q.
10th Jul.	3·8	99	101	0·98
5th Aug.	—	46	46	1·00
10th Aug.	—	101	97	1·04
25th Aug.	42	107	79	1·35
3rd Sep.	89	158	102	1·54
25th Sep.	155	105	91	1·15

Recalculated from Gerber (1897).

TABLE XVIII. Relation between oxygen uptake and utilization malic acid in Cabernet grapes[a]

Date	O_2 uptake (μlitres)	Decrease in malic acid (mmoles)	mmoles O_2 required for the oxidation of malic acid	O_2 uptake in mmoles
5th Jul.	3095	30	90	138
19th Jul.	2102	7	21	93
13th Sep.	417	7	21	18
15th Sep.	360	6	18	16
21st Sep.	715	8	24	32

From Gatet (1939).
[a] The results are expressed as g/fr. wt. for a 30-hour period at 30°C.

IX. FACTORS AFFECTING THE MATURATION AND QUALITY OF GRAPES

Several factors may influence the composition and quality of the grape at harvest. Some, such as the variety and root stock, the age of the vine, the nature and porosity of the soil, climatic conditions, etc. cannot be manipulated by the grower once the vines have been planted. These factors are called the "patrimoine" of the vineyard. Other factors, such as pruning, trimming, manuring and soil improvement are in the hands of the grower. To some extent also "accidental" factors such as exposure and microbial disease are capable of some degree of control (Table XIX).

On rich soils adequately supplied with water, the grape-vine yields heavy crops but the quality of wines is poor. Most of the best vineyards are generally encountered on poor and stony soils. Chemical analyses of soils of the most famous wine-growing regions in France indicate that the levels of inorganic nutrients are surprisingly low. However, since the grape-vine is able to send its roots into the soil to depths of 12–24 feet, the amounts of inorganic nutrients available to the roots may be considerable.

If the humidity of the soil at the root level is high during the maturation period, the grapes are richer in acids, particularly in tartaric acid, than if soil humidity is low. Clay soils produce more acidic, less delicate grapes, rich in pigments and tannins. Limestone soils give a particularly high content of odoriferous constituents to some varieties of grapes.

TABLE XIX. Influence of cultivation and climate on the composition of Merlot grapes

	Sugars (g/litre)	Acidity (meq/litre)	Optical density of pigments from 1 g of dr. wt. of skin (diluted to 500 ml)
Conditions of cultivation			
Control	190	104	3·27
Non-topped	202	91	2·87
Pruned	212	94	3·42
Humidity			
Control	190	104	3·27
Protected from rain	177	109	2·20
Control	176	116	2·82
Watered	171	108	3·12
Illumination			
Control	184	98	2·96
Protection from the sun	145	153	0·53
Control	204	90	3·64
Illuminated during the night	187	98	2·08

From P. Ribéreau-Gayon and G. Ribéreau-Gayon (1958).

TABLE XX. Comparison of three grapes harvests (Variety Merlot) in the same vineyard of the Bordeaux region under different climatic conditions

	Climatic conditions (April to September)			Composition/litre of must			
Year	Sum of mean daily tempera-tures (°C)	Rainfall (mm)	Sunshine (hours)	Sugars (g)	Acidity (meq)	Tartaric acid (meq)	Malic acid (meq)
1961	3295	231	1402	236	84	102	36
1966	3005	502	1058	182	160	131	64
1965	3164	359	1264	220	136	135	40

Table XX illustrates the effect of climate on the composition of grape juice. The amounts of sunshine increases sugars at the expense of acid. The year 1961 was a particularly good year for wines of Bordeaux.

The ripe grape is subject to attack by several types of fungus which rapidly spread over the surface of the fruit in hot, humid regions. Attack is particularly severe in compressed bunches of thin-skinned berries.

TABLE XXI. The composition of grapes as affected by *Botrytis cinerea* ("pourriture noble") (amounts per 1000 berries)

	Healthy grapes	Decaying grapes
Weight of 100 grape berries (g)	285	112
Sugars (g)	367	203
Acidity (meq)	158	67
Tartaric acid (meq)	100	15
Malic acid (meq)	105	69
Gluconic acid (meq)	0	3·7

The most common fungus to attack grapes is *Botrytis cinerea* which forms a grey felt, hence the name "grey rot". Damage can be very severe if there is a simultaneous attack by *Penicillium* and *Aspergillus*. Although fungal decay reduces the commercial value of black grapes, the attack on over-ripe white grapes by *Botrytis cinerea* has been utilized to produce special types of wine —Sauternes. The process is called "pourriture noble" in the wine industry. For optimal activity of the fungus a fairly humid atmosphere alternating with periods of sunshine is required. The effect of the fungal attack leads to a greater reduction of acid, particularly tartaric acid, than of sugar as will be seen in Table XXI; this results in a sweet wine.

X. GRAPE TECHNOLOGY

Grape technology is complex since grapes form the raw materials for several industries: fermentation into wines; production of alcohols by distillation of the fermented products; preparation of grape juice, and dehydration to produce raisins. There is also the problem of the storage of dessert grapes in order to extend the marketing period.

Each of these processes involves specific biochemical problems.

A. Vinification

This includes all the operations necessary to transform grapes and their juice into wine. These are many and varied and, apart from the modern development of special machinery for wine production, there remains the fact that grape growing for the production of wine still depends on craftsmanship and the maintenance of the distinctive quality of wines from the various regions. This makes rationalization of the industry difficult. Moreover, the wine grower is not entirely master of the quality of grapes, which is variable from year to year and from place to place. Neither can he control nor modify the nature of the active microflora to any great extent.

Since the production of wine forms the major industrial use of the grape it is pertinent here to discuss briefly the biochemical processes involved in the transformation of grape juice into wine.

The production of red wine involves the components of the skin and seeds as well as the pulp. These components, especially the phenolics, give the characteristic colour and flavour of the product. The first operations are the pre-fermentation processes which liberate the cell sap and allow the various enzymes and substrates to come into contact with one another. Pectolytic enzymes and phenolases are responsible for most of the visible changes but other enzyme systems also appear to be involved (Dourmichidzé, 1967). Under anaerobic conditions, the crushed grapes undergo an intracellular fermentation without the participation of yeasts which yields 2–3% of alcohol. The concentration of malic acid decreases and secondary productions of fermentation appear (Peynaud and Guimberteau, 1962). This is known as "carbonic maceration" and the characteristics of red wine depend on this process.

The control of subsequent alcoholic fermentation by yeasts is dependent on providing the best conditions for the complete fermentation of the sugars. This involves temperature control and the prevention of "abnormal" fermentation by unsuitable bacteria. The best red wines are not, as was thought from the work of Pasteur, due entirely to the action of the alcoholic fermentation. This must be followed by the fermentation of malic acid by lactic bacteria (the malolactic fermentation) which are able to multiply at a low pH and in the presence of alcohol. This process is particularly important since the quality of red wines is dependent on the decrease in acidity thus brought about.

White wines are the result of fermentation brought about by the fermentation of grape juice only. The process of production seeks to avoid direct or enzymic breakdown of the components of the skin, seed or stalk, and, therefore, pressing and filtration precede the fermentation processes. While red wines tend to have a relatively constant character, white wines exhibit a variety of characteristics—mildly or strongly aromatic, dry, medium dry or sweet. They may be "natural", self or artificially carbonated (sparkling). These characteristics are achieved by using unripe, ripe, over-ripe or grapes attacked by fungi (see above). The initial processes are carried through rapidly to reduce the time of contact between the juice and the solid parts of the grape. Malolactic fermentation is kept to a minimum so that white wines still contain malic acid.

Champagne and other sparkling wines may be carbonated under pressure or allowed to continue fermentation (to produce CO_2) in the bottle. Dessert wines may be fortified by added alcohol so that they retain some of the sugars present in the grapes. They develop their special quality by being allowed to undergo a prolonged, slow oxidation. Particularly dry wines are maintained under a surface layer of mycoderma yeasts which yield aldehydes.

Finally, the lees and marcs from wine-making are distilled to give various types of brandy.

B. Grape Juice

Non-alcoholic grape juice is a commercial product made from a whole variety of carefully selected, sound grapes to give a balance between acidity and sweetness. Treatment of the juice with SO_2 immediately after the crushing of the grapes is followed by a period of storage followed by filtration. This is followed by removal of SO_2 by heating and, finally, bottling.

A superior, but more costly type of grape juice is made under sterile conditions, thus eliminating the need for sulphiting. The final product is flash-pasteurized and stored in sterile tanks under an atmosphere of nitrogen. Bottling is then carried out under sterile conditions.

C. Storage of Dessert Grapes

Grapes are fruits which do not continue to ripen when detached from the vine but, owing to their fragile nature, the prevention of deterioration after harvesting is difficult. Senescence in grapes is mainly a question of withering and attack by fungi. Late varieties of large grapes firmly attached to the pedicels and having a sugar content of at least 20% are selected for preservation. The bunches of grapes are stored in bags to protect them against insects and fungal diseases. Fungistatic agents may be included in the bags and in this way the grapes may be stored for several weeks after harvesting. In the Thomery method, mainly used for the Chasselas variety, the cut ends of the bunches are placed in tanks of water followed by hanging in cold, well-aerated rooms.

The development of large co-operative stores has led to the use of cold storage (1°C) methods where the temperature must be kept constant to avoid condensation of water on the surface of the grapes which would lead to rapid fungal attack. To prevent shrivelling and loss of weight the humidity of the stores should be 85–95%. Solutions of SO_2 are sometimes used to sterilize the surface of grapes for storage. The preservation of grapes by controlled atmosphere storage has not as yet been investigated.

REFERENCES

Alquier-Bouffard, A. and Carles, J. (1963). *C. r. hebd. Acad. Sci., Paris* **256,** 3642.

Bertrand, A., Boidron, J. N. and Riberéau-Gayon, P. (1967). *Bull. Soc. chim. Fr.* **9,** 3149.

Biale, J. B. (1960a). *Adv. Fd Research.* **10,** 293.

Biale, J. B. (1960b). "Handbuch der Pflanzenphysiologie", vol. XII, p. 536. Springer Verlag, Berlin.

Bonastre, J. (1959). *Contribution à l'étude des matières minérales des vins.* Thèse Sciences, Bordeaux.

Bouard, J. (1960). *Recherches sur l'aoûtement des sarments de vigne.* Thèse de 3ème cycle, Bordeaux.

Bouard, J. (1966). *Recherches physiologiques sur la vigne et en particulier sur l'aoûtement des sarments.* Thèse Sciences Naturelles, Bordeaux.

Bouard, J. (1967). *In* "IIème Symposium intern. d'Oenologie", p. 53. 13–17 juin, Bordeaux. INRA, Paris.

Büchi, W. and Deuel, H. (1954). *Helvetica chim. Acta* **37,** 1392.

Carles, J. (1962). *C. r. hebd. Acad. Sci., Paris* **225,** 761.

Colagrande, O. (1959). *Annls Microbiol.* **9,** 62.

Cordonnier, R. (1956). *Annls Technol. agric.* **5,** 75.

Davies, D. D., Giovanelli, J. and Aprecs, T. (1964). "*Plant Biochemistry*'. Blackwell Scientific Publications, Oxford.

Dourmichidzé, S. V. (1967). *Question de biochimie appliquée au traitement du raisin. In* "IIème Symposium intern. d'Oenologie", p. 55. 13–17 juin, Bordeaux. INRA, Paris.

Drawert, F. and Stephan, H. (1965). *Vitis* **5,** 27.

Dupuy, P. (1960). *Annls Technol. agric.* **2,** 139.

Gatet, L. (1939). *Annls Phys. Physicochim. biol.* **15,** 984.

Geisler, G. and Radler, F. (1963). *Ber. dt. bot. Gesell.* **76,** 112.

Gerber, C. (1897). *Recherches sur la maturation des fruits charnus.* Thèse Sciences Naturelles, Paris.

Gore, H. C. (1911). *U.S. Dept. Agr. Bur. Chem. Bull.* **142,** 5.

Guichard, C. (1953). *Contribution à l'étude des glucides de la vigne et de certains fruits.* Thèse Sciences Naturelles, Bordeaux.

Gyr, J. (1960). *C. r. hebd. Acad. Sci., Paris* **251,** 263.

Hale, C. R. (1962). *Nature, Lond.* **195,** 917.

Hardy, P. J. (1968). *Pl. Physiol., Lancaster* **43,** 224.

Hawker, J. S. (1969). *Phytochemistry* **8,** 19.

Iwahori, S., Weaver, R. J. and Pool, R. M. (1968). *Pl. Physiol.* **43,** 333.

Karlson, P. (1964). "Biochimie." Doin, Paris.

Kliewer, W. M. (1964). *Pl. Physiol., Lancaster* **39,** 869.

Kliewer, W. M. (1966). *Pl. Physiol., Lancaster* **41,** 923.

Lafon-Lafourcade, S. (1962). Unpublished.

Lafon-Lafourcade, S. and Peynaud, E. (1959). *Vitis* **2,** 53.

Lefèbvre, A. (1968). *Recherches sur la biogénèse de l'acide tartrique chez Vitis vinifera* L. Thèse 3ème cycle, Bordeaux.

Luthra, J. C. and Chima, T. S. (1931). *Indian J. agric. Sci.* **1,** 695.

Moreau, L. and Vinet, E. (1932). *Annls agron.* p. 363.

Peri, C. (1967). *In* "IIème Symposium intern. d'Oenologie", p. 47. 13–17 juin, Bordeaux. INRA, Paris.

Peynaud, E. (1939). *Rev. Vitic.* **90,** 189.

Peynaud, E. (1946). *Contribution à l'étude biochimique de la maturation du raisin et de la composition des vins*. Thèse Ingénieur-Docteur, Bordeaux.
Peynaud, E. (1952). *Annls Falsif. Fraudes* **45**, 11.
Peynaud, E. and Guimberteau, G. (1962). *Annls Physiol. vég.* **4**, 161.
Peynaud, E. and Lafourcade, S. (1958). *Qualitas Pl. Mater. vég.* **3**, 405.
Peynaud, E. and Maurié, A. (1953). *Annls Technol. agric.* **2**, 15.
Peynaud, E. and Maurié, A. (1956). *Annls Technol. agric.* **4**, 111.
Peynaud, E. and Maurié, A. (1958). *Am. J. Enol. Vitic.* **9**, 32.
Puissant, A. and Leon, H. (1967). *Annls Technol. agric.* **18**, 217.
Radler, F. (1965a). *Nature, Lond.* **207**, 1002.
Radler, F. (1965b). *J. Sci. Fd Agric.* **16**, 638.
Radler, F. (1965c). *Am. J. Enol. Vitic.* **16**, 159.
Riberéau-Gayon, G. (1966). *Etude du métabolisme des glucides, des acides organiques et des acides aminés chez* Vitis vinifera L. Thèse Sciences Physiques, Paris.
Ribéreau-Gayon, G. (1968). *Phytochemistry*. **7**, 1471.
Ribéreau-Gayon, G. and Lefèbvre, A. (1967). *C. r. hebd. Acad. Sci., Paris* **264**, D, 1112.
Ribéreau-Gayon, J. and Peynaud, E. (1960). "Traité d'Oenologie", Vol. I. Dunod, Paris.
Ribéreau-Gayon, P. (1953). *C. r. hebd. Acad. Sci., Paris* **39**, 800.
Ribéreau-Gayon, P. (1963). *Ind. Alim. Agric., Paris* **80**, 1079.
Ribéreau-Gayon, P. (1964). "Les Composés phénoliques du Raisin et du Vin". Institut National de la Recherche agronomique, Paris.
Ribéreau-Gayon, P. (1968). "Les Composés phénoliques des Végétaux". Dunod, Paris.
Ribéreau-Gayon, P. and Ribéreau-Gayon, G. (1958). *Bull. Soc. Physiol. végét.* **4**, 57.
Saulnier-Blache, P. (1963). *Annls Physiol. vég.* **5**, 217.
Saulnier-Blache, P. and Bruzeau, F. (1967). *Annls Physiol. vég.* **9**, 179.
Stafford, H. A. and Loewus, F. A. (1958). *Pl. Physiol., Lancaster* **33**, 194.
Vickery, H. B. and Palmer, J. K. (1954). *J. biol. chem.* **207**, 275.
Webb, A. D., Kepner, R. E. and Maggiora, L. (1966). *Am. J. Enol. Vitic.* **17**, 247.

Chapter 5

Melons

HARLAN K. PRATT

Department of Vegetable Crops, University of California, Davis, California, U.S.A.

I Introduction 207
II Origins 209
III Botanical Aspects 209
IV Growth and Development 210
V Growth Regulation 212
VI Maturity and Quality 213
VII Composition 213
A. Carbohydrates 214
B. Flesh Pigments 217
C. Organic Acids 219
D. Amino Acids and Proteins 220
E. Miscellaneous Organic Compounds 220
VIII Enzymes 221
IX Respiration, Ethylene Production and the Initiation of Ripening .. 222
X Tissue Investigations 228
XI Summary 228
References 229

I. INTRODUCTION

The Cucurbitaceae have been treated extensively in a book by Whitaker and Davis (1962); all members of the family are discussed and the historical, botanical and genetic aspects of the group are treated in detail. The restriction for the purposes of this book to the group eaten as mature dessert fruits, the melons, is rather arbitrary, since available evidence makes it clear that the non-dessert fruit in this family (for example, cucumbers, squashes and pumpkins) are as comparable to other kinds of fleshy fruits in their biochemical and physiological phenomena as are the melons. Nevertheless we will confine our attention to two genera, *Cucumis* (muskmelon) and *Citrullus* (watermelon)—Smith and Welch (1964) have been referred to as our taxonomic authority.

Good illustrations of important American cultivars of muskmelon are provided by Davis *et al.* (1965), Doolittle *et al.* (1961) and (1962) for watermelon; these references also describe cultural and marketing practices, including production problems with insects and diseases. Pentzer *et al.* (1940) presented a comprehensive study of cantaloupe handling and shipping and Rattray (1938, 1939, 1940) studied Honeydew handling extensively. Lutz and Hardenburg (1968) summarize current recommendations for storage and shipping of all melons. Market diseases of melons have been treated by Ramsey and Smith (1961) and by Wiant (1938). It should be noted that a principal cause of market disease, especially in the Honeydew muskmelon, is chilling injury; the decay symptoms appear when the fruit has been weakened by an overlong period at chilling temperatures (Rattray, 1938). This cultivar is very resistant to decay organisms as long as it has been well handled.

The muskmelons have been relatively little studied biochemically which is unfortunate since they offer many advantages as biochemical and physiological research material, at least for investigators located in climates suitable for melon production. Well-defined yet very diverse cultivars are available with many simple genetic differences that could be biochemically investigated and exploited. Culture is relatively simple, if space consuming, and quite uniform material of known physiological age can be obtained by blossom tagging (McGlasson and Pratt, 1963) and can be made available by successive plantings over a long growing season. It is also unfortunate that, in common with comparable work on some other fruit, much detailed work on melons loses value because it was not done on taxonomically and physiologically authentic samples. For example, one worker who performed a very detailed analysis says, "the muskmelons used, which were the common orange-fleshed variety, were obtained in September at a local market". How much better it would have been if he had spent only a few hours more in co-operation with a horticulturist or physiologist and secured authentic samples!

American work purporting to be done on "muskmelon" or "cantaloupe", and generally that on "melon", probably refers to a cultivar (variety) of the *reticulatus* group of *Cucumis melo*, the netted, orange-fleshed melons, and this is the sense in which the term "cantaloupe" will be used in this chapter. The number of physiological and biochemical differences between the cultivars of this group is not known, but when we compared widely differing cultivars of *C. melo* in our laboratory, we found very striking differences among them in growth pattern and ripening habit as well as in the obvious differences in colour, flavour, shape, etc. (for example, PMR-cantaloupe, Honeydew, Persian, Crenshaw and Casaba).

Watermelons (*Citrullus lanatus* (Thunb.) Mansf.) show less diversity than do the muskmelons and have received less study, probably because they are a crop of relatively low unit value and show little obvious physiological change during marketing. Many cultivars exist with different adaptations to areas of production.

Because of the vast array of cultivars of the muskmelon and watermelon, many obviously of local interest only, it will be best generally to refer to findings about these two species as general classes. Only if sufficiently detailed work has been done on a cultivar of considerable commercial importance will that cultivar be designated, but work with various cultivars will be reported as needed to show the great diversity and contrasts which exist among what must be closely related types.

II. ORIGINS

In the summary of Whitaker and Davis (1962) it is suggested that watermelon is indigenous to tropical Africa, but it has been cultivated in the Mediterranean region for centuries and was introduced very early into Asia, a strong secondary centre of diversification developing in India. The origin of the muskmelons has not been resolved so clearly, but an African origin is likely; the species was introduced into Asia later than the watermelon, but even more striking secondary centres of variability developed in India, Persia, southern Russia and China. The extreme variability has led to naming of subspecies and botanical varieties.

III. BOTANICAL ASPECTS

The muskmelons share the genus *Cucumis* with cucumbers and gherkins. The species *Cucumis melo* L. is very diverse and includes many types of minor importance. Naudin (Whitaker and Davis, 1962) divided the species into several botanical varieties, but Smith and Welch (1964) considered that these do not exist in nature and reduced them to groups for horticultural convenience. The latter is undoubtedly valid since the cultivars of the two most important groups (*reticulatus* and *inodorus*) intercross freely. One might go even further and drop the *inodorus* group altogether, since the horticultural basis on which it has been retained (storability, lack of aroma) has no separate validity if the melons of the *inodorus* group are properly handled and ripened. Cultivars of the *cantaloupensis* group are not grown commercially in the United States and have been relatively little studied. The other groups recognized in *C. melo* are rare in commerce and are not used as fresh dessert fruit.

Cucumis and *Citrullus* are generally monoecious but may be andromonoecious. The flowering and fruiting habits have been well described by Jones and Rosa (1928) and McGlasson and Pratt (1963).

The fruits of the cultivated melons show a considerable range in size and shape, and the largest must be among the largest fruit known—watermelons may weight as much as 25 kg. The melon is classed as an inferior berry or "pepo"; it is indehiscent, with the fleshy floral tube adnate to the pericarp. Melon fruits may be spherical, oval, oblong, oblate and all possible intermediates; they may have solid flesh derived from the placentae as in the watermelon (*Citrullus lanatus*) or have a central cavity as in the muskmelons (*Cucumis melo*), the edible flesh being derived from the pericarp. The exterior may be smooth, netted or ribbed. Immature fruits are usually green; they may remain green or turn yellow or reddish-brown at maturity.

Very little has been published on the anatomy of melons. Partial cross-sections of the fruit pericarp have been presented by Winton and Winton (1935) and Matienko (1956). Kenny and Porter (1940) sought to relate relative rind toughness in watermelon to anatomical differences between cultivars.

IV. GROWTH AND DEVELOPMENT

The first published curve for muskmelon fruit growth is probably that of Gustafson (1926); the kinds of fruit he studied showed a simple sigmoid growth curve similar to those of other plant organs that had been investigated earlier. Other workers have confirmed this generality (Masuda and Kodera, 1953; Masuda and Hayashi, 1959; Liang, 1961; McGlasson and Pratt, 1963). In our laboratory, we have extended this work to cover some of the diverse cultivars important in North America. While the patterns of all the major cultivars are generally similar, there are intriguing differences that relate to the diverse growth and ripening physiology of these melons and have practical implications (Fig. 1).

For the normal growth pattern to be achieved, adequate pollination must be followed by fertilization and normal development of the ovules. Mann (1943) showed that for proper shape development of watermelon, all three parts of the stigma must be adequately pollinated as there is relatively little crossing from one stigma into an adjacent carpel. In muskmelon, ovaries that will finally fail to set may grow several-fold during a few days after anthesis; growth ceases in 4–8 days and the young fruit turns yellow and abscises within another 8–10 days (Mann and Robinson, 1950). McGlasson and Pratt (1963) give a diagram of a typical melon vine showing the pattern of fruit set and flower abortion, and Baksay (1967) discusses further the relations of vine growth and fruit set.

Of the types illustrated in Fig. 1, only cantaloupe regularly develops an abscission zone and falls from the vine, and it is interesting to note that this melon is still growing rapidly up to the time of abscission (Leeper, 1951; McGlasson and Pratt, 1963). Honeydew, in contrast, develops more rapidly at first and then almost ceases to grow until it ripens and senesces (see also

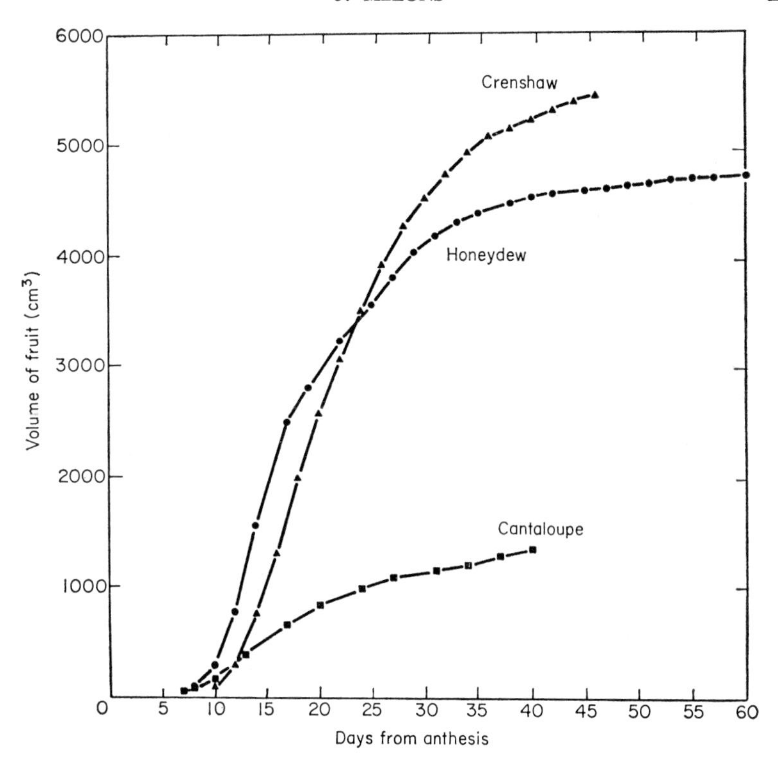

FIG. 1. Growth curves for three cultivars of muskmelon. Each point is the average of five fruit. (Pratt and Goeschl, unpublished.)

Figs. 2 and 7); no abscission zone develops in this and many other cultivars until the fruit is commercially over-ripe. Masuda and Hayashi (1959) compared several dimensions of growth in Honeydew; all give the same general pattern that can easily be obtained by a simple tape measurement of the equatorial circumference (McGlasson and Pratt, 1963). Avakyan (1957) weighed fruits of three cultivars twice daily and showed that the principal weight gain occurred at night during the 15 days of early, most rapid growth. Schroeder (1965) presented interesting data on diurnal temperature fluctuations in muskmelons and watermelons and showed large temperature gradients within these fruits when grown in a hot climate. The pattern of growth in watermelon is similar to that of the muskmelon types (Mizuno and Pratt, unpublished). Practical problems of sizing fruits for market have led to studies of weight, volume and density of muskmelons (Hoffman, 1940; Currence *et al.* 1944; Davis *et al.*, 1964; Kasmire, 1968) and watermelons (Showalter, 1961).

Masuda and Hayashi (1959) estimated that cell division finished in 3–5 days after anthesis in the cultivars they studied (Honeydew and New melon). During the period of most rapid growth, when about half the final size is reached, there is a transition stage before which the fruit is clearly immature and after which characteristics of maturity are assumed. In PMR-45 cantaloupe, the "net" develops at this time (McGlasson and Pratt, 1963) and biochemical developments occur (Rowan *et al.*, 1969).

That the quality and development of a given cultivar of muskmelon may vary from locality to locality is well known. Many authors have studied ripe melon fruits, comparing different cultivars and regions of production, and showing the effects of cultural practices (irrigation, soil salinity, fertilization, spacing, pesticides) on the content of sugar, vitamins and minerals, as well as yield. Many papers on these subjects are readily found in the abstracting journals. It is safe to say, as with most other kinds of crops, that the treatments which produce the highest yields from healthy plants are likely to produce the highest quality fruit. Davis *et al.* (1964, 1967) made a detailed statistical study of various growth phenomena and correlations in cantaloupe (PMR-45) fields from many areas of the Central Valley of California. They found that healthy melons reaching a length of 5–8 cm will almost certainly survive and develop, and in general the ultimate total time required for maturation, ultimate size and shape, and flesh thickness and proportion are established during this early growth stage.

V. GROWTH REGULATION

Surprisingly little has been done in this field with melons. Watermelon juice contains gibberellin-like factors (Maheshwari and Bhalla, 1966) and kinin-like activity (Maheshwari and Prakash, 1966). Ferenczy (1957) determined that watermelon extract contained indole-3-acetic acid and a "β-inhibitor" of *Avena* coleoptile growth.

Parthenocarpic fruit were obtained in several cultivars of watermelon by applying growth regulators to the cut styles (Wong, 1941). Cultivars varied in response, but some yielded seedless fruit of normal size and quality; seed coats developed to various degrees in fruits devoid of functional seeds. Bini and Raddi (1965) had less success; treatment with NAA, IAA, or 2,4-D reduced seed formation, but they got no seedless fruit.

The pattern of male and female flower production in the Cucurbitaceae can be altered toward increased femaleness by application of growth regulators. Halevy and Rudich (1967) showed that N,N-dimethylaminosuccinamic acid (Alan, B9) had this effect in muskmelon and we have found similar effects from field application of 2-chloroethylphosphonic acid (Ethrel) (unpublished). Davis *et al.* (1970) showed that Ethrel treatment of young plants affected the ultimate harvest date of cantaloupe fruits.

VI. MATURITY AND QUALITY

Maturity and quality in muskmelons are usually evaluated by sugar content; not only is adequate sugar content of paramount importance for good quality, but a particular sugar content may mark the physiological attainment of maturity, the state at which the fruit will ripen satisfactorily even after harvest. For example, the Agricultural Code of California requires that Honeydew melons shall have 10% sugar for legal sale; in that cultivar this concentration marks very closely the physiological or biochemical state which makes subsequent ripening possible (Davis *et al.*, 1965). Cantaloupe varieties generally develop an abscission layer at maturity which provides a good index for harvest (Doolittle *et al.*, 1961; Davis *et al.*, 1965). For example PMR-45 cantaloupe abscises from the vine at about the peak of the respiratory climacteric; the processes of ripening are well initiated, but the fruit is still at a satisfactory state for handling and shipping to distant markets. If such melons are to be stored (including any transit period), harvest at the optimum maturity is essential. Cantaloupes harvested before the abscission layer is fully developed will never develop the same flavour as those left for the full period. On the other hand, melons left too long in the field will have exhausted some of their storage life before harvest and the best flavour will be lost before completion of the marketing process (Hoover, 1955; Ogle and Christopher, 1957).

Attributes of muskmelon eating quality include sweetness (sugar content and sugar : acid balance), flavour (content of characteristic organic volatiles) and flesh texture (Gilbart and Dedolph, 1963). For commercial handling, other attributes are of interest: skin toughness, maturity changes that make selection easy to harvest time, and resistance to insects and disease.

In watermelon, the principal requirements for good eating quality are flesh crispness, good flesh colour and sweetness. Nip *et al.* (1968) have reviewed efforts to evaluate watermelon quality objectively. Specific gravity (Markov, 1956; Showalter, 1961) reflects maturity but is too variable for practical use. Colour of the ground spot is probably one of the best objective maturity tests for field use. For commercial handling of watermelons, rind toughness is an important attribute since the watermelon is usually its own "package". Objective tests of breakage resistance have been useful in breeding programmes (Ivanoff, 1954: Spurr and Davis, 1960).

VII. COMPOSITION

While a number of workers around the world have reported on melon composition, there have been relatively few studies providing the thorough and comprehensive standard that has characterized work on fruit such as the apple and pear. Numerous papers, often oriented toward human nutrition, have reported on sugar and vitamin contents, but unless careful attention has been paid to fruit age and cultivar, these data are of limited interest today and contribute almost nothing to our understanding of fruit biochemistry and physiology in general or that of melons in particular. Accordingly, most of these papers (to which reference is readily found in "Horticultural Abstracts" and "Chemical Abstracts") will not be cited in this chapter. In a number of works which have been quoted by others, especially older papers, the stated sugar contents have been so low that one must assume the melons studied were immature or of inferior quality, and the analyses of other components must therefore be equally suspect.

An exception is the study by Howard *et al.* (1962). In this study, Table I, reliable samples were taken fresh from the field or from sources of fresh supply, so that the time elapsed and the handling after harvest were known and satisfactory. This approach is necessary to increase reliability and to eliminate uncontrolled changes after harvest. Their values for watermelon confirm vitamin values reported earlier by French *et al.* (1951).

TABLE I. Composition of various melon cultivars[a]

	Cantaloupe	Casaba	Honeydew	Watermelon
Refuse (g)	45	45	45	50
Water (g)	90	92	87	90
Protein (g)	1·0	0·6	0·9	0 6
Fat (g)	0·1	0·1	0·1	0·1
Total sugar (g)	7·0	6·2	10·1	9·0
Other carbohydrate (g)	0·2	0·1	0·2	0·1
Vitamin A, I.U.	4200	tr.	500	300
Thiamine (mg)	0·06	0·06	0·06	0 08
Riboflavin (mg)	0·02	0·02	0·02	0·02
Niacin (mg)	0·9	0·4	0·6	0·2
Vitamin C (mg)	45	19	32	6
Minerals—Ca (mg)	10	5	6	5
Fe (mg)	0·4	0·4	0·2	0·2
Mg (mg)	17	8	10	11
P (mg)	39	7	14	9
K (mg)	330	210	330	130
Na (mg)	20	12	20	5
Calories	27–36	26	41	31–40

From Howard *et al.* (1962).
[a] Values are for 100 g fr. wt. of edible portion.

A. Carbohydrates

1. *Sugars*

Melons are among the sweetest of all fruits. For example, in the flesh of fully ripened Honeydew melons, sugars may comprise as much as 16% of the juice. Sugar accumulation during the development of melons is of especial interest because of the strong correlation between sugar content and subjective fruit quality in many diverse cultivars of muskmelon and in watermelon (Chace *et al.*, 1924; Rosa, 1928; Porter *et al.*, 1940; Currence and Larson, 1941; Hartman and Gaylord, 1941; Gilbart and Dedolph, 1963). This relation is important in enforcement of marketing regulations as well as in evaluations for research. For these purposes sugars have been specifically analysed or have been estimated from hydrometric or refractive index determination of soluble solids; useful information can be readily obtained by squeezing a few drops of juice onto the stage of a hand refractometer (Allinger *et al.*, 1940). It has been shown by several workers, notably by Scott and

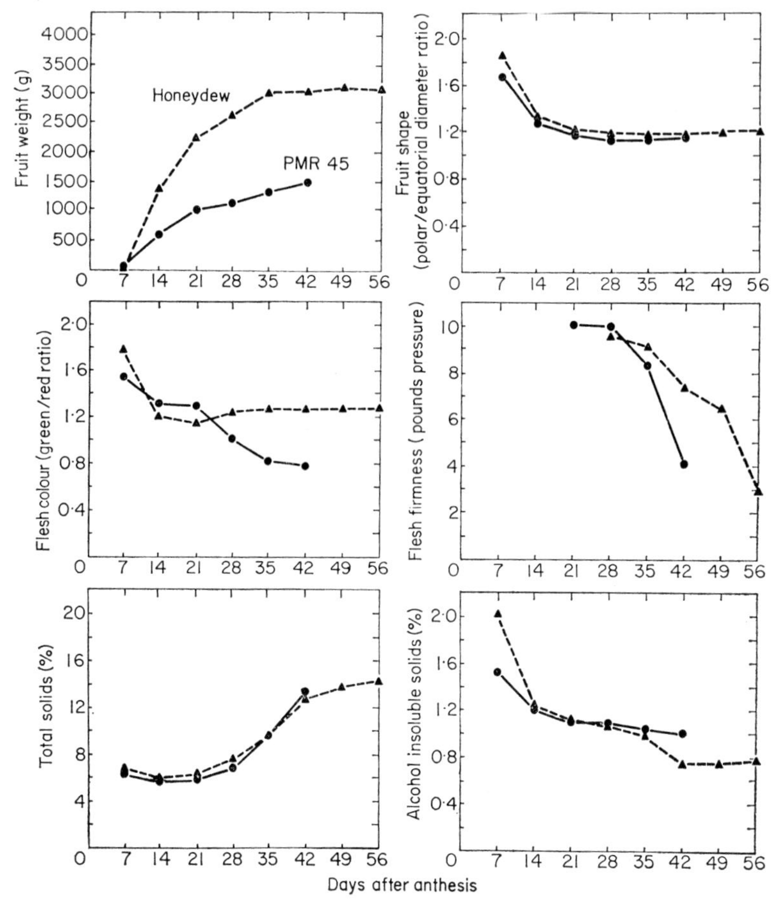

FIG. 2. Changes in characteristics of muskmelon development and ripening as a function of fruit age after pollination. Two cultivars were studied. Each point represents the average of ten fruits. (Bianco, 1969.)

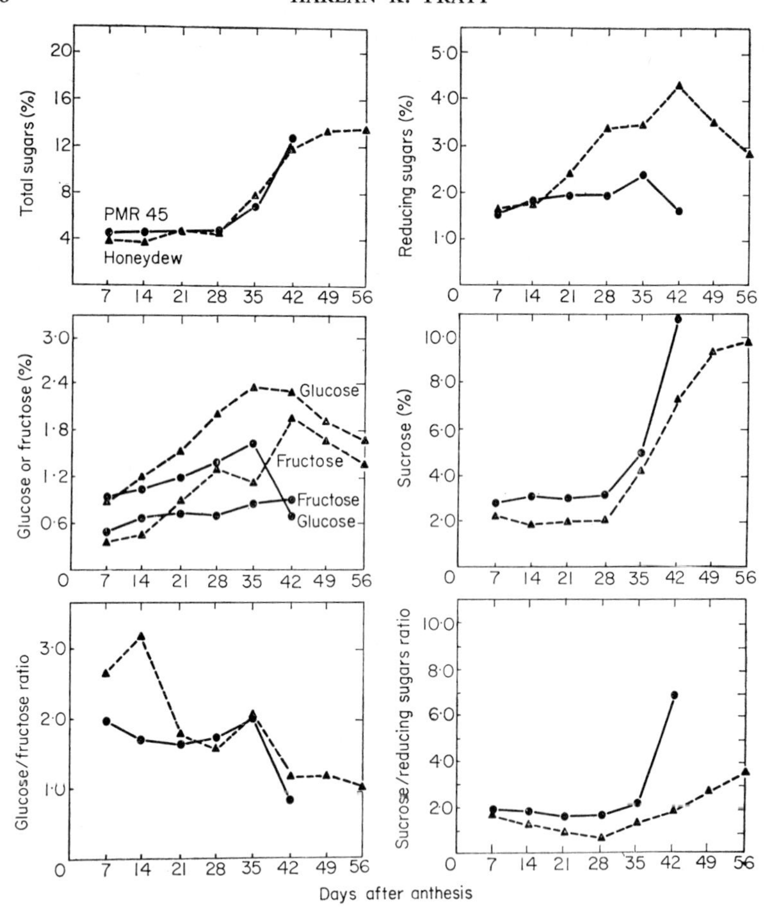

FIG. 3. Changes in sugar fractions in muskmelon as a function of fruit age after pollination. Two cultivars were studied. Each point represents the average of ten fruits. (Bianco, 1969.)

MacGillivray (1940) for cantaloupe, casaba, Honeydew and Persian, and by MacGillivray (1947) for watermelon, that differences exist in the sugar contents of different portions of these large fruits. We have found the best practice for convenient and relatively uniform sampling of melons is to cut a transverse slice and to take three plugs from the pulp between the points of placental attachment.

As pointed out above, many authors have shown relations between cultural factors and sugar content. We will confine our attention primarily to those who have reported sugar analyses in relation to stages of melon development.

The most comprehensive study is that of Bianco (1969) using two important cultivars of muskmelon. Growth and development of the fruit (weight, shape, softening and flesh colour changes) are compared with several characteristics of the sugar content (Figs. 2 and 3). The earlier work on muskmelons by Rosa (1928), Lü and Wang (1959), Lindner *et al.* (1963), Davis *et al.* (1964, 1967) and others is confirmed and extended over the total period of fruit development. The pattern of accumulation of glucose and fructose with subsequent conversion to sucrose in muskmelon is similar to that observed in watermelon (Iwanov *et al.*, 1929; Porter *et al.*, 1940; Lindner *et al.*, 1963) although fructose is relatively higher in the latter fruit. Biochemical aspects of the sugar transformations were investigated by Lü *et al.* (1962); the rapid synthesis of sucrose as the melons matured was correlated with high rates of formation of high-energy phosphate in the later stages of muskmelon development.

The question of whether ethylene treatment will increase the sugar content of melons occasionally arises when Honeydew melons have been harvested at less than 10% sugar, the legal limit in California. Since melons have essentially no starch reserve, there is no source of carbohydrate to convert to sugar, and the answer must be "no". Melons which are mature enough to harvest will respond to ethylene by ripening, but there is no significant change in the total sugar content (Bianco, 1969).

2. *Other carbohydrates*

Rosa (1928) examined changes in pectic substances during ripening and storage of muskmelons and watermelons. In muskmelons, total pectic substances ranged from 0·24 to 0·35% of the fresh weight, but watermelons contained only 0·09–0·10%. As the muskmelon fruits matured on the vine or ripened in storage, the total pectic substances decreased (confirmed by Tityapova, 1963), but the percentage in the soluble form increased; marked softening of the flesh is characteristic of most muskmelon cultivars as they ripen. Although there have been a number of more recent investigations of pectins in melons, they have been concerned primarily with melons as sources of commercial pectin, especially some non-dessert watermelon cultivars (Mel'nik and Arasimovich, 1962); little additional work has been done on pectic changes as biochemical or physiological phenomena of fruit growth and development in table varieties.

Tityapova (1963) reported on the cellulose content of a number of muskmelon cultivars. Carbohydrate contents of melons have been reviewed by Arasimovich (1966, muskmelon) and Arasimovich and Savchenko (1966, watermelon).

B. Flesh Pigments

The muskmelons are noted for their orange flesh, an important quality characteristic, and both red and orange flesh are common in watermelon cultivars. The predominant pigment of the orange-fleshed muskmelon cultivars is β-carotene (Vavich and Kemmerer, 1950) and Rabourn and Quackenbush (1953) identified phytofluene and phytoene. A detailed analysis of an unknown cultivar was presented by Curl (1966) as follows: Total carotenoid as β-carotene, 20·2 mg/kg fresh weight; β-carotene (84·7% of total carotenoids present), δ-carotene (6·8%), α-carotene (1·2%), phytofluene (2·4%), phytoene (1·5%), lutein (1·0%), violaxanthin (0·9%) and traces of other carotenoids.

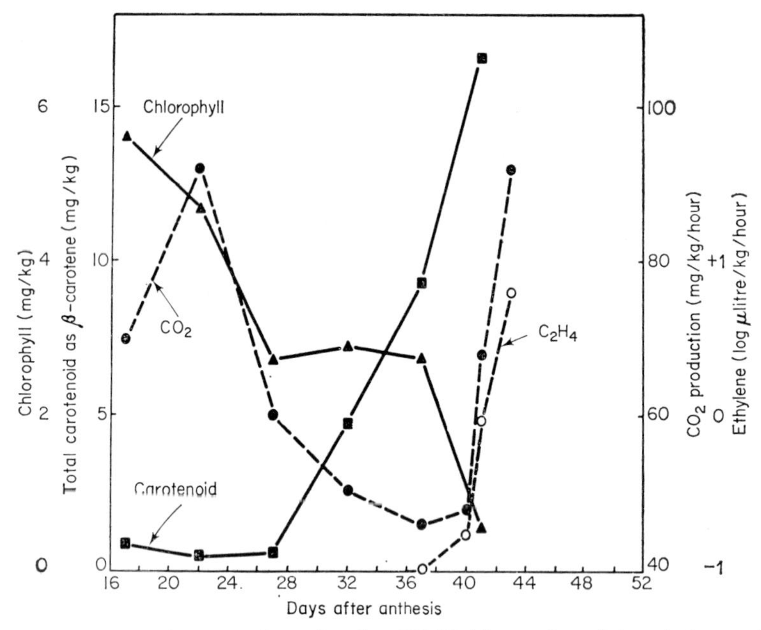

FIG. 4. Changes in pigments of muskmelon (PMR-45 cantaloupe) in relation to fruit age after pollination and to fruit ripening as shown by curves for respiration and ethylene production. Each point represents the average of five fruit; gas exchange values were determined at 20°C (Reid *et al.*, 1970.)

Reid *et al.* (1970) have followed the pigment content of several muskmelon cultivars of known physiological age in relation to other parameters of ripening (Fig. 4). Carotenoids begin to increase at least 10 days prior to the onset of the respiration climacteric. Pigmentation commences in the centre of the fruits and progresses outward through the pericarp until the flesh is uniformly orange at maturity. Hence, the rise in carotenoid content in developing fruits reflects primarily an increase in the amount of tissue containing high levels of carotenoids, rather than an increased concentration per cell. At the climacteric maximum, the carotenoid content of the Persian, PMR-45 cantaloupe and Crenshaw cultivars was shown to be 25, 16·5 and

8 mg/kg of fresh tissue respectively; there are only minor qualitative differences in their pigment composition. Chlorophyll content drops gradually as the fruits develop with a final rapid decline coinciding with ripening. Pigments of the green-fleshed muskmelons appear to have had little investigation, but Smith *et al.* (1944) found only 0·3 mg carotene/kg in Honeydew and 0·2 mg in casaba.

Zechmeister and Tuzson (1930) isolated lycopene and carotene from watermelon, showing that lycopene was responsible for the red colour. Zechmeister and Polgar (1941), with refined techniques, determined that 1 kg fr. wt. of pulp contained 1·0 mg of a complicated xanthophyll mixture, 6·1 mg lycopene, 0·06 mg γ-carotene, 0·46 mg β-carotene, 0·01 mg α-carotene and 0·16 mg unknown carotenoids. More recently Tomes *et al.* (1963) and Tomes and Johnson (1965) have made quantitative determinations on numerous cultivars of mature red- and orange-fleshed watermelons. Pigment contents of 9 red-fleshed cultivars were (in mg/kg fr. wt.): lycopene 12·5–52·4; phytofluene 0·8–2·9; β-carotene 0·4–6·0; ζ-carotene trace–2·2 and γ-carotene trace–0·7; neolycopene A and neurosporene were detected in one cultivar each. Visual red colour depended mainly on the lycopene content. An orange-flesh type was also examined; prolycopene (15·9 mg) was the major pigment and determined the visual flesh colour. Also present were phytoene (5·3), phytofluene (2·0), β-carotene (1·4), ζ-carotene (5·7), proneurosporenes (2·2), lycopene (1·5) and traces of other poly-*cis*-lycopenes. Morgan's results (1967) were similar to those for the red-fleshed type. Matienko (1967) reports that the red watermelon colours reside in chromoplasts containing a system of amorphous and crystalline pigments. He says that the various chromoplasts are generally the same colour, so the resulting flesh colour depends on the pigment mass and the number and distribution of the chromoplasts.

C. Organic Acids

Systematic and comprehensive studies of the organic acid complements of major, well-identified cultivars of melon appear to be lacking. Lindner *et al.* (1963) tested seven cultivars of muskmelon for malic and citric acid; all contained substantial amounts of citric acid but malic was present in only three. Ito and Sagasegawa (1952) and Jurics and Lindner (1965) found citric acid but no malic or tartaric in their cultivars. Jurics (1966) found caffeic acid to be absent from a cultivar of muskmelon. Mori *et al.* (1967) report their muskmelon cultivar to contain acetic, citric, formic, glycolic, malic and oxalic acids, as well as several unseparated or unidentified kinds; levulinic and tartaric acids were absent. In watermelon, six cultivars contained malic acid (Lindner *et al.*, 1963), agreeing with the earlier work of Ito and Sagasegawa (1952). Jurics and Lindner (1965) agreed and reported citric acid to be absent from watermelons. Caffeic acid (3 mg/kg) was found in watermelons by Jurics (1966).

D. Amino Acids and Proteins

Citrulline (δ-carbamido-orithine) was first discovered and identified in watermelon juice by Wada (1930); it has since been found in muskmelons (Inukai *et al.*, 1966). Shinano and Kaya (1957) reported another new amino acid from watermelon juice which they believed to be α-amino-β-(1-imidazolyl) propionic acid.

Orr and Watt (1957) reported that muskmelons contain (per 100 g fr. wt.) 600 mg protein and 1 mg tryptophane, 15 mg lysine and 2 mg methionine. Isawa *et al.* (1938) found arginine in watermelons. Muskmelon juice was shown to contain histamine by García-Blanco and Vento (1944). Hadwiger and Hall (1961) identified a number of amino acids from watermelon rind tissue; the pattern of occurrence was related to the colour pattern of the rind and possibly to disease susceptibility.

Rowan *et al.* (1969) followed protein and total nitrogen changes in cantaloupe fruits as they ripened after harvest; a net synthesis of protein occurred during the climacteric rise in respiration.

E. Miscellaneous Organic Compounds

While quality in melons is very highly correlated with sugar content, a really good melon also has a characteristic flavour which is a function of the organic volatiles that are produced or released as the fruit ripens. While the maximum sugar content will be present at the time of harvest, maximal flavour is a function of the state of ripeness, and in fact further changes in flavour volatiles, or their presence in unpleasant excess, often characterize the over-ripe fruit. Unfortunately very little is known of either the nature or the course of development of melon flavour constituents.

Rakitin (1935, 1945) identified acetaldehyde and ethanol in muskmelons and showed that the concentration of each increased during ripening. Serini (1957) measured the content of acetoin and 2,3-butylene glycol in a muskmelon cultivar ripened after harvest. Butylene glycol was present (5·5 mg/kg pulp) at harvest; the concentration rose to 28·5 mg at ideal eating ripeness and the substance disappeared when the fruit became slightly over-ripe. Acetoin was absent at harvest, first appearing (14·2 mg/kg) as the fruit passed the ideal eating stage, and rising to 748 mg/kg when the fruit was very over-ripe. Both compounds were absent from watermelons.

Rowan *et al.* (1969) detected the following nucleotides in extracts of mature cantaloupes, adenosine monophosphate, adenosine diphosphate, adenosine triphosphate, uridine monophosphate, uridine diphosphate, uridine triphosphate and uridine diphosphate-glucose.

As pointed out above, a large number of papers have reported on the vitamin content of melons. Chen and Schuck (1951), re-examined earlier analytical results and compared the diketogulonic acid, dehydroascorbic acid and ascorbic acid content of four cataloupe cultivars; a summary of their results is given in Table II.

TABLE II. Diketogulonic (DKGA), dehydroascorbic (DHAA) and ascorbic (AA) acid content of melons taken directly from the field (2–10 samples analysed)

Variety	Time	DKGA[a]	DHAA[a] (mg per 100 g)	AA[a]
Cantaloupe P44	Jul. and Sep.	2·46	8·17	33·3
Cantaloupe, Pride of Wisconsin	Aug.	2·60	7·74	41·8
Cantaloupe MR54	Aug.	2·03	7·23	25·4
Cantaloupe, Lafayette Austin	Sep.	2·04	9·69	22·6

[a] Average values—variation was considerable (see Chen and Schuck, 1951).

Bitter principles occur exclusively as glycosides in the fruits of watermelon, coinciding with the absence of the β-glucosidase, elaterase (Enslin *et al.*, 1956). Watermelon fruit can contain as much as 0·1% of α-elaterin, $C_{32}H_{44}O_8$ (Enslin *et al.*, 1957). Glycosides are absent or found only in very small amounts in the fruit of all species of *Cucumis*, all of which possess a high elaterase activity (Rehm *et al.*, 1957). Jurics (1967) examined cantaloupe and watermelon for the presence of (+)-catechol and (−)-epicatechol. She found chlorogenic acid to be absent from both muskmelon and watermelon (Jurics, 1966).

VIII. ENZYMES

Little work has been done on the enzymology of melon fruits. Melon tissue has been tested for the presence of a number of enzymes (Table III), but only Liang (1961) appears to have made a systematic study of enzyme activity during fruit growth and development. Using the cultivar "Bai-Lan" (Honeydew ?), he showed that activity of a flavoprotein and of copper and iron-containing metal enzymes are roughly correlated with the rate of fruit respiration, with peak activity corresponding to the climacteric respiratory peak.

Honeydew melons have an effect on the mucous membranes of sensitive persons, suggesting the presence of a proteolytic enzyme; Berkowitz-Hundert (1966) reported a proteinase of the trypsin type from a melon of unspecified cultivar. Mel'nik (1960) and Mel'nik and Arasimovich (1962) determined

TABLE III. Enzyme activities identified in melon tissue

Enzyme	Presence (+) or Absence (−)	Reference
Muskmelon		
Ascorbic oxidase	+	Aslanyan *et al.* (1960); Blesa and Pretel, (1967); Ito (1938)
Catalase	+	Matui (1940)
Elaterase	+	Enslin *et al.* (1956)
Glucose-6-phosphate dehydrogenase	+	Anderson *et al.* (1952)
Glutathione reductase	+	Anderson *et al.* (1952)
Invertase	+	Arasimovich (1939)
Malic enzyme	+	Anderson *et al.* (1952)
Peroxidase	+	Aslanyan *et al.* (1960)
Phosphoenolpyruvate carboxykinase	+	Mazelis and Vennesland (1957)
Phosphogluconic dehydrogenase	+	Barnett *et al.* (1953)
Polygalacturonase	−	Hobson (1962)
Polyphenol oxidase	+	Aslanyan *et al.* (1960)
Proteinase	+	Berkowitz-Hundert (1966)
"Oxidase" (flavo-enzyme)	+	Liang (1961)
(copper-enzyme)	+	Liang (1961)
(iron-enzyme)	+	Liang (1961)
"Oxidase"	+	Ezell and Gerhardt (1940)
Watermelon		
Ascorbic acid oxidase	+	de Farias and Magalhães Neto (1955)
Elaterase	−	Enslin *et al.* (1956)
Pyruvic carboxylase	+	Vennesland and Felsher (1946)
"Oxidase"	−	Ezell and Gerhardt (1940)

pectic and other enzymes in the "fodder" watermelon or "stock citron", an inedible type.

Recently the enzyme, alcohol dehydrogenase, has been partially purified from the flesh of the Honeydew melon (Rhodes and Wooltorton, 1970). The partially purified preparation shows activity towards both $NADH_2$ and $NADPH_2$ but attempts to resolve these two activities proved unsuccessful. Evidence was presented which showed that the $NADPH_2$-dependent rate of aldehyde oxidation was inhibited by very low concentrations of $NADH_2$.

IX. RESPIRATION, ETHYLENE PRODUCTION AND THE INITIATION OF RIPENING

Respiration of melons in relation to their ripening has been studied relatively little until recent years, perhaps because of the awkward size of the fruits. Gerber (1897) tabulated data for the variety Ananas; when his data are plotted, a respiration climacteric is revealed, corresponding with changes in colour and development of aroma. Gerber was, of course, not aware of the general significance of the climacteric phenomenon in relation to fruit ripening. Pratt (1953) discussed the climacteric in Honeydew melons and showed that it could be induced by ethylene treatment. Since then other workers have clearly established the occurrence of this pattern in melons (Masui *et al.*, 1954; Liang, 1961; Lyons *et al.*, 1962). Several melon cultivars, including watermelon, have been investigated in our laboratory; all are truly climacteric fruits (Fig. 5). Some respiration rates at various temperatures are presented in Table IV; Wardlaw and Leonard (1936) present data on the effect of temperature on the internal CO_2 concentration in watermelon.

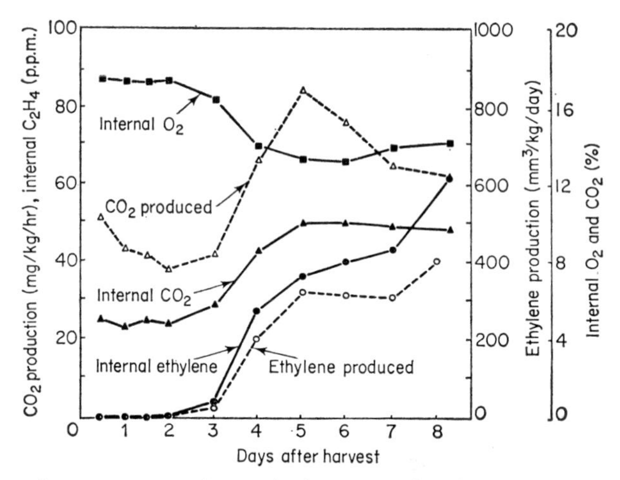

FIG. 5. Gas exchange patterns for a single muskmelon fruit (PMR-45 cantaloupe) harvested mature but unripe and held at 20°C. (Lyons *et al.*, 1962.)

TABLE IV. Approximate rates of respiration of melons at various temperatures (in mg CO_2/kg/hour)[a]

Temp. (°C)	Muskmelon Cantaloupe	Muskmelon Honeydew	Watermelon
32	89	33	44
27	67	30	—
21	48	23	21
16	36	14	—
10	15	8	8
4·5	9	4	4
1	5	—	—

[a] Calculated from data of Scholz *et al.* (1963). The rates were determined immediately after commercial harvest, hence the cantaloupes were probably near the climacteric peak, and the Honeydews were probably mature but unripe.

The first finding that melons produce ethylene must be attributed to Denny and Miller (1935), using a bioassay, and to Christensen *et al.* (1939) who made the first chemical determination of ethylene production by cantaloupe. Morris and Mann (1946) used a bioassay to demonstrate ethylene production by other muskmelon cultivars. Lyons *et al.* (1962) showed that the ethylene-forming mechanism was present in cantaloupe as young as 7 days after anthesis; other cultivars since studied reveal the same phenomenon. The rate of ethylene production remains below the critical level until the fruit is mature (Fig. 6). Some unknown event then brings about an increased rate of ethylene production, and when the internal concentration of ethylene in the fruit reaches the critical concentration (about 3 p.p.m.), the other phenomena of fruit ripening are induced (Pratt and Goeschl, 1968, 1969).

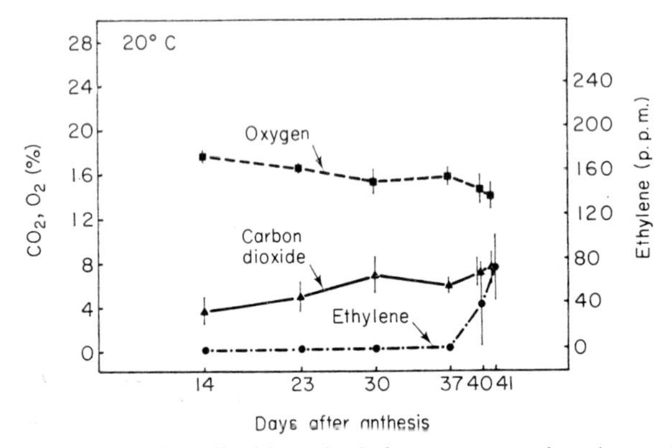

FIG. 6. Internal oxygen, carbon dioxide and ethylene concentrations in muskmelon fruit (PMR-45 cantaloupe) at six stages of maturity. Gas samples were taken from the central cavity shortly after harvest. Vertical lines through each point indicate the estimate of the standard derivation of the population of six fruit. (Lyons *et al.*, 1962.)

As in other fruit, ethylene treatment hastens ripening of pre-climacteric but mature muskmelons (McGlasson and Pratt, 1964a). The response to ethylene is a function of the physiological age of the fruit, tissue temperature, duration of treatment and concentration applied. Field application of 2-chloroethylphosphonic acid (Ethrel) leads to the same result (Kasmire *et al.*, 1970). Stimulation of ripening of most muskmelon cultivars is of no practical use since the fruits are adequately self ripening, and most commercial effort is devoted to delaying ripening by use of suitable low temperatures. However field treatment with Ethrel may find application in unifying ripening for mechanical harvest.

The potential diversity in the muskmelons is suggested by the observed variation in ripening behaviour among the cultivars most intensively studied. The cantaloupe (PMR-45) presents no surprises. The fruit grows steadily until it abscises from the vine (Figs 1 and 2) at which time it is approximately at the climacteric peak, having produced sufficient ethylene to induce self-ripening. In contrast, the Honeydew grows rapidly, reaching approximately its maximum size and mature appearance about 35 days after anthesis (Figs. 1, 2, 7). At this time it will typically have the required 10% soluble solids (Figs. 2 and 7) and will ripen into an acceptable product when ethylene is applied. However, the Honeydew does not develop the capability for self-ripening for several days more (Fig. 7). As a result melons which vary in

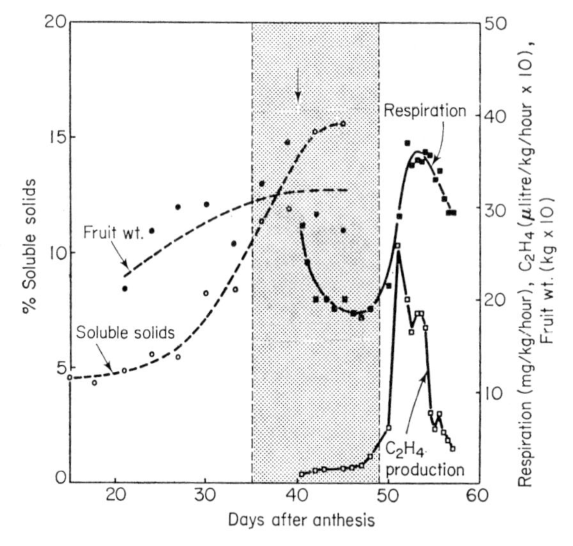

FIG. 7. Growth, development and ripening of the Honeydew muskmelon. For fruit weight and soluble solids content each point represents the harvest of a single fruit. Respiration and ethylene production curves are averages for five fruit harvested at 40 days after pollination and held at 20°C. The shaded area represents the ages of fruit that could appear in any one commercial harvest. (Reid, Murr and Pratt, unpublished.)

physiological age by 10 or more days will be harvested commercially; there is no readily identifiable change in appearance during this time. The physiologically younger melons are likely never to ripen unless given an ethylene treatment, while the physiologically older melons in the same lot may proceed to senescence. It is therefore commercially essential, both for uniformity of the pack and for the benefit of the ultimate consumer, to treat all mature but unripe Honeydews with ethylene before shipment (Pratt, 1953; Pratt and Goeschl, 1968). This has been generally practised in the United States for at least 35 years (although not always to an optimum degree), and one suspects that this cultivar and several other similar types would profit from the same treatment in European and other markets.

The casaba type of muskmelon is known to have exceptionally good keeping quality, albeit its eating quality is inferior. Recent studies (Reid and Pratt, unpublished) make it appear that this melon has an unusually long growing season and is generally harvested (as is the Honeydew) horticulturally mature but unripe. Like Honeydew it has a low ability to self-ripen at this stage, and it appears that the quality of this cultivar could be improved by ethylene treatment.

Both muskmelon and watermelon show an interesting phenomenon that contrasts with several other kinds of tree fruit. If fruit are harvested at various ages after flowering and held at a constant temperature (20°C), they tend to

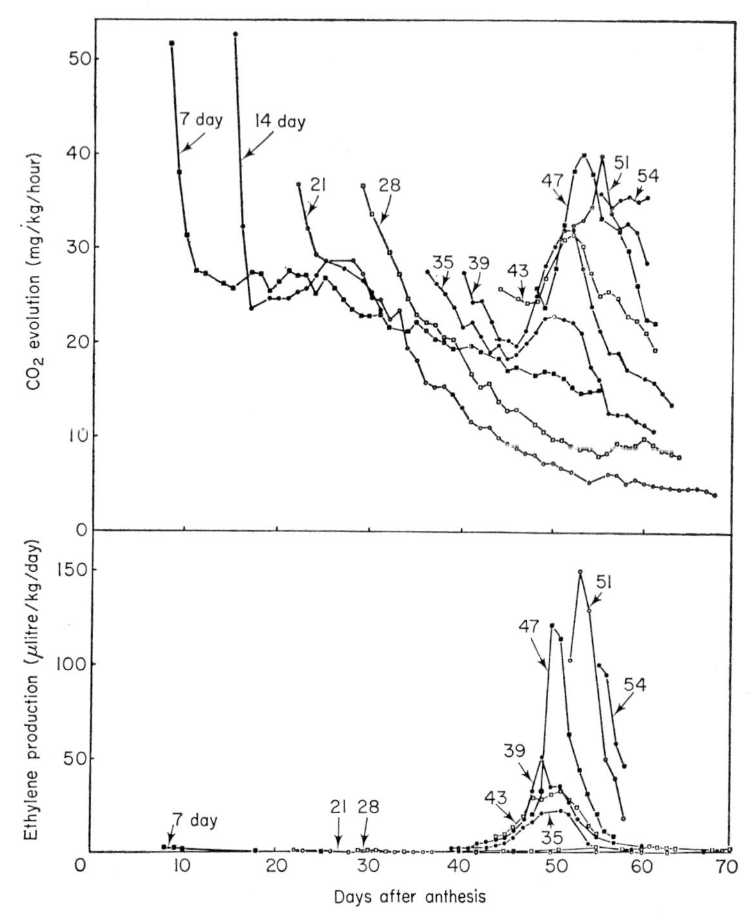

Fig. 8. Respiration and ethylene production by Honeydew muskmelons harvested at different ages after pollination and held at 20°C. Each point is the average of ten individually studied fruit. (Pratt and Geoschl, 1968.)

show increased ethylene production, the rise in respiration, and final senescence at about the same ultimate age after anthesis. This has been shown very clearly in muskmelon (Fig. 8) (McGlasson and Pratt, 1964a; Pratt and Goeschl, 1968) and watermelon (Mizuno and Pratt, unpublished). The phenomenon can also be seen in tomato (Lyons and Pratt, 1964), suggesting that there might be a fundamental difference between fruit from annual plants and from woody perennials (Pratt and Goeschl, 1969).

Little has been done on the biochemical fundamentals of respiration in melons, but Liang and Lü (1964) report that the pentose phosphate pathway of glucose utilization is most important in the early stages of muskmelon fruit development, but later the importance of the glycolytic pathway increases. Lü *et al.* (1962) state that the content of organic phosphates and the rate of high-energy phosphate formation are higher in fruits during their later stage of development. Rowan *et al.* (1969) followed cantaloupe fruits through the climacteric rise and showed that the concentration of adenosine triphosphate (ATP) in the tissue increased while that of adenosine diphosphate (ADP) remained essentially unchanged; thus, there was a net synthesis of adenosine pyrophosphate during this period of rapid increase in respiration.

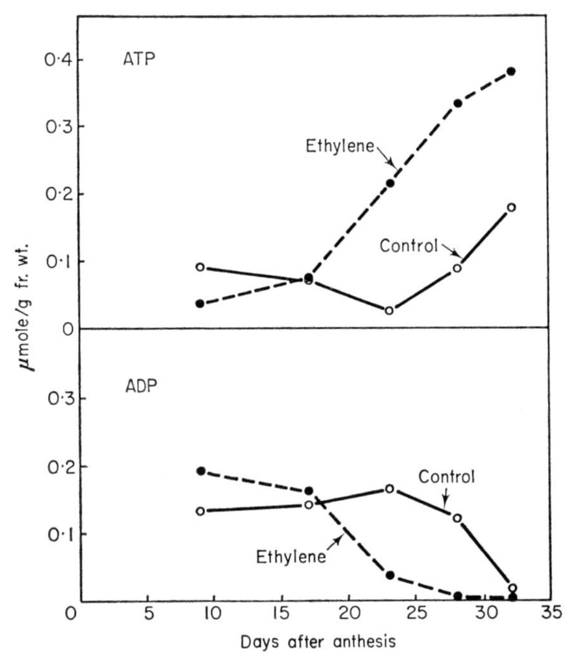

FIG. 9. Changes in concentration of ATP and ADP in muskmelon (PMR-45 cantaloupe) as a function of age and of ethylene treatment Each point represents an individual fruit; treated fruits received 1000 p.p.m. ethylene for 12 hours. (Rowan *et al.*, 1969.)

A net synthesis of protein was positively correlated with this increasing concentration of ATP. Fruits harvested at different ages showed little change in ATP or ADP content until 23 days after harvest; ATP then increased and ADP decreased (Fig. 9). Treatment with ethylene hastened these changes in fruits 23 or more days old, but ethylene had no effect on fruit at 17 days; the effect on younger fruits was in the opposite direction to that on the oldest fruits. The transition occurs during the period of development of the netted surface in cantaloupe, a critical stage in fruit development.

X. TISSUE INVESTIGATIONS

No published physiological or biochemical work on cells or cell organelles of melon fruit has come to my attention; there has been little work on tissue slices. McGlasson and Pratt (1964b) used cantaloupe (PMR-45) tissue slices to examine the effects of wounding on respiration, ethylene production and the effects of applied ethylene. Cutting caused an immediate increase in the respiration rate of flesh slices compared to that of intact fruits. The rates subsequently declined until the onset of tissue breakdown; only slices from fruits harvested more than 20 days after anthesis showed a climacteric respiratory pattern. Ethylene production by tissue slices was at least ten times that by intact fruits. Treating the slices with ethylene had relatively less effect compared to whole fruit, probably because of a high rate of endogenous ethylene production after injury. Lü *et al.* (1962) used tissue slices to investigate the effect of respiratory inhibitors and other substances on respiratory activity and sucrose synthesis in melon fruits. In further work, Liang and Lü (1964) compared the relative importance of the pentose phosphate and glycolytic pathways of glucose metabolism, and showed that the tricarboxylic acid cycle was operating in melon tissue respiration. VonAbrams and Pratt (1967) studied permeability changes in cantaloupe tissue slices in an attempt to explain the role of ethylene in the induction of fruit ripening.

XI. SUMMARY

The work on melons which has appeared in journals throughout the world, and which seems to represent real or potential contributions to our knowledge of these fruit, has been summarized. Many gaps and areas of imperfect knowledge are revealed; in fact there are more gaps than solid information. Biochemical and physiological work should be done on accurately named cultivars at precise physiological ages, and the potential for combining physiological and biochemical studies with genetic studies should be exploited.

REFERENCES

Allinger, H. W., Bisson, C. S., Roessler, E. B. and MacGillivray, J. H. (1940). *Proc. Am. Soc. hort. Sci.* **38,** 563.

Anderson, D. G., Stafford, H. A., Conn, E. E. and Vennesland, B. (1952). *Pl. Physiol., Lancaster* **27,** 675.

Arasimovich, V. V. (1939). *Biokhimiya* **4,** 251. (*Chem. Abstr.* **34,** 1694[6].)

Arasimovich, V. V. (1966). *Biokhim. Kul't. Rast. Mold.* No. 4, 12. (*Chem. Abstr.* **67,** 79607t.)

Arasimovich, V. V. and Savchenko, A. P. (1966). *Biokhim. Kul't. Rast. Mold.* No. 4, 31. (*Chem. Abstr.* **67,** 88285p.)

Aslanyan, G. Sh., Asmaeva, A. P., Avakyan, S. O. and Azatyan, S. A. (1960). *Izv. Akad. Nauk armyan. SSR, Biol. Nauki* **13**(7), 27.

Avakyan, A. G. (1957). *Soviet Pl. Physiol.* (AIBS trans.) **4,** 86.

Baksay, L. (1967). *Acta agron. hung.* **16,** 197.

Barnett, R. C., Stafford, H. A., Conn, E. E. and Vennesland, B. (1953). *Pl. Physiol., Lancaster* **28,** 115.

Berkowitz-Hundert, R. (1966). *Enzymologia* **31,** 281.

Bianco, V. V. (1969). M.S. Thesis, University of California, Davis.

Bini, G. and Raddi, P. (1965). *Riv. Ortoflorofruttic. ital.* **49,** 467. (*Hort. Abstr.* **36,** 2832.)

Blesa, A. C. and Pretel, A. (1967). *An. Edafol. Agrobiol.* **26,** 1049. (*Chem. Abstr.* **68,** 9720w.)

Chace, E. M., Church, C. G. and Denny, F. E. (1924). *U.S. Dep. Agr. Dep. Bull.* 1250, 27 pp.

Chen, S. D. and Schuck, C. (1951). *Fd Res.* **16,** 507.

Christensen, B. E., Hansen, E. and Cheldelin, V. H. (1939). *Ind Engng Chem., Analyt. Edn.* **11,** 114.

Curl, A. L. (1966). *J. Fd Sci.* **31,** 759.

Currence, T. M. and Larson, R. (1941). *Pl. Physiol., Lancaster* **16,** 611.

Currence, T. M., Lawson, R. E. and Brown, R. M. (1944). *J. agric. Res.* **68,** 427.

Davis, G. N., Whitaker, T. W., Bohn, G. W. and Kasmire, R. F. (1965). *Calif. agric. exp. Sta. Cir.* 536, 39 pp.

Davis, R. M., Jr., Baker, G. A. and Kasmire, R. F. (1964). *Hilgardia* **35,** 479.

Davis, R. M., Jr., Davis, G. N., Meinert, U., Kimble, K. A., Brown, L. C., May, D. M., May, G. E., Hendricks, L. C., Jr., Scheuerman, R. W., Schweers, V. H. and Wright, D. N. (1967). *Hilgardia* **38,** 165.

Davis, R. M., Jr., Fildes, R., Baker, G. A., Zahara, M., May, D. M. and Tyler, K. B. (1970). *J. Am. Soc. hort. Sci.* **95,** 475.

Denny, F. E. and Miller, L. P. (1935). *Contr. Boyce Thompson Inst. Pl. Res.* **7,** 97.

Doolittle, S. P., Taylor, A. L., Danielson, L. L. and Reed, L. B. (1961). *U.S. Dep. Agr. Handbook* 216, 45 pp.

Doolittle, S. P., Taylor, A. L., Danielson, L. L. and Reed, L. B. (1962). *U.S. Dep. Agr. Info. Bull.* 259, 31 pp.

Enslin, P. R., Joubert, F. J. and Rehm, S. (1956). *J. Sci. Fd Agric.* **7,** 646.

Enslin, P. R., Rehm, S. and Rivett, D. E. A. (1957). *J. Sci. Fd Agric.* **8,** 673.

Ezell, B. D. and Gerhardt, F. (1940). *J. agric. Res.* **60,** 89.

de Farias, L. V. and Magalhães Neto, B. (1955). *Revta Quím. Ind., Rio de J.* 1955, No. 274, 39. (*Chem. Abstr.* **50,** 1940d.)

Ferenczy, L. (1957). *Phyton, B. Aires* **9,** 47.

French, R. B., Abbot, O. D. and Townsend, R. O. (1951). *Fla agric. exp. Sta. Bull.* 482, 19 pp.

García-Blanco, J. and Vento, V. (1944). *Trabhs Inst. Nacl. Cienc. Méd.* (Madrid) **4,** 175. (*Chem. Abstr.* **43**, 328a.)

Gerber, C. (1897). *Annls. Sci. nat., Series* 8, *Bot.* **4,** 1.

Gilbart, D. A. and Dedolph, R. R. (1963). *Mich. agric. exp. Sta. Quart. Bull.* **45,** 589.

Gustafson, F. G. (1926). *Pl. Physiol., Lancaster* **1,** 265.

Hadwiger, L. A. and Hall, C. V. (1961). *Pl. Dis. Reptr.* **45,** 373.

Halevy, A. H. and Rudich, Y. (1967). *Physiologia Pl.* **20,** 1052.

Hartman, J. D. and Gaylord, F. C. (1941). *Proc. Am. Soc. hort. Sci.* **39,** 341.

Hobson, G. E. (1962). *Nature, Lond.* **195,** 804.

Hoffman, J. C. (1940). *Proc. Am. Soc. hort. Sci.* **37,** 836.

Hoover, M. W. (1955). *Proc. Fla Sta. hort. Soc.* **68,** 185.

Howard, F. D., MacGillivray, J. H. and Yamaguchi, M. (1962). *Calif. agric. exp. Sta. Bull.* 788, 44 pp.

Inukai, F., Suyama, Y., Sato, I. and Inatomi, H. (1966). *Meiji Daigaku. Nogakubu Kenkyu Hokoku* No. 20, 29.

Isawa, I. T., Takahashi, Y. and Togo, S. (1938). *Japan J. Med. Sci. VIII. Internal. Med., Pediat. Psychiat.* 5, *Proc.* 90 (*Chem. Abstr.* **35,** 521^2.)

Ito, N. (1938). *J. agric. Chem. Soc. Japan* **14,** 140.

Ito, S. and Sagasegawa, H. (1952). *Bull. hort. Div., Tokai-Kinki agric. exp. Sta.* **1,** 225. (*Chem. Abstr.* **50,** 1499i.)

Ivanoff, S. S. (1954). *J. Hered.* **45,** 155.

Iwanov, N. N., Alexandrowa, R. S. and Kudrjawzewa, M. A. (1929). *Biochem. Z.* **212,** 267.

Jones, H. A. and Rosa, J. T. (1928). "Truck Crop Plants." McGraw-Hill, New York.

Jurics, E. W. (1966). *Z. Lebensmittelunters. u. -Forsch.* **132,** 193.

Jurics, E. W. (1967). *Élelmiszerv. Közl.* **13**(3), 158. (*Chem. Abstr.* **67,** 115918f.)

Jurics, E. W. and Lindner, K. (1965). *Élelmiszerv. Közl.* **11**(1–2), 40. (*Chem. Abstr.* **64,** 4170h.)

Kasmire, R. F. (1968). *Calif. Agric.* **22**(5), 13.

Kasmire, R. F., Rappaport, L. and May, D. (1970). *J. Am. Soc. hort. Sci.* **95,** 134.

Kenny, I. J. and Porter, D. R. (1940). *Proc. Am. Soc. hort. Sci.* **38,** 537.

Leeper, P. W. (1951). *Proc. Am. Soc. hort. Sci.* **58,** 199.

Liang, H. K. (1961). *Acta bot. sin.* **9,** 219. (*Hort. Abstr.* **33,** 5068.)

Liang, H. K. and Lü, C. S. (1964). *Acta bot. sin.* **12,** 267.

Lindner, K., Hapka, S., Kramer, M. and Szoke, K. (1963). *Qualitas Pl. Mater. vég.* **9**(3), 203.

Lü, C. S., Lian, H. K., Wang, P. S., Wang, P. M., Li, M. C. and Shao, Z. P. (1962). *Acta bot. sin.* **10,** 317.

Lü, C. S. and Wang, P. H. (1959). *Acta bot. sin.* **8,** 221.

Lutz, J. M. and Hardenburg, R. E. (1968). *U.S. Dep. Agr. Handbook* 66, 94 pp.

Lyons, J. M., McGlasson, W. B. and Pratt, H. K. (1962). *Pl. Physiol., Lancaster* **37,** 31.

Lyons, J. M. and Pratt, H. K. (1964). *Proc. Am. Soc. hort. Sci.* **84,** 491.

MacGillivray, J. H. (1947). *Pl. Physiol., Lancaster* **22,** 637.

Maheshwari, S. C. and Bhalla, P. R. (1966). *Naturwissenschaften* **53,** 89.

Maheshwari, S. C. and Prakash, R. (1966). *Naturwissenschaften* **53,** 588.

Mann, L. K. (1943). *Bot. Gaz.* **105**, 257.
Mann, L. K. and Robinson, J. (1950). *Am. J. Bot.* **37**, 685.
Markov, V. M. (1956). *Sad Ogorod* 1956 (7), 27.
Masuda, T. and Hayashi, K. (1959). *Sci Rep. Fac. Agr. Okayama Univ.* **14**, 71.
Masuda, T. and Kodera, M. (1953). *Sci. Rep. Fac. Agr. Okayama Univ.* **2**, 38.
Masui, M., Fukushima, Y. and Suzuki, M. (1954). *Rep. Fac. Agr. Shizuoka Univ.* **4**, 24.
Matienko, B. T. (1956). *Bot. Zh., Kÿyiv* **41**, 558.
Matienko, B. T. (1967). *Bot. Zh., Kÿyiv* **52**, 229.
Matui, H. (1940). *J. agric. Chem. Soc. Japan* **16**, 1162. *Bull. agric. Chem. Soc. Japan* **16**, 173.
Mazelis, M. and Vennesland, B. (1957). *Pl. Physiol., Lancaster* **32**, 591.
McGlasson, W. B. and Pratt, H. K. (1963). *Proc. Am. Soc. hort. Sci.* **83**, 495.
McGlasson, W. B. and Pratt, H. K. (1964a). *Pl. Physiol., Lancaster* **39**, 120.
McGlasson, W. B. and Pratt, H. K. (1964b). *Pl. Physiol., Lancaster* **39**, 128.
Mel'nik, A. V. (1960). *Trudy l-oĭ (Pervoĭ) Nauch. Konf. Molodykh Uchenykh Moldavii, Kishinev* 1958, 313. (*Chem Abstr.* **55**, 27695h.)
Mel'nik, A. V. and Arasimovich, V. V. (1962). *Biokhim. Plodov i Ovoshchei, Akad. Nauk SSSR. Inst. Biokhim.* 1962 (7), 207. (*Chem. Abstr.* **57**, 12900g.)
Morgan, R. C. (1967). *J. Fd Sci.* **32**, 275.
Mori, T., Muraoka, N. and Shitomi, H. (1967). *Nippon Shokuhin Kogyo Gakkaishi* **14** (5), 187.
Morris, L. L. and Mann, L. K. (1946). *Proc. Am. Soc. hort. Sci.* **47**, 368.
Nip, W. K., Burns, E. E. and Paterson, D. R. (1968). *Proc. Am. Soc. hort. Sci.* **93**, 547.
Ogle, W. L. and Christopher, E. P. (1957). *Proc. Am. Soc. hort. Sci.* **70**, 319.
Orr, M. L. and Watt, B. K. (1957). *U.S. Dep. Agr. Home Econ. Res. Rep.* **4**, 82 pp.
Pentzer, W. T., Wiant, J. S. and MacGillivray, J. H. (1940). *U.S. Dep. Agr. Tech. Bull.* 730, 73 pp.
Porter, D. R., Bisson, C. S. and Allinger, H. W. (1940). *Hilgardia* **13**, 31.
Pratt, H. K. (1953). *In* "Proc. Conference on Transportation of Perishables, University of California, Davis, Feb. 5–6–7, 1953", pp. 104–110. Ass. Am. Railroads, Chicago.
Pratt, H. K. and Goeschl, J. D. (1968). *In* "Biochemistry and Physiology of Plant Growth Substances" (F. Wightman and G. Setterfield, eds), pp. 1295–1302. Runge Press, Ottawa.
Pratt, H. K. and Goeschl, J. D. (1969). *A. Rev. Pl. Physiol.* **20**, 541.
Rabourn, W. J. and Quackenbush, F. W. (1953). *Archs Biochem. Biophys.* **44**, 159.
Rakitin, Yu. V. (1935). *Dokl. Akad. Nauk SSSR*, n.s. **4**, 361.
Rakitin, Yu. V. (1945). *Biokhimiya* **10**, 373.
Ramsay, G. B. and Smith, M. A. (1961). *U.S. Dep. Agr. Handbook* 184, 49 pp. 24 pl.
Rattray, J. M. (1938). *S. Afr. Dep. Agr. For., Rep. Low Temp. Res. Lab., Capetown* 1936/37, 112.
Rattray, J. M. (1939). *S. Afr. Dep. Agr. For., Rep. Low Temp. Res. Lab., Capetown* 1937/38, 55.
Rattray, J. M. (1940). *S. Afr. Dep. Agr. For., Rep. Low Temp. Res. Lab., Capetown* 1938/39, 66.
Rehm, S., Enslin, P. R., Meeuse, A. D. J. and Wessels, J. H. (1957). *J. Sci. Fd Agric.* **8**, 679.

Reid, M. S., Lee, T. H., Pratt, H. K. and Chichester, C. O. (1970). *J. Am. Soc. hort. Sci.*, in press.
Rhodes, M. J. C. and Wooltorton, L. S. C. (1970), in preparation.
Rosa, J. T. (1928). *Hilgardia* **3,** 421.
Rowan, K. S., McGlasson, W. B. and Pratt, H. K. (1969). *J. exp. Bot.* **20,** 145.
Scholz, E. W., Johnson, H. B. and Buford, W. R. (1963). *J. Rio Grande Valley hort. Soc.* **17,** 170.
Schroeder, C. A. (1965). *Proc. Am. Soc. hort. Sci.* **87,** 199.
Scott, G. W. and MacGillivray, J. H. (1940). *Hilgardia* **13,** 67.
Serini, G. (1957). *Ann. Sper. agr.* **11,** 583.
Shinano, S. and Kaya, T. (1957). *Nippon Nôgei Kagaku Kaishi* **31,** 759. (*Chem. Abstr.* **52,** 15612a.)
Showalter, R. K. (1961). *Proc. Fla State hort. Soc.* **74,** 268.
Smith, M. C., Farrankop, H., Caldwell, E. and Wood, M. (1944). *Ariz. agric. exp. Sta. Mimeo. Rep.* 67, 11 pp.
Smith, P. G. and Welch, J. E. (1964). *Proc. Am. Soc. hort. Sci.* **84,** 535.
Spurr, A. R. and Davis, G. N. (1960). *Calif. Agric.* **14** (6), 5.
Tityapova, I. G. (1963). *Sb. Tr. Aspirantov i Molodykh Nauchn. Sotrudnikov Vses. Inst. Rastenievodestva* 1963 (3), 240.
Tomes, M. L. and Johnson, K. W. (1965). *Proc. Am. Soc. hort. Sci.* **87,** 438.
Tomes, M. L., Johnson, K. W. and Hess, M. (1963). *Proc. Am. Soc. hort. Sci.* **82,** 460.
Vavich, M. G. and Kemmerer, A. R. (1950). *Fd Res.* **15,** 494.
Vennesland, B. and Felsher, R. Z. (1946). *Archs. Biochem.* **11,** 279.
VonAbrams, G. J. and Pratt, H. K. (1967). *Pl. Physiol., Lancaster* **42,** 299.
Wada, M. (1930). *Biochem. Z.* **224,** 420.
Wardlaw, C. W. and Leonard, E. R. (1936). *Ann. Bot.* **50,** 621.
Whitaker, T. W. and Davis, G. N. (1962). "Cucurbits. Botany, Cultivation and Utilization". Interscience Publishers, New York.
Wiant, J. S. (1938). *U.S. Dep. Agr. Tech. Bull.* 613, 18 pp.
Winton, A. L. and Winton, K. B. (1935). "The Structure and Composition of Foods", Vol. II, pp. 451–464. John Wiley, New York.
Wong, C. Y. (1941). *Bot. Gaz.* **103,** 64.
Zechmeister, L. and Polgár, A. (1941). *J. biol. Chem.* **139,** 193.
Zechmeister, L. and Tuzson, P. (1930). *Ber. dt. chem. Ges.* **63B,** 2881.

Chapter 6

The Mango

A. C. HULME

A.R.C. Food Research Institute, Norwich, England

I Introduction 233
II Botanical Aspects 234
III Growth and Development 235
IV Biochemical Indices of Maturity 239
V Chemical Composition of Developing and Maturing Fruits 243
A. Carbohydrates, Organic Acids and Nitrogenous Compounds .. 243
B. Vitamins 245
C. Phenolic Compounds 247
D. Odoriferous Compounds 247
E. Mitochondrial and Other Enzymes 248
VI Biochemical Changes in Storage 250
References 253

I. INTRODUCTION

The mango is one of the oldest tropical fruits and has been cultivated by man for over 4000 years, originating apparently in the Indo-Burma region. During the Mogul Empire, 100,000 mango trees were planted in one area between A.D. 1556 and A.D. 1605. It is, at the present time, the most popular fruit among millions of people in the Orient where it is appreciated as the choicest of indigenous fruits, occupying, according to Singh (1960), relatively the same position in the tropics as the apple in Europe and North America. The mango is grown primarily in the sub-continent of India although it is now grown with increasing intensity in the south-east of the U.S.A., Central and South America, south-east Asia, Hawaii, the Philippines, Australia, the West Indies, East and South Africa, Egypt and Israel. The early history of the mango is elegantly described by Singh (1960) in his book on the fruit.

While results are available of a considerable amount of work on the growing of the mango, its diseases and its general food value, little systematic biochemical studies during growth, maturation and senescence appear to have been made, although an examination of certain aspects of mango biochemistry have appeared in the literature during the past decade. It is hoped that this brief chapter will point to deficiencies in our biochemical knowledge of this fruit and stimulate research to a level comparable with the growing commercial importance of the fruit in the West.

The average production of mangoes in India is 3,300,000 tons per annum (Cheema *et al.*, 1954).

II. BOTANICAL ASPECTS

As early as the sixteenth century, the name *mangos* (Tamil) was used for the mango in "Colloquies on Simples and Drugs of India" (Garcia da Orta, 1563, and quoted by Singh, 1960) and the common English term and the botanical name *Mangifera indica* L. originate from this ancient name. There are literally hundreds of varieties, and Singh (1960) gives an interesting insight into how many of them got their names.

The seed of the mango rapidly loses its viability and recently the method of vegetative propagation by budding and grafting has become normal practice. The importation of budsticks wrapped in polythene bags has, with modern rapid transport, been found to be the easiest and cheapest way of introducing mango varieties. Singh (1960) names 17 varieties of mango grown in India and 28 in Florida.

Mangifera indica belongs to the dicotyledenous family Anacardiaceae which consists of 64 genera, mostly trees and shrubs, some of which are poisonous. The leaves are exstipulate, usually alternate. In some species they are simple, in others, compound. The tree itself is evergreen and may attain a size of 50–60 feet. It is widely grown in the Indian continent and large mango trees are typical features of the Indian landscape. Its colour varies between green through yellow to red. The fruit is a laterally compressed, fleshy drupe. It varies considerably in size, shape, colour and flavour. The shape varies from rounded to ovate–oblong with the length varying from 2 to 30 cm in different varieties and the weight from several grams to more than a kilogram. The mesocarp provides the edible pulp which is firm containing a sweet, well-flavoured juice. In the mango, the endocarp develops into a tough leathery covering of the seed and is termed the husk. This is comparable to the strong endocarp present in mature fruits of the temperate fleshy drupes. The seed is exalbuminous. Details of the shape and specific names of the various outward physical features of the fruit are shown in Fig. 1.

Some allied species of *Mangifera*, while producing less valuable edible fruit, may be a good source of rootstocks for the common mango.

Horticultural aspects of mango growing up to 1960 are admirably dealt with by Singh (1960) in his book "The Mango". Although this book is primarily concerned with the mango as an important commercial fruit crop, Singh states that it is "intended to present in a comprehensive form all available information about this most popular of tropical fruits". It is, therefore, interesting that less than ten pages are devoted to the biochemistry of the mango. (The 1968 reprint does not differ from the 1960 edition.) The present author would emphasize, however, that this does not appear to be due to any deficiencies on the part of Dr. Singh, but simply a reflection of the general lack of detailed biochemical knowledge of the mango.

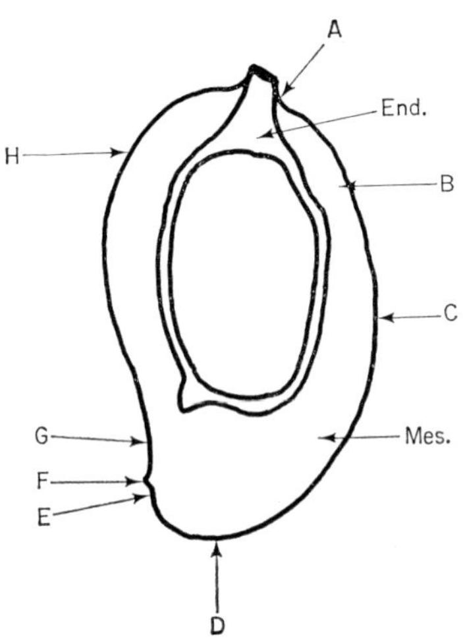

FIG. 1. A mango fruit. A, basal sinus; B, dorsal shoulder; C, back; D, apex; E, *nak*; F, beak; G, skin; H, ventral shoulder; End., endocarp (stone); Mes., mesocarp (fruit flesh).

As new techniques become available and as the biochemical research on other fruits points the way, the mango is now receiving more attention from biochemists. The earlier, and most of the more recent research work, was of a general nature and largely confined to the post-harvest changes. In general, the present chapter will be confined to work done in the past two decades, although reference to the pioneer work of Wardlaw and Leonard (1936) in the West Indies is inevitable.

III. GROWTH AND DEVELOPMENT

It is useful to divide the development of the mango fruit into four stages of development (Singh, 1960): (i) the juvenile stage, up to 21 days from fertilization (rapid cellular growth); (ii) stage of maximum growth, 21–49 days from fertilization; (iii) maturation (respiration climacteric and ripening), 49–77 days from fertilization; (iv) senescence.

The growth pattern of the mango, unlike many other stone fruits (see Chapter 12), appears to take the form of a simple, rather than a double, sigmoid curve. This will be seen from Fig. 2 which gives the growth curves of three Indian varieties studied by Mukerjee (1959).

Kennard and Winters (1956) sprayed Amini mango trees with 2,4,5-trichlorophenoxypropionic acid (2,4,5-TPP) at various concentrations (50, 200 and 800 p.p.m.) at intervals of 3, 6 and 12 weeks from flowering. Early applications decreased the final size of the fruit by approximately 15% (50 p.p.m. 2,4,5-TPP) and 40% (200 p.p.m. 2,4,5-TPP) and hastened maturity by about 2 weeks. Applications at 6 weeks from flowering also reduced the final size of the fruit and 200 p.p.m. 2,4,5-TPP hastened maturity by 1 week. An application at 12 weeks—one month before maturity—had little effect even at the highest concentration of 2,4,5-TPP. The two earlier applications at 800 p.p.m. seriously affected the growth of the fruit and the shape and appearance of the seeds. Early application of the growth regulator even

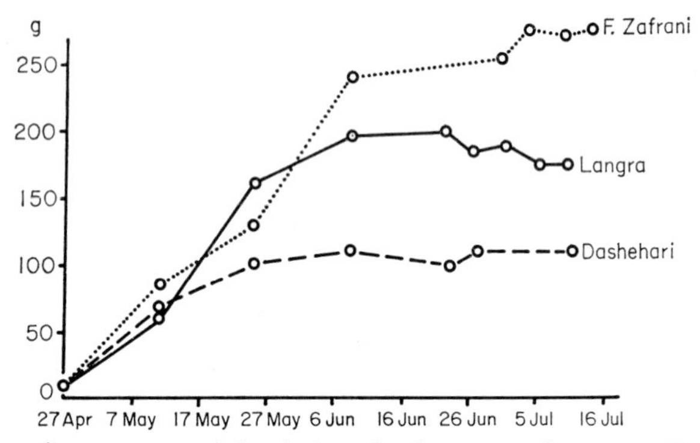

FIG. 2. Changes in average weight during development of mango. From Mukerjee (1959).

at 200 p.p.m. tended to increase fruit drop. The colour development of the fruit was unaffected by the application of 2,4,5-TPP which was, however found to decrease the total solids content by 3–5% and to increase the ascorbic acid content by as much as 45% (from 9·3 to 13·2 mg/100 g fresh weight) in the early applications at 50 p.p.m.

Most of the biochemical studies on mango to date appear to have been concerned with the mature fruit in the region of the time of harvest, and with subsequent post-harvest changes. Since so little is known of the biochemistry of the developing mango, any detailed assessment of the problem is impossible; all that can be done is to provide a catalogue of isolated observations made during this period.

Wardlaw and Leonard (1936) made a careful study of acid changes during the growth of the Julie variety. These authors describe in graphic detail the physical changes undergone by the developing fruit ". . . In the young fruit, the seed consists of the small white embryo with conspicuous cotyledons attached by a long stalk to the flesh and surrounded by a parchment-like integument. At this stage, the flesh is pale green but becomes white as the fruit enlarges, the first trace of yellow ripening colour appearing when the shoulders (that part of the fruit situated round the stem end) become level with the point of insertion of the stem instead of sloping away from it. Colouring and softening of the flesh is from the seed outwards; at this stage the latter has become surrounded by a cartilagenous and, finally, strong endocarp. If allowed to ripen on the tree the later growth developments include the raising of the shoulders, and the elevation of the stem on a small mound surrounded by a hollow. The flesh ultimately becomes deep orange throughout." As will be seen later, these readily observable changes have been used as a means of assessing the optimal picking date for immediate consumption or for storage.

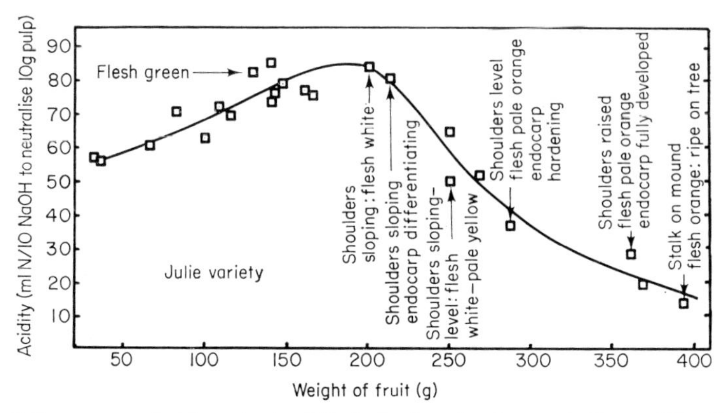

FIG. 3. Julie mango. Acid content of fruit during development. From Wardlaw and Leonard (1936).

During the 90 day period on the tree, starch accumulation is the main activity of the mango pulp tissue; Leley *et al.* (1943) found starch to increase from 1 to 13% in the fruit during growth. After picking, the starch was hydrolysed in 8 days. Wardlaw and Leonard (1936) made a careful study of changes in total acidity of the Julie mango during its development on the tree. Since fruit may be at different stages of development at the same time in the same orchard, they took fruit of different sizes taken *on the same day* to construct a graph of the relationship between acid content with development, irrespective of chronological age. This followed development on the tree right through to the ripe fruit and the results are shown in Fig. 3. The results show a rise and fall in acidity during development similar to that obtained for other fruits such as the apple (see Chapter 10). This pattern of acid change was also observed by several Indian workers, for example Singh *et al.* (1937) and Mukerjee (1959). Non-reducing and total sugars increase gradually showing a slight fall during ripening, while reducing sugars remain more of less constant throughout the period of development.

There is a continuous accumulation of starch up to maturity and this falls during ripening. Ash decreases during development with some increase near maturity while crude fibre remains more or less constant.

The ascorbic acid content is considerably greater in the green, young fruit than in ripe fruit, although the ripe mango is an excellent source of the vitamin. Mattoo and Modi (1969a) working with the Alfonso variety found 250 mg/100 g fresh pulp in the unripe fruit, 90 mg in the partially ripe and 165 mg in the ripe fruit. Soule and Hatton (1955) found 79 mg/100 g ascorbic acid in unripe and 25 mg in ripe Haden mango. It is interesting that Mattoo and Modi (1969a) found more ascorbic acid in the fruit injured by chilling (see later) than in healthy fruit at all three stages of maturity.

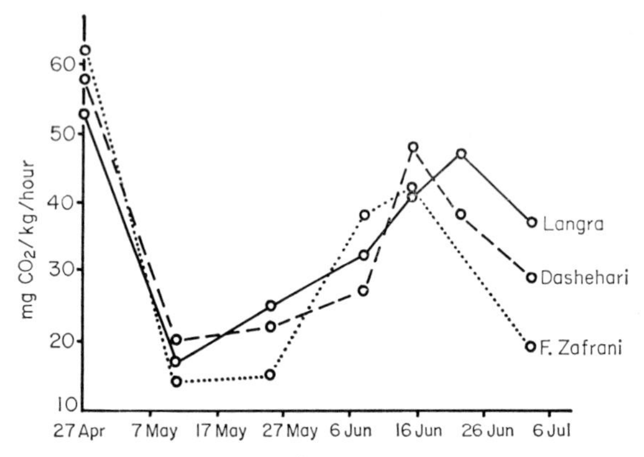

FIG. 4. The course of normal respiration of the mango. From Mukerjee (1959).

Cheema (1954, quoted by Singh, 1960) found that there were certain characteristic physiological attributes of the various stages of development mentioned at the beginning of this section. In stage (i) there was a high rate of respiration and a low carbohydrate : nitrogen (C/N) ratio; in stage (ii) the respiratory rate was steady and the C/N ratio was increasing; respiratory activity was low and the C/N ratio high in stage (iii), while in stage (iv) there was an increasing rate of respiration, an abrupt decline in sucrose with an increase in glucose and the C/N ratio was very high. Singh calls stage (iii) the "climacteric stage" although the respiration is low. This is a confusing use of "climacteric" since the mango is a climacteric fruit (see Volume 1, Chapter 17) and the respiration climacteric immediately precedes full maturity. This will be seen from Fig. 4 (Mukerjee, 1959).

Singh *et al.* (1937) found two maxima in the rate of respiration of developing mango, one in stage (i) and one near maturity—the true respiration climacteric. This early peak in respiration has also been observed in other fruits, for example the apple (Flood *et al.*, 1960). Singh *et al.* (1937) suggested that the changes in respiration paralleled changes in reducing sugars at all stages of development. This appears, at present, to be the sum total of our knowledge of biochemical changes in the mango up to the period when the fruit is ready for harvest.

IV. BIOCHEMICAL INDICES OF MATURITY

Numerous attempts have been made to establish the point at which to harvest mangoes for shipment over distances (short-term storage) or for immediate sale. In recent studies of this problem gross biochemical composition has been brought in to implement morphological and colour factors. Cheema and Dani (1934) put forward four different stages of maturity based on changes in colour and shape and size in an attempt to assess the optimal time for harvesting. These stages, designated A, B, C and D, appear in much of the subsequent papers dealing with this subject. At stage A, the fruits have their shoulders in line with the stem and the skin colour is olive green. At this stage, the fruit is not fully grown. Stage B, suggested as the best stage for export, occurs when the shoulders have outgrown from the stem end. At stage C, the colour has lightened towards yellow, while at stage D, the fruits are fully ripe with a typical flush developed on the skin. Singh (1960) points out, however, that this stem-end and shoulder relationship does not hold true for all varieties of mango. Nevertheless, Wardlaw and Leonard (1936) agreed, in effect, that the stages of maturity based on this relationship are similar in the Julie variety (West Indies) as in the Alphonso variety of Cheema and Dani, and that stage B was about the best time to pick for shipment or short-term storage at 7°C. Where such storage facilities are not available the fruit, they suggested, should be picked when the shoulders are level with the insertion of the stem end or slightly raised above it (that is, less ripe). What does this mean in biochemical terms? From Fig. 3, it is clear that acid changes are too gradual to provide a clear correlation with these morphological changes. The West Indian Julie variety shows little skin coloration but other varieties, especially Indian ones, show a marked colour change on ripening. The fruits are picked for market at this point which coincides with the time when fruits are beginning to drop (Jain, 1961). Jain (1961) remarks that these stages of maturity cannot, so far, be related to specific biochemical patterns. Studies of ripening at various temperatures abound in the literature but chemical changes have not been followed in any detail.

It is difficult to relate results of research carried out in Florida on chemical composition in relation to maturity with similar research in India owing to the wider range of physical attributes of maturity in the different varieties. It is unfortunate that, as yet, workers in this field have not investigated the

TABLE I. Changes in physico-chemical characteristics of mangoes during ripening after picking

Variety	Stage of maturity	Pressure test (lb)	Brix at 20°C	Water-insoluble solids	Total solids, %	pH	Acidity (anhy. citric acid) %	β-Carotene mg/100g	Glucose, %	Fructose, %	Sucrose, %	Total sugars %	Colour	Flavour
Badami	1	25·85	8·30	9·66	17·88	2·68	3·41	345	0·22	1·06	1·81	3·09	Pale white	No flavour
Badami	2	12·66	13·30	5 98	18·90	2·85	2·08	1413	0·70	1·79	6·50	8·99	Dull yellow	Mild
Badami	3	6·30	18·31	0·88	19·11	3·68	0·81	4883	1·91	2·95	11·25	16·11	Yellow	Marked flavour
Badami	4	3·90	19·32	0·22	19·37	4·04	0·38	6607	2·07	4·04	9·69	15·80	Orange yellow	Strong flavour
Raspuri	1	15·75	7·29	11·44	18·09	2·64	3·50	1231	0·25	0·75	1·33	2·33	Pale white	No flavour
Raspuri	2	7·37	15·31	6·31	19·98	2·94	2·12	1498	0·84	1·45	7·33	9·62	Dull yellow	Very mild
Raspuri	3	2·80	16·31	2·11	18·42	3·22	1·17	3094	2·13	3·11	8·13	13·37	Yellow	Mild
Raspuri	4	1·15	18·31	0·72	18·97	3·82	0·57	4952	2·06	3·44	10·52	16·02	Deep yellow	Marked
Totapuri	1	—	—	—	—	—	—	—	—	—	—	—	—	—
Totapuri	2	20·06	12·80	5·51	18·53	3·07	1·71	513	0·74	2·52	4·04	7·30	Pale white	No flavour
Totapuri	3	14·25	14·31	4·11	18·70	3·11	1·69	566	1·85	4·69	2·14	8·68	Dull yellow	Very mild
Totapuri	4	2·41	17·31	1·40	18·75	4·24	0·32	1976	0·55	3·23	9·52	13·30	Yellow	Mild
Neelam	1	30·00	12·80	5·82	17·81	3·06	1·50	130	1·79	2·71	1·40	5·90	Pale white	No flavour
Neelam	2	18·44	14·31	5·02	18·30	3·10	1·28	490	1·91	2·66	3·44	8·41	Dull yellow	Very mild
Neelam	3	10·08	16·31	3·65	18·84	3·46	0·88	844	2·46	2·68	4·95	10·09	Yellow	Mild
Neelam	4	3·47	18·31	1·30	18·90	4·74	0·16	2381	3·07	7·75	4·00	14·82	Slight orange yellow	Marked

From Krishnamurthy *et al.* (1960).

TABLE II. Chemical constituents of small and large mangoes[a] of 6 varieties analysed when hard and after ripening at 27°C.

(Each value is the average of 2–4 samples of 10–20 fruits each)

Variety and size[b]	% Starch in hard fruit	% Reducing sugars (as glucose)		% Sucrose (as invert sugar)		% Total sugars		% Total soluble solids		% Sugars–solids ratio		Phenolic compounds (as tannic acid) mg/100 g flesh	
		Hard fruit	Ripe fruit	Hard fruit	Ripe fruit	Hard fruit	Ripe fruit	Hard fruit	Ripe fruit	Hard fruit	Ripe fruit	Hard fruit	Ripe fruit
Haden													
small	5·1	4·2	3·8	1·6	6·7	5·8	10·5	7·9	14·1	0·73–1	0·74–1	40	59
large	6·8	4·3	3·6	1·7	7·8	5·9	11·4	7·9	14·2	0·75–1	0·80–1	44	49
Irwin													
small	6·0	5·2	6·1	2·7	5·4	8·0	11·5	8·7	14·2	0·92–1	0·81–1	38	59
large	6·1	5·3	5·6	2·5	5·3	8·0	11·0	8·9	12·4	0·90–1	0·89–1	42	53
Zill													
small	9·3	3·4	3·5	3·3	11·1	6·7	14·6	9·0	16·2	0·75–1	0·90–1	54	63
large	7·9	3·7	3·2	3·9	10·9	7·6	14·1	9·1	15·9	0·84–1	0·89–1	42	46
Sensation													
small	6·8	4·5	4·7	4·8	8·0	9·7	12·7	10·1	15·0	0·96–1	0·85–1	76	75
large	7·7	4·5	4·3	5·4	9·0	9·9	13·3	10·9	15·7	0·91–1	0·85–1	73	68
Kent													
small	10·6	4·5	4·3	2·9	7·9	7·1	12·2	9·2	14·2	0·77–1	0·86–1	37	50
large	9·9	4·4	4·3	2·8	8·4	7·1	12·7	9·3	14·5	0·76–1	0·88–1	31	45
Keitt													
small	4·2	4·2	3·9	0·6	5·7	4·8	9·8	6·9	11·0	0·70–1	0·89–1	52	39
large	8·4	4·6	3·8	0·7	7·5	5·3	11·2	6·7	12·4	0·79–1	0·90–1	39	31

From Soule and Harding (1956).

[a] Fruit was picked 10th May to 29th July, 1955, preserved by freezing at −18 to −23°C, and tested April and May 1956.

[b] Large is an average for medium and large fruits.

[c] Ripe fruit did not contain any starch.

10

use of the respiration climacteric as a yardstick of maturity in relation to chemical and physical properties associated with ripening. Popenhoe *et al.* (1958) suggested that for Haden and Zill mangoes grown in Florida, the point of maximum starch content was a good index of full (green) maturity for harvesting. Examples of various attributes, including acid and carbohydrate contents of some Indian mango varieties (Krishnamurthy *et al.*, 1960) and some varieties grown in Florida (Soule and Harding, 1956), are shown in Tables I and II, respectively. Stage 1 in the Indian fruit was "green and hard" and stage 4 was soft and well coloured and flavoured. These two stages of maturity may be compared with the hard (unripe) fruit and the ripe fruit (ripeness at 27°C) of the Florida samples. Thus we see that, averaging the results for the small and large fruits from Florida, the comparative values of reducing sugars, sucrose, soluble solids and sugars/solids ratios for the mango varieties grown in India and in Florida are as shown in Table III; calculating the reducing sugars for the Indian mangoes by adding the figures for fructose and glucose will not introduce any serious error into the comparison.

A comparison of the Indian Mango with those from Florida suggests that the green, unripe fruits of the Indian varieties contain much less sugar than the Florida fruits (giving a low sugars/solids ratio) but that, when ripe, the sucrose content of the Indian varieties is higher, with the exception of

TABLE III. Comparative values for mango varieties grown in India and in Florida

Variety and source	Stage	% Total reducing sugars	% Sucrose	% Soluble solids	Sugars/ solids
Badami (India)	1	1·28	1·81	8·22	0·38
	4	6·11	9·69	19·15	0·83
Raspuri (India	1	1·00	1·33	6·65	0·35
	4	5·50	10·52	18·27	0·88
Neelan (India)	1	4·50	1·40	11·99	0·50
	4	10·82	4·00	17·60	0·84
Haden	hard	4·15	1·65	7·9	0·74
	ripe	3·7	7·25	14·15	0·77
Irwin	hard	5·25	2·6	8·8	0·89
	ripe	5·85	5·35	13·3	0·84
Zill	hard	3·55	3·6	9·1	0·79
	ripe	3·35	11·0	16·1	0·89
Kent	hard	4·45	2·85	9·25	0·79
	ripe	4·3	8·15	14·35	0·87
Keitt	hard	4·4	0·65	6·8	0·74
	ripe	3·85	6·6	11·7	0·90

From Krisnamurthy *et al.* (1960) and Soule and Harding (1956).

the Neelan variety. It might be argued on the Indian data that a rise in the sugars/solids ratio to a value approaching unity might be a criterion of ripeness, but this suggestion falls down with the Florida mangoes where there is virtually no change in the ratio from hard to ripe. Presumably the increase in soluble solids during ripening reflects conversion of starch into sugars but little data appears to be available on "mobile" insoluble solids other than starch. Increase in sucrose and decrease in acid seem to be the only biochemical criteria available for the assessment of ripeness; an examination of enzyme changes during ripening accompanied by continuous recording of rate of respiration would appear to be long overdue for the mango.

Protein changes during development and ripening have scarcely received any attention. There are a number of examples of general analytical results comparable with those shown in Tables I and II but they are generally concerned with the later stages of maturity and ripening. Variation from variety to variety is considerable so that a collection of catalogues of composition would serve little useful purpose here.

Srivastara *et al.* (1962), in connection with a study of carotinogenesis in the Alphonso mango (see p. 250), picked fruit at stage A of Cheema and Dani (1934) and measured their respiration rate at "room temperature" (24–33°C). Although such measurements are, clearly, imprecise, the observed increase in CO_2-output from 35 mg/kg/hour after 3 days to 73 mg/kg/hour followed by a fall in rate confirms the climacteric nature of the mango fruit. By analogy with apples, stage B, being just before the climacteric rise in respiration, would appear to be a suitable stage of maturity for maximum transport and storage.

Burg and Burg (1964) first showed that mangoes are producing small amounts of ethylene at the time of the commencement of the respiration climacteric. They suggested that the rapid rate of ripening which commences as soon as the mature fruit is picked from the tree is due to a ripening inhibitor which becomes inactivated on picking, rendering the fruit more susceptible to the low concentrations of ethylene present in the fruit at this stage. When once the respiration climacteric commences autocatalytic production of ethylene proceeds apace as with other fruits in the climacteric class.

V. CHEMICAL COMPOSITION OF DEVELOPING AND MATURING FRUITS

A. Carbohydrates, Organic Acids and Nitrogenous Compounds

Singh (1960) has collected together in one table (p. 360 in his book) the average weight, seed, skin and pulp ratios, specific gravity, moisture, titratable acid, oil, protein, reducing and non-reducing sugars and ash contents of upwards of 30 samples of "mature" mango fruits of various varieties from various sources. Little more recent general information is available except on a few specific points. In a review of the chemistry and technology of the mango, Jain (1961) reported the presence of glucose, fructose and maltose; Sankar (1963) found xylose in ripening mangoes. The total sugar content of mangoes varies between 11·5 and 25% (fresh weight). Starch may attain to 15% of the fresh pulp of green, mature fruits. Jain (1961) states that pectin increases from the fifth week of fruit set until the stone is formed; thereafter the pectin content falls.

TABLE IV. The organic acids of mango fruits

Acid	Variety: Kent	Hsaing-Ien
Glycollic	0·061	0·026
Oxalic	0·036	0·008
Malic	0·074	0·045
Citric	0·327	0·194
Tartaric	0·081	0·051

From Fang (1965).

Leley *et al.* (1943) found the starch content of Alphonso mangos to be 14% of the fresh weight at harvest (Biale (1960) considers the mango to be one of the fruits to contain reserve polysaccharides at harvest). During subsequent ripening (21°C ?) of the fruit, sucrose rose from 5·8 to 14·2% of the fresh weight, while the pH rose from 3·0 to 5·2. In the post-climacteric stage, non-reducing sugars fell to 0·6% 10 days after the climacteric peak.

Total acidity varies from 0·13 to 0·71% (as citric). Jain *et al.* (1959) reported the presence of oxalic, citric, malic, malonic, succinic, pyruvic, adipic, galacturonic, glucuronic and mucic acids, together with two unidentified acids; Stahl (1935) noted the presence of tartaric acid. Fang (1965), using paper and silica-gel chromatography, investigated the non-volatile organic acids of Kent and Hsaing-Ien mangoes grown in Taiwan. He found the amounts of several acids to be as shown in Table IV; he also found traces of an unidentified acid. The predominance of citric acid is clearly indicated. Although a stage of "low temperature" (below 70°C) vacuum concentration was used in Fang's concentration of his mango extract, he considered that his unknown acid might be a "volatile organic acid".

The protein content of Indian mangoes varies between 0·5 and 1·0% while the Peruvian variety, La Molina, has an exceptionally high content of 1·57–5·42% (Jain, 1961). The skin of Java grown fruit contains 1–2% protein and the pulp 0·6–1%. Johnson and Raymond (1965) quote 0·9% protein for the skin of mangoes grown in Queensland. The principal free amino acids appear to be aspartic and glutamic acids, alanine, glycine, serine and α-amino butyric acid (Jain, 1961) but this is unlikely to be a complete list.

Shantha (1969) has recently made a thorough study of the sugar, acid, protein and amino acid changes in the ripening mango and has identified 11 amino acids.

B. Vitamins

Mangoes are a particularly rich source of vitamin C (ascorbic acid). Siddappa and Bhatia (1954) followed the vitamin C content of Raspuri mango fruits during growth from an average weight of 0·55 g to maturity (average weight 133 g). During the early stages of growth, the whole fruit (minus seeds) contained the strikingly large amount of over 300 mg of ascorbic acid/100 g fresh weight. This fell at maturity to 69·5 mg for medium-sized (108 g) and 39·1 mg/100 g fresh weight for large (158 g) fruit. In green but well developed fruits (58–90 g), the peel contained nearly 1·5 times as much ascorbic acid as the pulp; the peel constituted 20–25% of the total weight of the fruit. No ascorbic acid was present in the kernel

Singh (1960) lists the vitamin C content of more than 50 varieties of the fruit when ripe. The values vary between 13 and 178 mg/100 g. Many of the results quoted were obtained by Stahl (1935) and by Mustard and Lynch (1945) working with Florida-grown fruits. The range of ascorbic acid contents appears to be similar for mangoes grown in India. More recently, Iguina de George *et al.* (1969), in a study of mango varieties grown in Puerto Rica, gave a range of vitamin C contents from 6 to 63 mg/100 g for fully grown fruits, the highest value being recorded for the Julie variety. Ghosh (1960) found 36 mg of folic acid in 100 g of green mangoes.

Stahl (1935) gives the vitamin B1 (thiamine) content of 2 varieties of mangoes as 35 and 60 μg/100 g and the vitamin B2 (riboflavin) content of 3 varieties as between 45 and 55 μg/100 g fresh weight. Quinones *et al.* (1944) gave 0·057–0·060 mg/100 g as the thiamine content and 0·037–0·73 mg/100 g for the riboflavin content of 4 varieties of Philippine mangoes.

The mango is also rich in carotene and other carotenoid pigments (see p. 246) capable of being converted by the human body into vitamin A. Again, Singh (1960) lists the vitamin A equivalent (International Units, I.U.) of a number of mango varieties. These range between 1000 and 6000 I.U. and Singh points out that the vitamin A content of the mango may be equal to that of butter.

Iguina de George *et al.* (1969) examined the β-carotene content of 30 varieties of mangoes grown in Puerto Rico. They divided the varieties into three groups based on three ranges of β-carotene expressed in terms of I.U., namely, 400–2500 I.U., 2500–4000 I.U. and 5000–8000 I.U./100 g fresh weight. The Keitt, Kent and Haden varieties fell into the first group, the Julie into the second, while seven varieties, including Zill, occupied the third group with Carrie (7900 I.U.) occupying the premier place.

Several workers have studied the genesis and destruction of carotene in mango fruits and, since certain of the carotenoids function as provitamin A, it will be convenient to discuss this subject here. A qualitative and quantitative investigation of the carotenoids of flesh of fully ripe Alphonso mangoes was made by Jungalwala and Cama (1963). Sixteen different carotenoids were identified. β-Carotene was found to account for 60% of the total carotenoids. Of the oxycarotenoids, luteoxanthin and violaxanthin and *cis*-violaxanthin were present in significant amounts. All the oxycarotenoids were present as β-carotene derivatives, mostly as epoxides of zeaxanthin which was itself present only in trace amounts; lutein could not be detected and cryptoxanthin was the only monohydroxyxanthophyll present. Jain (1961) quotes the results of several workers on the carotenoids and xanthophylls present in several workers on the carotenoids and xanthophylls present in several varieties of mango. Modi and Patwa (1960) showed that ripe mangoes are ten times richer in carotene than partially ripe ones, while unripe, green mangoes do not contain even traces of carotene. Increase in carotene was accompanied by a decrease in acid content and an increase in sugar. These workers prepared cell-free extracts which increased in carotene content when incubated at 3°C for 18 hours in presence of either acetate or glucose. If acetate and glucose were omitted, there was a loss of carotene and this also occurred if carotene and extract were similarly incubated. This work was followed up by Modi *et al.* (1965) who found that mevalonic acid, a precursor of carotenoids (see Volume 1, Chapter 12) increased progressively during the ripening of mangoes. Mevalonate was isolated as the dibenzyl ethylene diamine derivative. With cell-free extracts as prepared by Modi and Patwa (1960)—essentially extraction at 0–3°C of ground mango with 0·5M sucrose and phosphate buffer at pH 7·0 followed by centrifugation at 12,800 g for 20 min—there was some increase in carotene content during incubation with 0·5 mmole mevalonate for 18 hours at 30°C. The amount of carotene formed was increased by about 8% if ATP, Mg^{2+}, reduced glutathione and $NADPH_2$ were added, but ATP alone (with the mevalonate) caused an increase of 5–6% so that the presence in the extracts of a carotene-synthesizing system was not altogether proven. Later, Modi and Reddy (1967) investigated further the factors relating to carotenogenesis in ripening Alphanso (this variety appears to be called Alphonso or Alphanso by different workers) mangoes. They studied changes in malic and citric acids, sucrose, glucose, fructose and pentose as well as malic enzyme (NADP dependent), glucose-6-phosphate and 6-phosphogluconate dehydrogenase during the ripening of the mango. They also showed that geraniol was present in ripe mangoes and that cell-free extracts of these fruits incubated with geraniol and farnesol showed an appreciable increase (7·2 mμmoles with geraniol) in carotene content. A five-fold increase in pentose, a three-fold increase in hexose sugars, a large decrease in malate and citrate and an increase in malic enzyme and the hexose monophosphate shunt (HMP) dehydrogenase all occurred during ripening. On this evidence, they suggest the formation during ripening of considerable amounts of $NADPH_2$ from the activity of malic enzyme and the HMP shunt which may be used in the reductive synthesis of isoprenic units. The evidence taken as a whole does indeed support the assertion by Modi and Reddy that carotene synthesis during the ripening of the mango follows the pattern of biosynthesis worked out for other plant material—see Volume 1, Chapter 12.

C. Phenolic Compounds

These compounds must make an important contribution, because of the astringency of tannins, to the flavour of the mango as they do to other fruits. Johnson and Raymond (1965) echo the view of Singh (1960) when they state "quantitative estimations of tannins in the mango do not yet seem to have been undertaken". Jain (1961) suggested that tannins are present only in small amounts—0·16 and 0·105% in the flesh and skin, respectively—while Singh (1960) states that according to other, much earlier workers, the mango contains "considerable amounts". Soule and Harding (1956) using the Folin–Denis reagent have, however, measured the total tannin content of a number of varieties grown in Florida, in the hard green and in the ripe state. Their results are seen in Table II (p. 241). The values are less than those quoted by Jain (above) but are still sufficient to impart astringency to the fruit; changes during ripening appear to be small.

The nature of the compounds contributing to the "tannin" content (in its widest context including phenolics not strictly tannins) have been little investigated. In the *leaf* and *bark* of the mango tree (varieties Taimour and Haram) El Sissi *et al.* (1965) identified quercetin and kaemferol and studied the sugars, presumably, combined with these compounds. They quoted earlier work on the detection of mangiferin in the *fruits*. The structure of this compound, isolated from the bark of the mango, has been studied in detail recently by Nott and Roberts (1967) using protonmagnetic resonance spectra. Their results confirm the suggestion of Billet *et al.* (1965).

D. Odoriferous Compounds

Preliminary studies in the extraction of odoriferous principles (see Volume 1, Chapter 10) from the mango have been made by Pattabhiraman *et al.* (1968) who found that chloroform and diethylether were the most satisfactory extractants for the compounds concerned. They used the pulp of ripe Bahami mangoes and employed column, thin layer and gas–liquid (GLC) chromatography to separate the compounds. They considered that carotene modified the overall aroma, and examination of individual fractions obtained from a column of silica gel G showed that the odoriferous principles travelled on the column along with the carotene; carbonyl compounds and alcohols appeared to be involved in the aroma but no definite identifications of individual compounds were made. Separation of odoriferous principles in a steam distillate of the tissue (which had a cooked aroma) by GLC yielded no definite results. Thus, the identification of the compounds responsible for the odour in the mango remains for the future.

E. Mitochondrial and Other Enzymes

Patwardhan (1965) isolated mitochondrial fractions from the pulp of seven varieties of mangoes using a medium containing 0·5M sucrose and 0·1M phosphate at pH 7·4. These preparations oxidized succinate and, less actively, citrate. They contained several of the usual mitochondrial dehydrogenases but required the addition of cytochrome c for maximum activity. They were, therefore, not entirely intact; since mangoes contain phenolics (see above) and since no precautions, apart from the presence of cysteine in the extracting medium, were taken to prevent the inhibiting effect of phenolics, it is not surprising that Patwardhan's preparations were fairly crude. Nevertheless they showed respiratory control (R.C.), although no R.C. values were recorded.

Mattoo *et al.* (1968) obtained active mitochondrial preparations from the pulp of Alphonso mangoes using a sucrose-tris-malate medium at pH 7·0 but, again, no precautions were taken to prevent interference by phenolics. They reported activity with pyruvate as substrate in the presence of glucose and hexokinase for preparations from mature, green (unripe) and from partly ripe (slightly yellow) fruit. They obtained P/O ratios of 2·88 for the green and 2·70 for the partly ripe fruit, suggesting no loss of oxidative phosphorylation during ripening. As with apples (see Chapter 10) mitochondrial activity in the extracted preparations increased as ripening proceeded.

As early as 1941, Banerjee and Kar showed that a relationship existed between the haem-iron content of mangoes and the activity of catalase and peroxidase. More recently Mattoo *et al.* (1968) found a four- to five-fold increase in the activity of catalase and peroxidase during the ripening of mangoes. They showed that the unripe fruit contained a nihydrin- and biuret-positive, heat-labile, non-dialysable inhibitor of both these enzymes—clearly not a phenolic compound. They partially purified this inhibitor using alumina gel followed by ammonium sulphate fractionation. The inhibition of catalase was linear in terms of the amount of inhibitor added. Later Mattoo and Modi Modi (1969a) claimed that ethylene (100–150 p.p.m. within the solution) passed through a solution of the partially purified inhibitor completely removed its inhibitory action on catalase and peroxidase. They showed that unripe bananas also contained a similar inhibitory substance similarly inactivated by ethylene.

In continuation of this line of investigation, Mattoo and Modi (1969b) showed that ethylene treatment of slices of pre-climacteric mango enhanced the catalase and peroxidase activity of the tissue. Since the ethylene treatment would be expected to induce the development of the respiration climacteric, this result might well mean that the activity of the two enzymes increases during the early stages of the climacteric. It could, of course, also be due, at least in part, to the ethylene rendering inactive the inhibitory substance discussed above.

Mattoo *et al.* (1968) found that the phosphatase activity of mangoes increased two-fold during ripening and that the enzyme activity could be doubled by the addition of β-carotene but not by vitamin A. They concluded that the increased carotene levels found during ripening might regulate carotenogenesis by promoting phosphatase activity—carotene synthesis involves phorphorylated compounds (see Volume 1, Chapter 12).

Changes in the activity of invertase and amylase during ripening were studied by Mattoo and Modi (1969b) who found that, in the early stages of ripening, invertase activity increased three-fold and amylase activity two-fold. This agreed with the observation that during the same period hexose and non-reducing sugars increased while starch fell from 6% to a trace. The disappearance of starch so early in ripening suggests that the use of maximum starch as an index of maturity by Florida growers (see p. 242) indicates that they prefer to harvest their fruit unripe; presumably this is to obtain maximum shelf-life, possibly at the expense of quality.

Mattoo and Modi (1969b) found a doubling in the activity of pectin esterase (PME) activity in the ripening mango, while Medina (1968), in a survey of the PME activity of several tropical fruits, found considerable activity in ripe mango pulp.

As mentioned earlier, Modi and Reddy (1967) observed an increase in the activity of NADP-dependent malic enzyme, glucose-6-phosphate and 6-phosphogluconic dehydrogenase during the ripening of the mango. De and Debnath (1966) also found an increase in "total" dehydrogenase activity during ripening measured, crudely, as the reducing power of the ground tissue towards methylene blue. Shantha (1969) has purified (66-fold) NADP-dependent malic enzyme.

Mattoo and Modi (1970) found an increase in ATP:citrate OAA-lyase (citrate-cleaving enzyme) activity during the ripening of mangoes and suggested that the acetyl-CoA and OAA formed in the reaction of the enzyme on citrate may contribute to synthetic processes taking place during the ripening period. They had previously found a large fall in the citrate content of mangoes during ripening. A crude fatty acid preparation from mango pulp when added to the enzyme preparations doubled their activity at the optimal pH of the enzyme. Mattoo and Modi suggest that the natural lipid breakdown products might regulate the activity of the enzyme *in vivo*. This suggestion is interesting in view of the fact that changes in fatty acid metabolism appear to be associated with the initiation of the respiration climacteric in apples, and that linolenic acid is involved in the production of ethylene (a ripening hormone) in these fruits (see Chapter 10). It is further circumstantial evidence in favour of the hypothesis that the products of the breakdown of lipid membranes are essential factors in the processes leading to ripening in fruits.

This somewhat disjointed account of the enzymes of the mango fruit illustrates that, while serious attempts are at last being made to investigate the complex biochemical changes occurring in the fruit, especially during the important ripening phase, there is still a long way to go; many of the results are conflicting.

VI. BIOCHEMICAL CHANGES IN STORAGE

Although many of the papers which have appeared on the storage of mangoes contain some account of gross chemical changes, few have been concerned with the exploration of modern, controlled atmosphere techniques such as those employed with temperate and sub-tropical fruits. Clearly, the mango being a tropical fruit and, therefore, very susceptible to low temperatures, lengthy storage at temperatures low enough to delay ripening and senescence (followed by fungal attack) is quite out of the question. As Wardlaw and Leonard showed as early as 1935, 7–8°C is the lowest "safe" temperature for mangoes and, even when harvested at the optimal state of maturity, a total storage life of 20–25 days is the best that can be expected. Although Jain (1961) quotes periods of up to 6 weeks for "satisfactory" storage of certain varieties at 7°C, using various fungistatic procedures, losses are likely to be high. As he pointed out, the unripe mango is too acid to be readily attacked by bacteria and fungi. However, as the fruit ripens, the acid content falls rapidly (see Table I, p. 240) and the fruit becomes more vulnerable to attack.

Srivastara *et al.* (1962) followed the respiration, carotenoid content and titratable acidity of Alphonso mangoes, harvested at stage B of Cheema and Dani (see above) after coating the fruit with a 6% sucrose wax (wax not specified) emulsion containing 0·5% sodium orthophenyl phenate and with mineral oil ("Prerox" D), during storage at 24–33·5°C. Both treatments had a similar effect in reducing the respiration rate (in terms of CO_2-production) and the rate of acid loss as compared with untreated fruits during storage up to 14 days. Loss of carotenoids was also reduced by the skin coating. The effects were so great (respiration rate after 14 days being 16·6 as compared with 70·5 mg CO_2/10 kg/hour in the controls; acidity as percentage of malic acid, 0·017 as compared with 0·30, and cartenoids 1·68 as compared with 6·20) that it seems probable that the skin coatings were causing a serious biochemical imbalance within the fruit.

Some biochemical changes occurring in Dusehri mangoes during storage at 32–38°C for 17 days were measured by Agnihotri *et al.* (1963). They found, rather surprisingly, little change in reducing sugars. Non-reducing sugars increased rapidly during the first 4–5 days and then remained constant; this was due, no doubt, to hydrolysis of starch. Total acidity remained constant for 11 days and then fell rapidly, while ascorbic acid showed a steady decline throughout the 17-day period from 22 to 10 mg/100 g fresh weight. Siddappa and Bhatia (1954), using mature green mangoes with a high ascorbic acid content (71 mg/100 g fresh weight), found a loss of only 10% of this acid during 6 days storage at 24–26°C.

The effect of post-harvest treatment of fully mature Badam mangoes with growth regulators, followed by storage at 18–29·5°C for 10 days, was studied by Date and Mathur (1960). Treatment with 2,4,5-T pp (1000 p.p.m.) caused a rise in total solids from 7·5 to 15% and a fall in titratable acid (expressed as citric acid) from 2·93 to 1·92%; the flavour of the fruit at the end of the 10 days was good. Maleic hydrazide at 1000 p.p.m. brought about a rise in total solids to 16·25% and a large fall in acid to 0·584%, the fruit, nevertheless, being somewhat tart; at 1500 p.p.m. the acidity fell less (to 1·13%) but the fruit developed an "off" flavour.

Saddappa and Bhatia (1954) found that the ascorbic acid content of slices of "tender" green mangoes steeped in 0·5% metabisulphite for 30 min, rinsed in water and then dried in the sun and stored for 6 months at 24–30°C retained more than 70% of the original ascorbic acid.

Teaotia *et al.* (1964) tested various packing media, such as various leaves, sawdust, paper cuttings, for short-term storage/ripening of mature green Gaurjeet, Bombay Green and Subul mangoes. Weight loss was least when double layers of fruit were stored in wooden boxes or in *Arusa* leaves for 1 week at, presumably, room temperature. Single layers in wooden boxes or leafy packings gave improved ripening results increasing the total soluble solids and total sugars. In most cases, starch had completely disappeared at the end of the week of storage. Few biochemical investigations of cool and controlled atmosphere storage have been undertaken.

Kapur *et al.* (1962) investigated the possibilities of refrigerated gas (controlled atmosphere) storage of Alphonso and Raspuri mangoes. Ventilated storage only was used so that the conditions of 5, 7·5 and 10% CO_2 also meant 16, 13·5 and 11% O_2 in the ambient atmosphere. The respiration rate of the fruit was, naturally, reduced by the CO_2 concentrations as compared with air; a distinct climacteric was shown by all the fruit, indicating that they were in the pre-climacteric state when placed in storage. Satisfactory storage with less than 10% wastage was achieved for 35 days at 8·5–10°C with the Alphonso variety and 49 days at 5·5–7°C with the Raspuri variety using 7·5% CO_2 in both cases. Under these conditions, loss of acid, including ascorbic acid (45·2 mg/100 g fruit in the controlled atmosphere stored fruit as compared with 20·1 mg/100 g in the "control" for Alphonso, and 27·4 as compared with 17·2 mg/100 g for the Raspuri at the end of the storage period), and total soluble solids were considerably reduced.

Storage of Keitt mangoes at 13°C in five different atmospheres for up to 40 days was studied by Hatton and Reeder (1966). They found 5% O_2 plus 5% CO_2 (90% N_2) to have the most beneficial result in terms of minimal loss of total solids combined with acceptable skin colour and flavour after ripening at 21°C. However, the small advantages obtained were of little practical value. When the O_2-tension was reduced to 1% "off" flavours and skin discoloration resulted.

Mattoo and Modi (1969b) studied changes in the activity of invertase and amylase in relation to chilling injury in storage. This they approached in two ways. First, they showed that the activity of amylase in cell-free extracts of the pulp from both unripe and ripe fruits, stored at 3°C until some injury appeared in the tissue, was two to three times lower in the chilled than in the healthy parts of the same fruit, whereas invertase activity was higher in the chilled than in the healthy tissue. It is not clear how much damage was present in the chilled tissue so that some retention of enzyme activity in the injured tissue could be a "post mortem" effect; the injured tissue was marked by "softening and darkening". Their second method of attack was to incubate for 12 hours at −10, 3 and 35°C the cell-free extracts from both healthy and chilled tissue and to measure the enzyme activity at the end of the incubation. Invertase from the injured tissue was more stable at all temperatures, but especially at 3°C. Amylase activity decreased at 3°C in extracts from chill-injured tissue, whereas it appeared actually to increase after incubation at −10 and 35°C. Pectin esterase activity was also higher in extracts of chilled tissue, especially in ripe fruit. This behaviour of isolated enzyme preparations is difficult to explain. Mattoo and Modi also measured the relative amounts of Mg^{2+}, Ca^{2+}, K^+ and Na^+ in injured and healthy tissue and suggested that an accumulation of Ca^{2+} and K^+ ions (they found K^+ and Ca^{2+} to inhibit amylase and to stimulate invertase activity) in injured tissue might contribute to the difference in the activity of the two enzymes in chilled tissue. They concluded that low temperature injury in mangoes may be a result of the impairment of cellular permeability resulting in an imbalance between various ions affecting the activity of the key enzymes in carbohydrate metabolism.

This work appears to be the first serious attempt to investigate the biochemistry of low temperature injury in mangoes and is, therefore, encouraging, but the present author feels that not enough attention was paid to the "post mortem" effects of injury or to differentiation between what precedes (and leads up to) injury and what is the immediate result of it.

In fruit in general, as will be seen from other chapters in this book, there are a number of varied biochemical changes associated with low temperature injury. In the mango, a tropical fruit, and, therefore, especially susceptible to injury at temperatures well above the freezing point of the tissue, the picture is likely to be particularly complex. In practice, West Indian mangoes carefully selected with regard to maturity (unfortunately not yet in biochemical terms) and freedom from mechanical damage, transported to the United Kingdom in fast banana boats (12 days) at 8–10°C with approximately 5% CO_2 in the hold, can be stored satisfactorily on arrival at the same temperature for 3–4 weeks. However, the relatively high price of these fruits in retail shops puts them into the luxury class. It is possible that an intensification of physiological and biochemical research—such as has been given to many other fruits—might well lead to methods of transport and storage which would enable the mango to enjoy in Europe the popularity it deserves and which it has had for centuries in the countries in which it is grown.

ACKNOWLEDGEMENT

The author wishes to thank Drs Caygill and Marriott of the Tropical Products Institute, London, for helpful suggestions during the preparation of this chapter.

REFERENCES

Agnihotri, B. N., Kapoor, K. L. and Srivastara, J. C. (1963). *Punjab hort. J.* **3,** 286.

Banerjee, H. K. and Kar, B. K. (1941). *Curr. Sci.* **10,** 289.

Biale, J. B. (1960). *Adv. Fd Res.* **10,** 293.

Billet, D., Massicot, G., Mercier, D., Anker, A., Matshenko, A., Mentzer, C., Chaigneau, G., Valdener, G. and Pacheo, H. (1965). *Bull. Soc. Chem.* 3006.

Burg, S. F. and Burg, E. (1964). *Pl. Physiol., Lancaster* **39,** (Suppl.), x.

Cheema, G. S. and Dani, P. G. (1934). *Bombay Dep. Agric. Bull.* **170.**

Cheema, G. S., Bhat, S. S. and Naik, K. C. (1954). "Commercial Fruit of India." Macmillan, London.

Date, W. B. and Mathur, P. B. (1960). *Fd Sci.* (*Mysore*) **9,** 248.

De, H. N. and Debnath, J. C. (1966). *Pakist. J. scient. ind. Res.* **9,** 57.

El Sissi, H. I., Saleh, N. A. M., El Sherbeiny, A. E. A. and El Ansary, M. A. I. (1965). *Qualitas Pl. Mater. veg.* **12,** 262.

Fang, T. T. (1965). *Mem. Coll. Agric. nat. Taiwan Univ.* **8** (2), 236.

Flood, A. E., Hulme, A. C. and Wooltorton, L. S. C. (1960). *J. exp. Bot.* **11,** 316.

Ghosh, S. (1960). *Sci. Cult.* **26,** 287.

Hatton, T. T. and Reeder, W. F. (1966). *Proc. Carribb. Reg. Am. Soc. hort. Sci.* **10,** 114.

Iguina de George, A. M., Collazo de Rivera, A. L., Benaro, J. R. and Pennock, W. (1969). *J. agric. Univ. P. Rico* **53,** 100.

Jain, N. L. (1961). *Rev. Fd Technol.* **3,** 131.

Jain, N. L., Krishnamurthy, G. V. and Gindhari, Lal (1959). *Fd Sci.* (*Mysore*) **8,** 115.

Johnson, R. M. and Raymond, W. D. (1965). *Trop. Sci.* **7,** 156.

Jungalwala, F. B. and Cama, A. R. (1963). *Indian J. Chem.* **1,** 36.

Kapur, N. S., Sarveshivara, R. and Srivastara, H. C. (1962). *Fd Sci.* (*Mysore*) **11,** 228.

Kennard, W. C. and Winters, A. F. (1956). *Proc. Am. Soc. hort. Sci.* **67,** 290.

Krishnamurthy, G. V., Jain, N. L. and Bhatia, B. S. (1960). *Fd Sci.* (*Mysore*) **9,** 277.

Leley, V. K., Narayana, N. and Daji, J. A. (1943). *Indian J. agric. Sci.* **13,** 291.

Mattoo, A. K. and Modi, V. V. (1969a). *Pl. Physiol., Lancaster* **44,** 308.

Mattoo, A. K. and Modi, V. V. (1969b). "Proc. int. Conf. subtrop. Fruits", London, 1969. p. 111.

Mattoo, A. K. and Modi, V. V. (1970). *Biochem. biophys. Res. Commun.* **39,** 895.

Mattoo, A. K., Modi, V. V. and Reddy, V. V. R. (1968). *Indian J. Biochem.* **5** (3) 111.

Medina, M. G. (1968). *Arch. Latinoamericados de Nutrit.* **18,** 401.

Modi, V. V. and Patwa, D. K. (1960). *Experientia* **16,** 352.

Modi, V. V. and Reddy, V. V. R. (1967). *Indian J. exp. Biol.* **5,** 233.

Modi, V. V., Reddy, V. V. R. and Shah, R. (1965). *Indian J. exp. Biol.* **3,** 145.

Mukerjee, P. K. (1959). *Hort. Adv.* (*India*) **3,** 95.

Mustard, M. J. and Lynch, S. J. (1945). *Fla agric. exp. Sta. Bull.* 406.

Nott, P. E. and Roberts, J. C. (1967). *Phytochemistry* **6,** 741.

Pattabhiraman, T. R., Rao, P. and Sastry, L. V. L. (1968). *Perfum. essent. Oil Rec.* **59,** 733.

Patwardhan, M. V. (1965). *Nature, Lond.* **207,** 983.

Popenhoe, J., Hatton, T. T. and Harding, P. L. (1958). *Ann. Soc. hort. Sci.* **71,** 326.

Quinones, V. L., Guernant, N. B. and Dutcher, R. A. (1944). *Fd Res.* **9,** 45.

Sankar, K. P. (1963). *Sci. Cult.* **29,** 51.

Shantha, H. S. (1969). Ph.D. Thesis, University of Mysore, India.

Siddappa, G. S. and Bhatia, B. S. (1954). *Indian J. Hort.* **11,** 104.

Singh, B. N., Seshagiri, P. F. V. and Gupta, S. S. (1937). *Indian J. agric. Sci.* **7,** 176.

Singh, L. B. (1960). "The Mango." Leonard Hill, London (second impression, 1968).

Soule, M. J. and Harding, P. L. (1956). *Proc. Fla Mango Forum* 13.

Soule, M. J. and Hatton, T. T. (1955). *Proc. Fla Mango Forum* 16.

Srivastara, H. C., Narsimhau, P., Kapur, N. S., Srecnivasan, A. and Subrahmanyan, V. (1962). *Abstr. Paps. 1st int. Congr. Fd sci. Technol.* 1962, p. 25.

Stahl, A. L. (1935). *Fla agric. exp. Sta. Bull.* 283.

Teaotia, S. S., Singh, L. P., Maurya, V. N. and Agnihotri, B. N. (1964). *Ind. J. Hort.* **21,** 136.

Wardlaw, C. W. and Leonard, E. R. (1936). Memoir No. 3, Low Temp. Res. Sta., Trinidad.

Chapter 7

The Olive

M. J. FERNANDEZ DIEZ

Instituto de la Grasa y sus Derivados, Sevilla, Spain

I	Introduction	255
	A. Brief History	255
	B. World Production and Importance	260
II	Fresh Olives	260
	A. Main Varieties of Commercial Interest	260
	B. Composition	260
	C. Changes during Fruit Development and Maturation	263
III	Processed Olives	267
	A. Methods	267
	B. Biochemical Changes during Processing and Fermentation	270
IV	Olives for Oil	274
	A. Post-harvest Problems	274
	B. Methods of Processing	274
	C. Olive Oil	276
V	Conclusion	276
	References	277

I. INTRODUCTION

A. Brief History

The olive tree (*Olea europaea*) is prehistoric. Its real origin is still a matter of conjecture and throughout history legend has involved Asia Minor, India, Africa and Europe. In fact, we may consider the Mediterranean and its surroundings as the home of the olive. Thence it was taken overseas to reach the New World with the Spanish colonists in the fifteenth century, finally reaching California with the Franciscan fathers at the end of the eighteenth century. Italian and Spanish emigrants and missionaries also carried the olive tree to Australia and South Africa. It is known that the fruit of this tree has been used since 2000 to 3000 B.C., mainly for oil extraction but also as a food and an "appetizer".

Those who are interested in the history of these products should consult the works of Kalogueria (1932), Molinari and Nicolea (1947), Garoglio

TABLE I. Olive varieties and their uses

Country	Variety	Use, quality, yield, etc. Oil	Table use
Algeria	Sigoise	good quality, medium yield (14–17%)	green or black
	Azeradj	yes	—
	Aghenfas }	—	yes, good quality
	Sevillano }		
	Chemlali }		
	Limli }		
	Rougette de Mitidja }	yes, good quality	—
	Hamra }		
Argentina	Arauco	yes, good quality and yield (22–24%)	large, good quality
	Empeltre }		
	Gordal }	—	yes, not so widely grown
	Sevillana }		
	Santa Catalina	—	good quality
	Ascolano	—	
	Cerignola	—	good quality, average to large fruits
	Maraiolo	yes	
	San Agostino	yes	
	Manzanilla }		yes
	Frantoio }		—
	Arbequina }	good oil yield, especially Arbequina and Frantoio	—
	Picudilla }		—
	Nevadillo blanco }		—
	Nevadillo negro }		—

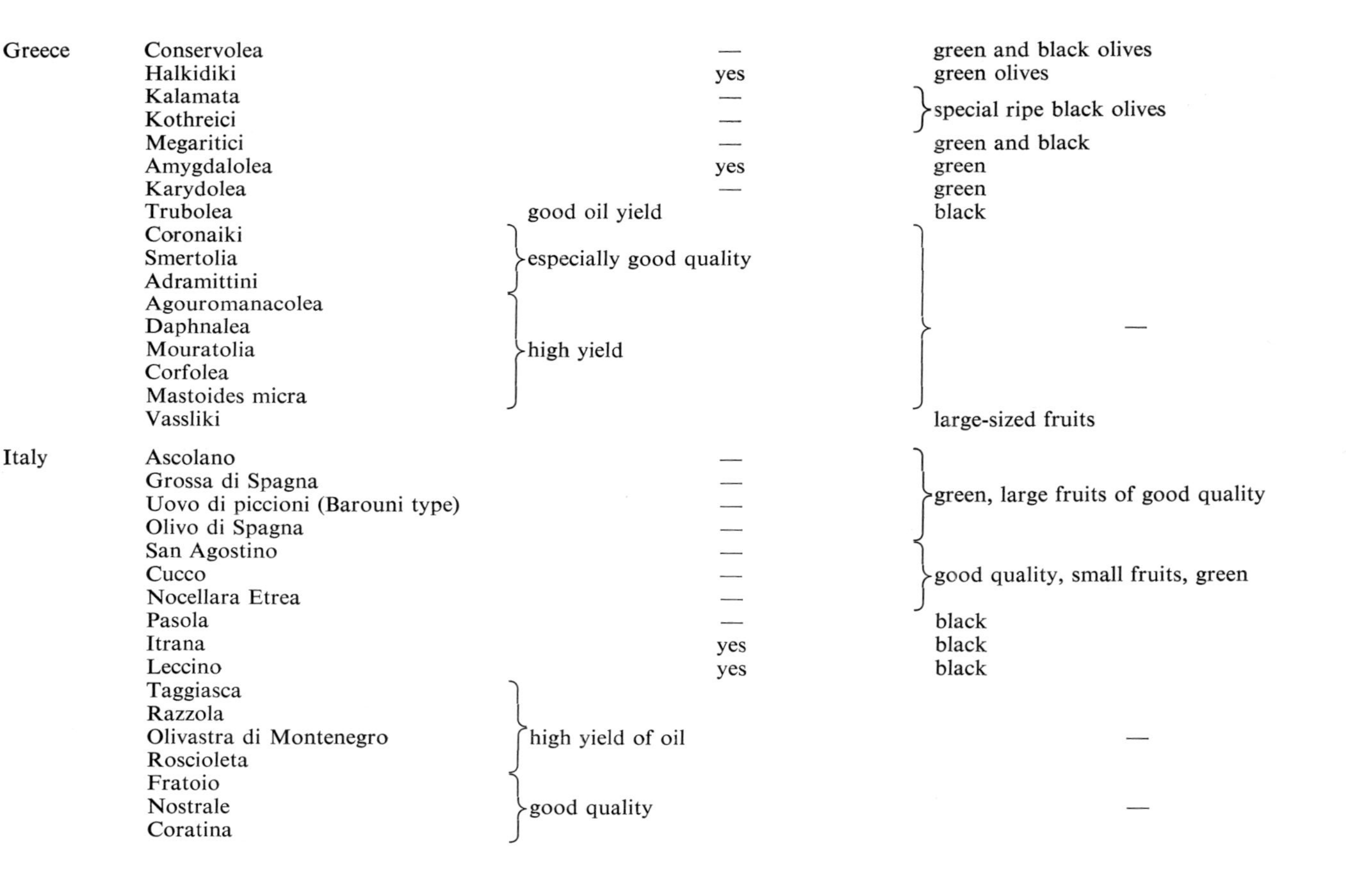

Greece	Conservolea		—	green and black olives
	Halkidiki		yes	green olives
	Kalamata		—	special ripe black olives
	Kothreici		—	
	Megaritici		—	green and black
	Amygdalolea		yes	green
	Karydolea		—	green
	Trubolea	good oil yield		black
	Coronaiki			
	Smertolia	especially good quality		
	Adramittini			
	Agouromanacolea			
	Daphnalea			—
	Mouratolia	high yield		
	Corfolea			
	Mastoides micra			
	Vassliki			large-sized fruits
Italy	Ascolano		—	
	Grossa di Spagna		—	
	Uovo di piccioni (Barouni type)		—	green, large fruits of good quality
	Olivo di Spagna		—	
	San Agostino		—	
	Cucco		—	good quality, small fruits, green
	Nocellara Etrea		—	
	Pasola		—	black
	Itrana		yes	black
	Leccino		yes	black
	Taggiasca			
	Razzola			
	Olivastra di Montenegro	high yield of oil		—
	Roscioleta			
	Fratoio			
	Nostrale	good quality		—
	Coratina			

TABLE I (*continued*)

Country	Variety	Use, quality, yield, etc. Oil	Table use
Morocco	Zitoun		
	Picholine marocaine		
	Soussia	for oil	pickling for eating
	Mesala		
	Gordal		
	Bon chonika		
	Beni	for oil	—
	Hojiblanca		
Portugal	Sevilhana	—	direct consumption
	Mancanilha	—	
	Negral		general purpose
	Galega		
	Redondil		—
	Carrasquenha	excellent oil	—
	Verdeal		—
	Cordobil		—
	Lentisca	high yield small quantities	—
Spain	Gordal	—	green, excellent quality, ferments readily, large
	Manzanilla	—	green, excellent quality
	Carrasqueña	—	green, good quality
	Morona	—	green, characteristic flavour
	Rapazalla	—	green, difficult to process
	Hojiblanca	yes	green, high percentage of flesh
	Picual	high yield	—
	Verdial	yes	pickling

	Lechin	good yield			—
	Zorzaleña				
	Negral	high quality			—
	Cornicabra				
Syria	Jlott		—	green, direct consumption	
	Kaiamani-Kachali		—	green, good yield	
	Sourani		yes		yes
	Khalklali	good yield			—
Tunisia	Zarazi		—	black	
	Meski		—		
	Various "Barouni" types		—		
	Cherullali	high yield			—
	Chetoni				
Turkey	Hurma		—	pickled	
	Tiribya		—		
	Edremit		—		
	Karamursel		—		
	Sam	large fruits			—
	Agvalik	excellent oil			—
	Girit	high yield			—
USA	Mission	small quantities used for oil; high yield		"California style" (see later), green	
	Manzanilla				
	Sevillano		—	excellent as "Spanish style" green olives difficult to process, large	
	Ascolano		—	California style, black, difficult to process, as ripe large	
	Barouni (introduced from Tunisia)		—	good quality	
	Nevadillo		—	limited quantity only grown	
	Redding Picholine		—		
	Chemlali		yes		—

(1950), Hartmann (1953), Patac *et al.* (1954) and Foytik (1960), for further information. Olive culture in the ancient Holy Land is the subject of a recent study by Goor (1966).

B. World Production and Importance

According to recent statistical data from the International Oil Council (1968), world production of fresh olives is about 7 million metric tons per annum, of which 450,000 tons are used as edible processed olives (table olives) and the rest for oil extraction amounting to about 1·3 million tons. The number of trees is estimated at 750 million on a cultivated surface of 9 million hectares. About two-thirds of this area is exclusively dedicated to olive tree culture, and approximately 98% of the planting is located in the Mediterranean area and its surroundings.

Table olives are consumed throughout the world, chiefly in American and European markets. They are either processed as green olives (p. 267) or as black olives (p. 268). Olive oil is mostly consumed (94%) by Mediterranean countries and the remainder of the production (6%) is exported to the United States, Brazil, Australia and other non-producing European communities.

II. FRESH OLIVES

A. Main Varieties of Commercial Interest

The main producers of table olives and/or olive oil are, in alphabetical order, Algeria, Argentina, Greece, Italy, Morocco, Portugal, Spain, Syria, Tunisia, Turkey and the United States of America. A very large number of varieties are grown for their specific qualities. For example, Garolio (1950) describes almost 300 varieties for Italy alone, of which at least 50 are used for sale as table olives (Baldini and Scaramuzzi, 1963). Further detailed information can be obtained from Molinari and Nicolea (1947), Hartmann (1953), Patac *et al.* (1954), Foytik (1960), Ortega Nieto (1963) and Balatsouras (1967).

In Table I some of the outstanding varieties are listed alongside the qualities for which they are valued as sources of oil or table olives.

B. Composition

The fruit of the olive tree is an edible, fleshy drupe, more or less oblong, of a green colour which changes to purple or black when mature and attains an average weight ranging from 10–12 down to 1·5–2 g, according to the variety. Exceptionally, the fruits can reach extreme values of 20 and 0·5 g respectively. The percentage of flesh, intensely bitter, especially when still green, comprises between 70 and 88% of the fruit. The pit or stone represents from 12 to 30% of the fruit, according to the variety, extent of growth and maturity. It is grooved and the seed it contains accounts for about 1·5% of the whole fruit and 7–7·5% of the pit.

Any attempt to give an average composition for the olive is fraught with difficulties since it varies widely from one variety to another, and even within the same variety, according to the state of development and maturity. Hence, different values found in the literature often appear to be contradictory. Table II shows some typical values, obtained by Borbolla *et al.* (1955) for

TABLE II. Composition of olive flesh (percentage of fresh flesh)

	Gordal variety		Zorzaleña variety	
	September	November	September	November
Water	73·73	68·10	60·55	59·75
Oil	6·25	18·60	17·70	25·20
Reducing sugars	5·96	3·66	5·87	3·40
Non-reducing sugars	0·27	0·03	0·15	0·13
Proteins (N × 6·25)	1·31	nd.	1·50	1·25
Crude fibre	3·40	2.05	2·60	1·15
Ash	0·68	0·84	1·10	0·82
Others	8·40	6·72	10·53	8·30

n.d. = non-determined.

the flesh of two Spanish varieties, Gordal and Zorzaleña, in two different seasons. The September figures are for the beginning of the month, when the fruits were ready to be harvested for Spanish-style pickled green olives. The results for November were obtained when harvesting for oil extraction was about to begin; colour change, characteristic of the last stage of maturity, had already set in.

The main constituents of the flesh are water and oil. In general, for the same degree of maturity, these two components show an inverse relationship, that is to say, the higher the oil content of a variety, the lower the water content (Cruess *et al.* 1939; Borbolla *et al.* 1961).

The main sugars found in the flesh are glucose, fructose and sucrose. The distribution of the hexoses varies with the variety. For example, the Manzanilla variety contains 88% glucose, while Gordal has 94% and Zorzaleña 100%. In the first two varieties, the remainder of the hexose content is largely made up of fructose.

From the processing viewpoint, another important component of the olive flesh is oleuropein, which is responsible for the bitter taste of the fruit, and has been identified as a compound of a glucosidic nature (Cruess and Sugihara, 1948). Simpson *et al.* (1961) considered that arabinose as well as glucose is involved. Its probable constitution is shown in Fig. 1. Mannitol and polysaccharides are also important as fermentable substances; other components are organic acids, salts and colouring matters.

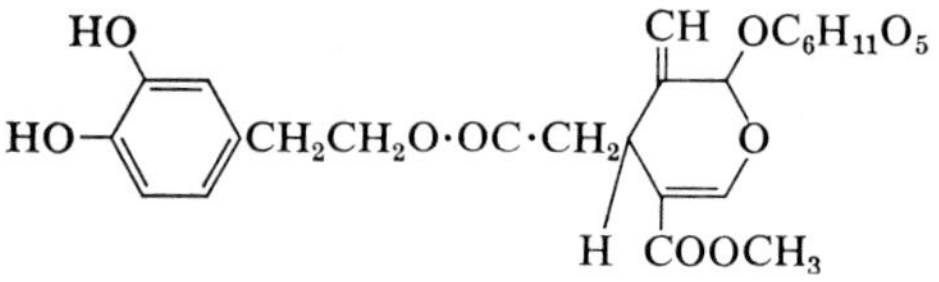

HO CH $OC_6H_{11}O_5$

HO— $CH_2CH_2O{\cdot}OC{\cdot}CH_2$ O

H $COOCH_3$

FIG. 1. Suggested constitution of oleuropein.

Mahecha (1967) identified six anthocyanins in Manzanilla olives, while Cantarelli (1961) found caffeic, ferulic and a 4-OH flavanone in ripe olives. Cruess *et al.* (1939), Borbolla *et al.* (1955, 1961), Fernandez Diez and Gonzalez Pellisso (1956), Sandret (1957) and Balatsouras (1964) give detailed information on the flesh composition of typical olive varieties from California, Spain, Greece, Italy, Morocco, Israel and Peru. Recently Vazquez Roncero (1963, 1964, 1965) extensively surveyed work on the different constituents of olive flesh.

A list of the amino acids found in the flesh of the olive is shown in Table III, while the citric, malic and oxalic acid content in the flesh of three Spanish

TABLE III. Amino acids found in the olive

Acid	Reference
Aspartic acid	a, c
Alanine	a, b, c
Arginine	b, c
Cystine	b
Glycine	a, b
Glutamic acid	a, b, c
Histidine	a, b, c
Isoleucine	a, b, c
Leucine	a, b, c
Lysine	a, b, c
Methionine	b, c
Phenylalanine	a, b, c
Proline	a, b, c
Serine	a, b
Tyrosine	a, b
Threonine	b
Tryptophane	b, c
Valine	a, b, c

[a] Narasaki and Katakura (1954).
[b] Ferrao (1957).
[c] Fernandez Diez (1958).

varieties as determined by Fernandez Diez and Gonzalez Pellisso (1956) is given in Table IV. Balatsouras (1964) found the pH of the flesh of several varieties of Greek grown olives to be in the range 4·1–5·4; the average value was 4·5.

TABLE IV. Organic acids found in the olive flesh for Gordal, Rapazalla and Zorzaleña Spanish varieties

	% dry flesh
Citric acid	0·05–0·38
Malic acid	0·09–0·19
Oxalic acid	0·09–0·24

Vasquez Roncero (1965) quotes various authors for the composition of the protein fraction of the fresh pulp of ripe olives: the protein content varied from 1·5 to 5% depending on the variety and source. The values given by Balatsouras (1964) average 1·4%. Tyrosine, tryptophane, phenylalanine and lysine were among the products of hydrolysis of the protein. Quantitatively, hydrolysis gave 14·8% amide and ammonia, 16·6% humin, 5·7% diamino amino acids, 27·6% mono amino acids and 35·1% imino acids. Among the enzymes detected were carbonic anhydrase, emulsin, lipase, peroxidase, phosphatase and polyphenolase.

Eisner *et al.* (1965) showed that squalene is the major hydrocarbon and campestrol and β-sitosterol were two sterols present in olives.

The woody part of the pit surrounding the seed is mostly composed, according to Mingo and Romero (1953), of cellulose (38%), other carbohydrates (41%), water (9%), proteins (3%), ash (4%) and a small percentage of oil (less than 1%). The components of the seed are: water (30%), oil (27%), various carbohydrates (27%), proteins (10%), cellulose (2%) and ash (1·5%). Seed proteins, the detailed study of which is considered to have the greatest importance in relation to genetic differences between varieties, have been further investigated by Fernandez Diez (1960a, b, c, 1961).

C. Changes during Fruit Development and Maturation

King (1938) implied that the growth curve of the olive was sigmoidal and this was confirmed later by Deidda (1965). The earliest research work on olives had as its object the study of oil formation within the fruit. According to Scurti and Tommasi (1910), De Lucca in 1861 was the first person to study this factor; besides the increase in oil in the fruit, he demonstrated the presence of a considerable amount of mannitol in olive leaves during the whole vegetative cycle of the plant. He pointed out that mannitol is also present in the fruit during the first stage of maturation, and considered it as the raw material for the formation of fatty acids.

Rousille (1878) also studied the maturation of olives, especially in relation to the possibility of a migration of oil from the leaf to the fruit. He found that the amount of oil in the leaves remains almost constant, while it increases continuously in the flesh. For this reason, he concluded that olive oil is elaborated *in situ*, from substances he was unable to isolate; Funaro (1880) reached the same conclusion.

Maxie *et al.* (1960) studied the respiration of olives and could find no clear cut autogenous climacteric related to ripening at temperatures between 15 and 35°C. They considered, from their results, that the olive may have a respiration climacteric while attached to the tree. At 20°C neither ethylene nor 2,4D had an appreciable effect on the rate of respiration. At 25°C ethylene caused a loss of green colour but no softening took place, that is, the fruit did not appear to ripen as a result of the treatment.

Gerber (1897) investigated the respiratory quotient of olives during growth, and found that during the period in which sugar content reaches its maximum value, and before oil is formed, the respiratory quotient is slightly below unity. However, during the period in which sugar content is decreasing and oil content increasing, the respiratory quotient is higher than one (1·46). From these values he concluded that oil is formed from carbohydrates. However, Scurti and Tommasi (1910) investigated changes during the maturation of fruits belonging to the Morella and Rosola varieties, taking samples every 15 days during the whole period of growth. Their determinations included average weight, crude fibre, ash, total nitrogen, proteins, cellulose and pentosans. They found that the ether extract in the early stages of development was composed almost exclusively of oleanol, a compound formerly discovered by Canzonery (1897), and obtained as a pure substance by Power and Tutin (1908). This alcohol is synthesized in leaves and migrates to the fruit. With increasing maturity, oleanol decreases and fatty acids are formed, followed by glycerides. Scurti and Tommasi concluded that oil formation within the mesocarp differs essentially from what happens in oil seeds where oil is formed from carbohydrates.

Nuccorini (1930) studied mannitol changes at different stages of maturity, as well as those of glucose, fructose and oleanolic acid. He believed that mannitol is produced by reduction of fructose and that it contributes to oleanolic acid formation from which glycerides are formed. Nichols (1930) made a detailed study of the changes in sugar content and related substances in olives, concluding that the ratio of reducing to total sugars decreases as the season progresses, and that the total sugar percentage increases early in the season and decreases later.

Leoncini and Rogai (1932, 1934) studied changes in the composition of olives at different stages of maturity, and found that oil and sugars increased during the growth. They did not, however, suggest any relation between the two substances but indicated that oil reaches a maximum some time before the fruit is mature. Cruz Valero *et al.* (1939) and Grado Cerezo (1943) also investigated the changes in oil content during fruit development and maturation, in an attempt to fix the best time for harvest. Later Borbolla *et al.* (1955, 1961) emphasized the importance of sampling in order to get significant and comparable results and studied the changes during fruit development and maturation of the main fruit characteristics (average weight, flesh-pit ratio, mean volume and density). They also studied the principal components of the flesh such as water, crude fat (ether-extract) crude fibre,

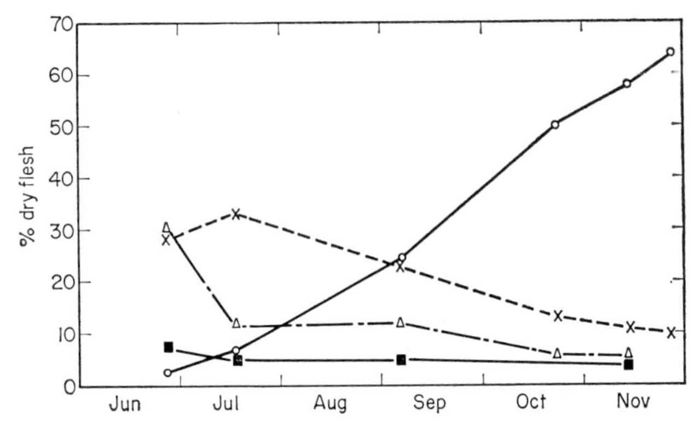

FIG. 2. Typical changes in olive flesh components for Gordal variety. ○, crude oil; ×, reducing sugars; △, crude fibre; ■, total N × 6.25.

total nitrogen and proteins, ash, reducing and non-reducing soluble sugars, starch, pentoses, mannitol and some oil characteristics. Changes in free and total acidity, pH and buffer capacity of the juice extracted from olive flesh, as well as its content in oxalic, citric and malic acids, were also determined by Fernandez Diez and Gonzalez Pellisso (1956). The main findings were as follows: average weight, mean volume, and flesh–pit ratio show increasing values during the whole period of growth and maturation. Fruit density decreases slightly near the time when colour starts changing. This fact has been used by Fernandez Villasante and Fernandez Diez (1968) to classify, before pickling olives of the same size but at different stages of maturity, by floating them on water and on brines of different concentrations. At the beginning of fruit development, when oil content is still low, the percentage of water reaches a maximum value and thereafter decreases up to the final period of growth and maturation, during which water content keeps approximately constant, except for slight fluctuations depending on rainfall. The decrease in water, parallel with the progressive increase in oil, seems to confirm Pitman's conclusion (1935) that the higher the oil content the lower the water percentage in the flesh. It will be seen from Fig. 2 (Borbolla *et al.*,

1955) that during the growing season for the Gordal variety, after an initial rise, reducing sugars show a continuous decline up to maturity. Mannitol, non-reducing soluble sugars, pentoses and starch follow the same declining trend.

Crude fibre decreases with growth and maturation and it appears that, as a structure support, it is formed only during the first stage of fruit development. Texture measurements on olives made recently by Fernandez Diez and Cordon Casanueva (1966) and Fernandez Diez and Vidal Sigler (1968) agree with this suggestion.

The total nitrogen content is low throughout the whole period of growth, a point also made by Narasaki and Katakura (1954), who indicated that proteins in olives are of a complex nature, probably being associated with other compounds such as lipids and carbohydrates. The difficulty experienced in extracting them, as indicated by Fernandez Diez (1958), supports this suggestion. Combination with phenolic compounds might also be a factor here. No attempt has been made to follow quantitative changes in amino acids during fruit development.

The ash content remains almost constant on a weight basis, indicating that fruits absorb mineral elements continuously during growth and maturation. A system having a maximum buffer capacity between 3 and 4 pH units, and constituted by a complex mixture of organic acids and their salts, maintains in the juice of the flesh a pH value between 4·40 and 5·05 depending on the variety, during the whole period of fruit development. The total acidity of the juice increases until the colour begins to change. At this point the value becomes stable or decreases slightly. The combined acid accounts for 60–70% of the total acidity. No significant variations have been found in the individual acids studied.

Sandret (1957) made a study of the chemical composition of olive flesh and its changes during fruit development and maturation for three varieties growing in Morocco, i.e. Picholine Marocaine or Zitoun, Hojiblanca and Gordal. He compared his results with those obtained by Borbolla *et al.* (1955, 1961) and Uzzan *et al.* (1956), and considered oil synthesis to be the most characteristic phenomenon commencing when the fruit is already well formed and sclerification of endocarp has almost finished, that is, at the end of July or the beginning of August. According to Sandret (1957), soluble sugars are responsible for oil formation. Sugars are continuously entering the fruit for the whole period of growth. Mannitol, which decreases regularly, appears to participate in oil synthesis after its oxidation to either fructose or mannose.

Sucrose disappears when olives become over-ripe. For the varieties studied by Sandret (1957), the fructose/reducing sugars ratios exhibited higher values (between 0·40 and 0·45) than those obtained by Borbolla *et al.* (1955). Other sugars, such as xylose, galactose and mannose, were also found in ripe and over-ripe olives, although they are present in very small proportions.

Changes found in the composition of the oil, mainly the increase in unsaturated fatty acids, were also in agreement with the earlier results of Borbolla *et al.* (1955) and Uzzan *et al.* (1956), and also with the later results of Rottini and Balestrieri (1964). A qualitative study of the variations in the most important fat-soluble olive components during fruit maturation has been made by Vazquez Roncero *et al.* (1965). Squalene does not appear until the beginning of August. Although triglycerides are present from the commencement of fruit development, significant amounts are not found before this date. The distribution of the various triglycerides having different degrees of unsaturation remains the same throughout the whole period of growth and maturation. Appreciable amounts of the 1,2-diglycerides always exist and increase in the same way as the triglycerides. Esters and crategolic acid a pentacyclic triterpene (see Volume 1, Chapter 12) remain constant, while oleanolic acid decreases and triterpenic alcohols increase during the development of the fruit.

In 1914, the Bureau of Chemistry of the U.S. Department of Agriculture recognized the need for standards of maturity for the Californian olive, and an investigation was conducted by Hilts and Hollingshead (1920). They found that of the various chemical changes occurring in the olive during ripening, the change in oil content is probably the best indication of maturity. Pitman (1935) came to the same conclusion, and considered that, for Californian varieties, the colour of the fresh fruit gave a good indication of oil content.

III. PROCESSED OLIVES

A. Methods

The olive industry has its roots in "home" pickling and other empirical methods of treatment. It is not surprising, therefore, that some of the procedures are still in the "pre-historic" stage of development. This brief description of processing methods will be confined to the three which are of particular importance either because of the high quality of the product or because of the extent of their commercial usage. These are, those used to produce Spanish-style pickled green olives, Californian-style ripe olives (black or green) and Greek naturally ripe olives (see Figs 3, 4 and 5).

1. *Spanish-style pickled green olives*

In this procedure, fresh olives of green to yellowish-green colour are treated with a diluted lye solution to eliminate most of the bitter glucoside. Lye solutions generally contain from 1·4 to 2·5% of sodium hydroxide, according to the olive variety and degree of maturation, and are allowed to penetrate the flesh up to two-thirds of the way to the pit. After the alkaline treatment, olives are water washed several times for variable periods of time to remove most of the lye, and, finally, put into a sodium chloride brine generally of from 9 to 11° Be in which they undergo a lactic fermentation for a length of time depending on the previous treatment, variety, temperature, microbial population, etc. Traditionally, lye treatment and water washing is done either in wooden tanks or, more frequently, in concrete vats. Fermentation used to be carried out in wooden barrels capable of holding 136–440 kg of olives, but nowadays there is a general tendency to perform the whole process (lye treatment, washing, brining and fermentation) in tanks of 10–20 tons' capacity, made from either paraffin or plastic coated concrete, glass fibre, plastics, stainless steel, etc. Different types of containers for this purpose have been tried mainly in the Argentine, Greece, Israel, Spain and the United States.

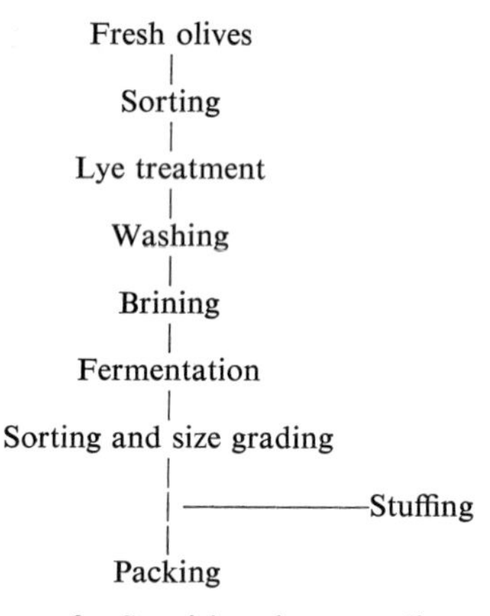

FIG. 3. Scheme for Spanish-style green olives.

2. *Californian-style ripe olives (black and green)*

For these two products, olives are picked when the colour is between green and cherry red. If the fruit is too green, the final product has a poor flavour; on the other hand, over-ripe olives become too soft during processing. In the preparation of black olives, an air oxidation process must be included to oxidize the phenolic pigments. The olives are treated 3–8 times with 0·5–2% NaOH with alternate dipping in water and aeration. The fruits are then washed vigorously with water several times to remove the lye. The black colour is fixed in a 0·1% aqueous solution of ferrous gluconate and the fruit is then put into 3% sodium chloride brine, pasteurized and canned in brine. Finally, the cans are sterilized for 50 min at 121°C or for 60 min at 116°C. For green olives, oxidation must be avoided and the fruit is processed immediately on reaching the factory. With black olives, processing is not always carried out immediately; the fresh fruits may be placed in tanks in a holding solution of brine, having an initial concentration of from 5 to 7% sodium chloride.

During the holding time, salt concentration is increased to 9–10% by successive additions of salt. A mild lactic fermentation is produced and helps to preserve the fruit until the time of processing.

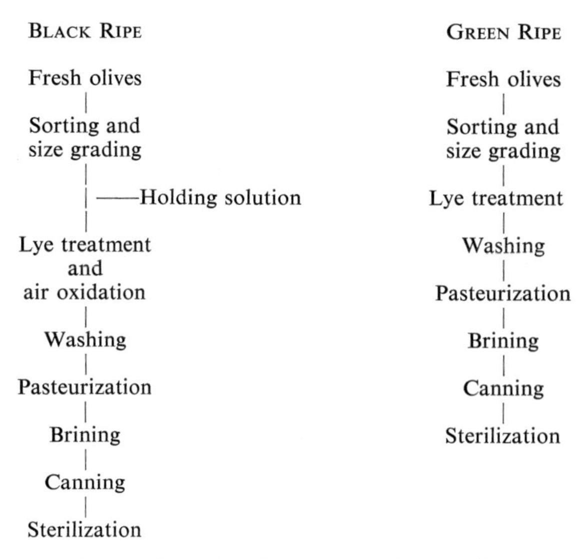

FIG. 4. Scheme for Californian-style ripe olives.

3. *Greek naturally ripe olives*

Greek ripe olives are generally prepared by three different methods. For naturally ripe olives in brine, the fruits, generally picked when they have changed to a purple or black colour, are put into either wooden or concrete tanks which are coated with paraffin or plastic paint and which have from 1 to 20 tons capacity. The olives are then covered with sodium chloride brine of about 10° Be. During this brine treatment—salt is added from time to time to give a final concentration of approximately 10%—an unknown type of fermentation occurs as a result of the action of lactobacilli, yeasts, etc. When the fermentation is finished, the olives are size and colour graded, and packed in fresh brine either in tin containers for 10–15 kg of olives or in paraffin-coated barrels of about 136 kg capacity. This type of product is used mostly as a food, rather than as an appetizer. Alternatively, the final product, especially with olives of the Kalamata variety, is packed in vinegar brine and good quality olive oil; this product is generally used as an appetizer.

A third method is frequently used in which the olives are allowed to become over-ripe on the tree, picked and placed in baskets and washed with water. After 2–3 days, the olives are removed to fresh baskets in which they are placed in alternate layers of solid salt. By this means, the natural wrinkles become more pronounced and the partially dried product keeps well due to the high salt concentration; it is used as a food chiefly for sale in local markets.

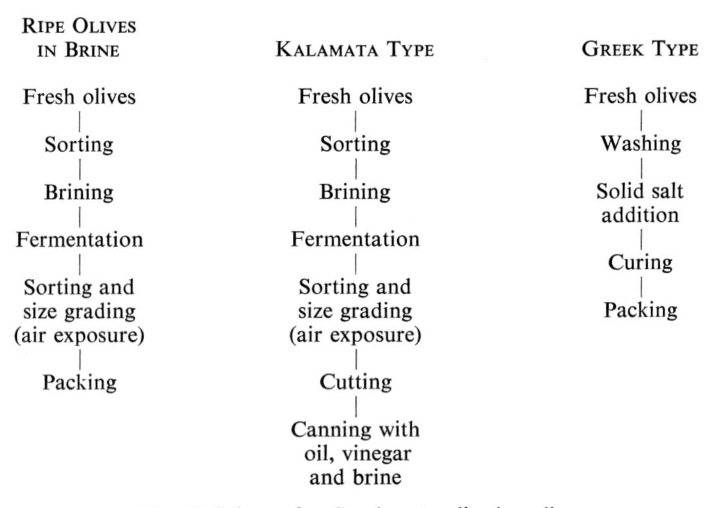

FIG. 5. Scheme for Greek naturally ripe olives.

Detailed descriptions of these methods, summarized in Figs 3, 4 and 5, are given by Borbolla *et al.* (1961) for Spanish-style, Vaughn (1954) and Cruess (1958, 1964) for Californian-style, Balatsouras (1967) and Fernandez Diez (1969) for Greek-style. Other commercial methods, less widely used, such as Sicilian-type olives, differing in one way or another from those previously mentioned, are also described by Vaughn (1954) and Cruess (1958). Those used in France and Italy are described by Savastano (1963).

B. Biochemical Changes during Processing and Fermentation

1. *Spanish-style pickled green olives*

One of the main purposes of lye treatment is to remove the bitter glucoside oleuropein, which, according to Cruess and Alberg (1934), is destroyed by cold sodium hydroxide solutions at pH values between 10·5 and 11·5. The degree of elimination is dependent on time and temperature. The action of the lye, however, is complex, and a number of chemical components of the olive, such as soluble sugars, mannitol, tannins, colouring matters, etc. are at least partially lost during the alkaline treatment, although those changes have not yet been accorded much study. Some of these changes are beneficial and responsible for the special characteristics of the final products; others, such as the loss of sugars, are unfavourable. The penetration rate of the lye as a function of its concentration, its absorption by various parts of the fruit, of permeability changes, sugar removal and conversion to organic acids have been studied by Borbolla *et al.* (1961) for different Spanish varieties mainly Gordal and Manzanilla.

As a general rule, the lye penetration factor is very important. Inadequate lye treatment gives rise to a bitter final product which does not ferment well. On the other hand, an excessive lye treatment can be the cause of a product which is soft, has a low rate of fermentation because of a lack of sugar and a high pH value. Washing operations, to eliminate excess of lye, may produce considerable loss of sugars, depending upon the number and length of these treatments. Table V, from Borbolla *et al.* (1961), illustrates the removal of reducing sugars as a consequence of lye treatment followed by different methods of washing.

TABLE V. Loss of reducing sugars from olives during lye treatment and washing

Variety	Lye treatment % NaOH	Washing Number	Washing Time (hours)	Reducing sugars % fresh flesh	Reducing sugars % in brine
Gordal	3·67	2	4	1·59	1·28
Gordal	3·67	5	25	0·78	0·18
Gordal	1·92	1	1	2·25	1·88
Gordal	1·92	4	24	1·18	0·39
Manzanilla	3·67	2	5	1·15	0·52
Manzanilla	3·67	5	26	0·55	0·10
Manzanilla	1·92	1	2	2·75	2·13
Manzanilla	1·92	4	25	1·47	0·43

Washing of the lye-treated olives has, therefore, to be carefully controlled since if it is inadequate, the fruit will be bitter; if carried too far, there will be a deficiency of sugars for satisfactory lactic fermentation.

The effect of alkaline treatment and washing on the composition of pickled green olives of the Conservolea variety has been studied recently by Polymenakos *et al.* (1967).

When the washed lye-treated olives are placed in the brine, a fermentation process is initiated as a result of which they acquire their special characteristic flavour, odour, texture, etc. This process, which is fundamentally a lactic fermentation of sugars, has been studied in detail by Vaughn *et al.* (1943) and Borbolla *et al.* (1961). Although lactobacilli (almost exclusively *L. plantarum*) are the main organisms responsible, a mixed flora is usually found; many of the micro-organisms present have been isolated and identified by Gonzalez Cancho (1956, 1960, 1963, 1965, 1966a, b). According to Borbolla *et al.* (1958a), the fermentation process can be divided into three well defined stages.

The first stage takes place during the first 48–72 hours of brine treatment. After a lag phase of 24–48 hours, lactobacilli begin to dominate the mixed flora and the pH falls to about 6. At the same time, the concentration of the brine decreases rapidly as it comes into equilibrium with the contents of the fruit. The yeasts present in the brine at this stage differ with the variety of the fruit.

In the second stage, only lactobacilli and yeasts persist after 10–15 days They grow rapidly when the pH has reached 6 and during this stage acidity increases at a rate which depends on the variety of olive. The pH values decrease but, because of buffering by the residual lye which continues to leach from the fruits for as long as 15–25 days, there is no direct correlation between pH and acidity.

The third stage persists until all the fermentable substances have been used up and, therefore, acid fermentation has ceased. When the process is carried out in the open air, weather conditions may stop fermentation. These three stages differ slightly from those reported by Vaughn *et al.* (1943).

At the end of the fermentation period, pH values are generally between 3·7 and 4·0 and the free acidity around 1·0–1·2%, expressed as g of lactic acid/100 ml of brine. The term "residual lye" is applied to the mixture of acetate (from the reaction between the sugars and the lye) and lactate present at the end of the process. It is of the greatest importance to reach the correct pH values at the end of the fermentation process. The correlation between free acidity and pH for different "residual lye" values has been investigated by Borbolla and Gonzalez Pellisso (1964).

When lactic fermentation of pickled green olives does not proceed normally, several types of microbial spoilage may occur during the different stages mentioned above. One of the most important from an economic point of view is that known in Spain as "alambrado" and "vejigas", and as "fish eyes" and "gas pockets" in the United States. The organisms responsible for this type of spoilage are several species of Gram-negative bacilli. It is characteristic of the first stage of fermentation and mainly affects certain olive varieties, such as Morona and Gordal. It can be avoided by acidification of the brine and control of pH (Borbolla *et al.*, 1959, 1960). Other types of spoilage which may lead to economically important losses include butyric fermentations and "zapateria" which have been investigated by Vaughn *et al.* (1943) and Borbolla *et al.* (1961). Other micro-organisms, such as several species of *Clostridium* (Gililland and Vaughn, 1943; Kawatovari and Vaughn, 1956), *Propionibacterium* (Plastourgos and Vaughn, 1957) and *Desulfovibrio* (Soriano, 1955; Levin and Vaughn, 1966), are associated with abnormal fermentations. The organic acid composition, acid–salt relations and sensory characteristics of spoiled brines are completely different from the normal pattern.

The thermal resistance of the main micro-organisms involved in both normal and abnormal fermentations has been studied by Fernandez Diez and Gonzalez Cancho (1964, 1966b, 1967) and Gonzalez Cancho and Fernandez Diez (1967).

2. *Californian-style ripe olives*

Cruess *et al.* (1939) reported that during preliminary storage in holding brine solutions no loss of oil occurred. During lye treatment and subsequent washing with water, a marked decrease was observed in certain constituents, such as sugars, mannitol, tannin and colouring matter, alcohol-precipitable matter and ash. Later Cruess (1958, 1964) reported that olives kept in holding solutions undergo lactic fermentation of sugars, which is influenced by variety, temperature, degree of maturity and salt concentration. As a general rule, not more than 0·4% of free acidity, expressed as g of lactic acid/100 ml of brine, is attained. When moulds and yeasts grow on the surface, softening problems may occur, the source of which is still unknown. Shrivelling can also occur if the initial salt concentration appropriate to the variety is not well chosen by the processor.

During the debittering–washing period, colour changes are important. Simpson *et al.* (1961) isolated and purified a phenolic compound, probably oleuropein, that contributes to colour formation in black olives. Acid hydrolysis yielded glucose and arabinose, and these workers consider that colour is probably due to an interaction between several compounds under the alkaline conditions prevailing. Although only three compounds giving rise to the colour changes were separated, they considered that other compounds in the olive could also be involved (see also p. 262). Pasteurization, after the removal of lye with water, helps to avoid quick fermentation and the softening effects of micro-organisms. The cooking effect of the final sterilization of canned olives probably also improves the characteristic flavour of the product.

Hartmann *et al.* (1959) related the characteristics of the Sevillano, Manzanillo and Mission varieties at harvest to the quality of the black-ripened, processed fruits. In general, the intensity of the dark colour in the processed olive decreased as picking was delayed.

3. *Greek-style naturally ripe olives*

According to Balatsouras (1967), the fermentation of Greek naturally ripe olives in brine is due to the activity of a mixed flora (Balatsouras, 1966a) composed of coliforms, yeasts, and possibly *Lactobacillus* species. The total acidity is always less than 0·5%. Sometimes when a skin composed of moulds, yeasts and bacteria is formed over the surface of the brine, this removes sugars and acids thereby increasing the pH of the brine. This produces a kind of spoilage due to the growth of *Clostridia*, propionic acid bacteria, and, possibly, sulphate-reducing organisms. Softening is also another type of undesirable change. As no lye treatment is used in the preparation of this product, the bitter principle and other fruit components are only partially and slowly leached into the brine. The degree of blackening depends upon, and is favoured by, high pH values and the author considers anthocyanins to be mainly responsible for the colour (see Chapter 16 and Chapter 21).

Changes in the chemical composition of the brine (pH, total acidity, salt, reducing substances, proteins, buffer capacity and tannins) were also investigated by Balatsouras (1966b). Under certain conditions (Balatsouras, 1967) black olives undergo complete lactic fermentation, develop a total acidity as high as 0·8–1·0% and can be kept in brines of moderate salt content. However, important colour changes occur and the fruits become cherry-red rather than black. Balatsouras and Polymenakos (1964) have also investigated a new product using a lye treatment similar to that used for green olives but applied to naturally ripe fruits.

IV. OLIVES FOR OIL

A. Post-harvest Problems

Harvesting of olives for oil has necessarily to be done within a relatively short period of time. In most of the olive-producing countries, depending mainly upon the size of the crop, not all the olives reaching the factory can be immediately extracted. As a consequence of the excess of production over processing capacity, it has become necessary to store a considerable quantity for a period of several months before they can be processed. Fruits so kept change more or less rapidly producing poor quality oils if good care is not taken to avoid spoilage. The main problems are flavour changes and an increasing free fatty acid content. In principle, there are three reasons for this acidification: (i) a simple chemical reaction favoured by the increase of temperature; (ii) lipolytic action of enzymes existing in the fruit; and (iii) lipolytic activity of the complex microflora which grow on the stored olives.

Different ways to avoid the increase in free fatty acids have been proposed, such as brining, the use of atmospheres of SO_2 or CO_2, low temperature and dehydration. The most important microbiological and chemical aspects of this subject have been studied and summarized by Borbolla *et al.* (1958b). Although these authors did not find any lipolytic action by fruit enzymes, recently Cantarelli (1964) isolated and identified two lipase-type enzymes from mesocarp and endosperm respectively.

B. Methods of Processing

Olives are first cleaned with water and then crushed and extracted by pressing. To separate the aqueous liquor from oil, decantation, centrifugation and

TABLE VI. Analyses of olive oils (%) from different countries

Identity of oil		Palmitic 16:0[a]	Palmi-toleic 16:1	Stearic 18:0	Oleic 18:1	Linoleic 18:2	Lino-lenic 18:3	19:1	Ara-chidic 20:0	Eico-sanoic 20:1
Greece	1	11·8	0·81	2·66	76·2	7·00	0·65	0·12	0·48	0·32
	2	12·2	0·70	2·67	75·8	6·99	0·61	0·15	0·50	0·36
	Av.	12·0	0·76	2·66	76·0	7·00	0·63	0·14	0·49	0·34
Italy	1	12·7	0·79	2·56	70·4	12·1	0·45	0·07	0·59	0·39
	2	12·9	0·69	2·56	70·6	11·9	0·45	0·06	0·51	0·36
	Av.	12·8	0·74	2·56	70·5	12·0	0·45	0·06	0·55	0·38
Tunisia	1	16·1	2·07	2·06	59·2	19·2	0·62	0·02	0·48	0·27
	2	15·9	2·04	2·06	59·8	19·2	0·45	tr.	0·40	0·20
	Av.	16·0	2·06	2·06	59·5	19·2	0·54	0·01	0·44	0·24
California, pressed	1	8·98	0·71	2·30	78·6	7·72	0·82	0·05	0·41	0·42
	2	9·38	0·71	2·36	78·1	7·59	0·84	0·07	0·42	0·49
	Av.	9·18	0·71	2·33	78·4	7·66	0·83	0·06	0·42	0·46
California, ext. hexane	1	10·8	0·72	2·49	74·3	9·59	0·70	0·41	0·46	0·50
	2	11·0	0·78	2·66	74·0	9·59	0·80	0·33	0·46	0·41
	Av.	10·9	0·75	2·58	74·2	9·59	0·75	0·36	0·46	0·46
Std dev. between duplicates		0·19	0·08	0·10	0·43	0·46	0·12	0·03	0·04	0·06

From Iverson *et al.* (1965).

[a] First figure refers to the number of carbons, second figure to number of double bonds.

filtration processes are used. If all these operations are carried out at room temperature, the oil obtained is called "virgin olive oil" (Gracian Tous, 1968). A second pressing is generally made to exhaust the cakes. Oils coming from either the first or the second pressing which are not of a sufficiently high quality to yield an edible product (a high free acidity, deficient organoleptic characteristics, etc.) need to be refined. This refining process consists of three steps, namely, alkaline neutralization, decoloration with bleaching earths and deodorization with water vapour under vacuum.

Cakes from the pressing of crushed olives are extracted with solvents producing an oil which is called "orujo oil" or "sulphur olive oil". When these oils are of a good quality after refining, they are used for edible purposes. The rest is used for soap making and other industrial purposes.

A summary of the reports on different types of industrial machinery for olive oil extraction was recently published (Martinez Suarez, 1967). Some modern systems, like Westphalia/Enfida (Buchmann, 1967), separate olive pits from the pulp before oil extraction.

C. Olive Oil

Glycerides comprise 95% of olive oil (Vazquez Roncero, 1963). They are accompanied by small amounts of lipo-soluble substances which come from the ripe pulp. Orujo oil contains higher percentages of these latter compounds due to the action of the solvents used in its preparation. Non-glyceride components are saponifiable (for example, chlorophyll and free fatty acids), or unsaponifiable (for example, hydrocarbons and fatty alcohols). Virgin olive oil contains between 0·5 and 1·5% of unsaponifiable matter, but solvent extracted oils may contain as much as 2·5%.

Among the compounds of great importance are those responsible for the aroma and flavour, and these impart special characteristics to olive oil. A research to identify these components by gas chromatography was recently initiated by Gutierrez *et al.* (1968). Organoleptic spoilage (rancidity) during the whole process of storage and marketing of oils has been reviewed by Gutierrez G. Quijano (1967).

Iverson *et al.* (1965) compared the fatty acid composition of olive oil from several countries; some of his findings are given in Table VI. Gracian (1968) has recently provided an excellent, up-to-date review of the chemistry and analysis of olive oil (see also Volume 1, Chapter 9, p. 229).

V. CONCLUSION

Several hundred papers have been published on olives, but those selected here are sufficiently comprehensive to give a picture of the research carried out to date, particularly on table olives. After reading this chapter, it is easy to understand that there is still much to do in the field. While it is true that an examination of the biochemical, biophysical and microbiological problems involved is being developed, it is clear that more fundamental studies are required for a complete understanding of the different methods of processing and fermentation at present in use. This would, undoubtedly, lead to better quality products especially with regard to table olives. Here is an exciting field of research open to the biochemist and microbiologist.

REFERENCES

Balatsouras, G. D. (1964). *Infs oleic. int.* **28,** 131.

Balatsouras, G. D. (1966a). "Contribution to the Study of the Chemical Composition and the Microflora for the Stored in Brine Greek Black Olives." National Printing Office, Athens, Greece.

Balatsouras, G. D. (1966b). *Grasas y aceitas* **17,** 83.

Balatsouras, G. D. (1967). "Processing the Naturally Ripe (Black) Olives." International Olive Oil Seminar Perugia-Spoleto, Italy.

Balatsouras, G. D. and Polymenakos, N. G. (1964). *Infs. oleic. int.* **27,** 153.

Baldini, E. and Scaramuzzi, F. (1963). *In* "Olive da Tavola" (E. Baldini and F. Scaramuzzi, eds,) pp. 61–111. Edagricole, Bologna, Italy.

Borbolla y Alcala, J. M. R. de la and Gonzalez Pellisso, F. (1964). *Grasas y aceitas* **15,** 233.

Borbolla y Alcala, J. M. R. de la, Fernandez Diez, M. J. and Gonzalez Pellisso, F. (1955). *Grasas y aceitas* **6,** 5.

Borbolla y Alcala, J. M. R. de la, Fernandez Diez, M. J. and Gonzalez Cancho, F. (1959). *Grasas y aceitas* **10,** 221.

Borbolla y Alcala, J. M. R. de la, Fernandez Diez, M. J. and Gonzalez Cancho, F. (1960). *Grasas y aceitas* **11,** 256.

Borbolla y Alcala, J. M. R. de la, Gomez Herrera, C., Gonzalez Cancho, F. and Fernandez Diez, M. J. (1958a). *Grasas y aceitas* **9,** 118.

Borbolla y Alcala, J. M. R. de la, Gomez Herrera, C., Gonzalez Cancho, F. and Fernandez Diez, M. J. (1958b). "Conservación de Aceitunas de Molino." Sindicato Nacional del Olivo, Madrid, Spain.

Borbolla y Alcala, J. M. R. de la, Gomez Herrera, C., Gonzalez Cancho, F., Fernandez Diez, M. J., Gutierrez G. Quijano, R., Izquierdo Tamayo, A., Gonzalez Pellisso, F., Vazquez Ladron, R. and Guzman Garcia, R. (1961). "El Aderezo de Aceitunas verdes." Consejo Superior de Investigaciones Científicas, Madrid, Spain.

Buchmann, E. (1967). *Infs oleic. int.* **37,** 47.

Cantarelli, C. (1961). *Ital. Sotanz. Grasse* **38,** 69.

Cantarelli, C. (1964). *Infs oleic. int.* **26,** 123.

Canzonery, F. (1897). *Gazz. chim. ital.* **27,** II, 1.

Cruess, W. V. (1958). "Commercial Fruit and Vegetable Products." McGraw-Hill, Inc., New York.

Cruess, W. V. (1964). *Infs oleic. int.* **26,** 91.

Cruess, W. V. and Alberg, C. L. (1934). *J. Am. chem. Soc.* **56,** 2115.

Cruess, W. V. and Sugihara, J. (1948). *Archs. Biochem.* **16,** 39.

Cruess, W. V., El Saifi, A. and Develter, E. (1939). *Ind. Engng Chem. ind. Edn.* **31,** 1012.

Cruz Valero, A., Grado Cerezo, A. de, and Gutierrez y Fernandez Salguero, A. (1939). *Bolm. Inst. Invest. agric.* **3,** 211.

Deidda, P. (1965). *Fruiticoltura* **27,** 443.
Eisner, J., Mozingo, A. K. and Firestone, D. (1965). *J. Ass. off. agric. chem.* **48,** 417.
Fernandez Diez, M. J. (1958). *Grasas y aceitas* **9,** 163.
Fernandez Diez, M. J. (1960a). *Grasas y aceitas* **11,** 19.
Fernandez Diez, M. J. (1960b). *Grasas y aceitas* **11,** 173.
Fernandez Diez, M. J. (1960c). *Grasas y aceitas* **11,** 220.
Fernandez Diez, M. J. (1961). *Grasas y aceitas* **12,** 67.
Fernandez Diez, M. J. (1969). *Grasas y aceitas* **20,** 12
Fernandez Diez, M. J. and Cordon Casanueva, J. L. (1966). *Grasas y aceitas* **17,** 88.
Fernandez Diez, M. J. and Gonzalez Cancho, F. (1964). *Microbiología esp.* **17,** 225.
Fernandez Diez, M. J. and Gonzalez Cancho, F. (1966). *Microbiología esp.* **19,** 119.
Fernandez Diez, M. J. and Gonzalez Cancho, F. (1967). *Rev. Ciencia apl.* **115,** 117.
Fernandez Diez, M. J. and Gonzalez Pellisso, F. (1956). *Grasas y aceitas* **7,** 185.
Fernandez Diez, M. J. and Vidal Sigler, A. (1968). *Grasas y aceitas* **19,** 199.
Fernandez Villasante, J. and Fernandez Diez, M. J. (1968). *Infs oleic. int.* **42,** 59.
Ferrao, J. E. M. (1957). *Bolm Jta nac. Azeite* **45-46,** 49.
Foytik, J. (1960). "California Olive Industry." *Calif. agric. exp. Sta. Circular* 493.
Funaro, A. (1880). "L'Agricoltura italiana" Vol. V, pp. 56, 57.
Garoglio, P. J. (1950). "Tecnologia de los Aceites vegetales. Vol. II. El Aceite de Oliva y su Industria." Ministerio de Educación, University of Cuyo, Mendoza, Argentina.
Gerber, C. (1897). *C. r. hebd. Séanc. Acad. Sci., Paris* **125,** 658.
Gililland, J. R. and Vaughn, R. H. (1943). *J. Bact.* **46,** No. 4,315.
Gonzalez Cancho, F. (1956). *Grasas y aceitas* **7,** 81.
Gonzalez Cancho, F. (1960). *Grasas y aceitas* **11,** 125.
Gonzalez Cancho, F. (1963). *Microbiología esp.* **16,** 221.
Gonzalez Cancho, F. (1965). *Grasas y aceitas* **16,** 230.
Gonzalez Cancho, F. (1966a). *Rev. Ciencia apl.* **108,** 24.
Gonzalez Cancho, F. (1966b). *Rev. Ciencia apl.* **109,** 125.
Gonzalez Cancho, F. and Fernandez Diez, M. J. (1967). *Microbiología esp.* **20,** 73.
Goor, A. (1966). *Econ. Bot.* **20,** 223.
Gracian Tous, J. (1968). *In* "Analysis and Characterization of Oils, Fats and Fat Products" (H. A. Boekenoogen, ed.), Vol. 2, pp. 315–606. Interscience, London.
Grado Cerezo, A. de (1943). *Bol. Inst. Invest. agric.* **8,** 101.
Gutierrez G. Quijano, R. (1967). *Grasas y aceitas* **18,** 68.
Gutierrez G. Quijano, R. and Nosti Vega, M. (1968). *Grasas y aceitas* **19,** 191.
Hartmann, H. T. (1953). "Olive Production in California." *Calif. agric. exp. Sta. Manual* 7.
Hartmann, H. T., Simone, M., Vaughn, R. H. and Maxie, E. C. (1959). *Proc. Am. Soc. hort. Sci.* **73,** 213.
Hilts, R. W. and Hollingshead, R. S. (1920). *U.S. Dep. Agric. Bull.* 803.
International Oil Council, 18th Meeting (1968). Madrid. (A summary in *Infs oleic. Int.* **42,** 13).
Iverson, J. L., Eisner, J. and Firestone, D. (1965). *J. Ass. off. agric. Chem., Wash.,* **48,** 1191.
Kalogueria, S. A. (1932). "Table Olives," Ermou, Athens, Greece.
Kawatovari, T. and Vaughn, R. H. (1956). *Fd Res.* **21,** 481.
King, J. R. (1938). *Hilgardia* **11,** 437.

Leoncini, G. and Rogai, F. (1932). *Boll. Inst. sup. agr., Pisa* **8,** 711.
Leoncini, G. and Rogai, F. (1934). *Boll. Inst. sup. agr., Pisa* **10,** 369.
Levin, R. E. and Vaughn, R. H. (1966). *Fd Sci.* **31,** 768.
Mahecha, G. (1967). M.S. Thesis, University of California, Davis.
Martinez Suarez, J. M. (1967). *Grasas y aceit* **18,** 33.
Maxie, E. C., Catlin, P. B. and Hartmann, H. T. (1960). *Proc. Am. Soc. hort. Sci.* **75,** 275.
Mingo, M. de and Romero, J. M. (1953). *Revta R. Acad. Cienc. exact. fís. nat. Madr.* **47,** 557.
Molinari, O. Ch. and Nicolea, H. G. (1947). "Tratado general de Olivicultura." El Ateneo, Buenos Aires, Argentina.
Narasaki, T. and Katakura, K. (1954). *Tech. Bull. Kagawa ken. agric. Coll.* **6,** 194.
Nichols, P. F. (1930). *J. agric. Res.* **41,** 89.
Nuccorini, R. (1930). *Annali chim. appl.* **20,** 533.
Ortega Nieto, J. M. (1963). "Las Variedades de Olivo cultivadas en España." Ministerio de Agricultura, Madrid.
Patac de las Traviesas, L. Cadahia Cicuendez, P. and Campo Sanchez, E. del (1954). "Tratado de Olivicultura." Sindicato Nacional del Olivo, Madrid.
Pitman, G. (1935). *J. Ass. off. agric. Chem., Wash.* **18,** 441.
Plastourgos, S. and Vaughn, R. H. (1957). *Appl. Microbiol.* **5,** 267.
Polymenakos, N. G., Balatsouras, G. D. and Balatsouras, V. D. (1967). "The Effect of the Type of Processing upon the Fermentability and the Chemical Composition of Green Olives." Division of Agricultural Industries. Athens College of Agriculture, Athens.
Power, F. B. and Tutin, T. (1908). *J. chem. Soc.* **93,** 891.
Rottini, C. T. and Balestrieri, G. (1964). *Infs. Oleic. Int.* **26,** 117.
Rousille, A. (1878). *Annls agron.* **4,** 230.
Sandret, F. G. (1957). "Contribution à l'Etude de la Composition chimique de la Pulpe d'Olive." Ecole Marocaine d'Agriculture, Meknes.
Savastano, G. (1963). *In* "Olive da Tavola" (E. Baldini and F. Scaramuzzi, eds), pp. 217–240. Edagricole, Bologna, Italy.
Scurti, F. and Tommasi, G. (1910). *Annali Staz. chim-agr. sper. Roma*, Series II, Vol. IV, p. 253.
Simpson, K. L., Chichester, C. O. and Vaughn, R. H. (1961). *J. Fd Sci.* **26,** 227.
Soriano, S. (1955). *In* "Primera Conferencia Nacional de Olivicultura" (Ministerio de Agricultura y Ganaderia), pp. 397–401. Buenos Aires, Argentina.
Uzzan, A., Pelle, C. and Mahdi, M. (1956). *Oleagineux* **11,** 705.
Vaughn, R. H. (1954). *In* "Industrial Fermentations" (L. A. Underkofler and R. J. Hickey, eds), Vol. II, pp. 417–478. Chemical Publishing Co. Inc., New York.
Vaughn, R. H., Douglas, H. C. and Gililland, J. R. (1943). "Production of Spanish-type Green Olives." *Calif. agric. exp. Sta. Bull.* 678.
Vasquez Roncero, A. (1963). *Grasas y aceitas* **14,** 262.
Vasquez Roncero, A. (1964). *Grasas y aceitas* **15,** 87.
Vasquez Roncero, A. (1965). *Grasas y aceitas* **16,** 292.
Vasquez Roncero, A. Vioque Pizarro, E. and Mancha Perello, M. (1965). *Grasas y aceitas* **16,** 17.

Chapter 8

The Persimmon

SABURO ITO

Horticultural Research Station, Ministry of Agriculture and Forestry, Okitsu, Shimizu, Shikuoka-Ken, Japan

I Introduction 281
A. Brief History 281
B. General Morphology 281
II Principal Varieties of Commercial Significance 283
A. Non-astringent Varieties 284
B. Astringent Varieties 284
III Constituents of the Fruit 288
A. Major Constituents 288
B. Changes during Growth and Maturation 294
C. Effects of Removal of Calyx Lobes 297
IV Post-harvest Treatments 298
A. Controlled Atmosphere Storage 298
B. Methods for Removal of Astringency 298
V Conclusion 300
References 301

I. INTRODUCTION

A. Brief History

Persimmons belong to the genus *Diospyros*. They are fairly common trees in warm regions of the world, mainly in Asia and North America, where nearly 190 species are known. Only four species have been used commercially for the production of fruit. They are *D. kaki* L., *D. lotus* L., *D. virginiana* L. and *D. oleifera* Cheng. *D. kaki* is known as the Japanese persimmon and is the most important species. It has numerous horticultural varieties and is thought to be a native of Japan where it occupies third place to mandarin oranges and apples. Recent studies (Ikegami, 1963), however, suggest that it was introduced from China during the Nara era (A.D. 710 to A.D. 794).

B. General Morphology

This is shown in Table I.

TABLE I. Main species of *Diospyros* producing fruit

Species	Main regions of cultivation	Use	Morphological details		Chromosome No.
D. kaki L.	Japan, China, Korea	fresh and processed	generally monoecious sometimes dioecious[a]	hexaploid	$2n = 90$[b]
D. lotus, L.	Asia	1. source of tannin and vinegar 2. rootstock	monoecious	diploid	$2n = 30$[b]
D. virginiana, L.	North America	1. edible but not normally eaten 2. rootstock		hexaploid, tetraploid	$2n = 60$ and 90[c]
D. oleifera, Cheng	China	tannin production			

[a] Yasui (1915); Namikawa *et al.* (1932).
[b] Namikawa and Higuchi (1928).
[c] Baldwin and Culp (1941).

II. PRINCIPAL VARIETIES OF COMMERCIAL SIGNIFICANCE

The Japanese persimmon has at least 1000 varieties. Based on the differences in flesh coloration as influenced by pollination, the kaki may be divided into at least two groups; (i) those which show no change in the colour of the flesh under the influence of pollination and (ii) those in which the flesh of the fruit is darkened as a result of pollination. Since the change in colour in the one case is directly due to pollination and in the other any change is independent of pollination, we shall refer to those varieties which undergo no change in colour as "pollination constant", and those which are light coloured when seedless and dark coloured when with seeds (pollinated) we shall call "pollination variant" (Hume, 1914).

There are two major groups, non-astringent and astringent. Also, both groups may be sub-divided into two further groups in relation to the effects of pollination as discussed in the previous paragraph. Thus, the Japanese persimmon can be classified into the four following groups (Mori, 1953).

1. Non-astringent and pollination constant (PCNA) varieties: Fuyu, Jiro, Gosho, Suruga. They usually have dark tannin spots.
2. Non-astringent and pollination variant (PVNA) varieties: Zenjimaru, Shogatsu, Mizushima, Amahyakume. They have dark tannin spots and may become astringent when seedless.
3. Astringent and pollination constant (PCA) varieties: Yokono, Yotsumizo, Shakokushi, Hagakushi, Hachiya, Gionbo. These varieties have no dark tannin spots.
4. Astringent and pollination variant (PVA) varieties: Aizumishirazu, Emon, Koshuhyakume, Hiratanenashi. Astringent when pollinated with some dark tannin spots around the seeds.

TABLE II. Production of persimmons in Japan

Year	Area (hectares)	Production (tons)
1966	38 500	419 400
1967	38 400	504 300
1968	37 800	450 000
1969	36 900	444 100

In Japan, non-astringent varieties are predominant and occupy about 55% of the total cultivated area (see Table II). Non-astringent varieties are eaten as fresh fruit, but astringent varieties are edible only after the removal of astringency or as a dried fruit.

A. Non-astringent Varieties

1. *Fuyu*

The most popular non-astringent variety and occupies more than 80% of the cultivated area of the sweet varieties. An average temperature of 15°C or higher throughout the year is desirable for its cultivation. The fruit keeps its quality well in storage and has a pale yellow flesh and sweet taste with 2–4 seeds per fruit. The harvesting season is November.

"Matsumoto Early Fuyu" is an early type of Fuyu and the time of harvest is about 20 days earlier than for Fuyu.

2. *Jiro*

Excellent quality and fine texture. The quality is retained under good storage conditions. The area of cultivation is second only to Fuyu. Average weight is 250–260 g. The harvesting season is late October to early November.

3. *Gosho*

An old variety with probably the best quality of all the varieties. It has a fine texture, an excellent, sweet taste and a beautiful appearance. The rind is orange-red when ripe. Average weight is about 150 g. Yield is poor because of fruit drop during maturation. It is a useful variety for breeding.

4. *Suruga*

A new and promising variety which appeared in 1959. It is derived from the Hanagosho and Okugosho varieties (Iikubo *et al.*, 1961). The tree is vigorous, gives a good yield of an average weight of approximately 200 g. It has an orange-red rind and is of excellent quality being sweeter than Fuyu and it stores well.

Examples of two non-astringent varieties are shown in Fig. 1.

B. Astringent Varieties

1. *Hiratanenashi*

The tree is vigorous. The fruit is flat in shape, seedless and of excellent quality. It has one defect, namely that the flesh becomes soft after treatment for removal of astringency and cannot be kept for long periods. The dried product is excellent. Harvest is from mid-October to November.

2. *Hachiya*

Generally yields well and has an oblong shaped fruit. The fruit may become dark after treatment for the removal of astringency. It is suitable for making the dried product. Average weight is 230–240 g. Few seeds or seedless. The harvesting season is after the middle of October.

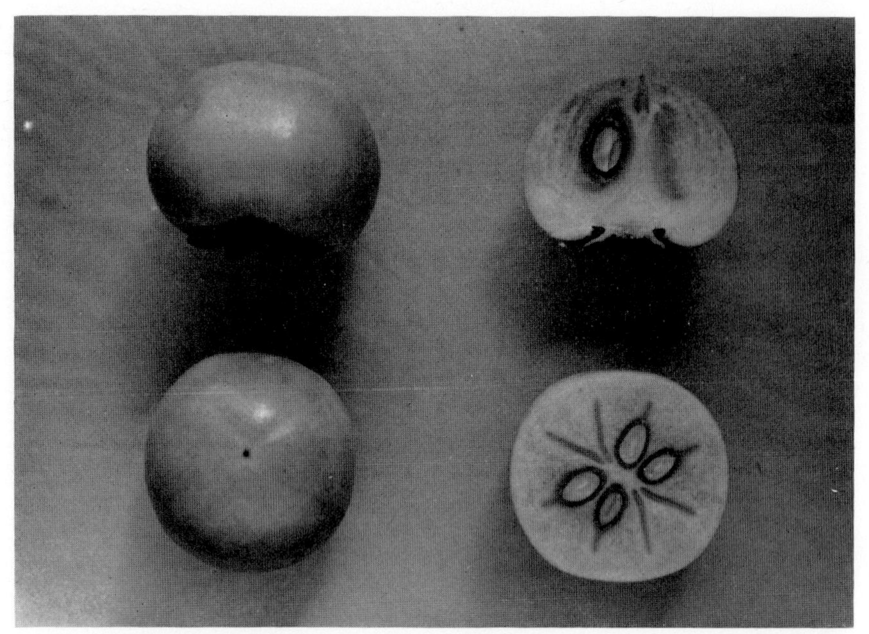

(a) 

(b)

FIG. 1. Two examples of the non-astringent variety. (a) Fuyu. (b) Jiro.

(a)

(b)

FIG. 2. Two examples of the astringent variety. (a) Hiratanenashi. (b) Gionbo.

TABLE III. Major constituents of persimmon fruit on a fresh weight basis

Variety	Av. weight (g)	Water(%)[a]	Sp. grav.(20°C)	Brix degree (BX)	pH	Protein (%)	Crude fibre (%)	Soluble pectin (%)[b]	Soluble tannin (%)[c]
Astringent									
Aizumishirazu (PVA)	243	79·00	1·074	18·0	5·5	0·46	0·38	0·74	0·92
Atago (PCA)	178	79·10	1·075	18·6	5·5	0·47	0·33	0·51	0·80
Hagakushi (PCA)	167	78·00	1·074	19·0	5·3	0·45	0·40	0·68	1·58
Hiratanenashi (PVA)	288	80·40	1·072	19·0	5·3	0·47	0·33	0·63	1·47
Schakokushi (PCA)	277	76·80	1·080	20·8	5·4	0·42	0·47	0·89	1·55
Yokono (PCA)	282	80·00	1·074	19·6	5·4	0·37	0·39	0·55	1·51
Yotsumizo (PCA)	125	79·20	1·078	20·3	5·3	0·36	0·28	0·80	1·68
Average	205	79·10	1·076	19·5	5·4	0·43	0·37	0·68	1·41
Non-astringent									
Fuyu (PCNA)	249	82·40	1·066	16·2	5·5	0·58	0·49	0·68	0

From Ito (1962).
[a] Water: by toluene distillation method.
[b] Pectin: by Carré-Haynes method, determined as Ca-pectate.
[c] Tannin: by one-tenth micro Lowenthal method.

TABLE IV. Sugar content of persimmon fruit on a fresh weight basis

Variety	Brix degree (Bx)	Total sugar (g)	Reducing sugar (g)	Sucrose (g)	Glucose (g)	Fructose (g)	Glucose : fructose
Fuyu	14·8	14·34	13·90	0·42	6·87	7·03	1 : 1·02
Jiro	16·7	14·38	13·78	0·57	6·40	7·38	1 : 1·15
Suruga	17·9	17·14	15·40	1·65	7·34	8·06	1 : 1·10
Gosho	16·4	14·91	13·26	1·57	6·56	6·70	1 : 1·02
Average	16·4	15·19	14·09	1·05	6·79	7·29	1 : 1·07

From Kakiuchi and Ito (1968).

3. *Aizumishirazu*

Non-alternating bearing variety and gives a good yield. It has a round shape, is of medium quality and has a rough texture. Black spots (insoluble tannin) appear around the seeds. It is suitable for the removal of astringency.

4. *Yotsumizo*

Yields well and the fruit has a good texture, sweet taste and is usually seedless. The astringency may be readily removed. The dried product is very small but has a good flavour.

5. *Yokono*

The tree is vigorous and gives a good yield of large fruit of good quality. It retains its quality in storage. When ripe, the rind is orange red. This variety is suitable for the removal of astringency although it sometimes darkens after the treatment. This is a late variety, harvested in November. It is subject to fruit drop during maturation. Average weight is 250–300 g (Kajiura, 1944).

Examples of two astringent varieties are shown in Fig. 2.

III. CONSTITUENTS OF THE FRUIT

Quantitative results for the main constituents of the flesh of the persimmon fruit are given in Table III.

A. Major Constituents

1. *Sugars*

The main sugars in the flesh of the mature fruit are fructose and glucose, the total amount of which is more than 90% of the total sugars. The amount of fructose is always rather greater than that of glucose. Sucrose is present as a minor component (Table IV). This suggests that most of the sucrose translocated from the leaves is enzymatically hydrolysed in the flesh.

2. *Pectins*

In mature fruits, the total amounts of pectic substances ranged from 0·52 to 1·07% in four non-astringent varieties tested (Table V). The pectic substances were divided into three groups on the basis of solubility in consecutive extractions, that is, water-soluble, 64–69%, aqueous sodium hydroxide-soluble, 20–29%, aqueous sodium hexametaphosphate-soluble, 5–10%.

3. *Tannins*†

The remarkable astringency of the persimmon fruit is due to water-soluble tannin present in the tannin cells. The soluble tannins are fluid and easily spread over cut surfaces. Coagulated (polymerized) tannin does not show astringency because of its water-insolubility. As the fruit ripens the tannin cells increase in size and number.

TABLE V. Pectin content of persimmon fruit per 100 g fresh weight[a]

Variety	Water soluble (g)	Na-hexameta-phosphate soluble (g)	Sodium hydroxide soluble (g)	Total pectin (g)
Fuyu	0·360	0·050	0·108	0·518
Jiro	0·370	0·040	0·140	0·555
Suruga	0·640	0·046	0·288	0·974
Gosho	0·740	0·084	0·244	1·070

From Ito and Tada (1969).
[a] Determined by the colorimetric method using carbazole.

The size and density of the tannin cells differ markedly in different varieties. Non-astringent and pollination constant varieties (PCNA) have particularly small tannin cells (Miyabayashi, 1941). The variety Yotsumizo (PCA) has several times as many tannin cells as the varieties Jiro or Fuyu (PCNA). Examples of the size and form of tannin cells are shown in Figs. 3 and 4.

In mature fruit of astringent varieties, the amount of soluble tannin is 0·80–1·94% (average 1·41) of the flesh weight. The main component of kaki tannin is a (+) leucodelphinidin-3-glucoside and has been named "diospyrin" (Ito and Oshima, 1962; Ito, 1962a). Leucocyanidin was also identified as a minor component, the amount present being about one-fifth of that of leucodelphinidin (Joslyn and Goldstein, 1964a, b). In the products of alkali fusion of the latter, phloroglucinol and gallic acid were detected (Fig. 5).

Nakayama and Chichester (1963) suggested that astringency is due to the presence of a *soluble* form of a delphinidin-generating leucoanthocyanin; the leucoanthocyanin itself appears to be present even in non-astringent varieties.

Chromatographically purified kaki tannin readily yields gallic acid, gallocatechin and gallocatechin gallate when heated with dilute hydrochloric acid (see Fig. 5). Treatment with stronger acid yields delphinidin (Ito and Joslyn, 1964). Kaki tannin is inferred to be complex compound, centreing around leucodelphinidin conjugated with, at least, gallic acid and gallocatechin.

†See also Volume 1, Chapter 11.

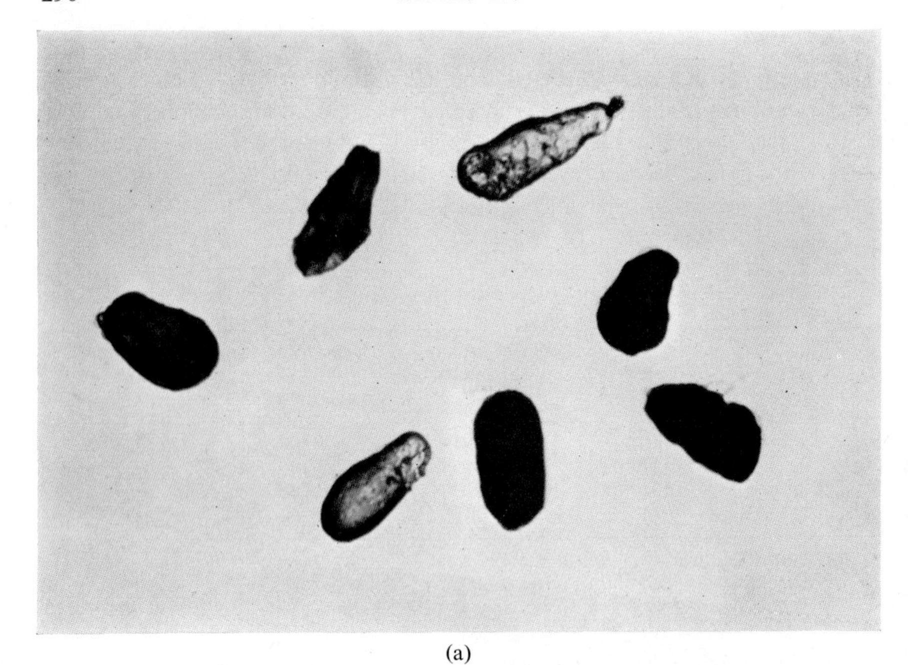

(a)

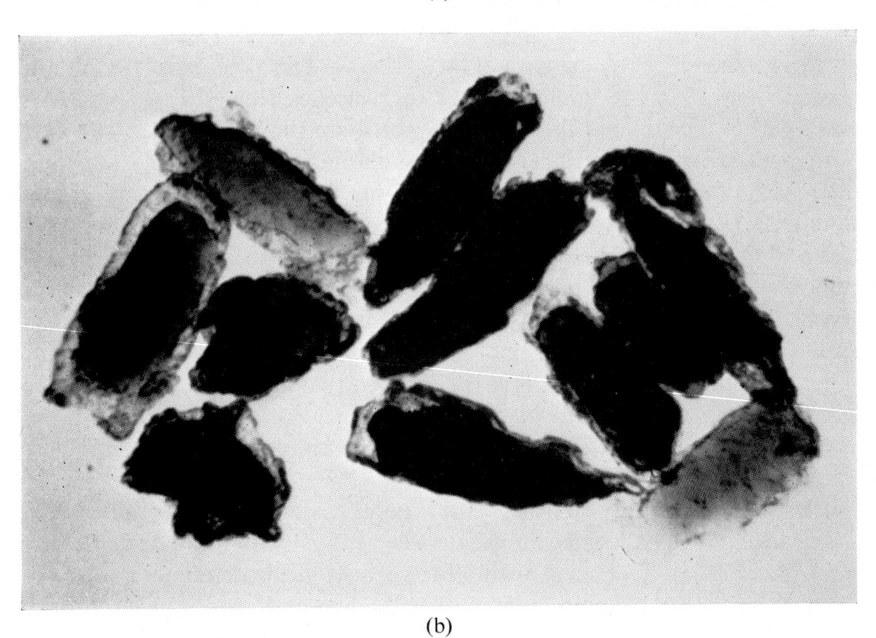

(b)

FIG. 3. Various sizes and forms of tannin cells, non-astringent varieties ($\times$ 230). (a) Jiro. (b) Fuyu. Photographs by N. Kakiuchi.

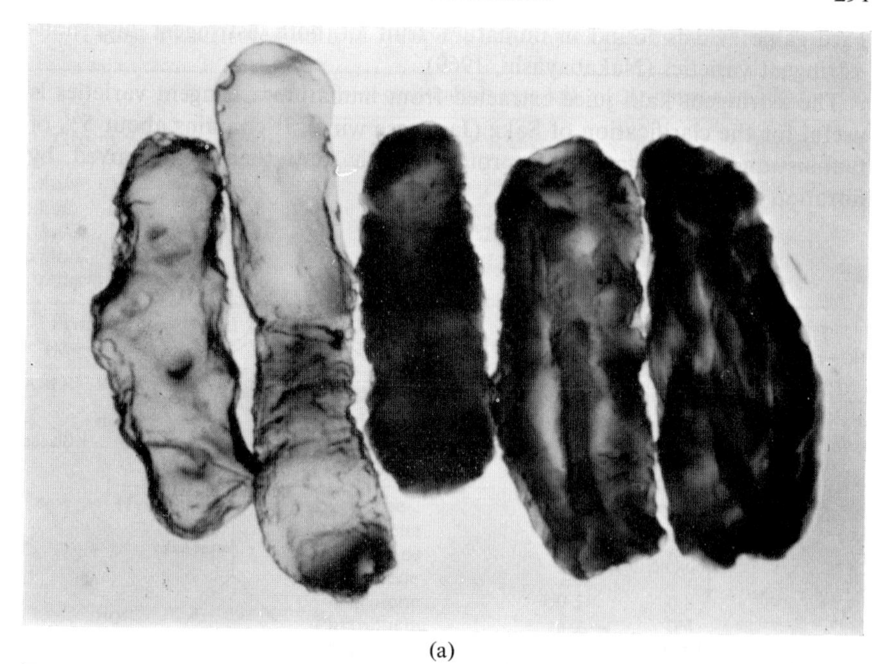

(a)

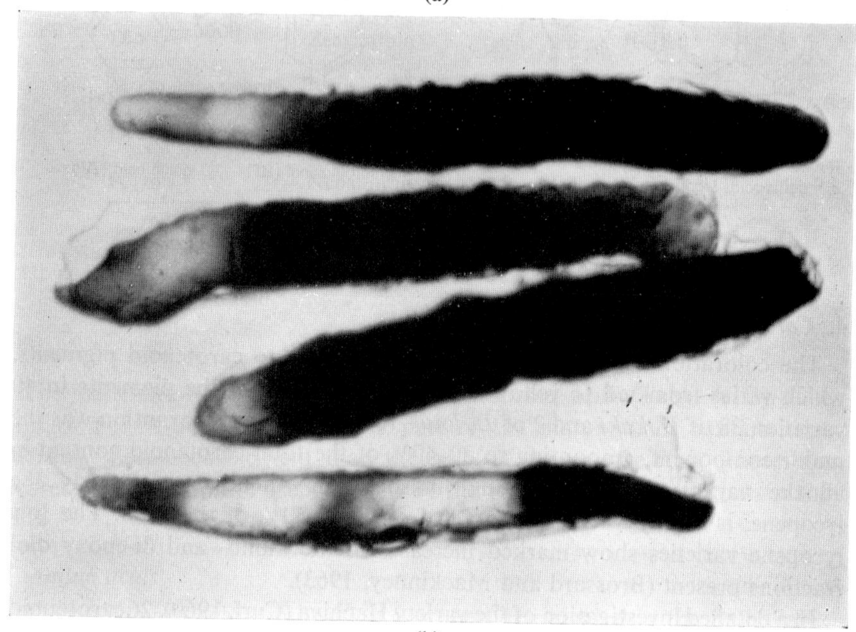

(b)

FIG. 4. Various sizes and forms of tannin cells, astringent varieties (× 230). (a) Yotsumizo. (b) Hachiya. Photographs by N. Kakiuchi.

relationship is unknown, the calyx lobes have high rates of assimilation and respiration and contain auxin, and it would appear that some compound present in the lobes plays an important part in fruit development (Nakamura, 1967; Maeda, 1968).

IV. POST-HARVEST TREATMENTS

A. Controlled Atmosphere Storage

The controlled atmosphere (CA) storage of persimmon fruit is not yet firmly established. The fruit used for storage is almost entirely limited to the non-astringent variety Fuyu. At present, the optimum conditions for CA storage are tentatively suggested as 8% carbon dioxide, 3–5% oxygen, 90–100% relative humidity and a temperature of 1°C. CA storage has the desirable effect of delaying the degreening of the calyx and results in a good quality, marketable product within 3 months.

The fruit may be stored at 0°C and packed in polyethylene bags (0·06 mm thick) giving a slight modification of CA storage condition, that is, a combination of 5% carbon dioxide with 5–8% oxygen, and almost 100% relative humidity (Tarutani, 1960).

B. Methods for Removal of Astringency

1. *Practice*

(a) *Warm water treatment.* The traditional method of removing the astringency from persimmon fruits consists of dipping in warm water at 40°C and storing for 15–24 hours before use. The product is not of very high quality.

(b) *Alcohol treatment.* Fruits treated by the ethyl alcohol method are pleasant to eat, but the method takes a comparatively long time. As the source of alcohol, Japanese Sake, white spirit (35% alcohol), or chemical ethanol is used.

(i) Sake-barrel. About 40–50 kg of fruit are put into a 70 litre sake-barrel. A half litre of the alcoholic source is sprayed on to the fruits. Then the barrel is sealed for 7–10 days at room temperature (20°C).

(ii) Polyethylene bag. About 20 kg of fruit are sprayed with 80 ml of 35% alcohol and covered with a polyethylene bag (0·02–0·03 mm thick). The bag is put into a corrugated board box, and left for 7–10 days. This method may be used during transport of the fruit.

(c) *Treatment with carbon dioxide.* Fruit treated with carbon dioxide gas remains firm and keeps its quality after the removal of astringency. The method is suitable for use on a large scale. The pressure of gas used ranges from 0·7 to 1·2 kg/cm². It requires 3–5 days under a comparatively high pressure at 20°C. Dry ice (solid CO_2) may be substituted for the gas, pieces of the dry ice being scattered throughout the containers. Sufficient dry ice is added to fill the vessel with gas which is then sealed and left for 3–5 days to complete the process.

(d) *Coating with chemicals*. This method is still being investigated. The dry surface of the fruit is coated with a fatty acid such as linoleic acid (2%), an ester of the acid or synthetic liquid resin. The fruit is allowed to stand for several days.

(e) *Freezing*. When fruit is frozen at −25°C for 10–90 days, soluble tannin decreases gradually, but the amount of loss of astringency varies considerably with the variety. As used at present, the method does not give complete removal of astringency (Nakamura, 1961).

(f) *Ionizing radiation*. Attempts have been made to use irradiation of the fruit with 0·15–0·25 Mrad of γ-rays from a ^{60}Co source as a means of removing astringency. Tannins decrease and there is some loss of astringency but the fruit becomes soft (Kitagawa *et al.*, 1964; Ogata *et al.*, 1968).

2. *Mechanism of removal of astringency*

The efficacy of many of the above-mentioned practices suggests that there may be some kind of mechanism in the fruit for decreasing astringency. As the result of inducing abnormal respiration, the carbon dioxide treatment results in an accumulation of appreciable amounts of acetaldehyde and ethanol in the fruit (Komazawa and Uchida, 1956). The acetaldehyde formed during the treatment may react with soluble tannin causing polymerization to an insoluble tannin (Tazaki and Matsuoka, 1924; Ito, 1962b). During the freezing treatment, however, the ethanol and acetaldehyde content does not increase. In this case removal of astringency may result from dehydration of the colloidal substances enclosing the soluble tannin (Nakamura, 1961).

Recently, Kitagawa (1969) has suggested that the removal of astringency during the warm water treatment is not due to chemical change in the tannins such as polymerization or condensation, but is due to the gelling of the protoplasm in the tannin cells.

3. *Physiological problems of astringency removal*

(a) *Partial softening*. "Partial softening" means an abnormal softening of the fruit which occurs around the blossom-end after removal of astringency by carbon dioxide treatment (see Fig. 9). The colour of the softened area changes to dark brown in one or two days. No relation has been found between the degree of partial softening and the stage of ripeness of the fruit. A practical method has not been found for controlling this partial softening, but it may be minimized by using the alcohol method for removing astringency (Sato *et al.*, 1962).

(b) *Darkening*. Dark speckles appear on the peel with the passage of time after removal of astringency by treatment with carbon dioxide. The degree of darkening varies with the variety. The varieties Yokono and Hachiya are most sensitive while Hiratanenashi and Yotsumizo are immune to the disorder.

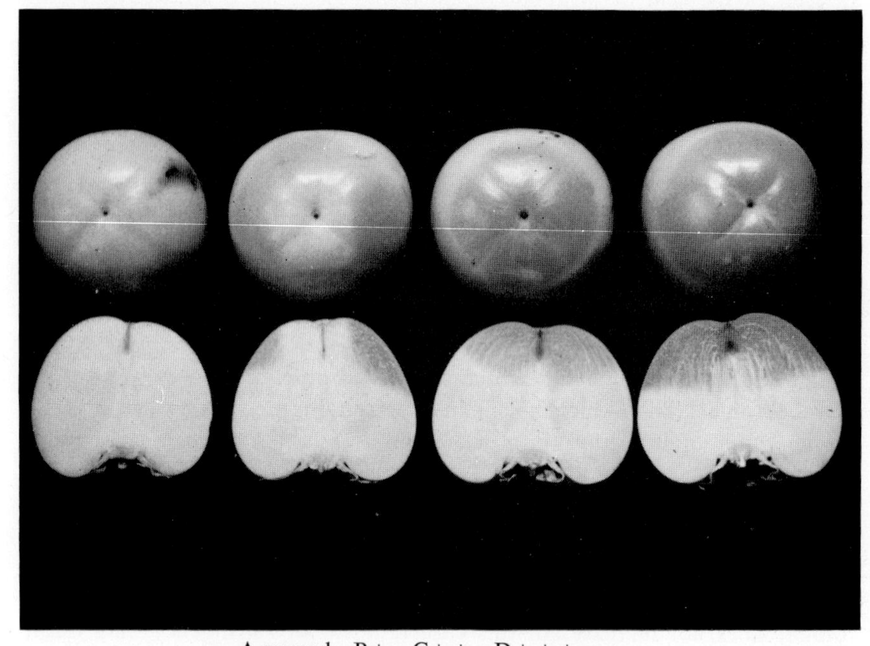

A normal B+ C++ D+++

FIG. 9. Partial softening of persimmon fruit (variety Yokono).

The mechanism of darkening has not yet been explained, although the polyphenol-oxidase and peroxidase activities of the peel have been found to increase gradually and to reach a maximum with the appearance of darkening (Tsukamoto, 1942).

V. CONCLUSION

The foregoing account of our present knowledge of the persimmon fruit emphasizes the paucity of the data relating to the biochemical changes taking place during the development of the fruit. Very little is known of the post-harvest physiology and biochemistry, especially the enzyme processes involved in ripening. A study of the respiratory changes taking place in the various variety-types during maturation would be interesting to determine whether the persimmon is a climacteric or a non-climacteric fruit.

REFERENCES

Baldwin, J. T. and Culp, R. (1941). *Am. J. Bot.* **28,** 942.
Brossard, J. and Mackinney, G. (1963). *Agric. Fd Chem.* **11,** 501.
Curl, A. L. (1960) *Fd Res.* **25,** 670.
Hume, H. H. (1914). *J. Hered.* **5,** 400.
Iikubo, S., Furuhashi, S., Hoshino, M. and Yamamoto, M. (1954). *Tokai-kinki Agr. Exp. Sta. Hort. Div. Bull.* No. 2, 42.
Iikubo, S., Sato, T. and Nishida, T. (1961). *Tokai-kinki Agr. Exp. Sta. Hort. Div. Bull.* No. 6, 33.
Ikegami, T. (1963). "Morphological Studies on the Origin of *Diospyros kaki* in Japan", pp. 1–57.
Ito, S. (1962a). *Abstract, Annual Meeting of Jap. Soc. hort. Sci.* p. 3.
Ito, S. (1962b). *Bull. Hort. Res. Sta. Japan*, Ser. B, No. 1, 1.
Ito, S. and Joslyn, M. A. (1964). *Nature, Lond.* **204,** 475.
Ito, S. and Oshima, Y. (1962). *Agric. Biol. Chem.* **26,** No. 3, 156.
Ito, S. and Tada, T. (1969). *A. Rpt Hort. Res. Sta. Okitsu* No. 5, p. 65.
Joslyn, M. A. and Goldstein, J. L. (1964a). *Agric. Fd Chem.* **12,** 511.
Joslyn, M. A. and Goldstein, J. L. (1964b). *Calif. Agric.* **18,** 13.
Kajiura, M. (1944). *J. Jap. Soc. hort. Sci.* **14,** No. 1, 1.
Kakiuchi, N. and Ito, S. Unpublished data.
Kitagawa, H. (1969). *J. Jap. Soc. hort. Sci.* **38,** 202.
Kitagawa, H., Yamane, H. and Iwata, M. (1964). *Am. Soc. hort. Sci.* **84,** 213.
Kitahara, M., Takeuchi, Y. and Matsui, M. (1951). *J. Jap. Soc. Fd Nutr.* **4,** 138.
Komazawa, T. and Uchida, I. (1956). *J. Utilization Agric. Prod.* **3,** 69.
Maeda, S. (1968). *Res. Bull. Tokushima Tree Fruit Exp. Sta.* No. 2, 1.
Matsushita, A. (1957). *J. agric. Chem. Soc. Japan* **31,** 921.
Miyabayashi, T. (1941). *J. Jap. Soc. Hort. Sci.* **12,** 143.
Mori, H. (1953). *Bull. natn. Inst. agric. Sci. Japan* **E 2,** 1.
Nakabayashi, T. (1968). *J. Jap. Soc. Brewing Ass.* **63,** 1149.
Nakabayashi, T. (1969). Symposium on Natural Phenolic Compounds, Japan, No. 5, 9.
Nakamura, R. (1961). *J. Jap. Soc. hort. Sci.* **30,** 73.
Nakamura, M. (1967). *Res. Bull. Fac. Agric. Gifu University* No. 23, 1.
Nakayama, T. O. M. and Chichester, C. O. (1963). *Nature, Lond.* **199,** 72.
Namikawa, J. and Higuchi, M. (1928). *Bot. Mag., Tokyo* **42,** 436.
Namikawa, J., Sisa, M. and Asai, K. (1932). *Jap. J. Bot.* **6,** 139.
Ogata, K., Yamanaka, H. and Chachin, K. (1968). *J. Fd Sci. Technol., Japan* **15,** 519.
Sato, T., Ito. S. and Shimura, I. (1962). *Bull. Hort. Res. Sta., Japan* Ser. B, No. 1, 48.
Tarutani, T. (1960). *J. hort. Ass. Japan* **29,** 212.
Tazaki, K. and Matsuoka, C. (1924). *Nogaku Kwai Ho.* (*J. Sci. Agric. Soc.*) No. 225, 119.
Tsukamoto, Y. (1942). *J. hort. Ass. Japan* **13,** 321.
Tsumaki, T., Yamaguchi, M. and Hori, F. (1954). *Sci. Rept Fac. Sci. Kyushu University* **2,** No. 1, 35.
Yasui, K. (1915). *Bot. Gaz.* **60,** 362.

Chapter 9A

The Pineapple: General

GERALD G. DULL

U.S.D.A., Southeastern Marketing and Nutrition Research Division, Athens, Georgia, U.S.A.

I Introduction .. 303
A. History .. 303
B. Cultural Practices .. 304
C. Fruit Development and Anatomy .. 304
II Composition .. 305
A. General .. 305
B. Carbohydrates .. 306
C. Lipids .. 307
D. Nitrogen .. 307
E. Inorganic .. 308
F. Miscellaneous .. 308
III Developmental Biochemistry .. 309
A. Definition of Ripeness .. 309
B. Composition Changes during Development .. 311
IV Post-harvest Physiology .. 316
A. Chemical Changes .. 316
B. Respiration .. 317
C. Ethylene Response .. 319
D. Oxygen and Carbon Dioxide Responses .. 320
E. Storage .. 321
F. Physiological Disorders .. 322
References .. 323

I. INTRODUCTION

A. History

The pineapple, sometimes called the King of Fruit, has its probable origin in Brazil or Paraguay. It is a member of the Bromeliaceae family, which is the horticultural source of a number of ornamental plants. Collins (1960) has recorded a detailed account of the history of the pineapple from both a botanical and an economic point of view. Commercial cultural practices in Guinea are thoroughly discussed by Py *et al.* (1957).

The major single area producing pineapples for commerce is the Hawaiian Islands. Substantial quantities of the fruit are produced in Taiwan, Malaysia, Australia, the Philippines, Thailand and several countries in Africa. Although South America is the ancestral home, the pineapple industry in this area is still relatively small in comparison to the other major areas.

The pineapple variety of commerce in Hawaii is *Ananas comosus* (var. Smooth Cayenne). In other areas the varieties of commerce include Red Spanish, Queen, Singapore Spanish, Selangor Green, Sarawak and Mauritius. Unless otherwise stated, the work presented in this chapter refers to the Smooth Cayenne variety. The author spent ten years at the Pineapple Research Institute of Hawaii and has drawn upon the Institute's published and unpublished† information for this chapter. It is felt that concentration on the biochemistry of a single variety grown in a single geographical location would give a more representative picture of the species.

B. Cultural Practices

The pineapple is, in general, cultivated by vegetative reproduction. In Hawaii the planting season is in the autumn and the first fruit are harvested between 18 and 22 months later. Each plant produces one fruit, the first fruit being called the plant crop fruit. Approximately 12 months later a second crop (called the ratoon crop) is harvested. Ratoon fruit are usually smaller than plant crop fruit.

C. Fruit Development and Anatomy

The pineapple is a composite or collective fruit, that is, it is a collection of small fruits called fruitlets. In ancestral species each fruitlet was borne separately on the fruit stem (peduncle). As a result of evolutionary processes the individual fruitlets became fused, thus forming a composite fruit. The general morphological structure of a pineapple fruit is illustrated in Fig. 1. A thorough study of the morphology and anatomy of the pineapple inflorescence and fruit has been carried out by Okimoto (1948).

The edible flesh of the fruit is comprised mainly of ovaries and the bases of sepals and bracts. The white pattern that is quite obvious in slices of many varieties is due to the nectary gland.

An important aspect of pineapple fruit development pertains to the flowering sequence. Flowering begins at the base of the inflorescence and

†The cooperation of the Pineapple Research Institute Board of Trustees in releasing unpublished research results is gratefully acknowledged. Credit should also be given for the research efforts of a large number of co-workers at the Pineapple Research Institute which provide the foundation for this chapter. Should the reader desire additional information regarding the unpublished data covered in this chapter, he should correspond with the author or the Director of the Pineapple Research Institute of Hawaii, Wahiawa, Hawaii 96786.

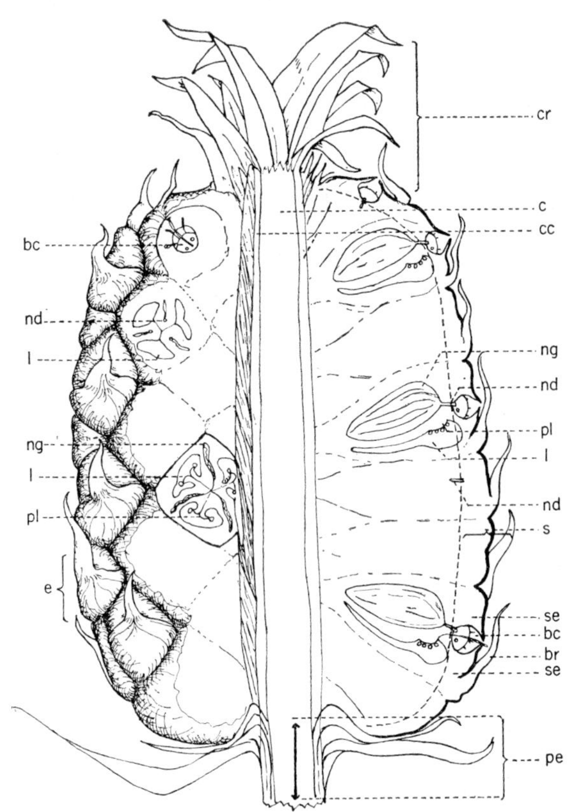

FIG. 1. The general morphology and anatomy of the pineapple fruit. s, shell; c, core; cc, core cortex, cr, crown; pe, peduncle; se, sepal; br, bract; bc, blossom cup; l, locule; pl, placenta; nd, nectary duct; ng, nectary gland. Drawing by Ruby Rice Little.

progresses in a spiral manner to the apex. The number of flowers that open each day varies from one to several, so that flowering can take from 3 to 4 weeks (Kerns and Collins, 1936). Therefore, at a given time, an entire pineapple fruit will represent many discrete stages of physiochemical development. This sets the pineapple aside as being considerably different from other commercial fruit. Since the pineapple is normally handled as a unit, it is necessary to consider the physiochemical properties of the total fruit. This range of physiochemical development in a single fruit contributes substantially to variation in results of experiments with pineapple.

II. COMPOSITION

A. General

The chemical composition of the pineapple fruit has been investigated and selected data on the composition of the edible flesh are presented in Tables

I to V. Unless otherwise stated, the data apply to the Smooth Cayenne variety. Ranges of composition are given whenever possible, illustrating the degree of variation encountered under commercial agronomic operations. It also illustrates clearly the wide range in fruit quality encountered by those who consume fresh pineapple. The data in Tables I to V are assumed to pertain to "ripe fruit". A ripe fruit is defined as a fruit that has attained its full development and its maximum aesthetic and edible quality. From a practical viewpoint this point of perfecıion is not often found in fresh pineapples of commerce in the United States. Therefore, the data on general composition in Table I reflect a broad range of ripeness and agronomic and environmental factors.

TABLE I. General analysis[a] of ripe pineapple fruit flesh

Analysis	Fresh weight (%)
Brix	10·8–17·5
Titratable acid (as citric)	0·6–1·62
Ash	0·30–0·42
Water	81·2–86·2
Fibre	0·30–0·61
Nitrogen	0·045–0·115
Ether extract	0·2
Esters (p.p.m.)	1–250
Pigments (p.p.m. of carotene)	0·2–2·5

[a] Extracted from numerous Pineapple Research Institute publications covering at least thirty years of research.

B. Carbohydrates

Representative carbohydrate composition ranges for ripe fruit are listed in Table II. Direct analyses for starch in fruit were essentially negative. The highest value found was 0·1% of the total dry matter at 60 days before ripeness. Pectin was determined by ethanol precipitation followed by drying in air.

TABLE II. Carbohydrate constituents of ripe pineapple fruit flesh

Constituent	Fresh weight (%)
Glucose	1·0–3·2
Fructose	0·6–2·3
Sucrose	5·9–12·0
Starch	<0·002
Cellulose	0·43–0·54
Hexosans	0·10–0·15
Pentosans	0·33–0·43
Pectin	0·06–0·16

C. Lipids

Relatively little work has been done with the lipid constituents of pineapple other than the yellow pigments. Singleton (1959) reported that sitosterol was probably the major sterol present and that it was present in trace amounts. Gardner (1966) made a preliminary thin-layer chromatography study of fruit phospholipids and reported the relative presence of phosphatidyl inositol (+2), phosphatidyl choline (+4), phosphatidyl glycerol (+1), phosphatidyl ethanolamine (+5) and phosphatidic acid (+5). He also noted a trace amount of monogalactosyldiglyceride.

D. Nitrogen

1. *General*

Since the amount and form of nitrogen fertilizer can affect the nitrogen distribution pattern in fruit, a nitrogen balance sheet for a representative ripe fruit composite sample is presented in Table III, including the free amino acids.

TABLE III. Nitrogen balance sheet for ripe pineapple fruit flesh

General analysis (% fresh weight basis)	
Total nitrogen	0·108
Soluble nitrogen	0·079
Protein	0·181
Ammonia	0·010
Amino acids (total)	0·331

Amino acids (p.p.m. of fresh weight)							
Alanine	407	Cystine[a]	20	Isoleucine	23	Proline	31
γ-Aminobutyric	124	Glutamic	90	Leucine	24	Serine	256
Arginine	46	Glutamine	256	Lysine	46	Threonine	78
Asparagine	1251	Glycine	65	Methionine	134	Tyrosine	58
Aspartic	293	Histidine	48	Phenylalanine	40	Valine	39

[a] Cystine was not measured in this series. The value given is an approximation based on later analyses.

The serotonin (5-hydroxytryptamine) concentration in canned and fresh fruit juice (variety not specified) was reported as 25 and 12 mg/litre respectively by Bruce (1960). The higher level in canned juice could be explained on the basis of different fruit sources. Foy and Parratt (1961) found that unripe fruit contain 50–60 μg of serotonin/g of tissue, while ripe fruit contained 19 μg. Canned juice contained an average of 2·8 μg/ml (variety not specified).

2. *Enzymes*

Four pineapple enzyme systems, bromelin, IAA oxidase, peroxidase and phosphatase, have been studied using the pineapple plant stem as the source material. For details the reader is referred to the work of Heinicke and Gortner (1957) on bromelin; Gortner and Kent (1953, 1958), Gortner *et al.* (1958) and Beaudreau (1964) on IAA oxidase and peroxidase; and Gortner (1957) on phosphatase.

The fruit enzymes have not been investigated as thoroughly as the stem enzymes. Purification and characterization of fruit bromelin was accomplished by Ota *et al.* (1964) (fruit variety not specified). Gortner (1965) reported a protease activity of six milk-clotting units/ml of ripe fruit juice. Nearly half of the protein in the fruit was found in the protease. Ota (1966) isolated and characterized a minor component which was designated fruit bromelin B, the major component being fruit bromelin A. Stem bromelin is basic (Heinicke and Gortner, 1957), while fruit bromelin is acidic (Ota *et al.*, 1964).

With regard to enzymes other than proteases, Gortner and Singleton (1965) found 5 peroxidase units (leucodichlorophenolindophenol method)/ml juice in ripe fruit. Hobson (1962) noted a polygalacturonase activity of 0·09 g galacturonic acid/100 g fresh weight/hour (variety not specified).

E. Inorganic

Representative concentrations of inorganic constituents of ripe pineapple fruit are tabulated in Table IV.

TABLE IV. Inorganic constituents of ripe pineapple fruit flesh

Constituent	mg/100 g fresh weight
Calcium	7–16
Chlorine	46
Iodine	0·006–0·107
Iron	0·3
Magnesium	11
Manganese	0·03
Nitrate	0–120
Phosphorus	6–21
Potassium	11–330
Silicon	11–69
Sodium	14
Sulphur	7

F. Miscellaneous

Miscellaneous constituents of ripe pineapple fruit, including organic acids, pigments, vitamins and phenolics are summarized in Table V.

TABLE V. Miscellaneous constituents of ripe pineapple fruit flesh

Organic acids (% of fresh weight)	
Citric	0·32–1·22
Malic	0·1–0·47
Oxalic	0·005
Total acid (as citric)	0·6–1·62
Pigments (mg/100 g fresh weight)	
Carotene	0·13–0·29
Xanthophyll	0·03
Vitamins (μg/100 g fresh weight)	
p-Aminobenzoic	17–22
Folic acid	2·5–4·8
Niacin	200–280
Pantothenic acid	75–163
A (as alcohol)	0·02–0·04
Thiamine	69–125
Riboflavin	20–88
B_6	10–140
Ascorbic acid	10–25 mg/100 g
Phenolics[a] (μg/g fresh weight)	
p–Coumaric acid	33–73
Ferulic acid	20–76

[a] Analyses for apical meristematic tissue in developing pineapple plant from planting to differentiation.

III. DEVELOPMENTAL BIOCHEMISTRY

A. Definition of Ripeness

Based upon the combined efforts of a number of workers at the Pineapple Research Institute of Hawaii, covering several crop years, Gortner *et al.* (1967) proposed a biochemical basis for horticultural terminology of development, maturation, ripening and senescence. The basis for this proposal is a series of physical and biochemical changes occurring in a very reproducible manner and at specific times during fruit development. A number of these changes are illustrated in Fig. 2. For pineapple, approximately 110 days elapse between the end of flowering and the attainment of ripeness. Marked changes in chemical composition occur when the shell is about half-yellow (ripe) as well as approximately three and seven weeks prior to the half-yellow stage. These points in time permit the delineation of four distinct stages in pineapple fruit development. The following definitions are offered and explained.

Development: the period from the end of flowering to and including the ripening of the fruit.

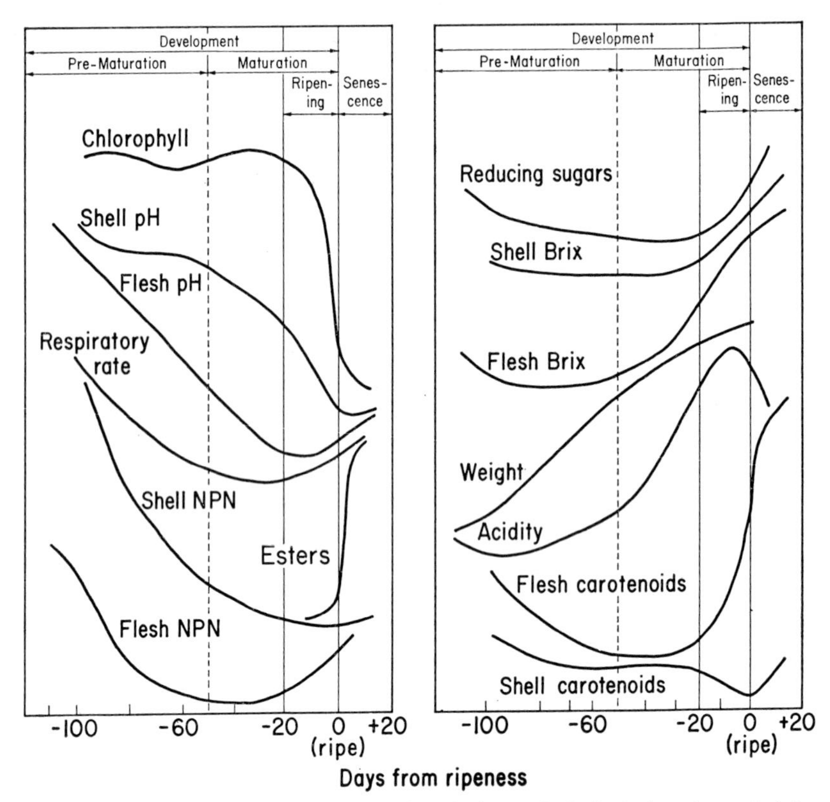

FIG. 2. Changes in physiochemical properties of pineapple fruit during the period from flowering to senescence. Reproduced from Gortner *et al.* (1967).

Pre-maturation: the period of fruit development after the completion of flowering up to 7 weeks prior to half-yellow shell.

Maturation: the 6- or 7-week period prior to half-yellow shell.

Ripening: the final 2- or 3-week period of maturation.

Senescence: the period following the ripening of the fruit.

Note that the onset of maturation is characterized by the start of an accelerated decline in shell pH (Fig. 2). The respiratory rate and the non-protein nitrogen (NPN) in flesh cease their marked decline. Flesh brix and titrable acid begin a marked increase. Flesh carotenoid pigments level off.

The start of ripening is indicated when there is a rapid loss of shell chlorophyll; the flesh pH reaches a minimum and begins to rise; the respiratory rate begins a slow rise after having reached a minimum; volatile esters accumulate rapidly; NPN and sugars in the flesh begin to increase; shell brix increases; flesh pigments increase rapidly while shell pigments decline; and titrable acid reaches a peak and begins to decline.

Chemical indicators reveal when the fruit passes from ripening to senescence, for example, shell chlorophyll is completely gone; shell pH and NPN begin to rise, flesh brix and pigment increase taper off and shell carotenoids begin to increase.

Based upon the preceding, Gortner *et al.* (1967) proposed the following definitions:

Development: the entire period during which new tissue is formed and brought to morphological completion, and perfective chemical changes take place. The period of fruit development covers the stages of pre-maturation and maturation, the latter of which includes ripening.

Pre-maturation: the developmental period prior to the onset of the maturation processes (see below), and generally including at least half the interval between blossoming and harvest. This stage is characterized by extensive cell enlargement.

Maturation: the stage of fruit development during which the fruit emerges from the incomplete stage to attain a fullness of growth and a maximum edible quality. Most of the maturation processes must take place while the fruit is still attached to the plant.

Ripening: the terminal period of maturation during which the fruit attains its full development and its maximum aesthetic and edible quality. Changes taking place during this period are primarily chemical. For some fruits, ripening may occur either before or after fruit harvest; for others, the fruit must be detached for ripening processes to proceed.

Senescence: the period following fruit development during which growth has ceased and the biochemical processes of ageing replace the perfective changes of ripening. Senescence may occur either before or after fruit harvest.

B. Composition Changes during Development

1. *Carbohydrates*

The major carbohydrate constituents in pineapple fruit are the simple sugars sucrose, glucose and fructose. These simple sugars have been studied extensively since taste is a primary quality factor in fresh fruit acceptance. The major change in carbohydrates occurs during the last 40 days of maturation when the total sugars increase dramatically. As long as the fruit is attached to the plant these sugars increase through senescence. At ripeness, the sucrose reaches a peak concentration and then declines, but the reducing sugars continue to increase (Singleton and Gortner, 1965).

It should be noted that there is no starch accumulation in the pineapple fruit and hence no reserve for major post-harvest, perfective, quality changes.

2. *Pectic substances*

Although texture is not considered a major quality problem in fresh pineapple, a situation arose in a canning operation that resulted in a major investigation of cell wall components. In this situation, fruit appeared to be normal in every sense of the word until the shell, core and ends of the fruit had been removed. At this time, the fruit cylinder appeared to sag. When such a cylinder was passed through the slicer, a pile of crushed fruit rather than whole slices was obtained. These fruit were called "fragile". In approaching this problem, the author first obtained developmental trends in cell wall constituents for normal fruit (Table VI). The general methods of Gee *et al.* (1958) were used in these analyses. The marc (alcohol insoluble materials) decreased throughout pre-maturation and maturation, reaching

TABLE VI. Cell wall composition of pineapple fruit flesh at different stages of development

Analysis	Days from ripeness					
	−103	−61	−26	−7	0	+7
Fruit (fresh weight)						
Moisture	91·9	92·5	91·9	84·5	83·9	82·5
Alcohol insoluble (%)	3·22	2·46	1·86	1·58	1·55	1·64
Alcohol insoluble (air dry)						
Moisture (%)	7·96	7·36	6·89	6·35	5·70	5·30
Ash (%)	5·37	4·69	3·19	2·24	2·30	2·83
Protein ($N_2 \times 6{\cdot}25$) (%)	13·0	10·6	8·6	9·3	9·5	10·5
Fibre (corrected for protein) (%)	52·4	58·4	59·8	59·0	56·6	55·2
Anhydrouronic acid (%)	7·23	6·19	5·78	5·43	5·17	4·22
Unaccounted for (%)	19·1	12·8	15·7	17·7	20·7	21·9

a minimum at ripeness, and then began to increase. Fibre appeared to reach a maximum at the onset of ripening and then decreased steadily through senescence. Because of the empirical nature of a fibre determination, the importance of small differences remains to be proved. The anhydrouronic acid exhibited a steady decline in concentration from flowering through senescence. The relative high percentage of unaccounted for material (assumed to be predominantly hemicellulose), coupled with the large increase during maturation, indicated that these substances could play a major role in fruit texture.

Samples of normal and fragile fruit of equivalent ripeness were analysed for cell wall characteristics. The fragile fruit contained slightly more marc than the normal fruit; the significance of this difference is not known. However, the composition of the two marcs was quite different (Table VII). The fragile fruit contained less fibre and more pectin and (presumed) hemicellulose than the normal fruit. The most dramatic difference was in the higher degree of esterification of pectin in normal fruit than in the fragile fruit. It was concluded that the problem of fragile pineapple fruit could be attributed to a decrease in fibre, an increase in total pectin and hemicellulose and a loss in esterification of pectin.

TABLE VII. A comparison of cell wall characteristics of normal and fragile pineapple

Analysis	Normal	Fragile
Marc (% of fresh weight)	1·52	1·79
Analysis of marcs		
Moisture (%)	6·71	7·79
Ash (corrected for CO_3^-) (%)	2·83	3·18
Protein (%)	10·2	10·8
Fibre (%)	62·9	56·7
Anhydrouronic acid (titration) (%)	5·51	7·48
Anhydrouronic acid (colorimetric) (%)	5·60	6·16
Unaccounted for (%)	11·8	15·4
Free acid (meq/g)	0·18	0·37
Total esters (meq/g)	0·76	0·59
Acetyl (meq/g)[a]	0·63	0·54
Esterification (%)[a]	42·4	6·16

[a] See Gee *et al.* (1958).

A corollary to this work was a problem encountered in analysing for pectin in pineapple. The titrimetric method was found to be less reliable than the colorimetric (carbazole) method. The interference was traced to the association of a phenolic moiety with marc samples. The phenolic substance was released by mild saponification. The substance was tentatively identified as ferulic acid.

Preston and Hepton (1959) found that IAA had an effect on both plastic and elastic extensibility in *Avena* coleoptiles. IAA is also effective in binding pectinmethylesterase in tobacco pith and Jerusalem artichoke (Glaziou, 1957). Gortner *et al.* (1958) showed that ferulic acid was a strong inhibitor of pineapple IAA oxidase. In view of these findings, it would appear that a polyphenolic system could play an important role in pineapple fruit texture.

3. *Acids*

The two major acids in pineapple are citric and malic. Citric acid levels during pineapple fruit development follow the total acid pattern (Fig. 2). Short-term environmental factors do not appear to affect the citric acid level (Singleton and Gortner, 1965). This is not true for malic acid, which exhibits an inverse relationship with water evaporation forces (Gortner, 1962).

Although ascorbic acid does not contribute substantially to fruit acidity, it should be noted that there is, apparently, a positive correlation between levels of ascorbic acid in pineapple fruit and the amount of solar radiation to which the developing plant and fruit are exposed (Singleton and Gortner, 1965).

The gross nature of pineapple fruit acids (pH and titratable acidity) has been investigated under widely varying agronomic conditions, but a detailed biochemical study has not been made. It is of special interest to note that the pineapple plant exhibits Crassulacean type acid metabolism (Sideris *et al.*, 1948; Joshi *et al.*, 1965).

4. *Nitrogen*

The major nitrogenous and enzyme constituents in pineapple vary with the stage of fruit development (Gortner and Singleton, 1965).

Several interesting trends in amino acid patterns during pineapple fruit development were reported by Gortner and Singleton (1965).

Glycine and alanine were low during maturation but high at other times (Table VIII).

Methionine was present in very low amounts until fruit ripening began; in the ripe fruit methionine was one of the major amino acids. The basic amino acids lysine, proline, histidine and arginine were present at relatively low levels throughout the entire period of fruit development.

In the same study, protease activity (bromelin) was shown to be very low in fruit just after flowering was complete but within two weeks it rose to a high level. This level was maintained up to the onset of ripening, at which time there was a marked decline in activity. The decrease in protease activity was not followed by an equivalent loss of nitrogen. At about the same time, the esters began to increase markedly. Gortner and Singleton (1965) speculated on an interrelationship of a loss in protease activity, an increase in volatile esters (and related enzymes) and an increase in methionine (one flavour component of pineapple is methyl-β methylthiopropionate).

In contrast to the protease activity, the peroxidase activity started at a high level and decreased steadily throughout the development period.

5. *Pigments*

Gortner (1965) has discussed pigment changes during fruit development. As expected, ripening is correlated with a loss of shell chlorophyll. This is true for fruit attached to the plant but not for the post-harvest situation.

In fruit flesh the carotenoids pass through a minimum concentration about 40 days before ripeness (Fig. 2) and then undergo an extremely rapid increase during the last three weeks of ripening. A key point to be kept in mind is that pineapple fruit carotenoids undergo rapid isomerization in tissue homogenates due to the high acidity (Singleton *et al.*, 1961).

TABLE VIII. Free amino acid concentrations (mM) in the juices of one lot of pineapple fruit during various stages of development

	Pre-maturation (−63 to −98 days)		Maturation (−28 to −48 days)		Ripening (−14 to 0 days)		Senescence (+7 to +14 days)	
Glycine	0·35	(0·08–0·70)[a]	0·10	(0·06–0·14)[a]	0·32	(0·2–0·41)[a]	0·98	(0·74–1·22)[a]
Alanine	0·48	(0·41–0·65)	0·09	(0–0·15)	0·66	(0·43–0·89)	1·29	
Valine	0·18	(0–0·37)	0		0·05	(0–0·16)	0·17	(0·15–0·19)
Methionine	tr	(0–0·05)	tr	(0–0·07)	0·49	(0·40–0·57)	0·66	(0·52–0·80)
Isoleucine	0·12	(0·05–0·24)	0		0·06	(0·04–0·08)	0·09	(0·08–0·10)
Leucine	0·39	(0·15–0·54)	0·05	(0·04–0·06)	0·11	(0·10–0·12)	0·14	(0·13–0·15)
Tyrosine	0·16	(0·06–0·30)	tr	(0–0·02)	0·05	(0·04–0·06)	0·08	(0·08–0·09)
Phenylalanine	0·17	(0·08–0·30)	tr	(0–0·03)	0·08	(0·06–0·10)	0·13	(0·12–0·14)
Lysine	0·18	(0·03–0·34)	0·04	(0·03–0·05)	0·10	(0·05–0·16)	0·09	(0·07–0·10)
Histidine	0·03	(0–0·06)	0·02	(0·02–0·03)	0·07	(0·03–0·13)	0·05	(0·04–0·06)

From Gortner and Singleton (1965).

[a] Values in parentheses are the ranges for the 5, 3, 3 and 2 samples analysed for each of the respective stages.

Gortner (1965) makes a strong point in observing that the "yellowing-up" of a ripe pineapple (shell) is really a degreening. Shell carotenoids actually decrease during ripening and then increase in senescence.

6. *Phenolics*

Gortner *et al.* (1958) showed the presence of p-coumaric and ferulic acids in stem-tip tissue. The p-coumaric acid or its ester serves as a co-enzyme for pineapple IAA oxidase, while ferulic acid is a strong inhibitor. Sutherland and Gortner (1959) identified quinyl-1,4-di-p-coumarate in pineapple stem tissue. Gortner (1963a) demonstrated an inverse relationship between soil moisture stress and the concentration of pineapple stem-tip phenolics.

In unpublished work, the author found a phenolic substance associated with cell wall fractions that interfered with pectin analyses. The material was tentatively identified as ferulic acid (see p.313).

To the best of the author's knowledge, no pineapple fruit quality problem such as tissue darkening has been positively linked with a phenolic substance Although Miller and Heilman (1952) proposed such a reaction in the case of chill injury during storage of several varieties, including Smooth Cayenne, they did not demonstrate its presence.

IV. POST-HARVEST PHYSIOLOGY

A. Chemical Changes

TABLE IX. Comparison of chemical changes in stored pineapple with chemical changes in fruit left on the plant for an equivalent time

Sample	Shell[a] colour	Esters as ethyl acetate (p.p.m.)	Pigment as carotene (p.p.m.)	Ascorbic acid (mg/100 ml)	Brix (°)	Titratable acid as citric (% w/v)
Original	2·00	4·2	0·98	12·4	16·8	1·02
Stored[b]						
2 days	2·70	8·0	1·00	12·0	16·7	0·92
4 days	3·81	11·2	1·12	12·5	16·1	0·93
7 days	5·00	16·6	1·18	13·3	16·6	1·03
10 days	5·00	31·6	1·46	13·0	15·7	0·95
On the plant						
2 days	2·64	8·8	1·06	13·8	16·4	0·85
4 days	3·47	7·0	1·27	11·9	17·3	0·85
7 days	4·74	13·6	1·39	11·3	17·5	0·81
10 days	4·92	53·6	1·73	9·1	17·4	0·67

From Singleton (1959).

[a] A subjective measure of percentage of shell showing yellow colour; 1 = <12%, 5 = >87%.

[b] Stored at ambient temperature; mean temperature was approximately 24°C.

In work carried out by Singleton (1959) at the Pineapple Research Institute of Hawaii, a comparison was made between chemical changes in stored fruit and the chemical changes in fruit left on the plant for an equivalent period of time (Table IX).

The fruit left on the plant underwent the natural ripening trends illustrated in Fig. 2. In the detached stored fruit, both ascorbic acid and Brix trends were opposite to the trends in normal fruit. The titratable acid change in stored fruit was relatively negligible in comparison with the decrease in acidity during the normal ripening process. The trends for pigments and esters were similar in the two sets of fruit except that the stored fruit never attained the levels of perfection of the normal fruit. Sensory evaluation work in later studies indicated that the flavour of stored fruit was not the same as fruit ripened on the plant. It is concluded from these data that the normal ripening process in pineapple fruit ripened on the plant does not take place in fruit harvested at the onset of ripening.

The effect of storage temperature on shell colour and acidity and Brix levels in the flesh was studied by Singleton (1957). From these data (Table X) it appears that high temperatures tend to decrease fruit acidity. The retardation of shell yellowing by refrigeration lessens markedly at 18·5°C. The Brix decreased under all storage temperatures.

TABLE X. Effect of storage temperature on quality of fresh pineapple

	Days	7°C	13°C	18·5°C	24°C	29·5°C	35°C
Shell colour	0	1	1	1	1	1	1
(% of area yellow)	5	2	2	8	57	27	32
	8	2	6	34	80	67	59
	15	4	11	82	98	98	88
Acidity	0	0·68	0·68	0·68	0·68	0·68	0·68
(% calculated on	5	0·89	0·77	0·75	0·70	0·53	0·44
original fresh	8	0·81	0·76	0·78	0·77	0·54	0·36
weight)	15	0·78	0·85	0·90	0·67	0·46	0·32
Brix(°)	0	14·5	14·5	14·5	14·5	14·5	14·5
	5	14·4	14·2	13·3	12·6	12·8	12·6
	8	12·8	14·0	13·4	13·7	13·0	12·7
	15	13·0	13·2	13·0	12·3	12·0	10·7

From Singleton (1957).

B. Respiration

1. *General*

Davies (1928) reported carbon dioxide production in an amount equal to the volume of the pineapple fruit (variety not specified) in five days at 7°C.

Gore (1911) found that the respiration rate of pineapple (variety not specified) increased 2·45- to 2·85-fold per 10°C rise in temperature.

2. *Respiration patterns*

The studies of Davies (1928) and Gore (1911) were apparently carried out with fruit harvested for commercial use. Dull *et al.* (1967) characterized the post-harvest respiratory patterns during the entire fruit development period. Six stages of development were studied: stage 1, one week after flowering; stage 2, six weeks after flowering; stage 3, eleven weeks after flowering; stage 4, less than 12% of the shell surface yellow; stage 5, 50% of the shell surface yellow; stage 6, 100% of shell surface yellow. Stages 1, 2 and 3 were in the pre-maturation phase, stages 4 and 5 were in the maturation phase, stage 6 was at the start of the senescence phase; stage 5 was ripe in accordance with the definitions presented earlier. A composite respiration rate for pineapple fruit was obtained, showing the initial respiration rate and the respiratory drift for each of the six stages of development (Fig. 3). The

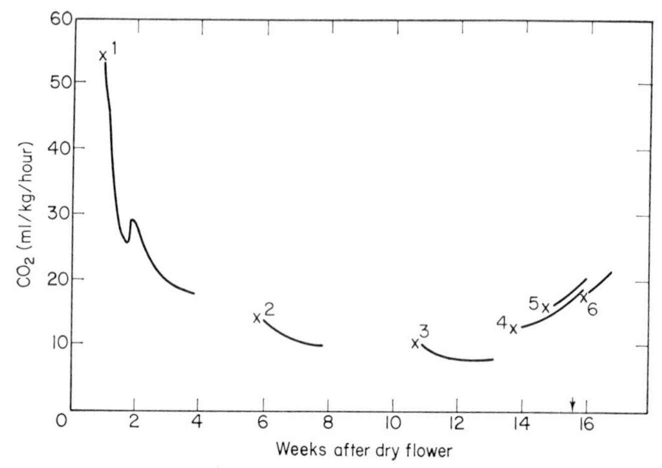

FIG. 3. A composite respiration pattern for pineapple (*Ananas comosus* var. Smooth Cayenne) based upon post-harvest respiration rates. Points marked (×) represent first measurement on equilibrated fruit ($t = 20$°C). Numbers denote stage of development. Arrow marks point when fruit are ripe. Reproduced from Dull *et al.* (1967).

initial respiration rate was approximately 50 ml carbon dioxide/kg/hour for stage 1, 10 ml (the minimum) for stage 3 and 15 ml for stage 6. There was a temporary increase in stage 1 during the second week but this was not associated with any ripening changes. The respiratory drifts for stages 2 through 6 showed no maxima. It was concluded that pineapple is a non-climacteric fruit. The term "climacteric" is used in the sense that dramatic biochemical changes occur during ripening. If respiration reaches a maximum shortly before, during, or shortly after these changes, the fruit is considered climacteric. If the respiration maximum does not coincide with the dramatic ripening changes, then the fruit is considered non-climacteric. This view represents the interpretation of the authors (Dull *et al.*, 1967) but does not necessarily have general acceptance. It appears that there is serious need for a clarification of the term climacteric as it applies to fruits (see Volume 1, chapter 17).

Since pineapple is a composite fruit, it could be argued that a climacteric maximum for each fruitlet is masked by the respiration of the total fruit. By using two other pineapple species (*Bromelia faustuoso* and *Bromelia pinguin*) in which individual fruitlets were not fused, it was possible to measure the respiration rate of individual fruitlets at different stages of development. Again, no climacteric maximum was found.

It should be pointed out that Bose *et al.* (1962) report obtaining a climacteric maximum for the Kew pineapple. The differences in the results of Dull *et al.* (1967) and Bose *et al.* (1962) are not readily explained.

C. Ethylene Response†

Ethylene production and internal ethylene concentrations in Smooth Cayenne pineapple were measured by Dull *et al.* (1967). In stage 4 fruit (the onset of ripening) the ethylene production ranged from 9 to 300 mμl/kg/hour; the internal concentration ranged from 80 to 1140 μl/litre. There was a trend of decreasing ethylene concentration from the base to the apex of the fruit, which indicates that ethylene increases as each fruitlet ages. These data on ethylene production and internal concentration agree with results of Burg and Burg (1962).

Stage 4 fruits were subjected to atmospheres in which ethylene concentrations varied from 0·01 to 1000 μl/litre. Maximum respiration response was obtained at the 100 μl/litre level and 1·0 μl/litre appeared to be the threshold level.

All six stages of fruit were placed in an air atmosphere containing 100 μl/litre of ethylene. The respiration patterns for these fruit are shown in Fig. 4. In stage 1 fruit, ethylene did give rise to a respiration maximum. At no time in the post-harvest life of this fruit were ripening changes detected. In stages 2, 3 and 4, ethylene stimulated respiration, the magnitude of stimulation decreasing as fruit development progressed. In these three stages, the respiration was maintained at a higher level (a plateau) until the experiment was terminated due to fruit spoilage. In stages 5 and 6 the application of ethylene had essentially no effect.

Ethylene appears to stimulate the respiration rate of pineapple fruit as long as there is some chlorophyll remaining in the shell. The rate of degreening is increased slightly but this effect would seem to be of little practical value.

† See also Volume 1, Chapter 10.

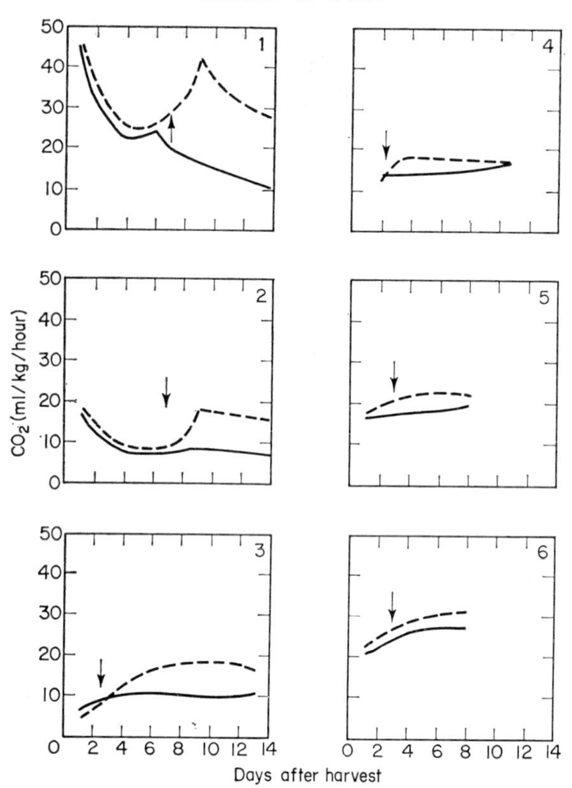

FIG. 4. Respiratory patterns during development of pineapple fruit (*Ananas comosus* var. Smooth Cayenne) from dry flower to senescence, showing the effect of treatment with atmospheres containing 100 μl ethylene/litre of air ($t = 20°C$). The numbers of the different diagrams denote the stage of fruit development. Stage 1: 1 week; stage 2: 6 weeks; stage 3: 11 weeks after dry flower; stage 4: <12% yellow shell; stage 5: 50% yellow shell; stage 6: 100% yellow shell. The arrow indicates the point of application of ethylene (broken curve); full-drawn curve = control. Reproduced from Dull *et al.* (1967).

D. Oxygen and Carbon Dioxide Responses

In view of the interest in controlled atmosphere storage, the effects of varying carbon dioxide and oxygen levels on respiration were studied by Dull *et al.* (1967). Stage 4 fruit were placed in atmospheres containing 21 (air), 10, 5 and 2·5% oxygen. Nitrogen made up the balance of the atmosphere. As the oxygen concentration decreased, so did the respiration rate (Fig. 5).

Stage 4 fruit were placed in atmospheres of 21% oxygen plus 0·03 (air), 5 and 10% carbon dioxide. The only noticeable effect of increased carbon dioxide concentrations was a slight suppression of the respiration rate (Fig. 6).

Although adequate numbers of fruit for evaluation of quality were not available in these experiments, it was the judgment of the author that decreased oxygen and increased carbon dioxide concentrations had no obvious effect on fruit quality. Thus, no major advantage of quality maintenance appears to be gained by manipulation of the concentration of these two environmental gases.

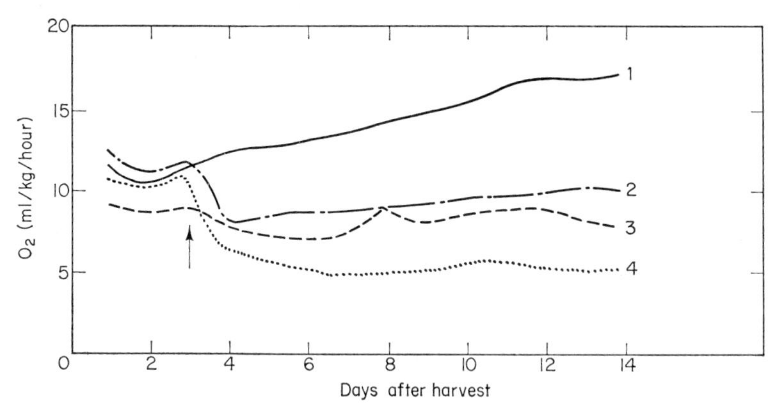

FIG. 5. Effect of varying oxygen concentrations on respiration of stage 4 fruit (*Ananas comosus* var. Smooth Cayenne). Arrow indicates change of O_2 level. 1 = air; 2 = 10% O_2; 3 = 5% O_2; 4 = 2·5% O_2. Reproduced from Dull *et al.* (1967).

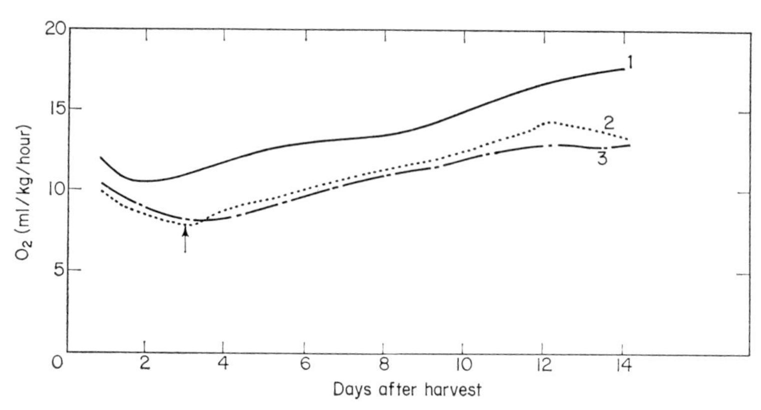

FIG. 6. Effect of varying carbon dioxide concentrations on respiration of stage 4 fruit (*Ananas comosus* var. Smooth Cayenne). CO_2 added at arrow. 1 = air; 2 = 5% CO_2; 3 = 10% CO_2. Reproduced from Dull *et al.* (1967).

E. Storage

The problem with fresh pineapple is not how to store the fruit so that it can be ripened for use by the consumer, but rather how best to store the fruit in order to minimize loss of quality in the fruit at the time of harvest.

As in the case of many fruits and vegetables, refrigeration delays quality deterioration in pineapple during shipping. For Smooth Cayenne, Akamine (1963) noted that one-half ripe fruit can be held for about two weeks at 7·5–12·5°C, and still have about one week of shelf life. Mature green fruit (fruit picked with no yellow colour showing on the shell) are very susceptible to chill injury when stored at temperatures less than 10°C. Py *et al.* (1957) reported similar observations for pineapple grown in Guinea, as did Bose *et al.* (1962) for the Kew variety grown in India.

Py *et al.* (1957) recommended 8°C for shipping ripe pineapple. Ginsburg (1953) recommended a storage temperature of 8·5°C for South African pineapple.

Gortner (1963b) showed that dipping a half-ripe Smooth Cayenne fruit in a 100 p.p.m. solution of 2,4,5-trichlorophenoxyacetic acid extended the shelf life from 6 to 14 days when the fruit was stored at ambient temperatures. The quality of the treated fruit at the end of the 14-day period was judged to be good, while the control was fair to poor. Measurements of quality indices, including sugar, acid, pigment and flavour, indicated that the hormone functioned as a senescence inhibitor. Bose *et al.* (1962) treated mature green Kew pineapples with 500 p.p.m. of 2,4,5-trichlorophenoxyacetic acid and extended the shelf life from 12–30 days at 21°C.

It has been the experience of workers at the Pineapple Research Institute of Hawaii that for each 6°C decrease in storage temperature for fruit showing approximately 25% shell yellowing at harvest, approximately one week additional storage life may be gained. At 7°C the maximum storage life was about four weeks.

F. Physiological Disorders

The pineapple is subject to a wide variety of pathological disorders. Of these only two are likely to be encountered frequently in the U.S. fresh fruit channels. Specifically, they are chill injury and endogenous brown spot.

Chill injury is characterized by failure of the green shell to turn yellow, the yellow shelled fruit turning a brown or dull colour, a wilting, drying and discoloration of crown leaves, and a breakdown of internal tissues, giving a watery appearance. Chill injury of Smooth Cayenne can be produced routinely by a one-week storage at 4·5°C, followed by one week at room temperature (21°).

Chill injury of pineapple (several varieties including Smooth Cayenne) was studied by Miller (1951), Miller and Heilman (1952), Miller and Hall (1953) and Miller and Marsteller (1953). These workers found a relationship between the physiological disorder and a decrease in pH and ascorbic acid.

Endogenous brown spot (EBS) frequently occurs simultaneously with chill injury. This is not always the case, since storage at 8°C for one week, followed by one week at 21°C, will frequently induce EBS but not chill injury. EBS does occur in fruit before harvest, where chilling occurs as a result of low field temperatures, such as in Australia. An early characteristic of the disease is the formation of water spots at the base of a fruitlet close to the core. As the severity of the disease increases, the spots enlarge and turn brown. In severe cases the entire centre of a fruit may be consumed, thus giving rise to another name for the malady, Black Heart.

REFERENCES

Akamine, E. K. (1963). *Hawaii Fm Sci.* **12,** 1.

Beaudreau, C. A. (1964). *Diss. Abstr.* **24,** 4385.

Bose, A. N., Lodh, S. B. and De, S. (1962). *First Intern. Congr. Fd Sci. Technol.* **2,** 117.

Bruce, D. W. (1960). *Nature, Lond.* **188,** 147.

Burg, S. P. and Burg, E. A. (1962). *Pl. Physiol., Lancaster* **37,** 179.

Collins, J. L. (1960). "The Pineapple—Botany, Cultivation, and Utilization". Leonard Hill, London.

Davies, R. (1928). *Union of S. Africa, Dept of Agric. Sci. Bull.* No. 71.

Dull, G. G., Young, R. E., and Biale, J. B. (1967). *Physiologia Pl.* **20,** 1059.

Foy, J. M. and Parratt, J. R. (1961). *J. Pharm Pharmac.* **13,** 382.

Gardner, H. L. (1966). *Pineapple Res. Inst. Hawaii, PRI News* **14,** 87.

Gee, M., McComb, E. and McCready, R. (1958). *Fd Res.* **23,** 72.

Ginsburg, L. (1953). *Fmg S. Afr.* **28,** 85.

Glaziou, K. T. (1957). *Aust. J. Biol.* **10,** 426.

Gore, H. C. (1911). *U.S. Dept. Agric., Bur. Chem. Bull.* No. 142.

Gortner, W. A. (1957). Pineapple-stem Phosphatase Essentially Free of Protease Activity. U.S. Patent No. 3,002,890.

Gortner, W. A. (1962). *J. Fd Sci.* **28,** 191.

Gortner, W. A. (1963a). *Nature, Lond.* **197,** 1316.

Gortner, W. A. (1963b). Delaying Senescence of Pineapple Fruit. U.S. Patent No. 3,346,397.

Gortner, W. A. (1965). *J. Fd Sci.* **30,** 30.

Gortner, W. A. and Kent, M. J. (1953). *J. biol. Chem.* **204,** 593.

Gortner, W. A. and Kent, M. J. (1958). *J. biol. Chem.* **233,** 731.

Gortner, W. A. and Singleton, V. L. (1965). *J. Fd Sci.* **30,** 24.

Gortner, W. A., Dull, G. G. and Krauss, B. (1967). *Hort Science* **2,** 141.

Gortner, W. A., Kent, M. J. and Sutherland, G. K. (1958). *Nature, Lond.* **181,** 630.

Heinicke, R. M. and Gortner, W. A. (1957). *Econ. Bot.* **11,** 225.

Hobson, G. E. (1962). *Nature, Lond.* **195,** 804.

Joshi, M. C., Boyer, J. S. and Kramer, P. J. (1965). *Bot. Gaz.* **126,** 174.

Kerns, K. R. and Collins, F. L. (1936). *New Phytol.* **35,** 305.

Miller, E. V. (1951). *Pl. Physiol., Lancaster* **26,** 66.

Miller, E. V. and Hall, G. D. (1953). *Pl. Physiol., Lancaster* **28,** 532.

Miller, E. V. and Heilman, A. S. (1952). *Science, N.Y.* **116,** 505.

Miller, E. V. and Marsteller, R. L. (1953). *Fd Res.* **18,** 421.

Okimoto, M. C. (1948). *Bot. Gaz.* **110,** 217.

Ota, S. (1966). *J. Biochem, Tokyo* **59,** 463.

Ota, S., Moore, S. and Stein, W. H. (1964). *Biochemistry* **3,** 180.

Preston, R. D. and Hepton, J. (1959). *J. exp. Bot.* **11,** 13.

Py, C., Tisseau, M. A., Oury, B. and Ahmada, F. (1957). "The Culture of Pineapples in Guinea." Institut Français de Recherches Fruitières d'Outre-Mer, Paris.

Sideris, C. P., Young, H. Y. and Chun, H. H. Q. (1948). *Pl. Physiol., Lancaster* **23**, 38.

Singleton, V. L. (1957). Pineapple Research Institute of Hawaii, Research Report No. 49.

Singleton, V. L. (1959). Pineapple Research Institute of Hawaii, Research Report No. 69.

Singleton, V. L. and Gortner, W. A. (1965). *J. Fd Sci.* **30**, 19.

Singleton, V. L., Gortner, W. A. and Young, H. Y. (1961). *J. Fd Sci.* **126**, 49.

Sutherland, G. K. and Gortner, W. A. (1959). *Aust. J. Chem.* **12**, 240.

Chapter 9B

The Pineapple: Flavour

R. M. SILVERSTEIN

State University College of Forestry, Syracuse, New York, U.S.A.

While some, possibly all, of the volatile compounds produced by the pineapple may be concerned in the flavour of the fruit, recent new appraisals of the subjective–objective relationships of taste and odour (Harper *et al.*, 1968; Guadagni, 1968) render it necessary to exercise extreme caution in allocating to any chemical substance or group of substances the role of flavour or aroma characteristic of a fruit. This subject is fully discussed in Volume 1, Chapter 10.

Studies of the volatile components of pineapple (see Nursten and Williams, 1967) have spanned the years that have witnessed spectacular developments of isolation and identification methodology. Chromatography, especially gas chromatography, has provided the chemist with a very effective tool for isolation of pure compounds from complex mixtures, and developments in spectrometry have furnished the means of identifying the small amounts of pure compounds made available by chromatography.

The pineapple studies during this period (1945 to the present day) may be described under three headings according to the methodology used:

1. *Classical.* Components are separated by fractional distillation, recrystallization and solubility differences before and after treatment with reagents designed to alter functional groups. Identification involves sodium fusion, boiling point, refractive index, solubility tests, functional group tests, derivative preparation, mixture melting points, combustion analysis, molecular weight and degradation with similar manipulations of the degradation products. The disadvantages are the time involved, the large sample size required, the inadequacies of the separation techniques and the very great hazard of unexpected chemical side reactions.

2. *Isolation and identification by gas chromatography.* Compounds separated by gas chromatography can be tentatively identified by comparing their retention data with that of likely authentic compounds. The inherent weakness of this method is that only those compounds whose presence can be surmised will be identified. Coincident retention times for an unknown and

an authentic sample, run as a mixture, should be obtained on several stationary phases of differing polarity to increase the certainty of identification. This requirement is frequently ignored.

3. *Isolation by chromatography and identification by spectrometry.* Chromatographic fractionation is carried out until each peak on the recorder (gas chromatograph) represents a pure compound. The sample is collected and identified from the complementary information afforded by mass, infrared, nuclear magnetic resonance and ultraviolet spectrometry (Silverstein and Bassler, 1967). Identification is routine at the milligram level and possible at the submilligram level. Of course not all molecules yield so easily and chemical manipulation may be necessary. Ozonolysis (Beroza and Bierl, 1967) and high temperature hydrogenation (Beroza and Sarmiento, 1963; Brownlee and Silverstein, 1968) followed by gas chromatography and mass spectrometry are particularly useful techniques at the microgram level.

The first study of pineapple flavour (Haagen-Smit *et al.*, 1945) was carried out using the classical techniques of organic chemistry. Fresh pineapple (1161 kg of trimmed summer fruit and 995 kg of trimmed winter fruit) of the Smooth Cayenne variety (Hawaii) was steam distilled ($<40°C/20$ mm), and a concentrate was obtained by saturating the distillate with ammonium sulphate. Components of the concentrate were separated by fractional distillation. Eight esters, one acid, one aldehyde, one ketone and a unique sulphur-containing ester were identified. Three other esters were partially identified, and concentrations were given in both summer and winter fruit

TABLE I. Volatile compounds

Compound	Fruit (mg/kg) Winter	Summer
Acetaldehyde	0·61	1·4
Acetic acid		0·49
Ethyl acetate	2·9	120
Ethyl acrylate		0·77
Ethyl alcohol		60
Ethyl caproate		0·77
Ethyl isovalerate		0·39
Ethyl ester of a C_5 unsaturated acid		1·1
Methyl caprylate	0·75	
Methyl isocaproate	1·4	
Methyl isovalerate	0·60	
Methyl propyl ketone		tr.
Methyl ester of a C_5 hydroxy acid	tr.	
Methyl ester of a C_5 unsaturated acid		0·68
Methyl ester of a C_5 keto acid		tr.
Methyl β-methylthiopropionate	1·1	0·88

From Haagen-Smit *et al.* (1945).

(Table I). The sulphur-containing ester, methyl β-methylthiopropionate, was identified by oxidation to the sulphone and determination of a mixed melting point of the sulphone with an authentic sample.

Gawler (1962) identified the following volatile compounds in canned pinapple juice prepared from the Singapore Canning variety: acetic acid, 5-hydroxymethylfurfural, furfural, formaldehyde, acetaldehyde and acetone. The juice was distilled at 20–30°C at 20 mm in a "circulating vacuum evaporator". An acid fraction from the distillate was subjected to paper chromatography (as ammonium salts) and acetic acid was identified as the major component. The carbonyl compounds were identified by matching R_f values on paper chromatography and ultraviolet spectra of their 2,4-DNP derivatives against those properties of 2,4-DNP derivatives of a number of simple aldehydes and ketones.

Connell (1964) obtained 1·4 g of a flavour concentrate by vacuum stripping 475 gallons of juice of fresh Queensland summer pineapples, extracting the distillate with ether and distilling the extract. Components were identified by matching gas chromatographic retention times against those of authentic samples. Transesterification was used effectively to locate peaks due to esters. In this way, Connell confirmed the presence of Haagen-Smit's sulphur-containing compound, methyl β-methylthiopropionate, and identified the

TABLE II. Volatile compounds

Compound	Connell Relative Amount	Mori
Acetic acid		present
Amyl caproate	++	
Biacetyl	+	
Ethanol	++++	
Ethyl acetate	++++	present
Ethyl butyrate	+	present
Ethyl caproate	+	
Ethyl caprylate		present
Ethyl lactate	+	
Ethyl β-methylthiopropionate	+	
Isobutanol	+	
Methanol	+	
Methyl acetate		present
Methyl butyrate		present
Methyl caproate	++	present
Methyl caprylate	+	present
Methyl isovalerate	+	
Methyl β-methylthiopropionate	+++	
Pentanol	++	
Propanol	+	

From R. Mori (1963). Period communication to D. W. Connell.

homologous ethyl ester. Connell's results are given in Table II. Results of unpublished work by R. Mori (personal communication to Connell, 1963) on Hawaiian fruit are included.

Howard and Hofman (1967) identified 16 compounds (Table III) by gas chromatographic analysis of headspace vapours of canned Malayan pineapple. Retention times on three columns of different polarities were matched against those of reference compounds. The object of this study was to develop rapid "fingerprint" chromatograms in order to compare canned pineapple from different sources. The problems of "peak matching" of authentic compounds with partially resolved peaks were discussed.

TABLE III. Volatile compounds

Acetaldehyde	Isobutyl acetate
Acetone	Isobutyl formate
Butyl formate	Isopropyl isobutyrate
Ethanol	Methyl acetate
Ethyl acetate	Methyl caproate
Ethyl formate	Methyl isobutyrate
Ethyl isobutyrate	Propyl acetate
Ethyl propionate	Propyl formate

From Howard and Hofman (1967).

In 1963, the Pineapple Research Institute of Hawaii furnished investigators at Stanford Research Institute (SRI) with a flavour concentrate prepared under mild conditions from 250 freshly picked pineapples (*Ananas comosus* (L.) Merr. var. Smooth Cayenne, winter harvest, field grown at 700–800 ft elevation in the Waipio region, Oahu, Hawaii). The fruit was one-half to fully yellow shell colour, and the flesh was moderately to highly translucent. The flavour concentrate was prepared by screw-pressing the cored fruit, adding sodium chloride and extracting the juice with peroxide-free ether. The dried ether was evaporated at 40–50 mm between room temperature and 40°C. The syrupy residue was subjected to short-path distillation at a bath temperature of 25–120°C between atmospheric pressure and 0·1 mm Hg. The distillate is the volatile flavour concentrate. All material was stored at −40°C during the course of the study. The final paper in the series resulted from collaboration between the SRI group and Dr. W. G. Jennings' laboratory at the University of California, Davis. This latter study utilized the volatile flavour concentrate from 1000 pineapples from a summer harvest. The hallmark of the investigation was gas chromatographic isolation of pure compounds and rigorous criteria for spectrometric identification.

The first paper of the series (Rodin *et al.*, 1965) describes the isolation and identification of 2,5-dimethyl-4-hydroxy-3(2H)-furanone:

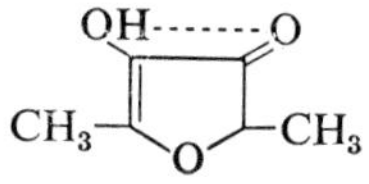

OH·······O

CH_3 O CH_3

This unique compound (a major component) was characterized by a remarkably intense aroma—described as "burnt pineapple" or "fruity caramel"—and a high degree of instability in air. A literature search proved negative, and this compound was described as a new structure. Shortly thereafter, the same structure was described by J. E. Hodge of Northern Regional Laboratory, U.S.D.A., at the International Symposium on Chemistry of Carbohydrates in Münster, Germany, July 13-17th, 1965. Exchange of spectra quickly established that Hodge's compound, prepared in the course of his intensive studies of caramelization by heating rhamnose with pyridine acetate, and the pineapple compound were indeed identical. Hodge, in fact, had formulated the structure in a review article in a rather obscure journal (Hodge *et al.*, 1963) and at an American Chemical Society Meeting (Hodge and Fisher, 1963), but the formula was not abstracted by "Chemical Abstracts". Several derivatives were prepared (Fisher and Hodge, 1965; Wilhalm *et al.*, 1965), and a synthesis was designed to establish the structure unequivocally (Henry and Silverstein, 1966). Hofman and Eugster (1966) identified it as a product from the hydrogenolysis of acetylformoin. Recently the same compound was found in a beef broth (Tonsbeek *et al.*, 1968) and in maple syrup (Underwood *et al.*, 1968). In both cases, "cooking" is required to generate the furanone. Its genesis in pineapple must be different. In a separate experiment, the ether extract of pineapple flavour was distilled below 40°C; the volatile flavour concentrate thus obtained had the same furanone content found in the material that had been heated to 120°C.

A gas chromatography "sulphur profile" of the volatile flavour concentrate was carried out (Rodin *et al.*, 1966) using a selective electrolytic conductivity detector for sulphur-containing compounds (Coulson, 1965). Peaks due to β-methylthiopropionate, ethyl β-methylthiopropionate and two minor unidentified compounds accounted for all of the volatile sulphur-containing material. In this connection, it may be noted that methionine, a likely biogenetic precursor, was found to be generally absent in pineapple until the onset of ripening, and present in considerable amounts thereafter (Gortner and Singleton, 1965).

Table IV presents the compounds identified in the volatile flavour concentrate (Rodin *et al.*, 1965, 1966; Silverstein *et al.*, 1965; Creveling *et al.*, 1968). There were at least five other compounds with infrared spectra very similar to those of the reported β-hydroxyl and β-acetoxyl esters. This series was a prominent feature of the composition of the volatile flavour concentrate. The fractions containing the common lower fatty esters were not examined.

TABLE IV. Volatile compounds

Compound	Pineapple flesh (mg/kg)
Acetoxyacetone	0·009[b]
γ-Butyrolactone	0·024[b]
γ-Caprolactone	0·12[a]
Chavicol	0·27[a]
2,5-Dimethyl-4-hydroxy-3(2H)-furanone	1·2[a]
Dimethyl malonate	0·06[b]
Ethyl-β-acetoxyhexanoate	0·006[b]
Ethyl-β-hydroxyhexanoate	0·03[b]
Ethyl-β-methylthiopropionate	0·09[b]
Methyl-β-acetoxyhexanoate	0·03[b]
Methyl-β-hydroxybutyrate	0·006[b]
Methyl-β-hydroxyhexanoate	0·021[b]
Methyl-β-methylthiopropionate	0·12[a]
Methyl *cis*-(4?)-octenoate	0·0009[b]
γ-Octalactone	0·3[b]
δ-Octalactone	0·3[b]
trans-Tetrahydro-α, α, 5-trimethyl-5-vinylfurfuryl alcohol	0·0009[b]

[a] Winter harvest.
[b] Summer harvest.

REFERENCES

Beroza, M. and Bierl, B. A. (1967). *Analyt. Chem.* **39,** 1131.

Beroza, M. and Sarmiento, R. (1963). *Analyt Chem.* **35,** 1353.

Brownlee, R. G. and Silverstein, R. M. (1968). *Analyt Chem.* **40,** 2077.

Connell, D. W. (1964). *Aust. J. Chem.* **17,** 130.

Coulson, D. M. (1965). *J. Gas Chromatog.* **3,** 134.

Creveling, R. K., Silverstein, R. M., and Jennings, W. G. (1968). *J. Fd Sci.* **33,** 284.

Fisher, B. E. and Hodge, J. E. (1965). Abstracts of Papers, p. 4D, 150th Meeting, Am. Chem. Soc., New York, September.

Gawler, J. H. (1962). *J. Sci. Fd Agric.* **1962,** 57.

Gortner, W. A. and Singleton, V. L. (1965). *J. Fd Sci.* **30,** 34.

Guadagni, D. G. (1968). ASTM 440. Am. Soc. for Testing and Materials 1968, pp. 36–48.

Haagen-Smit, A. J., Kirchner, J. G., Prater, A. N. and Deasy, C. L. (1945). *J. Am. chem. Soc.* **67,** 1646 and 1651.

Harper, R., Bate-Smith, F. C. and Land, D. G. (1968). "Odour Description and Odour Classification." Churchill, London.

Henry, D. W. and Silverstein, R. M. (1966). *J. org. Chem.* **31,** 2391.

Hodge, J. E. and Fisher, B. E. (1963). Abstracts of Papers, p. 3D, 145th Meeting, Am. Chem. Soc., New York, September.

Hodge, J. E., Fisher, B. E. and Nelson, E. C. (1963). *Am. Soc. Brewing Chemists Proc.* **1963,** 84.

Hofman, A. and Eugster, C. (1968). *Helv. Chim. Acta* **49,** 53.
Howard, G. E. and Hofman, A. (1967). *J. Sci. Fd Agric.* **18,** 106.
Mori, R. (1963). Personal communication to D. W. Connell.
Nursten, H. E. and Williams, A. A. (1967). *Chemy Ind.* **1967,** 486.
Rodin, J. O. Coulson, D. M. and Silverstein, R. M. (1966). *J. Fd Sci.* **31,** 721.
Rodin, J. E., Himel, C. M., Silverstein, R. M., Leeper, R. W. and Gortner, W. A. (1965). *J. Fd Sci.* **30,** 280.
Silverstein, R. M. and Bassler, G. C. (1967). "Spectrometric Identification of Organic Compounds", 2nd Ed. John Wiley and Sons, Inc., New York.
Silverstein, R. M., Rodin, J. O., Himel, C. M. and Leeper, R. W. (1965). *J. Fd Sci.* **30,** 668.
Tonsbeek, C. H. T., Koenders, E. B. and Losekoot, J. A. (1968). Abstracts of Papers, Paper AGFD 53, 156th Meeting, Am. Chem. Soc., Atlantic City, September. See also Tonsbeek, C. H. T., Planken, M. J. and Weerdhof, T. v. d. *J. agric. Fd Chem.* **16,** 1016.
Underwood, J. C., Filipic, V. J. and Bell, R. A. (1969). *J. Ass. Off. Anal. Chem.* **52,** 717.
Wilhalm, B., Stoll, M. and Thomas, A. F. (1965). *Chemy Ind.* **1965,** 1629.

Chapter 10

Pome Fruits

A. C. HULME AND M. J. C. RHODES

A.R.C. Food Research Institute, Norwich, England

I Introduction 333
II History and Distribution 334
III Morphology of the Fruit 335
IV The Biochemistry of Pome Fruits 336
A. Introduction 336
B. Changes Occurring during Growth 337
C. Changes Occurring during Ripening 343
V Conclusion 368
References 369

I. INTRODUCTION

Pome fruits are produced by members of the family Rosaceae and include the apple, pear, quince and medlar. Only the commercially important apple and pear will be considered here. The quince, closely related to the pear, has some small commercial use in the preparation of jellies and jams.

World production of apples and pears, including cider and perry varieties, has risen from 17·7 million tons per annum for the years 1951 to 1955 to 25·7 million tons in 1964; pear production was just over one-third of apple production. In addition to North America, Australia and New Zealand have been sending increasing quantities of apples and pears to Europe. Nevertheless, Western Europe is becoming increasingly self-supporting in these fruits and in a good cropping year some form of control of imports is exercised. Particularly striking is the recent expansion of apple production in Hungary, Italy, France and Japan. It is not surprisingly, therefore, that the total world exports of apples and pears was, in 1964, running at little more than 2·2 million tons, less than 10% of the total world production.

According to the Commodities Division of the (British) Commonwealth Secretariat ("Fruits", 1967) there is the possibility of world overproduction of apples and pears and it is suggested that, among other factors engendered by this possibility, increasing emphasis should be placed on the need for improvement of quality. As will become abundantly clear from the contents of this chapter, improvement in quality both at harvest and throughout storage is dependent on a better knowledge of the biochemistry of the fruit.

II. HISTORY AND DISTRIBUTION

This subject will only be briefly dealt with here, with special reference to the more commercially important apple, since several text books provide full coverage of the field, notably the classic work of Hall and Crane (1933) and the more recent "Apples and Apple Products" by Smock and Neubert (1950). These latter authors state, "The original home of the apple (*Malus sylvestris*) is not known but it is thought to be indigenous to the region south of the Caucasus, from the Persian province of Ghilan on the Caspian Sea to Trebizond on the Black Sea". It has probably existed from prehistoric times in both the wild and cultivated states in Europe from the Caspian Sea to the Atlantic Ocean. Apples were available as early as 100 B.C. but "Pearmain" appears to be the first variety recorded in history appearing in 1204 in a deed relating to the lordship of Runton in Norfolk. In America, there are records as early as 1647 of apples having been grafted on seedling rootstocks in Virginia. By 1773 apples from America were found in the London markets.

Most of the early orchards were of seedling trees, that is trees growing on their own rootstocks, although there is evidence of grafting of superior varieties onto seedling rootstocks while apple growing was still a small side line on a mixed farm. A multiplicity of seedling rootstocks, often incorrectly designated, gradually developed. In recent years, however, as the demand for authentic, reproducible stocks grew with the development of large, specialized commercial orchards, the situation has been systematically clarified, notably by the East Malling Research Station in England. East Malling rootstocks, numbered according to the vigour they impart to a scion graft, are now used all over the world. Of particular interest are the dwarfing and semi-dwarfing stocks yielding trees of a size which greatly facilitate the harvesting of the fruit. Little is yet known of the physiology of the influence of stock on scion development, although incompatibility between stock and scion (this is common especially with quinces) has been variously attributed to cyanogenitic glycosides and phenolic compounds (see Mosse, 1962). As the apple industry expanded, varieties were selected and new ones bred with a view to high average yield, absence of biennial bearing, improved appearance and texture, resistance to disease, winter hardiness, etc. This process of selection has resulted in the world markets becoming dominated by less than 20 dessert and culinary varieties of apples of which, perhaps, the most widely distributed are Bramley's Seedling (largely confined to the British Isles), Cortland, Cox's Orange Pippin, Delicious, Golden Delicious, Granny Smith, Jonathan, McIntosh, Newtown Pippin and Winesap. From time to time there is a shift in the importance of some varieties and some of the varieties listed are unsuitable for growing in certain parts of the world. For example, the Cox's Orange Pippin is the most important dessert variety in the United Kingdom, but is little grown in America or Australasia, where the fruit develops a thick skin, has an inferior texture and little of the characteristic flavour of the English grown fruit. While many varieties of dessert pears are grown in small quantities in Europe, the main varieties of commerce are the Bartlett (also known as the Williams) the Comice (Doyenné du Comice) and the Conference.

The countries now growing apples on a considerable scale are Argentina, Australia, Austria, Bulgaria, Canada, France, Hungary, Italy, Japan, Netherlands, Poland, Spain, United Kingdom, United States and West Germany. France, the United States and Italy are the largest producers of apples, producing two to three times as many as the next largest producers, West Germany and Japan (1·1 million tons each in 1964). Poland and the United Kingdom produce about 0·7 million tons each while the remainder of those countries quoted produce between 0·3 and 0·4 million tons each ("Fruits", Commonwealth Secretariat, 1967).

III. MORPHOLOGY OF THE FRUIT

Some knowledge of the morphology of the apple fruit is necessary for the appreciation of the physiological and biochemical problems which form the main body of the present chapter. Only essential points will, however, be examined—the subject is dealt with very fully by Smock and Neubert (1950).

By botanical definition a fruit is a mature or ripened ovary or ovaries together with any closely associated parts. Some fruit such as peaches and plums involve only the ripened ovary of a single carpel (stigma, style and ovary). In pome fruits more than just the ovaries are involved. The five ovaries of the apple flower are imbedded in tissue which, as well as the carpel tissue, becomes fleshly and edible. This outer structure is interpreted by one school of anatomists as being the "receptacle" (in which case the apple fruit of commerce is not a "true" fruit but a swollen receptacle). Another view, and the one most generally accepted (McDaniels, 1940; Professor A. J. Eames, personal communication), is that the fruit consists of the fused base of the calyx (sepals), corolla (petals) and stamens. Figure 1 shows in diagramatic form the apple flower and the mature fruit. The development of the pear fruit is similar to that of the apple.

The growth curve of pome fruits is sigmoidal, the main phase of steady, rapid growth commences as cell division ceases and continues until the period of the respiration climacteric (see Flood *et al.*, 1960).

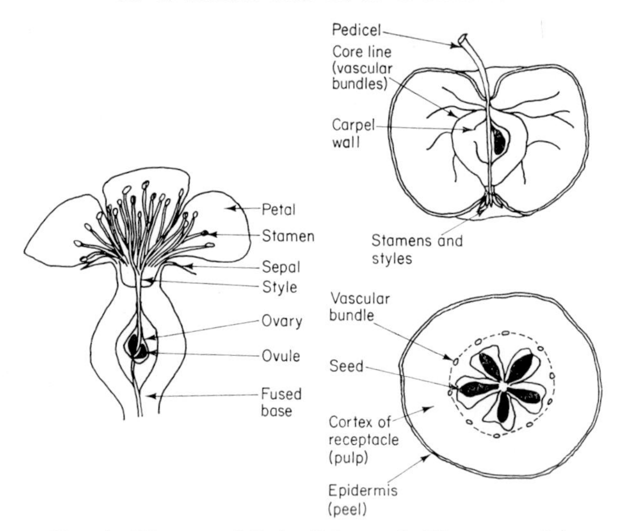

FIG. 1. Flower and fruit of the apple (diagrammatic).

IV. THE BIOCHEMISTRY OF POME FRUITS

A. Introduction

Work on the physiology and biochemistry of pome fruits up to 1957 was comprehensively reviewed by Hulme (1958). Most of the early research will, therefore, not be dealt with here except where no more recent work on important aspects of the subject has been reported.

Since 1957 the two most important advances appear to the authors to have been in the modifications to the metabolism of the fruit brought about by pre-harvest spray treatments of the fruit with "growth regulators" such as N-dimethylaminosuccinamic acid (B9, B995, Alar), and in the detailed knowledge of the biochemical changes occurring in the early stages of ripening. The modifications induced by pre-harvest sprays are far from being fully understood, but the principle involved may well be as important to an improvement in storage behaviour as was the development of Controlled Atmosphere Storage. With regard to present storage techniques, the realization that there is a short period during which the fruit should be picked for successful storage, and the use of controlled atmospheres in cold storage are already firmly established in commercial practice. Perhaps the most serviceable criterion for picking for storage is still the number of days from full bloom or petal fall, but there is every reason to believe that soon the physiologist and biochemist will be able to improve on this rule-of-thumb method. In the field of Controlled Atmosphere Storage, results obtained within the last decade suggest that the use of an atmosphere of low O_2

(2·5–3·0%) with no CO_2 is the ideal for many varieties of apples especially those susceptible to core flush; core flush is a type of injury which appears to be similar to the CO_2 injury (brown heart) which develops in stores in which the CO_2 concentration has been allowed to rise to too high a value.

Biochemical investigations during the past decade have gone a long way towards explaining why the storage techniques laboriously developed over the past 40 years have proved successful, and a fruit biochemist might be tempted to feel that if the modern knowledge of biochemistry and the analytical techniques now available had been to hand 20 or 30 years ago he could have pointed the way to improved storage techniques much earlier. This is, however, only partly true since the conditions of storage found most suitable through *ad hoc* trials have themselves influenced the direction the biochemical investigations have taken.

Although events taking place during the development of the fruit on the tree affect the condition of the fruit at maturity, it has become clear that the later stages of development (the early stages of ripening) are crucial to the subsequent storage behaviour of the fruit. In the present chapter, while some attention will be given to the developmental stages of the fruit (this is also dealt with in Volume 1, Chapter 14), the main preoccupation will be with post-harvest changes in the mature fruit.

B. Changes Occurring during Growth

The growth and development of pome fruits is not only interesting as a basic problem in plant physiology, but it is also of economic importance since the development of the fruit will affect its post-harvest behaviour. Much of the earlier work on biochemical changes occurring during the development of the fruit has been reviewed by Hulme (1958) so that the present discussion will concentrate mainly on specific changes affecting particularly the behaviour of the fruit at maturity.

During the first few weeks after pollination a phase of rapid cell division ensues. The length of this period varies considerably with the variety and the growing conditions. For Cox's Orange Pippin apples grown in England, Denne (1960) has shown that intensive cell division ceases abruptly 30–40 days from full blossom and that this phase coincides with the period of exponential growth of the whole fruit. With the Australian apple, Granny Smith, cell division ceases about 4 weeks after full blossom (Bain and Robertson, 1951) although this variety has a long overall growing period. The major period of growth of the fruit succeeds the period of cell division and is almost exclusively dependent on cell expansion. During this period the meristematic cells differentiate into the various tissues of the fruit. Auxins produced by the seeds in waves during growth appear to influence fruit set and growth; the exact nature of this influence is still uncertain and synergism between the auxins and gibberellins appears to be involved (Luckwill, 1953). Spraying the trees with gibberellin can overcome the effects of frost damage on the fruitlets and result in a reasonable crop of parthenocarpic fruits (Luckwill, 1961). This subject is dealt with in Volume 1, Chapter 15.

The length of the period of growth of the apple on the tree from petal fall (the point at which 90% of the petals have fallen) to commercial harvest varies considerably between varieties. For the major English varieties growing in the southern half of England (Cox's Orange Pippin, Worcester Pearmain, Bramley's Seedling) the growth period is between 105 and 140 days. Cox's Orange Pippins grown in quite different environmental conditions, for example in Tasmania, have a somewhat longer growth period—160 days instead of the 120 days in England. Other Australian varieties, however, take much longer to reach commercial maturity, Granny Smith, 170–190 days, Sturmer Pippin, 180 days and Democrat, 200 days (Bain and Robertson, 1951; Lewis and Martin, 1965). The length of the cell division and cell enlargement phases have an important bearing on the size of the apple at harvest. This can be important both from the point of view of total crop size and of keeping properties since, in general, smaller fruits have smaller cells and generally have a longer storage life (Smith and Hulme, 1935; Martin *et al.*, 1964).

Changes in chemical constituents during growth have been studied by various workers (see Hulme, 1958). Expressed on the basis of fresh weight, both reducing sugars and sucrose rise throughout the period of growth while starch rises to a peak in July and August (Northern Hemisphere) and then falls with considerable amounts of starch still being present at harvest. The time of the peak in starch is variable but the decline does not appear to be directly related to the onset of the respiration climacteric. Malate accumulates in the fruit during the early stages of growth on the tree and then slowly declines. It is interesting that Clijsters (1969) has shown that Jonathan apples grown in light fix CO_2 by photosynthesis and that during the period of cell enlargement the photosynthetic CO_2-fixation is almost balanced by respiratory CO_2-production. When fruits were grown on trees in continuous darkness (black bags), the distribution of malate in the apple differed from normally grown control fruits. There was no marked effect of continuous darkness on sugar content and fruit growth.

The respiration per unit weight of the fruit is high during the cell division stage and immediately after its cessation. It then declines as maturity is approached and remains relatively steady until it rises again as the respiration climacteric develops. As might be expected, the activity of mitochondria per unit weight isolated from the fruit decreases during the growing period (Hulme *et al.*, 1966).

Phenolics, which cause so much difficulty in the isolation of active enzyme preparations from the fruit, are present in high concentration during the early stages of growth, then decline and finally reach a constant level from the end of August, with little further change during ripening (Fig. 2). Figure 2 shows changes in total phenolics as well as changes in chlorogenic acid, one of the more important phenolic compounds of pome fruits. Once again when results are expressed on the basis of fresh weight, there is a fall followed by a steady state and, in this instance, a small rise associated with the ripening phase. The expression in terms of fresh weight has little meaning when the whole organ is in a state of rapid expansion. Change in anthocyanin in Fig. 3 illustrates this point well; although the concentration of the fraction is falling throughout most of the growth period, its synthesis (in terms of amount per fruit) is continuing until maturity is reached.

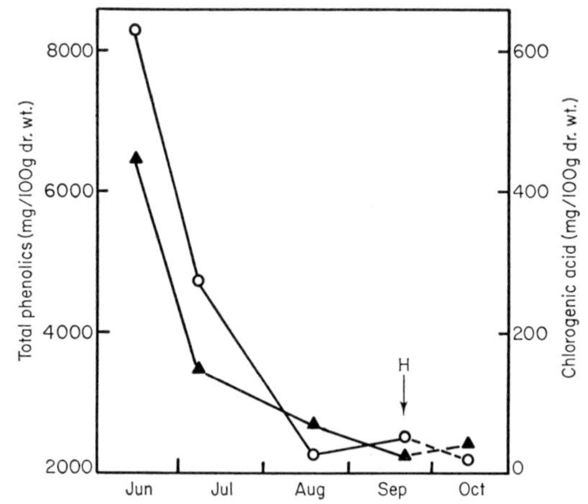

FIG. 2. Changes in total phenolics (○) chlorogenic acid (▲) in Cox's Orange Pippin apples during growth on the tree and in storage at 12°C. 'H' indicates where the fruit was picked for storage.

The nitrogen content of the fruit, expressed on the basis of fresh weight, falls as the fruit grows, rapidly at first, then more slowly, finally reaching a steady level. If the fruit is left on the tree, as the respiration climacteric develops, the nitrogen content increases mainly as protein nitrogen (Hulme, 1954).

Since the total nitrogen of the mature apple is small, seldom more than 80 mg/100 g fresh weight (as nitrogen) the amino acid complement of the mature fruit, while varied in nature, is also small. While the young fruit just after petal fall may contain 150 mg nitrogen/100 g fresh weight in the form of free amino acids (the peel generally contains twice as much of its nitrogen in the form of protein than the pulp), the pattern of amino acids

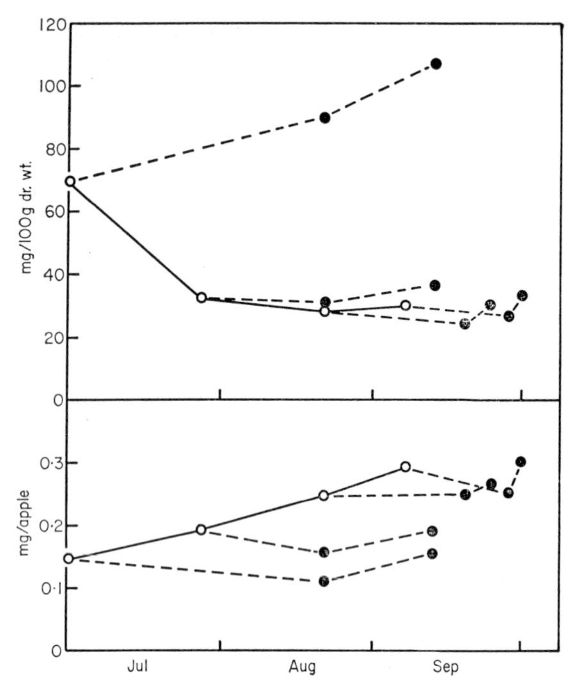

FIG. 3. Changes in anthocyanidins in the peel of Cox's Orange Pippin apples during development on the tree and during storage at 12°C (○ = on tree; ● = in storage at 12°C).

scarcely changes from this period up to maturity; the only significant change is a reduction in the glutamine content as starch disappears (McKee and Urbach, 1953). The principal amino acids present are aspartic acid, asparagine and glutamic acid with moderate amounts of serine, α-alanine, γ-amino-butyric acid, valine, the leucines, phenylalanine, piperidine-2-carboxylic acid and smaller amounts of β-alanine, the basic amino acids arginine and histidine, tyrosine, tryptophane and various proline derivatives. There is little difference in the pattern with different varieties of apples. These are the usual amino acids found in plants, the only unusual ones so far found are 4-hydroxymethylproline in apples and 1-aminocyclopropane-1-carboxylic acid found in pears by Burroughs (1957). The amino acids of apples and pears have been discussed in detail by Hulme (1958); nothing further of significance in this field has appeared since this review.

Many factors affect the growth and development of fruit but only those which have proved or may prove significant in the commercial exploitation of apples and pears will be considered. Manurial treatments have marked effects on the condition of the fruit at maturity and on its subsequent behaviour in storage (Perring, 1968). Owing, however, to the complexity of the interaction of the environment with manurial treatments, precise knowledge on this subject is still not available. Much of the data obtained over the years is conflicting. The effect of manurial treatments on storage disorders is discussed in Volume 1, Chapter 18. In one aspect of this field, the effect of nitrogenous fertilizers, the evidence seems to agree with the feeling among growers that fruit from trees supplied with high levels of nitrogen have storage properties inferior to those of fruit from similar trees fed with normal levels of nitrogen (Hulme, 1958). While the manurial treatments can cause a wide variation in total nitrogen content at maturity, the protein nitrogen content shows much less variation (Hulme, 1956a). Some storage disorders, for instance, bitter pit, have been traced to nutritional deficiencies during growth of the apple. Spraying the trees with calcium nitrate has been found, in some cases, to reduce the incidence of bitter pit (see Volume I, Chapter 18), but in other cases boron deficiency appears to be involved.

In recent years, interest has developed in the use of chemical tree sprays as a means of modifying the development of the fruit with a view to the enhancement of desirable properties. Among the materials tested are growth regulators such an apthalene acetic acid (NAA) and 2:4-dichlorophenoxyacetic acid (2:4 D) and growth retardants such as N-dimethylaminosuccinamic acid (Alar, B995, B9) and 2-(chloroethyl)-trimethylammoniumchloride (CCC). Unfortunately many of these growth regulators have more than one effect and some of these "subsidiary" effects may be detrimental to the commercial usage of the fruit. For example, 2,3,5-trichlorophenoxypropionic acid, which is used to control fruit drop and to improve the development of anthocyanin pigment in the fruit, also advances the maturity of the fruit and thus reduces its effective storage life (Walter, 1967). The indole compounds, indole acetic, butyric and propionic acids, another series of growth regulators have long been used as a means of preventing pre-harvest drop, although they are less effective in this respect than NAA. One of the most effective groups of these "drop stoppers" are the phenoxy compounds and especially 2:4 D (Edgerton and Hoffman, 1951). Again it is unfortunate that most of these growth promoters also advance the maturity of the fruit. Satisfactory results may, however, often be obtained by the choice of a concentration of the compound which will achieve a reduction in fruit drop without a serious advancement in the maturity of the fruit. For instance, Batjer *et al.* (1948) achieved this effect by using 2:4 D at a concentration of 20 μmoles/litre. Somewhat paradoxically the growth retardant (B9) when sprayed onto trees at a concentration of 1000 p.p.m. during the first 70 days after full bloom significantly reduced pre-harvest drop of apple fruits (see Cooper *et al.*, 1968). This compound has other effects which appear to be advantageous. It increases blossom yield, reduces shoot elongation, enhances fruit colour and delays some aspects of maturity (Dilley and Austin, 1967).

The effect on maturity does, however, appear to be complex, and, as yet, far from fully understood. The authors have found that B9 sprayed onto Cox's Orange Pippin trees about 58 days after petal fall at a concentration of 2500 p.p.m. delays the onset of the respiration climacteric by about 10 days. In addition the ethylene production of the fruit was reduced ten-fold and the respiration rate at the climacteric peak was appreciably lower than in comparable control fruit (Rhodes *et al.*, 1969). The effect on fruit drop was striking, a large proportion of the crop remaining on the tree in mid-November, several weeks after all the fruit had fallen from the control trees. The treated fruit was smaller in size and extremely highly coloured. Surprisingly, the treatment with B9 had little or no effect on the development of the malate effect (see p. 352) in the fruit during the period of the respiration climacteric. In the authors' experience, fruit from B9 treated trees does not appear to develop such good flavour as fruit from comparable untreated trees. This growth regulator is, therefore, having a differential effect on various aspects of ripening while B9 may not itself, in the long term, be the ideal compound for the control of growth and ripening, its discovery and application have at least demonstrated the potential value of chemical sprays in the control of the metabolism of fruit trees (see Luckwill and Child, 1967; Child, 1968).

Finally, as an example of the complexities of the situation, we may cite the suggestion of Reed (1965) that B9 *in vivo* is converted to 1,1-dimethylhydrazine, a powerful inhibitor of diamine oxidase which controls the important step in the biosynthesis of indolyacetic acid, namely the conversion of tryptamine to indole acetaldehyde. Some evidence that this might be possible, has been provided by the work of Martin *et al.* (1964) who applied ^{14}C N-dimethylaminosuccinamic acid to apple trees and found that although most of the B9 remained unaltered over long periods, a significant amount was broken down; the nature of the breakdown products was, however, not determined. Other workers have suggested an indirect relation between the action of B9 and that of the gibberellins (Moore, 1967). The effect of B9 on ethylene production by the fruit at the time of the climacteric has already been mentioned, and ethylene is itself an important regulator of growth and maturation. Looney (1968) has provided evidence that the delayed onset of the respiration climacteric and ripening of the McIntosh apple brought about by B9 may be reversed by the application of ethylene to the fruit. He suggests that B9 curtails ethylene production by reducing the indolyacteic acid in the fruit. However, the present authors have found that with apple peel tissue (a non-growing tissue as opposed to growing tissues where IAA does undoubtedly stimulate ethylene production—Abeles, 1966), treatment with IAA does not stimulate ethylene production (see p. 358). Ashby and Looney (1968) have also shown that early application of B9 sprays to apple trees modifies the mineral content of the mature fruit. Clearly the effect of B9 on the metabolism of the fruit is complex.

The importance of the physiological condition of the fruit in relation to storage behaviour in the immediate pre-harvest phase is the main theme of this chapter, and it will become obvious that treatments which modify the metabolism of the growing fruit will repay further study. Growth regulators such as those mentioned and others, for example the gibberellins, perhaps offer the best hopes of achieving a regulated modification of the biochemical processes during growth in directions favourable to good quality and lengthy storage.

C. Changes Occurring during Ripening

1. *Introduction*

Apples and pears, unlike some other fruits, will ripen while still attached to the tree, although pears tend to become "mealy" if ripened on the tree. With dessert varieties of apples, optimal eating quality is often best obtained in this way. Apple fruits ripened on the tree are, however, seldom encountered in practice. Storm and bird damage would render this practice uneconomic in many apple growing countries and, in addition, it has been found that ripe fruit has a short storage and shelf-life. Research on transport and storage has shown that the best results are obtained if the fruit is picked before the onset of ripening and this has become general commercial practice. Most storage practices are, in fact, aimed at delaying or prolonging the first stages of ripening (the respiration climacteric, see later). A comparison of the physiological and biochemical changes taking place in apples ripening on the tree and, under approximately the same conditions, in storage shows general similarities but some differences which will be discussed later.

The English word "ripen" is derived from the Old Saxon *ripi* meaning to reap or gather. It has, therefore, no precise scientific meaning, but the biochemical and physical changes which are associated with the visible process of ripening are embraced by changes in colour, texture, sweetness and astringency and the development of the characteristic flavour of a given variety. With pome fruits these readily perceived changes do not give the complete picture. Before *visible* ripening sets in, the respiration of the fruit begins to rise and during this rise profound changes take place which initiate the ripening processes.

The characteristic "flavour" of apples and pears is probably centred in a range of volatile compounds which will be discussed later in this chapter and which tend to increase during the ripening process. However, the most important volatile compound from a physiological viewpoint, and one which first appears in readily detectable quantities as the respiration of the fruit begins to rise, is the olefine, ethylene. This compound is, in fact, the initiator of ripening and its physiology will be discussed in the next section.

"Taste" is centred in the balance between sugar and acid; phenolic compounds contribute both to flavour and taste, the astringency of many of these compounds influencing the latter.

2. *Ethylene production*

In 1932, Kidd and West (1933) showed that the vapours produced by ripe apples when passed over unripe apples would stimulate their respiration into the climacteric phase and cause these unripe apples to ripen. Gane (1935) demonstrated that the active component in the vapours was ethylene and the gas is now firmly established as a "hormone of ripening" (see Volume I, Chapter 16). Ethylene introduced into the atmosphere surrounding unripe apples will cause them to ripen and it has been shown that endogenous production of ethylene immediately precedes the climacteric rise in respiration and the onset of ripening (Meigh *et al.*, 1967). As will be seen later, this endogenous ethylene production initiates an increase in many of the enzyme processes associated with ripening (Hulme *et al.*, 1968).

Hansen (1942) established that the production of ethylene is a part of the metabolism of mature pears. The rate of production of ethylene and of respiration followed a similar trend but Hansen could find no indication that ethylene production was *directly* related to the rate of respiration in a quantitative sense. He found wide differences in the CO_2/ethylene rates for different varieties of pear. In O_2-free atmospheres ethylene production was inhibited. Hansen (1942) could not, in effect, decide whether or not ethylene was a direct stimulator of the burst of respiration which immediately precedes visible ripening. Incidentally, Hansen (1946) later showed that the synthetic auxin 2:4 D at 1000 p.p.m. hastened the ripening of pears by increasing their rate of production of ethylene and appears to be one of the first investigators to link auxins with ethylene production.

Since ethylene induces ripening it is important to know what initiates its production, how it is synthesized and what is the mode of its action in promoting ripening. Little is yet known about the first of these problems, but Galliard *et al.* (1968a) showed that synthesis of RNA and protein appeared to be directly involved in the development of the capacity of the tissue to produce ethylene. Indolylacetic acid appears to increase ethylene production in some growing plants (Burg and Clagett, 1967), but it is not certain whether this is an effect on the growth responses of the tissue or a direct effect on ethylene biosynthesis. Indolylacetic, at low concentrations which stimulate growth (10^{-6} M) in vegetative tissue has little or no effect on the ethylene production of discs of apple peel. Two possible precursors of ethylene, namely methionine and its derivative methional (Lieberman *et al.*, 1966; Burg and Clagett. 1967; Mapson and Wardale, 1968) and linolenic acid (Galliard *et al.*, 1968b) have been undergoing intensive study. It now seems likely that methionine or 4-methylmercapto-2-oxobutyric acid serves as the precursor in a number of tissues including the apple (Mapson *et al.*, 1970). Linolenic acid and lipoxidase, which increase in activity during the climacteric in apples (Wooltorton *et al.*, 1965), appear to be involved in some way in the synthetic process. No conclusive evidence has yet been presented on the mechanism by which ethylene initiates the ripening process. It is likely, however, that it acts, as do some other plant hormones, by modifying the pattern of RNA synthesis (Holm and Abeles, 1967).

3. *The respiration climacteric*†

Kidd and West (1922) observed a characteristic rise in the rate of respiration, measured as output of CO_2, when Bramley's Seedling apples are detached from the tree in the region of the normal commercial harvesting time and stored at ripening temperatures. This phenomenon was first called the "climacteric" by Kidd and West in 1925. They also showed that a similar rise in respiration occurred in fruit left to ripen on the tree. The Respiratory Quotient (R.Q.—the output of CO_2 divided by the uptake of O_2) during this rise in respiration rose from 1·02 to 1·4 (see section on acid metabolism, p. 350). A typical set of respiration data for apples picked at several stages of maturity, and for fruit ripening on the tree are shown in Fig. 4. The results for the "on the tree" fruit were obtained by measuring the CO_2 output of the fruit (at 12°C) *immediately* after picking at each point. It will be seen that the climacteric on the tree proceeds more slowly and rises to a higher peak value than in detached fruit.

In view of the importance in the respiration of living cells of certain cell organelles, the mitochondria, their activity has been widely studied in relation to the respiration climacteric. Pearson and Robertson (1954) were the first to obtain a crude mitochondrial fraction from apples. Although their particles had less than one-tenth the activity of preparations obtained by later workers using the more sophisticated methods developed for plant mitochondria generally, their results established that there was a marked increase in mitochondrial activity as the fruit passed through the climacteric. This work of Pearson and Robertson together with later work by Lieberman (1958) and by Hatch *et al.* (1959) established that a Krebs Tricarboxylic Acid cycle (TCA cycle) involving the cytochrome oxidase system operated in apples, and that the activity of the enzymes concerned increased over the climacteric. This work has been confirmed by Hulme *et al.* (1964a), who obtained mitochondrial fractions from the peel of apples having 30 times the activity of the original preparations of Pearson and Robertson and comparable with those shown by mitochondria from leaves. These preparations

† See Volume 1, Chapter 17.

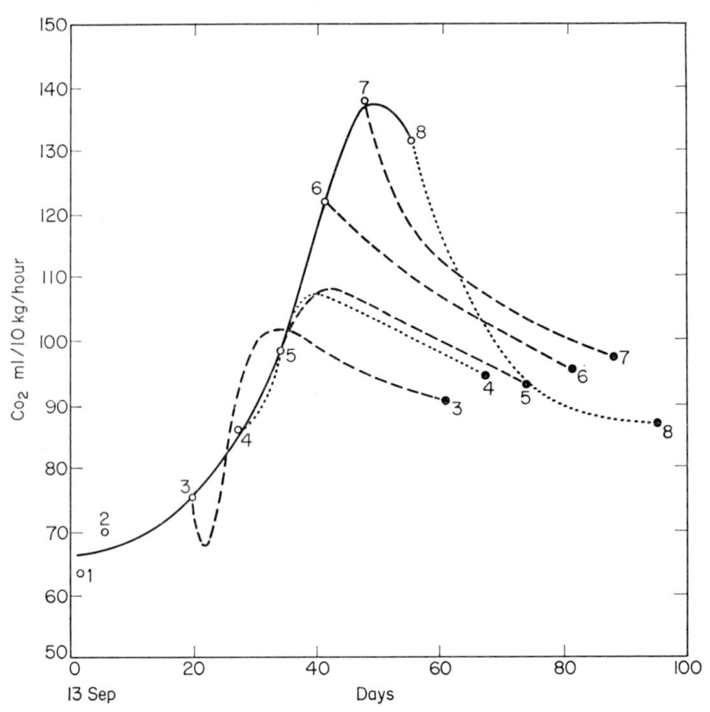

FIG. 4. Respiration rate of Cox's Orange Pippin apples 'on the tree' (continuous line) and after detachment at various stages of development and storage at 12°C (dotted line).

showed the typical structure of plant mitochondria in the electron microscope. The method of preparation owed its success to complete but gentle disintegration of the tissue cells in the presence of polyvinylpyrrolidone (PVP) which "protected" the mitochondria by preventing the oxidation of phenolic compounds in which the apple tissue is rich. Such oxidized phenolic compounds belong to a class of compounds (the tannins) which complex with proteins to form insoluble aggregates. It is probable that these oxidized phenolic compounds not only directly inactivate the mitochondrial enzymes (Hulme and Jones, 1963) but also cause considerable contamination of the mitochondrial preparations by co-sedimentation of non-mitochondrial protein–tannin complexes.

The apple mitochondria prepared by Hulme and his associates showed active coupling between O_2-uptake and esterification of P_i giving P/O ratios with succinate and malate substrates close to the "theoretical" values. Furthermore, as activity increased over the climacteric, no uncoupling of oxidation and phosphorylation occurred, the P/O ratio only decreasing as the fruit became over-ripe (Jones *et al.*, 1964). In intact mitochondria an important factor controlling respiration rate is the amount of ADP available in the system (see Lehninger, 1964) for combination with the P_i to give ATP (a reservoir of energy in the form of the high-energy phosphate bond). This control of respiration by added ADP (respiratory control) is a criterion of mitochondria which are structurally and functionally "intact". The Respiratory Control Index (R.C. value) is calculated as the rate of respiration (O_2-uptake) in the presence of ADP divided by the rate when it has been exhausted. Although R.C. values of 5 or 6 and higher have been obtained for some plant mitochondria, the highest value so far found for mitochondria from apples is 2·0 (Wiskich, 1966). Values up to 3 have been obtained in this laboratory for mitochondria prepared from the pulp of pears. Even higher values (up to 8·5) have been obtained by Romani *et al.* (1969) from pulp of pears using particularly gentle methods of tissue maceration. The relatively low values could be due to the difficulty in preventing completely the damaging effect on the mitochondrial preparations of the large amount of phenolics present in the apples (Hulme and Jones, 1963). There is some evidence that respiratory control by ADP might be changing in mitochondria prepared at successive stages during the climacteric; Hulme *et al.* (1967a) showed that the addition of glucose and hexokinase (which would cause a regeneration of ADP from ATP) had an increasingly stimulatory effect on the activity of the mitochondria as the climacteric progressed.

The increased activity of mitochondrial fractions isolated from the apple (especially the peel; the change in activity is less in the pulp, see Hartmann (1962b)) as the climacteric develops, whether the fruit is "on" or "off" the tree (Hulme *et al.*, 1965; Jones *et al.*, 1965), poses the question as to whether there is an actual increase in the number of mitochondria during this period or merely an increase in the activity of the mitochondria present just before the climacteric begins. Frenkel *et al.* (1968) in experiments in which the protein synthesis inhibitor, cycloheximide, was injected into unripe pears, showed that although the cycloheximide inhibited subsequent ripening as judged by changes in ethylene production, chlorophyll breakdown and flesh softening, it did not inhibit the climacteric rise in respiration. This would suggest that the rise in respiration (and the associated rise in mitochondrial activity) was independent of new protein synthesis and hence of increase of mitochondrial in number. However, cycloheximide tended to *stimulate* the rate of respiration of the fruit at all stages of the climacteric (see Frenkel *et al.*, 1968, Fig. 5), and it may be that the cycloheximide was having a secondary effect on respiration, possibly by uncoupling oxidative phosphorylation in addition to affecting protein synthesis directly (MacDonald and Ellis, 1969). The suggestion that the rise in respiration is independent of protein synthesis may thus not be conclusive.

Preliminary electron microscope studies in our laboratory suggest that there is a phase early in the climacteric phase during which there is a relative abundance of dumb-bell shaped profiles which might indicate mitochondrial multiplications. On the other hand, it could be that the number of mitochondria does not increase and we must look for a stimulation of the activity of the existing mitochondria through, for example, an increase within the tissue of essential co-factors, especially the pyridine nucleotides. It has been shown, however, that the concentration of these nucleotides (NAD and NADP) is relatively high; NAD shows no major change during the period of the climacteric (Rhodes and Wooltorton, 1968). Regulation could take place through an increase in the concentration of ADP in relation to the concentration of ATP as suggested by Pearson and Robertson (1954). No direct measurements of the ADP/ATP ratio have been made over the climacteric (there are technical difficulties involved here peculiar to tissues of pome fruits), but this suggestion, as well as that advanced earlier for other fruits, of the appearance of endogenous uncouplers of oxidative phosphorylation have been discounted by the work of Young and Biale (1967) with the avocado pear. It seems most likely, therefore, that the increased activity of the mitochondria during the climacteric is due to an increase in their number. Lance and Bonner (1968) showed for a number of tissue (not, unfortunately, including fruits) that the number of mitochondria per cell is the controlling factor in the intensity of tissue respiration.

It has been shown that the susceptibility of mitochondrial oxidation of succinate to inhibition by oxaloacetic acid (see Lehninger, 1964, p. 146) increases as the climacteric developes (Hulme *et al.*, 1967b) so that there could be a change in the balance of enzyme activity within the mitochondria during the climacteric period (see Volume 1, Chapter 17).

4. *Protein and RNA metabolism*

Interest in the metabolism of proteins in relation to ripening was first aroused by the findings of Hulme (1937, 1954) that there was an increase in the 80% ethanol insoluble nitrogen fraction expressed as a percentage of the total nitrogen at the time of the climacteric rise in respiration in the apple. When the respiration climacteric was induced in young apples by treatment with ethylene, a rise in protein nitrogen was also induced (Hulme, 1948). Conditions such as relatively high CO_2 in the storage atmosphere which delay the onset of the climacteric also delay the rise in "protein" (Kidd *et al.*, 1939). Although the ethanol insoluble fraction estimated by Hulme would not consist entirely of protein, subsequent work has shown that the original conclusions were correct in general terms. The relatively small increase in protein nitrogen may well considerably underestimate the degree of underlying redistribution of enzymic proteins during the climacteric. For example, as we shall see later (Section 9a) lipo-protein membranes within the chloroplasts are being broken down during the early stages of the climacteric and may provide an additional pool of amino acids for the synthesis of "new" enzymic proteins. Thus there are likely to be major changes in the pattern of protein during the early stages of ripening; this has recently been confirmed by electrophoretic studies of protein patterns in pre- and post-climacteric peas (Frenkel *et al.*, 1968; see Volume 1, Chapter 7).

In addition to the increase in net protein nitrogen during the climacteric, there appears, in some apples at least, to be a secondary rise to a peak during senescence according to Lewis and Martin (1965). These authors suggest that this increase in protein may hasten the onset of senescent breakdown by accelerating the decline of the ATP supply to a level below that required for the maintenance of cellular organization.

Both Hulme (1951) and Pearson and Robertson (1953) showed that from the end of the period of active cell division, during the growth of the apple until the onset of the respiration climacteric the rate of respiration per unit protein nitrogen (the R/P ratio) remained remarkably constant. During the climacteric, however, the R/P ratio rose considerably suggesting that there is no simple relationship between protein synthesis and the respiratory rise. For instance, it rules out the possibility that the respiration rate is determined by ADP made available by protein synthesis. Richmond and Biale (1966), using slices of ripe avocado tissue, found that inhibitors of protein syntheis did not affect the rate of respiration even when protein synthesis was inhibited completely. The increase in and redistribution of protein is, however, evident in the number of enzymic processes that increase at the time of the climacteric.

It has been demonstrated that there is an increase in the activity of a variety of mitochondrial and soluble enzymes during the climacteric in apples and pears. The isolation of these enzymes is dependent upon the development of suitable extraction techniques which protect the enzymes during isolation from damage due to the presence of phenolics and the high acidity of the sap. Another problem in studying changes during ripening is the change in the texture of the tissue which may lead to differences in the extractability of enzymes. The overcoming of the first of these difficulties has already been discussed when considering the mitochondria; the same methods are applicable to soluble enzymes. There is no reason to believe that the second of these difficulties leads to serious errors at least during the climacteric although it may be significant when senescent disintegration of the tissue has commenced.

Among the enzymes showing increased activity during the climacteric phase and beyond are: NADP-malic enzyme (Neal and Hulme, 1958; Dilley, 1962; Hulme and Wooltorton, 1962), pyruvated ecarboxylase (Hartman, 1962a; Hulme *et al.*, 1963), lipoxidase (Wooltorton *et al.*, 1965), chlorophyllase (Rhodes and Wooltorton, 1967; Looney and Patterson, 1967), acid phosphatase and ribonuclease (Rhodes and Wooltorton, 1967). A review of some of the enzyme changes in relation to fruit ripening has been made recently by Ulrich and Hartmann (1967).

The work of Meynhardt *et al.* (1964), Frenkel *et al.* (1968) and Hulme *et al.* (1968) on the capacity of apple and pear tissue to incorporate labelled amino acids into protein at various stages of the climacteric showed that there is a phase in the early stages of the climacteric when protein synthesis is stimulated and reaches a peak value. By the time the climacteric maximum is reached the rate of protein synthesis is falling. Using polyacrylamide gel electrophoresis technique Frenkel *et al.* (1968) found that only a relatively specific group of proteins was synthesized at this early stage and that in this group malic enzyme was particularly important.

Although detailed data on the subject are still scarce, the work of Rhodes *et al.* (1968) and Frenkel *et al.* (1968) indicates that nucleic acid synthesis is involved in the ripening of both apples and pears. Looney and Patterson (1967) showed that there was an increase in the total level of RNA associated with the ripening of Yellow Transparent apples; there was no change in the base composition of the RNA synthesized. They showed that the RNA of apple pulp tissue is relatively rich in guanylic (30%) and adenylic (26%) acids.

It is clear from the foregoing that the early stages of ripening (the climacteric phase) is a period of rapid change in enzyme synthesis and activity. There are, no doubt, many important enzyme changes other than those already studied—for example, those responsible for the synthesis of anthocyanins (see later). For a study of the pattern of enzyme changes over the climacteric the use of the disc electrophoresis technique with polyacrylamide gel developed by Clements (1965) for apple material may prove to be a useful starting point.

5. *Metabolism of organic acids*

The major organic acids in pome fruits are malate and citrate but, as will be seen from Table I, a variety of other acids occur in much smaller amounts. In apples malate predominates, but in some varieties of pears citrate and malate occur in approximately equal amounts (see Hansen, 1966). Succinate appears in more than trace amounts only in fruit stored in Controlled Atmosphere Storage under relatively high CO_2 conditions. Such fruit subsequently develops CO_2 injury (Hulme, 1956b). Oxaloacetic acid is always present in fruit in small amounts (it is an intermediate in the Krebs cycle) but a rapid *increase* in the concentration of this acid in cold storage generally presages the onset of the physiological disorder known as low temperature breakdown (Hulme *et al.*, 1964b).

Kidd *et al.* (1951), in a study of changes in the total titratable acids (which they regarded as at least 95% malate) in detached apples passing through the

TABLE I. Organic acids known to be present in apple and pear fruits

Apples			Pears		
Whole fruit or juice of whole fruit	Pulp	Peel	Whole fruit or juice of whole fruit	Pulp	Peel
Quinic	Malic	Malic	Quinic	Malic	Malic
Glycolic	Citric	Citric	Glycolic	Citric	Citric
Succinic	Quinic	Quinic	Succinic	Quinic	Quinic
Lactic	Shikimic	Shikimic	Lactic	Shikimic	Shikimic
Galacturonic	Succinic	Citramalic	Galacturonic	Glyceric	Glyceric
Citramalic	Glyceric	Glyceric		Mucic	Citramalic
Mucic	α-Oxoglutaric	α-Oxoglutaric			
	Pyruvic	Pyruvic			
	Oxalacetic	Oxalacetic			
	Glyoxylic				
	Isocitric				

From Hulme (1958).

the climacteric, found no change in the rate of loss of acid as the respiration rose. As a result of more careful investigation, using separation on ion exchange resins, of changes of malic acid in detached fruits taken at various stages of maturity, Flood *et al.* (1970) have challenged this finding. Some of their results are shown in Fig. 5, and the general conclusion from them is that a marked increase in malic acid utilization in both peel and pulp of apples commences during the period of the climacteric. In fruit picked when the climacteric had already begun on the tree, the fall in malate had already commenced.

Hansen (1967) has shown that in Bartlett pears, in which the climacteric was induced by the application of ethylene, parallel with the increase in respiration, there was an increase in utilization of malate. Hulme and Wooltorton (1958) studied changes in malic, citramalic, citric, quinic and shikimic acids during the ripening at 15°C of Bramley's Seedling apples. Citramalate appeared in the peel only and rose during storage to a maximum of 25 mg/100 g fresh weight in extreme senescence. There was a rise in citrate in the pulp (from 6·5 to 10·0 mg/100 g fresh weight) but not in the peel where it remained constant at less than 2 ml/100 g fresh weight. Quinic acid rose to a maximum in both pulp (80 mg/100 g fresh weight) and peel (320 mg/100 g fresh weight) and subsequently fell, the peak in both tissues occurring about half-way through the 100 day storage period. Finally, shikimic acid, present only in the peel, began to rise as the quinic acid fell and reached a value of 8 mg/100 g fresh weight towards the end of the storage period. Hansen (1967) also found an increase in shikimic acid during the ripening of Bartlett pears. The physiological significance of these changes in the "minor" acids of the fruit has not been explored.

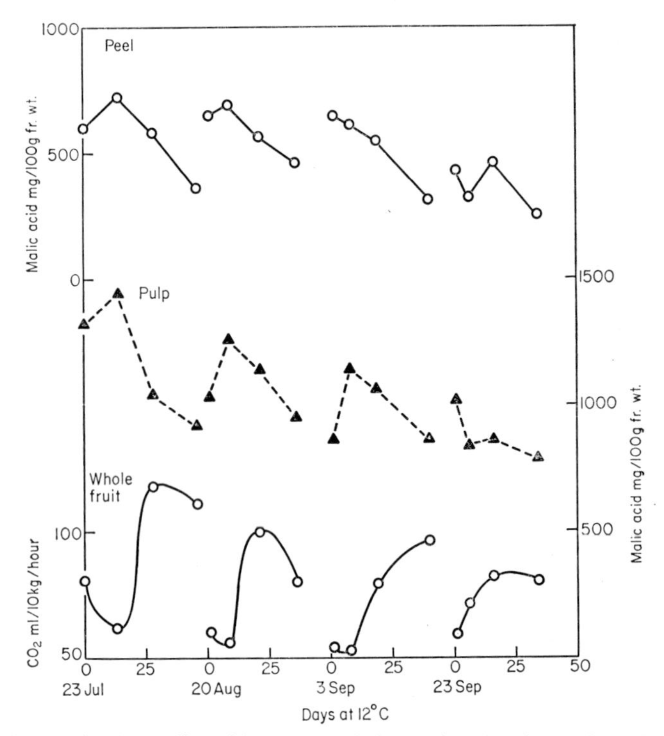

FIG. 5. Changes in the malic acid content of the peel and pulp, and respiration rate (at 12°C) of the whole fruit of Cox's Orange Pippin apples harvested at various dates during growth.

Together with sugars, malic acid is the main substrate for respiration in apples. Undoubtedly a major pathway of metabolism of the acid is via the Krebs (TCA) cycle. The respiratory quotient (R.Q.) for the complete oxidation of malate via the cycle is 1·33. The R.Q. of a system metabolizing both malate and sugar should be between 1·0 and 1·33. Kidd and West (1938) showed that the R.Q. of whole Bramley's Seedling apples ripening at 22·5°C rose from 1·02 to 1·25 at the climacteric peak and subsequently to 1·4. Neal and Hulme (1958) variously found rises in the R.Q. of discs of apple peel taken over the climacteric period of 1·39–1·62 and 1·08–1.36. In view of this rising R.Q. it appeared likely that a system was developing during the climacteric which had a relatively high R.Q. This possibility led Neal and Hulme (1958) to a study of the oxidation of various acids of the Krebs cycle by apple peel discs taken at various stages of the climacteric. They found that if malate or pyruvate were added to discs taken near the climacteric peak there was a large increase in CO_2-production without a correspondingly large increase in O_2-uptake. Further work established that the malate decarboxylation system developed in the tissue during the climacteric; the pyruvate decarboxylation system also increased in activity but to a lesser extent. Hartmann (1962a) also found an increased decarboxylation of pyruvate over the climacteric in pears. Decarboxylation of other TCA cycle acids could be traced to the development of the malate decarboxylation system. This system at present appears to be characteristic of apples and pears; it is not for example found in ripening bananas. Neal and Hulme (1958) suggested that the enzyme responsible for this "malate effect" might well be NADP-dependent malic enzyme. The presence of this enzyme in apple tissue was confirmed by Okamoto (1961) and by Dilley (1962) who isolated a crude, NADP and Mn^{+2}-dependent malic enzyme preparation from apples. This he later further purified and characterized. Neal and Hulme (1958) showed that the R.Q. of this "malate respiration" was considerably in excess of 1·33 and that acetaldehyde was a major product; they proposed the following series of reactions for the overall process:

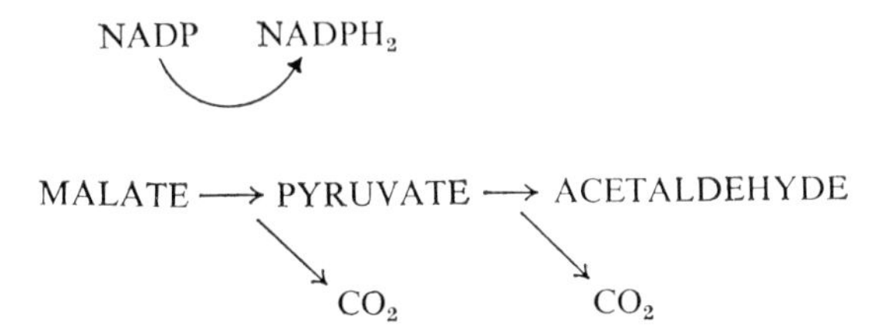

They studied the stoichiometry of the reaction and found that the experimental results approximated closely to the theoretical value 1 : 2 : 1 for malate utilization, CO_2 production and acetaldehyde formation. Subsequently Hulme *et al.* (1963) established that there was an increase in malic enzyme and pyruvate decarboxylase (the enzyme responsible for the conversion of pyruvate to acetaldehyde) during the climacteric period. The development of this combined malate and pyruvate decarboxylating would explain the rise in the R.Q. over the climacteric. The R.Q. of the system converting malate to acetaldehyde is 4·0. Table II shows the change in the malate effect (the increased CO_2-output following addition of malate) and in the basic R.Q. and the R.Q. after addition of malate for discs of apple peel taken from Cox's Orange Pippin apples from the pre-climacteric minimum to just past the climacteric peak (at 12°C). It will be seen from the table that, as the climacteric proceeds, the tissue becomes more active in decarboxylating malate, its basic R.Q. rises while the response to exogenous malate is a large increase in the malate induced R.Q. often in excess of the "theoretical" value of 4·0. This would suggest that towards the climacteric peak the malic enzyme system is making an increased contribution to the respiration of the tissue. It almost certainly does not mean that, because the malate-induced

TABLE II. Respiratory quotients for apple peel discs taken from Cox's Orange Pippin apples passing through the respiration climacteric (Fruit stored at 12°C)

Days at 12°C	Respiratory quotient[a] Basic	Malate	Δ Malate
0	1·05	1·96	7·7
5	1·11	1·85	3·2
8	1·25	2·22	5·0
14	1·25	2·45	8·4
22 } [b]	1·36	3·02	9·0
27 }	1·32	3·30	17·5
34	1·33	3·74	16·9

[a] Basic R.Q.s are for tissue alone; malate R.Q.s are for tissue given malate, and Δ R.Q.s are calculated from the *increased* O_2-uptake and CO_2-output after adding malate.
[b] Climacteric peak is between these days.

R.Q. approaches 4·0, the *whole* of the malate (endogenous) respiration of the tissue is via the malic enzyme system. As Lips and Beevers (1966) point out in a study of the effect of adding malate to corn root tissue, malate is likely to be present in the tissue in more than one pool, each pool having a different location within the cell. Probably the added malate will not permeate into the mitochondria so that the contribution to the respiration of this added malate will be almost entirely through the action of the (cytoplasmic) malic enzyme system. This will emphasize any increase in the activity of the system in relation to utilization of malate by the mitochrondria. In the tissues, of course, the endogenous malate will be available to mitochondria and malic enzyme (mitochondria in the apple do not contain any NADP-dependent malic enzyme) only in relation to the size and position in the cell of the malate pools. Hence in the fruit itself increased activity of malic enzyme will not necessarily have a dramatic effect on the R.Q. of the fruit. The efficiency in terms of availability of energy from the utilization of malate via malic enzyme is considerably less than via the TCA cycle. Nevertheless, the action of malic enzyme in the direction of decarboxylation of malate leads to an increase in $NADPH_2$, a requisite of many synthetic processes. Whether this is the role of the enzyme in the apple is a matter of speculation. As Sanwal *et al.* (1968) have stated when considering and, to a considerable extent delineating, its function in *E. coli* that the function of malic enzyme in diverse organisms remains largely enigmatic.

Rhodes *et al.* (1968) showed that there is no general increase in the permeability of apple peel discs to malate during the climacteric so that the rise in malate respiration cannot simply be due to an increase in the accessibility of the added malate to the enzyme site.

The acetaldehyde formed under the combined action of malic enzyme and pyruvate decarboxylase would be metabolized (Fidler, 1968). In Neal's experiments on the stoichiometry of the system (see above), he swept the acetaldehyde formed from pyruvate from his flasks into bisulphite as a preliminary to the determination of the aldehyde.

Although Neal and Hulme (1958) found that malate loss was exactly balanced by production of CO_2 and acetaldehyde with no formation of ethanol, this may have been due largely to the removal of acetaldehyde. Clijsters (1965a, b) has shown that discs of both peel and pulp of Jonathan apples produce ethanol as well as acetaldehyde when supplied with malate. Rhodes *et al.* (1969) confirmed this for Cox's Orange Pippin apples and showed that, while the acetaldehyde plus ethanol was equal to half the CO_2 produced, about one-third of the acetaldehyde was converted to ethanol. Rather surprisingly there appeared to be two alcohol dehydrogenase (ADH) involved, one NADP-dependent and one NAD-dependent (the usual ADH present in higher plants). In Cox's Orange Pippin apples the proportion of the two ADH enzymes appears to be equal whereas in experiments with the Bramley's Seedling variety there appears to be considerably more of the NADP-dependent enzyme. These facts provide a possible explanation of the low O_2-uptake (in relation to CO_2 production) and R.Q. value in excess of 4 of discs fed with malate; some of the $NADPH_2$ formed in the reaction malate to pyruvate may be utilized in reducing the acetaldehyde to alcohol. We have obtained recently (unpublished data) an extract from ripe Cox's Orange Pippin apples which produced ethanol from added malate in the presence of NADP but not in the prescence of NAD.

In an attempt to overcome the difficulty in studying the biochemistry of the climacteric in whole fruits that the fruits are only available in a suitable physiological state for a short period of the year, coupled with the fact that at physiological temperatures the climacteric period extends to 10–15 days, Hulme and his co-workers have looked for a system (a "model" system) in which at least some of the biochemical changes occurring in whole fruit will develop in a matter of hours. This would enable many more experiments to be done in each growing season. Furthermore, if such a model system would operate in small pieces of tissue, then the way is open to investigate the mechanisms of any biochemical changes by the use of labelled substrates and the many general and specific modifiers (inhibitors and accelerators) of enzyme processes which are now becoming available. It is difficult to make such an approach if whole, intact fruit must be used. Such a model system has materialized, insofar as certain phenomena characteristic of the climacteric in whole fruits are concerned, in the form of discs of peel tissue, prepared from apples in the pre-climacteric state, and "aged" at 25°C over periods up to 24 hours. (It should be noted that discs of peel tissue cannot be compared with thin slices of the cortical tissue of fruits inasmuch as there is much less damaged tissue since the inner side of the peel disc is the only significant cut surface. This statement is supported by the fact that peel discs taken from the pre-climacteric apples do not immediately (i.e. within the first 2–3 hours of preparation) produce any ethylene. Also these peel discs follow the general physiological trends of the fruit from which they have been taken (Rhodes *et al.*, 1970)). Using this system, the factors associated with the development of the malate decarboxylating system (malate effect) have been examined. Rhodes *et al.* (1968) showed that the development of this malate effect was inhibited by the presence of inhibitors of RNA and protein synthesis such as actinomycin D, azauracil, cycloheximide, puromycin and fluorophenylalanine. With, for example, cycloheximide (an inhibitor of protein syntheis) the inhibition of the development of the malate effect was removed by the removal of the inhibitor. The inhibition by nucleoside and amino acid analogues was partially reversible when an equal concentration of the normal nucleoside or amino acid was also included in the medium. The growth retardant B9 (see p. 341) when included in the ageing medium inhibits the development of the malate effect. Ageing in 3% oxygen (the ripening of intact fruits is delayed in an atmosphere of 3% oxygen) also inhibits the phenomenon. These effects are shown in Fig. 6. Rhodes *et al.*

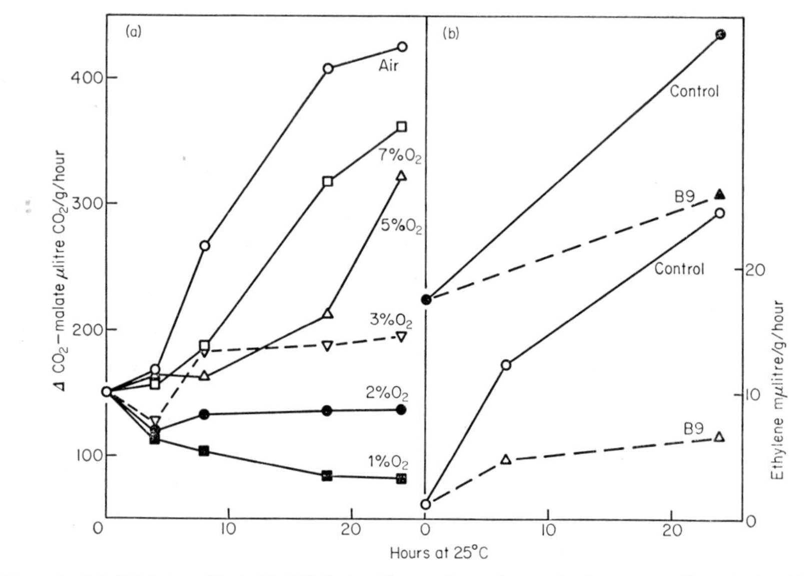

FIG. 6. (a) Malate effect (Δ CO_2) in discs of apple peel, from preclimacteric Cox's Orange Pippin apples, during ageing in various concentrations of O_2. (b). Malate effect (●, ▲) and ethylene production (○, △) in discs from pre-climacteric apples and aged in the presence and absence of B9 (2 mg/ml).

(1968) concluded that the increase in this malate decarboxylating system requires the synthesis of new RNA and protein. The similarity in increased rate of respiration and the development of the malate effect between the model system and the climacteric is seen in Fig. 7 which the two parameters are compared in ageing pre-climacteric discs and in discs taken from whole fruits as they passed through the climacteric.

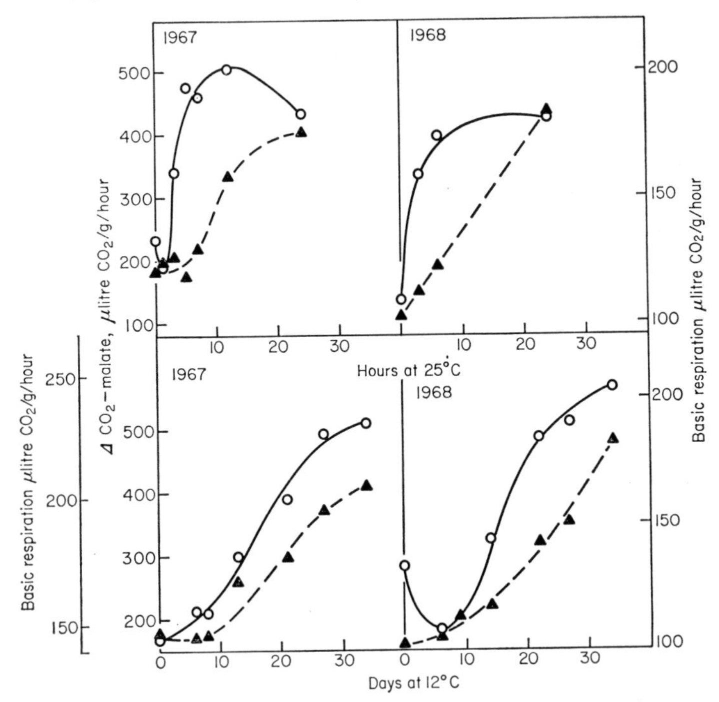

FIG. 7. A comparison of the basic CO_2-output (○) and extra CO_2-production on addition of malate (▲) in initial discs taken from apple fruits as they pass through the respiration climacteric (lower curves), and during the ageing of discs from pre-climacteric fruit (upper curves).

Galliard *et al.* (1968b), studying the ageing of discs of peel from pre-climacteric apples, showed that just prior to the development of the malate effect the discs began to produce ethylene, while Rhodes *et al.* (1968) found that ageing such discs in an atmosphere containing ethylene accelerated the development of the malate effect (see Rhodes *et al.*, 1968, Fig. 6). Indolyl-acetic acid (10^{-4} M) applied to the discs during ageing inhibited both ethylene production and the development of the malate effect (Fig. 8).

It is clear from the foregoing discussion that the development of the malate decarboxylating system is intimately concerned in the respiration climacteric although its significance in quantitative terms is still not clear.

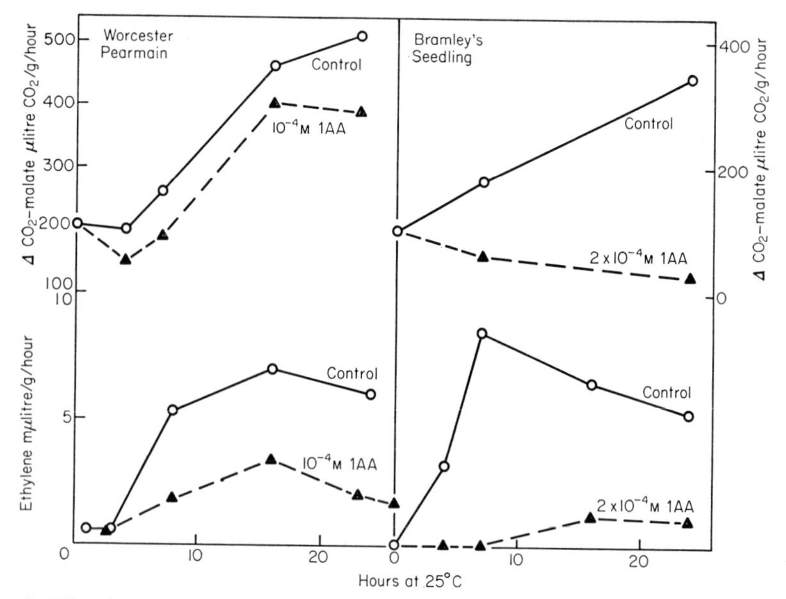

FIG. 8. The changes in ethylene production (lower curves) and malate decarboxylation capacity (upper curves) of discs of two varieties of apples, Worcester Pearmain (left and Bramley Seedling (right), aged in the absence (○——○) and presence (▲——▲) of various concentrations of IAA.

6. *Metabolism of carbohydrates*

In the early biochemical work concerned with the respiration climacteric in pome fruits, attention was focused on carbohydrates as the respiratory substrate since in those days the importance of acid metabolism via the TCA cycle was not fully appreciated. Inevitably, no direct relationships were found between the climacteric and loss of starch or changes in sucrose and reducing sugars. Much of this earlier work has been reviewed by Hulme (1958) and will not be considered here. In recent years more emphasis has been placed on the *pathways* of metabolism of sugars and possible changes in the activity of the glycolytic and the pentose phosphate pathways during ripening. Some of the work on this subject has been reviewed recently by Ulrich and Hartmann (1967). Meynhardt *et al.* (1965) established that the pentose phosphate pathway is present in ripening Bartlett pears, but did not measure quantitative changes in the enzymes involved and were unable to decide whether there was an increased activity of the pathway during ripening. Faust *et al.* (1966), using specifically labelled glucose, concluded that the pentose phosphate pathway plays an increasingly important role in the respiratory mechanism of Cortland apples during ripening and senescence. Hartmann (1966) studied changes in the activity of the enzymes involved in the two pathways as well as the transformations of specifically labelled glucose and came to the conclusion that the contribution of the pentose phosphate pathway decreased during ripening largely because of an increase in glycolysis; the present authors have confirmed this for Cox's Orange Pippin apples. Neither the use of glucose labelled in the C_1 or C_6 position nor the measurement of changes in enzyme activity provide rigorous proof of quantitative changes in the relative importance of the two pathways and this remains an open question during the ripening of fruits.

7. *Metabolism of lipids*

Markley and Sando (1931) were the first to show that the waxy surface coating of apple skin increased during maturation on the tree and in subsequent storage. By virtue of differential solubilities in organic solvents these coatings could be divided into four crude fractions, an oil, a wax, the triterpenoid, ursolic acid and cutin. Chibnall *et al.* (1931, 1934) showed that the wax constituents contained mostly n-nonacosane together with smaller amounts of heptacosane and a range of long-chain alcohols. Huelin and Gallop (1951) found that when Granny Smith apples were stored for 27 weeks at 1°C, the oil fraction trebled, but if the apples stored at 1°C were transferred to 20°C the fraction fell rapidly. By comparison, the changes in the wax and ursolic acid fractions was small. Davenport (1956, 1960), working with the same variety of apple, found that the saturated acids comprising 10% of the oil from the cuticle of ripe apples consisted principally of stearic and arachidic acids with smaller amounts of palmitic, behenic and acids of higher molecular weight. Mazliak (1960) found that the wax fraction was rich in the saturated C_{20} to C_{30} acids while the oil fraction contained a high proportion of unsaturated C_{18} acids. Subsequently he showed that during the ripening of apples both at 4°C and 15°C the total wax content rose to a maximum at the time of the climacteric peak (Mazliak, 1963). Huelin (1959) showed that polymerization of cutin in Granny Smith apples occurred only in the early stages of storage at 1°C; he purified the cutin by treatment with snail gut extract. The cutin fraction increased by about 20% during the storage period. Meigh and Hulme (1965) and Meigh *et al.* (1967) showed that there was a rise in both free and esterified fatty acids during the early stages of the climacteric rise followed by a fall. A discussion of the fractionation of the cuticular components of the apple is given by Mazliak in Volume 1, Chapter 9.

Recently interest has moved from the cuticular waxes and oils to the lipids involved in lipoprotein membranes such as galacto- and phospholipids in the peel and pulp of apples which have been investigated in some detail by Mazliak (1967) and Galliard (1968). The major components in the peel are phosphatidyl-choline, phosphatidyl-ethanolamine, phosphatidyl-inositol and

TABLE III. Quantitative composition of lipids in pulp of pre- and post-climacteric apples

Lipid	μmoles liquid/1000g fr. wt.		Weight lipid mg/1000g fr. wt.		% by weight of total lipid	
	Pre-climacteric	Post-climacteric	Pre-climacteric	Post-climacteric	Pre-climacteric	Post-climacteric
Phosphatidyl choline	240	272	189	214	21·5	22·7
Phosphatidyl ethanolamine	168	136	124	101	14·4	10·7
Phosphatidyl inositol	62	69	53	59	6·0	6·2
Phosphatidyl glycerol	35	10	27	7·7	3·1	0·8
Diphosphatidyl glycerol	4	3	5·8	4·4	0·7	0·5
Phosphatidyl serine	5	5	4·0	4·0	0·4	0·4
Phosphatidic acid	4	9	2·8	6·0	0·3	0·7
Unknown phospholipid	2	2	—	—	0·2[a]	0·2[a]
Total phospholid	520	506	405·6	396·1	46·6	42·2
Monogalactosyl diglyceride	54	15	42	12	4·8	1·3
Digalactosyl diglyceride	114	52	107	49	12·2	5·3
Polygalactosyl diglyceride	tr.	tr.	—	—	—	—
Unidentified galactolipid	tr.	tr.	—	—	—	—
Total galactolipid	168	67	169	61	17·0	6·6
Esterified steryl glucoside	14	18	12	15	1·4	1·6
Steryl glucoside	89	89	51	51	5·8	5·4
Sterol	314	460	129	190	14·7	20·1
Sterol esters	27	32	18	22	2·0	2·1
Total steryl lipids	444	599	210	278	23·9	29·2
Sulpholipid	tr.	tr.	—	—	1·0	1·0
Glucocerebroside	47	69	34	49	3·9	5·2
Triglyceride	50	55	44	50	5·0	5·3
Chlorophyll	1	0·3	0·9	0·3	0·1	0·03
Combined neutral lipids	—	—	208[b]	272[b]	23·1	28·7
Combined polar lipids	—	—	672	675	76·9	71·3
Total lipids	—	—	880[b]	947[b]	—	—

Reproduced from Galliard (1968).
[a] Estimated value.
[b] Results from direct weighing.

a little phosphatidic acid. Galliard (1968) showed that the main change in the lipid components of apple pulp over the climacteric period was a fall in the galactolipids and phosphatidyl-glycerol (see Table III). Both these lipids are rich in linolenic acid and are associated with chloroplast membranes (a rudimentary form of chloroplast is present in the pulp as well as the true chloroplasts which appear in the peel). Most other components of the lipid membranes show no marked changes. Galliard concludes from these results and other unpublished results for peel tissue, that there is a selective breakdown in plastid membranes during ripening. Hulme *et al.* (1968) have shown that the capacity of discs taken from apples undergoing the climacteric to incorporate labelled acetate into lipids, particularly phospholipids, rises rapidly during the early part of the respiratory climacteric (see Galliard *et al.*, 1968a). Although the picture is still far from clear, it would appear that the metabolism of lipids plays an important part in the changes occurring during the climacteric. Further work is needed, particularly to answer the question whether or not the biosynthesis of ethylene is indeed a result of the breakdown of lipids involving lipase and lipoxidase (see later).

8. *Volatile products of the ripening apple*†

The advent of gas–liquid chromatography (GLC) has given new life to the search for the volatile compounds given off by apples. Minute amounts (for example 0·001 μg) can be detected by this technique. At present it might almost be said that the more sophisticated the GLC column used the more peaks are obtained. The identification of the compounds represented by the peaks is much more difficult. Drawert *et al.* (1968) have detected as many as 120 compounds allegedly involved in apple aroma from various varieties of apples at various stages of maturity. Pailliard (1967a, b) fractionated apple volatiles into 100 peaks by GLC of which 80 were identified as alcohols, aldehydes, ketones and esters of formic, acetic, propionic, butyric, valeric and caproic acids. Quantitatively the most important were acetates, butyrates and caproates and the aldehydes, 2-hexanal and hexanal. Flath *et al.* (1967) suggests that the main component responsible for the characteristic apple flavour are these two aldehydes together with ethyl-2-methyl butyrate. However, the pitfalls in trying to relate flavour and aroma to specific substances or groups of substances will be seen from Volume 1, Chapter 12. This subject of the development of flavour in relation to an increased production of volatile constituents has exercised fruit biochemists since the early work of Power and Chesnut in 1920. No doubt in time the problem will be solved through a combination of GLC and mass spectrometry and the more sophisticated methods employed in the assessment of

† See also Volume 1, Chapter 10.

flavour and aroma by carefully organized taste panels (Harper *et al.*, 1968). This subject has been dealt with for the apple in an excellent article entitled "Requirements for Coordination of Instrumental and Sensory Techniques" by D. G. Guadagni (1968). As an example of recent work attempting to correlate production of volatiles with the physiological state of the fruit, the results of Brown *et al.* (1966) indicated an increase in the more volatile compounds to just past the climacteric peak; the compounds were, however, not identified. It has been suggested (Somogyi *et al.*, 1964) that nitrogenous fertilizers especially when used in conjunction with potassium and phosphorus increase, both qualitatively and quantitatively, the amount of volatile compounds produced by the harvested apples. Angelini and Pflug (1967) investigated by GLC and mass spectrometry the volatile compounds absorbed on activated carbon from McIntosh apples in Controlled Atmosphere Storage. They identified 26 hydrocarbons, 11 alcohols, 8 esters and 6 aldehydes.

The physiological function, if any, of these volatile emanations is at present uncertain. The sesquiterpenoid farnesene has been detected in the air passed over apples and it is now thought that this compound or its oxidation products are connected with the storage disorder known as "superficial scale" (Huelin and Murray, 1966). This is discussed in more detail in Volume 1, Chapter 19.

The volatile compound given off by ripe apples which is the most important both quantitatively and physiologically is not directly involved in the aroma or flavour of the fruit. This is the olefine, ethylene, which may appear in amounts more than twice that of all the other volatiles added together (Fidler, 1950). The function and biosynthesis of ethylene has already been discussed (Section 2).

9. *Colour changes*

(a) *Chlorophyll.* The peel of apples contains chloroplasts and, although these chloroplasts are smaller than those found in leaves, they have a similar structure and, of course, they contain chlorophyll which gives the green ground colour to the fruit. The pulp tissue also contains rudimentary plastids, which also contain chlorophyll.

The relative chlorophyll content of the peel and pulp of mature apples is in the region of 100 : 1, and there is a rapid breakdown of chlorophyll in both tissues as ripening proceeds. The fall commences at about the time of the onset of the climacteric and continues throughout the period of ripening and senescence. The chlorophyll content of the tissue varies with the variety. Francis *et al.* (1955) have shown that in the variety McIntosh the chlorophyll content of the flesh fell during ripening at 1°C from 2·2 μg/g to 1·5 μg/g.

For Cox's Orange Pippin apples stored at 12°C, Rhodes and Wooltorton (1967) found that the chlorophyll content of the peel fell during the climacteric period from 105 μg/g to 25 μg/g. Hansen (1955) found that the failure of the peel of Anjou pears to become yellow during ripening at various temperatures was due to incomplete decomposition of chlorophyll which tended to mask the carotenoids present.

The fate of chlorophyll during senescence is not known in any detail. Phaeophytin appears as an early product of breakdown and from this stage oxidation to much smaller molecules appears to occur; no large molecular intermediates have been found (Chichester and Nakayma, 1965). It has been suggested that the removal of the phytol side group converting chlorophyll to chlorophyllide precedes further breakdown to phaeophytin. However, the function of the enzyme concerned, chlorphyllase, is the subject of some dispute. It has generally been thought to catalyse the *removal* of phytol (Granick, 1967), but Holden (1961) has suggested that it may rather be concerned with the synthetic addition of the phytol group. It cannot, however, be without significance that during the climacteric period in apples there is an increase in the activity of chlorophyllase (Rhodes and Wooltorton, 1967; Looney and Patterson, 1967). Virtually nothing else is known of the enzymology of chlorophyll breakdown.

Associated with the loss of chlorophyll commencing in the early stages of ripening, there is a disorganization of the chloroplast lamellae which contain the chlorophyll (Bain and Mercer, 1964; Rhodes and Wooltorton, 1967). The granal and inter-granal membranes disintegrate and osmophilic granules appear between the membranes. At the end of the ripening period many of the chloroplasts have lost most of their lamellar structures while other cell organelles, such as the mitochondria, appear to be still structurally intact. As the chloroplast membranes break down, so do the chloroplast proteins and lipids of which these membranes are composed. Galliard (1968) has shown that the major lipid undergoing breakdown at the time of the respiration climacteric is the chloroplast-bound galactolipid fraction which contains a high proportion of linolenic acid. This breakdown of protein and lipid may well supply the building materials to be used in synthetic processes involved in ripening; the former belief that ripening is entirely a downgrade process is no longer held. The possibility suggested by Galliard *et al.* (1968b, c) that linolenic acid from the breakdown of galactolipids may be involved in the synthesis of ethylene is of particular interest. Another facet of chloroplast breakdown which does not appear to have been given attention is the influence of the termination of photosynthetic activity of the peel tissue might have on the subsequent metabolism of the fruit. Thus, all in all, the breakdown of chloroplasts and of chlorophyll may well have a considerable bearing on the whole ripening processes.

(b) *Carotenoids.* Goodwin (1966) has described the transformation of chloroplasts into chromatoplasts as a fundamental process in fruit ripening. Generally as the chlorophyll is lost there is a build up of carotenoids (Goodwin, 1952), but this may not always be so, and there is no real evidence that carotenoids arise directly from the breakdown products of chlorophyll. Workman (1963) showed that during the ripening of Golden Delicious apples at 20°C the chlorophyll content fell by 75% while at the same time the xanthophylls increased five-fold. During this period, however, the carotene concentration decreased while there was no change in the concentration of quercitin glycosides which had been thought to contribute to the yellow colour of the fruit.

(c) *Anthocyanins.* The red colour of the skin of apples is due mainly to the flavonoid compound idaein (cyanidin-3-galactoside) as first shown by Sando (1937). Cyanidin-7-arabinoside and cyanidin-3-arabinoside have been shown to be minor anthocyanin components of 74 varieties of apples (Bonnie and Francis, 1967). The extent of anthocyanin formation is very variable from variety to variety and the "reddening" of the fruit caused by the development of the pigment may play an important part in the commercial acceptability of many varieties of apples. Colour is much less encountered and is much less important in pears; light of a wavelength between 600 and 750 mμ is necessary for the formation of anthocyanins in apple peel (Siegelman and Hendricks, 1958). Consequently various post-harvest light treatments have been used to increase the reddening of apples. Spraying with water when the sun is shining just before harvest has also been employed, but the most effective environmental factor in the development of anthocyanin pigments appears to be a combination of cold nights and warm sunny days in the immediate pre-harvest period. Smock (1963), using discs of peel and whole fruits dipped or immersed in various solutions and illuminated, showed that sucrose, glucose, sorbitol, mannitol and various carbonates were effective in promoting an increase in anthocyanins in the peel. It is interesting to note that Smock also found that peel discs floated on solutions of N-dimethyl-aminosuccinamic acid (commercially known as B9 or Alar) inhibited anthocyanin formation. This is the opposite effect obtained when trees are sprayed early in the growing season with solutions of B9; for example, if Cox's Orange Pippin trees are so treated, abnormally red fruits result. This is a good example of the difference a growth regulator may have when applied to a complex growing system such as a tree compared with its direct effect on a more specialized organ such as a fruit. A general review of factors affecting anthocyanin formation in apples has recently been made by Walter (1967).

The factors controlling the biosynthesis of anthocyanins in the skin of apples has received little detailed study, although it is tempting to link the finding of Siegelman and Hendricks (1958) that in the dark (when anthocyanin formation is inhibited) acyl compounds accumulate with that of Faust (1965a) who found that metabolic activity through the pentose phosphate pathway increased as anthocyanins were being formed. As Siegelman and Hendricks pointed out, incorporation of acyl compounds, as acetate, could be responsible for the "A" ring of anthocyanins. Then shikimic acid, a known precursor of the "B" ring of anthocyanins can arise from erythrose 4-phosphate (formed in the pentose phosphate pathway) and phosphoenol-pyruvate (formed in glycolysis). In a subsequent paper, Faust (1965b) made the interesting suggestion that anthocyanin formation is regulated by a balance between the activity of the pentose phosphate pathway and protein synthesis through changes in synthetic functions of the RNAs. This work was, however, based on varying responses to a range of protein and nucleic acid synthesis inhibitors applied to the skin of apples and must be considered as somewhat speculative. It does, nevertheless, serve as a pointer to the future work in this field.

10. *Texture*

The texture of apples in biochemical terms derives from the pectin,† hemicelluloses, cellulose, pentosans and hexosans and their interactions and interconnections. The chemistry of these compounds is dealt with in Volume 1, Chapters 3 and 4. Texture will also be affected by changes in the tugor of the cells and by changes in cell elasticity and adhesion. Cell elasticity will be influenced by changes in the composition of the cell wall and in lipoprotein membranes bordering the cells. It is generally thought that pectins and their combinations with calcium play a central role in the softening of the tissues during ripening. It does not appear that, in apples, there is a shortening of the chains of the polygalacturonic acid of which pectin is composed during the ripening process but rather that the softening is due to an enzymic breakdown of protopectin. Protopectin consists of chains of polygalacturonic acid cross-linked in various ways with metals (Ca^{2+}, Mg^{2+}, and possibly Fe^{3+}), and hydrogen bonding between hydroxyl groups and, possibly, by methylene bridges (Neukom, 1949). This protopectin may also embrace polysaccharides, lignin and even proteins linked within it by ether-like linkages (Joslyn, 1962). It seems likely that the various constituents and linkages within this complex will vary from fruit to fruit and during the development of individual fruits. Some form of "protopectinase" must exist but very little appears to be known about it. Since softening in apples appears to be concerned primarily with changes in protopectin, it is not surprising

† For a comprehensive definition of the various forms of pectin, see Kertesz (1951). The subject of "pectin" changes in apples and pears is considered in detail by Hulme (1958).

that polygalacturonase (P.G.), the enzyme responsible for reducing the basic unit of pectin, polygalacturonic acid, to smaller units, appears to be absent from the apple. In extreme senescence, polygalacturonic acid is degraded, but in spite of some speculative attempts to explain this fact nothing definite is known of the mechanism of this breakdown. The enzyme responsible for removing methoxy side groups from the polygalacturonic acid, pectin methyl esterase (PME), is present in small amounts in the apple (Hulme and Wooltorton, unpublished results) but shows little change in activity during ripening. In any case removal of the methoxy groups has no appreciable effect on softening.

In pears, where softening during ripening is carried further than in apples, P.G. is present (Hulme and Wooltorton, unpublished results) although active preparations cannot be obtained at all stages of maturity (Weurman, 1952, 1954). Weurman (1953) explained earlier failures to obtain active enzyme from pears as due to the presence of two inhibitors of the enzyme, one being thermolabile, which appear in the fruit at different stages of development. The importance of metal binding in the control of texture is amply demonstrated by the fact that EDTA, a metal chelator, readily causes separation of cells when many fruit tissues are incubated in its presence. Wiley and Stembridge (1961) suggest that starch and starch-like substances may be intimately associated with pectinic acids and that change in these substances contributes to changes in apple texture. This may be another facet of the protopectin story. Knee (1970) was able to separate satisfactorily several polysaccharide fractions from the apple using DEAE cellulose and fractional elution by phosphate at pH 6·8. He obtained an arabinogalactan neutral fraction, a true poly galacturonic fraction and a polyuronide fraction containing also neutral sugars. This last fraction Knee considered as an artifact arising from the cell wall during the extraction procedures.

It is clear from what has been said that the softening of pome fruits is a complicated process, the mechanism of which is still not fully understood.

11. *Phenolic compounds and enzymes involved in their oxidation*

In reviewing the biochemistry of apples, Hulme (1958) suggested that "if this review had been written five years hence the largest and most exciting section would be concerned with . . . phenolic substances . . . ". This was at a time when leucoanthocyanins were being considered as possible modulators of growth. Hulme has proved to be a false prophet and the role of phenolics and phenolase (the enzyme responsible for the oxidation of the hydroxyl groups of phenolics) in plants, and especially fruits, remains obscure. There are, however, indications that, for example, phenolic compounds and phenolase are involved in the biosynthesis of ethylene in the cauliflower (Mapson and Mead, 1968) and in the apple (authors' laboratory, unpublished results). Anthocyanin pigments have already been considered (Section 9c).

(a) *Phenolic constituents.* Williams (1960) lists the following compounds occurring in apple fruits: Chlorogenic acid; p-coumaryl-quinic acid; catechin; epicatechin; leucoanthocyanin; the 3-galactosides, 3-glucosides, 3-rhamnosides, 3-arabinosides and 3-xylosides of quercetin and kaempferol, and idaein. Phloridzin occurs in the seeds only. In pears the phenolics are chlorogenic acid, traces of catechins and leucoanthocyanins, isohamnetin glycoside (Nortje and Koeppen, 1965), with arbutin in the seeds only. The leucoanthocyanins of pome fruits are complex and their exact configuration and degree of polymerization are uncertain. In cider and perry varieties the concentration of these leucoanthocyanins is relatively high and accounts for the intense astringency of these fruits.

Hulme and Edney (1960), in a study of the possible fungistatic effect of phenolics in the peel of Cox's Orange Pippin apples, found little change in the phenolic pattern of the peel of the fruits during storage at 12°C. The chlorogenic acid content, however, fell from 20 mg/100 g dry weight to 8 mg/100 g dry weight in 24 days in fruit picked in the immature state (about 100 days from petal fall). They suggested that one of the factors rendering the fruit progressively more susceptible to fungal attack during senescence could be a fall in the concentration of chlorogenic acid since this acid appeared to inhibit the germination of spores of the fungus, *Gloeosporium perennans.*

(b) *Oxidizing enzymes.* The oxidation of chlorogenic acid and, to a lesser degree, catechins, by the action of phenolase (polyphenolase, polyphenol oxidase, catechol oxidase) in apple and pear fruits leads to the browning of the tissue when cut. This reaction is also probably responsible for the darkening of bottled fruit juices, although some oxidation of chlorogenic acid may be due to non-enzymic oxidation by metals (see Chapter 19). Mayer and his colleagues (Mayer *et al.*, 1964; Harel *et al.*, 1965) have studied the phenolase of chloroplast preparations from the peel of apples and mitochondrial preparations from peel and pulp. They used surface active agents to isolate the enzymes from the organelles and found Triton X100 most satisfactory for the chloroplasts and digitonin for the mitochondria. They suggested that different catechol oxidases were involved for each organelle and separated four distinct enzymes by gel electrophoresis, three from the chloroplasts, having an optimum pH of 5·1, and one from the mitochondria, having an optimum of 7·3. They did not regard these various enzyme preparations as isozymes and suggested that each enzyme is bound to a specific site on the subcellular structures "where it fulfils a definite physiological role". Later Harel and Mayer (1968) further purified their enzyme preparations from the chloroplast fraction of apples and came to the conclusion that the three enzymes from the chloroplasts resulted from various degrees of aggregation of sub-units of the same enzyme. Walker and Hulme (1966) obtained several phenolase fractions from crude mitochondrial preparations from the peel of apples, but regarded them as different forms of the same enzyme since they oxidized both chlorogenic acid and catechin at similar rates. They suggested that the enzymes were intimately associated with lipid material in the cell. It could be that an affinity for lipids accounts for the association of phenolase with isolated cell organelles, and that as suggested by Mason (1955) this association occurs during the extraction procedures with no close association being present in the living cell.

Finkle (Nelson and Finkle, 1964; Kelly and Finkle, 1969) has shown that the darkening of pome fruit tissue on damage may be prevented by the addition of enzymes which destroy (protocatechuate-3,4-dioxygenase) the phenolic substrates or modify them (*o*-methyltransferase) and suggests that such treatments might be used to prevent the darkening of fruit juices.

Both catalase and peroxidases are present in pome fruits but their function remains obscure. Ascorbic oxidase appears to be absent and this is a common occurrence with tissues rich in phenolase.

12. *Vitamins*

The only vitamin present in appreciable amounts in pome fruits is ascorbic acid (Vitamin C) and this is largely confined to the peel of the fruit. This subject is dealt with in detail in Volume 1, Chapter 13.

V. CONCLUSION

Commercially the most important phase in the life of a pome fruit is the period just before and just after harvesting. There are as yet no detailed biochemical criteria of what is desirable in an apple of a given variety at the time it is harvested; such a possibility seems to be still quite remote and not necessarily of any great commercial value. It seems that, for transport and storage, the ideal situation would be to take fruit harvested in the pre-climacteric state and then to be able to maintain it in this pre-climacteric state throughout a period of storage regulated to meet market requirements. When removed from the store the fruit would be "switched on" to develop, in a reasonably short time, a normal climacteric and the subsequent process of ripening. This ideal has almost been obtained with the banana since these fruits, picked green (pre-climacteric), can be stored in 3–5% oxygen for long periods even at quite high ambient temperatures (20–25°C). Unfortunately this does not happen with pome fruits; in low oxygen (down to levels not sufficiently low to give anaerobiosis) ripening is delayed but not prevented. The low oxygen atmosphere with bananas prevents them producing ethylene and with apples it delays ethylene production, so that in biochemical terms the ideal situation is one which cuts off ethylene production and the sequence of enzymic reactions triggered off by ethylene. This is particularly satisfactory from a commercial viewpoint since all that would be necessary to ripen the fruit at the end of the storage period would be to expose it to low concentrations of ethylene. It would be easier to devise means of controlling the developing of the ethylene-producing system if the mechanism of the system was known. Already some progress has been made in this direction and research in the field is very active. Meanwhile there appear to be three possible ways of controlling ethylene production. Firstly by modifying the metabolism of the fruit through treatment of the tree, secondly by treating the fruit itself either just before or just after harvesting, and thirdly by maintaining permanently in the store conditions which prevent the production of ethylene. Spraying the trees with the growth retardant B9 is already known to reduce ethylene production in apples so that the first of these methods is a distinct possibility; research will be necessary to find a spray that has no concomitant "adverse" effects. The second way of inhibiting ethylene production remains only a possibility, while the third method has already been achieved for bananas in the form of a low oxygen tension in the store.

Finally, when more is known of the way in which ethylene initiates the development of the enzymes of ripening it may be possible to block the development of ripening in storage at a later stage.

A subject which has not, in the view of the authors, received enough attention is the question of quality—not only does one want to produce an acceptable product at harvest (immediate consumption) or at the end of storage, but one wants to have a highly desirable product in terms of appearance, flavour and texture.

REFERENCES

Abeles, F. B. (1966). *Pl. Physiol., Lancaster* **41,** 585.

Angelini, P. and Pflug, T. J. (1967). *Fd Technol.* **21,** 1643.

Ashby, D. L. and Looney, N. E. (1968). *Can. J. Plant Sci.* **48,** 422.

Bain, J. M. and Mercer, F. V. (1964). *Aust. J. biol. Sci.* **17,** 78.

Bain, J. M. and Robertson, R. N. (1951). *Aust. J. scient. Res.* **B4,** 75.

Batjer, L. P., Thompson, A. H. and Gerhardt, F. (1948). *Proc. Am. Soc. hort. Sci.* **51,** 71.

Bonnie, H.-Sun and Francis, F. J. (1967). *J. Fd Sci.* **32,** 647.

Brown, D. S., Buchanan, J. R. and Hicks, J. R. (1966). *Proc. Am. Soc. hort. Sci.* **88,** 98.

Burg, S. P. and Clagett, C. O. (1967). *Biochem. biophys. Res. Commun.* **27,** 125.

Burroughs, L. F. (1957). *Nature, Lond.* **179,** 360.

Chibnall, A. C., Piper, S. H., Pollard, A., Smith, J. A. B. and Williams, E. F. (1931). *Biochem. J.* **25,** 2095.

Chibnall, A. C., Piper, S. H., Pollard, A., Williams, E. F. and Sahai, P. N. (1934). *Biochem. J.* **28,** 2189.

Chichester, C. O. and Nakayama, T. O. M. (1965). Pigment changes in senescent and stored tissue. *In* "Chemistry and Biochemistry and Plant Pigments" (T. W. Goodwin, ed.) Chapter 12. Academic Press, London.

Child, R. D. (1968). *Rep. Long Ashton Res. Stn for 1967*, 95.
Clements, R. L. (1965). *Analyt. Biochem.* **13,** 390.
Clijsters, H. (1965a). Ph.D. thesis, University of Louvain.
Clijsters, J. (1965b). *Physiologia.* **18,** 85.
Clijsters, H. (1969). *Qualitas Pl. Mater Vég.* **19,** 129.
Cooper, W. C., Rasmussen, G. K., Rogers, B. J., Reece, P. C. and Henry, W. H. (1968). *Pl. Physiol., Lancaster* **43,** 1560.
Davenport, J. B. (1956). *Aust. J. Chem.* **9,** 416.
Davenport, J. B. (1960). *Aust. J. Chem.* **13,** 411.
Denne, M. P. (1960). *Ann. Bot.* **24,** 397.
Dilley, D. R. (1962). *Nature, Lond.* **196,** 387
Dilley, D. R. and Austin, W. W. (1967). Michigan State Hort. Soc. 96th Ann. Rep. 102.
Drawert, F., Heimann, W., Emberger, R. and Tressl, R. (1968). *Phytochem.* **7,** 881.
Egerton, L. J. and Hoffman, M. B. (1965). *Proc. Am. Soc. hort. Sci.* **86,** 28.
Faust, M. (1965a). *Proc. Am. Soc. hort. Sci.* **87,** 1.
Faust, M. (1965b). *Proc. Am. Soc. hort. Sci.* **87,** 10.
Faust, M., Chase, B. R. and Massey, L. M. Ir. (1966). *Pl. Physiol., Lancaster* **41,** 1610.
Fidler, J. C. (1950). *H. hort. Sci.* **25,** 81.
Fidler, J. C. (1968). *J. exp. Bot.* **19,** 41.
Flath, R. A., Black, O. R., Guadagni, D. G., McFadden, W. H. and Schultz, T. H. (1967). *J. agric. Fd Chem.* **15,** 29.
Flood, A. E., Hulme, A. C. and Wooltorton, L. S. C. (1960). *J. exp. Bot.* **11,** 316.
Flood, A. E., Hulme, A. C. and Wooltorton, L. S. C. (1970). In preparation.
Francis, F. J., Harvey, P. M. and Bulstrode, P. C. (1955). *Proc. Am. Soc. hort. Sci.* **65,** 211.
Frenkel, C., Klein, I. and Dilley, D. R. (1968). *Pl. Physiol., Lancaster* **43,** 1146.
"Fruits" (1967). A Review by Commodities Div. Commonwealth Secretariat, London,
Galliard, T. (1968). *Phytochem.* **7,** 1915.
Galliard, T., Rhodes, M. J. C., Wooltorton, L. C. S. and Hulme, A. C. (1968a). *Phytochem.* **7,** 1453.
Gailliard, T., Rhodes, M. J. C., Wooltorton, L. S. C. and Hulme, A. C. (1968b). *Phytochem.* **7,** 1465.
Galliard, T., Hulme, A. C., Rhodes, M. J. C. and Wooltorton, L. S. C. (1968c). *FEBS letters* **1,** 283.
Gane, R. (1935). *J. Pomol.* **13,** 351.
Goodwin, T. W. (1952). "The Comparative Biochemistry of the Carotenoids", p. 39. Chapman & Hall, London.
Goodwin, T. W. (1966). The Carotenoids. *In* "Comparative Biochemistry", (T. Swain, ed.), Chapter 7. Academic Press, London.
Granick, S. (1967). The haem and chlorophyll biosynthetic chain. *In* "The Biochemistry of Chlorophyll", Volume II (T. W. Goodwin, ed.), p. 373. Academic Press, London.
Guadagni, D. G. (1968). *A.T.S.M. Special Tech. Bull. No. 440*, 36.
Hall, A. D. and Crane, C. (1933). "The Apple". Martin and Hopkinson, London.
Hansen, E. (1942). *Bot. Gaz.* **103,** 543.
Hansen, E. (1946). *Pl. Physiol., Lancaster* **21,** 588.
Hansen, E. (1946). *Pl. Physiol., Lancaster* **21,** 588.
Hansen, E. (1955). *Proc. Am. Soc. hort. Sci.* **66,** 118.
Hansen, E. (1966). *Ann. Rev. Pl. Physiol.* **17,** 459.

Hansen, E. (1967). *Proc. Am. Soc. hort. Sci.* **91,** 863.
Harel, E. and Mayer, A. M. (1968). *Phytochem.* **7,** 199.
Harel, E., Mayer, A. M. and Shain, Y. (1965). *Phytochem.* **4,** 783.
Harper, R., Bate-Smith, E. C. and Land, D. G. (1968). "Odour Description and Odour Classification". J. & A. Churchill, London.
Hartmann, C. (1962a). *Rev. Gen. Botanique* **69,** 26.
Hartmann, C. (1962b). *Compt. Rend.* **255,** 996.
Hartmann, C. (1966). *Compt. Rend.* **263,** 1718.
Hatch, M. D., Pearson, J. A., Millerd, A. and Robertson, R. N. (1959). *Aust. J. biol. Sci.* **12,** 167.
Holden, M. (1961). *Biochem. J.* **78,** 359.
Holm, R. E. and Abeles, F. B. (1967). *Pl. Physiol., Lancaster* **42,** 1094.
Huelin, F. A. (1959). *Aust. J. biol. Sci.* **12,** 175.
Huelin, F. A. and Gallop, R. A. (1951). *Aust. J. scient. Res. B.* **4,** 526, 533.
Huelin, F. A. and Murray, K. E. (1966). *Nature, Lond.* **210,** 260.
Hulme, A. C. (1937). *Rep. Fd Invest. Bd. for* 1936, 128.
Hulme, A. C. (1948). *Biochem. J.* **43,** 343.
Hulme, A. C. (1951). *J. hort. Sci.* **26,** 118.
Hulme, A. C. (1954). *J. exp. Bot.* **5,** 159.
Hulme, A. C. (1956a). *Nature, Lond.* **178,** 218.
Hulme, A. C. (1965b). *J. hort. Sci.* **29,** 142.
Hulme, A. C. (1958). *Adv. Fd Res.* **8,** 297.
Hulme, A. C. and Edney, K. L. (1960). Phenolic substances in the peel of Cox's Orange Pippen apples. *In* "Phenolics in Plants in Health and Disease" (J. B. Pridham, ed.), p. 87. Pergamon Press, Oxford.
Hulme, A. C. and Jones, J. D. (1963). Tannin inhibition of plant mitochondria. *In* "Enzyme Chemistry of Phenolic Compounds" (J. B. Pridham, ed.), p. 97. Pergamon Press, Oxford.
Hulme, A. C. and Wooltorton, L. S. C. (1958). *J. Sci. Fd Agric.* **9,** 150.
Hulme, A. C. and Wooltorton, L. S. C. (1962). *Nature, Lond.* **196,** 388.
Hulme, A. C., Jones, J. D. and Wooltorton, L. S. C. (1963). *Proc. Roy. Soc. B.* **158,** 514.
Hulme, A. C., Jones, J. D. and Wooltorton, L. S. C. (1964a). *Phytochem.* **3,** 173.
Hulme, A. C., Smith, W. H. and Wooltorton, L. S. C. (1964b). *J. Fd Sci. Agric.* **15,** 303.
Hulme, A. C., Jones, J. D. and Wooltorton, L. S. C. (1965). *New Phytol.* **64,** 152.
Hulme, A. C., Jones, J. D. and Wooltorton, L. S. C. (1966). *J. exp. Bot.* **17,** 135.
Hulme, A. C., Rhodes, M. J. C. and Wooltorton, L. S. C. (1967a). *Phytochem.* **6,** 1343.
Hulme, A. C., Rhodes, M. J. C. and Wooltorton, L. S. C. (1967b). *J. exp. Bot.* **18,** 277.
Hulme, A. C., Rhodes, M. J. C., Galliard, T. and Wooltorton, L. S. C. (1968). *Pl. Physiol., Lancaster* **43,** 1154.
Jones, J. D., Hulme, A. C. and Wooltorton, L. S. C. (1964). *Phytochem.* **3,** 201.
Jones, J. D., Hulme, A. C. and Wooltorton, L. S. C. (1965). *New Phytol.* **64,** 158.
Joslyn, M. A. (1962). *Adv. Fd Res.* **11,** 1.
Kelly, S. H. and Finkle, B. J. (1969). *J. Sci. Fd Agric.* **20,** 629.
Kertesz, Z. I. (1951). "The Pectic Substances". Interscience Publishers, New York.
Kidd, F. and West, C. (1922). *Rep. Fd Invest. Bd* for 1921, 17.
Kidd, F. and West, C. (1933). *Rep. Fd Invest. Bd* for 1932, 55.
Kidd, F. and West, C. (1938). *Rep. Fd Invest. Bd* for 1937, 101.

Kidd, F., West, C. and Hulme, A. C. (1939). *Rep. Fd Invest. Bd* for 1938, 119.
Kidd, F., West, C., Griffiths, D. G. and Potter, N. A. (1951). *J. hort. Sci.* **27,** 179.
Knee, M. (1970). *J. exp. Bot.* In press.
Lance, C. and Bonner, W. D. (1968). *Pl. Physiol., Lancaster* **43,** 756.
Lehninger, A. L. (1964). "The Mitochondrion". W. A. Benjamin Inc., New York.
Lewis, T. L. and Martin, D. (1965). *Aust. J. biol. Sci.* **18,** 1093.
Lieberman, M. (1958). *Science, N.Y.* **127,** 189.
Lieberman, M., Kunishi, A., Mapson, L. W. and Wardale, D. A. (1966). *Pl. Physiol., Lancaster* **41,** 376.
Lips, S. H. and Beevers, H. (1966). *Pl. Physiol., Lancaster* **41,** 713.
Looney, N. E. (1968). *Pl. Physiol., Lancaster* **43,** 1133.
Looney, N. E. and Patterson, M. E. (1967). *Nature, Lond.* **214,** 1245.
Luckwill, L. C. (1953). *J. hort. Sci.* **28,** 14 and 25.
Luckwill, L. C. (1961). *Rep. Long Ashton Res. Stn* for 1961, 61.
Luckwill, L. C. and Child, R. D. (1967). *Rep. Long Ashton Res. Stn* for 1966, 74.
McDaniels, L. W. (1940). *Cornell Univ. agt. exp. Sta.* Mem. 230.
Macdonald, I. R. and Ellis, R. J. (1969). *Nature, Lond.* **222,** 791.
McKee, H. S. and Urbach, G. E. (1953). *Aust. J. biol. Sci.* **6,** 369.
Mapson, L. W. and Mead, A. (1968). *Biochem. J.* **108,** 875.
Mapson, L. W. and Wardale, D. A. (1968). *Biochem. J.* **107,** 433.
Mapson, L. W., March, J. F., Rhodes, M. J. C. and Wooltorton, L. S. C. (1970). *Biochem. J.* **117,** 473.
Markley, K. S. and Sando, C. E. (1931). *J. agr. Res.* **42,** 708.
Martin, D., Lewis, T. L. and Czerny, J. (1964). *Aust. J. agr. Res.* **15,** 905.
Martin, G. C., Williams, M. W. and Batjer, L. P. (1964). *Proc. Am. Soc. hort. Sci.* **84,** 7.
Mason, H. S. (1955). *Adv. Enzymol.* **16,** 105.
Mayer, A. M., Harel, E. and Shain, Y. (1964). *Phytochem.* **3,** 447.
Mazliak, P. (1960). *Compt. Rend.* **250,** 182.
Mazliak, P. (1963). La Cire cuticulaire des Pommes. Doctorate Thesis, Paris.
Mazliak, P. (1967). *Phytochem.* **6,** 687.
Meigh, D. F. and Hulme, A. C. (1965). *Phytochem.* **4,** 863.
Meigh, D. F., Jones, J. D. and Hulme, A. C. (1967). *Phytochem.* **6,** 1507.
Meynhardt, J. T., Maxie, E. C. and Romani, R. J. (1964). *S. African J. agr. Sci.* **7,** 485.
Meynhardt, J. T., Romani, R. J. and Maxie, E. C. (1965). *S. African J. agr. Sci.* **8,** 691.
Moore, T. C. (1967). *Pl. Physiol., Lancaster* **42,** 677.
Mosse, Barbara (1962). "Graft Incompatibility in Fruit Trees." Common. Bur. hort. and plant Crops, Tech. Commun. No. 28. C.A.B. Farnham Royal, England.
Neal, G. E. and Hulme, A. C. (1958). *J. exp. Bot.* **9,** 142.
Nelson, R. F. and Finkle, B. J. (1964). *Phytochem.* **3,** 321.
Neukom, J. (1949). Dissertation, Eidg. Tech. Hodschule, Zurich, Switzerland.
Nortje, B. K. and Koeppen, B. H. (1965). *Biochem. J.* **97,** 209.
Okamoto, T. (1961). *Nippon Nogei Kagaku Kaishi* **35,** 1355.
Paillard, N. (1967a). *Physiol. Vég.* **5,** 85.
Paillard, N. (1967b). *Fruits (Paris)* **22,** 141.
Pearson, J. A. and Robertson, R. N. (1953). *Aust. J. biol. Sci.* **6,** 1.
Pearson, J. A. and Robertson, R. N. (1954). *Aust. J. biol. Sci.* **7,** 1.
Perring, M. A. (1968). *J. Sci. Fd Agric.* **19,** 186.

Reed, D. J. (1965). *Science, N.Y.* **148,** 1097.
Rhodes, M. J. C. and Wooltorton, L. S. C. (1967). *Phytochem.* **6,** 1.
Rhodes, M. J. C. and Wooltorton, L. S. C. (1968). *Phytochem.* **7,** 337.
Rhodes, M. J. C., Wooltorton, L. S. C., Galliard, T. and Hulme, A. C. (1968). *Phytochem.* **7,** 439.
Rhodes, M. J. C., Harkett, P. J., Wooltorton, L. S. C. and Hulme, A. C. (1969). *J. Fd Technol.* **4,** 377.
Rhodes, M. J. C., Wooltorton, L. S. C. and Hulme, A. C. (1969). *Qualitas Pl. Mater. Vég.* **19,** 167.
Rhodes, M. J. C., Wooltorton, L. S. C., Galliard, T. and Hulme, A. C. (1970). *J. exp. Bot.* **21,** 40.
Richmond, A. and Biale, J. B. (1966). *Archs Biochem. Biophys.* **115,** 211.
Romani, R. J., Yu, I. K. and Fisher, L. K. (1969). *Pl. Physiol., Lancaster* **44,** 311.
Sando, C. E. (1937). *J. biol. Chem.* **117,** 45.
Sanwal, B. D., Wright, J. A. and Smando, R. (1968). *Biochem. biophys. Res. Commun.* **31,** 623.
Smith, A. J. M. and Hulme, A. C. (1935). Ditton Laboratory Interim Record, No. 828.
Smock, R. M. (1963). *Proc. Am. Soc. hort. Sci.* **83,** 162.
Siegelman, H. W. and Hendricks, S. B. (1958). *Pl. Physiol., Lancaster* **33,** 188.
Smock, R. M. and Neubert, A. M. (1950). "Apples and Apple Products". Interscience Publishers, New York.
Somogyi, L. P., Childers, N. F. and Chang, S. S. (1964). *Proc. Am. Soc. hort. Sci.* **84,** 51.
Ulrich, R. and Hartmann, C. (1967). *Annls Nutr. Aliment.* **21B,** 161.
Walker, J. R. L. and Hulme, A. C. (1966). *Phytochem.* **5,** 259.
Walter, T. E. (1967). *Rep. East Malling Res. Stn* for 1966, 70.
Weurman, C. (1952). *Publ. centr. Inst. Voedingsonderg, T.N.O., Utrecht* **147,** 1.
Weurman, C. (1953). *Acta bot. neerl.* **2,** 107.
Weurman, C. (1954). *Acta bot. neerl.* **3,** 100.
Wiley, R. C. and Stembridge, G. E. (1961). *Proc. Am. Soc. hort. Sci.* **77,** 60.
Williams, A. H. (1960). The distribution of phenolic compounds in apple and pear trees. *In* "Phenolics in Plants in Health and Disease" (J. B. Pridham, ed.), p. 3. Pergamon Press, Oxford.
Wiskich, J. T. (1966). *Nature, Lond.* **212,** 641.
Wooltorton, L. S. C., Jones, J. D. and Hulme, A. C. (1965). *Nature, Lond.* **207,** 999.
Workman, M. (1963). *Proc. Am. Soc. hort. Sci.* **83,** 149.
Young, R. E. and Biale, J. B. (1967). *Pl. Physiol., Lancaster* **42,** 1357.

Chapter 11

Soft Fruits

AUDREY GREEN

Beecham Products (U.K.) Ltd., Coleford, England

I	Introduction	375
II	Characteristics of Soft Fruits and their Changes with Maturation	375
	A. General and Physical	375
	B. Chemical	380
	C. Some Effects of Cultural Practice	405
III	Post-harvest Changes	405
	A. Holding Fruit	405
	B. Transportation	406
	References	407

I. INTRODUCTION

The term soft fruits includes a number of botanically unrelated fruits which have become associated, rather through their culinary qualities than for any morphological similarity. The group includes berry fruits, currants and false fruits or achenes. These fruits are popular as dessert, but they are also widely used in processing where they may be canned, frozen, used as pie fillings or made into jams, jellies or conserves, or their juices may be expressed. The juices are often syruped, either for use diluted as beverages or as pour-over sauces for use with ice cream. Juices of soft fruits are also used in the manufacture of the European drinks known as "Süssmost", for which there would appear to be no really satisfactory English translation. Some idea of the importance of the processing outlets for soft fruits can be gauged by the fact that in 1966 as much as 47% of the English fruit was used for processing.

II. CHARACTERISTICS OF SOFT FRUITS AND THEIR CHANGES WITH MATURATION

A. General and Physical

There are a number of botanical differences in the various soft fruits, and as these have an important bearing on the representative sampling for biochemical investigations, they will be considered briefly. The structures of the fruit are shown in Fig. 1.

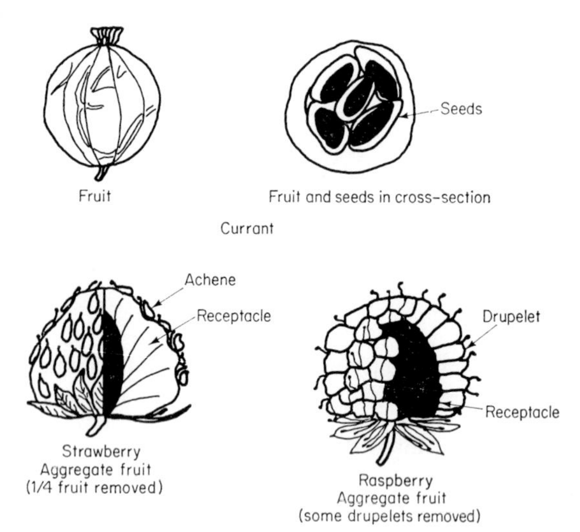

FIG. 1. Structure (diagrammatic) of various types of soft fruit (not to scale).

1. *Berry fruits of the genus* Rubus

These are the fruit commonly called brambles and consist of a number of single-seeded drupelets surrounding a pulpy "plug" or core (the receptacle) which cannot readily be removed prior to maturity; the ease with which this can be accomplished varies with the species. The number and arrangement of the drupelets are characteristic in the different berry fruit.

(a) *Blackberries*. There are numerous species of blackberry, varying in size and shape, of which the common bramble *R. fructicosus* forms a good example. Variations in character are shown by *R. allegheniensis* which has many small drupelets, *R. recurvans* in which these are fewer and much longer, *R. glandicaritis* with its long-shaped fruit and *R. frondosus* round with comparatively few drupelets.

Closely related to the blackberries are *R. procumbens* the wild dewberry and *R. villosus* the cultivated dewberry. These differ from blackberries in that they produce new plants by rooting at the tops of runners instead of producing suckers.

The blackberries and dewberries are normally picked with the plug *in situ*, and hence analysis will include the receptacle.

(b) *Raspberries*. *Rubus idaeus* is the common European raspberry, but it has largely been replaced in America by *R. strigosus*. In addition, the black raspberry *R. occidentalis* and to a lesser extent the yellow raspberry developed from the American red raspberry are also cultivated. Leinbach *et al.* (1951) found 68–108 seeds per berry in red raspberries, indicating the number of drupelets on a plug. In contrast to the blackberry, the berries of raspberry can be readily separated from the receptacle when ripe.

(c) *The loganberry. Rubus loganbaccus* is often considered to be a blackberry-raspberry hybrid, but Wilkinson (1945) regarded this as unlikely, considering it more likely to be a red fruited sport of the blackberry. It is rarely eaten raw and the berries remain tightly adhering to the receptacle even when the fruit is fully mature. It is mainly used for canning and in the manufacture of jam.

(d) *The boysenberry.* This is a cross of the blackberry, raspberry and loganberry. It is larger and more vigorous than its parents. The receptacle is small and there are few seeds, hence it provides a more suitable fruit for processing than many of the berries.

The berry fruits frequently have flowers and fruit on the plants simultaneously and in general the fruit tends to ripen randomly on the stems. It is thus possible to obtain fruit at different stages of maturity at a single picking; on the other hand, commercial fruit will not vary widely in maturity as a number of pickings are taken from the same bushes as the fruit ripens.

(e) *The mulberry.* (*Morus*) may perhaps be mentioned with the berry fruits. In contrast to the preceding berries it grows on trees which may reach a considerable size. The fruits, which resemble large loganberries in appearance, tend to be too sweet for cooking or canning and also present harvesting problems. They are, however, sometimes used for their rich pigment. Their main interest in the past was in their leaves as a source of food for silk-worms.

2. *Currants*

This group includes the Ribes and Vaccinium families. The seeds are enclosed in a fleshy pericarp, the number and size of the seeds being characteristic of the fruit.

(a) *The gooseberry.* Those of cultivation are mainly derived from *R. gossularia.* The fruit develops singly or in small clusters and the fruit may thus be picked at a single stage of maturity if desired.

(b) *Currants.* The currants are composed of the red and white currants derived from *R. sativum* and the blackcurrant from *R. nigrum.* In contrast to gooseberries, currants are arranged on short stems or "strigs", with a single currant adjacent to the strig on the main stem. The fruit ripens in order along this strig, the single fruit ripening first and then the fruit nearest the branch with the terminal fruit last.

Under normal harvesting conditions whole strigs are picked and not individual currants. Even at optimum picking periods, there will almost certainly be some under-ripe and some over-ripe fruit in any sample of fruit examined at a particular date, unless special further grading is performed.

The length and character of the strigs of these fruit is of considerable importance in the selection of varieties for commercial use. Blackcurrants with well-developed long strigs (for example, the Boskoop Giant) are considerably easier to pick than those with short strigs (for example, the Amos Black). The weight of strig can vary considerably even within a variety; in Baldwin the range is from 0·5 to 2·0% of the fruit as picked.

The number of seeds in the currant type of fruit varies widely as can be seen from Table I. There is a marked difference here between the redcurrant and the blackcurrant, the redcurrant having fewer larger seeds. These values may give a guide to the fruit contents of some whole-fruit jams and conserves.

TABLE I. Some typical numbers and sizes of seeds in soft fruit[a]

Fruit	Number of seeds/100 g	Weight of seeds/100 g	Average seed wt. (mg)
Bilberry	13,200	2·31	0·17
Blackberry	3180	6·75	2·12
Raspberry	4190	4·74	1·13
Gooseberry	600	1·22	2·03
Currant (black)	4450	4·66	1·03
Currant (red)	970	4·13	4·26

[a] Mainly from Macara (1931).

(c) *The Vaccinium (blueberry) family.* This family includes the cowberry or red whortleberry (*V. vitis idaea*), the low bush blueberry (*V. myrtillis*), the late low bush blueberry (*V. vacillans*) also called bilberry and whortleberry and the high bush blueberry (*V. corymbosum*) which is also sometimes referred to as bilberry or whortleberry. While the low bush varieties may be only a few inches in height, the high bush varieties may reach a height of 10–12 ft. These fruit mature over a period and several pickings may be made. It is to this family that the American and European cranberries belong (*V. macrocarpon* and *V. oxycoccus*). This plant is a slender creeping vine up to 120 cm long. The fruit, which turns red in the early autumn, may remain on the vine throughout the winter when protected by snow. Picking can thus be carried out over a considerable period.

3. *Achenes (false fruit)*

The false fruit or achenes are represented by the Fragaria—the strawberry family. These consist of an enlarged pulpy receptacle surmounted by seeds. There is no fleshy mesocarp surrounding the seeds. The fruits grow individually and several pickings are made. Fruit of similar maturity can be readily sampled or a range of maturities. Culpepper *et al.* (1935) discuss the implications of ripening rates on the selection of samples.

4. *Growth changes*

As berry fruits mature there is a gradual colour change from green to the characteristic colour of the respective fruits, which, in the case of raspberries, may be yellow, red or black depending on the variety. During this ripening period an increase occurs in the pericarp area which pushes the seeds farther apart and there is a gradual loosening of the drupelets from the core. According to Rohrer and Luh (1959) the increase in fruit weight in boysenberries from under-ripe to over-ripe progresses steadily throughout the developing period.

Ripening currant fruits also show an increase in the pericarp area as the colour changes from green to that of the mature fruit.

Attention has been drawn previously to the systematic ripening in these fruits, and the problem of differing maturity and fruit size in blackcurrant samples was considered by Olliver (1938) with regard to ascorbic acid analysis. Figure 2 shows data obtained by the present author for changes in mean currant weight in two seasons for blackcurrants of the Baldwin variety.

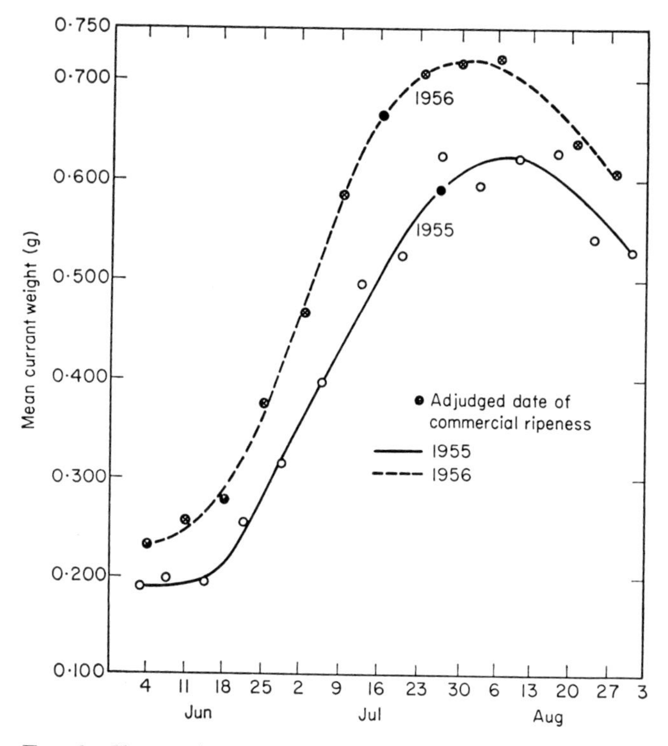

FIG. 2. Changes in the weight of blackcurrants with maturity.

As strawberries mature the receptacle enlarges and the fertilized ovules (achenes) become further apart. Abbott and Webb (1970) have shown that fresh berry weight is highly correlated to achene spacing and achene number, and have developed a formula for berry weight based on the achene number and spacing. As in any individual berry the number of achenes is an unalterable constant—fresh weight × achenes/cm². It was found with the variety Red Gauntlet that the limit to the expansion was 6 achenes/cm². It thus became possible from achene-spacing measurements to calculate short fall in yield relating to sub-maximal receptacle development, and thus to distinguish between those fruits which were inherently small and those whose size could have been increased if conditions had been more favourable.

The author suggests that a similar approach might also prove practical with berry and currant type fruit.

B. Chemical

1. *Total solids and moisture*

While the major constituent of all soft fruit is water, it is more common for analysts to determine the total solid content and to obtain the moisture by difference. When drying methods are used there are two sources of error which may occur. The most obvious is that volatile compounds are included in the value for water, but the greatest source of error arises in the caramelization of sugars when high temperatures are used. Typical values for the solids content of mature soft fruits are shown in Table II. In berries the seeds constitute a relatively high proportion of the fruit and hence the increase in moisture and soluble solids during growth is less for these fruit than for currants or strawberries.

TABLE II. The solids content of soft fruit

Fruit	Solids % w/w			Reference
	Soluble	Insoluble	Total	
Blackberry				
Cultivated	8·9	6·3	15·2	Money and Christian (1950)
Wild	9·9	9·2	19·1	Money and Christian (1950)
Blueberry			16·8	Macara (1931)
Boysenberry			13·7	Rohrer and Luh (1959)
Cranberry			13·0	McCance and Widdowson (1960)
Currant				
Black	13·8	5·9	19·7	Hughes and Maunsell (1934)
Red	10·5	5·5	16·0	Hughes and Maunsell (1934)
White			15·7	McCance and Widdowson (1960)
Gooseberry	8·9	2·2	11·1	Money and Christian (1950)
Loganberry	10·0	6·3	16·3	Money and Christian (1950)
Mulberry			15·0	McCance and Widdowson (1960)
Raspberry	8·8	5·1	13·9	Money and Christian (1950)
Strawberry	7·8	2·4	10·2	Money and Christian (1950)

Experiments were carried out on Baldwin blackcurrants in the author's laboratory in 1956 over a 14-week period (28 May–27 August) during which the fruit ranged from extremely small and green to over-ripe. The total solids remained practically constant for the first 6 weeks. There was a relatively sharp increase in solids 2 weeks before the fruit was judged ripe for processing into juice. Fig. 3 shows the soluble and insoluble solids together with the percentage of non-green currants (i.e., red and black) from which it can be seen that the changes in solids are accompanied by the development of colour in the fruit.

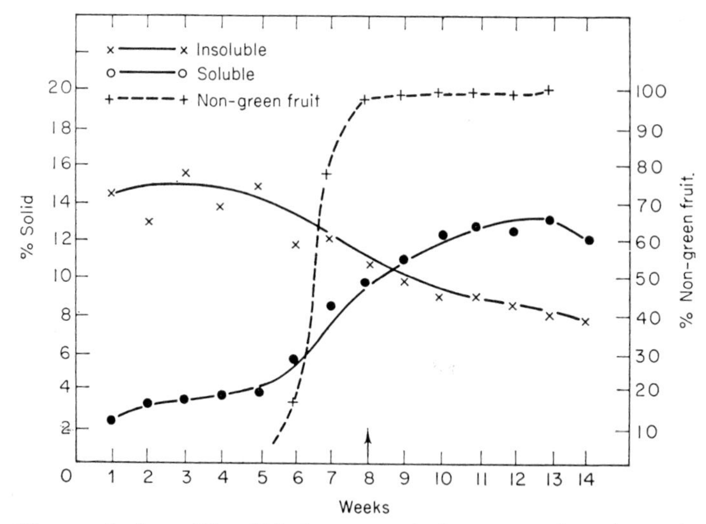

FIG. 3. Changes in the solids of blackcurrants during maturation. Arrow indicates commercial ripeness.

Kushman and Ballinger (1963) found that with blueberries, which fruit over a period and may be harvested at intervals, the date of harvesting had a greater effect on the moisture content of the fruit than the interval between harvestings. Culpepper *et al.* (1935) found a marked increase in the moisture content of strawberries in the early stages of development with a tendency to level off as the fruit reached maturity. The moisture content of the mature fruit of the variety Missionary was found by Kimbrough (1930) to be influenced more by the rainfall in the 3–6 days prior to picking than by the fertilizer treatments studied.

2. *Insoluble solids*

The "insoluble solids" of soft fruits are closely linked to the size and weight of the seeds. In blackberries, for example, the average weight of seeds found by one author has exceeded the average value for insoluble solids reported by another (cf. Tables I and II). This is no doubt due to varietal differences but serves to indicate the high contribution that the seeds make to the insoluble solids of berry fruits; the remaining insoluble matter consists mainly of cell wall constituents.

In the currant family the seeds constitute a smaller proportion of the insoluble matter.

In blackcurrants the parenchymatous tissue forming a network in the fruit contributes about 14% of the insoluble solids.

In strawberries the seeds though numerically high contribute a still smaller amount to the solid matter, as the edible portion of the fruit consists essentially of the swollen receptacle.

3. *Pectins and other cell wall materials*

The general chemical characteristics of the pectic substances have been considered in Volume 1, Chapter 3. Table III gives the levels of total pectic

TABLE III. The pectin content of soft fruit

Fruit	Pectin as calcium pectate (% w/w) Min.	Max.	Average	Soluble	Reference
Blackberry					
Cultivated	0·4	1·19	0·93		Money and Christian (1950)
Wild	0·35	1·10	0·70		Money and Christian (1950)
Dewberry			0·72	0·51	Conrad (1926)
Loganberry	0·50	0·68	0·59		Money and Christian (1950)
Raspberry	0·10	0·88	0·40		Money and Christian (1950)
	0·68	0·97			Leinbach *et al.* (1951)
Currant					
black	0·55	1·79	1·13		Money and Christian (1950)
	0·98	1·38		0·56	Beecham Products (1959)
Strawberry	0·13	0·90	0·54		Money and Christian (1950)

substances in soft fruit expressed as calcium pectate. Although calcium precipitation has disadvantages as a means of assessment of pectin, the majority of the published values for soft fruit have been obtained by this method, so that the results given in Table III have been expressed in terms of calcium pectate. Ahmed and Scott (1958) compared the calcium pectate method with the decarboxylation and carbazole methods for hydrochloric acid (N/50) extracts of a number of fruits including blackberry and bilberry, and found considerably higher values by the calcium pectate method. The present author, working on blackcurrants, found that the high values for calcium pectate were largely associated with the material extracted with EDTA and hydrochloric acid, and did not occur when true (water-soluble) pectin was estimated (see p. 384).

Soft fruit pectins have been less extensively studied than those of citrus and pome fruits. Swindells, in previously unpublished work, obtained a molecular weight of 45,000–46,000 for ripe blackcurrant pectin determined by the method of Luh and Phaff (1954). Kertesz (1951) quotes a similar value of 42,000 for the redcurrant.

The degree of methoxylation reported for the pectin of raspberries by Rohrer and Luh (1959) was 11·7–14·0%. Swindells (1960) found only 6·4% methoxyl, corresponding to 57% esterification of a pectin isolated from mature blackcurrants, compared with a value of 82·7% reported by Kieser *et al.* (1957) in *fresh* pulp. The degree of esterification fell when the pulp was allowed to stand for longer than 20 minutes. Worth (1967) quotes a value of only 0·2% methoxyl for strawberries. Seegmiller *et al.* (1955) studied the synthesis of pectin in boysenberries, using radioactive tracers. They found glucose to be converted into pectin and into arabinose without splitting of the carbon chain. The splitting of the chain into two triose molecules did not appear to play an important part in this synthesis. Subsequently, Seegmiller *et al.* (1956) showed that similar synthesis occurred in strawberries and that in this fruit both glucose and galactose could be utilized.

The saline-insoluble constituents of ripening strawberries of the varieties Cambridge Favourite and Huxley were examined by Wade (1964). These varieties were chosen because of the marked difference in behaviour on processing into jam, the former tending to break up whilst the latter remain intact. It was found that, as the fruit ripened, a decrease took place in the percentage of cell wall recovered and in the anhydrouronic content of this material. The main differences in composition of the isolated cell walls of the two varieties appeared to lie in the greater fall in galactan and araban content in ripening of Cambridge Favourite as compared with Huxley. Neal (1965) made both microscopical and chemical investigations into the changes occurring during maturation of these two varieties of strawberry and found that ripening was accompanied by a splitting of the middle lamella of the cell walls of large cortical parenchyma cells. He found a relationship between the degree of methylation of the pectinous middle lamella and the ability of divalent cations to "firm" the tissue. Ripening was accompanied by a change in the cationic stabilization of the pectinous material.

Woodruff *et al.* (1960) followed the soluble pectin in blueberry fruit during ripening and found that after red coloration there was a continued decrease in soluble pectin which was associated with an increase in methyl esterase activity.

The present author, working with blackcurrants, found that during ripening there was a decrease in the total pectin, expressed as a percentage by weight; in the Baldwin variety it reached a maximum per currant just prior to ripeness. These experiments were made by immersing samples of fruit in boiling alcohol to inactivate the pectin esterase and to precipitate the pectic matter which was then extracted successively, in centrifuge bottles, into cold water, hot water, EDTA and dilute (N/50) hydrochloric acid at 80°C. The first three extractions were continued until the viscosity of the supernates was not more than 1·1 relative to water, the individual extracts being bulked

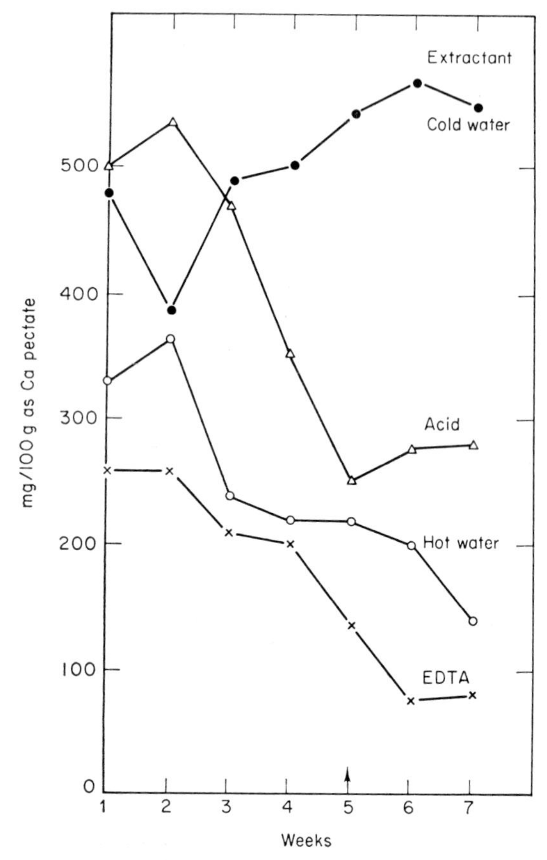

FIG. 4. Changes in blackcurrant pectin (as calcium pectate) during ripening. Arrow indicates commercial ripeness.

for analysis. Two 2-hour extractions with acid were made and these were examined separately. The pectic matter was determined both as calcium pectate, by a modification of the Carré and Haynes (1922) method and by the photometric method of Lawrence and Groves (1954). There were considerable changes in the character of the pectic constituents. As the fruit ripened, the soluble pectin increased while the insoluble (pectate and protopectin) decreased. The results which were obtained are shown in Figs. 4 and 5 in which the values given are the mean of duplicate determinations. The soluble pectin increased at the expense of the insoluble fractions. A statistical analysis of the results from the two methods showed that those of the water extracts were not significantly different, while those of the EDTA and acid extracts differed significantly. It appeared that extractions with these reagents liberated non-uronide material which precipitated with calcium. When the

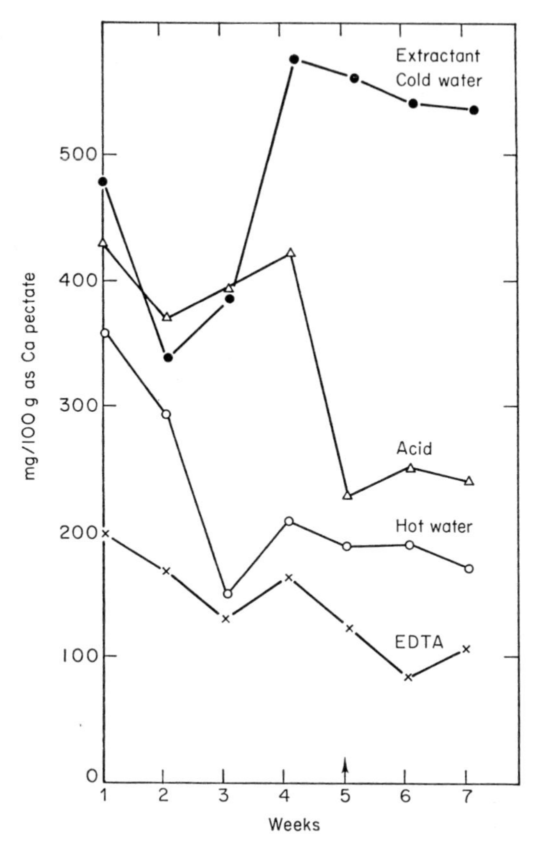

FIG. 5. Changes in blackcurrant pectin during ripening—photometric assay. Arrow indicates commercial ripeness.

viscosities of the extracts were plotted against the concentration found, a decrease during ripening was observed for the pectin (cold-water-soluble) fraction, indicating a decrease in molecular weight as the fruit matured. In contrast the material extracted with hot water appeared to be of a similar character throughout the season.

4. *Sugars*

The sugars constitute one of the major soluble components of soft fruits. Table IV gives a general indication of the sugars found in ripe fruits, together with the sugar to acid ratios. The latter is of importance in determining the level of flavour acceptance. Minimum and maximum values have been included for raspberries as the variation is large for this fruit.

Rohrer and Luh (1959) determined the sugars of boysenberries at different

TABLE IV. The sugar content of mature soft fruit

Fruit	Sugars (% w/w) Reducing	Sucrose	Total	Sugar/Acid	Reference
Blackberry					
Cultivated			4·3	2·8	Crang *et al.* (1950)
Wild			5·0	7·1	Money and Christian (1950)
Blueberry	12·4	1·4	13·8	31·0	Woodruff *et al.* (1960)
Boysenberry	4·20	1·14	5·34	3·5	Rohrer and Luh (1959)
Cranberry			3·5	1·2	McCance and Widdowson (1960)
Currant					
Black	7·0	0·9	7·9	2·1	Green (unpublished)
Red			4·4	2·3	McCance and Widdowson (1960)
White			5·6	2·7	McCance and Widdowson (1960)
Gooseberry			4·6	2·3	Money and Christian (1950)
Loganberry			3·4	1·3	McCance and Widdowson (1960)
Mulberry			8·1		
Raspberry					
Min.	1·44	0·06	1·57	0·9	Knight (1932)
Max.	4·25	1·21	5·34	0·9	Knight (1932)
Strawberry	4·13	0·87	5·00	5·3	Culpepper *et al.* (1935)

stages of maturity. In over-ripe fruit an increase in sugar was observed compared with under-ripe and medium ripe fruit. Sucrose represented 30% of the total sugars in over-ripe fruit compared with 20% in under-ripe fruit.

In blackcurrants (Baldwin variety) the rate of increase in total sugars has been followed in the author's laboratory over a 3-year period. The fruit was obtained from two locations. The sucrose was not determined in these experiments but separate determinations were made for glucose and the sugar acid ratio was also measured. Fig. 6 shows the values obtained at one location together with the percentage of non-green currants. Week 8 (end of July) coincided with commercial harvesting of this fruit. The rise in total sugar, glucose and sugar : acid ratio commenced at about the same time as the colour from green to red was observed.

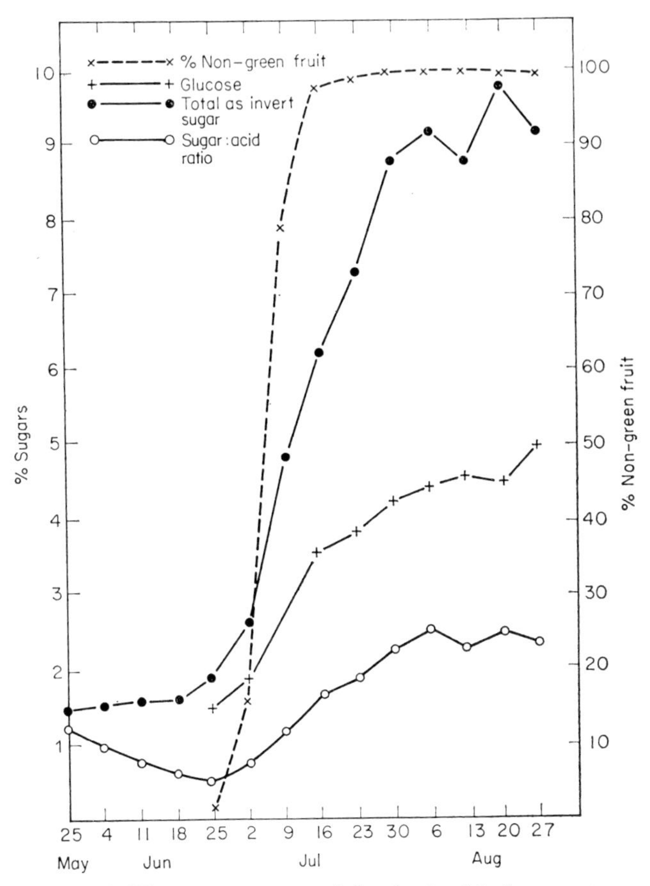

FIG. 6. The sugar content of developing blackcurrants.

TABLE V. The major acids of soft fruit

Fruit	pH	Typical Ripe fruit	Acidity calculated as % w/w: Range covered Min.	Range covered Max.	Individual acids: Citric	Malic	Iso-citric and lactone	Reference
Blackberry								
Cultivated	3·0	1·5	0·68	1·84	Tr.	0·82	0·81	Nelson (1925), Mehlitz and Matzik (1957b), Josysch (1962)
Wild		0·7	0·68	1·51	—	0·57	0·94	Whiting (1958), Woidich and Langer (1959), Crang and Sturdy (1953)
Blueberry		0·3	0·22	0·52				Woodruff *et al.* (1960)
Boysenberry	2·91	1·51			1·24	0·21	0.06	Rohrer and Luh (1959)
Cranberry		3·0	2·90	3·17	Main			Fellers and Esselen (1955), Money and Christian (1950)
Currant								
Black	3·1	3·1	0·85	4·55	3·16	0·44	+	Nelson (1925), Mehlitz and Matzik (1957b), Whiting (1958), Bryan (1946)
Red	2·9	2·5	1·93	3·46	2·02	0·26	+	Nelson (1925)
White	2·6	2·31			1·72	0·60	+	Nelson (1925)
Dewberry	3·53	0·98						Pederson and Beattie (1942)
Gooseberry	3·1	1·99	1·21	2·91	0·88	0·73	+	Money and Christian (1950), Woidich and Langer (1959), Bryan (1946)
Loganberry	2·9	2·63	1·02	3·12	2·10	0·53		Money and Christian (1950), Whiting (1958), Crang and Sturdy (1953)
Raspberry								
Red	3·47	2·8	0·74	3·62	2·06	0·80		Leinbach *et al.* (1951), Josysch (1962)
Purple	3·95	1·55						Josysch (1962)
Black	3·56	0·9	0·84	1·01				Josysch (1962)
Strawberry	3·26	1·01	0·57	2·26	0·92	0·09		Pederson (1942), Hulme and Wooltorton (1958), Charley *et al* (1937), Money and

5. *Acids*

Quantitatively the second largest contribution to the soluble solids of soft fruits is made by the organic acids. The individual fruits show marked qualitative as well as quantitative differences. Table V shows some of the main acids which have been found in the more common soft fruits and Table VI the trace acids. In the majority of the fruits citric acid is the predominant acid, but a notable exception is the blackberry where isocitric acid and its lactone predominate (Whiting, 1958). The cranberry contains an appreciable amount of benzoic acid (Fellows and Esselen, 1955). Ascorbic acid is considered separately because of its biological importance. In the blackcurrant, at approximately 0·2% w/w, it contributes significantly to the total acidity.

Hane (1962) studied in detail the acid metabolism in strawberries; from his results he considered that malic acid is formed in the tricarboxylic acid cycle.

TABLE VI. Trace acids of mature soft fruit (% w/w)

Fruit	Shikimic	Quinic	Oxalic	Others	Reference
Bilberry		+		Benzoic Salicylic	Mehlitz and Matzik (1957)
Blackberry	+	+		Mucic Lactoisocitric	Anet and Reynolds (1954) Whiting (1958) Whiting (1958)
(cultivated)				Galacturonic	Mehlitz and Matzik (1957)
Boysenberry				*Cis*-aconitic	Rohrer and Luh (1959)
Loganberry				Nine detected	Whiting (1958)
Raspberry				Ten detected	Mehlitz and Matzik (1957) Whiting (1958)
Cranberry		+		Benzoic, 0·08	Mehlitz and Matzik (1957) Fellers and Esselen (1955) Butkus (1963)
Currant black			0·02	Glycollic	Mehlitz and Matzik (1957)
Gooseberry	0·10	+	0·01		Whiting (1958)
Strawberry	tr.	0·02		Succinic, 0·04 Glyceric, trace Glycollic, trace Oxaloacetic	Hulme and Wooltorton (1958) Hane (1962)

Rohrer and Luh (1959) determined the acids in developing boysenberries and found a decrease in total titratable acidity as the fruit developed; the proportion of citric and isocitric acids increased while that of malic acid decreased from 17·9% of the total acid in unripe fruit to 8·3% when the fruit was over-ripe.

Whiting (1957) followed the shikimic acid content of gooseberries and found that the acid remained practically constant per berry over a 6-week ripening period.

Woodruff *et al.* (1960), following the chemical changes associated with ripening blueberries, found the percentage of acid to decrease after coloration, the changes in titratable acidity being greater than for other constituents during ripening. They showed that the sugar : acid ratio may be correlated with maturity for purposes of harvesting and that for the Jersey variety a sugar : acid ratio below 12 is unacceptable for fresh consumption.

The change in acidity of blackcurrants was followed by the present author at two farms in 1955 and 1956. Both sets of results were similar in the two years and those for 1956 are given in Table VII.

TABLE VII. Changes in the acidity of Baldwin blackcurrants during ripening

Date	Titratable acidity % w/w	Titratable acidity mg/berry	Sugar : Acid ratio
28th May	1·17	1·57	1·32
4th June	1·50	3·45	1·05
11th June	2·17	5·58	0·79
18th June	2·32	6·46	0·69
25th June	3·04	11·40	0·65
2nd July	3·25	15·20	0·82
9th July	3·85	22·59	1·26
16th July	3·64	23·84	1·71
23rd July	3·76	26·70	1·94
30th July	3·82	27·50	2·29
6th August	3·59	26·10	2·57
13th August	3·84	22·75	2·28
20th August	3·86	24·79	2·54
27th August	3·94	24·10	2·33

Line indicates date of picking commercially.

6. *Vitamins*

Quantitatively the most important vitamin in soft fruit is vitamin C (ascorbic acid), which ranges from the negligible level in some whortleberries to around 200 mg/100 g in blackcurrants.

Typical average values are given in Table VIII together with the range encountered in sound mature fruit. As will be seen from the minimum and maximum values there is a wide variation between varieties of a fruit.

The routes of synthesis of ascorbic acid have been described by Mapson in Volume 1, Chapter 13. The synthesis in strawberries has been studied by a number of workers (Isherwood *et al.* 1954; Loewus *et al.*, 1956, 1958; Finkle *et al.*, 1960).

Olliver (1938) examined the ascorbic acid levels in developing blackcurrants, gooseberries and strawberries. In blackcurrants, variety Westwick Choice, she found the ascorbic acid per berry increased very rapidly in the early stages and then remained constant; as the fruit continued to increase in weight there was a decrease per unit weight of fruit. The present author,

TABLE VIII. The ascorbic acid content of soft fruits

Fruit	Minimum	Maximum	Typical Average Value
Blackberry			20 (McCance and Widdowson, 1960)
Blueberry	1 (Watt and Merril, 1963)	52 (Butkus, 1963)	15 (Olliver, 1967)
Boysenberry			13 (Rohrer and Luh, 1959)
Cranberry	11 (Watt and Merrill, 1963)	33 (Watt and Merrill, 1963)	12 (Watt and Merrill, 1963)
Currant			
Black	106 (Barker *et al.*, 1961)	297 (Fore, 1957)	210 (Olliver, 1967)
Red			40 (McCance and Widdowson, 1960)
White			40 (Olliver, 1967)
Dewberry			21 (Watt and Merrill, 1963)
Gooseberry	20 (Watt and Merrill, 1963)	50	40 (McCance and Widdowson, 1960)
Mulberry			15 (Olliver, 1967)
Raspberry			
Black			18 (Watt and Merrill, 1963)
Red	19	38	25 (Watt and Merrill, 1963)
Strawberry		89	60 (Olliver, 1967)

working with Baldwin fruit, found the maximum concentration per berry when the fruit was fully ripe. Later there was a decrease and the levels per unit weight dropped more steeply for this variety than for the one used by Olliver. The changes in the gooseberry (also a member of the Ribes family) was found by Olliver to be similar to that in blackcurrants. In contrast to this, the curve of development of ascorbic acid in strawberry was markedly different. Here the total ascorbic acid per berry remained low prior to colour development when there was a rapid increase in ascorbic acid followed by a decrease at the end of the season.

The soft fruits cannot be regarded as good sources of carotene (provitamin A). The pigment has, however, been determined in a number of soft fruits; Curl (1964) fractionated blackberry, blueberry, cranberry and strawberry carotenoids and Galler and Mackinney (1965) strawberry carotenoids (see also Volume 1, Chapter 12).

The level of carotene found is shown in Table IX. Some of these figures have been calculated from Vitamin A values, taking one I.U. of Vitamin $A \equiv 0{\cdot}6\gamma$ carotene. Table X shows the distribution of carotenoids between hydrocarbons, monols, diols and polyols.

Vitamins of the B group have been determined in a number of soft fruits and their concentration is shown in Table XI. In general, soft fruits are nutritionally poor sources of these vitamins.

TABLE IX. Carotene content of soft fruits

Fruit	Carotene (mg/100 g)	Reference
Blackberries	0·10, 0·12, 0·59	McCance and Widdowson (1960) Watt and Merrill (1963) Curl (1964)
Raspberries		
Black	Trace	Watt and Merrill (1963)
Red	0·05, 0·08	Watt and Merrill (1963) McCance and Widdowson (1960)
Blueberries	0·27	Curl (1964)
Currants		
Black	0·14, 0·20	Watt and Merrill (1963) McCance and Widdowson (1960)
Red	0·07	Watt and Merrill (1963)
Cranberries	0·02, 0·58	McCance and Widdowson (1960) Curl (1964)
Gooseberries	0·18	McCance and Widdowson (1960)
Strawberries	0·03, 0·064, 0·15	McCance and Widdowson (1960) Curl (1964) Galler and Mackinney (1965)

TABLE X. Distribution of carotenoids from solvents

Fruit	I Hydrocarbons	II Monols	III Diols-polyols	Reference
Blackberry	12·2	3·2	84·7	Curl (1964)
Blueberry	11·0	2·2	86·7	Curl (1964)
Cranberry	7·9	4·0	88·1	Curl (1964)
Strawberry	14·1	1·5	84·3	Curl (1964)
	13·8	5·2	81·0	Galler and Mackinney (1965)

TABLE XI. Vitamin B content of soft fruits[a]

Fruit	Thiamine (mg/100 g)	Riboflavin (mg/100 g)	Nicotinic acid (mg/100 g)
Blackberries	0·03	0·034–0·038	0·4
Boysenberries[b]	0·02	0·13	1·0
Loganberries		0·026–0·029	
Raspberries red	0·02–0·03	0·03–0·09	0·4–0·9
Cranberries	0·03	0·02	0·1
Currants			
Black	0·03–0·05	0·05–0·06	0·3
Red	0·04		0·1
White	0·04		0·1
Gooseberries		0·024–0·03	0·3
Strawberries	0·03	0·027–0·07	0·6

[a] From James (1944); McCance and Widdowson (1960); Watt and Merrill (1963)
[b] Frozen.

7. *Mineral constituents*

Whilst there is some variation in mineral content depending on cultural conditions, the amounts found in soft fruits do not vary as widely as might be anticipated. Cultivation effects do, however, show up in the leaves and these, rather than the fruits, are used to assess the mineral status of bushes and trees. Table XII gives the mineral levels found in ripe soft fruit by a number of workers from different countries. While there are certain noticeable differences in values such as the phosphate and calcium in raspberry, gooseberry and strawberry there is in general a considerable measure of agreement between the values for the various fruits reported by authors from different countries.

While the intake of mineral constituents is influenced by cultural conditions, experiments carried out on Baldwin blackcurrants indicated considerable changes in the level of minerals during growth. Fig. 7 shows the ash and polyphenol content of Baldwin blackcurrants over a 14-week period. The polyphenol was determined by the permanganate titration method (Ayres *et al.*, 1962) and the ash by dry ashing at 600°C. There was a considerable degree of correlation between the concentrations of these constituents which might have arisen as a result of the chelating properties of the flavonoids present in the "polyphenol" fraction.

8. *Nitrogenous constituents*

Although soft fruits are not rich sources of nitrogenous substances they do contain proteins, polypeptides and amino acids, and Boland *et al.* (1968) have determined the protein and amino acid content of a number of soft

TABLE XII. Typical average values for the mineral content of soft fruit

Fruit	Ash (% w/w)	P (mg/100 g)	K (mg/100 g)	Na (mg/100 g)	Ca (mg/100 g)	Mg (mg/100 g)	Fe (mg/100 g)	Cu (mg/100 g)	S (mg/100 g)	Cl (mg/100 g)	Reference
Blackberry	0·5	23·8	208	3·7	63·3	29·5	0·85	0·18	9·2	22·1	Watt and Merrill (1963), McCance and Widdowson (1960)
		30·0	—	—	60						Strimiska (1967)
Loganberry	0·5	17·0	170	1·0	35·0		1·2				Watt and Merrill (1963)
		24·3	257	2·5	35·1	25·0	1·37	0·14	18·1	15·8	McCance and Widdowson (1960)
Raspberry											
Red	0·5	22·0	168	1·0	22·0		0·9				Watt and Merrill (1963)
		28·7	221	2·5	40·7	21·0	1·2	0·21	17·3	22·3	McCance and Widdowson (1960)
		40·0	—	—	50·0	—	—	—	—	—	Strimiska (1967)
Black		—	—	—	24·0	22·0	—	—	—	—	Leinbach *et al.* (1951)
	0·6	22·0	199	1·0	30·0		0·9				Watt and Merrill (1963)

Cranberry	0·2	11·2	119	1·8	14·7	8·4	1·11	0·14	11·1	1·0	Watt and Merrill (1963)
Currant											
Black	0·8	43·2	327	2·7	60·3	17·1			33·1	14·8	McCance and Widdowson (1960)
Red	0·6	29·5	227	2·3	35·8	12·8	1·12	0·12	28·6	14·6	McCance and Widdowson (1960), Strimiska (1967), Watt and Merrill (1963)
White	0·6	28·0	291	1·5	22·4	12·7	0·93	0·14	23·6	14·7	McCance and Widdowson (1960), Watt and Merrill (1963)
Gooseberry	0·4	19·0	170	1·2	18·5	8·6	0·58	0·15	13·5	10·7	McCance and Widdowson (1960)
		30·0	—	—	50·0	—	—	—	—	—	Strimiska (1960)
		15·0	115	1·0	18·0	—	0·5	—	—	—	Watt and Merrill (1963)
Strawberry	0·5	23·0	161	1·5	22·0	11·7	0·71	0·13	13·4	17·5	McCance and Widdowson (1960)
		40·0	—	—	30·0	—	—	—	—	—	Strimiska (1960)
		12·0	164	1·0	21·0		1·0	—	—	—	Watt and Merrill (1963)

fruits. Their findings are shown in Table XIII. With the advent of chromatography it has become possible to obtain quantitative and semi-quantitative assessments of the amino acids and using this method Burroughs (1960) obtained the results shown in Table XIV for a number of British-grown fruits.

Tinsley and Bochain (1959) and Swindells, in previously unpublished work, chromatographically examined the juices of strawberry and blackcurrant respectively. Swindells' findings are shown in Table XV. The values are given as mg/100 ml and for comparative purposes have been put on a 1–10

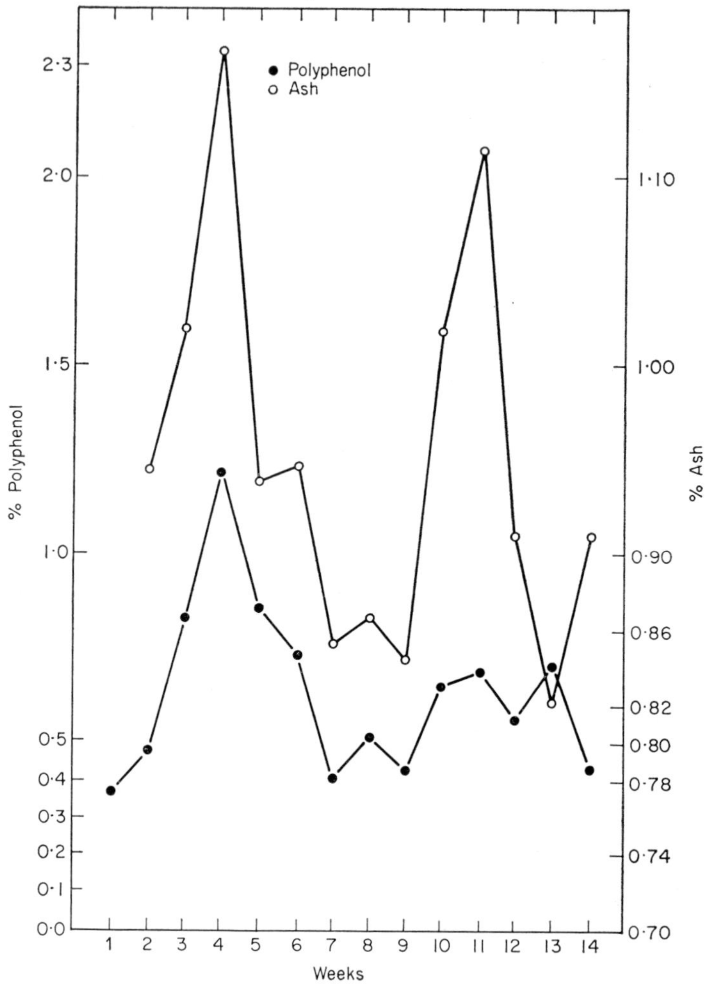

FIG. 7. The ash and polyphenol content of developing blackcurrants.

scale. Barker *et al.* (1961), reporting on the examination of blackcurrant juices over a 9-year period, obtained a mean nitrogen content for juice of 53 mg/100 ml.

TABLE XIII. Nitrogenous constituents of soft fruit[a]

Fruit	Origin of fruit	Protein %	Amino Acid (meg/100 g)
Blackberry	Washington and Oregon	0·56	2·25
Boysenberry	Washinton and Oregon	0·51	2·42
Raspberry			
Black	Washington and Oregon	0·39	1·41
Red (1965)	Washington and Oregon	0·50	1·99
(1963)	Washington and Oregon	0·42	2·14
Cranberry	Several	0·39	3·63
Strawberry (1963)	Washington and Oregon	0·23	0·82
(1964)	Tenessee	0·23	—

[a] From Boland *et al.* (1968).

The values obtained on juices show certain differences when compared with the values given for whole fruit, particularly a higher glutamic acid, proline and valine content of blackcurrant juice. It would seem possible that these have been derived from the action of pectolytic enzymes used to aid in the expression of the juice.

In certain instances chromatographic amino acid maps may be of value in ascertaining the authenticity of fruit juices, although, as pointed out by Pollard (1959), they do little to distinguish between similar fruits such as red and blackcurrants, raspberries and blackberries. It will be seen from Table XV that there are, however, marked differences between the juices of blackcurrant and those of fruits such as the strawberry.

There appear to have been few studies on the changes in the amino acids of soft fruits during maturation.

9. *Pigments*†

(a) *Carotenoids* have been considered under vitamins.

(b) *Flavonoids.* In most instances the anthocyanins have been examined qualitatively rather than quantitatively. For some fruit products

† See also Volume 1, Chapters 11 and 12.

TABLE XIV. Amino acids of soft fruit[a]

	Strawberry	Gooseberry (green)	Blackcurrant	Redcurrant	Raspberry	Blackberry	Loganberry
			(mg of nitrogen 100 g of fruit)				
Total N	148	256	290	183	177	181	278
Alcohol-soluble N	44	165	53	15	44	49	94
Alcohol-insoluble N	104	91	237	168	133	141	184
Amino acids							
(*Visual assessments on arbitrary scale*	0-10)						
Aspartic acid	3	2	2	3	2	3	2
Asparagine	7[b]	3	3	1	3	6[b]	6[b]
Glutamic acid	5	5	3	5[b]	3	5	2
Glutamine	8[b]	10[b]	5	5	3	5	2
Serine	2	5	4	3	5[b]	5	3
Glycine	tr.	1	1	1	0	0	tr.
Threonine	1	2	2	1	1	1	1
α-Alanine	3	8[b]	7[b]	8[b]	8[b]	5	6[b]
β-Alanine	tr.	tr.	2	1	tr.	0	0
γ-Aminobutyric acid	1	3	3	3	3	4	1
Valine	tr.	2	2	1	2	2	2
Leucine	tr.	2	2	1	1	1	1
Proline	0	A	A	A	tr.	tr.	tr.
Arginine	0	tr.	1	tr.	tr.	tr.	1
Lysine	0	tr.	tr.	tr.	tr.	0	0
Tyrosine	0	tr.	0	0	tr.	tr.	tr.

[a] From Burroughs (1960).
[b] Dominant amino acid.
tr. = trace; A = small amount.

TABLE XV. The amino acid content of blackcurrant and strawberry juices

Fruit	Blackcurrant	Strawberry	Blackcurrant	Strawberry
	(mg/100 ml)		(on 1–10 scale)	
Aspartic acid	2·3	2·8	1	1
Asparagine	8·7	59·4	3	8
Glutamic acid	21	7·7	6	1
Glutamine	29	14·5	9	2
Serene	6	2	2	1
Glycine	—	—	—	—
Threonine	1·7	2	1	1
α-Alanine	20	12·1	5	2
β-Alanine	1·2	—	1	—
γ-Amino butyric	—	—	—	—
Valine	17	2	5	1
Leucine/isoleucine	5·6	2	2	1
Proline	9·8	—	3	—
Arginine	3·5	—	1	—
Lysine	—	—	—	—
Tyrosine	—	—	—	—
Norvaline	1·7	—	1	—
Cysteic acid	—	2	—	1

anthocyanins have been used for identification purposes and for the detection of adulteration by other, usually cheaper or more readily available fruit (Woidich and Langer, 1959; Mehlitz and Drews, 1960a, b). Table XVI shows the anthocyanins which have been reported in the different soft fruits. In 1965, Luh *et al.* reported the presence of 3-rutinosides and 5-glucosides in the boysenberry but more recently Nybom (1968), reinvestigating the problem because of the differences in the claims of Luh *et al.* and of Harborne and Hall (1964), found, in agreement with the latter workers, only the 3-glycosides. There are still discrepancies in the results which have been obtained with black raspberry, but in general it would appear that these berries, which have been bred by crossing, contain the pigments of both parent plants.

In the bilberry, Casoli *et al.* (1967) found, on hydrolysis of the eleven anthocyanin pigments, glucose, galactose and one unidentified sugar. Considerable interest has been taken in Europe in the anthocyanins of blackcurrants and redcurrants as a means of detecting the adulteration of blackcurrant juice with that from the redcurrant. This is a situation unlikely to occur in England where the redcurrant is only grown in small quantities. Mehlitz and Matzik (1957a), isolating the pigments by an ion-exchange technique, found seven anthocyanins in redcurrant, three generally present and four in small quantities or not at all depending on the sample examined; blackcurrants also contained seven pigments of which three were generally distributed. Two of the redcurrant pigments were absent from the blackcurrant. These authors subsequently developed a scheme of identification

TABLE XVI. The anthocyanins reported as present in the different soft fruits

Fruit	Pigments	Reference
Berry fruits		
Blackberry	3-Glucosidyl cyanidin chloride	Huang (1955)
Boysenberry	Cyanidin-3-glucoside; cyanidin-3-rutinoside; cyanidin-3-sophoroside; cyanidin-3-glucosylrutinoside; Cyanidin-3-rutinoside, -5-glucoside	Nybom (1968)
Loganberry	As for blackberry and raspberry	Nybom (1968)
Mulberry	Cyanidin-3-monoside	Bate-Smith (1954)
Raspberry		
American black	As for boysenberry	Chandler and Harper (1962)
	Cyanidin-3-rhaminoglucoside; 3,5-diglucoside	Daravingas and Cain (1966), Daravingas and Cain (1968)
	3-Rhaminoglucoside-5-glucoside; 3-diglucoside; cyanidin-3-glucoside; cyanidin-3-rutinoside; cyanidin-3-xyloglucoside; cyanidin-3-xylosylrutinoside	Co and Markakis (1968)
Red	Cyanidin-3-bioside; 4 pigments	Harris and Brown (1956), Lamport (1959)
	Cyanidin-3-glucoside; cyanidin-3-rutinoside; cyanidin-3-diglucoside; cyanidin-3-diglucosyl rhamnoside	Nybom (1960)
Purple	As for black and red raspberry	Nybom (1968)
Currants		
Bilberry	Eleven pigments including: 2 delphinidins; 2 cyanidins; 2 pelargonodins; 1 peonidin derivative; 1 malvidin derivative	Casoli *et al.* (1967) Soumalainen and Kevanen (1961)
Currant		
Black	2 cyanidin glycoside; 2 delphinidin glycosides	Fouassin (1956)
	Cyanidin; delphinidin; cyanidin-3-glucoside; delphinidin-3-glucoside; cyanidin-3-rutinoside; delphinidin-3-rutinoside	Chandler and Harper (1962)
	Seven pigments	Melhitz and Matzik (1957)
Red	Seven pigments two absent from blackcurrants; cyanidin derivatives	Mehlitz and Matzik (1957)
Red (*Ribes betraeum*)	A branched triglycoside containing 1 mol. each of glucose rhamnose and xylose	Harborne and Hall (1964)
Cranberry	Cyanidin-3-monogulactoside; cyanidin-3-arabinoside; peonidin-3-monogalactoside; peoniden-3-arabinoside	Fuleki and Francis (1968)
Achenes		
Strawberry	Pelargonidin-3-monoglucoside + trace of cyanidin	Co and Markakis (1968), Sondheimer (1968), Akuta and Koda (1954) Casoli *et al.* (1967)
	Pelargonidin-3-monoglucoside; diglucoside; cyanidin-3-glucoside	Lukton *et al.* (1955)

based on two-dimensional chromatography or one-dimensional chromatography on specially prepared paper (Mehlitz and Matzik, 1958). However, they found difficulty in obtaining constant R_f values and developed a position constant for comparisons. In blackcurrants, the present author found no evidence of preferential production of either cyanidin or delphinidin pigments when samples were taken over a 14-week period. In contrast, Soumalainen and Keränen (1961) found that the first pigments to appear during the ripening of blueberries were both cyanidin derivatives although the glycosides of malvidin, delphinidin and petunidin were also present. While the citrus fruits are the usual commercial sources of flavonoid material some of the soft fruits, notably blackcurrant, contain appreciable quantities of biologically active bioflavonoids (see Volume 1, p. 299). Flavonoids other than anthocyanins can also be used to identify products from different fruits:

(i) Blackberry and raspberry. These fruits contain flavonoids in sufficient concentration to have been classified as moderate sources of vitamin P activity. The flavonoid of raspberry fruit have not been extensively examined, but Haskell and Garrie (1966) used two-dimensional chromatograms of the flavonoid compounds (without identification of the individual members) of the leaves as "finger-prints" for identification of various cultivars.

(ii) Blackcurrant. This berry possesses the highest flavonoid content of any soft fruit; high biological activity was reported in 1943 by Baccharach and Coates. Pollard (1945) isolated biologically active flavonoid material equivalent to 0·04% of the fresh weight of blackcurrants, obtaining the most potent extract then known. In 1952 Williams *et al.* isolated quercetin and isoquercetin from the fruit, while Kajanne and Stén (1958a, b) isolated the three aglycones myricetin, quercetin and kaempherol from blackcurrants at a level of 0·045% on a wet basis of juice press cake. Gleisberg and Aumann (1958) also found the same three aglycones together with their respective 3-glycosides. Their total yield amounts to 0·31% of the dry weight, which, with a solids content of 16·6%, would have been equivalent to 0·02% of the fresh weight of the fruit. The present author found, in addition to the above, some evidence of rutin and, possibly, dihydroquercetin. Morton (1968) confirmed the aglycones together with rutin and isoquercetin. In addition, working on juice, he found the presence of citrin and the 5-methyl ether of quercetin. His yield of total phenolics was 69 mg/100 ml of which 24 mg were flavonols.

The redcurrant is not a rich source of flavonoid compounds.

(iii) Elderberry. While not a commercial fruit, it may be of interest to note that this fruit contains a relatively high level of rutin (Mayer and Cook, 1943).

(iv) Cowberry and cranberry. In 1958 Kajanne and Stén extracted the flavonoids from cowberry press waste, and obtained 0·013% flavonoid (on wet basis) of which the main portion was quercetin. They also obtained a small amount (about 2–3% of the total of material) which was probably kaempferol. Recently, Puski and Francis (1967) have examined the American cranberry, and have found the major pigment to be quercetin-3-galactoside; quercetin-3-rhamnoside and quercetin-3-arabinoside and free quercetin were also present. In addition, small amounts of the 3-arabinoside and 3-digalactoside of myricetin were found.

(v) Strawberry. Williams and Wender (1952) isolated and identified quercetin and kaempferol from strawberries and this was confirmed by Hörhammer and Muller (1954), Co and Markakis (1968) identified quercetin and kaempferol-3-glucoside as well as catechin.

10. *Tannins (polyphenols) and other phenolic compounds*†

The phenolic compounds contributing to the colour of soft fruits have already been dealt with under flavonoids in Section IIB 9 on p. 397. Table XVII shows the percentage of tannin material in soft fruits, reported in the literature and the major phenolic compounds present. Special reference is advised to the excellent review of fruit tannins by Herrmann (1960). The term "tannin" is used here as it has been in use in the fruit juice industry for a number of years. In practice it represents the material oxidizable by potassium permanganate under standard conditions. The term includes complex polyphenolic materials capable of combining with proteins, as well as the leucoanthocyanins. These compounds are not necessarily capable of tanning hide.

There are considerable changes in the phenolic compounds as fruit mature and these are closely linked with the oxidative enzyme systems. Timberlake (1960) has considered the importance of the phenolics in this connection (see Chapter 17). The changes affecting different types of fruit only will therefore be considered here:

(a) *Berry fruits.* Rohrer and Luh (1959) found that the highest tannin content in boysenberries occurred at ripeness with lower levels in either immature or over-mature fruit.

(b) *Currants.* In blackcurrant the present author has observed a very sharp rise in total phenolic material approximately 35 days before commercial ripeness (Fig. 7). Prior to colour development in the fruit, there was a correlation between the total phenolics determined from potassium permanganate titration and the light absorption at 270 nm. After colour development this relationship no longer existed. When the permanganate titration was performed on fractions separated by water on a chromatogram, it was found that one of the anthocyanin pigments did not react, while both pigments contributed to the absorption at 270 nm. The fraction containing

† See Volume 1, Chapter 11.

leucoanthocyanins also titrated and this would appear to account for the discrepancy between the methods in mature fruit. A chromatographic examination of the fruits during ripening showed that there was a decrease in leucoanthocyanin-type material as anthocyanins appeared. This inverse relationship between the two types of compound, suggests that the anthocyanin pigment was formed from the leuco-compound.

(c) *Strawberries.* Co and Markakis (1968) found that glycosylated leucoanthocyanidins constituted the largest fraction of the "tannin" in both ripe and unripe strawberries.

TABLE XVII. The tannin content and compounds associated with "tannins" in soft fruit

Fruit	Tannin (%)	Hydroxy acids	Catechin and leucoanthocyanins
Blackberry	—	Chlorogenic	+Catechin
		Ferulic	−Epicatechin
		Neochlorogenic	
Boysenberry	0·09		
Blackcurrant	0·21–0·37[a]	Caffeic	+Catechin
	0·34	Chlorogenic	−Catechin
		Cinnimic	−Gallocatechin
		Dactylifric	leucoanthocyanin
		Ferulic	
		Gentisic	
		Neochlorogenic	
		p-Coumaric	
		Protocatechuic	
		Syringic	
		trans-chlorogenic	
Gooseberry	0·06–0·1[a]	Chlorogenic	+Catechin
		Ferulic	+Gallocatechin
		Neochlorogenic	leucoanthocyanin
		p-Coumaric	
Loganberry	0·19[a]		
Raspberry	0·10–0·14[a]	Chlorogenic	+Catechin
		Ferulic	−Epicatechin
		Neochlorogenic	
Strawberry	0·11–0·15	Chlorogenic	+Catechin
		Ester	−Epigallocatechin
		Neochlorogenic	−Gallocatechin
		p-Coumaric	

[a] Juice values.

11. *Volatile constituents*

Nursten has described in detail the volatiles of the more important soft fruits in Volume 1, Chapter 10 and the reader is referred there for a comprehensive report on this important field. Kieser and Pollard (1962) comparing loganberries with raspberries found fewer alcohols and aldehydes in the loganberry which, however, contained a muscatel aroma absent from raspberries.

Kuusi *et al.* (1966) found a good correlation between the ethanol content of blackcurrants and the total volatiles, while methanol showed some correlation with the degree of methylation of the pectin. They found that the peaks obtained by gas chromatography were influenced by the weather to which the crop had been subjected.

Croteau and Fugerson (1968) examined the volatiles of cranberry and found that 42 compounds accounted for 95% of the aroma complex, while the remaining 5% contained over 200 compounds in very small concentration. These compounds present in low concentration were, however, important in the overall aroma.

12. *Enzymes*†

Weurman (1961) isolated an enzyme from the core of mature raspberries which was capable of producing typical raspberry volatile material when added to a substrate obtained from immature fruit. This was accompanied by the production of a raspberry odour, but the relation between the odour and the volatiles was not investigated further.

In the handling of fruit, the enzymes of prime importance are those concerned with oxidation and degeneration. The ascorbic acid oxidation in soft fruits is mainly due to the direct action of ascorbic acid oxidase rather than through a polyphenol oxidase. In blackcurrant juice, Timberlake (1960) showed that very little enzymic oxidation occurred providing the juice was uncontaminated. The enzymes concerned with fruit juice production have been covered in Chapter 17. Because of their influence on soft fruit development, however, the pectic enzyme are also considered here. Pectin methyl esterase was reported in gooseberries, redcurrants and blackcurrants by Kertesz (1951). In the latter fruit, Kieser *et al.* (1957), using the variety Mendip Cross, found the highest activity in the cell debris with no activity in the skin ($0{\cdot}28 \times 10^{-4}$ PE‡ units in whole fruit and $3{\cdot}64 \times 10^{-4}$ units in the cell debris). The present author found for the Baldwin variety that the skin fraction separated on a commercial sieve was capable of causing gelation in a pectin solution. Leinbach *et al.* (1951) found levels of 0·0012–0·0030 PE units/g in red raspberries.

Arakji and Yang (1969) examined the pectic enzymes of the McFarlin cranberry and found the esterase to be low, relative to strawberry and tomato, and with an optimum pH of 7·5. These authors also found endopolymethylgalacturonase in this fruit. Gizis (1964), studying external changes

† See also Volume 1, Chapters 3 and 8.

‡ One PE unit = 1 meq, methanol liberated/min/g tissue.

in strawberries, concluded that these fruits contained an endopolymethyl-galacturonase capable of hydrolysing pectins, pectates and protopectins with a pH maximum of 4·5–5·0. This enzyme showed the same rate of action on both sodium pectate and citrus pectin.

Changes in the enzyme pattern during the maturation of soft fruits has been little studied. Rohrer and Luh (1959) found that pectin esterase increased during the ripening of boysenberries, while the present author found that in blackcurrants, different varieties produced their maximum pectin esterase activity at different stages of ripeness. The varieties Mendip Cross and Wellington showed increased pectin esterase activity up to ripeness followed by a decrease, while Baldwin fruit continued to show an increase throughout the period under test. Woodruff *et al.* (1960), in a study of ripening in blueberries, found that the pectin methyl esterase activity remained low until about 6 days after coloration following which there was a steep rise in activity.

C. Some Effects of Cultural Practice

Attention has been drawn earlier to the effect of rainfall on crop yield. With blackcurrants, the present author has found that mulching promotes greater water retention and leads to larger fruit. Irrigation can now be used to control water supply and similar enhanced yields have been obtained by Greenham (1961) and Goodes (1963) on irrigated plots.

Certain manurial treatments have a considerable influence on the quality as well as quantity of the resultant fruit. In blackcurrants, a low nitrogen status has been found (Kieser *et al.*, 1950) to produce fruit with a higher vitamin C content. The work of Bould (1958, 1963) has shown that for optimum growth, yield and vitamin content there is a considerable variation in the requirement from soil to soil and that a balance must be struck between the requirements of nitrogen for general growth and yield and for good quality in the fruit.

III. POST-HARVEST CHANGES

A. Holding Fruit

Unlike citrus and pome fruits, soft fruits degrade very rapidly after harvest. The period in which such fruit, if untreated, can be regarded as sound rarely extends beyond 72 hours after picking, and, probably because of the short times involved, relatively little work has been carried out on the post-harvest changes in these fruits.

Woodruff *et al.* (1960) found the shelf life of the blueberry to be highly correlated with the sugar : acid ratio at harvest. The deterioration increased with rising sugar : acid ratios regardless of the temperature at which the fruit was stored. This emphasizes the importance of picking at optimum maturity for the market rather than merely at optimum gross yield.

The present author found that blackcurrants held for 24 hours continued to ripen as judged by colour change. During this period, a weight loss of approximately 3% occurred and the ascorbic acid fell by approximately 7%, expressed on a fresh weight basis. There were also changes in the pectic substances of the fruit involving a reduction of the methylated component and an increase in pectate indicating that pectin esterase was active during this period.

Controlled atmospheres have been used successfully for the short-term storage of soft fruit. In 1932 Brooks *et al.* (1932) examined the use of carbon dioxide in controlling transit diseases of fruits and vegetables. Later (Brooks *et al.*, 1936) found that blackberries retained their flavour well when cooled and treated with up to 40% CO_2 for 2 days, but the flavour of strawberries and raspberries could be impaired. Subsequently, Thornton (1937) reported on the use of CO_2 for storage of strawberries and found that at 15% CO_2, ripe strawberries could be held at 0·4 or 10°C without noticeable injury. A higher concentration of CO_2 (28%) was, however, found to result in a rapid softening, and removal to warm air accelerated softening. He also found, when studying the effects of CO_2 concentrations ranging from 0 to 60%, that the ascorbic acid could be seriously affected. Duvekot (1958) showed that strawberries kept for 5 days at 3°C in an atmosphere of 30% CO_2 remained firm and had a better texture after canning than untreated fruits; no off flavour was detected. Smith (1957) investigated the storage of blackcurrants in various concentrations of CO_2 and recommended storage at 2°C with 1 week in 50% CO_2 followed by 25% CO_2 for the remainder of the storage period. Hall (1968) has recently reviewed the work on the use of controlled atmosphere in the storage of fruit and vegetables and considers this form of storage likely to be useful in the short term-holding of many kinds of fresh produce particularly where optimum refrigeration is not possible, or to extend the life of very perishable items such as berry fruits. It has been found that strawberries can be held at 0°C in 99–100% N_2 for up to 10 days while maintaining their flavour but tending to lose texture.

B. Transportation

Smith (1958) studied the handling and transport of raspberries and strawberries and found that the fruits ripened appreciably during long journeys of 200–500 miles. He found considerable advantages in pre-cooling this type of fruit before transport and preferably transporting in insulated vehicles. Where cooling was not possible, minimum sheeting for weather protection was recommended to enable free air circulation and natural cooling while the vehicle is in motion.

Some changes occurring in the blackcurrant during the normal commercial transport were examined by the Research Development of Beecham Products. It was found that over an 18–22-hour period from the farm to the factory there was a 2–3% loss of total fruit weight. This appeared to be accounted for both by evaporation and respiration since the increase in the percentage of total solids of the fruit after transport was less than would have been anticipated if all the weight loss was due to evaporation.

Hall (1968) has considered the value of using controlled atmospheres in the transport of fresh soft fruit; it is, however, difficult to assess the economy of these measures at present, and such fruit must still be regarded as a highly perishable commodity.

ACKNOWLEDGEMENT

The author wishes to thank the Directors of Beecham Products (U.K.) for permission to publish some hitherto unpublished data.

REFERENCES

Abbott, A. J. and Webb, R. A. (1970) *Nature, Lond.* **225,** 663.

Ahmed, E. and Scott, L. E. (1958). *Proc. Am. Soc. hort. Sio.* **71,** 376.

Akuta, S. and Koda, R. (1954), *Fermentat Technol. Japan* **32,** 257; from Fuleki, T. (1969). *J. Fd Sci.* **34,** 365.

Anet, E. F. L. J. and Reynolds, T. M. (1954). *Nature, Lond.* **174,** 930.

Arakji, O. A. and Yang, H. V. (1969). *J. Fd Sci.* **34,** 340.

Ayres, A. D., Charley, V. L. S. and Swindells, R. (1962). *Fd Process. Packag.* Jan. p. 13.

Baccharach, A. L. and Coates, M. E. (1943). *J. Soc. chem. Ind., Lond.* **62,** 85.

Barker, A. G., Billington, A. E. and Charley, V. L. S. (1961). *Fd Process. Packag.* Sept. p. 325.

Bate-Smith, E. C. (1954). *Adv. Fd Res.,* **5,** 261.

Beecham Products (U.K.) (1959). Previously unpublished work of the Product Research Department.

Boland, F. E., Blonquist, H. V. and Estrin, B. (1968). *J. Ass. off. agric Chem.* **51,** 1203.

Bould, C. (1958). *A. Rep. L. Ashton Res. Sta.* p. 82.

Bould, C. (1963). *A. Rep. L. Ashton Res. Sta.* pp. 79, 83.

Brooks, C., Bratley, C. O. and McCollock, L. P. (1936). *U.S. Dept. Agr. Tech. Bull.* No. 519.

Brooks, C., Miller, E. V., Bratley, C. O., Cooley, J. S., Moor, K. and Johnson, H. B. (1932). *U.S. Dept. Agr. Tech. Bull.* No. 318.

Bryan, J. D. (1946). *Ann. Rep. L. Ashton Res. Sta.* p. 138.

Burroughs, L. F. (1960). *J. Sci. Fd Agric.* **11,** 14.

Butkus, V. (1963) Tr. l-Oi (Pervoi) Naucha Konf. Po Issled i Obogashck. Rastit. Resursov—Pribalt. Resp. i Belorussii Viln. 183; *Chem. Abs.* (1965) **62,** 4530.

Carré, M. H. and Haynes, D. (1922). *Biochem. J.* **16,** 60.

Casoli, K., Cultrera, R. and Dall'aglro, G. (1967). *Industria Conserve* **42,** 1, 111.

Chandler, B. V. and Harper, K. A. (1962). *Aust. J. Chem.* **15,** 114.

Charley, V. L. S., Curtis, R. C. and Sills, V. E. (1937). *A. Rep. L. Ashton Res. Sta.* p. 186.
Co, H. and Markakis, P. (1968) *J. Fd Sci.* **33,** 281.
Conrad, C. M. (1926). *Am. J. Bot.* **13,** 531; Kertesz (1951). "The Pectic Substances." Interscience, New York and London.
Crang, A., Kendall, A. and Sturdy, M. (1950). *A. Rep. L. Ashton Res. Sta.* p. 197.
Crang, A. and Sturdy, M. (1953). *J. Sci. Fd Agric.* **4,** 449.
Croteau, R. J. and Fugerson, I. S. (1968). *J. Fd Sci.* **33,** 386.
Culpepper, C. W., Caldwell, J. S. and Moon, H. H. (1935). *J. agric. Res.* **50,** 645.
Curl, A. L. (1964). *J. Fd Sci.* **29,** 5, 241.
Daravingus, G. and Cain, R. F. (1966). *J. Fd Sci.* **31,** 927.
Daravingus, G. and Cain, R. F. (1968). *J. Fd Sci.* **33,** 138.
Duvekot, N. S. (1958). Ann. Rep. Inst. Voor Bewaring Vewerkin von Turnboun Prod. Netherlands.
Fellers, C. R. and Esselen, W. B. (1955). *Agr. Expt. Sta. Univ. Mass. Bull.* No. 481, 13.
Finkle, B. J., Kelly, S. and Loewas, F. A. (1960). *Biochim. biophys Acta* **38,** (2), 332.
Fore, H. (1957). *Flussig. Obst.* **24,** (XI), 14.
Fouassin, A. (1956). *Revue Ferment. Ind. aliment.* **11,** 173.
Fuleki, T., and Francis, F. J. (1968). *J. Fd Sci.* **33,** 471.
Galler, M. and Mackinney, G. (1965). *J. Fd Sci.* **30,** 393.
Gizis, E. J. (1964). *Diss. Abstr.* **1,** 394.
Gleisberg, W. and Aumann, H. (1958). *Gartenbauwissenschaft* **23** (5), 4.
Goodes, J. E. (1963). *A. Rep. East Malling Res. Sta.* p. 152.
Greenham, D. W. P. (1961). *A. Rep. East Malling Res. Sta.* p. 20.
Hall, E. G. (1968). *Fd Res. Quarterly* **28** (1–2), 2–8.
Hane, M. (1962). *Gartenbauwissenschaft* **27,** 453.
Harborne, J. B. and Hall, E. (1964). *Phytochem.* **3,** 453.
Harris, A. T. and Brown, H. D. (1956). *Proc. Am. Soc. hort. Sci.* **68,** 482.
Haskell, G. and Garrie, J. B. (1966). *J. Sci. Fd Agric.* **17,** 189.
Herrmann, von K. (1960). *Fruchtsaft-Ind.* **5,** 87.
Hörhammer, L. and Müller, K. H. (1954). *Arch. Pharm. Berl.* **287,** 488.
Huang, H. T. (1955). *J. agric. Fd Chem.* **3,** 141.
Hughes, E. B. and Maunsell, J. (1934). *Analyst* **59,** 231.
Hulme, A. C. and Wooltorton, L. S. C. (1958). *Chemy Ind.* 659.
Isherwood, F. A., Chen, Y. T. and Mapson, L. W. (1954). *Biochem. J.* **56,** 1.
James, W. E. (1944). *Soap. Perfum. Cosm.* **17,** 673.
Josysch, D. (1962). *Fd Technol.* **16,** 90.
Kajanne, P. and Sten, M. (1958a). *Suomen kemistilehti* **B.31,** 149.
Kajanne, P. and Sten, M. (1958b). *Suomen kemistilehti* **B.31,** 211.
Kertesz (1951). "The Pectic Substances," Interscience, New York and London.
Kieser, M. E., Pollard, A., Timberlake, C. F. (1950). *A. Rep. L. Ashton Res. Sta.* p. 194.
Kieser, M. E., Pollard, A. and Sissons, D. J. (1957). *A. Rep. L. Ashton Res. Sta.* p. 134.
Kieser, M. E. and Pollard, A. (1962). "Symposium on Volatile Fruit Flavour." Berne Int. Fruit Juice Union.
Kimbrough, W. D. (1930). *Proc. Am. Soc. hort. Sci.* **27,** 184.
Knight, L. D. M. (1932). *A. Rep. L. Ashton Res. Sta.* p. 32.
Kushman, L. J. and Ballinger, W. E. (1963). *Proc. Am. Soc. hort. Sci.* **83,** 395.

Kuusi, T., Siiria, A. and Kuusi, T. (1966). *J. Sci. Agric. Soc. Finland* **38,** 162.
Lamport, C. (1959). *Rev. Ferment. Ind. aliment.* **13,** 153.
Lawrence, J. M. and Groves, K. (1954). *J. Agric. Fd Chem.* **2,** 882.
Leinbach, L. R., Seegmiller, C. G. and Wilbur, J. S. (1951). *Fd Technol.* **5,** 51.
Loewus, F. A. and Jang, R. (1958). *J. biol. Chem.* **232,** 305, 521.
Loewus, F. A., Finkle, B. J. and Jang, R. (1958). *Biochim. biophys. Acta* **30,** 629.
Loewus, F. A., Jang, R. and Seegmiller, C. G. (1956). *J. biol. Chem.* **222,** 649.
Loewus, F. A., Jang, R. and Seegmiller, C. G. (1958). *J. biol. Chem.* **232,** 533.
Luh, B. S. and Phaff, H. J. (1954). *Archs Biochem. Biophys.* **51,** 102.
Luh, B. G., Stachowicz, K. and Hsia, C. L. (1965). *J. Fd Sci.* **30,** 300.
Lukton, A., Chichester, C. O. and Mackinney, G. (1955). *Nature, Lond.* **176,** 790.
Macara, T. (1931). *Analyst* **56,** 43.
McCance, R. A. and Widdowson, E. M. (1960) M.R.C. Spec. Rep. Ser. No. 297. H.M.S.O.
Mayer, F. and Cook, A. H. (1943). "The Chemistry of Natural Colouring Matters." Reinhold Publ. Co., New York.
Mehlitz, A. and Drews, H. (1960a). *Flussig. Obst.* **27,** 28.
Mehlitz, A. and Drews, H. (1960b). *Ind. Obst. Gemüseverwert.* **45** (8), 174.
Mehlitz, A. and Matzik, B. (1956). *Fruchtsaft-Ind.* **1** (3), 130.
Mehlitz, A. and Matzik, B. (1957a). *Ind. Obst. Gemüseverwert.* **42,** 127.
Mehlitz, A. and Matzik, B. (1957b). *Die Industrielle Obst—und Gemuseverwertung,* No. 6, Verlag Dr. Serger Hempel, Braunschweig; via Pollard, A., 1st Rep. 5th Int. Fruit Juice Congress at Vienna, 3–6 June 1959.
Mehlitz, A. and Matzik, B. (1958). *Ind. Obst Germüse verwert.* **43,** 3.
Money, R. W. and Christian, W. A. (1950). *J. Sci. Fd Agr.* **1,** 8.
Morton, A. D. (1968). *J. Fd Technol.* **3,** 269.
Neal, G. E. (1965). *J. Sci. Fd Agric.* **16,** 604.
Nelson, E. N. (1925). *J. Am. chem. Soc.* **47,** 568.
Nybom, N. (1960). *A. Rep. Balsgrad Fruit Inst. Sweden,* p. 31.
Nybom, N. (1968). *J. Chromat.* **38,** 382.
Olliver, M. (1938). *Analyst,* **63,** 2.
Olliver, M. (1967). *In* "The Vitamins", 2nd ed., Vol. 1 (Sebrell, W. H. and Harris, R. S., eds.) Academic Press, London.
Pederson, C. S. and Beattie, H. G. (1942). *Fruit Products J.* **22,** 260.
Pollard, A. (1945). *Fruit Prod. J. Am. Fd Mfr,* **24,** 139.
Pollard, A. (1959). 1st Rep. 5th Int. Fruit Juice Congress, Vienna.
Puski, G. and Francis, F. J. (1967). *J. Fd Sci.* **32,** 527.
Rohrer, D. E. and Luh, B. S. (1959). *Fd Technol.* **13,** 645.
Seegmiller, C. G., Axelrod, B. and McCready, R. M. (1955). *J. biol. Chem.* **217,** 765.
Seegmiller, C. G., Jang, R. and Mann, W. (1956). *Archs Biochem.* **61,** 422.
Smith, W. H. (1957). *Nature, Lond.* **179,** 876.
Smith, W. H. (1958). "The Handling, Precooling, Transport and Storage of Strawberries and Raspberries. A.R.C. leaflet, London.
Sondheimer, E. (1953). *J. Am. Chem. Soc.* **75,** 1507.
Sondheimer, E. and Kertesz, Z. I. (1948). *J. Am. chem. Soc.* **70,** 3476.
Soumalainen, H. and Keränen, A. J. A. (1961). *Nature, Lond.* **191,** 498.
Strimiska, F. (1967). *Ind. Obst. Gemussevertwert.* **52,** 397.
Swindells, R. (1960). Previously unpublished work, Beecham Products Research Department.
Thornton, N. C. (1937). *Contr. Boyce Thompson Inst. Pl. Res.* **10.**

Timberlake, C. F. (1960). *J. Sci. Fd Agric.* **11,** 268.
Tinsley, I. S. and Bochain, A. H. (1959). *Fd Res.* **24**, 410.
Wade, P. (1964). *J. Sci. Fd Agric.* **15,** 51.
Watt, B. K. and Merrill, A. L. (1963). *U.S. Dept. Agr. Handbook* No. 8. Rev. ed.
Weurman, C. (1961). *Fd Technol.* **15,** 531.
Whiting, G. C. (1957), *Nature, Lond.* **179,** 531.
Whiting, G. C. (1958). *J. Sci. Fd Agric.* **9,** 244.
Wilkinson, A. E. (1945) "The Encyclopedia of Fruits, Berries and Nuts, and How to Grow Them." The New Home Library, Blakiston, Philadelphia.
Williams, B. L. and Wender, S. H. (1952). *J. Am. chem. Soc.* **74,** 5919.
Williams, B. L., Ice, B. L. and Wender, S. H. (1952). *J. Am. chem. Soc.* **74,** 4566.
Woidich, H. and Langer, T. (1959). *Fruchtsaft Ind.* **4,** (6) 234.
Woodruff, R. E., Dewey, D. H. and Sell, H. M. (1960). *Proc. Am. hort. Sci.* **75,** 387.
Worth, H. G. J. (1967). *Chem. Rev.* **67,** 465.

Chapter 12

Stone Fruits

R. J. ROMANI AND W. G. JENNINGS

University of California, Davis, California, U.S.A.

I	Introduction	411
II	Botanical Aspects	412
III	Origins	412
IV	Growth and Development	413
	A. Patterns of Growth	413
	B. Responses to Growth Regulators	414
	C. Endogenous Hormones	415
V	Maturation, Ripening and Senescence	416
	A. Effects of Growth Regulators	416
	B. Response to Cultural Practices	417
	C. Composition of Fruit and Biochemical Changes	417
	D. Response to Modified Atmospheres	420
	E. Post-harvest Abnormalities	420
VI	Maturity Indices	421
	A. Conventional	422
	B. Light Transmission and Delayed Light Emission	424
VII	The Respiratory Climacteric	424
VIII	Flavour and Quality	425
IX	Enzymes and Organelles	429
X	Summary	430
	References	431

I. INTRODUCTION

Stone fruits have long been appreciated for their gustatory and aesthetic qualtities. In spite of these characteristics, they have not been highly favoured as specimens in biochemical research. A major portion of this brief and necessarily cursory treatment is devoted to the chemical and physiological changes which have been shown to occur in developing and ripening stone fruits. It is our hope that this presentation may encourage further investigation of these changes, and efforts to understand how they relate to aspects of fruit development and underlying biochemical phenomena.

II. BOTANICAL ASPECTS

Stone fruits are characterized by highly lignified endocarp (pit or stone), fleshy mesocarp (pulp) and a thin epicarp (skin). All fruits discussed in this chapter are members of the family *Rosaceae*, genus *Prunus*. The olive, *Olea europaea*, is the only species of the family Oleaceae to produce edible fruits of commercial value (Chandler, 1957). This fruit is dealt with in Chapter 7.

All stone fruits or "drupes" (from *druppa*, the Latin for an overripe olive) are derived from a superior ovary and therefore lack the crown of dried floral parts characteristic of pome fruits. Fruit of the various *Prunus* species have a characteristic indentation or ventral suture. In analogy with a leaf, the fruit's dorsal edge corresponds to the midrib and the ventral suture to adhered outer edges of a folded leaf. In some peach and nectarine varieties the ventral suture is especially pronounced in the pit. Inadequate adhesion along the suture plus internal strains as the stone hardens can result in undesirable "split pits" (Chandler, 1957).

The mesocarp or edible portion of stone fruit consists of large parenchyma cells, smaller cells sometimes containing oxalate rosettes, and vascular bundles with accompanying fibres. Plums, peaches and cherries have thin-walled sclerenchyma cells and fibres. Similar cells in the apricot are thick-walled, thus contributing to typical textural characteristics.

III. ORIGINS

De Candolle (1959) concluded that the cherry first grew wild in northern Persia and the Russian provinces south of the Caucasus. From there it spread rapidly because of its attractiveness to birds, and hence the name *Prunus avium* L., or bird cherry. The sour or tart cherry, *Prunus cerasus* L., is characterized both by the taste of its fruit and the spreading characteristics of the tree. The sour cherry is also thought to have originated in Asia and was known in Europe at the beginning of Greek civilization.

Innumerable varieties of plums are found in nearly all temperate zones. Among the more common species are *Prunus domestica* L. or the domestic plum, *Prunus insititia* L., also known as the damson plum, and *Prunus salicina* Lindl. or salicina plum. The latter is thought to be native to China (Chandler, 1957). Fruit of the salicina varieties are elongate and somewhat pointed at the apex. In contrast to cherries and plums, the apricot had a later introduction into the western world. Greeks and Romans knew the apricot at about the beginning of the Christian era.

The botanical name for the apricot, *Prunus armeniaca* L., implies an Armenian origin just as *Prunus persica* L. suggests a Persian origin for the peach. However, it is now generally thought that both the apricot and the peach were first grown in ancient China.

IV. GROWTH AND DEVELOPMENT

For the purposes of this discussion fruit development has been divided into four phases: growth, maturation, ripening and senescence. Growth and maturation occur while the fruit is attached to the tree; ripening may take place either before or after harvest. However, post-harvest ripening is generally thought to occur only if the fruit is sufficiently mature (horticulturally mature) when picked. Senescence is generally restricted to changes occurring after the fruit has reached optimum (edible) ripeness. As a biological phenomenon, however, senescence is coincident with ripening; indeed, all aspects of development are part of a continuum and are not separate phenomena.

A. Patterns of Growth

A double sigmoid growth curve (Fig. 1) is characteristic of all the stone fruit discussed in this chapter. The growth patterns of peaches (Conners, 1920; Blake *et al.*, 1931; Lilleland, 1932; Tukey, 1934), apricots (Lilleland, 1930) and cherries (Tukey and Young, 1939) have been well described.

Phase (period, stage) I of the growth curve results principally from the enlargement of all parts of the ovary with the exception of the endosperm and embryo. Lignification of the endocarp takes place during phase II and growth is confined principally to the endosperm and embryo. It is during phase III that expansion of the mesocarp (edible portion), leading to the mature fruit (Crane, 1964), is resumed.

Ryugo (1962) has shown that 95% of the cell wall lignin is deposited in the peach pit before onset of phase III. The coincidence of pit hardening with depressed fruit growth had led to the suggestion that competition for nutrients suppressed growth of the mesocarp. However, the theory was questioned (Lilleland, 1930) when a double sigmoid growth curve was observed even for fruits grown on heavily thinned trees and thus having an abundant supply of nutrients. A time-dependent interaction between the pit and mesocarp influencing growth of the latter has been demonstrated by Tukey (1936). Destruction of the embryo early in growth phase II caused an abrupt check in development, whereas destruction of the embryo during the transition between phase II and phase III had little or no effect on subsequent fruit growth.

B. Responses to Growth Regulators

Hormonal interactions between seed and mesocarp were implied in studies by Gustafson (1939). Nitsch (1953) attributed the onset of mesocarp growth in phase III to the availability of auxin resulting from the cessation of embryo growth. However, this hypothesis was questioned by Crane and Punsri (1956) who found that 75 p.p.m. 2,4,5-trichlorophenoxyacetic acid (2,4,5-T), applied at the beginning of pit hardening, resulted in increased size of apricot fruit (growth of mesocarp) with no effect on the relative rates of endosperm and embryo development.

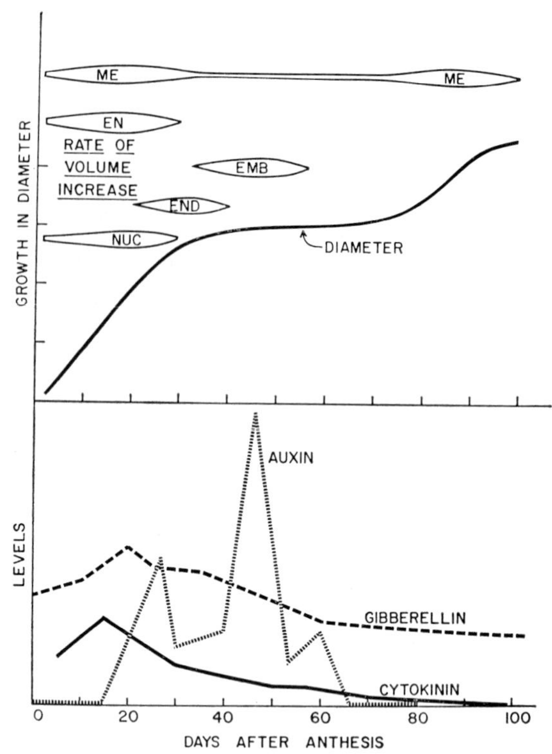

FIG. 1. *Upper graph:* Sigmoidal growth curve characteristic of stone fruit and the rates of increase in volume of component parts. ME=mesocarp; EN=endocarp; NUC=nucellus; END=endosperm and EMB=embryo. *Lower graph:* Changing levels of hormones in developing stone fruit (levels of hormones bear no relationship to one another) Taken from a recent review by Crane (1969) and including data presented by Jackson and Coombe (1966), Letham (1964) and Powell and Pratt (1966).

1. *Parthenocarpy*

Gibberellic acid (GA) at 500 p.p.m. will induce parthenocarpy in the apricot and peach (Crane *et al.*, 1960). Growth characteristics of the seedless fruit are the same as for pollinated and seeded counterparts thus providing evidence against the theory that suppression of mesocarp growth in phase II is due to competition for auxin by the growing seed. Bradley and Crane (1962) confirmed that number and size of cells in the mesocarp of GA-induced parthenocarpic peaches were the same as in seeded fruit. Parthenocarpy of cherry fruit has also been induced with combined applications of GA and

2,4-dichlorophenoxyacetylmethionine, or with GA plus other auxins, applied 2 days after full bloom (Reibeiz and Crane, 1961; Crane and Hicks, 1968).

2. *Metabolic responses*

The metabolic responses to hormone application are conditioned by the state of fruit development. For instance, respiratory rates of auxin-treated apricot fruit, determined 24 hours after picking, decline rapidly during growth phase I, remain relatively constant during phase II and decline again during phase III (Catlin and Maxie, 1959). Effects of phytohormones on the respiratory climacteric are discussed in Section VII.

Auxin (Catlin and Maxie, 1959) and other growth regulators (Carlone, 1965a b) stimulate continued growth during phase II, although the effect is somewhat variable. Auxin stimulated growth results from increased cell enlargement and not cell multiplication (Bradley and Crane, 1955). Varietal differences in the response of apricots to 2,4,5-T have been shown to be attributable to the different stages of development achieved by these fruits at any given date (Crane and Punsri, 1956).

C. Endogenous Hormones

Changes in endogenous auxin levels in developing peach ovules were first reported by Stahly and Thompson (1959). Phinney and West (1960) found gibberellic acid activity in both plums and apricots. A natural kinin was reported (Lavee, 1963) to be present in 10- to 15-day old peach fruit. Meanwhile, Letham and Bollard (1961) and Letham (1963) reported the presence of both growth stimulants and inhibitors in developing plum but deemed it unlikely that inhibitors act as controlling factors. Jackson and Coombe (1966) found the concentration of gibberellin-like substances in apricot seed, endocarp and mesocarp to correlate with the growth rates of these tissues.

Cytokinins have been extracted from peach embryo, endosperm and flesh (Powell and Pratt, 1964, 1966). The cytokinins were present at higher concentrations coincident with development of the nucellus and endosperm, with embryo growth, and with abscission of early and mid-season peaches. However, a peak in growth substances also occurred in late peaches well before they abscissed. Preliminary evidence suggests that abscisic acid (1–2 μg/g dr. wt.) is present in young peach and cherry fruits (Davison and Bukovac, 1968). Lipe and Crane (1966) have confirmed the presence of dormin (abscisic acid) or a dormin-like substance in resting peach seeds.

Jackson (1968) recently concluded that increased hormone levels in specific parts of fruit occur only with the growth of that particular tissue, thus detracting from the concept of a strong hormal interrelationship between seed growth, pit hardening and expansion of the mesocarp. However, there is little question that knowledge of the phytohormones and their modes of action is vital to an understanding of the growth and development of stone fruits, and the interrelationship of fruit parts. The present status of this vital and complex subject is well summarized by Crane (1969), who notes that "the interactions and interrelationships among the growth promoting substances are obscure to say the least". A graphic representation (Fig. 1) prepared by Crane from the data of several laboratories summarizes the current understanding.

Both Crane (1964) and Jackson (1968) evoke the concept, credit to Osborne and Hallaway (1964), that hormones increase the ability of a tissue to compete for metabolites thus resulting in increased growth. Though the concept could explain the effects of phytohormones, it does not indicate the biochemical mechanisms involved. Recent evidences for interactions between phytohormones and nucleic acid metabolism (Fox, 1968) offer possibilities of a fundamental, molecular mechanism in hormonal control. Stone fruits, having a distinct separation of tissues each with distinct developmental phases, should provide an excellent source of materials with which to study the relationship among phytohormones, nucleic acid metabolism and cellular development.

V. MATURATION, RIPENING AND SENESCENCE

The later phases of fruit development have been studied extensively, largely with reference to horticultural problems encountered in fruit harvesting, storage and marketing. A limited number of these studies will be touched upon in drawing attention to specific biochemical responses or "changes" characteristic of maturing, ripening and senescent stone fruits.

A. Effects of Growth Regulators

Apricot maturity can be advanced by approximately 1 week with an application of one of various auxins (Crane, 1956). In contrast, a pre-harvest spray of 250 p.p.m. cytokinin or gibberellic acid will cause a definite and, in some instances, marked retardation in maturation of apricots (Abdel-Gawad and Romani, 1967, 1968). A later application of the hormones (4 weeks as opposed to 6 weeks before harvest) was more effective. Treatment with GA caused a significant (2–4 days) delay in the climacteric of fruits subsequently harvested and held at 20° C. In contrast, cytokinin did not cause a temporal shift in the climacteric, but effected a slight decrease in respiratory rate. A marked effect of an application of 1000 p.p.m. kinin or 250 p.p.m. GA was the temporary reversal of the normal changes in light transmission characteristics known to accompany the ripening of apricots (Romani *et al.*, 1963b).

Preliminary reports have described the effects of Alar (B9,N-dimethyl-amino succinamic acid) in promoting the growth and ripening of sweet (Chaplin and Kenworthy, 1968) and tart (Unrath and Kenworthy, 1968) cherries. Pre-harvest spraying of Italian prunes with 100 p.p.m. GA is reported to decrease internal browning and to extend market life (Proebsting and Mills, 1966).

B. Responses to Cultural Practices

Cultural practices can be expected to influence, directly or indirectly, all aspects of development with effects on maturation, ripening and ultimate quality of the fruit being most notable. Practices tending to increase vigour of the tree will, in turn, affect fruit ripening and quality (Addoms *et al.*, 1930; Kenworthy and Mitchell, 1952; Schneider *et al.*, 1958; Claypool *et al.*, 1966). The above and many similar studies document in detail what has been implicitly understood for centuries.

Among the somewhat more definitive experiments are those of Albrigo *et al.* (1966) demonstrating that high nitrogen conditions prolong cell division thus protracting development and resulting in delayed ripening. Curwen *et al.* (1966) report that high nitrogen or potassium may affect the levels of water and pectic substances in cherries. Fruit colour of Elberta peaches was somewhat improved with higher potassium fertilization, but diminished by magnesium (Cummings, 1965). Boron deficiency resulted in pronounced suberization of peach parenchyma cells (Kamali and Childers, 1967).

Whether accepted cultural practices or deviations therefrom significantly affect the biochemistry of maturation and ripening will depend upon the husbandry as practised in a given region, the soil characteristics, the climacteric peculiarities, and other properties inherent to a region or to the varieties grown therein. Descriptions of the end results, e.g. changes in soluble solids, colour, firmness, etc. of the fruit are of interest and value to the management of orchards. However, the understanding and control of underlying biochemical phenomena to the full benefit of horticulture are likely to come only with detailed enquiries at the cellular and/or molecular levels.

C. Composition of Fruit and Biochemical Changes†

Extensive data on the constituents of various stone fruits have been recorded (Winton and Winton, 1932a, b, Pienaar *et al.*, 1965). Only those changes in constituents which are related to specific developmental or "quality" aspects of stone fruits will be discussed in this section.

Apricots: Treatment with 2,4,5-T not only affects the rate of apricot growth

† See also Volume 1, Chapters 5, 6, 11 and 12.

but can result in higher concentrations of niacin, biotin and pantothenic acid (Catlin, 1959). In allied studies, applied 2,4,5-T accumulated in the exocarp and mesocarp of apricot fruits (Maxie *et al.*, 1962; Crane *et al.*, 1965).

Citric acid is the principal acid in both fresh and canned apricots. Agarwal and Date (1966) found no malic acid in these fruit but did report an unidentified acid closely related to citric. Polyphenolic compounds in canned apricots were analysed by El-Sayed and Luh (1965). Using two-dimensional paper chromatography, they found a predominance of the three isomers of chlorogenic acid and two *p*-coumaric acid derivatives (probably *p*-coumaryl quinic acids) with smaller amounts of quercetin, isoquercitin, an unidentified quercetin glucoside, rutin, catechin and epicatechin.

Cherries: In discussing cherries and cherry products Marshall (1954) concluded that few physiological or chemical changes occur in the cherry after it is harvested. Allen (1941) suggested that increased sweetness in harvested cherries results from losses of acids and tannins and not increases in sugars or soluble solids. Extensive chemical analyses of both tart and sweet cherries are provided by Boland and Blomquist (1965). Constantinides and Bedford (1964) reported that sugars comprised 50–60% of total dry matter in red tart cherries. Fructose comprised 99% of the total sugars; sorbitol and mannitol have also been detected in cherry pulp (Haeseler and Misselhorn, 1966). Maleic acid has been found to represent 75–95% of the total non-volatile acid in red tart cherries, and to decrease in concentration with maturity (Das *et al.*, 1965).

Cyanidin-3-monoglucoside is reported as a minor pigment in tart cherries (Schaller and Elbe, 1968), along with antirrhinin and mesocyanin (Li and Wagenknecht, 1956). Carotenoids are present in some cherries (Galler and Mackinney, 1965), and range in concentration from 1·2 to 9·5 μg/g tissue.

The vitamin C content of cherries was found by Jaquin (1965) to range from 13 to 56 mg/100 g edible fruit, depending upon variety. Olliver (1967) in reviewing published results, reported lower average values for the edible portions of other stone fruits, i.e. (in mg ascorbic acid/100 g) peach, 7; apricot, 10; nectarine, 15 and plum, 3.

Perhaps the chemical changes most nearly related to the acceptability of cherries are those associated with polyphenols (flavour and colour) and pectins (texture). In a detailed study, Wolf (1958) found a seven-fold decrease in chlorogenic acid from pit hardening to full ripeness. Though total water soluble and insoluble pectins decrease with maturation of sour cherries, these constituents are not markedly changed by freezing or by frozen storage (Al-Delaimy *et al.*, 1966). Van Buren (1967) found protopectin as well as the intrinsic viscosity of the pectic materials to decrease during ripening of sweet cherries.

Peaches: The anthocyanin pigments of peaches have received considerable attention because of discolorations in the canned product. Van Blaricom and Senn (1967) analysed anthocyanin pigments from nine varieties of peaches and found the main constituent to be cyanidin-3-monoglucoside. The same pigment increased markedly as peaches ripened on the tree (Hsia *et al.*, 1965). Leucoanthocyanins (Hsia *et al.*, 1964) and proanthocyanins (leucoanthocyanins) (Joslyn and Dittmar, 1967) have also been reported in Elberta peaches. Curl (1959) found peach carotenoids to be a complex mixture with xanthophylls being the predominant species.

Organic acid levels in peaches and nectarines reach a maximum and then decrease as the fruit approach harvest maturity (David *et al.*, 1956; Ryugo and Davis, 1958). Ryugo (1964b) compared the concentrations of malic, citric and total titratable acidity in peaches and nectarines. He found the total titratable acidity was less than the sum of citric and malic acids, and concluded that a portion of the acids existed as salts. It should be noted, however, that the lactones, which are major flavour components of the peach, would also account for a portion of the titratable acidity. Similar findings on the gross constituents of peaches were reported by Salunkhe *et al.* (1968).

In 1898, Storer reported on the cellulose, lignic acid and xylan present in peach pits. Other constituents of peach and apricot endocarp include vanillin, syringaldehyde, coniferaldehyde and sinapaldehyde (Ryugo, 1962). Ryugo (1964a) noted a constant increase in the methoxyl content of the pectin of the endocarp over a 50-day period following initiation of lignification.

A well-described relationship exists between the pectic constituents of cling and freestone peaches and their flesh texture. Whereas the pectins of cling peaches vary little with fruit maturity, the opposite is true for freestone, in which ripening is accompanied by conversion from insoluble to soluble pectic material (Postlmayr *et al.*, 1956; Sterling and Kalb, 1959; Shewfelt, 1966; Souty *et al.*, 1967).

Histochemical changes in ripening peaches were assessed in detail by Reeve (1959a, b, c). Methyl esterification of cell wall pectins remained constant (75–80%) during fruit growth, increased to near 100% in hard-ripe fruits and declined again as the fruits ripened and softened. Tannin content per cell also increased ninety-fold from the beginning of cell enlargement to half fruit size and was followed by an approximate eight-fold decrease as ripening progressed. However, as pointed out by Reeve, there may be a tendency to oversimplify the relationship between cell wall thickness or composition and textural quality of ripened fruits.

No significant difference in total polyphenols of tree-ripened as compared with stored-ripened peaches was found by Craft (1961). The major polyphenols in order of prominence in the Elberta peach are reported to be leucoanthocyanins, chlorogenic acids, catechins and flavonoids. Total phenolic content, higher at pit-hardening stage, decreases as fruits enlarge and mature. However, the decrease in concentration (expressed on a fresh weight basis) is accompanied by an actual increase in total phenolics. An analysis of polyphenolic compounds in canned cling peaches (Luh *et al.*, 1967) revealed four chlorogenic acid isomers, five leucoanthocyanidin isomers and catechin. A clear relationship between tannins and astringency of the canned fruits was shown by Guadagni and Nimmo (1953).

Plums and prunes: Malic acid is reported to be the principal non-volatile acid in prunes, with the exception of sweet Italian prunes which contain an almost equal amount of quinic acid (De Moura and Dostal, 1965).

D. Response to Modified Atmospheres

Refrigerated and controlled atmosphere storage (low O_2 and/or high CO_2 has been used most successfully with apples. However, early studies by Allen (1939) and Claypool and Allen (1948), and the recent report by Anderson *et al.* (1967) indicate the feasibility of using Controlled Atmosphere Storage to extend the life of various stone fruits. The observation by Claypool and Allen (1948) that, contrary to common experience, Wickson plums had an increased CO_2 production at low O_2 tensions was later (Claypool and Allen, 1951) related to the onset of fermentative reactions.

Specific (and varietal) differences in the response of plums to modified atmospheres (Controlled Atmosphere Storage) were noted by Couey (1960). He suggested that a suitable atmosphere (7% O_2, 7% CO_2, 86% N_2) which delayed ripening and reduced the loss of soluble solids in Eldorado plums stored for several weeks at 0–1°C, with no impairment of texture. Eldorado plums stored well for up to 6 weeks, whereas Santa Rosa plums did not. Varietal differences have also been observed in response to storage at 32°C (Uota, 1955). Plum varieties that normally would not ripen at that temperature could be induced to do so with ethylene treatment. Porritt and Mason (1965) reported various combinations of controlled atmospheres to be ineffective for storage of cherries.

Controlled atmosphere storage, perhaps more than any other manipulation, emphasizes the "living" nature of fruit cells. In retrospect, it is remarkable that fruit cells can exist for months under such marginal environments, often without serious impairment of the normal developmental (ripening) processes.

E. Post-harvest Abnormalties

Many types of damage can befall harvested fruits. Only a few are listed to illustrate the biochemical phenomena that underly some of the abnormalties observed in stone fruits.

Scald in cherries is described as the localized translocation of pigment from the skin to the flesh as a consequence of bruising (Dekazos, 1966). Buch *et al.* (1961) found bruising of red tart cherries to be accompanied by demethylation of pectin. Pollack *et al.* (1958a, b) determined that scald development in bruised fruit was accompanied by an increase in respiration and a change in R.Q. from 0·9 to 1·1. In related studies, Peng and Markakis (1963) defined the conditions under which mushroom phenolase would discolour the anthocyanin pigments of red tart cherries.

"Cracking" of sweet cherries is temperature dependent with a Q_{10} of approximately 1·5 between 15 and 25°C (Bullock, 1952). Direct relationships between metabolic rates and deterioration of harvested cherries are implicit in the studies by Micke *et al.* (1965) and Mitchell and Micke (1966) in which rapid post-harvest cooling was shown to be a requisite for maintaining product quality.

Pentzer and Allen (1944) determined that for various plum varieties a temperature of 0°C was required to suppress ripening sufficiently to permit transcontinental shipment in the United States. Under these conditions, however, low-temperature breakdown occurred in the less mature plums.

Smith (1940, 1947) noted that while potential for injury is acquired at low temperatures, injury symptoms develop most rapidly at higher temperatures, a strong indication of their metabolic origins. A minimum temporal requirement for development of low-temperature injury is implicit in the discovery that an interim period of 2 days at 18°C after the 15–16th day of cold storage (at −0·5°C) markedly reduced the incidence of injury in Victoria plums (Smith, 1967). Equilibrium and kinetic factors in the phenomena of injury at low temperature were defined by van der Plank and Davies (1937). The relationship between degree of immaturity and chilling injury of Italian prunes was clearly shown by Gerhardt *et al.* (1943) and Gerhardt and English (1945). Based on recent experiments, Smith (1967) recommends storage of Victoria plums at −0·5°C in 1% O_2 with a 2-day interruption at 18°C in air on the 16th day.

At the opposite extreme, heat injury in harvested prunes was attributed to limiting oxygen tensions in the inner mesocarp of fruits exposed to warm temperatures (Maxie and Claypool, 1957). Several factors affecting post-harvest behaviour of fruits including information on stone fruits have been reviewed by Pentzer and Heinze (1954).

Cellular senescence and an accompanying loss of the ability to compensate for stress (Romani *et al.*, 1968) may account for many of the detrimental changes observed in ripening fruit. However, and as noted below, other changes during ripening are vital to the ultimate flavour and quality of the fruit. Taking the view that all transitions are part of biochemically inter-related events, one is led to search for the inter-connecting, controlling aspects of development rather than biochemical changes *per se*.

VI. MATURITY INDICES

The search for maturity indices stems from the recurrent need to decide when fruits should be harvested. The goals are not always compatible: e.g. marketing flexibility as opposed to optimum fruit quality. Moreover, assessments of maturity are inherently complex, especially for stone fruits where significant biochemical changes affecting fruit characteristics occur after harvest, but are predestined by the stage of fruit development (maturity) at the time of harvest. Estimates of maturity are therefore estimates of performance under known (or assumed) storage and marketing conditions. The final decision of when to harvest fruits, or on what basis to separate an array of mechanically harvested fruits into maturity categories, will necessitate an integration of criteria and, ultimately, a compromise.

A. Conventional

Many gross physiological and biochemical changes in maturing stone fruits have been assessed as indices to the final, ripened product. The observed

TABLE I. Firmness, soluble solids and acid content associated with different stages of matuity in some stone fruits

Fruit	Variety	Maturity	Firmness	Soluble solids (%)	Malic acid (%)	Reference
			Pounds			
Plum	Beauty	Green	13·2	11·5	1·10	Allen (1932)
		SM[a]	9·0	11·3	1·43	
		Mature	4·9	13·4	0·94	
Peach	Elberta	Green	17·6	11·7	0·63	Allen (1932)
		SM	12·4	12·1	0·62	
		Mature	3·7	13·1	0·49	
	Phillips Cling	Green	12·0	11·5	0·77	
		SM	8·8	12·4	0·63	
		Mature	8·4	12·6	0·57	
Apricot	Derby Royal	Green	16·6		1·40	Allen (1932)
		SM	13·8		1·29	
	Royal	Green	19·0		1·41	
		SM	10·2		1·19	
		Harvest period	Compression (%)			
Sour cherry	Montmorency	Early	23·53	16·1		Taylor and Mitchell (1953)
		Mid	25·52	18·9		
		Late	22·16	19·5		

[a] SM = shipping maturity.

changes have included an increase in soluble solids, decrease in firmness, loss of chlorophyll, increase in specific pigments (where they occur), some decrease in acidity and a decrease in electrical impedance. These observations were well summarized by Allen in 1932. A few of the more commonly used measurements of fruit maturity are illustrated in Table I. Results from recent

TABLE II. Correlating fruit firmness and soluble solids at time of harvest with organoleptic acceptability of ripened stone fruit[a]

Fruit	Estimated field maturity[b]	At time of harvest: Average soluble solids	At time of harvest: Average firmness	Percentage acceptable when ripe[c]
Apricots				
Derby	1	11·1	13·4	36
	2	15·7	5·1	73
	3	19·1	3·2	92
Tilton	1	8·6	14·1	10
	2	10·0	11·3	56
	3	11·3	8·5	85
Peaches and Nectarines				
Rio Oso Gem	1	10·2	12·8	75
	2	11·0	11·9	84
	3	12·9	7·8	89
Early Sun Grand	1	10·1	12·7	58
	2	10·6	12·0	73
	3	12·0	9·0	87
Plums				
Laroda		0 to 12		0
		12 to 14		69
		14 to 16		93
		16+		100
Queen Ann		0 to 13		25
		13 to 15		5
		15 to 16		66
		16+		100

[a] Data from unpublished studies conducted at the University of California, Davis, by Gordon Mitchell and Roger Romani.

[b] All fruit were picked at time of the normal commercial harvest. The apricots, nectarines and plums were separated into 3 maturity classes on the basis of subjective evaluation, prior to the analyses for soluble solids, firmness and other indices not shown here. The plums were prepared for analyses in the same manner except that the data shown summarize the results of 3 years' work.
Maturity 1 = least mature of the fruit picked; maturity 2 = average maturity; maturity 3 = most mature (but still unripe) of the fruit picked.

[c] Fruit were evaluated by a taste panel utilizing a 10-point hedonic scale. A sample of fruit in which 40% of the fruit rated 3 or less was classified as unacceptable. All fruits were brought to uniform ripeness prior to evaluation.

studies attempting to directly relate maturity at harvest to ultimate organoleptic acceptability of the fruit are summarized in Table II.

Among published reports are those with reference to maturing and ripening peaches (Haller, 1952; Bedford and Robertson, 1955; Rood, 1957; Kott, 1965), prunes (Jouret and Maugenet, 1967) and cherries (Taylor and Mitchell, 1953; Kott, 1965; La Belle, 1968). In most instances these studies define, or extend to a particular growing region or fruit variety, the general types of chemical or physical analyses already well known. We have found no study that relates any of the gross physiological changes to specific biochemical developments or, more important, to their control.

B. Light Transmission and Delayed Light Emission

Light transmission techniques, based on the absorption spectrum of light passing through an intact fruit, were developed by Norris (1956) for the non-destructive measurement of changes in pigmentation. When these methods were applied to peaches, nectarines (Romani *et al.*, 1962) and apricots (Romani *et al.*, 1963b), a reasonable correlation was noted between certain light transmission ratios at time of harvest and quality of the ripened fruit.

In the course of light transmittance measurements, it was found that the fruit themselves emit minute amounts of energy after illumination (Romani *et al.*, 1963a). The delayed light emission (DLE) has characteristics affected by excitation energy, chlorophyll content and physiological state of the fruit (Jacob *et al.*, 1965). As opposed to light transmission, DLE is eminently suited to rapid, mechanized grading of fruits, a necessity if the full potential of mechanical harvesting is to be realized. Moreover, DLE may have unexplored discriminatory potential based on selective excitation and emission spectra. A further understanding of mechanisms of energy absorption, transmittance by pigments, energy pools and reconversion of chemical energy into light are requisites of a successful application of this technique.

VII. THE RESPIRATORY CLIMACTERIC

Described by Kidd and West in 1930, the respiratory climacteric (see Volume 1, Chapter 17) is a well-documented phenomenon characteristic of most, but not all, ripening fruits. Apricots have a distinct climacteric that is affected by pre-harvest spray with GA or kinetin (Abdel-Gawad and Romani, 1967), and by a post-harvest dip in an auxin solution (Maxie and Crane, 1956; Abdel-Gawad and Romani, 1967).

Moore and Scott (1966) did not detect a climacteric in carbon dioxide production by ripening peaches although a peak in ethylene production occurred after approximately 7 days' storage at 18°C. This is in contrast to the evidence of Lim and Romani (1964) which indicated a definite climacteric pattern for both carbon dioxide and ethylene evolution. In an earlier work

Roux (1940) reported the occurrence of a climacteric in both peaches and plums.

Hartmann (1957) concluded that cherries, as well as apricots, go through a typical respiratory climacteric, but that the phenomenon is difficult to discern in cherries because of their susceptibility to fungus. His findings support earlier data (Ulrich, 1946) indicating a slight rise in the respiratory activity of cherries as the fruit begin to colour. Others (Pollack *et al.*, 1961; Romani *et al.*, 1961), however, did not observe a "climacteric" in ripening cherries, but rather a steadily decreasing rate of respiration (Fig. 2). The

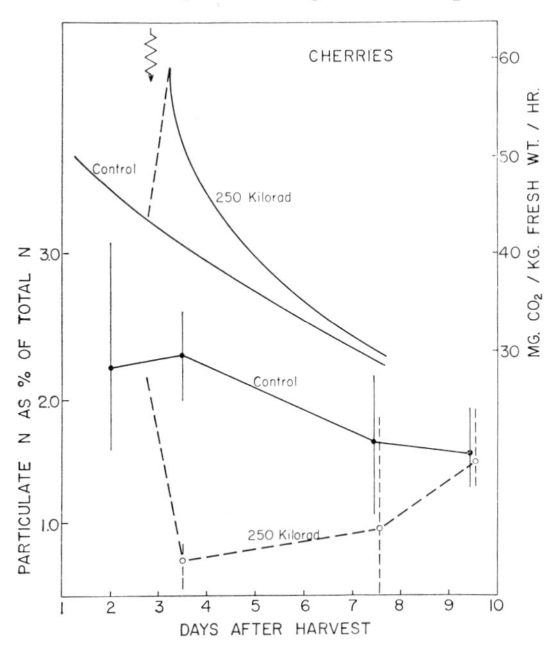

FIG. 2. Respiration rates and mitochondrial yield of Bing cherries with and without prior absorption of 250 Krad of ionizing radiation. From Romani *et al.* (1961) and Romani and Van Kooy (1962).

question is unresolved. Future investigators should give careful attention to the possibility of fungal infection, wide varietal differences and the occurrence of a climacteric while the cherries are still on the tree.

Maxie *et al.* (1960) found no autogenous climacteric in ripening olive fruits (see Chapter 7); however, a marked respiratory response to ethylene treatment was observed at 25°C or higher.

VIII. FLAVOUR AND QUALITY†

Many of the gross changes accompanying ripening (Section V) relate indirectly to the quality of the final product. Certainly one of the more direct

† See also Volume 1, Chapter 10.

and important criteria of quality is the flavour of the fruit, but comparatively little work has been done on the flavour of stone fruits; this is particularly true for the cherry. Nelson and Curl (1939) did some early work, and more recently Spanyar *et al.* (1964) reported additional compounds. Volatile components which have been reported in cherry now include methanol, ethanol, butanol, pentanol, octanol, and geraniol, ethyl acetate and benzaldehyde.

A little more is known about the behaviour of apricot. Salunkhe *et al.* (1962) and Deshpande and Salunkhe (1964) recorded gross constituent changes in ripening apricots and peaches, including increases in volatile reducing substances. Fideghelli *et al.* (1967) studied natural and artificial ripening of apricots and reported that apricots harvested green and kept at 19°C for 24–48 hours in air containing 0·1% ethylene ripened in just over 3 days but lacked the aroma, flavour and red tinge of those ripened naturally in 6 or 7 days on the tree. The volatile constituents of apricot have been shown to include (Tang and Jennings, 1967, 1968) myrcene; limonene; *p*-cymene; terpinolene; trans-2-hexenol; α-terpineol; geranial; geraniol; benzyl alcohol; 2-methyl butyric, hexanoic and acetic acids; linalool; three and probably four isomers of epoxydihydrolinalool; and the γ-capro-, γ-octa, δ-octa-, γ-deca-, δ-deca- and γ-dodeca-lactones. The "epoxydihydrolinalool" nomenclature was used (Stevens *et al.*, 1966; Tang and Jennings, 1968) because these compounds are derived via epoxidation of linalool (Klein and Rojahn, 1964). The isomers (I, II, III, IV) might more precisely be termed *trans*- and *cis*-α,α,5-trimethyl-5-vinyl tetrahydrofurfuryl alcohol (Creveling *et al.*, 1968), and *trans*- and *cis*-1,1,5-trimethyl-2-hydroxy-5-vinyl tetrahydropyran, respectively.

Peaches (and nectarines) have received considerably more attention than other stone fruits. Bedford and Robertson (1955) made a detailed study of peach ripening, and found that the flavour of the canned and frozen products was markedly influenced by the method of artificial ripening. Better flavoured products were obtained from fruit ripened at lower temperatures 24°C than from fruit ripened in a relatively short period at higher temperatures 29°C. Peaches requiring longer ripening times (harvested somewhat immature) tended to develop a stale 'off' flavour.

Power and Chesnut (1921, 1922) reported that formic acetic, pentanoic and oetanoic esters of linalool, plus traces of acetaldehyde, were the major aroma constituents of peach. Daghetta *et al.* (1956) analysed Hale peach volatiles and found ethyl alcohol, acetaldehyde, pentyl acetate and acetic acid. Lim (1963) found that the Red Globe variety produced more volatiles than did other varieties, and on the basis of gas chromatographic retentions identified ethylene, ethanol, acetaldehyde and ethyl and hexyl acetate. In a continuation of these studies, Lim and Romani (1964) compared the production of volatiles from peaches harvested at three different maturities. Although all fruit ripened normally (in appearance and CO_2 production) the less mature fruit had a dramatically lower volatile production.

Fideghelli *et al.* (1967) found that peaches harvested when turning from green to very pale yellow and kept at 19°C in air containing ethylene or 50% O_2 compared unfavourably in weight and size with tree-ripened fruit.

Ethylene treatment failed to accelerate ripening of the Earlihale variety. Oxygen treatment of Fairhaven and ethylene treatment of Pulchra advanced ripening by 3-4 days (see also Section V D).

In a study concerned with qualitative identification of peach volatiles, Jennings and Sevenants (1964) and later Sevenants and Jennings (1966) found a number of compounds, which they characterized by retentions on two gas chromatographic columns and infrared spectroscopy of the isolated components. The compounds identified included ethanol; n-hexanol; *trans*-2-hexanol; benzyl alcohol; acetaldehyde; benzaldehyde; acetic, isopentanoic and n-hexanoic acids; hexyl formate; methyl, ethyl, n-hexyl, trans-2 hexenyl and benzyl acetate; ethyl and n-hexyl benzoate; an α-pyrone, and a series of lactones—γ-hexa-, γ-hepta-, γ-octa-, γ-nona-, γ-deca- and δ-deca-lactones. They suggested that while the lactones were probably most responsible for the characteristic peach aroma, the full flavour resulted from the integrated response to a spectrum of volatiles. Broderick (1966) confirmed the identification of the lactones, and postulated that these compounds were of primary importance to peach flavour.

In recent studies (Do *et al.*, 1969; Salunkhe, 1968) of volatile compounds in Gleason Early Elberta peach fruit as a function of maturity, the presence of many of these compounds has been confirmed, including the lactones. In addition, γ-pentalactone and γ- and δ-dodecalactone have been identified. These workers also found that tree-ripened fruit possessed a higher level of volatiles than artificially ripened peaches, and reported that the most striking differences occurred in the lactones; artificially ripened peaches possessed about one-fifth the quantity of lactones as the tree-ripened fruit. The concentration of γ-decalactone was particularly low, and the dodecalactones were absent in the artificially ripened fruit. They suggested that this large difference in lactone concentration might account for the inferior flavour of artificially ripened peaches.

Apricots differ from other stone fruits in that they possess a variety of terpenes among their volatiles—these are apparently completely lacking in the peach. In view of the early reports by Power and Chesnut (1921, 1922), Sevenants and Jennings (1966) searched specifically for terpenoids in peach essence and concluded that they were absent.

While lactones undoubtedly are major contributors to the flavour of peaches, this may not always be a positive sense. The typical bakery-produced peach pie possesses a rather overpowering lactone flavour. Many such products utilize sliced peaches preserved by freezing, and frozen peaches also possess this strong lactone flavour (Jennings, personal observation). Whether this is due to an increase in the concentration of lactone(s), or to a decrease in the amount of some contributory flavour compound that masked this strong flavour, remains to be elucidated.

The occurrence of lactones in both ripened peach and apricot suggests that these fruit possess some biochemical pathways in common. Tang and Jennings (1968) suggested that some general mechanism would be required to account for the wide variety of lactones observzd. Scheuerbrandt *et al.* (1961) isolated from certain eubacteria an enzyme system capable of random hydration–dehydration of unsaturated moieties. If peach and apricot differed from other fruits in possessing such an enzyme(s), a variety of hydroxy acids could result. In efforts to stimulate interest in this direction, Tang and Jennings (1968) suggested that hydroxy acids might occur in sub-detectable quantities, but whenever the hydroxyl group assumed the γ- or δ-position, lactonization would occur, and the compound formed would no longer participate in reactions and would accumulate:

XXXXX

$$\downarrow\uparrow$$

$$R{-}CH{=}CH{-}CH_2{-}CH_2{-}CH_2{-}COOH$$

$$\downarrow\uparrow$$

$$R{-}CH_2{-}\underset{\displaystyle OH}{\underset{|}{C}H}{-}CH_2{-}CH_2{-}CH_2{-}COOH \longrightarrow R{-}CH_2{-}(\delta\text{-lactone ring, O, }{=}O)$$

$$\downarrow\uparrow$$

$$R{-}CH_2{-}CH{=}CH{-}CH_2{-}CH_2{-}COOH$$

$$\downarrow\uparrow$$

$$R{-}CH_2{-}CH_2{-}\underset{\displaystyle OH}{\underset{|}{C}H}{-}CH_2{-}CH_2{-}COOH \longrightarrow R{-}CH_2{-}CH_2{-}(\gamma\text{-lactone ring, O, }{=}O)$$

$$\downarrow\uparrow$$

XXXXX

A valid objection to this suggestion is the relative ratios of γ-and δ-lactones which occur in these fruits. Random hydration–dehydration of a normal fatty acid would proceed from deeper in the molecule toward the carboxyl group, and one would expect higher concentrations of the δ- and lower concentrations of the corresponding γ-lactones. The γ-lactones are actually dominant (Tang and Jennings, 1967, 1968).

This argument, however, could be countered if the fatty acids were being degraded at the same time that the hydration–dehydrations were occurring. Results of Heinz and Jennings (1966) and Creveling and Jennings (1969) indicate that a β-oxidation cycle is operative in the maturing Bartlett pear, and results observed by Tressl (1967) and Emberger (1967) indicate that a similar fatty acid degradation occurs in apple.

It has also been pointed out (Heinz and Jennings, 1966) that the roles of oxidases—which are widely distributed in fruits—have never been explained, nor do we fully understand the respiratory changes that occur in ripening fruit. As the fruit ripens, cell walls soften and become more permeable, various reactants may gain access to compounds from which they were formerly restricted. The increased tissue permeability may result in enzyme-substrate interactions leasing to compounds found exclusively during this phase of the life cycle. Oxidases, together with the higher levels of oxygen which can diffuse through the more permeable tissues, may account for the oxygenated compounds we associate with fruit flavour. A loss of differential permeability may also result in internal oxygen concentrations well above normal and lipoxidases may be pressed into service as a compensatory response thus producing the oxygenated flavour compounds.

Other plausible but equally unsupported mechanisms could be suggested, but much more work utilizing labelled compounds will be required to help us understand this aspect of fruit biochemistry.

IX. ENZYMES AND ORGANELLES

Enzymes. The enzyme content of various stone fruits has been examined in relation to injury, softening, discoloration or other undesirable developments. Onslow in 1920 noted the presence of various oxidases in *Prunus* species and correctly attributed injury-induced browning to the oxidation of aromatic substances, namely polyphenols. Subsequently, Kertesz (1933) found that Sunbeam peaches did not discolour when injured or sliced because of the absence of catechol tannins rather than from a lack of oxidizing enzymes. Guadagni *et al.* (1949) also correlated browning in peaches with the amount of oxidizable tannins present and found that changes in pH markedly affected the rate of browning. Motawi (1967) noted, however, that browning in peach varieties having intermediate to high tannin contents can be directly correlated with enzyme levels. Polyphenolase and peroxidase in peaches, characterized by Reyes and Luh (1966), act primarily on the orthodihydroxy configuration and are inhibited by sodium diethyldithiocarbamate.

The inhibition of polygalacturonase in brined cherries is extremely important for the maintenence of texture. Yang *et al.* (1960) found the enzyme to be stable over the pH range 3·6–6·0, with a pH of 1·6 required for satisfactory inhibition. Since high acidity is detrimental to the brining process, 0·01–0·025% alkyl aryl sulphonate is added to inhibit the enzyme. The mechanism leading to sulphite inhibition of polyphenol oxidase has been studied by Embs and Markakis (1965). Time, pH and temperature relationships for apricot phenolase have been reported by Soler Martinez *et al.* (1965).

Pectinmethylesterase (PME) has been found in peaches (Shewfelt, 1966), sweet cherries (Somogyi and Romani, 1964) and tart cherries (Al-Delaimy *et al.*, 1966) where it is reported to increase with maturation.

Organelles. Though organelles undoubtedly exist in stone fruit cells, there have been few reports describing their isolation from these fruit. Catlin (1963) obtained cytoplasmic particles from developing apricots and demonstrated the ability of the organelles to oxidize various Krebs cycle intermediates. However, he very appropriately describes some of the difficulties encountered in correlating intracellular activity with tissue development.

A considerable decrease in mitochondrial protein was found to occur as Bing cherries ripened over a period of 7–9 days at 15°C (Romani *et al.*, 1965) and quantitative changes in cherry mitochondria, in response to massive doses of ionizing radiation (Fig. 2), were also noted by Romani and Van Kooy (1962). As demonstrated for other fruit (Romani *et al.*, 1968), the ability of fruit cells to recover or compensate for radiation-induced losses of mitochondria and mitochondrial integrity is indicative of the continued vitality of ripening cells. However, the ability to compensate for stress terminates as fruit reach their climacteric peak. There is thus established a clear age-relationship between critical "life" functions of fruit cells and their capacity to withstand stress be it radiation or the rigours of harvesting, storage and distribution.

X. SUMMARY

Innumerable and often well-defined biochemical changes—some subtle as in the case of volatiles and some a little more obvious as exemplified by the pectins—occur in growing, maturing and ripening stone fruits. Implicit but still largely unexplored is a dynamic biochemistry that must underly these diverse cellular transitions.

Hormones or growth regulators clearly play important roles in modulating cellular events. Changing levels of endogenous hormones have been detected in the various tissues of stone fruits coincident with periods of development. Significantly, the responses to applied growth regulators often vary with fruit age reflecting altered biochemical propensities in developing, maturing fruit cells.

This aspect, the sequential expression of inherent cellular traits, comprises much of the research in current molecular biochemistry. Undoubtedly, "molecular" systems can be isolated from stone fruits as they have been from other fruit species (Ku and Romani, 1966; Richmond and Biale, 1967). Most of the significant horticultural problems associated with fruits derive from cellular functions (or malfunctions). There would appear to be promise, therefore, in studying the genetically programmed events that regulate growth, development and ripening of cells and in turn predetermine the ultimate nature and quality of fruits.

ACKNOWLEDGEMENT

We express our appreciation to Drs Julian Crane, George Martin, Kay Ryugo and Silviero Sansavini who have read portions of the manuscript and offered valuable comments.

REFERENCES

Abdel-Gawad, H. A. and Romani, R. J. (1967). *Pl. Physiol., Lancaster* **42** (suppl.), 43.
Abdel-Gawad, H. A. and Romani, R. J. (1968). *65th Ann. Meeting: Amer. Soc. hort. Sci.*, Abstr. p. 122, No. 236.
Addoms, R. M., Nightingale, G. T. and Blake, M. A. (1930). *New Jersey Agric. Exp. Sta. Bull.* **507**, 1.
Agarwal, J. D. and Date, W. B. (1966). *J. Fd Sci. Technol.* (Mysore) **3**, 70.
Albrigo, L. G., Claypool, L. L. and Uriu, K. (1966). *Proc. Am. Soc. hort. Sci.* **89**, 53.
Al-Delaimy, K. A., Borstrom, G. and Bedford, C. L. (1966). *Mich. agric. Exp. Sta. Quart. Bull.* **49**, 164.
Allen, F. W. (1932). *Hilgardia* **6**, 381.
Allen, F. W. (1939). *Proc. Am. Soc. hort. Sci.* **37**, 467.
Allen, F. W. (1941). *Calif. agric. Exp. Sta. Memo.*
Anderson, R. E., Parsons, C. S. and Smith, W. L. (1967). *Agric. Res., Wash.* **15**, 7.
Bedford, C. L. and Robertson, W. F. (1955). *Mich. agric. Exp. Sta. Tech. Bull.* **245**, 31 pp.
Blake, M. A., Davidson, O. W., Addoms, R. M. and Nightingale, G. T. (1931). *New Jersey agric. Exp. Sta. Bull.* **535**.
Boland, F. E. and Blomquist, V. (1965). *J. Ass. off. agric. Chem., Wash.* **48**, 523.
Bradley, M. V. and Crane, J. C. (1955). *Amer. J. Bot.* **42**, 273.
Bradley, M. and Crane, J. C. (1962). *Bot. Gaz.* **123**, 243.
Broderick, J. J. (1966). *Am. Perf. Cosm.* **81**, 2, 43.
Buch, M. L., Satori, K. G. and Hills, C. H. (1961). *Fd Tech.* **15**, 526.
Bullock, R. M. (1952). *Proc. Am. Soc. hort. Sci.* **59**, 243.

Carlone, R. (1965a). *Annali Accad. Agric. Torino*, 1961-1962, **104,** 16.
Carlone, R. (1965b). *Annali Accad. Agric. Torino*, 1961–1962, **104**, 36.
Catlin, P. B. (1959). *Proc. Am. Soc. hort. Sci.* **74,** 174.
Catlin, P. B. (1963). *Proc. Am. Soc. hort. Sci.* **83,** 217.
Catlin, P. B. and Maxie, E. C. (1959). *Proc. Am. Soc. hort. Sci.* **74,** 159.
Chandler, W. H. (1957). "Deciduous Orchards". Lea and Febiger, Philadelphia.
Chaplin, M. H. and Kenworthy, A. L. (1968). *65th Ann. Meeting: Am. Soc. hort. Sci.*, Abstr. p. 125, No. 257.
Claypool, L. L. and Allen, F. W. (1948). *Proc. Am. Soc. hort. Sci.* **51,** 103.
Claypool, L. L. and Allen, F. W. (1951). *Hilgardia* **21,** 129.
Claypool, L. L., Uriu, K. and Albrigo, L. G. (1966). *In* "Proceedings of the XVII International Horticulture Congress" (R. E. Marshall, ed.), Vol. I, Abstr. 372. Michigan State University, Michigan.
Conners, C. H. (1920). *New Jersey agric. Exp. Sta. Ann. Rept.*, **1919,** 82.
Constantinides, S. M. and Bedford, C. L. (1964). *J. Fd Sci.* **29,** 804.
Couey, M. (1960). *Proc. Am. Soc. hort. Sci.* **75,** 207.
Craft, C. C. (1961). *Proc. Am. Soc. hort. Sci.* **78,** 119.
Crane, J. C. (1956). *Proc. Am. Soc. hort. Sci.* **67,** 153.
Crane, J. C. (1964). *Ann. Rev. Pl. Physiol.*, **15,** 303.
Crane, J. C. (1969). *HortScience* **4,** 108.
Crane, J. C. and Hicks, J. R. (1968). *Proc. Am. Soc. hort. Sci.* **92,** 113.
Crane, J. C. and Punsri, P. (1956). *Proc. Am. Soc. hort. Sci.* **68,** 96.
Crane, J. C., Primer, P. E. and Campbell, R. C. (1960). *Proc. Am. Soc. hort. Sci.* **75,** 129.
Crane, J. C., Erickson, L. C. and Brannaman, B. L. (1965). *Proc. Am. Soc. hort. Sci.* **87,** 123.
Creveling, R. K. and Jennings, W. G. (1968). Unpublished data.
Creveling, R. K., Silverstein, M. S. and Jennings, W. G. (1968). *J. Fd Sci.* **33,** 284.
Cummings, G. A. (1965). *Proc. Am. Soc. hort. Sci.* **86,** 133.
Curl, A. L. (1959). *Fd Res.* **24,** 413.
Curwen, D., McArdle, F. J. and Ritter, C. M. (1966). *Proc. Am. Soc. hort. Sci.* **89,** 72.
Daghetta, A., Forti, G. and Monzini, A. (1956). *Annali Sper. agr.* **10,** 321.
Das, S. K., Markakis, P. and Bedford, C. L. (1965). *Mich. agric. Exp. Sta. Quart. Bull.* **48,** 81.
David, J. J., Luh, B. S. and Marsh, G. L. (1956). *Fd Res.* **21,** 184.
Davison, R. M. and Bukovac, M. J. (1968). *65th Ann. Meeting: Am. Soc. hort. Sci.*, Abstr. p. 91, No. 18.
De Candolle, A. (1959). "Origin of Cultivated Plants." Hafner Publishing Co., New York.
Dekazos, E. D. (1966). *J. Fd Sci.* **31,** 226.
De Moura, J. and Dostal, H. C. (1965). *J. agric. Fd Chem.* **13,** 433.
Deshpande, P. B. and Salunkhe, D. K. (1964). *Fd Tech.* **18,** 1195.
Do, J. Y., Olson, L. E. and Salunkhe, D. K. (1969). *J. Fd Sci.* **34,** 618.
El-Sayed, A. R. S. and Luh, B. S. (1965). *J. Fd Sci.* **30,** 1016.
Emberger, R. (1967). "Beitrage zur Biogenese pflanzlicher Aromastoffe." Doctoral thesis, Technische Hochschule Karlsruhe.
Embs, R. J. and Markakis, P. (1965). *J. Fd Sci.* **30,** 753.
Fideghelli, C., Cappellini, P. and Monastra, F. (1967). *Progr. Agric.*, Bologna **13,** 405.

Fox, J. E. (1968). "Molecular Control of Plant Growth." Dickenson Publishing Co., Belmont, California.
Galler, M. and Mackinney, G. (1965). *J. Fd Sci.* **30,** 393.
Gerhardt, F. and English, H. (1945). *Proc. Am. Soc. hort. Sci.* **46,** 205.
Gerhardt, F., English, H. and Smith, F. (1943). *Proc. Am. Soc. hort. Sci.* **42,** 247.
Guadagni, D. G. and Nimmo, C. C. (1953). *Fd Technol.* **7,** 59.
Guadagni, D. G., Sorber, D. G. and Wilbur, J. S. (1949). *Fd Technol.* **3,** 359.
Gustafson, F. G. (1939). *Am. J. Bot.* **26,** 189.
Haeseler, G. and Misselhorn, K. (1966). *Z. Lebensmitt. Untersuch.* 129, 71–75. Abstr., *J. Sci. Fd Agric.* (1966), **17,** ii–139.
Haller, M. H. (1952). *U.S. Dept. Agric. Biblio. Bull.* **21.**
Hartmann, C. (1957). *Fruits d'Outre Mer* **12,** 45.
Heinz, D. E. and Jennings, W. G. (1966). *J. Fd Sci.* **31,** 69.
Hsia, C., Claypool, L. L., Abernethy, J. L. and Esau, P. (1964). *J. Fd Sci.* **29,** 723.
Hsia, C. L., Luh, B. S. and Chichester, C. O. (1965). *J. Fd Sci.* **30,** 5.
Jackson, D. I. (1968). *Aust. J. biol. Sci.* **21,** 209.
Jackson, D. I. and Coombe, B. G. (1966). *Science, N.Y.* **154,** 277.
Jacob, F. C., Romani, R. J. and Sprock, C. M. (1965). *Transactions Am. Soc. agric. Eng.* **8,** 18, 19, 24.
Jaquin, P. (1965). *Ann. Technol. Agric.* **14,** 157; *Hort. Abstr.* (1966), **36,** 403.
Jennings, W. G. and Sevenants, M. R. (1964). *J. Fd Sci.* **29,** 796.
Joslyn, M. A. and Dittmar, H. F. K. (1967). *Mitt. Klosterneuburg* **17,** 227. *Hort. Abstr.* (1968), **38,** 2715.
Jouret, C. and Maugenet, J. (1967). *C. R. Acad. Agric., France*, **53,** 231. *Hort. Abstr.* (1968), **38,** 435.
Kamali, A R. and Childers, N. F. (1967). *Proc. Am. Soc. hort. Sci.* **90,** 33.
Kenworthy, A. L. and Mitchell, A. E. (1952). *Proc. Am. Soc. hort. Sci.* **60,** 91.
Kertesz, Z. I. (1933). *Tech. Bull. New York St. agric. Exp. Sta.* **219.**
Kidd, F. and West, C. (1930). *Proc. Roy. Soc.*, **B106,** 93.
Klein, E. and Rojahn, W. (1964). *Tetrahedron* **20,** 2025.
Kott, V. (1965). *Prun. Potravin.* 16, 336. Abstr. from *J. Sci. Fd Agric.* (1966), **17,** i–35.
Ku, L. L. and Romani, R. J. (1966). *Science, N.Y.* **154,** 408.
La Belle, R. L. (1968). *65th Ann. Meeting: Am. Soc. hort. Sci.*, Abstr. p. 99, No. 83.
Lavee, S. (1963). *Science, N.Y.* **142,** 583.
Letham, D. S. (1963). *New Zealand J. Bot.* **1,** 336.
Letham, D. S. (1964). *Coll. Int. Centr. Nat. Res. Sci.*, No. 123, Paris.
Letham, D. S. and Bollard, E. G. (1961). *Nature, Lond.* **191,** 1119.
Li, K. C. and Wagenknecht, A. C. (1956). *J. Am. chem. Soc.* **78,** 979.
Lilleland, O. (1930). *Proc. Am. Soc. hort. Sci.* **27,** 237.
Lilleland, O. (1932). *Proc. Am. Soc. hort. Sci.* **29,** 8.
Lim, L. (1963). M.Sc. Thesis, University of California, Davis.
Lim, L. and Romani, R. J. (1964). *J. Fd Sci.* **29,** 246.
Lipe, W. N. and Crane, J. C. (1966). *Science, N.Y.* **153,** 541.
Luh, B. S., Hsu, E. T. and Stachowicz, K. (1967). *J. Fd Sci.* **32,** 251.
Marshall, R. E. (1954). "Cherries and Cherry Products." Interscience Publishers, New York.
Maxie, E. C. and Claypool, L. L. (1957). *Proc. Am. Soc. hort. Sci.* **69,** 116.
Maxie, E. C. and Crane, J. C. (1956). *Proc. Am. Soc. hort. Sci.* **68,** 113.
Maxie, E. C., Catlin, P. B. and Hartmann, H. T. (1960). *Proc. Am. Soc. hort. Sci.* **75,** 275.

Maxie, E. C., Bradley, M. V. and Robinson, B. J. (1962). *Proc. Am. Soc. hort. Sci.* **81,** 137.

Micke, W. C., Mitchell, F. G. and Maxie, E. C. (1965). *Calif. Agric.* **19,** 12.

Mitchell, F. G. and Micke, W. C. (1966). *West. Fruit Grow.* **20,** 25.

Moore, M. D. and Scott, L. E. (1966). *In* "Proceedings of the XVII International Horticulture Congress" (R. E. Marshall, ed.), Vol. I, Abstr. 376. Michigan State University, Michigan.

Motawi, K. El-Din H. (1967). *Diss. Abstr.*, Sect. B. **27,** 2403B.

Nelson, E. K. and Curl, A. L. (1939). *J. Am. chem. Soc.* **61,** 667.

Nitsch, J. P. (1953). *A. Rev. Pl. Physiol.* **4,** 199.

Norris, K. H. (1956). *Ag. Eng.* 30, 640.

Olliver, M. (1967). *In* "The Vitamins." Vol. I. (Sebrell, W. H. and Harris, R. S., eds). Academic Press, London.

Onslow, M. W. (1920). *Biochem. J.* **14,** 546.

Osborne, D. and Hallaway, M. (1964). *New Phytol.* **63,** 334.

Peng, C. V. and Markakis, P. (1963). *Nature, Lond.* **199,** 597.

Pentzer, W. T. and Allen, F. W. (1944). *Proc. Am. Soc. hort. Sci.* **44,** 148.

Pentzer, W. T. and Heinze, P. H. (1954). *A. Rev. Pl. Physiol.* **5,** 205.

Phinney, B. O. and West, C. A. (1960). *A. Rev. Pl. Physiol.* **11,** 411.

Pienaar, W. J., Bartel, E. E., Gurgen, K. H. and Schutte, C. E. (1965). *Tech. Commun. Dept. Agric. Tech. Serv.*, South Africa, No. 43.

Pollack, R. L., Ricciuti, C., Woodward, C. F. and Hills, C. H. (1958a). *Fd Technol.* **12,** 102.

Pollack, R. L., Whittenberger, R. T. and Hills, C. H. (1958b). *Fd Technol.* **12,** 106.

Pollack, R. L., Hoban, N. and Hills, C. H. (1961). *Proc. Am. Soc. hort. Sci.* **78,** 86.

Porritt, S. W. and Mason, J. L. (1965). *Proc. Am. Soc. hort. Sci.* **87,** 128.

Postlmayr, H. L., Luh, B. S. and Leonard, S. J. (1956). *Fd Technol.* **10,** 618.

Powell, L. E. and Pratt, C. (1964). *Nature, Lond.* **204,** 602.

Powell, L. E. and Pratt, C. (1966). *J. hort. Sci.* **41,** 331.

Power, F. B. and Chesnut, V. K. (1921). *J. Am. chem. Soc.* **43,** 1725.

Power, F. B. and Chesnut, V. K. (1922). *J. Am. chem. Soc.* **44,** 2966.

Proebsting, E. L., Jr. and Mills, H. H. (1966). *Proc. Am. Soc. hort. Sci.* **89,** 135,

Reeve, R. M. (1959a). *Am. J. Bot.* **46,** 210.

Reeve, R. M. (1959b). *Am. J. Bot.* **46,** 241.

Reeve, R. M. (1959c). *Am. J. Bot.* **46,** 645.

Reibeiz, C. and Crane, J. C. (1961). *Proc. Am. Soc. hort. Sci.* **78,** 69.

Reyes, P. and Luh, B. S. (1966). *Fd Technol.* **14,** 570.

Richmond, A. and Biale, J. B. (1967). *Biochem. biophys. Acta.* **138,** 625.

Romani, R. J. and Van Kooy, J. (1962). *Second Intern. Congr. Rad. Res.*, August, 1962, Harrogate, Yorkshire, England, Abstr. p. 138.

Romani, R. J., Van Kooy, J. and Robinson, B. J. (1961). *Fd Irrad.* **2,** A11.

Romani, R. J., Jacob, F. C. and Sprock, C. M. (1962). *Proc. Am. Soc. hort. Sci.* **80,** 220.

Romani, R. J., Breidenbach, R. W. and Van Kooy, J. (1965). *Pl. Physiol., Lancaster* **40,** 561.

Romani, R. J., Jacob, F. C., Sprock, C. M. and Mitchell, F. G. (1963a). *In* "Proceedings of the XVI International Horticulture Congress" (A. Lecrenier and Ir. P. Goeseels, eds), Vol. 3, pp. 337–342. J. Duculot, S.A., Gembloux, Belgique.

Romani, R. J., Jacob, F. C. Mitchell, F. G. and Sprock, C. M. (1963b). *Proc. Am. Soc. hort. Sci.* **83,** 226.
Romani, R. J., Yu, I. K., Ku, L. L., Fisher, L. K. and Dehgan, N. (1968). *Pl. Physiol., Lancaster* **43,** 1089.
Rood, P. (1957). *Proc. Am. Soc. hort. Sci.* **70,** 104.
Roux, E. R. (1940). *Ann. Bot.* (N.S.) **4,** 317.
Ryugo, K. (1962). *Proc. Am. Soc. hort. Sci.* **81,** 180.
Ryugo, K. (1964a). *Proc. Am. Soc. hort. Sci.* **84,** 110.
Ryugo, K. (1964b). *Proc. Am. Soc. hort. Sci.* **85,** 154.
Ryugo, K. and Davis, L. (1958). *Proc. Am. Soc. hort. Sci.* **72,** 106.
Salunkhe, D. K. (1968). Personal communication.
Salunkhe, D. K., Cooper, G. M., Dhaliwal, A. S., Boe, A. A. and Rivers, A. L. (1962). *Fd Technol.* **16,** 119.
Salunkhe, D. K., Deshpande, P. B. and Do, J. Y. (1968). *J. hort. Sci.* **43,** 235.
Schaller, D. R. and Elbe, J. H. von (1968). *J. Fd Sci.* **33,** 442.
Scheuerbrandt, G., Goldfine, H. Baronowski, P. E. and Block, K. (1961). *J. biol. Chem.* **236,** PC70.
Schneider, G. W., Jones, I. D. and McClung, A. C. (1958). *Proc. Am. Soc. hort. Sci.* **71,** 110.
Sevenants, M. R. and Jennings, W. G. (1966). *J. Fd Sci.* **31,** 81.
Shewfelt, A. L. (1966). *J. Fd Sci.* **30,** 573.
Smith, W. H. (1940). *J. Pomol.* **18,** 74.
Smith, W. H. (1947). *J. Pomol.* **23,** 92.
Smith, W. H. (1967). *J. hort. Sci.* **42,** 223.
Soler Martínez, A. Sabater García, F. and Lozano, J. A. (1965). *Rev. Agroquím. Technol. Aliment.* **5,** 353. Abstr., *J. Sci. Fd Agric.* (1965), **17,** i–249.
Somogyi, L. and Romani, R. J. (1964). *J. Fd Sci.* **29,** 366.
Souty, M., Perret, A. and André, P. (1967). *Ann. technol. Agric.* **16,** 55. *Hort. Abstr.* (1967), **37,** 6384.
Spanyar, P., Kevei, E. and Blazovich, M. (1964). *Industr. Clim. Agric.* **81,** 1063.
Stahly, E. A. and Thompson, A. H. (1959). *Maryland agric. Exp. Sta. Bull.* A–104.
Sterling, C. and Kalb, A. J. (1959). *Bot. Gaz.* **121,** 111.
Stevens, K. L., Bomber, J., McFadden, W. H. (1966). *J. agric. Fd Chem.* **14,** 249.
Storer, F. H. (1898). *Bull. of the Busseg. Inst.* No. 36B.
Tang, C. S. and Jennings, W. G. (1967). *J. agric. Fd Chem.* **15,** 24.
Tang, C. S. and Jennings, W. G. (1968). *J. agric. Fd Chem.* **16,** 252.
Taylor, O. C. and Mitchell, A. E. (1953). *Proc. Am. Soc. hort. Sci.* **62,** 267.
Tressl, R. (1967). "Über die Bildung und enzymatische Veränderung einiger Frucharomastoffe." Doctoral Thesis, Technische Hochschule Karlsruhe.
Tukey, H. B. (1934). *Proc. Am. Soc. hort. Sci.* **30,** 209.
Tukey, H. B. (1936). *Bot. Gaz.* **98,** 1.
Tukey, H. B. and Young, J. O. (1939). *Bot. Gaz.* **100,** 723.
Ulrich, R. (1946). *Bull. Soc. Bot. France* **93,** 248.
Unrath, C. R. and Kenworthy, A. L. (1968). *65th Ann. Meeting: Proc. Am. Soc. hort. Sci.*, Abstr. p. 125, No. 258.
Uota, M. (1955). *Proc. Am. Soc. hort. Sci.* **65,** 231.
Van Blaricom, L. O. and Senn, T. L. (1967). *Proc. Am. Soc. hort. Sci.* **90,** 541.
Van Buren, J. P. (1967). *J. Fd Sci.* **32,** 435.

Van der Plank, J. E. and Davies, R. (1937). *J. Pomol.* **15,** 226.
Winton, A. L. and Winton, K. B. (1932a) "The Structure and Composition of Foods", Vol. I, pp. 587–597. John Wiley and Sons, New York.
Winton, A. L. and Winton, K. B. (1932b) "The Structure and Composition of Foods", Vol. II, pp. 635–636. John Wiley and Sons, New York.
Wolf, J. (1958). *Planta* **51,** 547.
Yang, H. Y., Steele, W. F. and Graham, D. J. (1960). *Fd Technol.* **14,** 644.

Chapter 13

The Tomato

G. E. HOBSON AND J. N. DAVIES

A.R.C. Glasshouse Crops Research Institute, Littlehampton, Sussex, England

I Introduction and General Survey 437
II Fruit Composition and Chemical Changes during Growth and Maturation 440
A. Carbohydrates 440
B. Organic Acids 443
C. Amino Acids 447
D. Proteins 449
E. Volatile Components 450
F. Steroids 452
G. Pigments other than Flavonoids 453
H. Flavonoids 457
I. Other Phenolic Compounds 457
J. Cell Wall Constituents 459
K. Lipids and Long-Chain Components 459
L. Mineral Constituents 460
III Aspects of Metabolism during Maturation and Senescence 462
A. The Pre-climacteric Fruit 464
B. The Climacteric Fruit 465
C. The Senescent Fruit 466
D. The "Blotchy" Fruit 467
IV Factors Affecting Tomato Fruit Ripening 468
A. Effects of Temperature and Controlled Atmospheres during Storage 468
B. The Function of Ethylene 469
C. Ripening Disorders 471
D. Sprays and Dips 473
V The Control of Tomato Fruit Ripening—A Summary 473
References 475

I. INTRODUCTION AND GENERAL SURVEY

The tomato is a member of the potato family Solanaceae, belonging to the relatively small genus *Lycopersicon*, detailed accounts of which are available (Muller, 1940; Luckwill, 1943). It is indigenous to the lower western slopes of the Andes in South America and is said to have been brought to Europe by Columbus in 1498. It was recorded as growing in Italy in 1554 and in England in 1576. Almost all the commercially important varieties grown in the world belong to the species *L. esculentum* (Mill.). A survey carried out in 1958 showed that 70 varieties of tomato were being grown commercially in Great Britain at that time.

The fruit is a fleshy berry, or, in botanical terms, a swollen ovule. The body of the fruit, developed from the ovary wall which surrounds and encloses the seed, is known as the pericarp and consists of outer, radial and inner walls (Fig. 1). The locular cavities occur as gaps in the pericarp and contain the seeds embedded in a jelly-like parenchymatous tissue originating from the placenta. The number of locules in normal fruit varies from two upwards, and is more or less characteristic for each variety. Prior to pollination, as well as during a relatively short period after anthesis, the growth of the fruit is mainly by cell division (Spurr, 1959), after which cell enlargement is responsible for the growth of the fruit. During the latter phase, vacuoles appear in the cells and differentiation in composition becomes evident. Growth-promoting and growth-inhibiting substances moving into and out of the fruit no doubt play a major part in the maturation process, while respiratory and fermentative mechanisms assume a dominant role in ripening. The habit of growth of most varieties under glass is indeterminate, but for outdoor tomato production determinate varieties are largely used.

In this chapter, the meaning of the terms development, maturation, ripening and senescence are similar to those used by Gortner *et al.* (1967) in connection with the pineapple (see Chapter 9). In the tomato, a climacteric fruit (see Volume 1, Chapter 17), ripening, which is immediately preceded by the onset of the respiration climacteric, is the terminal period of maturation. It may occur before or after harvest with little loss in edible quality provided that the mature green stage (pre-climacteric minimum respiration rate) is reached before harvesting.

Both Biale (1960) and Maxie and Abdel-Kader (1966) discussed the climacteric respiration rise in fruits in general terms and stressed that the overall climacteric phenomenon should be regarded as a composite coordinated event in which the biochemical and physiological changes associated with ripening occur simultaneously. This subject is considered in detail in Volume 1, Chapter 17.

The economic importance of the tomato is considerable and world production of the fruit at 24 million metric tons was only surpassed by grapes at 51 million, citrus fruit at 31 million and pome fruits at 26 million (F.A.O., 1968). The commercial value of the crop coupled with the many virtues of the tomato as an experimental plant have together led to a voluminous literature. The present account attempts to summarize the biochemical

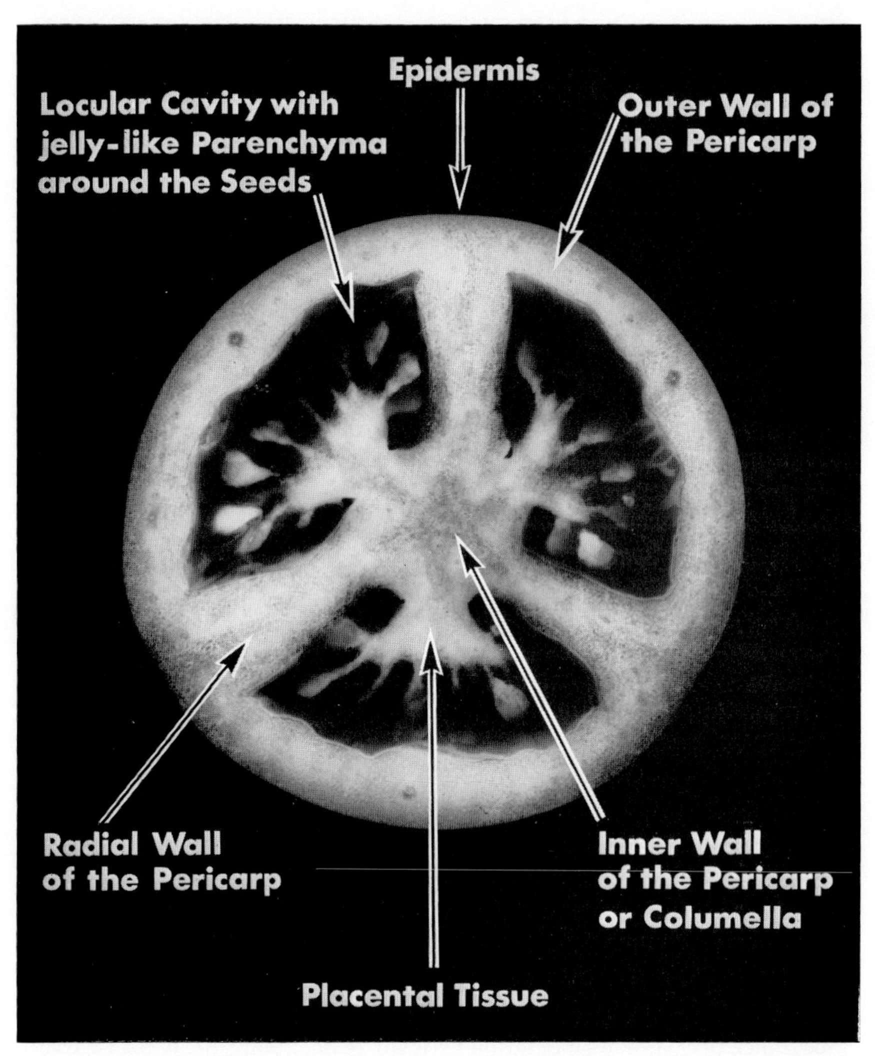

FIG. 1. Transverse section of mature tomato fruit. (Photograph by courtesy of Glasshouse Crops Research Institute.)

status of tomato fruit in relation to the more recent work on its physiology, nutrition and composition.

The increasing adoption of precision growing techniques in the horticultural industry, particularly in the closed and controllable environment of the glasshouse, is necessitating more detailed studies of the tomato with a view to obtaining a better understanding of the factors controlling fruit ripening, quality and shelf life. Interest is increasing in the major and minor components of the fruit which influence taste and flavour, particularly with regard to the effects of variety, nutrition, post-harvest storage and processing. In addition, the mechanical harvesting of outdoor crops is beset by its own problems such as the damage inflicted on the fruit and contamination with soil.

II. FRUIT COMPOSITION AND CHEMICAL CHANGES DURING GROWTH AND MATURATION

The moisture content of immature green tomatoes increases from about 91% to 93% as the fruit develops. Good quality ripe fruits have an average moisture content of 94–94·5%, the range being 92·5–95%.

A. Carbohydrates

1. *Sugars*

The soluble carbohydrates of the fruits of commercial varieties of tomato are almost entirely reducing sugars (Hamner and Maynard, 1942; Winsor *et al.*, 1962a; Lambeth *et al.*, 1964). Since the sugars constitute 1·5–4·5% of the fresh weight, equivalent to some 65% of the total soluble solids (Winsor, 1966), they have an important effect on the taste of the ripe fruit (see Tables I and III, pp. 444 and 446). The free sugars consist of glucose and fructose present in approximately equal amounts with a slight preponderance of the latter; sucrose, when found at all, rarely exceeds 0·1% of the fresh weight (Davies, 1958; Freeman, 1960; Simandle *et al.*, 1966; de Bruyn *et al.*, 1968). Traces of other sugars have been reported, notably a "ketoheptose" (Williams and Bevenue, 1954) and raffinose (Airan and Barnabas, 1953). In contrast, fruit of some *Lycopersicon* species, widely used as sources of genetic material, contain large quantities of sucrose and relatively small amounts of glucose and fructose (Davies, 1966a).

The sugar content increases progressively throughout maturation and ripening (for surveys of the earlier literature see Winsor *et al.*, 1962a; Lambeth *et al.*, 1964), a particularly pronounced rise occurring with the appearance of yellow pigmentation (Winsor *et al.*, 1962a, b). A post-harvest decline in the sugar content of ripe fruit during storage at room temperature has been observed (Winsor *et al.*, 1962a). Shading has an adverse effect on the sugar content (McCollum, 1946; Davies and Winsor, 1959), as does the removal of leaves (Winsor, 1966).

Other changes in sugar content, reflecting differences in the intensity and duration of light, are typified by seasonal trends. In America, Forshey and Alban (1954) reported a decrease in sugar content with each successive harvest during an autumn crop and an increase during the spring crop. Yamaguchi *et al.* (1960) also found a decrease during September and October, while Lambeth *et al.* (1964) showed an increase during July with little change during August. In England, spectacular increases in the sugar content of early crops have been observed between April and early June (Winsor and Davies, 1966). Subsequently a maximum is reached in late July or early August, followed by a decline towards the end of the season (Winsor, 1966).

In general, the levels of the major nutrients have little effect on the sugar content of tomato fruit, but high nitrogen fertilization has an adverse effect (Neubert, 1959; Davies and Winsor, 1967).

2. *Starch*

While immature tomatoes contain considerable amounts of starch, it is only a minor constituent of ripe fruit. Thus Sando (1920) found that the starch content decreased progressively with age from just over 1% of the fresh weight in immature fruit 14 days old to between 0·10 and 0·15% in red-ripe fruit, depending on the ripening conditions. However, the analytical method used would include both starch and hemicellulose. Rosa (1925, 1926) found smaller quantities of starch (0·012–0·052% fr. wt.) in ripe fruit, and similar values were reported by Saywell and Cruess (1932). Beadle (1937) was unable to detect starch in young fruit up to about 10 days from fertilization. He found that the starch content subsequently increased to a maximum at 8 weeks old, shortly before the fruit began to colour, and then disappeared rapidly as ripening proceeded. Davies and Cocking (1965) substantiated these results. Further studies by Hobson (1967b) showed that the starch content of green fruit immediately prior to ripening was about 0·07%, falling to half this value at ripeness. Fruit showing blotchy ripening (see p. 471) had an enhanced starch content in the affected locule walls, and bruising the fruit also appeared to delay starch breakdown. Yu *et al.* (1967) found a maximum of 1·22% starch in immature fruit of one variety and 0·7% in another, while the amylopectin–amylose ratio at first decreased as the fruit began to swell but then increased during the ripening period.

3. *Ascorbic acid*

Results indicating the importance of tomato fruit as a valuable source of vitamin C have been comprehensively reviewed (Maclinn and Fellers, 1938; Hamner and Maynard, 1942; Aberg, 1946; Murneek *et al.*, 1954). The discovery, chemistry and distribution of the vitamin have been surveyed by Bicknell and Prescott (1946), and its possible functions summarized by Mapson (1958).

Although a typical average value of 25 mg/100 g fr. wt. has been given by Olliver, (1967) published values vary from 16 to 25 mg/100 g fr. wt. for English (Bicknell and Prescott, 1946), from 18 to 36 mg for Canadian (McHenry and Graham, 1935), and from 5 to 60 mg for American varieties

(Maclinn *et al.*, 1936; Maclinn and Fellers, 1938). Baudisch (1963) found 16–70 mg/100 g in German mutants induced by X-rays.

Much of this wide variation in the ascorbic acid content is probably attributable to differences in light intensity during growth. Thus, fruits grown outdoors were found to contain more ascorbic acid than fruits grown under glass (Currence, 1939; Brown and Moser, 1941; Crane and Zilva, 1949; Daskaloff and Ognjanowa, 1962) while those grown in full sun contained significantly more than those grown in shade (McCollum, 1946; Crane and Zilva, 1949; Murneek *et al.*, 1954). Attempts to show a quantitative relation between sunshine hours and total ascorbic acid content were, however, unsuccessful (Pollard *et al.*, 1948; Crane and Zilva, 1949). The amounts of ascorbic acid varied from side to side and from top to bottom of individual fruit, depending on the relative exposure to sunlight (McCollum, 1946). Other distribution studies have shown a fall in concentration of the vitamin as the distance from the skin increases (McCollum, 1946; Ward, 1963). Sowinska (1966) confirmed these observations, and found, in addition, that the jelly surrounding the seeds contained the highest concentration of ascorbic acid.

A rise in ascorbic acid content with ascending truss position, due possibly to increasing light intensity, has been reported (Hassan and McCollum, 1954; Fryer *et al.*, 1954). Frazier *et al.* (1954), however, could find no consistent increase in the ascorbic acid content of fruit from plants receiving supplementary lighting.

Data concerning the ascorbic acid content during the development and ripening of tomato fruit are inconsistent. Some earlier investigators (e.g., Maclinn and Fellers, 1938; Wokes and Organ, 1943; Kaski *et al.*, 1944) reported little change, while more recent work has indicated an increase in ascorbic acid concentrations during maturation (Georgiev and Balzer, 1962), with either a continuing rise (Fryer *et al.*, 1954; Dalal *et al.*, 1965) or a slight fall (Dalal *et al.*, 1966) during the final stages of ripening. Inconclusive results have been obtained in attempts to relate the size of the fruit and the ascorbic acid content (Hassan and McCollum, 1954).

Any of the foregoing factors may outweigh varietal differences (Maclinn *et al.*, 1936; Reynard and Kanapaux, 1942; Pollard *et al.*, 1948) but faster ripening varieties contain more ascorbic acid than those which take longer to ripen (Clutter and Miller, 1961).

Hester (1940, 1941) found that the addition of potassium and manganese to certain soils resulted in both increased yield and in ascorbic acid content, and Murneek *et al.* (1954) showed that additional soil nitrogen depressed the ascorbic acid in the tomato, particularly in good light conditions. The influence of mineral nutrition was considered by Hamner *et al.* (1942) to be of minor importance compared with climatic factors.

This wealth of often contradictory data concerning factors influencing the ascorbic acid content of the tomato serves to stress the need for the utmost care in the control of experimental and analytical conditions if reliable results are to be obtained.

B. Organic Acids

The acids of tomato fruit have been the subject of considerable investigation. Not only are they important as major taste components, but total acidity plays an important part in the satisfactory processing of tomato products (Lambeth *et al.*, 1964).

It is generally agreed that the predominant acid of ripe tomato fruit is citric, with malic the next most abundant (Carangal *et al.*, 1954; Rice and Pederson, 1954; Bradley, 1960; Davies, 1964; Sakiyama, 1966a). Other acids reported include formic and acetic (Carangal *et al.*, 1954; Bradley, 1960; Villarreal *et al.*, 1960), and *trans*-aconitic (Bulen *et al.*, 1952; Carangal *et al.*, 1954; Anderson, 1957) with traces of lactic and fumaric acids in processed tomato products (Rice and Pederson, 1954; Bradley, 1960). Hamdy and Gould (1962) found three alpha-oxo acids (α-oxoglutaric, pyruvic and dihydroxytartaric) in small amounts (0·0002–0·002%) in canned tomato juice, while galacturonic acid has been shown to occur in ripe tomatoes (McClendon *et al.*, 1959; Bradley, 1960; Simandle *et al.*, 1966).

Pyrrolidonecarboxylic acid (PCA) is not a normal constituent of the fresh fruit but is found in processed tomato products (Rice and Pederson, 1954; Mahdi *et al.*, 1959; Villarreal *et al.*, 1960; Bradley, 1960) and in alcoholic extracts of tomato fruit (Davies, 1964). PCA may be formed from either glutamic acid or glutamine, but under normal conditions of processing or alcohol extraction it is formed entirely from glutamine (Mahdi *et al.*, 1959; Davies, 1964).

According to Rosa (1925) maximum acidity during ripening coincides with the first appearance of pink colour, and this has been largely substantiated by much of the subsequent work (see Winsor *et al.*, 1962a). During the ripening of whole fruit from mature green to red, acidity increases initially to a maximum value coinciding approximately, but not always precisely, with the first appearance of yellow pigment, followed by a progressive decrease in acidity (Winsor *et al.*, 1962a). The acidity of the locular contents decreases from the green or green-yellow stage onwards, but consistent trends in the acidity of the fruit walls after the green-yellow stage are difficult to establish (Winsor *et al.*, 1962b). Various reports have agreed that the titratable acidity of the outer fruit walls is relatively low compared with that of the locular contents (Bohart, 1940; Anderson, 1957; Winsor *et al.*, 1962b; Sakiyama, 1966a). Changes in titratable acidity have been attributed either to changes in citric acid alone (Bradley, 1962) or to changes in both citric and malic acid (Davies, 1964).

A highly significant negative correlation between fruit weight and titratable acidity has been established for some English varieties of tomato (Davies and Winsor, 1969). This is consistent with the observations that small tomatoes have a lower proportion of fruit walls and placentae than large fruit (Shafshak and Winsor, 1964), and that the fruit walls are lower in acidity than the locular contents (see above). Thompson *et al.* (1964), however, concluded that differences in acidity within large and small fruited breeding lines could not be attributed solely to differences in locular content.

Malic acid concentration falls as the tomato ripens (Mollard, 1953; Carangal *et al.*, 1954; Anderson, 1957; Davies, 1966b; Belluci and Aldini, 1967), while citric acid increases up to the green-yellow stage of ripeness and then either falls (Carangal *et al.*, 1954; Villarreal *et al.*, 1960; Belluci and Aldini, 1967) or shows no subsequent significant change (Davies, 1966b). On the other hand, Dalal *et al.* (1965, 1966) reported that both citric and malic acids increased steadily throughout maturation and ripening, while the titratable acidity rose to a peak as the fruit changed colour and then declined somewhat.

TABLE I. Average composition of seven varieties of tomato fruit grown as a main crop (1958–61)[a]

Variety	Titratable acidity (meq/100 ml)	Reducing sugars (g/100 ml)	Sugar : acid ratio	Total solids (g/100 ml)
Potentate	7·71	3·19	0·42	4·82
Radio	7·87	3·07	0·40	4·80
E.S.5	7·99	3·21	0·41	4·83
Moneymaker	8·09	3·13	0·39	4·80
Delicious	9·13	2·91	0·33	4·75
Ailsa Craig	9·34	3·28	0·36	5·13
L.M.R.1	10·38	3·04	0·30	5·02
Significance (*P*)	0·001	—	0·001	0·05
L.S.D. ($P = 0{\cdot}05$)	0·26	—	0·02	0·11

[a] From Davies and Winsor (1969).

Tomato varieties can vary markedly in acidity (Lambeth *et al.*, 1964; Lambeth *et al.*, 1966; Simandle *et al.*, 1966), and English varieties grown under comparable conditions differ far more in acidity than in sugar content (Davies and Winsor, 1969) (see Table I). Koch (1960) found that early varieties had a high malic : citric acid ratio whereas later varieties contained more citric acid. Davies (1965b) concluded that the malic : citric acid ratio was a varietal attribute. His results are given in Table II. Little information is available regarding the heritable basis for variation in acidity. Walkof and Hyde (1963) reported that acidity was under monofactorial control, but Thompson *et al.* (1964) in breeding studies indicated that acidity levels are under polygenic control (see also Lower and Thompson, 1967).

TABLE II. Malic and citric acids in the fruit of tomato varieties[a]

Variety	Acid in fresh tissue (μ-*equiv*/10 g)		Percentage of total acidity		Malic : citric acid ratio
	Malic	Citric	Malic	Citric	
Moneymaker	239	458	26	51	0·52
Radio	313	502	28	45	0·62
Ailsa Craig	288	612	24	51	0·47
Potentate	254	463	26	48	0·55
E.S.5	304	526	26	45	0·58
Ware Cross	313	696	24	54	0·45
L.M.R.1	192	736	16	60	0·26
Delicious	146	649	13	60	0·23
Immuna	90	878	7	66	0·11
Lycopersicon pimpinellifolium	158	978	10	64	0·16
Neverripe	124	1367	6	66	0·09

[a] From Davies (1965b).

Successive harvest dates are not always reflected in consistent acidity trends (Saywell and Cruess, 1932; Massey and Winsor, 1957; Anderson, 1957), but Lee and Sayre (1946) found that acidity was highest at the beginning of the season, declined as the season progressed and rose again at the end of the season, while Forshey and Alban (1954) reported a decrease in acidity with each successive harvest. Table III shows changes in sugar and acid content of tomatoes during ripening.

The relationship between potassium and acidity in tomato fruit is very close, and highly significant positive correlations have been reported between potassium content and the total, titratable and combined acidities (Davies, 1964; Sakiyama, 1966b; Davies and Winsor, 1967). The juices of the tomato constitute a weak acid/strong base buffer system in which the anions are mainly citrate and malate and the cations are mainly potassium. Bradley (1962) stated that potassium accounts for 86% of the total cations in the fruit while Davies (1964) reported an average of 90%, with values as low as 73% where high nitrogen and phosphate fertilization were combined with low potassium (see also Section IIL). Some 50% of the total acidity is neutralized in the normal fruit, and wide variations in the free acid and potassium contents can be found with only a minor effect on pH. Any factor

TABLE III. Composition of juices from whole fruit and locules (Variety Potentate) at six stages of ripeness[a]

	No. of samples	Stage of ripeness 1 Green	2 Green–yellow	3 Yellow–orange	4 Orange	5 Orange–red	6 Red	Significance, P^b	L.S.D. ($P = 0{\cdot}05$)
Whole fruit									
Titratable acidity (meq/100 ml)	3	7·47	9·50	9·19	8·36	8·27	7·14	0·01	0·97
Reducing sugars (g/100 ml)	3	2·93	3·31	3·37	3·43	3·57	3·60	0·01	0·27
Total solids (g/100 ml)	3	4·46	5·00	5·08	5·00	5·18	5·10	0·01	0·33
Locules									
Titratable acidity (meq/100 ml)	3	14·5	17·1	15·8	14·2	13·0	11·2	0·01	2·4
Reducing sugars (g/100 ml)	3	2·13	2·55	2·59	2·75	2·95	3·24	0·001	0·23
Total solids (g/100 ml)	2	4·54	4·99	5·06	4·91	4·93	5·13	—	—

[a] From Winsor *et al.* (1962a).
[b] Significance of *F*-test.

which increases the potassium content of the fruit will produce a corresponding increase in organic acids in order to maintain a constant pH, which normally ranges between 4·0 and 4·5 (Hamner and Maynard, 1942). A positive logarithmic relation between tomato acidity and soil potassium levels has been reported (Davies and Winsor, 1967), and there is also a positive correlation between leaf potassium and fruit acidity (Winsor and Davies, unpublished data).

The acidity of the tomato is also increased by nitrogen and decreased by phosphate (Davies and Winsor, 1967); this is true also for malic and citric acid concentrations (Davies, 1964). The form in which nitrogen is applied has a marked influence on the acidity. With the ammonium form the fruit is less acid than when nitrate-nitrogen only is supplied (Carangal *et al.*, 1954; Sakiyama, 1967). The combined effect of high nitrogen and high potassium is particularly favourable not only to the acidity of the tomato (Davies, 1964; Davies and Winsor, 1967) but also to pectic enzyme activities (Hobson, 1963b), high yields of fruit (Winsor *et al.*, 1967) and reduced incidence of ripening disorders (Winsor and Long, 1967). Calcium and magnesium have been reported to have little effect on tomato acidity (Bradley, 1962; Davies and Winsor, 1967), but Sakiyama (1966b) found that high calcium decreased acidity when combined with high potassium.

C. Amino Acids†

Reports on the amino acid composition of tomato fruit differ considerably and the limited quantitative data available are listed in Table IV. In addition, small amounts of indole derivatives, notably tryptamine, 5-hydroxytryptamine and tyramine, have been detected (Udenfriend *et al.*, 1959; West, 1959a, b). It is generally agreed that glutamic acid is the dominant amino acid of the ripe fruit. Much of the diversity in the various reports is probably attributable to varietal and nutritional differences.

During ripening the total free amino acid content remains relatively constant, but glutamic acid concentrations rise sharply and aspartic acid increases to a lesser extent (Freeman and Woodbridge, 1960; Yu *et al.*, 1967). Davies (1966b) found an increase in glutamic acid during ripening from 28 (green) to 272 (red) mg/100 g fr. wt. Larger quantities of both aspartic and glutamic acids were found in fruit ripened in store rather than on the plant (Freeman and Woodbridge, 1960). Apart from serine and threonine, which reach maxima before the fruit is fully ripe, the other amino acids all tend to decrease during ripening, and it has been suggested that the utilization of free amino acids for protein synthesis could account for this decrease (Freeman and Woodbridge, 1960; Yu *et al.*, 1967).

† See also Volume 1, Chapter 5.

TABLE IV. Amino acid content of ripe tomato fruit

Constituent	Concentration range in mg/100 g fr. wt.	Source	Reference
Alanine	3–7	Walls	Carangal *et al.* (1954)
	3·7	Whole fruit	Freeman and Woodbridge (1960)
	3·3	Whole fruit	Yu *et al.* (1967)
β-Alanine	0·6	Whole fruit	Freeman and Woodbridge (1960)
γ-Aminobutyric acid	22	Whole fruit	Freeman and Woodbridge (1960)
	48	Whole fruit	Burroughs (1960)
Arginine	tr.–9	Walls	Carangal *et al.* (1954)
	58	Whole fruit	Freeman and Woodbridge (1960)
	3·9	Whole fruit	Yu *et al.* (1967)
Asparagine	30	Whole fruit	Freeman and Woodbridge (1960)
Aspartic acid	2–32	Walls	Carangal *et al.* (1954)
	26	Whole fruit	Freeman and Woodbridge (1960)
	31	Whole fruit	Davies (1966b)
	50	Whole fruit	Yu *et al.* (1967)
Glutamic acid	9–125	Walls	Carangal *et al.* (1954)
	10–166	Pulp	Carangal *et al.* (1954)
	77	Whole fruit	Freeman and Woodbridge (1960)
	272	Whole fruit	Davies (1966b)
	240[a]	Whole fruit	Yu *et al.* (1967)
Glycine	2·3	Whole fruit	Yu *et al.* (1967)
Histidine	3–44	Walls	Carangal *et al.* (1954)
	3·3	Whole fruit	Yu *et al.* (1967)
Leucine(s)	tr.–9	Walls	Carangal *et al.* (1954)
	8	Whole fruit	Freeman and Woodbridge (1960)
	6·3	Whole fruit	Yu *et al.* (1967)
Lysine	2–16	Walls	Carangal *et al.* (1954)
	4·2	Whole fruit	Yu *et al.* (1967)
Methionine	1·6	Whole fruit	Yu *et al.* (1967)
S-Methyl methionine	1·6–3·5	Whole fruit	Wong and Carson (1966)
Phenylalanine	7·2	Whole fruit	Yu *et al.* (1967)
Serine	7–21	Walls	Carangal *et al.* (1954)
	13	Whole fruit	Freeman and Woodbridge (1960)
	5·7	Whole fruit	Yu *et al.* (1967)
Threonine	7	Whole fruit	Freeman and Woodbridge (1960)
	6·5	Whole fruit	Yu *et al.* (1967)
Tryptophane	1–56	Walls	Carangal *et al.* (1954)
	0·1	Pulp	West (1959b)
Tyrosine	3·8	Whole fruit	Yu *et al.* (1967)
Valine	tr.–7	Walls	Carangal *et al.* (1954)
	2	Whole fruit	Freeman and Woodbridge (1960)
	1·9	Whole fruit	Yu *et al.* (1967)

[a] Including proline.

Higher concentrations of the free amino acids have been observed in the gelatinous pulp of the tomato than in the fruit walls (Carangal *et al.*, 1954; Davies, 1966b). The total amino acid content has been found to decrease with ascending truss position (Freeman and Woodbridge, 1960).

With the exception of mineral nutrition, little or no information appears to be available for the effect of environmental factors on the amino acid composition of the tomato fruit. Carangal *et al.* (1954) showed that increased nutritional levels of ammonium-nitrogen were reflected in higher concentrations of tryptophan and glutamic acid. Saravacos *et al.* (1958a) found that applications of nitrogen brought about increases in the content of glutamic and aspartic acids, glutamine, asparagine, alanine and valine in canned tomato juice but had little effect on serine, threonine and leucine content. In extracts of the whole fruit, Davies (1964) showed that glutamic and aspartic acids were increased by applications to the plant of nitrogen and decreased by phosphate. In addition, aspartic acid was significantly decreased by high potash levels in the soil. Palmieri (1965) found a positive linear relation between the titratable acidity and the content of dicarboxylic amino acids for ten tomato varieties.

Little attention has so far been given to the possible significance of the amino acids in tomato flavour. High nitrogen fertilization has, however, been reported to have an adverse effect on the taste of canned tomato juice (Saravacos *et al.*, 1958b), while Hamdy and Gould (1962) suggested that the ratio of α-amino to citric acid could be used, *inter alia*, to assess the flavour of processed tomatoes. The ratio of dicarboxylic amino acids to soluble carbohydrates has also been shown to be specific for eleven varieties of tomato (Taverna, 1965). Subsequent work by Yu *et al.* (1968a, b) has suggested that certain amino acids may serve as precursors for the synthesis of volatile aroma components in the tomato.

D. Proteins

The total nitrogen content of tomato fruit during ripening has been variously reported to decrease (Sando, 1920; Neubert, 1959), increase (Rosa, 1925), or show no change (Vendilo, 1964). According to Yu *et al.* (1967) the total nitrogen content fell during development from an initially high value in small green fruit to a minimum value near incipient ripeness, then rose again to a peak near the red stage and finally decreased towards over-ripeness. Both Rosa (1925) and Yu *et al.* (1967) showed that the alcohol-soluble (non-protein) nitrogen increased with advancing maturity.

Both the total and soluble nitrogen contents of ripe fruit were enhanced by increasing applications of nitrogenous fertilizers, but the protein content was only slightly increased (Neubert, 1959). The alcohol-soluble nitrogen was increased by nitrogen fertilization according to Davies (1965a), while phosphate had the opposite effect. The alcohol-insoluble fraction (which is approximately equivalent to protein) was unaffected.

An increase in protein synthesis during the climacteric has been reported in a number of fruits (see Richmond and Biale, 1966), but convincing evidence for tomatoes is very limited. Rowan *et al.* (1958) showed that the ratio of protein-N : total-N increased just before the respiration peak and then decreased progressively as the fruit ripened. Woodmansee *et al.* (1959) found a small increase in protein during ripening in two varieties, but the data given by Yu *et al.* (1967) are inconsistent. Davies and Cocking (1965) examined the changes in protein levels in tomato fruit locule tissue on a "per cell" basis during development and maturation and found a marked decrease in protein from 10 weeks after fertilization onwards, which probably included most of the ripening stages. They suggest, however, that although the evidence points towards a breakdown rather than synthesis during the period immediately before ripening, this does not preclude the selective synthesis of pectic enzymes involved in the process of ripening (Hobson, 1963a, 1964). Subsequent work (Davies and Cocking, 1967) indicated that the plastids are the major site of protein synthesis.

E. Volatile Components†

The significance of the four basic tastes, sweet, sour, bitter and salt, in tomato flavour have been discussed in general terms by Winsor (1966), while attempts to correlate chemical analyses with organoleptic characteristics have been made (Simandle *et al.*, 1966). The presence of minute traces of volatile components make a marked contribution to the typical aroma of a freshly picked tomato and consequently to its flavour. Much of the aroma appears to be associated with the calyx, however, and is unfortunately lost before the fruit reaches the consumer via the markets. Spencer and Stanley (1954), in one of the first studies of tomato flavour components, concluded that a typical tomato odour fraction contained alcohols, carbonyl compounds and unsaturated compounds modified by the presence of traces of terpenes. This statement has been confirmed by later workers.

In recent years the development of gas chromatography as an analytical tool has facilitated research into the volatile flavour constituents of tomato fruit. In 1960, Bidmead and Welti illustrated its use with, *inter alia*, a tomato flavour concentrate, but the identity of the various components was not stated. The complexity of the problem may be judged by the qualitative approach adopted by the majority of workers in this field, and only recently have quantitative studies been made. A summary of the data available for the tomato is given in Table V. Prior to 1965 the volatile compounds investigated were principally aldehydes and ketones identified by means of their 2,4-dinitrophenylhydrazones. By these means, several workers established the presence of acetaldehyde and isovaleraldehyde, together with some higher aldehydes particularly *n*-hexanal. Unsaturated aldehydes such as glyoxal, diacetyl and citral have also been encountered. Of the ketones, 2-propanone

† See also Volume 1, Chapter 10.

TABLE V. Volatile components of tomato fruit

Compound	Reference														
	A	B	C	D	E	F	G	H	I	J	K	L	M	N	O
Aldehydes															
Acetaldehyde	+		+	+				+	+	+	+				
Isovaleraldehyde	+			+						+		+			
n-Hexanal				+				+		+		+			
Furfural			+									+			
Benzaldehyde								+				+			
Others[a], including unsaturated aldehydes	+				+	+			+	+	+				
Ketones															
Acetone		+	+						+	+	+				
Other ketones						+		+		+				+	+
Alcohols															
Methanol		+					+				+				
Ethanol							+			+	+				
Propanols								+		+	+				
Butanols								+			+	+			
Pentanols							+			+	+	+	+		
Hexanols							+	+			+	+	+		
Miscellaneous															
Methyl salicylate								+				+			
Other esters								+		+	+	+			
α-Pinene					+							+			
Other terpenes	+				+										

[a] For example, glyoxal, diacetyl and citral.

A. Spencer and Stanley (1954).
B. Jacquin and Tavernier (1955).
C. Matthews (1960).
D. Schormüller and Grosch (1962).
E. Hein and Fuller (1963).
F. Schormüller and Grosch (1964).
G. Schormüller and Grosch (1965).
H. Pyne and Wick (1965).
I. Hills and Collins (1966).
J. Giannone and Baldrati (1967a, b, c).
K. Katayama *et al.* (1967).
L. Dalal *et al.* (1967); Dalal *et al.* (1968).
M. Johnson *et al.* (1968).
N. Buttery and Seifert (1968).
O. Buttery *et al.* (1969).

(acetone), 2-butanone and 2-pentanone are those most frequently detected. According to Spencer and Stanley (1954), acetaldehyde accounts for 70% of the total carbonyl compounds in fresh tomato distillates. Early work by Rakitin (1945) showed that acetaldehyde and ethanol increased in concentration during the ripening of several fruits, including the tomato. Isovaleraldehyde reached a peak (about 6×10^{-3} p.p.m.) as the fruit began to colour and declined with further ripening (Dalal *et al.*, 1968).

The presence of alcohols other than methanol and ethanol, especially certain substituted pentanols and hexanols, has been increasingly recognized. For example, relatively large quantities (4–30 p.p.m.) of *cis*-3-hexenol-1 are reported to occur (Pyne and Wick, 1965; Johnson *et al.*, 1968). Such compounds have pleasant odours variously described as "green" or "green leafy" (Pyne and Wick, 1965; Dalal *et al.*, 1967).

Dalal *et al.* (1965, 1966) found that the total volatile reducing substances increased continuously as tomatoes matured and ripened, reaching maximum concentrations of around 0·3 μg and 0·5 μg/100 g fr. wt. in fruit grown under glass and outdoors respectively. In a subsequent paper, Dalal *et al.* (1968) studied the individual volatile components during ripening. Except for isovaleraldehyde and hexanol the concentration of the volatiles increased with maturity and was higher in field-grown tomatoes than in those grown under glass.

Variety of fruit does not appear to affect the qualitative composition of the volatiles of tomato, but quantitative differences have been established (Katayama *et al.*, 1967; Nelson and Hoff, 1969). Johnson *et al.* (1968) in a study of ten varieties found that the sum of the concentrations of isoamylol, *n*-pentanol and *cis*-3-hexenol-1 (calculated as hexanol) ranged between 7 and 40 p.p.m. They also found considerable variation between weekly harvests. Processing reduced the amounts of the three alcohols, particularly that of *cis*-3-hexenol-1.

One of the major changes in the volatiles of canned tomato juices during heat processing is the production of methyl sulphide (Miers, 1966) attributable to the thermal decomposition of a salt of S-methyl methionine present in the fruit (Wong and Carson, 1966). Guadagni *et al.* (1968) showed that the methyl sulphide content could provide a measure of the relative aroma intensity of canned tomato juice.

The existence in ripening tomatoes of an enzyme system capable of producing carbonyl compounds and alcohols from amino acid substrates was demonstrated by Yu *et al.* (1968a, b). This mechanism may be important in the synthesis of aroma components.

At present, by the use of suitable varieties, fertilizer régimes and environmental conditions, the sugar and acid contents of tomato fruit can be manipulated in order to improve the taste. Our knowledge of the volatile components is far from complete; it will no doubt be possible in the future to influence them in a similar manner.

F. Steroids†

A number of species in the Solanaceae contain alkaloidal glycosides (Blaim, 1963). These are bitter to the taste (Borchers and Nevin, 1954), and

† See also Volume 1, Chapter 12.

on acid hydrolysis, yield alkaloids possessing a steroidal structure and a mixture of sugars.

Following the suggestion by Irving (1947) that tomatine might play a role in the resistance of tomatoes to *Fusarium* wilt, this steroidal glycoalkaloid has been investigated extensively. The structure of tomatine is given in Fig. 2.

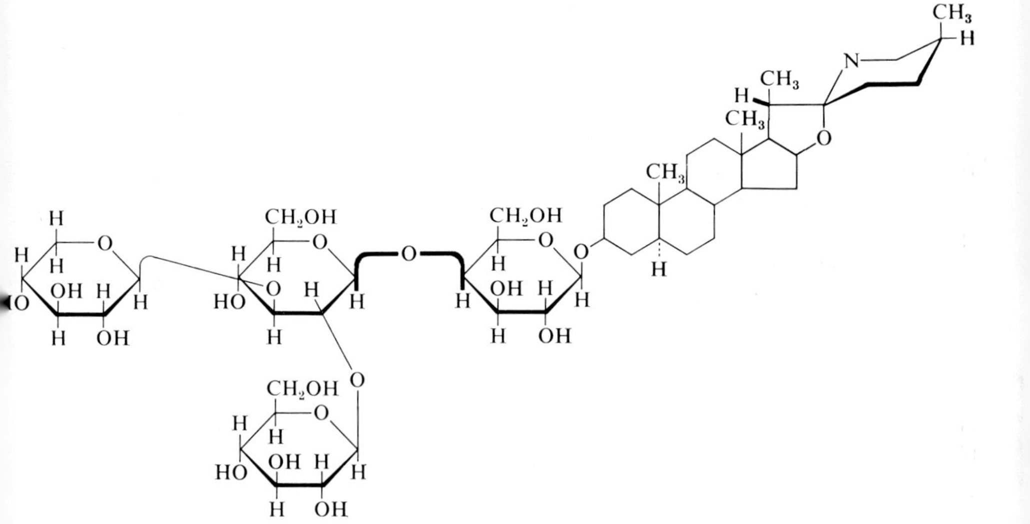

FIG. 2. Tomatine.

Much of the work has been confined to parts of the plant other than the fruit (see Arneson and Durbin, 1967) and, according to Tukalo (1956), the highest concentrations of tomatine are found in the flowers. The growing shoot is the principal site for the formation of tomatine, while degradation occurs mainly in the fruit (Sander, 1956).

G. Pigments other than Flavonoids

The green colour of immature tomato fruit is attributable to the presence of a mixture of chlorophylls, which seem to perform a definite photosynthetic role during maturation (Boe and Salunkhe, 1967a). With incipient ripening, yellow pigments (β-carotene and xanthophylls) are produced and become more apparent as the chlorophyll content decreases. Subsequently the rapid accumulation of the red pigment lycopene influences the fruit colour despite the reinforcement of the yellows by lutein and lycoxanthin (Fig. 3). Chichester and Nakayama (1965) have reviewed the pigment changes in senescent and stored tissue in general (see also Chapter 21).

Both Ramirez and Tomes (1964) and Edwards and Reuter (1967) found

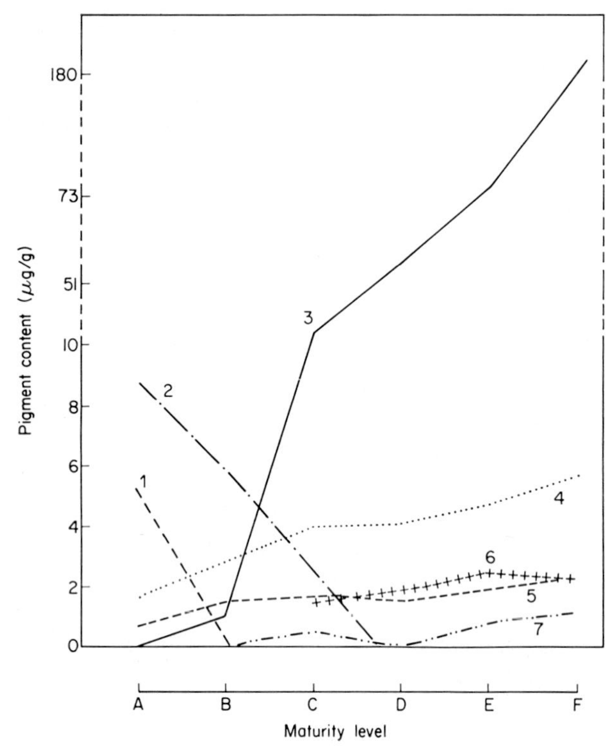

FIG. 3. Changes in the concentration of tomato pigments during maturation (variety San Marzano). 1, chlorophyll b; 2, chlorophyll a; 3, lycopene; 4, β-carotene; 5, lutein 5,6-epoxide; 6, lycoxanthin; 7, lycophyll. Maturity levels. A, mature green; B, green-yellow; C, yellow-orange, some trace of green; D, orange-yellow, no trace of green; E, orange-red; F, red. (Revised from Edwards and Reuter, 1967.)

more chlorophyll "a" than "b" in green tomato fruit, but Dalal *et al.* (1965, 1966) reported that the ratio of chlorophyll "b" to "a" was consistently about 2 : 1 during maturation and ripening. All agree, however, that with incipient ripeness the chlorophylls decrease rapidly. Ramirez and Tomes (1964) found that during ripening of normal tomatoes the chlorophyll content and the activity of chlorophyllase changed concurrently. Vogele (1937) showed that chlorophyll decomposition in tomatoes is prevented at 40°C or above, while exposure to ethylene accelerates breakdown as also does exposure of light. Reduction in oxygen supply did not, however, inhibit chlorophyll decomposition at temperatures between 24 and 36°C.

There is evidence for the involvement of chlorophyll in the synthetic reactions of tomato fruit but its contribution is of necessity limited in time by its loss during ripening. The amount of phytol which could be liberated during the disappearance of chlorophyll appears to be insufficient to account for the amount of lycopene formed, so that some independent system for its synthesis is indicated.

The occurrence and identity of the major carotenes in various species and strains of *Lycopersicon* were established largely by Porter and Zscheile (1946) and by Trombly and Porter (1953) (see also Volume 1, Chapter 12). The alicyclic hydrocarbon β-carotene is an important contributor to the colour of half-ripe fruit, and together with α-carotene is present in small quantities in mature green fruit (Meredith and Purcell, 1966). Several workers (Biswas and Das, 1952; Tarnovska, 1961; Edwards and Reuter, 1967; Czygan and Willühn, 1967) have reported that β-carotene slowly increases in concentration throughout the ripening of tomatoes. Meredith and Purcell (1966) observed a small decrease after the light pink stage of ripeness, but Dalal *et al*. (1965, 1966) reported a very rapid fall in concentration after this stage such that the amount left in fully ripe fruit was very low. No non-cyclic carotenes were found until the fruit began to colour, but subsequently phytoene, phytofluene, ζ-carotene and γ-carotene increased in concentration throughout ripening (Meredith and Purcell, 1966). Generally similar results were reported by Edwards and Reuter (1967).

Lycopene constitutes the main red pigment of tomatoes, and its concentration increases steadily throughout ripening. Denisen (1951b), Imparato-Gargano and Violante (1963) and Edwards and Reuter (1967) all quote figures for the lycopene content of specific varieties, and the last authors give details of six lycopene derivatives during ripening. Tomes (1963) studied the inhibiting effect of temperature on carotene synthesis in various strains of tomato and used the results to shed light on the biosynthetic pathways in normally and abnormally pigmented varieties. There is some evidence for a common connection between the amounts of chlorophyll and carotene in some varieties, but the two pathways are not inseparably linked. Denisen (1951a), Tarnovska (1961) and Tomes (1963) showed that lycopene synthesis is markedly temperature dependent and is virtually inhibited above 32°C (Tomes, 1963). Nettles *et al*. (1955) observed that light improved the colour of harvested mature green tomatoes. Although Vogele (1937) showed that lycopene formation was dependent upon the presence of oxygen, illumination of green fruit at suitable temperatures will act as a substitute, presumably through the formation of enough oxygen by photosynthesis (McCollum, 1954). A comprehensive summary of the evidence for interrelations between the commonly occurring pigments has been given by Goodwin (1965).

Oxygenated carotenoids or xanthophylls also occur in tomato fruit to the extent of about 6% of the carotenoid content (Curl, 1961). The main constituents according to this author were lutein, violaxanthin and neoxanthin, while Edwards and Reuter (1967) found principally lutein 5,6-epoxide, lycoxanthin, lycophyll and lutein. The diol and polyol xanthophylls only increase slowly during ripening but they become more obvious because of the disappearance of chlorophyll. Possible mechanisms for the formation of xanthophylls were discussed by Goodwin (1965), but concentrations of these pigments are very low in tomato fruit and seem to be of minor consequence.

The chlorophyll content of tomato fruit appears to be as much under genetic control (Ramirez and Tomes, 1964) as are many of the other pigments (Kargl *et al.*, 1960; Baker and Tomes, 1964). Changes in the relative amounts of the commonly occurring pigments following various genetic manipulations have been widely used to deduce some of the more likely biosynthetic pathways.

Hall (1961, 1964a, b) studied the storage of ripened fruit in terms of colour. McCollum (1954) showed that between 21 and 24°C lycopene could be increased by light until after the chlorophyll had disappeared. Carotene also was increased in ripe fruit by illumination at any stage during maturation. Takahashi and Nakayama (1959) studied the effect of shading and found that chlorophyll, carotene and xanthophyll were higher in exposed fruit while the lycopene content was lower. Shewfelt and Halpin (1967) concluded that the quality of the light received influenced the rate of colour development of picked fruit. Boe and Salunkhe (1967b) re-affirmed that light increased the amounts of β-carotene and lycopene. McCollum (1955) and Tarnovska (1961) examined the distribution of carotenoids in the various parts of tomato fruit, and McCollum (1954) showed that the pattern of pigment formation is light dependent.

In a factorial manurial trial, Denisen (1951b) found that nitrogen (in combination with potassium and phosphate for maximum effect) increased the lycopene content. Takahashi and Nakayama (1962) came to the opposite conclusion with regard to nitrogen fertilization, but agreed that phosphate was beneficial for good colour.

During ripening the tomatine content falls rapidly from around 0·09% fr. wt. in green fruit to less than 0·01% in red fruit (Kajderowicz-Jarosinska, 1965). Very low concentrations of the alkaloid were also observed in ripe fruit by Sander (1956) and Kolankiewicz *et al.* (1956). Recently Czygan and Willühn (1967) concluded that, contrary to the suggestion by Sander (1958), tomatidine is not the precursor of lycopene in the tomato.

Solanine has been stated to be responsible for the bitter taste of the fruit of some *Lycopersicon* species (Luckwill, 1943), and instances of its occurrence in "wild type" tomatoes in considerable quantities are quoted by Wehmer (1931). Further work, however, indicates that commercial varieties contain only traces, at least by the time the fruits are mature (Kolankiewicz *et al.*, 1954).

Bennett *et al.* (1961) would appear to have been the first to detect stigmasterol in tomatoes; they showed that mevalonic acid was one of its precursors. Later, Yamamoto and Mackinney (1967) concluded that stigmasterol and β-sitosterol are the two principal sterols of the tomato fruit. Brieskorn and Reinartz (1967a) found β-sitosterol and stigmasterol in tomato skins in the ratio 4 : 1, together with much smaller quantities of a campesterol.

H. Flavonoids†

The term "flavonoid" embraces all those compounds whose structure is based on that of flavone (2-phenylchromone) and includes the anthoxanthin (yellow) and anthocyanin (orange, red and blue) pigments. They may occur both in the free form and in combination with sugars as glycosides.

The formation of these pigments in the cuticular layer of the epidermal cell walls affects the appearance of tomato fruit. Piringer and Heinze (1954) studied the light-controlled production of a yellow pigment in tomato skin which they considered to be a flavonoid. Subsequently Wu and Burrell (1958) isolated naringenin and quercitrin from the skin of three varieties of tomato fruit, but no flavonoids could be detected in the flesh. Rivas and Luh (1968) have identified rutin, naringenin (which was thought to have come from the skin) and a possible isoflavone in canned tomato paste. They found that naringenin made up about 28% of the total polyphenolic content and rutin about 12%.

I. Other Phenolic Compounds

Surprisingly little information has been published about other phenolic compounds in tomato fruit despite their probable importance. Tissue affected by blossom-end rot contains a considerable amount of a dark brown or black material which may well be a polymer resulting from the oxidation and condensation of simple phenols, while the browning of tissue affected by virus disease or physiological disorders could result from the interaction of phenolic substrates with the enzyme phenolase (Hobson, 1967a). The tomato possesses little or no astringency although phenolic compounds present may take part in flavour development; there is an increase in their concentration from the mature green stage at least until mid-ripeness (Rubatzky, 1965; Davies, 1967).

Phenolic acids have long been recognized as potent germination inhibitors, and Akkerman and Veldstra (1947) showed that "blastocholine" isolated from tomatoes consisted of a mixture of ferulic and caffeic acids. These compounds may serve both to keep the seeds in a dormant state and as fungistatic agents.

Herrmann (1957) in a survey of the phenolic compounds of fruit and vegetables found p-coumaric, ferulic, caffeic and chlorogenic acids in tomatoes, and Walker (1962) subsequently confirmed the presence of the same acids in

† See also Volume 1, Chapter 11.

the fruit walls. Caffeic and chlorogenic acid concentrations in whole tomatoes of 0·6 and 1·8 mg/100 g fr. wt. respectively were reported by Jurics (1966). Rivas and Luh (1968) investigated the polyphenols of canned tomato paste and produced evidence for the presence of caffeic, ferulic and *cis*- and *trans*-chlorogenic acids together with two apparent derivatives of caffeic acid and a further three unidentified phenolic compounds. They could not, however, find p-coumaric acid which has been detected in tomato skins (Brieskorn and Reinartz, 1967a) and might be present in the rest of the fruit only in small quantities.

Walker (1962) showed that ferulic, caffeic and chlorogenic acids in tomato fruit wall tissue all increased with ripening but the level of p-coumaric acid was unchanged. Affected areas of the walls from blotchy fruit contained levels of these four acids similar to those in mature green tomatoes although this is not always so for other constituents (Winsor *et al.*, 1962c).

Very young tomatoes contain considerable amounts of polyphenolic substances but these decrease in concentration towards maturity (see Table VI). During ripening there is a tendency for a small increase in total phenolics as the fruit becomes red. This will be seen from Table VII; in both Tables VI and VII phenolics were determined by the relatively unspecific Folin-Denis method as modified by Swain and Hillis (1959).

TABLE VI. Polyphenols in immature green tomatoes (variety Potentate)

Diameter (in.)	mg/100 g fr. wt.
<0·5	156
0·5–0·75	97
0·75–1·00	64
1·00–1·75	50

TABLE VII. Total phenolics in whole tomato fruit (mg/100 g fr. wt.)

Variety	No. of series	Stage of ripeness: Green	Green–yellow	Yellow–orange	Orange	Red
Potentate	3	27	36	44	48	57
Immuna	4	27	34	40	43	48
Moneymaker	3	34	32	43	43	45
Overall means	10	29	34	42	45	50

Varieties did not differ significantly from each other but stages of ripeness were significant at $P<0{\cdot}001$; L.S.D. at $P=0{\cdot}05$: $\pm 4{\cdot}8$.

Tomato fruit of species other than *L. esculentum* contain higher amounts of polyphenols, although no published values for the levels of individual components are available.

J. Cell Wall Constituents†

The main constituents of the cell walls of tomato fruit are pectic substances, hemicelluloses, cellulose and some protein. The progressive loss of firmness with ripening is the result of a gradual solubilization of protopectin in the cell walls to form pectin and other products. Concomitantly, either erosion of the cell wall or stretching of the wall due to enlargement of the cell, or both, may be responsible for the thinness of the cell walls seen in electron micrographs of ripe fruit.

The cell wall fraction is typically that insoluble in 70 or 80% ethanol; this constitutes between 1·0 and 3·0% of the fresh weight and was found, for ripe fruit, to contain 6–7% of ash and 16–20% of protein according to Woodmansee *et al.* (1959) and Young and Davison (1968). The remainder was assumed by the latter authors to be carbohydrate and, on analysis, yielded pectic substances, hemicelluloses and cellulose in the ratio of 11 : 6 : 3.

The insoluble pectic compounds in the middle lamella act as intercellular binding material and contribute largely to the firmness of the young fruit while at the same time allowing some plasticity accommodating to the changing size and shape of the protoplast during growth (Davies and Cocking, 1965). The pectic substances undergo transformation and degradation by enzymic mechanisms during the later stages of maturation and ripening (Hobson, 1963a, 1964) accompanied by an almost complete loss of tissue cohesion (Hobson, 1965b). A discussion of the pectic enzymes concerned is given in Volume 1, Chapter 3. Although a cellulase-like enzyme is present in tomato fruit (Hobson, 1968), evidence at the moment points towards its having a very limited effect on cellulose in the primary and secondary walls.

K. Lipids and Long-Chain Components

Until recently little attention has been given to the lipid composition of tomato fruit. In view of the small quantities present (about 10–20 mg saponifiable lipids/g dry matter in the fruit pericarp according to Kapp (1966)), this is perhaps not surprising. The possible influence of the fatty acids on tomato flavour and the availability of gas–liquid chromatography will no doubt stimulate further work.

Kapp (1966) could find no evidence of an association between colour development and the total lipids in the pericarp of tomatoes. He detected 33 saturated and unsaturated fatty acids mainly with a chain length of 18 or

† See also Volume 1, Chapters 2 and 3.

less carbon atoms. Quantitative studies were confined to linolenic, linoleic, oleic, stearic, palmitic and myristic acids which together constituted about 75% of the total lipids. All increased during ripening of the fruit, but during this time the percentage of linoleic and palmitic acids in the total decreased. The concentration of individual acids varied with the maturity of the fruit at harvest, the duration of storage and the temperature.

The saponifiable fraction of tomato skins was found by Brieskorn and Reinartz (1967a) to contain palmitic, stearic, oleic and linoleic acids, with a small amount of linolenic acid, while the unsaponifiable matter contained five hydrocarbons in the range n-$C_{29}H_{60}$ to n-$C_{34}H_{70}$ together with α- and β-amyrin. The alkali-soluble fraction (Brieskorn and Reinartz, 1967b) contained a variety of long-chain organic acids, principally 10,16-dihydroxy-decanoic acid.

L. Mineral Constituents

In contrast to the extensive literature on the chemical, morphological and physiological effects of mineral nutrition on tomato leaf tissue, information on the fruit is relatively limited. Early work on the mineral constituents was reviewed by Lewis and Marmoy (1939). Subsequently other summaries have been included in comprehensive surveys of the composition of fruit and vegetables (see, for example, Beeson, 1941). Selected results for both macro- and micro-nutrients from more recent investigations are listed in Table VIII.

A detailed study of the major nutrient composition of tomato fruit was made by Arnon and Hoagland (1943) who concluded that the inorganic composition is relatively little influenced by fluctuations in external nutrient concentrations short of outright deficiencies. Andreotti (1956) found that during ripening the mineral content, calculated on either a fresh or dry weight basis, decreased slightly initially, but increased towards the end of ripening. Expressed on a "per fruit" basis, however, it showed a progressive increase.

As discussed in Section IIB, potassium plays a dominant role in tomato fruit composition and quality; together with nitrate and phosphate it constitutes some 93% of the total mineral matter (Ansiaux, 1960). The specific role of potassium is still unknown, but there is little doubt that, in addition to being required to maintain cell organization and permeability, it can act as an activator for various systems such as pyruvic kinase, and is associated with protein metabolism (see Nason and McElroy, 1963; Evans and Sorger, 1966).

The concentration of other metals in the fruit is low. Sodium may partly substitute for the potassium requirement in some plants, and under conditions of incipient potassium deficiency tomato fruit tend to accumulate sodium (Davies and Winsor, 1967). The physiological significance of sodium is also

TABLE VIII. Mineral constituents of whole tomato fruit

Constituent	Basis	Range	Reference
Potassium	g/100 g dr. wt.	1·88–5·88	Beeson (1941)
	g/100 g fr. wt.	0·126–0·492	Hopkins and Eisen (1959)
	g/100 g fr. wt.	0·078–0·312	Davies (1964)
	g/100 g fr. wt.	0·087–0·190	Sakiyama (1966b)
Sodium	g/100 g dr. wt.	0·04–0·66	Kidson (1963)
	g/100 g fr. wt.	0·001–0·003	Hopkins and Eisen (1959)
Calcium	g/100 g dr. wt.	0·10–0·24	Geraldson (1957)
	g/100 g dr. wt.	0·12 (normal fruit) 0·08 (fruit with blossom-end rot)	Wiersum (1966)
	g/100 g fr. wt.	0·008–0·024	Sakiyama (1966b)
Magnesium	g/100 g dr. wt.	0·13–0·59	Beeson (1941)
	g/100 g dr. wt.	0·18–0·24	Kidson (1963)
	g/100 g fr. wt.	0·007–0·016	Bohart (1940)
	g/100 g fr. wt.	0·007–0·025	Hopkins and Eisen (1959)
Phosphorus	g/100 g dr. wt.	0·29–0·84	Beeson (1941)
	g/100 g fr. wt.	0·011–0·031	Bohart (1940)
	g/100 g fr. wt.	0·013–0·041	Hopkins and Eisen (1959)
Boron	p.p.m. dr. wt.	3–14	Wilkinson (1957)
	p.p.m. dr. wt.	24–27	Eaton (1944)
	p.p.m. dr. wt.	3–12	Messing (1957)
Manganese	p.p.m. fr. wt.	0·2–1·1	Bohart (1940)
	p.p.m. fr. wt.	0·5–3·0	Hopkins and Eisen (1959)
Copper	p.p.m. dr. wt.	5–9	Beeson (1941)
	p.p.m. fr. wt.	0·6–2·6	Bohart (1940)
	p.p.m. fr. wt.	0·1–1·6	Hopkins and Eisen (1959)
Iron	p.p.m. dr. wt.	48–800	Beeson (1941)
	p.p.m. fr. wt.	0·9–5·6	Hopkins and Eisen (1959)
	p.p.m. fr. wt.	2·8–11·1	Bohart (1940)
Zinc	p.p.m. dr. wt.	12–67	Beeson (1941)

obscure. On the other hand, fruit calcium levels in excess of about 0·12% dry matter are required in order to ensure freedom from blossom-end rot (Geraldson, 1957; Wiersum, 1966). Studies with ^{45}Ca have shown that calcium chloride sprays are readily absorbed by both tomato leaves and fruit, but translocation away from the leaves is negligible (Barke, 1968). Magnesium has a beneficial effect on tomato ripening disorders particularly at low levels of potassium (Winsor *et al.*, 1965; Winsor and Long, 1967). Although Williams (1957) noted a marked reduction in the carboxylic acid content of magnesium-deficient tomato leaves, soil magnesium levels have little effect on the acidity of the fruit (Bradley, 1962; Davies and Winsor, 1967).

Comprehensive accounts of the effects and modes of action of the minor elements in plants have been included in reviews by Hewitt (1963) and Nason and McElroy (1963). The influence of elements such as manganese, copper and zinc on tomato fruit composition has been comparatively little investigated and few unequivocal results are available. Deficiency of boron is, however, associated with striking fruit abnormalities and a lowering of the dry matter content (Gum *et al.*, 1945; Messing, 1957), together with a reduction in the concentrations of ascorbic acid (Ruszkowska and Zinkiewicz, 1953; Jelenic *et al.*, 1958) and sugar (Gum *et al.*, 1945; Darkanabaev *et al.*, 1964). Boron toxicity levels also have the effect of lowering the amounts of sugars in the fruit (Govindan, 1952). Sugar in the fruit is also reduced by manganese deficiency (Gum *et al.*, 1945; Ruszkowska, 1960), but increased by zinc deficiency (Govindan, 1952).

III. ASPECTS OF METABOLISM DURING MATURATION AND SENESCENCE

The respiratory activity of most fruits towards the end of maturation is a distinctive and peculiar feature of their physiology. As the tomato is a climacteric fruit (see Volume 1, Chapter 17), there is, under normal conditions, a rapid rise in respiration more or less coincident with the first signs of red colour (see Winsor *et al.*, 1962a; Pratt *et al.*, 1965). Although the dynamic aspects of the changes in metabolism of tomato fruit during the climacteric cycle have not been investigated in detail, the present state of knowledge, with particular emphasis on respiration and mitochondrial activity, is summarized below.

Mitochondria are readily separated from tomato fruit by a variety of techniques (Dickinson and Hanson, 1965; Drury *et al.*, 1968a; Hobson, 1969) although for non-routine use the most satisfactory method so far described is probably that of Ku *et al.* (1968). It is essential to break open the cells of the fruit with the minimum of mechanical force, and to neutralize the highly acid contents with the least possible delay. In these circumstances it is then necessary to examine the particles both for physiological activity and by electron microscopy to demonstrate that relatively intact particles reasonably similar to those in the whole fruit are indeed being produced. Electron micrographs kindly supplied by Professor A. R. Spurr of the University of California are shown in Fig. 4, and other examples are discussed by Ku *et al.* (1968).

FIG. 4. Mitochondria in the outer region of the pericarp of tomato fruit at different stages of maturity. (a) Mature green stage. The matrix of the mitochondria is relatively dense. ×26,720. (b) Green-yellow stage near the climacteric respiration peak. The matrix of the mitochondria is less dense ×26,330. (c) Red ripe stage. The matrix of the mitochondria is least dense and the cristae appear less numerous. ×26,720. Glutaraldehyde-osmium fixed, lead citrate contrasted, except (c) which is without lead. CW= cell wall; V =vacuole; LD =lipid droplet; D =dictyosome; Chr =portions of chromoplasts. (Electron micrographs by kind permission of Professor A. R. Spurr.)

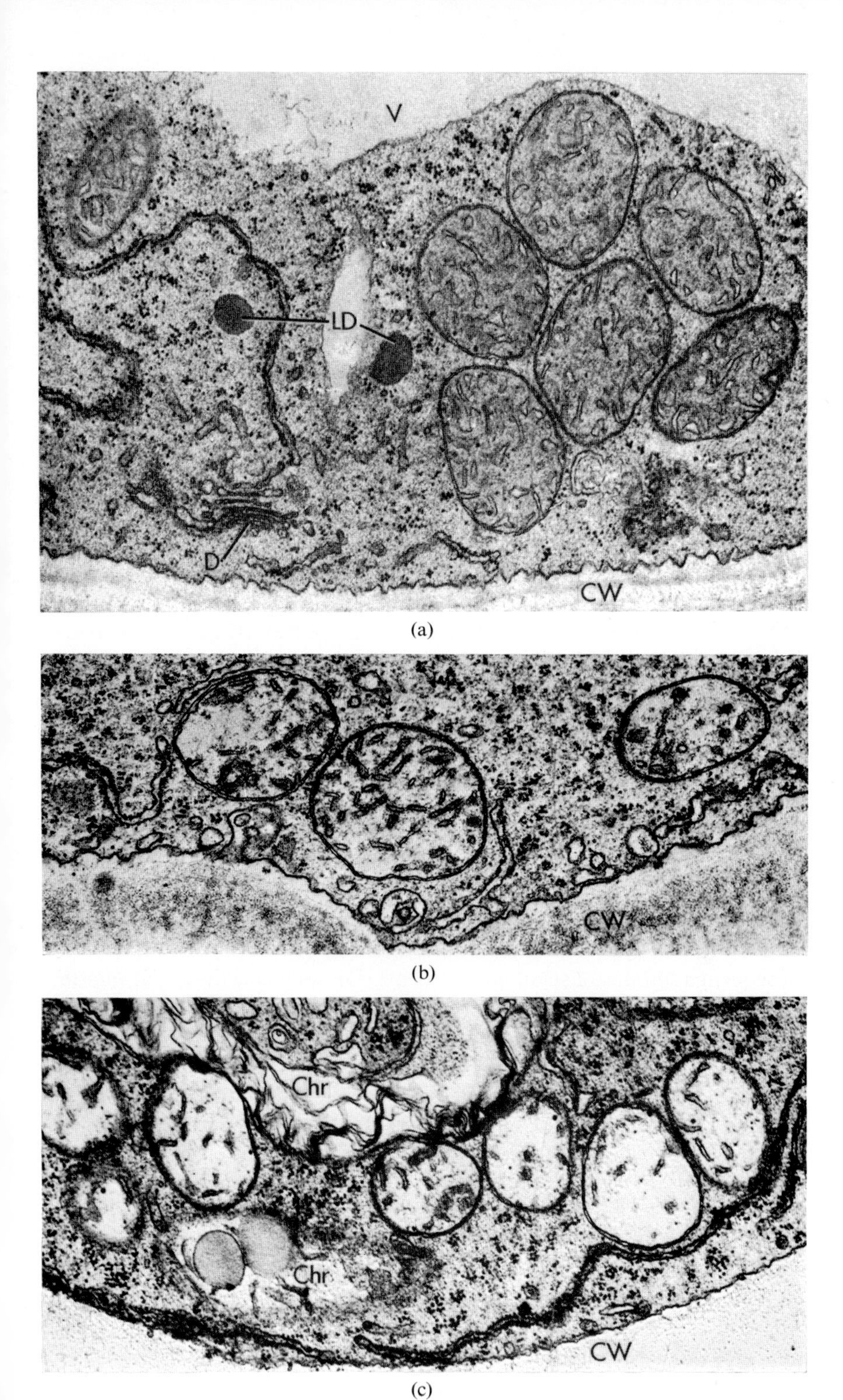

(a)

(b)

(c)

FIGURE 4

A Clark oxygen electrode is often used for polarographic measurement of mitochondrial oxidation rates. With intact particles, the response to succinate is markedly stimulated by adenosine diphosphate (ADP) (mitochondria in "state 3" according to Chance and Williams (1956)) and suppressed as ADP becomes limiting (mitochondria in "state 4"). Respiratory control (RC) ratios are calculated from the state 3 to state 4 rates of oxidation, a high figure indicating good mitochondrial integrity based on the degree of coupling of electron transport with the phosphorylation of ADP to produce adenosine triphosphate (ATP). Breakdown in one or more of the sites of phosphorylation, the presence of ATPase, the identity of the substrate and a number of other factors may influence the degree of control observed.

A. The Pre-Climacteric Fruit

The cuticle is generally thought to form a gas-tight envelope around the tomato fruit so that gaseous exchange for photosynthesis and respiration takes place exclusively through the stem end of the fruit (Clendenning, 1942). Respiration falls quickly from a high initial value during the first few weeks after fertilization, and continues to decline slowly throughout growth until a pre-climacteric minimum is reached. Changes in various constituents during this period have been investigated by Davies and Cocking (1965), but very few studies of the composition of pre-climacteric fruit have been published.

Mitochondria separated from tomatoes during maturation oxidize succinate at increasing rates and efficiencies up to a maximum at the mature green stage (RC ratios from 2·1 to 2·4; ADP : O ratios from 1·2 to 1·4). Malate oxidation shows trends similar to those with succinate (RC ratios from 1·8 to 2·2; ADP : O ratios from 1·3 to 2·2) but the rates diminish with subsequent additions of small quantities of ADP following the initial addition Exogenous supplies of thiamine pyrophosphate (TPP), which is a co-factor in the oxidative decarboxylation of α-oxo acids, not only stimulates the oxidation of malate by some 150% but also induces constant oxidation rates after the first ADP addition. It seems likely that with mitochondria derived from young fruit up to the mature green stage, oxaloacetate accumulation is sufficient to cause some inhibition of malate oxidation (Drury *et al.*, 1968a; Hobson, 1969).

Small green fruit yield mitochondria that are virtually incapable of oxidizing α-oxoglutarate but particles from more mature fruit metabolize the substrate increasingly well. Thus the state 3 oxidation rate gradually improves with time following the first addition of ADP and thereafter the rates in state 3 and 4 become constant. The oxidation rates with α-oxoglutarate are always increased by the presence of external sources of TPP and the co-factor also facilitates a display of respiratory control. Thus, for example, the original RC and ADP : O ratios for mitochondria from mature green fruit were 1·5 and 1·7 respectively and were increased to 1·7 and 2·3 when TPP was added in non-limiting amounts (Hobson, 1969).

During the pre-climacteric part of maturation the RC and ADP : O ratios for mitochondria appear to improve continuously when succinate, malate or α-oxoglutarate is being oxidized. Thus this period appears to be a time for growth and development in preparation for the peak in activity that accompanies the climacteric respiration rise.

A study of the major metabolic pathways in tomato fruit has been made by Wang and his group, using radiochemical tracer techniques. Acetate-1-^{14}C was shown to be widely metabolized in mature green fruit and possible pathways for the synthesis of citric and malic acids were examined (Wang *et al.*, 1953; Buhler *et al.*, 1956). Glyoxylate was also extensively incorporated into organic acids especially into malic acid (Doyle *et al.*, 1960). The relative importance of the Embden–Meyerhof glycolytic pathway and the hexose monophosphate pathway was deduced from studies with a number of labelled compounds. The latter pathway was shown to play an increasingly important part in the catabolic breakdown of hexoses in the fruit during ripening (Ramsey and Wang, 1962). In mature green tomatoes it was calculated that 73% of glucose degradation took place through the Embden–Meyerhof processes and 27% through the alternative oxidative pathway, which functions mainly as a mechanism for the conversion of glucose to various intermediates for biosynthesis (Wang *et al.*, 1962) and, possibly, to provide $NADPH_2$ for this process.

B. The Climacteric Fruit

Incipient ripening is accompanied by increased respiratory activity and rapid changes in a wide range of chemical constituents, many of which are detailed in Section II. Lycopene and carotene progressively replace chlorophyll, and often make their first appearance in the semi-liquid material surrounding the seeds. When this occurs the fruit is close to the climacteric peak.

Mitochondria appear to be at their most active when obtained from fruit shortly before the climacteric peak (more or less equivalent to the mature green stage). The rate of oxidation of succinate is high with good respiratory control ratios (see p. 464). This property is also shown when malate and α-oxoglutarate are used as substrates, both in the presence and absence of TPP, although in the former case the rates of oxidation are considerably increased. Particles separated from fruit at selected stages up to full ripeness showed a declining ability to oxidize succinate, malate and α-oxoglutarate, but with the last two substrates the presence of TPP continued to stimulate the oxidation rate (Hobson, 1969). Whether the co-factor is removed during the isolation of the mitochondria or is present in the particles only in limited quantities is, at present, an open question.

One of the first studies of the activity of particles from tomatoes between the mature green stage and over-ripeness was made by Abdul-Baki (1964). Using a mixed pyruvate–malate substrate in the presence of TPP, it was shown that there was a progressive decrease in the oxidative capabilities of the particles during the 3 weeks or so over which the fruit was stored. Similar results were obtained by Lyons *et al.* (1964) in the presence of succinate and α-oxoglutarate, but the reverse situation obtained when β-hydroxybutyrate was used. Pratt *et al.* (1965) found evidence that the respiration climacteric was, unlike the apple (see Chapter 10), not accompanied by greatly increased mitochondrial activity. More detailed work by Dickinson and Hanson (1965) confirmed that the oxidative capacity of the particles decreased considerably with increasing ripeness from the mature green stage, but their phosphorylative efficiency declined much more slowly, while the specific activity of malic dehydrogenase was maintained. Drury *et al.* (1968a) have shown that mitochondria from mature green fruit (probably near the beginning of the climacteric respiration) were more than twice as active as those from red-ripe fruit when oxidizing succinate. Particles from both sources were remarkably stable during storage. These investigators also studied the mechanism of the delay in the oxidation of succinate brought about by oxaloacetate (Drury *et al.*, 1968b).

An increasing amount of evidence suggests that the capacity for phosphorylation and synthesis continues beyond the climacteric respiration rise in tomato fruit, and the mitochondria are capable of providing a source of energy for the formation of additional enzymes required for the furtherance of ripening, although with diminishing efficiency. The enhanced rate of ripening of unpicked tomatoes when injected with small quantities of chemicals traditionally active in uncoupling oxidation from phosphorylation (Hobson, 1965a) is not necessarily inconsistent with the continuing provision of energy for the ripening processes.

Of the non-mitochondrial enzymes, Pratt *et al.* (1965) found a small increase in the activity of NADP-dependent malic enzyme during the climacteric on the vine but no consistent increase off the vine when stored at 15–20°C. Pyruvate decarboxylase activity was low throughout. Isocitric dehydrogenase (NADP-dependent) activity was low but increased slightly over the climacteric. Lipoxidase activity was high immediately before and during the early stages of the climacteric but fell rapidly as the fruit became over-ripe.

C. The Senescent Fruit

Once the fruit is fully ripe, tissue disorganization becomes increasingly dominant. Cell walls become very thin and the organized cytoplasmic units largely disintegrate. The degradation of cellulose, as well as most of the pectic components, leads to a progressive loss of tissue cohesion. Respiration continues to fall slowly concurrently with ethylene production. Mitochondrial protein as a percentage of cytoplasmic protein decreases markedly during ripening (Dickinson and Hanson, 1965), and this continues into over-ripeness. Nevertheless, those mitochondria that do survive the degenerative processes are still capable of oxidation and phosphorylation although with reduced efficiency. Rowan *et al.* (1958) showed that high energy phosphates were present throughout the ripening of tomatoes, and the continuing ability of mitochondria to incorporate inorganic phosphate into nucleotides substantiates this observation. However, one of the main points that emerges from mitochondrial studies is that the upsurge in respiratory activity during the climacteric rise is not wholly reflected in the oxidative behaviour of the particles *in vitro*, although the restraints imposed *in vivo* may well have been altered during isolation. Furthermore, the bruising of fruit produces an immediate increase in total respiration but this is not recognizable in isolated particles (Abdul-Baki, 1964). Much is still obscure about the structural and biochemical organization of the fruit cell, particularly with regard to the mechanism that triggers the onset of the climacteric rise in respiration.

D. The "Blotchy" Fruit

A major contribution to knowledge about the gross differences in composition between evenly ripened fruit and those showing blotchy ripening (see Section IV C) was made by Winsor and Massey (1958, 1959), and Winsor *et al.* (1962a, b, c). Their main findings were that blotchy areas of the fruit walls contain less acids, sugars, nitrogenous compounds and total solids than the adjacent red areas. Since the analytical results were such as to preclude blotchy regions from being comparable with normal fruit at any stage during ripening, these areas should be regarded as abnormal.

The disorder causes an increase in both the total and the insoluble pectic substances in blotchy fruit as a whole (Hobson, 1963a). There is also a small but significant reduction in the activity of pectinesterase, and a much more marked fall in polygalacturonase activity in the abnormal regions (Hobson, 1964). The incidence of blotchy ripening is affected to a greater degree by potassium than nitrogen nutrition (Winsor *et al.*, 1961; Winsor and Long, 1967; Ozbun *et al.*, 1967), but the opposite result was found with regard to the total activities of the pectic enzymes (Hobson, 1963b). It is possible that the abundant supplies of nitrogen were not utilized because protein synthesis was impaired by potassium deficiency. Likely mechanisms by which potassium exerts its strong effect on tomato fruit quality have been briefly discussed by Ozbun *et al.* (1967). It should not be thought, however, that blotchy ripening is exclusively a nutritional problem, and the extensive work that has been done on the influence of various environmental factors on the incidence of tomato ripening disorders has been reviewed by Cooper (1957) and again by Woods (1966).

A comparative study was made by Davies (1966c) of some of the more important non-volatile organic acids occurring in the walls of blotchy fruit. The titratable and total acidities of the green areas were lower and the combined acidity higher than those of the red areas. Evidence was presented to show that both the red and green areas of abnormal fruit were exceptional; some of the characteristics suggest that blotchy fruit as a whole would appear to be at a post-climacteric stage, while other characteristics, particularly of the green areas, imply that the tissue is more akin to a juvenile stage of development.

An enzyme system leading to the condensation of polyphenolic acids was considered by Kidson (1958) to be highly active in immature tomato fruit. Although the activity decreased as normal fruit ripened, it did not do so in the green areas of blotchy fruit. Subsequently, Hobson (1967a) showed that phenolase activity increased with maturity up to incipient ripeness after which the activity decreased. The red areas of blotchy fruit contained an abnormally high activity whereas in the green areas it was even higher.

Mitochondria derived from the red and green parts of blotchy fruit appear to be biochemically similar to particles isolated from the correspondingly coloured parts of healthy fruits (Hobson, 1969). Garrison (1967) also found that mitochondria from blotchy tissue retain many of the characteristics of particles from mature green tissue although the RC and ADP : O ratios were often somewhat lower in the former tissue than in the latter.

IV. FACTORS AFFECTING TOMATO FRUIT RIPENING†

Much of the tomato crop is picked at the green mature stage and most of the remainder, for the fresh market at least, is harvested before the fruit reaches full ripeness. Many of the biochemical changes so far investigated occur in the fruit whether the climacteric rise in respiration takes place naturally or is artificially stimulated by ethylene, although in the latter case inadequate pigmentation and poor flavour development may occur.

A. Effects of Temperature and Controlled Atmospheres During Storage

Conditions for the successful ripening and storage of tomatoes have been investigated for many years; early work has been summarized by Wright *et al.* (1931). Only recently has the ability of certain varieties to withstand prepacking and prolonged storage been fully appreciated. Handling and

† See Volume 1, Chapters 16 and 17.

storage damage is minimized by the use of firm varieties which ripen slowly, and such varieties have long been popular with the producers. Factors in the storage life of tomatoes and their resistance to infection were studied by Tomkins (1963), while Okubo and Maezawa (1966) followed the build-up of carbon dioxide by tomatoes stored in sealed plastic bags which caused a slowing down of fruit colour development. Haber (1931) considered that 10°C was the best temperature to store both green and red tomatoes in order to maintain quality, while Herregods (1963) established minimum storage temperatures ranging from 16°C for green fruit to 8°C for red fruit, allowing them to be kept for from 3 to 4 weeks. Truscott and Brubacher (1964) sought the best conditions for maximum length of storage without loss of quality. They recommended storing mature green fruit for 10 days at 10°C, ripening at 21°C (2–6 days) followed by holding at 10°C for a further 8–10 days. Truscott and Warner (1967) subsequently reiterated that 21°C was the best storage temperature where maximum uniformity of ripening, high quality and minimum loss from rot were required. Low temperature storage tended to inhibit the usual decline in acidity during ripening (Hall, 1966, 1968), while a decrease in firmness and an increase in colour occurred as the storage time at 7°C before ripening was extended (Hall, 1966). Adverse effects of low temperature storage on the firmness, colour and susceptibility to decay have also been reported (Hall, 1961, 1966). In contrast, high air temperatures during tomato harvesting in Florida have prompted investigations into the effects of exposure to high temperatures for short periods (Hall, 1964a, 1965). The best colour was obtained between 27 and 32°C, and a significant variety × temperature interaction was established with respect to fruit firmness. Parsons *et al.* (1964) found that storage in nitrogen at 16°C for 7 days retarded ripening. On removal to air at 21°C, tomatoes previously held in 99% nitrogen ripened slowly without marked deleterious effects on either flavour or appearance, but fruit stored for longer than 4 days in 100% nitrogen subsequently acquired abnormal flavours and ripened atypically.

As the underlying causes of the ripening processes become better established, storage techniques will become more sophisticated and precise. Among recent developments, delays in ripening by lowering the air pressure surrounding stored tomatoes, with consequent reduction in oxygen tension and sensitivity to ethylene (Burg and Burg, 1966) and the development of stabilizing substances affecting basic metabolism (Hurst, 1967), may be cited. The delay in ripening brought about by ethylene oxide (Lieberman *et al.*, 1964) may prove to be a further method of slowing down the natural sequence of events. Storage investigations should be allied to analytical and organoleptic tests to ensure that the quality of the fruit does not diminish too rapidly.

B. The Function of Ethylene†

Three reviews (Burg, 1962; Biale, 1964; Hansen, 1966) have summarized current opinion on the role of ethylene in fruit ripening, and only recent advances in connection with the tomato are included in this section.

Spencer (1956) showed that from a pre-climacteric minimum rate of respiration an initial rise in ethylene production was closely related to carbon dioxide evolution, and this was followed by a series of peaks for each of the two gases towards ripening and beyond. She went on to produce evidence for a close relation between respiration, ripening and ethylene production. Another study of ethylene evolution during ripening of tomatoes was made by Workman and Pratt (1957), who indicated that production followed a sigmoid curve. More recently, Basham (1965) gave further results which included examples of both wave and sigmoid production curves.

In normal circumstances, exogenous supplies of ethylene enable tomatoes to ripen more quickly and with more uniformity by initiating an earlier climacteric rise (Pratt and Workman, 1962), although this treatment is ineffective if applied at too low or too high a temperature or in atmospheres poor in oxygen (see Burg, 1962). Burg and Burg (1962, 1965) presented evidence that amounts of ethylene sufficient to stimulate ripening were present in several fruits prior to the climacteric respiration rise and that the concentration increased towards the respiration peak. Severance of the fruit from the plant altered the sensitivity towards ethylene and it was suggested that this might follow from cutting off supplies of a ripening inhibitor (Burg and Burg, 1964). The same authors went on to show that when fruit was subjected to unnatural atmospheres such as occur in gas storage chambers (C.A. storage) then carbon dioxide may displace ethylene at a receptor site in the tissue.

Evidence for the synthesis of ethylene by mitochondria has been critically reviewed by Hansen (1966). An early report of ethylene evolution by sub-cellular particles from tomatoes was later shown to be ethane (Lieberman and Mapson, 1962a). Ku and Pratt (1968) confirmed previous findings (Meigh *et al.* 1960; Ram Chandra and Spencer, 1962) that intact mitochondria do not produce significant quantities of ethylene. Even in the presence of light and adequate supplies of substrates and co-factors, disruption of isolated particles by blending, sonication or ageing, effectively led to ethylene evolution (Ram Chandra and Spencer, 1962), the action of various substances on the rate of production has been investigated (see Meheriuk and Spencer, 1967a, b, c). It is evident that the major source of ethylene *in vivo* is a soluble enzyme system, but contributions from aged mitochondria and non-enzymic sources cannot at present be discounted.

† See also Volume 1, Chapter 16.

The mechanism by which ethylene carries out its action is, as yet, by no means clear. The olefin appears to have an effect upon membrane permeability (Lyons and Pratt, 1964), which may, in turn, increase the respiration rate, leading to the initiation of the climacteric rise. However, Burg (1968) argued that ethylene at biologically active concentrations did not affect the permeability. Alterations in compartmentation were suggested by Sacher (1967) to lead to increased ethylene synthesis the effect of which, through altered membrane properties, then spreads throughout the tissue. Basham (1965) emphasized that whereas ethylene production and respiration rates were closely correlated up to the onset of the climacteric rise, once this had begun there was no longer a consistent association. However, the relation between ethylene production and red colour formation is a continuing one and both processes have similar temperature coefficients. Ethylene could, therefore, both initiate the onset of the respiratory rise and control the rate of red pigment formation. However, Frenkel *et al.* (1968) showed that when protein synthesis in pears was blocked by cycloheximide, ripening and ethylene synthesis were both inhibited whereas the respiration rate was unaffected or increased.

A report by Lieberman and Mapson (1962b) indicated that ethylene oxide delayed the ripening of tomatoes, and subsequent work (Lieberman *et al.*, 1964) demonstrated an antagonism between ethylene oxide and ethylene in their respective effects upon green tomatoes. While ethylene oxide undoubtedly retards ripening, it does not appear that the inhibition is competitive (Burg and Burg, 1965). Dostal and Leopold (1967) reported that ethylene stimulated the pigment changes associated with the normal ripening of tomatoes, but prior applications of gibberellin eliminated this effect. However, the usual respiratory behaviour of normal tomatoes and those treated with ethylene, was unaffected by the presence of gibberellin; this constitutes an interesting instance of the disturbance of the normal coordinated processes of ripening.

C. Ripening Disorders

Ripening disorders are a serious problem in tomato production in many parts of the world. Symptoms range from minor pigmentation blemishes to pronounced tissue discoloration and other abnormalities, rendering the fruit both unmarketable and inedible. Lack of adequate descriptions of the various disorders encountered in different tomato-growing areas, together with a multiplicity of names, has led to extreme confusion as yet largely unresolved.

One of the first disorders to be described adequately was "blotchy ripening" (Bewley and White, 1926), characterized by the appearance of green, yellow or translucent irregularly shaped hard patches of tissue often occurring near the calyx (Winsor, 1960) on an apparently normal red background. Kidson and Stanton (1953a, b, c) use the word "cloud" and a more severe form of the disorder has been called "waxy patch". Numerous histological investigations (e.g. Seaton and Gray, 1936; Kidson and Stanton, 1953a; Gigante, 1954; Sadik and Minges, 1966) demonstrated breakdown of the parenchyma near the vascular bundles in the outer walls of the fruit; necrotic areas often surround the vascular bundles but seldom invade them. These areas may well be formed when the spatial separation of phenolase and its substrates breaks down under stress (Hobson, 1967a). Parenchymatous cells adjacent to the bundles are less well expanded and somewhat disorganised (Fogleman, 1966). Sadik and Minges (1966) distinguished two types of abnormal tissue, one, "white tissue," consists of hard, lignified opaque cells containing considerable amounts of gas and starch. The other, "brown tissue", contains lignified parenchymatous cells occuring in strands or bands surrounding the vascular bundles. In severe cases, collapse, necrosis and breakdown of the parenchyma lead to the formation of lacunae containing brown particulate material which sometimes occludes the system. Possible causes of blotchy ripening are discussed in Section IIID.

A disorder of green tomatoes in which necrosis of the fruit wall is visible externally was termed "vascular browning" by Conover (1949). In mild cases the browning is restricted to tissue immediately adjacent to the vascular bundles, but in severely affected fruits the entire outer walls are involved. Haenseler (1949) called a similar, but more extensive, disorder "internal browning". This disease has also been investigated by Wharton and Boyle (1957) and by Smith *et al.* (1965). Nothing appears to be known of the biochemical changes involved in this type of disease which appears to be caused by infection with certain strains of tobacco virus. Various other virus infections leading to "pitting", "bronzing", etc. have been described by Broadbent (1963) and Jenkins *et al.* (1965). Some virus infections of the vine may not cause permanent injury and the fruits may even be of high quality (Broadbent and Winsor, 1964).

"Graywall", another disease of the tomato possibly akin to blotchy ripening, has been investigated by Conover (1949), Stoner (1951), Murakishi (1960), Kidson and Stanton (1953a) and Minges and Sadik (1965). Tobacco virus could be involved here although Hall and Stall (1967) have suggested bacterial infection as the source of graywall.

The symptom of another type of injury which goes under a number of names, is a hard green or yellow band of tissue encircling the calyx area of the fruit; this disease appears to be under genetic control and has been examined by Hester and Hall (1942), Winsor (1960), Sadik and Minges (1966) and Woods (1966). Here again nothing appears to be known of the biochemistry of the disorder.

"Blossom-end" rot is a widespread disorder; its development and histology have been studied in detail by Spurr (1959). Brown proteinaceous inclusions occur both in the epidermis and in the associated cells of the pericarp at the distal end of the fruit. Necrosis of the tissues quickly follows and in severe cases this can affect the whole fruit. It seems likely that these symptoms stem from extensive condensation of polyphenolic compounds to form dark-coloured pigments through the action of phenolase (Hobson, 1967a). Calcium deficiency appears to be the basic cause of the disorder (see Wiersum, 1966), with water stress as a possible secondary factor. Work by Van Goor (1968) has shown that disorganization of the cell membranes is associated with the disorder.

D. Sprays and Dips

A number of chemicals, usually growth inhibitors or promoters, affect the rate at which mature green tomatoes ripen after harvest. Ammonium thiocyanate and 2,4,6-trichlorophenoxyacetic acid were shown by Hartman (1959) to shorten the ripening time of green fruit. Boe *et al.* (1968) found that both 2-naphthoxyacetic acid and 2,4-dichlorophenoxyacetic acid also accelerated ripening, although previous workers indicated that the latter compound was inactive in this respect (Abdel-Kader *et al.*, 1966).

The methyl ester of indole-3-acetic acid was used by Mathur (1963) to reduce the decay rate of irradiated tomatoes, but at the same time it slowed the ripening of both control and irradiated fruit. The ethyl ester of gibberellic acid was also used in similar circumstances (Mathur, 1965), and although it appeared to hasten the ripening of all treated tomatoes it also increased the incidence of decay.

Other substances extend the ripening period, such as 1-naphthylacetic acid (Hartman, 1959). Gibberellin itself and kinetin are also active in this respect together, possibly, with indole-3-acetic acid and its ethyl ester (Abdel-Kader *et al.*, 1966). Gibberellin and kinetin are particularly effective in extending the shelf life of tomatoes; their action, within limits, is inversely related to concentration. Dostal and Leopold (1967) also showed that gibberellin was able to defer ripening, although ethylene-stimulated respiration was not affected, as discussed in Section IVB. These active substances probably enter the fruit by absorption through the calyx or the stem scar, but their mode of action is not known.

V. THE CONTROL OF TOMATO FRUIT RIPENING—A SUMMARY

The tomato has been widely used experimentally because it is easily and rapidly grown, is readily responsive to experimental treatment and is of world-wide importance commercially. Changes in the main constituents during the ripening and senescence of the fruit have attracted considerable attention, and much information is available on the nutritional and environmental conditions required to give high yields combined with good marketable quality. Only recently has the statistical design and evaluation of experiments become more commonplace and the validity of some of the earlier work must be questioned on these grounds. The nomenclature of tomato ripening disorders is still extremely confused; unless clear unequivocal descriptions are given and rigorous standards adopted, the value of work in this field is seriously diminished. Explanations in biochemical terms of the major ripening disorders should follow in the next few years. The effect of temperature on the ripening processes is far from simple and it is not merely a question of accelerating or delaying the changes. Further information on storage phenomena both at the cellular and sub-cellular levels is needed.

The nature of the stimulus initiating ripening is, as yet, unknown; indeed, it is still not clear whether changes in the susceptibility of the fruit to internal or external stimuli is the primary cause or whether the parent plant provides a diminishing supply of a ripening inhibitor. In many other fruits, ethylene appears to be the initiator of the processes leading to ripening. However, in the tomato it has not been established whether the respiration changes, possible increases in phosphate acceptor groups, alterations in membrane permeability, ethylene production, or shifts in metabolic pathways directed by selective protein synthesis, are primary or secondary effects during fruit ripening. The increasing use of electron microscopy, coupled with detailed studies of the enzyme systems operating, may help to clarify some of these problems.

As Spencer (1965) has pointed out, the processes controlling the development of climacteric and non-climacteric fruit may have much in common except for differences in the time scales. Even within the class of climacteric fruit wide variations exist; thus an intact tomato will ripen in a few days on the plant whereas an avocado in similar circumstances takes several months. On the other hand, mitochondria from both these fruits are markedly influenced by the presence of TPP, and this co-factor probably plays some central role in controlling the rate of oxidation of Krebs cycle intermediates *in vivo*.

Ethylene synthesis by fruit is intimately concerned with ripening although its precise role is still inadequately defined. Basham (1965) obtained evidence that, as with other climacteric fruits, ethylene production is directly correlated with the respiration rate up to the onset of the climacteric rise but not thereafter. The development of red colour, however, which occurs mainly in the post-climacteric period, does appear to be related to the rate of ethylene production. Ethylene in tomato fruit is probably produced by a soluble enzyme system (Ku and Pratt, 1968) and such a system in the tomato has been investigated by Mapson and Wardale (1971). The high volatility of ethylene militates against its involvement in blotchy ripening where a sharp demarcation between the red and green areas exists. It seems more likely that damage associated with the vascular bundles in this disorder results from a stress reaction, and that this leads to a partial or complete isolation of the affected tissue, the composition of which is consequently influenced by its separation from the mainstream of development. However, alterations in composition extend beyond the limits of the green areas and blotchy fruit are abnormal as a whole.

As the effects of the interactions between ethylene, growth regulators and other physiologically active substances on the ripening of tomato fruit become better understood, an integrated picture of the life of the tomato is being built up. However, the enjoyment to be gained from eating a freshly picked fully ripe tomato having a high acid and sugar content cannot yet be measured in biochemical terms.

REFERENCES

Abdel-Kader, A. S., Morris, L. L. and Maxie, E. C. (1966). *HortScience* **1,** 90.
Abdul-Baki, A. A. (1964). Ph.D. Thesis, University of Illinois.
Aberg, B. (1946). *K. LantbrHogsk. Annl.* **13,** 239.
Airan, J. W. and Barnabas, J. (1953). *J. Univ. Bombay* **22,** 29.
Akkerman, A. M. and Veldstra, H. (1947). *Rec. trav. chim.* **66,** 411.
Anderson, R. E. (1957). Ph.D. Thesis, University of Illinois.
Andreotti, R. (1956). *Industria Conserve* **31,** 305.
Ansiaux, J. R. (1960). *Annls Physiol. veg. Brux.* **5,** 19.
Arneson, P. A. and Durbin, R. D. (1967). *Phytopathology* **57,** 1358.
Arnon, D. I. and Hoagland, D. R. (1943). *Bot. Gaz.* **104,** 576.
Baker, L. R. and Tomes, M. L. (1964). *Proc. Am. Soc. hort. Sci.* **85,** 507.
Barke, R. E. (1968). *Qd J. agric. anim. Sci.* **25,** 179.
Basham, C. W. (1965). Ph.D. Thesis, University of Maryland.
Baudisch, W. (1963). *Kulturpflanze* **11,** 244.
Beadle, N. C. W. (1937). *Aust. J. exp. Biol. med. Sci.* **15,** 173.
Beeson, K. C. (1941). *Misc. Publs U.S. Dep. Agric.* No. 369
Belluci, G. and Aldini, R. (1967). *Industria Conserve* **42,** 99.
Bennett, R. D., Heftmann, E., Purcell, A. E. and Bonner, J. (1961). *Science, N.Y.* **134,** 671.
Bewley, W. F. and White, H. L. (1926). *Ann. appl. Biol.* **13,** 323.
Biale, J. B. (1960). *In* "Handbuch der Pflanzenphysiologie" (W. Ruhland, ed.), Vol. 12, pp. 536–592. Springer Verlag, Berlin.
Biale, J. B. (1964). *Science, N.Y.* **146,** 880.
Bicknell, F. and Prescott, F. (1946). "The Vitamins in Medicine." William Heinemann Medical Books Ltd., London.
Bidmead, D. S. and Welti, D. (1960). *Research, Lond.* **13,** 295.
Biswas, T. D. and Das, N. B. (1952). *Sci. and Cult.* **18,** 91.
Blaim, K. (1963). *Postepy Nauk roln.* **10,** 99.
Boe, A. A. and Salunkhe, D. K. (1967a). *Experientia* **23,** 779.
Boe, A. A. and Salunkhe, D. K. (1967b). *Econ. Bot.* **21,** 312.

Boe, A. A., Do, J. Y. and Salunkhe, D. K. (1968). *Econ. Bot.* **22,** 124.
Bohart, G. S. (1940). *Fd Res.* **5,** 469.
Borchers, E. A. and Nevin, C. S. (1954). *Proc. Am. Soc. hort. Sci.* **63,** 420.
Bradley, D. B. (1960). *J. agric. Fd Chem.* **8,** 232.
Bradley, D. B. (1962). *J. agric. Fd Chem.* **10,** 450.
Brieskorn, C. H. and Reinartz, H. (1967a). *Z. Lebensmittelunters. u. -Forsch.* **133,** 137.
Brieskorn, C. H. and Reinartz, H. (1967b). *Z. Lebensmittelunters. u. -Forsch.* **135,** 55.
Broadbent, L. (1963). *Proc. XVIth int. hort. Congr. 1962* **2,** 350.
Broadbent, L. and Winsor, G. W. (1964). *Ann. appl. Biol.* **54,** 23.
Brown, A. P. and Moser, F. (1941). *Fd Res.* **6,** 45.
de Bruyn, J. W., van Keulen, H. A. and Ferguson, J. H. A. (1968). *J. Sci. Fd Agric.* **19,** 597.
Buhler, D. R., Hansen, E., Christensen, B. and Wang, C. H. (1956). *Pl. Physiol., Lancaster* **31,** 192.
Bulen, W. A., Varner, J. E. and Burrell, R. C. (1952). *Analyt. Chem.* **24,** 187.
Burg, S. P. (1962). *A. Rev. Pl. Physiol.* **13,** 265.
Burg, S. P. (1968). *Pl. Physiol., Lancaster* **43,** 1503.
Burg, S. P. and Burg, E. A. (1962). *Pl. Physiol., Lancaster* **37,** 179.
Burg, S. P. and Burg, E. A. (1964). *Pl. Physiol., Lancaster* **39,** x.
Burg, S. P. and Burg, E. A. (1965). *Science, N.Y.* **148,** 1190.
Burg, S. P. and Burg, E. A. (1966). *Science, N.Y.* **153,** 314.
Burroughs, L. F. (1960). *J. Sci. Fd Agric.* **11,** 14.
Buttery, R. G. and Seifert, R. M. (1968). *J. agric. Fd Chem.* **16,** 1053.
Buttery, R. G., Seifert, R. M. and Ling, L. C. (1969). *Chemy Ind.* 238.
Carangal, A. R., Jr. Alban, E. K., Varner, J. E. and Burrell, R. C. (1954). *Pl. Physiol., Lancaster* **29,** 355.
Chance, B. and Williams, G. R. (1956). *Adv. Enzymol.* **17,** 65.
Chichester, C. O. and Nakayama, T. O. M. (1965). *In* "Chemistry and Biochemistry of Plant Pigments" (T. W. Goodwin, ed.), pp. 439–457. Academic Press, New York.
Clendenning, K. A. (1942). *Can. J. Res.* **20C,** 197.
Clutter, M. E. and Miller, E. V. (1961). *Econ. Bot.* **15,** 218.
Conover, R. A. (1949). *Pl. Dis. Rptr* **33,** 283.
Cooper, A. J. (1957). *Rep. Glasshouse Crops Res. Inst. 1956* 39.
Crane, M. B. and Zilva, S. S. (1949). *J. Pomol.* **25,** 36.
Curl, A. L. (1961). *J. Fd Sci.* **26,** 106.
Currence, T. M. (1939). *Proc. Am. Soc. hort. Sci.* **37,** 901.
Czygan, F. C. and Willühn, G. (1967). *Planta Med.* **15,** 404.
Dalal, K. B., Salunkhe, D. K., Boe, A. A. and Olson, L. E. (1965). *J. Fd Sci.* **30,** 504.
Dalal, K. B., Salunkhe, D. K. and Olson, L. E. (1966). *J. Fd Sci.* **31,** 461.
Dalal, K. B., Olson, L. E., Yu, M.-H. and Salunkhe, D. K. (1967). *Phytochem.* **6,** 155.
Dalal, K. B., Salunkhe, D. K., Olson, L. E., Do, J. Y. and Yu, M.-H. (1968). *Pl. Cell Physiol., Tokyo* **9,** 389.
Darkanabaev, T. B., Lysenko, M. K. and Niretina, N. V. (1964). *Trudy Inst. Bot., Alma-Ata* **20,** 144.
Daskaloff, C. and Ognjanowa, A. (1962). *Arch. Gartenb.* **10,** 193.

Davies, J. N. (1958). *Rep. Glasshouse Crops Res. Inst. 1957* 67.
Davies, J. N. (1964). *J. Sci. Fd Agric.* **15,** 665.
Davies, J. N. (1965a). *Rep. Glasshouse Crops Res. Inst. 1964* 64.
Davies, J. N. (1965b). *Rep. Glasshouse Crops Res. Inst. 1964* 139.
Davies, J. N. (1966a). *Nature, Lond.* **209,** 640.
Davies, J. N. (1966b). *J. Sci. Fd Agric.* **17,** 396.
Davies, J. N. (1966c). *J. Sci. Fd Agric.* **17,** 400.
Davies, J. N. (1967). *Rep. Glasshouse Crops Res. Inst. 1966* 65.
Davies, J. N. and Winsor, G. W. (1959). *Rep. Glasshouse Crops Res. Inst. 1958* 23.
Davies, J. N. and Winsor, G. W. (1967). *J. Sci. Fd Agric.* **18,** 459.
Davies, J. N. and Winsor, G. W. (1969). *J. hort. Sci.* **44,** 331.
Davies, J. W. and Cocking, E. C. (1965). *Planta* **67,** 242.
Davies, J. W. and Cocking, E. C. (1967). *Planta* **76,** 285.
Denisen, E. L. (1951a). *Iowa St. Coll. J. Sci.* **25,** 549.
Denisen, E. L. (1951b). *Iowa St. Coll. J. Sci.* **25,** 565.
Dickinson, D. B. and Hanson, J. B. (1965). *Pl. Physiol., Lancaster* **40,** 161.
Dostal, H. C. and Leopold, A. C. (1967). *Science, N.Y.* **158,** 1579.
Doyle, W. P., Huff, R. and Wang, C. H. (1960). *Pl. Physiol., Lancaster* **35,** 745.
Drury, R. E., McCollum, J. P. and Garrison, S. A. (1968a). *Pl. Physiol., Lancaster* **43,** 248.
Drury, R. E., McCollum, J. P., Garrison, S. A. and Dickinson, D. B. (1968b). *Phytochem.* **7,** 2071.
Eaton, F. M. (1944). *J. agric. Res.* **69,** 237.
Edwards, R. A. and Reuter, F. H. (1967). *Fd Technol. Aust.* **19,** 352.
Evans, H. J. and Sorger, G. J. (1966). *A. Rev. Pl. Physiol.* **17,** 47.
F.A.O. (1968). "Production Yearbook, 1967", Vol. 21. Food and Agricultural Organization of the United Nations, Rome.
Fogleman, M. E. (1966). Ph.D. Thesis, Iowa State University.
Forshey, C. G. and Alban, E. K. (1954). *Proc. Am. Soc. hort. Sci.* **64,** 372.
Frazier, J. C., Ascham, L., Cardwell, A. B., Fryer, H. C. and Willis, W. W. (1954). *Proc. Am. Soc. hort. Sci.* **64,** 351.
Freeman, J. A. (1960). Ph.D. Thesis, Washington State University.
Freeman, J. A. and Woodbridge, C. G. (1960). *Proc. Am. Soc. hort. Sci.* **76,** 515.
Frenkel, C., Klein, I. and Dilley, D. R. (1968). *Pl. Physiol., Lancaster* **43,** 1146.
Fryer, H. C., Ascham, L., Cardwell, A. B., Frazier, J. C. and Willis, W. W. (1954). *Proc. Am. Soc. hort. Sci.* **64,** 360.
Garrison, S. A. (1967). Ph.D. Thesis, University of Illinois,
Georgiev, H. P. and Balzer, I. (1962). *Arch. Gartenb.* **10,** 398.
Geraldson, C. M. (1957). *Proc. Am. Soc. hort. Sci.* **69,** 309.
Giannone, L. and Baldrati, G. (1967a). *Industria Conserve* **42,** 86.
Giannone, L. and Baldrati, G. (1967b). *Industria Conserve* **42,** 176.
Giannone, L. and Baldrati, G. (1967c). *Industria Conserve* **42,** 252.
Gigante, R. (1954). *Boll. Staz. Patol. veg. Roma* **12,** 127.
Goodwin, T. W. (1965). *In* "Chemistry and Biochemistry of Plant Pigments" (T. W. Goodwin, ed.), pp. 127–173. Academic Press, New York.
Gortner, W. A., Dull, G. G. and Krauss, B. H. (1967). *HortScience* **2,** 141.
Govindan, P. R. (1952). *Curr. Sci.* **21,** 15.
Guadagni, D. G., Miers, J. C. and Venstrom, D. (1968). *Fd Technol., Champaign* **22,** 1003.
Gum, O. B., Brown, H. D. and Burrell, R. C. (1945). *Pl. Physiol., Lancaster* **20,** 267.

Haber, E. S. (1931). *Iowa St. Coll. J. Sci.* **5,** 171.
Haenseler, C. M. (1949). *Pl. Dis. Rptr* **33,** 336.
Hall, C. B. (1961). *Proc. Am. Soc. hort. Sci.* **78,** 480.
Hall, C. B. (1964a). *Proc. Am. Soc. hort. Sci.* **84,** 501.
Hall, C. B. (1964b). *Proc. Fla St. hort. Soc. 1963* **76,** 304.
Hall, C. B. (1965). *Proc. Fla St. hort. Soc. 1964* **77,** 252.
Hall, C. B. (1966). *Proc. Fla St. hort. Soc. 1965* **78,** 241.
Hall, C. B. (1968). *HortScience* **3,** 37.
Hall, C. B. and Stall, R. E. (1967). *Proc. Am. Soc. hort. Sci.* **91,** 573.
Hamdy, M. M. and Gould, W. A. (1962). *J. agric. Fd Chem.* **10,** 499.
Hamner, K. C. and Maynard, L. A. (1942). *Misc. Publs U.S. Dep. Agric.* No. 502.
Hamner, K. C., Lyon, C. B. and Hamner, C. L. (1942). *Bot. Gaz.* **103,** 586.
Hansen, E. (1966). *A. Rev. Pl. Physiol.* **17,** 459.
Hartman, R. T. (1959). *Pl. Physiol., Lancaster* **34,** 65.
Hassan, H. H. and McCollum, J. P. (1954). *Bull. Ill. agric. Exp. Stn* No. 573.
Hein, R. E. and Fuller, G. W. (1963). Conference on Advances in Flavour Research, Southern Utilization Research and Development Division, U.S. Dept. of Agric., New Orleans.
Herregods, M. (1963). *Tuinbouwberichten* **27,** 434
Herrmann, K. (1957). *Z. Lebensmittelunters. u. -Forsch.* **106,** 341.
Hester, J. B. (1940). *Am. Fertil.* **93**(11), 5–7, 24, 26.
Hester, J. B. (1941). *Science, N.Y.* **93,** 401.
Hester, J. B. and Hall, H. F. (1942). *Campbell Soup Co., Bull.* No. 4.
Hewitt, E. J. (1963). *In* "Plant Physiology" (F. C. Steward, ed.), Vol. 3, pp. 137–360. Academic Press, New York.
Hills, H. G. and Collins, R. P. (1966). *Phyton, B. Aires* **23,** 91.
Hobson, G. E. (1963a). *Biochem. J.* **86,** 358.
Hobson, G. E. (1963b). *J. Sci. Fd Agric.* **14,** 550.
Hobson, G. E. (1964). *Biochem. J.* **92,** 324.
Hobson, G. E. (1965a). *J. exp. Bot.* **16,** 411.
Hobson, G. E. (1965b). *J. hort. Sci.* **40,** 66.
Hobson, G. E. (1967a). *J. Sci. Fd Agric.* **18,** 523.
Hobson, G. E. (1967b). *Rep. Glasshouse Crops Res. Inst. 1966* 130.
Hobson, G. E. (1968). *J. Fd Sci.* **33,** 588.
Hobson, G. E. (1969). *Qualitas Pl. Mater. vég.* **19,** 155.
Hopkins, H. and Eisen, J. (1959). *J. agric. Fd Chem.* **7,** 633.
Hurst, H. (1967). *Grower* **67,** 457.
Imparato-Gargano, E. and Violante, P. (1963). *Annali Fac. Sci. agr. Univ. Napoli* **28,** 229.
Irving, G. W. (1947). *J. Wash. Acad. Sci.* **37,** 293.
Jacquin, P. and Tavernier, J. (1955). *Bull. Soc. scient. Hyg. aliment.* **43,** 214.
Jelenic, D., Vajnberger, A. and Rasic, J. (1958). *Zemlj Biljka* **7,** 387.
Jenkins, J. E. E., Wiggell, D. and Fletcher, J. T. (1965). *Ann. appl. Biol.* **55,** 71.
Johnson, J. H., Gould, W. A., Badenhop, A. F. and Johnson, R. M., Jr. (1968). *J. agric. Fd Chem.* **16,** 255.
Jurics, E. W. (1966). *Z. Lebensmittelunters. u. -Forsch.* **132,** 193.
Kajderowicz-Jarosinska, D. (1965). *Acta Agr. Silvestria, Ser. Roln.* **5,** 3.
Kapp, P. P. (1966). Ph.D. Thesis, Louisiana State University.
Kargl, T. E., Quackenbush, F. W. and Tomes, M. L. (1960). *Proc. Am. Soc. hort. Sci.* **75,** 574.

Kaski, I. J., Webster, G. L. and Kirch, E. R. (1944). *Fd Res.* **9**, 386.
Katayama, O., Tsubata, K. and Yamato, I. (1967). *Nippon Shokuhin Kogyo Gakkaishi* **14**, 444.
Kidson, E. B. (1958). *N.Z. Jl agric. Res.* **1**, 896.
Kidson, E. B. (1963). *N.Z. Jl agric. Res.* **6**, 376.
Kidson, E. B. and Stanton, D. J. (1953a). *N.Z. Jl Sci. Technol.* **A34**, 521.
Kidson, E. B. and Stanton, D. J. (1953b). *N.Z. Jl Sci. Technol.* **A35**, 1.
Kidson, E. B. and Stanton, D. J. (1953c). *N.Z. Jl Sci. Technol.* **A35**, 368.
Koch, B. (1960). *Agrobotanika* **2**, 115.
Kolankiewicz, J., Mlodecki, H. and Szymczyk, F. (1954). *Roczn. panst. Zakl. Hig.* **5**, 214.
Kolankiewicz, J., Mlodecki, H. and Szymczyk, F. (1956). *Roczn. panst. Zakl. Hig.* **7**, 537.
Ku, H. S. and Pratt, H. K. (1968). *Pl. Physiol., Lancaster* **43**, 999.
Ku, H. S., Pratt, H. K., Spurr, A. R. and Harris, W. M. (1968). *Pl. Physiol., Lancaster* **43**, 883.
Lambeth, V. N., Fields, M. L. and Huecker, D. E. (1964). *Bull. Mo agric. Exp. Stn* No. 850.
Lambeth, V. N., Straten, E. F. and Fields, M. L. (1966). *Bull. Mo agric. Exp. Stn* No. 908.
Lee, F. A. and Sayre, C. B. (1946). *Tech. Bull. N.Y. St. agric. Exp. Stn* No. 278.
Lewis, A. H. and Marmoy, F. B. (1939). *J. Pomol.* **17**, 275.
Lieberman, M. and Mapson, L. W. (1962a). *Nature, Lond.* **195**, 1016.
Lieberman, M. and Mapson, L. W. (1962b). *Nature, Lond.* **196**, 660.
Lieberman, M., Asen, S. and Mapson, L. W. (1964). *Nature, Lond.* **204**, 756.
Lower, R. L. and Thompson, A. E. (1967). *Proc. Am. Soc. hort. Sci.* **91**, 486.
Luckwill, L. C. (1943). *Aberd. Univ. Stud.* No. 120.
Lyons, J. M. and Pratt, H. K. (1964). *Archs Biochem. Biophys.* **104**, 318.
Lyons, J. M., Wheaton, T. A. and Pratt, H. K. (1964). *Pl. Physiol., Lancaster* **39**, 262.
McClendon, J. H., Woodmansee, C. W. and Somers, G. F. (1959). *Pl. Physiol., Lancaster* **34**, 389.
McHenry, E. W. and Graham, M. (1935). *Biochem. J.* **29**, 2013.
Maclinn, W. A. and Fellers, C. R. (1938). *Bull. Mass. agric. Exp. Stn* No. 354.
Maclinn, W. A., Fellers, C. R. and Buck, R. E. (1936). *Proc. Am. Soc. hort. Sci.* **34**, 543.
McCollum, J. P. (1946). *Proc. Am. Soc. hort. Sci.* **48**, 413.
McCollum, J. P. (1954). *Fd Res.* **19**, 182.
McCollum, J. P. (1955). *Fd Res.* **20**, 55.
Mahdi, A. A., Rice, A. C. and Weckel, K. G. (1959). *J. agric. Fd Chem.* **7**, 712.
Mapson, L. W. (1958). *A. Rev. Pl. Physiol.* **9**, 119.
Mapson, L. W. and Wardale, D. A. (1971). *Phytochem.* **10**, 29.
Massey, D. M. and Winsor, G. W. (1957). *Rep. Glasshouse Crops Res. Inst. 1956* 52.
Mathur, P. B. (1963). *Nature, Lond.* **199**, 1007.
Mathur, P. B. (1965). *Nature, Lond.* **207**, 212.
Matthews, R. F. (1960). Ph.D. Thesis, Cornell University; *Dissert. Abstr.* (1961) **21**, 1693.
Maxie, E. C. and Abdel-Kader, A. (1966). *Adv. Fd Res.* **15**, 105.
Meheriuk, M. and Spencer, M. (1967a). *Phytochem.* **6**, 535.
Meheriuk, M. and Spencer, M. (1967b). *Phytochem.* **6**, 545.

Meheriuk, M. and Spencer, M. (1967c). *Phytochem.* **6,** 551.
Meigh, D. F., Norris, K. H., Craft, C. C. and Lieberman, M. (1960). *Nature, Lond.* **186,** 902.
Meredith, F. I. and Purcell, A. E. (1966). *Proc. Am. Soc. hort. Sci.* **89,** 544.
Messing, J. H. L. (1957). *Rep. Glasshouse Crops Res. Inst. 1954/55* 79.
Miers, J. C. (1966). *J. agric. Fd Chem.* **14,** 419.
Minges, P. A. and Sadik, S. (1965). *Proc. Fla St. hort. Soc. 1964* **77,** 246.
Mollard, J. (1953). *Annls Inst. natn. Rech. agron., Paris* (E) **2,** 27.
Muller, C. H. (1940). *Misc. Publs U.S. Dep. Agric.* No. 382.
Murakishi, H. H. (1960). *Phytopathology* **50,** 408.
Murneek, A. E., Maharg, L. and Wittwer, S. H. (1954). *Bull. Mo agric. Exp. Stn* No. 568.
Nason, A. and McElroy, W. D. (1963). *In* "Plant Physiology" (F. C. Steward, ed.), Vol. 3, pp. 451–536. Academic Press, New York.
Nelson, P. E. and Hoff, J. E. (1969). *J. Fd Sci.* **34,** 53.
Nettles, V. F., Hall, C. B. and Dennison, R. A. (1955). *Proc. Am. Soc. hort. Sci.* **65,** 349.
Neubert, P. (1959). *Arch. Gartenb.* **7,** 29.
Okubo, M. and Maezawa, T. (1966). *J. Jap. Soc. hort. Sci.* **35,** 277.
Olliver, M. (1967). *In* "The Vitamins" (Sebrell, W. H. and Harris, R. S., eds), p. 359. Academic Press, New York.
Ozbun, J. L., Boutonnet, C. E., Sadik, S. and Minges, P. A. (1967). *Proc. Am. Soc. hort. Sci.* **91,** 556.
Palmieri, F. (1965). *Annali Fac. Sci. agr. Univ. Napoli* **30,** 579.
Parsons, C. S., Gates, J. E. and Spalding, D. H. (1964). *Proc. Am. Soc. hort. Sci.* **84,** 549.
Piringer, A. A. and Heinze, P. H. (1954). *Pl. Physiol., Lancaster* **29,** 467.
Pollard, A., Kieser, M. E. and Bryan, J. D. (1948). *J. Soc. chem. Ind.* **67,** 281.
Porter, J. W. and Zscheile, F. P. (1946). *Archs Biochem.* **10,** 537.
Pratt, H. K. and Workman, M. (1962). *Proc. Am. Soc. hort. Sci.* **81,** 467.
Pratt, H. K., Hulme, A. C., Jones, J. D. and Wooltorton, L. S. C. (1965). *Rep. Ditton and Covent Garden Laboratories 1964–65* 44.
Pyne, A. W. and Wick, E. L. (1965). *J. Fd Sci.* **30,** 192.
Rakitin, Y. V. (1945). *Biokhimiya* **10,** 373.
Ram Chandra, G. and Spencer, M. (1962). *Nature, Lond.* **194,** 361.
Ramirez, D. A. and Tomes, M. L. (1964). *Bot. Gaz.* **125,** 221.
Ramsey, J. C. and Wang, C. H. (1962). *Nature, Lond.* **193,** 800.
Reynard, G. B. and Kanapaux, M. S. (1942). *Proc. Am. Soc. hort. Sci.* **41,** 298.
Rice, A. C. and Pederson, C. S. (1954). *Fd Res.* **19,** 106.
Richmond, A. and Biale, J. B. (1966). *Archs Biochem. Biophys.* **115,** 211.
Rivas, N. and Luh, B. S. (1968). *J. Fd Sci.* **33,** 358.
Rosa, J. T. (1925). *Proc. Am. Soc. hort. Sci.* **22,** 315.
Rosa, J. T. (1926). *Proc. Am. Soc. hort. Sci.* **23,** 233.
Rowan, K. S., Pratt, H. K. and Robertson, R. N. (1958). *Aust. J. biol. Sci.* **11,** 329.
Rubatzky, V. E. (1965). Ph.D. Thesis, Rutgers University, New Brunswick.
Ruszkowska, M. (1960). *Acta Soc. Bot. Pol.* **29,** 553.
Ruszkowska, M. and Zinkiewicz, J. (1953). *Roczn. Nauk. roln., Ser. A* **66,** 29.
Sacher, J. A. (1967). *In* "Aspects of the Biology of Ageing" (H. W. Woolhouse, ed.), pp. 269–304. University Press, Cambridge.
Sadik, S. and Minges, P. A. (1966). *Proc. Am. Soc. hort. Sci.* **88,** 532.

Sakiyama, R. (1966a). *J. Jap. Soc. hort. Sci.* **35,** 36.
Sakiyama, R. (1966b). *J. Jap. Soc. hort. Sci.* **35,** 260.
Sakiyama, R. (1967). *J. Jap. Soc. hort. Sci.* **36,** 399.
Sander, H. (1956). *Planta* **47,** 374.
Sander, H. (1958). *Naturwissenschaften* **45,** 59.
Sando, C. E. (1920). *Bull. U.S. Dep. Agric.* No. 859.
Saravacos, G., Luh, B. S. and Leonard, S. J. (1958a). *Fd Res.* **23,** 329.
Saravacos, G., Luh, B. S. and Leonard, S. J. (1958b). *Fd Res.* **23,** 648.
Saywell, L. G. and Cruess, W. V. (1932). *Bull. Calif. agric. Exp. Stn* No. 545.
Schormüller, J. and Grosch, W. (1962). *Z. Lebensmittelunters. u. -Forsch.* **118,** 385.
Schormüller, J. and Grosch, W. (1964). *Z. Lebensmittelunters. u. -Forsch.* **126,** 38.
Schormüller, J. and Grosch, W. (1965). *Z. Lebensmittelunters. u. -Forsch.* **126,** 188.
Seaton, H. L. and Gray, G. F. (1936). *J. agric. Res.* **52,** 217.
Shafshak, S. A. and Winsor, G. W. (1964). *J. hort. Sci.* **39,** 284.
Shewfelt, A. L. and Halpin, J. E. (1967). *Proc. Am. Soc. hort. Sci.* **91,** 561.
Simandle, P. A., Brogdon, J. L., Sweeney, J. P., Mobley, E. O. and Davis, D. W. (1966). *Proc. Am. Soc. hort. Sci.* **89,** 532.
Smith, P. R., Stubbs, L. L. and Sutherland, J. L. (1965). *Aust. J. exp. Agric. Anim. Husb.* **5,** 75.
Sowinska, H. (1966). *Przem. spozyw.* **20,** 404.
Spencer, M. S. (1956). *Can. J. Biochem. Physiol.* **34,** 1261.
Spencer, M. S. (1965). *In* "Plant Biochemistry" (J. Bonner and J. E. Varner, eds), pp. 793–825. Academic Press, New York.
Spencer, M. S. and Stanley, W. L. (1954). *J. agric. Fd Chem.* **2,** 1113.
Spurr, A. R. (1959). *Hilgardia* **28,** 269.
Stoner, W. N. (1951). *Proc. Fla St. hort. Soc.* 1950, **63,** 129.
Swain, T. and Hillis, W. F. (1959). *J. Sci. Fd Agric.* **10,** 63.
Takahashi, T. and Nakayama, M. (1959). *J. hort. Ass. Japan* **28,** 165.
Takahashi, T. and Nakayama, M. (1962). *J. Jap. Soc. hort. Sci.* **31,** 151.
Tarnovska, K. (1961). *Izv. tsent. nauchnoizsled. Inst. Rast., Sof.* **12,** 343.
Taverna, G. (1965). *Annali Fac. Sci. agr. Univ. Napoli* **30,** 495.
Thompson, A. E., Lower, R. L. and Hepler, R. W. (1964). *Proc. Am. Soc. hort. Sci.* **84,** 463.
Tomes, M. L. (1963). *Bot. Gaz.* **124,** 180.
Tomkins, R. G. (1963). *J. hort. Sci.* **38,** 335.
Trombly, H. H. and Porter, J. W. (1953). *Archs Biochem. Biophys.* **43,** 443.
Truscott, J. H. L. and Brubacher, L. (1964). *Rep. hort. Exp. Stn, Ont. 1963* 61.
Truscott, J. H. L. and Warner, J. (1967). *Rep. hort. Res. Inst., Ont. 1966* 96.
Tukalo, E. A. (1956). *Sb. nauch. Rab. dnepropetr. gos. med. Inst.* **1,** 347.
Udenfriend, S., Lovenberg, W. and Sjoerdsma, A. (1959). *Archs Biochem. Biophys.* **85,** 487.
Van Goor, B. J. (1968). *Physiologia Pl.* **21,** 1110.
Vendilo, G. G. (1964). *Vest. mosk. gos. Univ. Ser. VI, Biol. Pochv.* **19** (2), 52.
Villarreal, F., Luh, B. S. and Leonard, S. J. (1960). *Fd Technol., Champaign* **14,** 176.
Vogele, A. C. (1937). *Pl. Physiol., Lancaster* **12,** 929.
Walker, J. R. L. (1962). *J. Sci. Fd Agric.* **13,** 363.
Walkof, C. and Hyde, R. B. (1963). *Can. J. Pl. Sci.* **43,** 528.
Wang, C. H., Hansen, E. and Christensen, B. E. (1953). *Pl. Physiol., Lancaster* **28,** 741.
Wang, C. H., Doyle, W. P. and Ramsey, J. C. (1962). *Pl. Physiol., Lancaster* **37,** 1.

Ward, G. M. (1963). *Can. J. Pl. Sci.* **43,** 206.
Wehmer, C. (1931). "Die Pflanzenstoffe", Vol. 2, p. 1099 Fischer, Jena.
West, G. B. (1959a). *J. Pharm. Pharmac.* **11,** 319.
West, G. B. (1959b). *J. Pharm. Pharmac.* **11,** 275T.
Wharton, D. C. and Boyle, J. S. (1957). *Phytopathology* **47,** 208.
Wiersum, L. K. (1966). *Acta Bot. Neerl.* **15,** 406.
Wilkinson, B. (1957). *Nature, Lond.* **180,** 666.
Williams, K. T. and Bevenue, A. (1954). *J. Agric. Fd Chem.* **2,** 472.
Williams, A. H. (1957). *J. Sci. Fd Agric.* **8,** Suppl., S33.
Winsor, G. W. (1960). *Rep. Glasshouse Crops Res. Inst. 1959* 83.
Winsor, G. W. (1966). *Scient. Hort.* **28,** 27.
Winsor, G. W. and Davies, J. N. (1966). *Rep. Glasshouse Crops Res. Inst. 1965* 74.
Winsor, G. W. and Long, M. I. E. (1967). *J. hort. Sci.* **42,** 391.
Winsor, G. W. and Massey, D. M. (1958). *J. Sci. Fd Agric.* **9,** 493.
Winsor, G. W. and Massey, D. M. (1959). *J. Sci. Fd Agric.* **10,** 304.
Winsor, G. W., Davies, J. N. and Long, M. I. E. (1961). *J. hort. Sci.* **36,** 254.
Winsor, G. W., Davies, J. N. and Massey, D. M. (1962a). *J. Sci. Fd Agric.* **13,** 108.
Winsor, G. W., Davies, J. N. and Massey, D. M. (1962b). *J. Sci. Fd Agric.* **13,** 141.
Winsor, G. W., Davies, J. N. and Massey, D. M. (1962c). *J. Sci. Fd Agric.* **13,** 145.
Winsor, G. W., Messing, J. H. L. and Long, M. I. E. (1965). *J. hort. Sci.* **40,** 118.
Winsor, G. W., Davies, J. N. and Long, M. I. E. (1967). *J. hort. Sci.* **42,** 227.
Wokes, F. and Organ, J. G. (1943). *Biochem. J.* **37,** 259.
Wong, F. F. and Carson, J. F. (1966). *J. agric. Fd Chem.* **14,** 247.
Woodmansee, C. W., McClendon, J. H. and Somers, G. F. (1959). *Fd Res.* **24,** 503.
Woods, M. J. (1966). *Potass. Symp.* **8,** 313.
Workman, M. and Pratt, H. K. (1957). *Pl. Physiol., Lancaster* **32,** 330.
Wright, R. C., Pentzer, W. T., Whiteman, T. M. and Rose, D. H. (1931). *Tech. Bull. U.S. Dep. Agric.* No. 268.
Wu, M.-.A and Burrell, R. C. (1958). *Archs Biochem. Biophys.* **74,** 114.
Yamaguchi, M., Howard, F. D., Luh, B. S. and Leonard, S. J. (1960). *Proc. Am. Soc. hort. Sci.* **76,** 560.
Yamamoto, M. and Mackinney, G. (1967). *Nature, Lond.* **213,** 799.
Young, R. and Davison, P. K. (1968). Private communication.
Yu, M.-H., Olson, L. E. and Salunkhe, D. K. (1967). *Phytochem.* **6,** 1457.
Yu, M.-H., Olson, L. E. and Salunkhe, D. K. (1968a). *Phytochem.* **7,** 555.
Yu, M.-H., Olson, L. E. and Salunkhe, D. K. (1968b). *Phytochem.* **7,** 561.

Part II

The Biochemistry of Fruit Processing

Chapter 14

General Introduction

MAMIE OLLIVER

Consultant, Histon, Cambridge, England

I Introduction 485
II The Preservation of Foods 486
A. General Principles 486
B. The Preservation of Fruits 487
III Fruit for Processing 488
A. Factors Influencing Fruit Quality and Crop Yields 488
B. The Biochemistry of Fruit Quality and Crop Yields 493
IV Processes for Fruit Utilization and Preservation 496
A. The Biochemistry of Fruit Processing 496
B. Fruit Components as Products 499
V Developments in Biochemical Studies of Fruits 500
A. Collection and Utilization of Data 500
B. Selection of Projects for Improved Processing Control 504
References 505

I. INTRODUCTION

In 1793 a brief description of a curious African fruit was given in "The History of Dahomy" by Dalzel (1793). This "miraculous berry", which resembled a dusky red olive, had the extraordinary quality, on being chewed, of making acid foods taste sweet, so that "a glass of vinegar will taste . . . like sweet wine; a lime will seem to have the flavour of a very ripe China orange". Dalzel emphasized that this effect was quite different from that of neutralization of the acid to a saline flavour, or from that of direct sweetening since no change was observed in food or drink which did not contain acid. An even earlier record is found in the journal of the Chevalier des Marchais during a voyage to Guinea between 1725 and 1727. Labat (1730), whose work deals with this journal, chides des Marchais for not having collected more information about the miraculous fruit, since the remarkable sweetening properties he described were likely to benefit those, who, in Labat's words, "ont tant de peine à prendre les remedes amers & desagréables que les Medecins ordonnent, contre lesquels la nature se revolte, sans que la raison puisse y apporter du remède". Nevertheless it was not until 1844 that the plant was botanically described, under a section of *Sideroxylon*, as *Synsepalum dulcificum* (De Candolle, 1844). Daniell (1852) made *Synsepalum* a full genus and at the same time referred to reports confirming the unusual flavouring effects of the fruit.

Little interest in the miracle fruit, as it became known, was, however, shown by chemists and it was not until 1965 that concentrates containing the active principle were obtained and examined. Inglett and his co-workers who carried out these investigations tentatively identified the substance as a glycoprotein. Purification was subsequently effected by Kurihara and Beidler (1968), identification of a basic glycoprotein was confirmed, and a molecular weight of the order of 44,000 was determined. Almost simultaneously Brouwer *et al.* (1968) published their method for the isolation and purification of the sweetness-inducing principle, which they named miraculin. However, whilst the nature of this taste-modifying protein is becoming established, the mechanism of its action is still almost completely unknown, although results of recent psychological studies designed to clarify the position (Kurihara and Beidler, 1969) are of considerable interest.

The relatively small amount of research carried out on this interesting flavour property during the 250 years following the original observations is in striking contrast to that afforded in recent years to other aspects of fruit flavours, especially to the analysis of their volatile constituents (see Volume 1, Chapter 10). This comparative neglect is probably to a large extent due to the difficulties of obtaining freshly harvested samples of the miracle fruit. Nevertheless, this example has been selected not only to illustrate the gaps that exist in our understanding of the causative factors of biochemical phenomena which may be important when utilizing fruits for human consumption, but also to demonstrate the need for and possibilities of instilling new and imaginative approaches into studies for the elucidation of these phenomena; many other examples may be found throughout this volume. The purpose of this chapter is to evaluate the position as regards our present knowledge of the biochemistry of fruits in relation to the science and technology of fruit processing, and to consider possible lines of future research in this field.

II. THE PRESERVATION OF FOODS

A. General Principles

The principles of food preservation are essentially based upon the destruction or inhibition of activity of micro-organisms which would otherwise adversely affect, through changes in flavour, colour and texture, the palatability of foodstuffs during transit or storage, and which could produce toxic effects as well as altering nutritional value. Because of the susceptibility of micro-organisms to extremes of temperature, one of the major controlling principles applied is that of high temperatures, as adopted in canning, or of low temperatures, as in freezing. Other processing methods depend upon the limitation of available water, as in the various dehydration processes or in those in which high concentrations of sugar or salt are used. Control may be effected in suitable products by high acidity; low pH values aid in heat sterilizing. A limited number of chemical preservatives, such as sulphur dioxide, sorbic acid and benzoic acid, are permitted in specified low level quantities in some foodstuffs according to the legal food standards of different countries. Metabolites of fermentation may help in preservation, and in curing processes use is made of preservatives in smokes. Atomic irradiation has received much attention experimentally, although relatively little commercial application has so far been made, and the results are of considerable biochemical interest (see Chapter 20).

Conditions of all these preservation processes have to be so devised that inactivation is ensured not only of contaminating micro-organisms but also of those enzymes naturally occurring in the foodstuff and which could, during storage, have deleterious effects, similar to those of the microbial enzymes, on palatability. The processing technologist is, however, faced with the dilemma that these requisite conditions are usually such that they also alter the organoleptic qualities of the food. Sometimes the effects are adverse, sometimes they may be advantageous as, for instance, with heat-treated vegetables, meat and fish which can thereby be made more acceptable, but, even in such cases, variation in quality of the final product may be anticipated according to the characteristics of the particular sample treated. To achieve standard optimum quality of any processed food, therefore, the effects must be known not only of the conditions of processing on the components of the food, but also of the possible variations in the material to be processed. By reason of the relative inflexibility of the former the latter becomes of paramount importance.

B. The Preservation of Fruits

Because fruits have lower pH values than the majority of other foods, milder conditions of processing are effective in their preservation. Thus, while pressure cooking is essential for sterilization in the canning of vegetables, meat and fish, atmospheric pressure is adopted in fruit canning. Provided precautions are taken with the less acid fruits, e.g. pears and tomatoes, to ensure that the effective pH value of 4·5 is not exceeded, bacterial spoilage occurs only in exceptional cases at the temperatures and times adopted in approved commercial fruit canning although ascospores of fungi of the genus *Byssochlamys*, *B. fulva* (Olliver and Smith, 1933) and *B. nivea* (Westling, 1909), may survive. However, even these comparatively low temperature conditions may be sufficient to destroy the delicate balance of the flavour components of the fruit and to have deleterious results on texture by reason of disintegration of cellular structure. Indeed, the fragility of the cell walls of many fruits introduces problems of damage to texture not only in this and in other methods of processing but also during harvesting, transport and pre-processing storage, through the consequent lowered resistance to fungal enzyme attack. Enzymes liberated from the fruit itself by such cellular damage may, in addition, rapidly affect both colour and flavour. The activities of such enzymes are normally destroyed during the preservation process, but non-enzymic changes in organoleptic qualities of the processed fruit are liable to take place on storage, especially at elevated temperatures, in the presence of oxygen, light and/or metal.

Scientists and technologists concerned with fruit processing are, therefore, faced with many difficulties in the control of flavour, texture and colour when applying known methods. The nature of some of these problems and the extent to which they have been resolved are described in the following chapters.

III. FRUIT FOR PROCESSING

A. Factors Influencing Fruit Quality and Crop Yields

Selection of fruit for processing is based to a large extent upon experience combined with an understanding of the possible effect of known variables, such as those listed below, on fruit composition and behaviour. Because of the limitations in knowledge of these effects, fruit of untried quality, e.g. from a new source, is normally assessed through small batch trials. The practical difficulties on an industrial scale of this empirical approach, especially when handling short-season types, are great and underline the need for improved information on fruit qualities.

1. *Variety of fruit*

A range of textures, of fullness or freshness of flavour, of colour depths and brightness, and of stability of these characteristics to specific processing methods is found amongst different varieties of the same type of fruit. Nutritional value as indicated by ascorbic acid content may similarly be related to varietal differences.

The grower of fruit for the fresh market concentrates on those crop varieties which produce a high yield and with quality determined by visual criteria such as shape, size, colour and freedom from blemishes and disease. Standards for fruit composition are rarely used although most countries exporting citrus fruit specify sugar (soluble solids): acid ratios for oranges and grapefruit. The processor is in line with the grower in favouring high croppers on economic grounds, but his requirements of fruit quality are more precise than those demanded by the fresh market. Thus, although high visual quality is also desirable for most fruit processing methods, the composition of the fruit in relation to the major organoleptic qualities of flavour, texture and colour and to nutritional value is of paramount importance. In addition, these qualities should be such that they are impaired as little as possible during the specified process.

The interests of grower and processor may, in consequence, not always be compatible. For instance, in strawberry cultivation in the United Kingdom the variety Cambridge Favourite is widely grown commercially at the expense of other varieties because it is a relatively heavy cropper. The fruit is also readily detached from the plant, thus aiding picking. Unfortunately, the fruit tissue readily breaks down in acid, as when preserved by sulphur dioxide or made into jam, and for some time therefore this strawberry had limited use in preserve manufacture. Addition of calcium salts, however, materially assists in overcoming this defect and provides an example of the value of the biochemical approach to such problems (see p. 498).

The breeding of new varieties of fruits, which is being actively pursued, is also subject to the danger that selection for processing quality may be subordinated to economic considerations. Increasingly, these demand not only high crop yield and resistance to disease and pests but also, because of rising labour costs, ease in pre-processing handling, e.g. in the removal of inedible portions of the fruit such as plugs (calyx and core) of strawberries or strigs (stalks) of blackcurrants. Similarly, as discussed, for instance, in a comprehensive report edited by Cargill and Rossmiller (1969), mechanical harvesting is now being widely extended and is encouraging the breeding of plants with well-spaced trusses and with the fruit readily separated from the stalk. The possibility that less desirable qualities for processing such as tougher fruit or skins may be simultaneously introduced with these cultivar developments cannot be overlooked. Reference has already been made to the relatively limited flexibility in conditions of processing as compared with choice of fruit. Nevertheless, economic pressures such as those here indicated may impel the manufacturer either to modify existing conditions of processing or to devise new processes in order to meet the characteristics of the new varieties. The importance of increasing biochemical understanding of underlying principles is evident.

2. *Agronomy and environment*

Biochemical properties of fruits as evinced by fruit quality and yield are fundamentally influenced by rootstock, by soil and by climatic conditions, e.g. solar radiation and incidence of wind and rain. Little is known, however, of the biochemical mechanisms of these effects; attention has mainly centred upon means of qualifying some of these conditions. For instance, there is an interesting and potentially valuable trend towards the general control of environment as demonstrated by phytotrons, growth chambers or growing rooms. The use of chemicals for control has received major attention in recent years. In soil treatment they provide means of increasing nutrients to the plant, of improving the physical condition of the soil and of removing unwanted plants. Marked effects on yield and quality of fruit have been found in many instances from the use of specific fertilizers. Growth characteristics can be modified by growth regulators or hormones†; those affecting abscission are especially valuable, either in preventing premature fruit-dropping or, conversely, in enhancing ease of detachment as required in mechanical harvesting or in fruit thinning. Pesticide chemicals improve yields and, despite causing occasional spray injury, generally reduce the incidence of blemishes. There is little experimental evidence to show direct effects of pesticides on fruit growth and development under conditions of adequate moisture and nutrition (National Academy of Sciences, 1968) but biochemical problems can be created in that spray residues may have toxic properties not necessarily destroyed during processing, or harmful metabolites may be produced. Flavour taints have been found in processed fruit containing pesticide residues. The importance of using conditions of washing the fruit which will help in the elimination, before processing, of such hazards is evident. This is emphasized by the difficulties of determining these residues, either qualitatively or quantitatively, in deliveries of fruit as received at the processing factory. Indeed, the main responsibility must lie with the grower in ensuring correct application of the pesticides so that negligible amounts remain on the fruit when harvested.

Recent advances in research into the systemic fungicides have resulted in the production of several compounds which are likely to have increasing application to fruit growing and to overcome many of the difficulties associated with surface spraying whilst, at the same time, increasing yield and extending the range of control of fungal diseases. An interesting facet of these developments is their possible impact upon plant breeding in which the disadvantage exists that disease resistance may eventually be broken down by mutation. The biochemistry of the systemic fungicides in relation to the phytoalexins, the antifungal compounds produced by the plant when invaded, is another approach which is being watched with interest by the fruit technologist. In particular, however, he will be concerned with the absence of toxicity in fruit from plants treated with the fungicides and with the effect on quality of fruit to be processed. These points cannot, however, be even predicted until more information is available on the mode of action of systemic fungicides and their natural counterparts.

† See Volume 1, Chapter 15.

3. *Maturity and ripeness*

Because of the rapidity with which biophysical and chemical changes take place in fruit tissue during maturation and ripening, close contact between grower and processor is essential in order to determine the stage of ripeness optimum for use in a specified type of process and at which the fruit should be harvested. In apples for storage the length of time from petal fall has been used as an indicator of the suitability of ripeness of the fruit. The processor mainly relies upon empirical tests of organoleptic qualities, and decisions on harvesting time are usually based upon experience reinforced by the results of small batch processing. The biochemist may determine the stage at or near which maximum respiration coincides with marked biochemical changes, i.e. the respiration climacteric,† to indicate the approach of ripening but this pointer may either be absent from or not sharply defined in some types of fruit, e.g. citrus fruits and the pineapple. Specified changes in composition have been proposed as the basis of assessment of the onset of ripening in climacteric as well as non-climacteric fruits. Since detection of these changes by chemical assay is often too time consuming and too demanding of skilled manipulation for field tests, physical methods are receiving increased attention, e.g. the use of DLE (delayed light emission) for colour measurement of stone fruits, density tests for the olive and refractive index determinations for the avocado.

Timing of the harvesting of fruit for processing is further complicated by the great differences which occur in the rate of development and maturation of individual fruits even on the same tree, bush or plant, or, as with black-currants, on the same stalk. This problem of uneven ripening becomes accentuated by mechanical harvesting where the personal discrimination exercised in hand picking is absent. The use of hormones in promoting more even maturation (e.g. the ethylene-producing 2-chloroethanephosphonic acid, Ethrel, used for inducing abscission of currant fruits) is likely to become of increasing value, provided other quality characteristics are not adversely affected. Meanwhile, in order to maintain the quality of the processed fruit, resort has to be made to sorting and to the simultaneous working in the factory of different processes requiring fruit of different quality standards, such as canning, for which firm fruit is normally required, alongside preserve manufacture for which riper fruit may not only be acceptable but even preferable from the flavour angle.

The shortness of season at which many fruits are at their optimum quality for processing constitutes a major problem of handling, particularly in the case of soft fruits. An attempt to extend the season of ripe fruit is seen in the breeding of varieties which overlap in their times of ripening. Growth regulators have also been used in, for example, the apricot where fruit

† See Volume 1, Chapter 17.

maturity can be advanced by auxins but retarded by gibberellic acid. Of particular interest is the observation that high nitrogen conditions in cultivation prolong cell division in stone fruits and consequently delay ripening (Albrigo *et al.*, 1966).

4. *Handling before processing*

Skin abrasion and tissue bruising of the ripened fruit during handling are major causes of cell breakdown and hence of the effect on fruit quality of enzymes derived either from the fruit or from invading micro-organisms. The introduction of mechanization into harvesting has increased the risk of such damage, although improved machinery design, not only for gathering the fruit but also for removing extraneous material such as stalks and leaves, and precautionary measures such as immediate cooling of the fruit after harvesting, are being implemented. Further mechanical damage to the fruit, with correspondingly adverse effects on quality, may be anticipated during sorting, grading and washing.

More use is being made of refrigerated transport of fruit from field to factory with consequent reduction in loss of quality. Holding of ripe fruit for convenience in production management can usually be carried out for short periods in cold rooms, but the time of satisfactory storage depends upon the type of fruit. Firm fruits, such as citrus, can normally be stored in this way for several weeks without general gross deterioration in quality although changes in composition may be detected in, for instance, the pectinous materials. Soft fruits usually suffer rapid changes, sometimes within a few hours of harvesting. However, storage in CO_2 has been advocated for out-of-season use, as in the case of blackcurrants (see Chapter 11).

It may be necessary to apply some form of preliminary process, such as freezing, for preserving the fruit before the required process itself is used, as, for example, with soft fruits which are ultimately to be made into jam. Canning and sulphiting are other examples of this pre-processing, the former being universally applied to fruits which are to be transported for long distances, the latter to soft fruits in some European countries. The effects of double processing on fruit quality are likely to be cumulative and, when freezing is used, subsequent thawing of the tissue, if not controlled, may introduce further deleterious changes.

An extremely interesting line of research is that of determining conditions under which fruit harvested at the mature but firm unripe stage could be induced by artificial means during storage to ripen normally. Such procedures have been worked out using ethylene and have been successfully applied commercially to the banana, citrus fruits and melons. If suitable controlled conditions could be devised for all types of fruits, many of the present difficulties related to the harvesting and handling of ripe fruit could undoubtedly be resolved with great practical benefits.

B. The Biochemistry of Fruit Quality and Crop Yields

1. *Organoleptic qualities*

Whilst the effects of many of the influencing factors on the organoleptic qualities of fruits have already been established as outlined above, the related biochemical mechanisms are still little understood. To a large extent this is because the qualities themselves are so difficult to assess with precision. Thus, although the majority of the chemical components of acidity or sweetness can be determined with accuracy, both qualitatively and quantitatively, the more delicate physical aspects of flavour and aroma can usually only be estimated by the palate and by the nose. Results of this sensory method of assessment by trained panels will vary with the rating scales adopted and will also be qualified by considerations of other physical characteristics such as fruit colour and texture. Similarly, with texture measurements the physical apparatus available do not always give results in line with those obtained from palate-feel, which is still the ultimate criterion although, as with flavours, this is subject to personal variation in judgement. Instruments based upon such physical principles as those of delayed light emission help to evaluate colour, but the bloom on a peach still retains its indeterminable beauty.

Another difficulty in studying the biochemistry of the organoleptic properties of fruits is that, in order to determine chemical composition, the cellular structure of the fruit tissue must be disintegrated. Consequently, interaction or loss of components or liberation of interfering substances may result, thus further complicating the estimation of the low levels in which flavour or colour constituents normally occur. Even if these difficulties can be overcome by suitable precautionary measures, only the average level of overall composition of the fruit is obtained; no indication is given of the distribution of the components within the cellular structure. The microscope helps in combatting this problem and can be of special value when used in conjunction with the chemical approach. Thus, the importance of the cell wall in relation to fruit texture has been demonstrated histologically, histochemically and by chemical determinations of the composition of the isolated cell walls. (See, for example, Reeve (1959) for the peach and Wade (1964), Neal (1965) and Armbruster (1968) for the strawberry.) The biophysical characteristics and organization of the microfibrils of cellulose and the nature of the pectins probably have a marked influence on plasticity of the cell walls and hence on fruit texture during, for instance, maturation and ripening. The degree of esterification of the carboxyl groups of the polygalacturonic acid units of the pectinous materials is closely associated with the biochemical mechanisms involved in these changes, but the roles of associated polysaccharides, and of the various pectin-enzymes and of ions such as calcium are still only partly understood. Cell turgor is also undoubtedly involved in properties of texture and imparts crispness and juiciness to the fruit. Although this turgor, which reaches a pressure of many atmospheres in plant tissues, probably plays a major part in several aspects of fruit biochemistry, few figures appear to be available for fruits.

Recent work on flavours in fruits has largely been concerned with the volatile constituents, mainly because of the recognized importance of aroma. Gas–liquid chromatographic separation combined with mass spectrometric identification of the components has proved an especially useful technique in these studies, but the preliminary extraction systems, by causing loss of volatiles and by permitting enzyme action, have introduced serious difficulties. Some of these are undoubtedly the reason for lack of sensory identity between the fruit flavours tested and mixtures of the analytically determined constituents. Nevertheless, several non-volatile components are intricately concerned with flavour effects. Thus the sugar : acid ratio is a valuable indicator to the processor for specific requirements, the bitterness of grapefruit relates to naringin content and the solubility of tannins determines the astringency of persimmons. Reference has already been made to the non-volatile miraculin. Another substance of interest, which appears to be a complex glycoside although final classification has not yet been achieved, and which is one of the sweetest materials known (it has been estimated to be 800 to 1500 times as sweet as sucrose), has recently been isolated from the serendipity berry, *Dioscoreophyllum cumminsii* (Inglett and May, 1969). Very little is known of the biochemical mechanisms involved in the production of either volatile or non-volatile flavours.

Similar lack of information is found with the three classes of fruit pigments, i.e. the anthocyanins, carotenoids and chlorophylls. These are composed of a diversity of compounds and although biochemists have accomplished much in elucidating their nature, structure and occurrence, the biochemical mechanisms controlling the formation of some fruit colours are still unknown. For instance, although the disappearance of chlorophyll from fruits is a useful guide to ripeness in post-harvest fruit, it has not yet been determined whether chlorophyllase is active in the pathway of biosynthesis or of degradation and consequently how far this enzyme operates in the colouring of fruits. Browning in fruit tissue after bruising or other damage is mainly the result of enzymic oxidation, but the mechanisms operating during the translocation of anthocyanin which occurs in cherry scald are not fully understood. Clarification of the relation of the fruit pigments to colour quality continues to be limited, not only by the problems associated with the small amounts in which the operating substances occur, but also by the difficulty of accurate assessment of colour due to the complication of light effect such as translucency and reflectance.

2. *Nutritional value*

Fruits are of dietetic importance for the variety, freshness and flavour they provide. Their main nutritional value, however, lies in the position they hold with vegetables in being, with the exception of milk and some animal internal organs used as food, the only source of natural Vitamin C in human diet. For this reason, and also because of the development of simple, rapid and reliable chemical methods for estimation of the vitamin, an extensive amount of information is available on the occurrence and distribution of ascorbic acid in different fruits and the changes in concentration during fruit development, maturation and ripening and with variety. The biochemical mechanisms causing these differences and changes are not, however, understood nor are the seasonal fluctuations which have been observed in the concentration of ascorbic acid in, for instance, blackcurrants. Knowledge of the probable pathways of synthesis of the vitamin in fruits, the related enzymic activities and the nature of stabilizers is, however, increasing and may help to elucidate these problems (See Volume 1, Chapter 13).

Other nutritionally important components of fruits such as vitamins other than ascorbic acid and mineral salts have received comparatively little attention. The use of extracted pectins in therapeutics, especially in the treatment of intestinal disorders, is well known, and their antibacterial properties have been reported. Essential oils of citrus fruits have been found to have strong antimicrobial activity and extension of shelf life of milk products by their addition has recently been claimed (Dabbah *et al.*, 1970). Saponins and allied substances may be responsible for the surface activity found with some extracted fruit juices and the consequent strengthening of pectin gels (Olliver *et al.*, 1957); their possible role in, for instance, fat digestion merits further attention. Some of the natural components of fruits, e.g. oxalic acid, are known to have toxic properties, but it seems unlikely that any are present in sufficient quantities to be harmful in human diet. More information is, however, needed about this aspect of the subject as well as a greater understanding of the biochemistry of the beneficial nutrients.

3. *Crop yield*

It is significant that the tissue cells of most fruits cease to divide before or at the onset of maturation, the fruit subsequently increasing in size through cell enlargement and not through cell multiplication. An outstanding exception is the avocado. The possibility of an underlying relationship between cell size and cell turgor is further indicated in that larger fruits may develop as the result of increased mulching or irrigation, as demonstrated, for instance, with blackcurrants (see p. 405). Using achene-spacing measurements as a means of calculating the short fall in yield of strawberries, a possible increase of 50–100%, provided that berry expansion could be promoted after the onset of ripening, has recently been calculated in one series of tests (Abbott and Webb, 1970). The plasticity of the cell wall in relation to the pectinous materials, the action of auxins and the effects of increased water retention are important aspects of means of achieving such increases in yields and demand improved understanding of their biochemical mechanisms.

IV. PROCESSES FOR FRUIT UTILIZATION AND PRESERVATION

A. The Biochemistry of Fruit Processing

The biochemistry of the processes themselves and of storage of the processed fruit is little understood, due to the lack of precise information on many aspects of the biochemistry of fruits and also because the technology of fruit processing is largely based upon experience of long-established practices. Resort in some instances has been made to model systems, but direct application of the results of such tests may be misleading. Knowledge has been gained mainly from observations of chemical, physical and, to a more limited extent, histological changes occurring under specified conditions of processing. These changes may vary according to the characteristics of the fruit, as already discussed but, in general, the following aspects of processing conditions apply.

1. *Temperature*

Exposure of fruits to temperatures of the order of 100°C for a short period, as adopted in blanching, inactivates the fruit enzymes and reduces the number of micro-organisms, whilst sterilization can be effected if the temperature is maintained for longer periods, as in canning, provided conditions of pH and heat penetration are suitable.

The extent to which changes take place in fruits during heat treatment depends upon the type and quality of the fruit and the conditions of processing, e.g. the time of treatment and rate of heat penetration and the content of oxygen, water, acid, sugar, salt or metal ions. Some leaching of soluble constituents is, however, generally to be anticipated in the presence of liquids. Ascorbic acid is readily soluble in water and is partially extracted from the tissue during stewing or canning of fruit. Such solubilization does not significantly affect nutritional value since the liquor is normally consumed with the fruit, but other quality characteristics can be considerably altered. Thus, solubility of the pectinous materials of the middle lamella and cell walls may be induced, with consequent softening and eventual breakdown of the fruit tissue. Colour and flavour of the fruit may be altered by selective extraction or by absorption of additives such as sugar or acid. Volatile constituents of aroma can be lost and even relatively stable non-volatiles may be affected under more stringent heat conditions, such as those persisting in processes carried out in the absence of added water, e.g. drying. Degradation of other components is accelerated under the more extreme forms of treatment, and complete breakdown of anthocyanins is possible together with caramelization of sugars affecting both colour and flavour. Texture changes indicated by reduced efficiency of rehydration are seen in dried fruits.

Similar effects of elevated temperatures are found with storage of the fruit product, as, for instance, in the non-enzymic Maillard type browning reactions in colour deterioration and in the breakdown of flavour substances. The shelf life of processed fruits as measured by quality criteria is consequently to a large extent determined by the temperature of storage.

Problems of a different nature arise from low temperature processes, e.g. freezing, where the chemical substances of the fruit are stabilized but damage to cell structure occurs mainly through ice crystal formation and change in solute concentration. The extent of cytological change is dependent upon the conditions of freezing and becomes evident on thawing of the processed fruit, when "drip" of cell contents, with consequent loss of colour, flavour and nutrients and shrinkage of the tissue may take place. These effects are discussed in Chapter 19.

Whilst texture changes in frozen fruit are mainly directly attributable to cellular damage, darkening of colour arises in unblanched fruits from the action of enzymes, e.g. phenolase, liberated during this breakdown. Loss of aroma followed by the development of off-flavours may occur during the storage of frozen fruits, temperature and oxygen being determining factors, but little is known of the mechanisms involved. The greater resistance to change of flavour during low temperature storage of some fruits compared with others is an aspect that demands further consideration.

2. *Enzymes*

The extent to which quality changes will occur as the result of enzyme action during processing depends upon the type of enzyme, the degree of cell fracture, the time and conditions of the process as related to those required for inactivation of the enzyme, and the presence of enzyme inhibitors. Thus polygalacturonase may survive the mild processing conditions adopted for the brining of cherries and so cause fruit breakdown. A major loss of ascorbic acid in fruit juice manufacture results from phenolases but natural inhibitors, possibly flavanoids, stabilize the vitamin. Control of enzyme activity in many processes is exercised by steam or hot-water blanching. Adjustment of pH to values outside the range for enzyme activity is another means of control. More information is, however, needed for determining conditions of pH, temperatures and times essential for inactivation of fruit enzymes whilst allowing only minimum changes in fruit quality. Greater knowledge is also required of the occurrence and location of specific enzymes within different fruits, not only as regards type of fruit but also taking variety and degree of maturity and ripeness into consideration.

Enzyme preparations may be added in processing control as, for example, in juice manufacture where pectin enzymes are used to reduce viscosity in order to increase ease of pressing and for aiding clarification. An interesting finding is the restoration of flavour to heat-processed bananas by the addition of preparations of the enzymes of the raw fruit (see Chapter 2, p. 81).

3. *Oxygen*

Oxidation of fruit constituents on processing or during subsequent storage may be either enzymic or non-enzymic, the required oxygen being derived from the atmosphere, the intercellular spaces of the fruit tissue or, in the case of canned fruit, from dissolved air in the can syrup. Carotenoids are readily oxidized to a diversity of compounds which are not easily identified. Terpenes and other fruit components produce off-flavours; natural anti-oxidants to combat such changes appear to be present in citrus fruits (see Chapter 16, p. 554). Ascorbic acid is oxidized to dehydroascorbic acid and other degradation products. Many of these reactions are accelerated by light and by metal ions and consequently their effects during product storage can be controlled by suitable packaging. Chilling of the fruit, deaeration and the use of chemical anti-oxidants are other measures adopted in processing for minimizing oxidation.

4. *Metal ions*†

Iron, copper and tin ions may be introduced into processed fruits through factory equipment or through packaging and, even in the trace amounts normally found, can seriously affect quality. In addition to immediate metallic flavours, the production of off-flavours may be catalysed during storage. Fruits with high anthocyanin content may cause can corrosion with consequent formation, by chelating, of blue or purple colours; "pinking" of canned fruit is possibly dependent upon dissolved tin. Catalytic oxidation of ascorbic acid is also induced by metallic ions, notably iron and copper.

Calcium ions are of undoubted importance in the texture of fruits, possibly through ionic bonding of partly esterified polygalacturonic acid in the cell walls and middle lamella. Toughening of blackcurrant skins can occur, for instance, when the fruit is heated with unsoftened water. Calcium salts are used as

† See also Chapter 17.

additives when firming is required, as with sulphited strawberries where breakdown of fruit during storage would otherwise result from acid demethylation and depolymerization of the pectinous material.

5. *Additives*

Sugar is the most common additive in fruit processing where it is used for sweetening and, in concentrations above about 65%, as a method of preservation. The rate and degree of penetration of sugar into whole fruit or into large sections of fruit are of considerable importance for organoleptic qualities as well as for ensuring preservation. Information on the mechanism of its action is limited, although there is experimental evidence to suggest that sugar in canned fruits is adsorbed by the cell walls against a concentration gradient (Sterling and Chichester, 1960). The texture of fruits may be affected by cell shrinkage due to osmosis.

Sulphur dioxide is used in solution as a preservative for fruits for subsequent processing as preserves; as a gas it is widely applied in the prevention of enzymic and non-enzymic discoloration during drying and subsequent storage of fruits, but the biochemical reactions involved are still not fully understood. Reference has already been made to the use of anti-oxidants, texture-firming agents and enzyme inhibitors in the processing of fruit. Synthetic ascorbic acid, apart from its function as an anti-oxidant, is sometimes incorporated into fruit products to enhance their nutritive value. Other chemical additives, such as ethylene glycol and glycerine, have been used for protection against cell damage during freezing. Colour may be improved by selected artificial pigments and flavour fortified by synthetic flavouring materials or modified by acids and sweeteners. The permitted identities and quantities of these various additives are dependent upon the legal requirements of individual countries. These in turn are based upon stringent biochemical and biological experimentation to ensure safety for human consumption.

B. Fruit Components as Products

In the majority of fruit-processing methods the fruit, with inedible portions removed, is preserved whole or after division into portions. With many fruit juice products, however, only the soluble components are used, the bulk of the cellular material being eliminated. For satisfactory separation of the juice by pressing, the fruit tissue must be easily ruptured but with insufficient soluble pectin to raise the viscosity of the press material (see Chapter 17). Varieties of fruit and stages of ripeness showing these characteristics coincident with optimum flavour are not necessarily those suitable for whole fruit preservation, e.g. the presence of lignified stone cells in perry pear tissue aids pressing for juice but makes the texture of the fruit granular. Disintegration of fruit tissue during pressing releases enzymes which accelerate changes in organoleptic and nutritional qualities similar to those occurring in processes where the fruit is not pressed.

Whilst juices are used as foods in themselves, other fruit components can be extracted individually and used as additives mainly to foods but occasionally for non-food purposes. Pectinous materials extracted as by-products in the apple and citrus fruit processing industries are widely used in preserves and in sugar confectionery; they are also of value in medicine. Pectins are being applied increasingly to other industrial uses due to the diversification of their properties according to degree of esterification and equivalent weight.

Olives are economically important not only for processing as fruit but also for their oil content. Variety of fruit is a determining factor in suitability for both quality and yield of oil as also is the stage of ripeness at which the fruit is harvested. Essential oils are other fruit constituents which are extracted and used as additives in many foodstuffs. Volatile flavours can be concentrated from fruits and their juices, e.g. by extraction followed by evaporation or, as more recently claimed, by concentration to the order of 100,000-fold (without deleterious heat changes) by means of liquid CO_2 extraction (Schultz and Randall, 1970). Dihydrochalcones with exceptional sweetening powers have recently been prepared from flavanones of citrus fruits, e.g. neohesperidin and naringin (Krbechek *et al.*, 1968; Horowitz and Gentili, 1969). The possibility that other fruit constitutents of economic importance will be found through new and more sensitive techniques cannot be ignored, especially when these are applied to some of the less known tropical and sub-tropical fruits; this aspect of the biochemistry of fruits is likely to be afforded increasing attention.

V. DEVELOPMENTS IN BIOCHEMICAL STUDIES OF FRUITS

A. Collection and Utilization of Data

1. *Sample selection*

The value of biochemical data on fruits to the horticulturist and processor is essentially determined by the reliability and scope of the given information. Unfortunately, precautions for ensuring reliability are not always taken in the laboratory and misleading interpretations may consequently be made. In addition, sample selection may be so restricted that data are collected in isolation without related information being available for comparison and for ensuring full value of application. Thus, for instance, the diversity of botanical structures which may exist between different types of fruit and, consequently, the possible individuality of their biochemical systems is not always recognized. It is unusual, for example, to find the strawberry, where the edible portion is the receptacle swollen by the auxins induced by the true fruits, the achenes, investigated alongside a true fruit such as the plum. Even more striking is the relative lack of attention given to comparative research with fruit and vegetable tissues, e.g. concurrent studies of delicate textured, low pH value fruits and the more robust, high pH value vegetables.

In sample selection the marked influence of varietal, agronomic, seasonal and maturity factors on the biochemistry of fruits is also sometimes ignored. This applies not only to the presentation of data but also to their collection. Experiments with material of undescribed and sometimes even unknown origin, and of unstated time and conditions of post-harvest storage before examination, can be wasteful of effort and harmful in interpretation. Special care needs to be taken in determining varietal designation where the complexity in many fruits today is often confusing and creates difficulties in segregation for both the fundamental biochemist and the technologist.

Because of these influencing factors, accuracy in sampling is one of the most important features of biochemical experimentation. Structure variation has to be considered, e.g. separate component parts taken for individual assay or samples representative of the whole fruit determined. Degree of fruit development creates especially vexed problems in this context; small fruits, where several may need to be taken to obtain sufficient material for experimentation, may be at widely different stages of maturity or ripeness, even on the same plant or stalk. In larger fruits this difficulty may exist within individual fruits; ripening, for instance, may proceed more quickly on the side exposed to light compared with the leaf-sheltered tissue.

2. *Chemical assay*

To ensure complete extraction from the fruit of the desired constituents for chemical assay, the tissue is first disintegrated. In order to avoid change in the constituents during extraction, enzymic action must be controlled alongside non-enzymic oxidation induced by occluded air or by catalysing trace metals. This can usually be reasonably effected by choice of disintegrator and solvent and by working at a low temperature, but the possibility of some biochemical changes taking place during this stage of the process can rarely be completely excluded. Satisfactory separation of soluble from insoluble constituents is sometimes difficult to achieve; this applies particularly in the examination of cell wall components when the viscosity of solubilized pectin may interfere. Another example is found in methods of determination of specific pesticides on fruits or in their products where various "clean-up" procedures have to be adopted. Failure to ensure reliability of sample extracts is likely to be as damaging to data as failure to ensure representative sampling.

The availability of new and sophisticated physico-chemical methods allied to well-tried traditional chemical techniques of measurement means that determinations of constituents in extracts of fruit tissue can usually be carried out with a high degree of accuracy and specificity. Automatic analyses are not only eliminating the personal element but are also increasing the rate at which results can be obtained, a feature of considerable importance to the processor. The possibility cannot always be excluded that the technique adopted will induce changes in the substance to be estimated, as in the determination of fruit volatiles by gas chromatography, or that interfering substances may react similarly to the required component. Since even slight changes in sample preparation or in measurement of constituents may significantly affect results, the importance of using standard procedures and of recording essential modifications is evident.

One determination which is often omitted in fruit analysis is water content. Understanding of biochemical systems might possibly be enhanced in some instances if dry weight figures were available so that changes in amounts of fruit components could be attributed to synthesis or degradation on the one hand or to water addition or extraction on the other. Improved methods for the determination of the water content of fruits and their products are needed. pH is another determination frequently neglected, yet this value can be helpful to the processor when considering both enzymic and non-enzymic changes in organoleptic qualities.

3. *Histological and cytological studies*

It is unfortunate that the tools of the botanist have not received attention comparable with those of the chemist in biochemical studies of fruit. The relatively limited use which has been made of the light microscope is possibly because of lack of specificity of stains for characterizing and localizing fruit components. More information is needed on these and related techniques, including those of enzyme cytochemistry. Accurate methods for the assessment of cell size and of cell wall thickness are required for studies in cell turgor which appear necessary for a greater understanding of fruit quality, particularly in relation to texture, and of fruit crop yields. For the scientist concerned with processing, improved equipment is needed whereby fruit tissue, whilst being examined under the microscope, can be submitted to conditions simulating those adopted in processing. Increasing use is being made of the electron microscope in studies of fruit tissue, and recent developments in freeze etching and ion etching together with the scanning electron microscope hold great promise. Attention to preparation techniques such as fixation and embedding of fruit tissue is required together with wider and more realistic interpretations of microscope images as, for instance, by computer analysis.

Preliminary experimental studies of growth *in vitro* of intact cells of fruit by tissue culture methods have been beset by many difficulties, but techniques based on this principle may ultimately prove to be of the greatest value to fruit biochemists and they merit persistent investigation. An especially interesting and promising new approach has developed through the successful isolation by Cocking and his co-workers of intact protoplasts from tomato fruit, using controlled enzymic breakdown of the cell walls. The protoplasts can be prepared in sufficient numbers for biochemical and physiological experiments (Gregory and Cocking, 1965) and cell wall regeneration can be induced (Pojnar *et al.*, 1967). It is possible that this technique might prove especially valuable in studies of the effects of conditions of processing on fruits. Potentialities in plant breeding have already been demonstrated by the formation of hybrid cells from protoplasts of desired genetic characteristics (Power *et al.*, 1970). Of fruit cell organelles, mitochondria have received attention for many years, particularly in respect to respiration studies. Studies of the activity of other cell organelles such as peroxisomes (see Volume 1, Chapter 8) will, no doubt, eventually provide data of value to the fruit processor.

4. *Utilization of data*

Although extensive data on amounts of chemical components of fruits are available for demonstrating the results of biochemical mechanisms controlling fruit quality, the paths of metabolism and function of the various constituents are but little known. Classification of data, as for instance through the reviews presented in these volumes, is essential for helping to elucidate the position. More comparative tables are also needed to determine the effects of variety, agronomic and seasonal factors and stage of fruit development. In addition, comparison is required between data available for different types of fruit, especially in relation to the more practical aspects of fruit biochemistry such as growth curves and optimum conditions for controlled atmosphere storage. In particular, however, the value of using comparative biochemical data from vegetable and fruit tissues as a means of understanding differences in characteristics between these two classes of plant foodstuffs should be recognized. It is, however, not only the evidence of the work of plant biochemists that should be considered in fruit research; the results of investigations into animal tissues may prove of potential value (see Chapter 19).

Computer analysis of reliable data might bring out differences which at present are extremely difficult to extricate from attendant detailed information. This could profitably lead to effort being transferred from well-tried lines into new and exciting research approaches. Nevertheless, the full value of available information and of future research is unlikely to be realized until a more unified approach is shown by those working in the disciplines of agriculture, bioengineering, biophysics, botany, chemistry, cytology, food science and technology, microbiology, nutrition and physiology.

In the collection of new data the contributions of the developing countries will undoubtedly become important, for many tropical and sub-tropical fruits are already known to be of special interest from the biochemical standpoint, e.g. the high contents of lipids in the olive and the avocado and of ascorbic acid in the acerola cherry and the guava, the inert flavour-enhancing material of the miracle fruit and the retention of quality of the serendipity fruit after post-harvest storage for several weeks at room temperature (Inglett *et al.*, 1969). However, as evinced at a recent symposium on tropical fruits (Tropical Products Institute, 1970), some laboratories in tropical countries are adopting conventional methods of approach and are restricting their selection of study to limited spheres whilst tending to omit measures for ensuring the maximum value of the presented data. Imagination blended with precaution is needed for the fullest opportunities to be taken.

B. Selection of Projects for Improved Processing Control

Whilst a greater knowledge of fundamental biochemical principles is essential before the problems of processing control of fruits can be fully resolved, the fruit industry cannot await the outcome of extensive research and choice will have to be made of those projects likely to result most rapidly in improvements in the economics of the process and in the quality of the product.

The selection of new varieties of fruit for breeding continues to be actively carried out but, here again, greater cooperation and interchange of information is required in this work between fruit growers, basic research workers and fresh storage and processing technologists. Such collaboration also needs to be extended to studies of agronomic factors on fruit quality in relation to processing methods. To effect this, more joint experimental facilities, both on and off farm, in the form of nurseries and greenhouses and farm and processing equipment are required. In the laboratory, in addition to the more conventional approaches, it will probably be found profitable to study at cytological level the products of this collaborative experimentation, with the microscope taking rightful place alongside the other tools of the biochemist. Such studies will undoubtedly relate not only to fruit quality but also to yield. In particular, investigations into cell turgor and cell wall plasticity are likely to be needed for determining conditions under which uptake of water for maximum fruit yield can be ensured.

One of the most rewarding lines of research will probably be the determination of storage conditions under which mature but firm, unripe fruits can be artificially ripened before sale or processing. The practical advantages of handling the firmer fruit would reverse the present costly trends towards the use of refrigerated transport between farm and factory or the transfer of processes from factory to farm. In addition, as already discussed, many fruit quality problems associated with mechanical harvesting and the uneven ripening and short season of fruit could probably be overcome by early harvesting and long-term storage, with consequent improvement in quality and processing economics. These studies will demand precise means of identifying stages of fruit maturity and ripeness, a need which is indeed already evident for the substantiation of parameters of organoleptic assessment, the climacteric and chemical composition as used at the present time.

Of the various principles adopted in food preservation, drying has considerable economic advantages from the viewpoints of both storage and transport. Microbiological and chemical spoilage of quality can be significantly reduced in the absence of water. In principle, therefore, dehydration of fruits is a most desirable process. In practice, however, the changes in quality induced by present processes and the difficulty of rapid and complete rehydration of the dried tissue limit the application of this principle. It seems possible, however, that improved bioengineering techniques will eventually allow water to be drawn from fruit tissue with the minimum of damage so that quality on rehydration will be reasonably comparable with that of the fresh material. Advances in freeze drying and in the new techniques of reverse osmosis already lend encouraging support to this idea. The results of biochemical studies at cell level for determination of optimum uptake of water, as mentioned above, could undoubtedly be of value in this context.

The possibility of the development of processes based upon the *in vitro* culture of fruits is attractive to both fundamental biochemist and food scientist or technologist. Attainment of this objective as regards whole fruit may be remote, but the production of fruit components such as flavours, or of fruit tissue for preserves and similar products may become an earlier commercial proposition.

The validity of these predictions of trends in biochemical research in the interests of fruit cultivation and processing may be disputed at the present time and will certainly be tested by the future. Nevertheless, to those engaged in studies of the biochemistry of fruits the wisdom of the advice "as the Scripture saith, that we make a stand upon the ancient way, and then look about us, and discover what is the straight and right way, and so to walk in it" still persists.

REFERENCES

Abbott, A. J. and Webb, R. A. (1970). *Nature, Lond.* **225**, 663.

Albrigo, L. G., Claypool, L. L. and Uriu, K. (1966). *Proc. Am. Soc. hort. Sci.* **89**, 53.

Armbruster, G. (1968). *Proc. Am. Soc. hort. Sci.* **91**, 876.

Brouwer, J. N., van der Wel, H., Francke, A. and Henning, G. J. (1968). *Nature, Lond.* **220**, 373.

Cargill, B. F. and Rossmiller, G. E. (editors) (1969). "Fruit and Vegetable Harvest Mechanization. Technological Implications", Rural Manpower Center Report No. 16. Michigan State University, East Lansing, U.S.A.

Dabbah, R., Edwards, V. M. and Moats, W. A. (1970). *Appl. Microbiol.* **19**, 27.
Dalzel, A. (1793). *In* "The History of Dahomy", Introduction, p. iv. London.
Daniell, W. F. (1852). *Pharm. J. and Trans.* (Bell, J., ed.), **11**, 445.
De Candolle, A. P. (1844). *In* "Prodromus Systematis Naturalis Regli Vegetabilis", Vol. 8, p. 183.
Gregory, D. W. and Cocking, E. C. (1965). *J. Cell Biol.* **24**, (1) 143.
Horowitz, R. M. and Gentili, B. (1969). *J. Agric. Fd Chem.* **17**, 696.
Inglett, G. E. and May, J. F. (1969). *J. Fd Sci.* **34**, 408.
Inglett, G. E., Dowling, B., Albrecht, J. J. and Hoglan, F. A. (1965), *J. Agric. Fd Chem.* **13**, 284.
Krbechek, L., Inglett, G., Holik, M., Dowling, B., Wagner, R. and Riter, R. (1968). *J. Agric. Fd Chem.* **16**, 108.
Kurihara, K. and Beidler, L. M. (1968). *Science, N.Y.* **161**, 1241.
Kurihara, K. and Beidler, L. M. (1969). *Nature, Lond.* **222**, 1176.
Labat, le R. Pere (1730). *In* "Voyage du Chevalier des Marchais en Guinée", Vol. II, p. 255. Paris.
National Academy of Sciences (1968). "Principles of Plant and Animal Pest Control", Vol. 6, p. 36.
Neal, G. E. (1965), *J. Sci. Fd Agric.* **16**, 604.
Olliver, M. and Smith, G. (1933). *J. Bot., Lond.* **71**, 196.
Olliver, M., Wade, P. and Dent, K. P. (1957). *J. Sci. Fd Agric.* **8**, 188.
Pojnar, E., Willison, J. H. M. and Cocking, E. C. (1967). *Protoplasma* **64**, 460.
Power, J. B., Cummins, S. E. and Cocking, E. C. (1970). *Nature, Lond.* **225**, 1016.
Reeve, R. M. (1959). *Am. J. Bot.* **46**, 241.
Schultz, W. G. and Randall, J. M. (1970). *Fd Technol.* **24**, 1282.
Sterling, C. and Chichester, C. O. (1960). *Fd Res.* **25**, 157.
Tropical Products Institute (1970). Proceedings of the 1969 Conference on Tropical and Subtropical Fruits. Ministry of Overseas Development, London.
Wade, P. (1964). *J. Sci. Fd Agric.* **15**, 51.
Westling, R. (1909). *Svensk bot. Tidskr.* **3**, (2), 134.

Chapter 15

Canned Fruits other than Citrus

J. B. ADAMS AND H. A. W. BLUNDSTONE

Fruit and Vegetable Preservation Research Association, Chipping Campden, England

I Introduction 507
II Organoleptic Changes during the Canning Processes 508
A. Preparation of Fruit for Canning 508
B. Processing and Storage of Canned Fruit 513
III Vitamin Changes during the Canning Processes 526
A. Preparation of Fruit for Canning 526
B. Processing and Storage of Canned Fruits 527
IV Interaction between the Fruits and their Containers 530
A. Factors which Influence the Corrosivity of the Pack 530
B. Prevention of Undesirable Container–product Interactions .. 534
V Selection of Varieties of Fruit for Canning 535
A. Selection for Organoleptic Qualities 535
References 537

I. INTRODUCTION

In this chapter we hope to give the reader some idea of the range and nature of the biochemical changes which may take place during the preparation of fruit for canning and its subsequent processing and storage. Detailed processing conditions such as levels of enzyme inhibitors and processing times and temperatures have not generally been included, but an attempt has been made throughout to relate biochemical reactions with commercial practice. As very little work appears to have been published on the selection of fruit at the "correct stage of maturity for canning" (from a biochemical as distinct from an empirical viewpoint), we have not attempted a discussion on this subject.

Production of canned fruit represents an important part of the total fruit production in some countries. Thus, in the U.S.A. and Australia canned fruit amounts to approximately 20–25% of the total fruit production. In other countries, however, canned fruit is not so important. For instance, in the U.K. and Canada, it represents approximately 10% of total fruit production, and in

Italy, where a large amount of fruit is grown, only about 1% is converted into the canned product. It would appear, from the limited data available, that approximately 10% of the total fruit production of the world is canned.

II. ORGANOLEPTIC CHANGES DURING THE CANNING PROCESSES

From the consumer's point of view, alterations in the organoleptic properties of the fruit are the most important changes which take place during canning.

A. Preparation of Fruit for Canning

1. *Colour changes*

Any discoloration which arises during preparation of the fruit will be carried through to the final product and will result in a loss of consumer acceptability. It is thus vitally important to minimize such changes wherever possible.

(a) *Enzymic browning*. Many fruits undergo rapid browning as a result of cellular disruption and access of oxygen during the peeling and slicing operations prior to canning, if preventative measures are not taken. This applies to apples, pears, bananas, apricots, peaches, quinces, medlars and persimmons.

Plums, strawberries, grapes, olives and cherries will also brown if injured during the preparation, but only in the latter two cases are any measures, except careful handling, taken to prevent browning. Other fruits such as redcurrants and blackcurrants, gooseberries, raspberries, loganberries, blackberries, pineapples, tomatoes and melons do not become discoloured so readily so that no preventative action is required.

The browning reaction is mainly caused by the enzyme(s) polyphenol oxidase acting on a suitable phenolic substrate in the presence of oxygen (see Fig. 1) (Joslyn and Ponting, 1951).

All three reactants are required for browning to take place and so control is based on rendering at least one of these inactive. This is generally done by partially or totally inactivating the enzyme(s) either by heat (blanching) or by chemical inhibition. Examples of the use of the first method are the preparation of apricot purée (Ponting *et al.*, 1954) and the preparation of Bartlett pear halves for canning (Tate *et al.*, 1964). Steam at 93°C (200°F) or higher is used in these cases to give a rapid blanch. Low-temperature, short-time blanch may destroy only a fraction of the total oxidase present in the fruit. Thus, when halved, peeled peaches are sliced after a short hot water blanch, unattractive crescent-shaped brown areas are often observed about 1 to 2 mm below the surface of the fruit. This occurs at the line of demarcation between the flesh that has been heated sufficiently to destroy the oxidase, and the unheated flesh near the centre of the half (Cruess, 1958).

The most effective permitted inhibitor is sulphur dioxide, especially when used before enzymic oxidation has taken place. The use of sulphur dioxide for canned fruits is, however, limited because of its corrosive effect on the can.

Acids can sometimes be used as inhibitors, as in the acid rinsing of cling peaches after lye-peeling. If the pH is lowered sufficiently all enzyme activity will be inhibited.

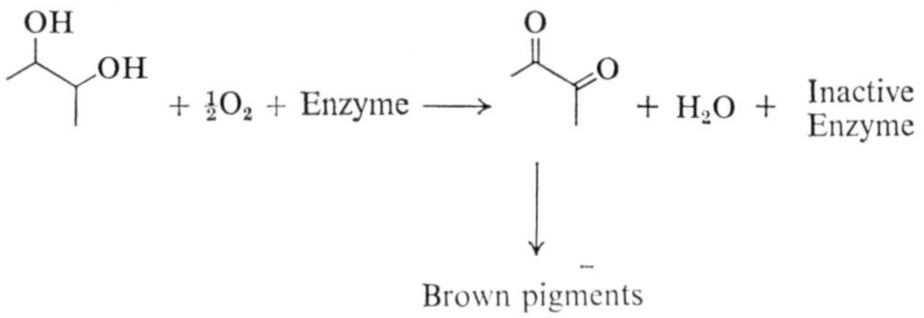

OH, OH + $\frac{1}{2}O_2$ + Enzyme ⟶ O, O + H_2O + Inactive Enzyme

↓

Brown pigments

FIG. 1.

Ascorbic acid has been used extensively to control enzymic browning in fruits. Unlike sulphur dioxide it does not inhibit the enzyme directly, but reduces the quinones first formed. This process is accompanied by a gradual decrease in enzyme activity and is known as reaction inactivation (Ponting, 1960). An example of the use of ascorbic acid is the prevention of browning in the preparation of apple halves for canning (Hope, 1961).

Sequestrants synergize the antioxidant activity of ascorbate. Thus, while browning of canned apple halves persists when less than 1000 p.p.m. ascorbate is employed, the addition of only 250 p.p.m. of the disodium salt of ethylenediaminetetracetic acid (EDTA) permits the effective use of 250 p.p.m. ascorbate. In the absence of ascorbate, levels as high as 1000 p.p.m. EDTA citrate or phosphate do not inhibit browning (Furia, 1968).

Sulphydryl inhibitors are very effective but usually leave undesirable tastes or odours. Cysteine, however, has been recommended instead of ascorbic acid for use in preventing browning of apple slices (Walker and Reddish, 1964). It reacts with the enzymically formed quinones to give a colourless cysteine–phenolic complex.

Many salts have been shown to inhibit polyphenol oxidase but sodium chloride is the most widely used. It has found application in holding solutions for apples, pears and quinces prior to blanching, as well as for ripe olives prior to pickling (see also Chapter 7). The strengths of brines used for this inhibition are such that only partial enzyme inactivation is achieved. Complete inactivation of the enzyme would require very high salt concentrations which would render the product unpalatable. The mechanism of the inhibition does not appear to be known.

Other chemicals, such as adenosine triphosphate (ATP), have been investigated; ATP seems to be unique in that it can prevent or diminish colour development in apples, avocados and peaches, without affecting the intrinsic activity of the polyphenoloxidase enzyme(s) (Makower and Schwimmer, 1957). This method has not, however, been applied commercially.

The most widely used methods for inhibiting enzyme activity (blanching, brining and ascorbic acid treatment) also lead to a reduction of oxygen in the fruit tissue. This is useful both in reducing the amount of enzymic browning which can take place, and in giving a potentially less corrosive product. No attempt is usually made to control directly the phenolic substrate required for the browning reaction.

A phenomenon which is often associated with enzymic browning is that of cherry scald. This arises as the result of bruising and is characterized by migration of the red anthocyanin pigment from the skin to the flesh of the fruit, followed by the loss of pigment and the appearance of a brown discoloration. It is important to differentiate between scald and the subsequent browning of cherries. The latter is believed to be due to enzymic oxidation, whereas the former is typified by a loss of anthocyanin whether browning occurs or not. It has been suggested that the effect of bruising is to disrupt the cells of the cherry peel and the flesh beneath the peel, leading to anthocyanin migration from the peel to the flesh followed by decolorization of the anthocyanin by anthocyanase from the peel (see also Chapter 12). However, more recent evidence appears to show that anthocyanase is not involved in the scalding of red tart cherries, at least when held in an atmosphere of 100% nitrogen (Dekazos, 1966a). There is some conflict in the literature on the effect of oxygen on the scald observed in cherries soaked in cold water prior to processing. Some authors (Dekazos, 1966b) state that lack of oxygen causes bruised cherries to scald, whereas others (Whittenberger *et al.*, 1968) maintain that there is no benefit in aerating the cherry soak tanks either at the orchard or at the factory. The present state of knowledge indicates that there is a definite risk in using aeration to reduce scalding since both anthocyanase (Wagenknecht *et al.*, 1960) and polyphenol oxidase require oxygen for their activity and both these enzymes may possibly contribute to the discoloration of cherries during soaking.

The best methods of reducing cherry scald at the moment appear to be to reduce bruising at all stages of handling, to cool orchard tanks promptly to 16°C, (60°F) and to maintain soak tanks at 10°C (50°F) or below.

(b) *Non-enzymic.* An example of non-enzymic browning is found in the preparation of olives for canning, where ripe olives are browned intentionally by pickling in dilute sodium hydroxide, removing the lye and exposing the fruit to the action of air (see Chapter 7).

2. *Texture changes*

If the texture is soft before the fruit goes into the can it will be even worse after heat processing. Control of texture breakdown can be achieved by limiting the activity of pectic enzymes naturally present in the fruit, and by the application of calcium salts.

(a) *Enzymic changes in texture.* The loss in firmness of texture which occasionally occurs in brined sweet cherries from which Maraschino-type cherries are prepared, is either due to the action of a polygalacturonase (PG) or to pH if this is outside the range 3·1–3·5 (Van Buren, 1967). As texture is of prime importance in this product the PG activity must be inhibited. This can be done effectively by adding 0·01–0·025% "Nacconal" (a commercial preparation of sodium alkyl aryl sulphonate) to the brine, only a trace of which remains in the finished product. The enzyme can also be rendered inactive by using a brine with a pH of 1·5 or lower but this tends to produce cracks in the cherries and encourages skin injuries (Yang *et al.*, 1960).

Failure to inactivate pectic enzymes early in the procedure used to process tomatoes during paste manufacture results in a rapid destruction of pectic substances in the macerate and a low viscosity product. If, however, the macerates are rapidly preheated to 85°C (185°F) pectin-methyl esterase (PE) is inactivated. This heat process may not be sufficient to destroy PG but this is unimportant as its substrate, pectic acid, will not have formed in the macerate (McColloch *et al.*, 1950) because of the inactivation of the esterase. It would appear that the pectic enzymes in tomatoes occur in greatest quantity near the surface of the fruit so that a large proportion of such enzyme activity may be destroyed prior to maceration by passing the whole fruit through a steam or hot water blanch (McColloch *et al.*, 1952).

Recently (Wagner and Miers, 1967) the effect of chemical inhibitors on the pectic enzymes in tomatoes has been investigated, and it has been shown that strong acids are most effective inhibitors. Large increases in product consistency were reported when acidification below pH 2·3 was combined with heat treatment. Nacconal was also effective, acting, presumably, by blocking PE. This enzyme may not, however, always lead to a loss of texture in canned fruits. Thus it has been shown (Leinbach *et al.*, 1951) that PE is not related to the tendency of some raspberry varieties to "mush" during processing. Furthermore, it has been shown, (Buch *et al.*, 1961) that when raw, red tart cherries are allowed to stand in air or water prior to canning, a small amount of demethylation by PE gives an increased firmness in the canned product. This is probably due to the action of calcium salts present in the fruit causing cross-linking of the partially demethylated pectin molecules. A similar observation has been made (Hsu *et al.*, 1965) in the canning of whole tomatoes. Here, selective demethylation resulting from a 30 sec blanch at 100°C and a 10 min hold, increased the firmness of the canned tomatoes. Calcium chloride was added to the cans before processing to ensure that the cross-linking of the partially demethylated pectins was complete.

(b) *Effect of calcium salts on texture.* Much attention has been paid to the effect of various calcium salts and their manner of application (Doesburg, 1965). The main salts used for firming canned fruits are the hydroxide (for example in cherry brining), the chloride (in tomato firming) and the lactate (in firming apple slices). Care must, however, be taken in their use. For instance, calcium chloride gives an undesirably tough texture of low crispness when used for brining cherries, and gives a bitter after-taste when used for firming apple slices. Effectiveness of calcium salts on improving fruit texture depends on the presence of partially demethylated pectin molecules and also on the absence of calcium complexing agents, such as oxalate and citrate ions. The tissue-firming compound appears to be calcium pectate and this has been identified in tomatoes treated with calcium chloride (Loconti and Kertesz, 1941).

3. *Flavour and related changes*

Very little information is available in the literature concerning flavour deterioration (enzymic or non-enzymic) during preparation of fruits for canning. Some of the compounds responsible for flavour are volatile to a greater or lesser degree, and so can be lost by evaporation from the fruit during peeling, cutting and pulping operations without any action of degrading enzymes. This is especially true when canning purées or sauces. Thus, in the conventional canning of Bartlett pear purées and apple sauces the volatiles are driven off to the atmosphere, yielding products almost entirely lacking in flavour. The problem can be partially overcome by processing in specially designed closed systems, thus obtaining an aqueous essence with the characteristic aroma of the fruit, which is used for fortifying the purée or sauce before canning (Buch *et al.*, 1956; Jennings *et al.*, 1960). If this should prove economically impossible, the least that can be done is to give a minimum heat treatment to purées and sauces so that degradation and evaporation are limited.

Prolonged soaking or brining of fruit may affect flavour by leaching of the free acids and sugars into the soaking liquid. Thus it has been reported (Peterson, 1938) that cherries, soaked in water at 10–12°C (50–55°F) for 12–24 hours, show losses of flavour which increase with soaking time, and which correspond to losses of acids. Similarly, the blanching of figs, a necessary step required to remove their "raw" taste, is carried out in the cans in steam to prevent loss of the natural fruit sugars (Cruess, 1958).

Sometimes, however, soaking can improve flavour as in olive processing where lye treatment removes the bitter principle (see Chapter 7, p. 270).

It has also been shown that puréeing persimmons results in a loss of astringency. This would seem to be due to adsorption by cellular materials, during blending and storage, of the soluble phenolics from the broken tannin cells, rather than to depolymerization or oxidation of the phenolic compounds (Joslyn and Goldstein, 1964).

B. Processing and Storage of Canned Fruit

1. *Colour changes*

The first change which takes place as soon as the hot syrup makes contact with the fruit involves a redistribution of the water-soluble pigments between fruit and syrup. This continues rapidly during processing and at a slower rate during storage until equilibrium is attained, usually within a few weeks of storage. The second change which accompanies redistribution involves the chemical interaction between pigment, fruit and syrup constituents. These interactions lead to alterations in the nature and type of pigment present in the fruit, and it is these changes which will be considered in this section.

(a) *Pigment degradation.* (i) Anthocyanins.† Much of the research carried out on the degradation of anthocyanins, the red pigments in red fruits (other than tomatoes), has been on fruit juice and model systems so that extrapolations have to be made between anthocyanin breakdown under these conditions and that actually occurring in the canned fruit.

It is evident from the work on liquid systems that the factors which will accelerate anthocyanin degradation most rapidly are high temperature processes of long duration, high concentrations of oxygen in the can contents and headspace, high ascorbic acid concentration, appreciable amounts of sugar degradation products, such as furfural and hydroxymethyl furfural, and high temperature during a long period of storage. Daravingas and Cain (1965) showed, for example, that prolonged storage at high temperatures in the presence of O_2 caused considerable breakdown of the anthocyanins in canned raspberries and Tinsley and Bockian (1960), using a model system, showed that the pelargonidin-3-glucoside of strawberries is more stable in N_2 than in O_2.

The reactions by which anthocyanins are degraded have yet to be established but it has been suggested that the breakdown in pure pelargonidin-3-glucoside (principal strawberry anthocyanin) solutions is due to a hydrolytic opening of the pyrylium ring with the formation of a sugar-substituted chalcone (see Fig. 2). Further degradation of this ketone would eventually lead to the brown precipitate which has constantly been observed as an end-product of the

† See Chapter 21.

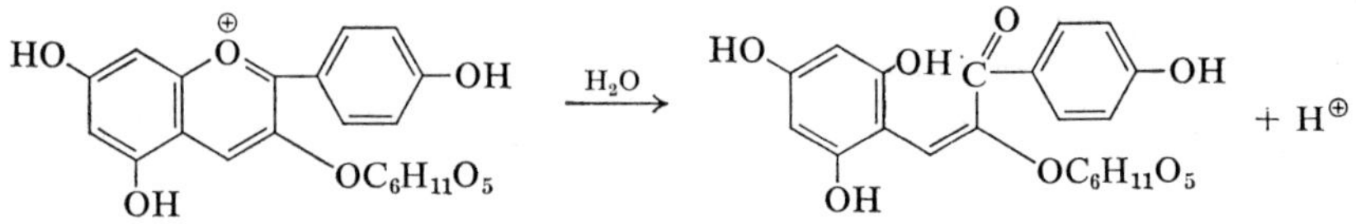

FIG. 2.

degradation of the pigment. The alternative mechanism for anthocyanin degradation, consisting of hydrolysis to the aglycone followed by conversion directly or through intermediates, to the brown precipitate would seem unlikely since extraction with ethyl acetate of pure pigment solutions and juices, which had been heated for various times and temperatures, failed to reveal the presence of an aglycone (MacKinney *et al.*, 1955; Markakis *et al.*, 1957). The effects of the various accelerating factors mentioned above on such a pathway, if in fact it occurs in canned fruits, are unknown. Work is at present being carried out in our own laboratory on the effects of processing and of conditions of storage on canned red fruits in an attempt to gain a better understanding of the anthocyanin degradation, and to devise means of preventing it. In previous work, various additives and antioxidants have been tried in an attempt to reduce % anthocyanin degradation, but generally these have proved to be of little value because of their relative ineffectiveness and of their toxicity. For example, it has been shown that out of some nineteen different additives tested for their effect on colour retention in strawberry juice, only thiourea, propyl gallate and quercetin showed any beneficial effects (Markakis *et al.*, 1957). Other workers have found that rutin in combination with small amounts of sulphur dioxide has a positive influence on pigment retention in processed strawberries and cherries (Kyzlink *et al.*, 1962).

At present the best means of preventing the breakdown of anthocyanins in cans of fruit is to store the cans at as low a temperature as possible. This was recommended in the early 1930's (Kohman, 1931) and is supported by more recent studies (Whittenberger and Hills, 1956; Dalal and Salunkhe, 1964; Adams, 1968).

(ii) Carotenoid changes.† Carotenoids play an important part in the colour of canned apricots, peaches, mangoes, pineapples, jack-fruit and tomatoes. However, the amount of work reported on the changes which take place in the carotenoids of these fruits during heat processing is small compared with that for changes in anthocyanins. It has been shown that peaches lose approximately 50% of their total carotenoids during can exhausting and processing (Mitchell *et al.*, 1948). Apricots lose between 2 and 34% of their carotene on canning, whereas this loss is increased to 50% if the fruit is puréed before canning (Moyls, 1951). These results may, in fact,

† See also Chapter 21.

reflect apparent losses due to isomerization of highly coloured 5,6-epoxides to the less intensely coloured 5,8-epoxides rather than actual destruction of carotenoids (Borenstein and Bunnell, 1966). A similar change has been shown to occur on canning pineapple where the carotenoid pigments of the fruit undergo complete isomerization from the original 5,6-epoxide into their 5,8-furanoid forms as shown in Fig. 3 (Singleton *et al.*, 1961).

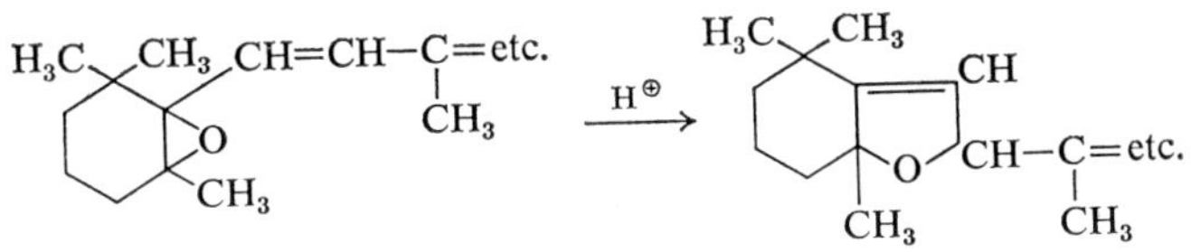

H_3C CH_3 CH=CH—C=etc.
O CH_3 CH_3
$H^{\oplus}$ →
H_3C CH_3 CH
O CH—C=etc.
CH_3 CH_3

FIG. 3.

Cis-trans isomerization on canning and storing may also lead to colour deterioration, and it has been shown that considerable changes in absorption characteristics of the carotenoid pigments take place during the canning of mango pulp (Ranganna and Siddappa, 1961).

Apart from changes due to isomerization, and the suggested conversion of protein-bound, water-insoluble carotenoid into a lipid-soluble form via a non-protein water-insoluble form (Hanson, 1954), little appears to be known about carotenoid breakdown. It has been observed that prolonged heating causes lycopene to become brown, presumably by oxidation. It follows that care must be taken not to overheat tomato purées during their preparation.

The effect of storage on carotenoids depends on the product being stored and the temperature of storage. Thus 52% of the original β-carotene is preserved after 12 months storage of canned papayas at 25–30°C (77–86°F), whereas in canned jack-fruit, under the same conditions, 97% remained after 6 months (Siddappa and Bhatia, 1956). After 4 months' storage, cans of Moorpark apricots kept at 21°C (70°F) showed approximately one-quarter of the amount of carotenoids found in those kept at 4·4°C (40°F) (Dalal and Salunkhe, 1964).

(iii) Changes in other pigments. Little work has been done on the changes, during canning and in storage, in the yellow pigments (anthraquinones, flavones, etc.) present in some fruits.

Chlorophyll, the green pigment of unripe tomatoes, turns brown during cooking, presumably due to pheophytin formation, so that use of too great a proportion of green fruit during manufacture of tomato preserves will lead to a brown or brownish-red product. The chlorophyll present in rhubarb appears to be quite stable during canning and subsequent storage (even at 35°C, 95°F) and does not lead to appreciable quantities of brown substances (Adams, 1968).

The dark brown or black colour of canned ripe olives is well retained on canning provided that the pH is maintained in the range 8·0–9·5 and that lacquered cans are used (Cruess, 1952). Further information on this topic is given in Chapter 7.

(b) *Discolorations*. (i) Purple discolorations. These occur in cans of fruit containing anthocyanins, and are caused by polyvalent metal ions, such as tin and iron, which may be formed by corrosion of the can. These ions react with the anthocyanins to form unattractive purplish-blue "lakes" (as in canned blackcurrants), which will be considered in more detail in the section on the interaction of the fruit with the container. This effect can be reduced by using lacquered cans to minimize corrosion.

(ii) Pink discolorations. These may occur in canned gooseberries, apples, peaches, pears, bananas and guavas. They are caused by colourless "leuco-anthocyanins" being converted into anthocyanins when heated under acid conditions. The problem seems to be of a spasmodic nature with some varieties of fruit, e.g. Bartlett pears, whereas with other varieties it occurs with great regularity, e.g. Packam and Bon Chrétien pears. The main processing factors affecting the formation of the pink colour are excessive heating and delayed cooling (Luh *et al.*, 1960). The most important factors in cultivation of the fruit appear to be maturity of fruit, locality, rootstocks, and exposure to direct sunlight; the conditions required in the fruit for appreciable pinking are low pH and high leucoanthocyanin concentration (Nortjé, 1966). The presence of oxygen in the canned fruit may also have some bearing on the problem (Adam, 1949). It has been suggested that tin is required for the pinking reaction (Anthistle and Dickinson, 1959). As the fruits mentioned are normally packed in plain cans, the pink discoloration may become a purple discoloration by reaction with the metal ions resulting from can corrosion.

The pink discoloration may be prevented in canned guavas by using a syrup containing 0·06% citric acid and 0·125% ascorbic acid (Ranganna *et al.*, 1966), and can be minimized in canned banana purée by acidifying to pH 4·2 (Guyer and Erickson, 1954). It has recently been reported that sulphur dioxide at a level of 200 p.p.m. prevented pinking in pear purée even after 2 hours boiling regardless of the leucoanthocyanin content of the purée (Clegg, 1967).

By treating pears with 100–500 p.p.m. of the disodium salt of ethylene-diaminetetracetic acid, citrate or phosphate prior to processing, pink discoloration can be significantly inhibited in the canned product (Furia, 1968). This is presumably due to traces of copper, iron, tin and zinc being complexed, and is a good indication that these metals are required, probably as catalysts, for the conversion of leucoanthocyanins to anthocyanins in canned pears. Prevention of pinking in canned peaches can also be achieved by thorough cooling after processing (Mahadeviah, 1966).

Conversely, in canned quinces, where the pink colour is considered a desirable characteristic a high-temperature, long-time process is used followed by slow cooling of the cans in air (Thompson, 1950).

(iii) Brown discolorations (enzymic). If the correct procedure for eliminating the effects of oxidative enzymes is not taken (see Section A, 1(a)), then they may act during the preparation and processing to give a browned product. Thus, because of the active enzyme system in bananas, a brown discoloration may be obtained in canned banana purée unless it is blanched at 85°C (185°F) and maintained at or above this temperature until the final cooling of the canned product (Guyer and Erickson, 1954). Canned cling peaches occasionally show a brown discoloration which is presumed to be due to polyphenoloxidase activity during the preparation and canning processes (Luh *et al.*, 1967); greengages are particularly prone to enzymic browning during exhausting and processing due to an apparently heat-resistant enzyme system. It has also been shown that in normal processes of fruit bottling, oxidase enzyme systems may not be inactivated (Crang and Sturdy, 1953), and peroxidase systems persist in some fruit tissues with consequent browning.

(iv) Brown discoloration (non-enzymic). Non-enzymic browning is a common cause of discoloration in canned fruits. The reactions which lead to this type of browning are little understood, but it is quite clear that no simplified mechanism which will cover all cases of discoloration can be proposed for the creation of the brown polymers. The connection between the Maillard reaction and deterioration of colour in canned fruits is not clear. This is because the pH of the fruit is generally acid, and because ascorbic and other organic acids (together with much sugar and relatively small quantities of amino compounds) are believed to be involved in the browning.

During storage of canned red fruits, brown polymers arise concomitantly with, or as a direct result of, pigment degradation (see section on anthocyanin degradation).

When gooseberries and apples are given long processes at high temperatures in lacquered cans, brown colours form in the fruit and the syrup, compared with pink coloration when plain cans are used, which is thought to be due to polymerization of the colourless "leucoanthocyanins" into brown "phlobaphenes" (Anthistle and Dickinson, 1959). A similar reaction may occur during the canning of bananas in acidified syrup. Here the discoloration is described as "fine pink to brown lines concentrated mainly in the carpel walls of the banana" (Board and Seale, 1954).

Lycopene browning on prolonged heating of tomato products has already been mentioned, but discolorations have been observed in tomato paste and ketchup which are not caused by this reaction. Thus, browning on storage of these products has been attributed to the presence of fructose in the tomato (Luh, 1960), whereas the "blackneck" discoloration has been shown to be caused by the reaction between iron salts, arising from equipment or the bottle cap, and the spice tannins. Ferrous tannate is formed first and this is oxidized by the oxygen in the headspace to give black ferric tannate (Davis and Kefford, 1955). Blackneck can be eliminated by using air-tight bottle caps, deaeration of the ketchup prior to filling, using stainless steel equipment, controlling headspace and by selection of spices. A similar blackening has also been observed during the processing of Maraschino-type cherries. In this case prevention may be achieved by using calcium phytate to precipitate the iron (Cohée and Nelson, 1951).

In an attempt to ensure that red fruits appear red when the can is opened, artificial dyes are commonly used. Generally, these are more stable than the natural red pigments, but on long storage they may break down to colourless end-products by reaction with metal ions liberated by corrosion (Yang *et al.*, 1960; Raven, 1962).

Storage temperature is an important factor in many non-enzymic browning reactions. Two examples of this are to be found in canned red fruits and in tomato products. Thus, canned red fruits will retain most of their natural red pigment if stored at less than 5°C (41°F) for periods of over a year (Adams, 1968), whereas they lose all their natural colour in less than 6 months at 35°C (95°F). High storage temperatures accelerate the blackneck phenomenon in tomato ketchup. It is therefore important to cool these products adequately before storing, and to keep the storage temperature as low as possible.

2. *Texture changes*

The application of heat, either during processing or storage of canned fruits, may lead to irreversible alterations in texture caused by loss of the semi-permeability of the cell membranes and by breakdown of pectic substances.

(a) *Destruction of the semi-permeability of cell membranes.* The first important effect of heat on the texture of fruits containing whole cells is the destruction of the selective permeability of the cell membranes. As a result of this most of the cell distension is lost and the crispness associated with fresh fruit is lost with it. Another consequence of this loss of permeability is that cell sap can be exuded into the extracellular spaces and vascular system. The gases originally present in these spaces are ultimately displaced by the fluids and thus the texture is affected. There appears to be no known method for preventing the loss of selective permeability of the cell membranes, (Matz, 1962).

(b) *Changes in pectic substances.* (i) Enzymic. Generally, the temperature of the fruit is increased so rapidly during the heat processing that the time available for any enzyme activity to take place before it is denatured is quite short and no effect relatable to the increased speed of reaction occurs. Thus it has been shown that pectic enzymes are unlikely to play a part in the disintegration of canned Bulida apricots (Van Der Merwe *et al.*, 1966), and that pectinesterase activity is not in itself related to firmness-retention in canned apricots although it may be involved indirectly in demethylation for cationic-binding (Shewfelt, 1965).

(ii) Non-enzymic. Of major importance for the maintenance of good texture in canned fruits is protopectin, the water-insoluble material which is to be found mainly in the intercellular layer and primary cell wall of the fresh fruit. Thus it has been shown that a partial conversion of protopectin into water-soluble pectin occurs during heat processing of both clingstone and freestone peaches. Since, however, the fresh ripe clingstone peach has a firm texture due to the high retention of protopectin in thick cell walls, this conversion is relatively unimportant to the final texture of the canned product. During the ripening of the freestone peach, however, protopectin is converted to water-soluble pectin, giving a low retention of the former in a thin cell wall. Any further changes in this direction during heat processing will have a profound effect on the final texture of the canned fruit. It is for this reason that stage of ripeness at the time of canning is an important factor in maintenance of good texture in canned freestone peaches (Postlmayr *et al.*, 1956). The suggestion has also been made that it is advantageous to heat apricots to temperatures no higher than 80–85°C (176–185°F) so that protopectin decomposition is reduced. After such moderate heat treatment the apricots can be firmed by application of calcium salts (Joux, 1957; Mohammadzadeh–Khayat and Luh, 1968). Selection of less ripe fruit does not ensure the avoidance of "mushiness" in canned appricots since the higher content of acid accelerates breakdown of protopectin when the fruit is heated (Patron, 1956). Acidification may, however, lead to a firmer canned product in some less acid fruits. Thus, addition of citric acid to pimentos, especially when used in conjunction with calcium salts, results in a significant increase in firmness of the canned product. Here, however, no correlation could be found between the differences in firmness and the major pectic fractions (Powers *et al.*, 1961).

It has been shown that high storage temperatures are detrimental in maintaining a good texture in canned clingstone peaches; recently this was attributed to a more rapid transformation of protopectin to water-soluble pectin under these conditions (Kanujoso and Luh, 1967). Protopectin breakdown does not always seem to be involved during the canning of fruits, as is shown by the lack of any marked transformation of water-insoluble pectin to water-soluble pectin during the heat processing of cherries (Al-Delaimy *et al.*, 1966). The main factors which appear to be involved in protopectin stability and, therefore, in the maintenance of a good texture are, the type of fruit, its acidity at the time of canning, the length and temperature of processing and the length and temperature of storage.

The role of other components of the cell wall in texture changes has not been finally established, but there are indications that, at least in the case of strawberries (var. Cambridge Favourite), the changes in texture occurring on canning are not due to large changes in the composition of the insoluble cell wall polysaccharides (Wade, 1964).

It would seem that, as for non-enzymic browning, it is impossible to give a general mechanism for non-enzymic texture deterioration and that each fruit must be treated individually.

(c) *Effect of sugar*. The following theory has been proposed for the effect of sugar on the texture of canned fruit. When a fruit is canned in sugar syrup, it shrinks because water leaves the fruit faster than syrup solutes can move in to equalize solute concentrations. The cells walls then absorb sugar, probably by hydrogen bonding of the latter with polysaccharides in the walls, and this causes cessation of water movement from the fruit followed by a reverse in its direction. Depending on the relative sugar concentration in the canning syrup, a greater or lesser amount of water may be withdrawn from the cell walls of the fruit. The effect of this is to bring the molecules of the cell wall polysaccharides together, permitting a greater degree of polymer–polymer hydrogen bonding. This, in turn, decreases the number of hydroxyl groups available for sugar bonding and produces some deformation of the cell walls. Because of this dehydration effect, fruit canned in heavier syrup will be firmer than fruit canned in a lighter one (Sterling, 1959).

If larger polysaccharide molecules, such as pectinic acids and maltosaccharides, are present in the syrup solution, these may compete with the wall polysaccharides for sugars and water, and the equilibrium distribution of simpler sugars between fruit and syrup will be modified correspondingly. Thus, it has been demonstrated that a syrup containing high molecular weight sugars shows less sugar translocation into the fruit and more water translocation into the syrup (Ross, 1955).

3. *Flavour and related changes*

The information available on flavour deterioration during canning is extremely sparse, and this is no doubt because the subject of flavour, as a whole, is not well understood. It is therefore difficult to recommend methods for avoiding flavour loss except the obvious one of not overprocessing canned fruit.

(a) *Changes in volatiles*. Many volatile compounds which may contribute towards the flavour of fresh fruit may be lost during heat processing and storing, either by evaporation into the headspace or by decomposition. For instance, the major carbonyl component of fresh banana, 2-hexenal, is not detected in the heat-processed purée. This is probably due either to inactivation during purée manufacture, of the enzyme required for its formation, or, if formed, to decomposition during processing and storage of the purée. The concentration of acetic acid esters in banana purées was also found by Hultin and Proctor (1961) to be very low, probably due to hydrolysis at the elevated temperatures used for puréeing. This explanation is supported by the fact that the purées contained a greater concentration of acetic acid than the fresh fruit.

Many of the volatiles present in fresh peaches are destroyed by canning. Of these, acetaldehyde and methyl octanoate suffer the largest losses. The result is a cooked peach flavour (Li, 1966). One of the fruits most affected by heat processing is the strawberry. However, no information appears to be available for this fruit on the changes in volatiles which occur in canning. It has, however, been suggested that changes in unsaturated volatile compounds may be involved in the deterioration of the flavour of stored tomato products (Spencer and Stanley, 1954). Recent work has shown that, in general, the concentration of carbonyls, such as acetaldehyde, acetone, isovaleraldehyde and hexanal, in canned tomato products decreases with storage time, while alcohols, such as methanol and ethanol, increase (Nelson and Hoff, 1969). In plain tin cans this is probably due to reduction of aldehydes and ketones to primary and secondary alcohols at the tin–acid interface. Due to oxidation, the concentration of acetaldehyde and hexanal may be lower in lacquered than in plain cans. Nelson and Hoff found that, as storage time increased, the aldehyde concentration continued to fall with an increase in primary alcohols, indicating that the initial oxidation was followed by reduction.

New volatile compounds may be formed during the storage of processed fruits. Thus, apple sauce develops a rank flavour on storage. This has been attributed to the formation of n-caproic acid. The source of this acid is not known, but it is clear that it does not arise from esters of caproic acid which are only present in very low concentrations (Mattick *et al.*, 1958).

(b) *Changes in non-volatiles.* Off-flavours in some canned fruits can be attributed to ammonium pyrrolidonecarboxylate formed by the cyclization of glutamine (Fig. 4). It is theoretically possible for pyrrolidonecarboxylic acid

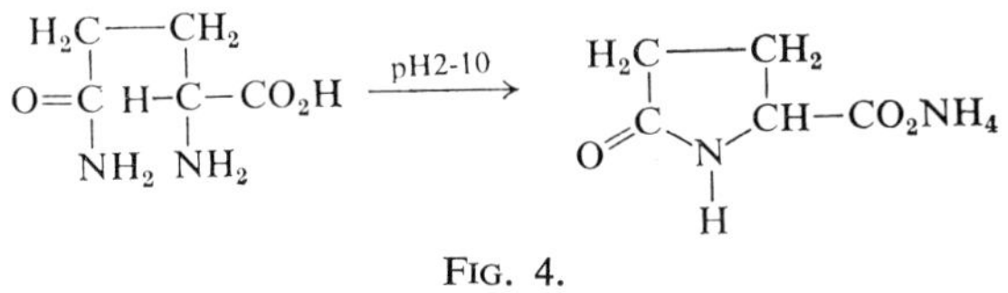

$$H_2N{-}C(=O){-}CH_2{-}CH_2{-}CH(NH_2){-}CO_2H \xrightarrow{\text{pH2-10}} \text{(cyclic: } H_2C{-}CH_2,\ O{=}C{-}N(H){-}CH{-}CO_2NH_4\text{)}$$

FIG. 4.

(PCA) to be formed from the cyclization of glutamic acid, but it has been shown that, in most products, glutamic acid is stable during processing and storage, whereas glutamine is converted to ammonium PCA quite rapidly during processing and, more slowly, during storage (Mahdi *et al.*, 1959). Ammonium PCA has been found in canned raspberries, cherries, pineapple, peaches, pears and prunes, but the concentration required to give a detectable off-flavour varies from product to product (Mahdi *et al.*, 1961).

During the processing of tomatoes, methyl sulphide is formed in an amount which exceeds the reported flavour threshold (Amerine *et al.*, 1965; Miers, 1966). It is thought that the precursor of this compound is an S-methyl methionine sulphonium salt, which, on heating, yields homoserine and methyl sulphide (Wong and Carson, 1966).

One of the artificial colours used in canned fruits, erythrosine, may lead to an off-flavour since it tends to lose iodine on prolonged storage of the product (Raven, 1962).

(c) *Enhancing and masking of natural flavours.* (i) Use of sugars. The sugar : acid balance of canned fruits has an important bearing on the acceptability of the product. In England the average sugar : acid ratio required is 15 : 1. This accounts for the different strengths of syrups used for different fruits. Thus, gooseberries and raspberries (more acid fruits) are normally packed in 45° Brix syrup, whereas sweet cherries (less acid fruit) are normally canned in 35° Brix syrup. The syrup also helps to retain, and may even enhance, the natural fruit flavours but will mask these if used in too high concentrations.

(ii) Flavour enhancers. The most commonly used flavour enhancer in the food industry, monosodium glutamate, has found no use in canned fruits because of the low equilibrium concentrations of the flavour-active ionic form at pH values below 4·0 (Fagerson, 1954). Flavour-producing enzymes may find some application in enhancing the flavour of canned fruit. These enzymes convert flavour precursors, which have not been destroyed during heat processing or storing, into flavoured compounds (Hewitt, 1963).

(d) *Enzyme activity.* The flavour producing enzymes present in the fresh fruit are destroyed during heat processing, but other enzymes, which can modify the flavour of canned fruit, may not be inactivated. Thus, a β-glucosidase system has been isolated from the kernels of canned Victoria plums which will break down amygdalin into glucose, benzaldehyde and hydrocyanic acid (Haisman and Knight, 1967a).

β-Glucosidases are not particularly heat-resistant enzymes, but there are two main reasons for the β-glucosidase stability in canned plums. The first is the favourable physical environment of the plum kernel which lies in an air pocket inside the stone of the plum and is, therefore, to some extent thermally insulated and is also isolated from the acid constituents of the plum flesh. The second is the favourable chemical environment of the kernel. The pH of the latter is between 6·0 and 6·5, where the enzyme has the greatest heat stability. At pH 3·0, the pH of plum flesh, the β-glucosidase can be rapidly inactivated even at relatively low temperatures (in about 6 hours at 30°C). However, only very few plum stones are penetrated by the syrup during processing and so the pH remains high. The kernel is also a relatively dry medium (50–60% water) and contains appreciable amounts of the enzyme substrate, amygdalin, both factors helping to stabilize the enzyme.

Haisman and Knight (1967a) showed that at least 9 min at 100°C (212°F) was required totally to inactivate the enzyme. However, if the plums are processed for only 6 min at 100°C (212°F), most of the amygdalin in the kernels is decomposed during storage and the cyanide content of the syrup rises to about 2 p.p.m. and then declines. At this level it does not present a health hazard.

(e) *Changes in astringency.* We have observed that quinces lose all their astringency on canning. This is probably due to leucoanthocyanins which are related to astringency (Swain, 1962), being converted to anthocyanins. Similar changes may occur during the canning of other fruits.

(f) *Pesticide residues.* Traces of some pesticide and fungicide residues may remain on fruit and may lead to bitter or musty flavours in the canned product (Gilpin *et al.*, 1954; Hope, 1964).

4. *Microbial spoilage*

The type of spoilage organisms which can grow in canned foods depends to a large extent on the pH of the food in question. The influence of pH value must be considered at two stages—during processing and during storage, the former being concerned with the survival of spores of spoilage micro-organisms, and the latter with the control of either the germination and out-growth of any spores which may have survived the heating process, or of the growth of vegetative organisms which may have gained entry into the can after processing, through leaks in the seams. The principal demarcation in the acidity classification of canned foods (Cameron and Esty, 1940) lies at pH 4·5, i.e. between the low and medium acid groups on the one hand and the acid and high acid groups on the other. It is commercial canning practice to pressure process foods in the former groups whilst atmospheric pressure is effective under the conditions of time and temperature adopted for the latter groups. At the pH persisting in canned fruit, i.e. generally within the range 3·0–4·5, bacterial spores are usually destroyed under such atmospheric processing conditions, whilst the growth of any surviving spores is inhibited by the low pH. Subdivision of fruits into acid and high acid groups is necessary because of the incrimination of acid-tolerant, spore-forming bacteria as spoilage agents in canned fruit with pH values between 3·7 and 4·5. In cases where there is a risk of the final pH being greater than 4·5, e.g. in canned tomatoes, pears, mangoes and figs, it is recommended that small amounts of citric or malic acid be added to the syrup prior to processing. Since the time of processing can often be greatly reduced as a result of acidification, the final quality of the product is also much improved.

Instances of sporing and non-sporing aciduric bacteria, yeasts and moulds which may occasionally be encountered in canned fruit as a result of inadequate sterilizing or of can leakage have been reported and are summarized below.

(a) *Spore-forming bacteria.* (i) The growth of the highly pathogenic *Clostridium botulinum*, the most heat-resistant of food-poisoning organisms, is generally regarded as being inhibited below pH 4·5 and therefore does not normally present a health hazard in canned fruits. An exceptional case has, however, been recorded of the presence of this bacillus and its toxin in canned pears having a pH of 3·9. The ability to grow was attributed to the presence of active yeasts and lactobacilli in the can (Adam and Dickinson, 1959).

(ii) *Clostridium pasteurianum.* Spore-forming saccharolytic anaerobes of the *Cl. pasteurianum* type have been found responsible for gaseous spoilage of canned pears, figs and solid-pack tomatoes (Townsend, 1939) and a similar organism was found in canned pineapple (Spiegelberg, 1940). The latter strain was not as heat resistant, however, or as acid tolerant as the former strains. During investigations of the growth of *Cl. pasteurianum* in canned whole tomatoes and tomato purée (Bowen *et al.*, 1954) it has been shown that at pH 4·3–4·4, can-centre temperatures of 93·3–95·6°C (200–204°F) were required to prevent spoilage, but, at pH 4·0, centre temperatures as low as 83·9°C (183°F) gave satisfactory results.

(ii) *Clostridium butyricum.* An organism closely related to *Cl. butyricum* has been isolated from swelled canned tomatoes (Clark and Dehr, 1947) and, in Australia, the C.S.I.R.O. Division of Food Preservation have noted spoilage of canned fruits, such as pears, by organisms of this type.

(iv) *Bacillus coagulans.* This organism is a thermophile with the general characteristics of the flat–sour type, but is tolerant of an acid environment, growth occurring in the region of pH 4·2. It is of considerable importance in canned tomato juice in which it produces off-flavours and souring. The organism may survive heating in tomato juice (pH 4·4–4·5) for 22 min at 100°C (212°F) and may, therefore, survive the thermal processing of canned juice (Anderson *et al.*, 1949).

(v) *Bacillus macerans* (*Bacillus polymyxa* group). Several outbreaks of gaseous spoilage of commercially canned fruits attributable to this group have been reported (Vaughn *et al.*, 1952). These organisms were able to grow at pH 3·8–4·0; however, circumstantial evidence pointed to infection by post-process leakage in the cooling water.

(b) *Non-sporing bacteria.* The most significant group of organisms of this type in the spoilage of acid packs are the Gram-positive, lactic-acid producing bacteria. These organisms, some of which will produce gas, are widely distributed and develop best under conditions of reduced oxygen tension.

(i) *Lactobacillus brevis.* This organism is gas-forming and can cause vigorous fermentation in tomato ketchup and similar products (Pederson, 1929).

(ii) *Leuconostoc pleofructi*. These bacteria have been incriminated as spoilage agents in fruit juices (Savage and Hunwicke, 1923) and tomato products.

(iii) *Leuconostoc mesenteroides*. This organism has caused gaseous spoilage of canned pineapple (Spiegelberg, 1940) and ropiness in canned peaches (Fabian and Henderson, 1950).

(c) *Yeasts*. Spoilage by yeasts in canned fruits is nowadays a rare occurrence as the heat resistance of both vegetative forms and ascospores is low. Outbreaks of understerilization spoilage are generally associated with gross underprocessing, except in the case of solid fruit packs where the very slow rate of heat penetration may give problems. Leaker spoilage is also uncommon, as the large cell size of yeasts, compared with bacteria, appears to prevent their entry into the can.

(d) *Moulds*. Generally speaking moulds are insignificant spoilage agents in canned fruits because of their low heat resistance but there are two notable exceptions, viz. *Byssochlamys fulva* and *B. nivea*.

(i) *Byssochlamys fulva*. This mould was isolated from canned and bottled fruits and described by Olliver and Smith (1933). The ascospores are exceptionally resistant to heat and survived after heating for 30 min at 84–88°C in the syrups from a number of different canned fruits (Olliver and Rendle, 1934). Gillespy and Thorpe (1962) found resistance times of 30 min at 85°C and 10 min at 87·7°C in canned strawberries. The mould causes complete disintegration of fruit due to breakdown of pectic substances and has been the cause of much spoilage in the fruit canning and bottling industry. Infected cans may swell due to CO_2 production (Hersom and Hulland, 1969). *B. fulva* can tolerate a low O_2 tension and will grow in a vacuum of 20 in.

The degree of softening of the fruit caused by this species is unrelated to the amount of visible mycelium; complete disintegration may take place when mycelium is not apparent to the naked eye.

(ii) *Byssochlamys nivea*. This mould has been reported in England (Gillespy and Thorpe, 1962) and Holland (Put, 1964), and its ascospores have the same order of heat resistance as those of *B. fulva*.

(iii) Other types. A heat-resistant species of *Penicillium* has been isolated from canned blueberries (Williams *et al.*, 1941) which has sclerotia which are considerably more heat-resistant than the ascospores. Growth occurred in a vacuum of more than 25 in.

Another heat-resistant mould, a species of *Aspergillus*, found in canned strawberries, has ascospores which will survive a temperature of 100°C (212°F) for more than 60 min when suspended in distilled water.

As these moulds are usually found in the soil, a good degree of control of infection and spoilage can be obtained by thorough washing of the raw material. Agitated processes may also be used to increase the rate of heat penetration into the can without adversely affecting the quality of the pack.

III. VITAMIN CHANGES DURING THE CANNING PROCESSES

A. Preparation of Fruit for Canning

Little research has been carried out on the influence of washing and handling operations on the vitamin content of canned fruits, but the suggestion has been made (Cameron *et al.*, 1955) that for the purposes of discussion of nutrient retentions, fruit may be divided into that which is canned without peeling and that which is peeled prior to canning.

1. *Vitamin changes in fruit canned without peeling*

Fruit which is canned without peeling may have the pits removed either by hand or by mechanical means, but, apart from this operation, the fruit is given a minimum of handling and, in general, retention of nutrients is high. In six samplings of red sour pitted (Montmorency) cherries an average retention of 96% of the ascorbic acid originally present was obtained. Retention of carotene was practically complete (Guerrant *et al.*, 1946). It has also been shown (Lamb *et al.*, 1947) that the ascorbic acid retention of canned apricots of various degrees of maturity varied from 76 to 97% with an average of 87%. Carotene retention in this case varied from 78 to 98%, with an average of 89%. It is noteworthy that the riper fruit was higher in total solids, ascorbic acid and carotene, thus emphasizing the advantages from a nutritional standpoint of allowing the fruit to obtain optimum ripeness before canning.

2. *Vitamin changes in fruit peeled prior to canning*

Lye peeling of clingstone peaches grown in California resulted in the loss of approximately 20% of the ascorbic acid, 15% of the thiamine and 5% of the nicotinic acid, while carotene appeared to be completely retained (Lamb *et al.*, 1947). Further losses on holding the fruit prior to processing raised the figures to 37%, 20% and 10% respectively. Later work (Mitchell *et al.*, 1948) showed that losses of ascorbic acid and carotene from peaches due to lye-peeling varied considerably with the variety and the year of picking. The retention of carotene and ascorbic acid found in one experiment was approximately 85% and 74% respectively. The exact conditions used in lye-peeling (lye-strength, temperature of peeling solution, etc.) may also be of importance here; the minimum time to loosen the skin followed by washing with water and immediate transfer to the cans should be employed. Lye-peeling has also been found to give considerable losses of Vitamin C in plums (Hargrave and Hogg, 1946). Steam peeling causes as much destruction of Vitamin C as does lye-peeling (Lamb *et al.*, 1947). Vitamins may be lost by the mere fact that the rejected portions contain a high proportion of the vitamin content of the fruit. Thus, apple and peach peel and pineapple core contain higher levels of Vitamin C than the flesh which is processed (see Volume 1, Chapter 13).

It is essential for the maintenance of a high level of vitamins, especially those sensitive to oxygen, such as B_1, C, E and K, that as much oxygen as possible is removed from the fruit tissue prior to canning.

When blanching is a necessary step in the preparation of the fruit for canning then steam is preferred to hot water (Powers *et al.*, 1958). Loss of water-soluble vitamins by leaching will then be minimized. Considerable losses in Vitamin C have been shown at this stage during the blanching of bananas and guavas (Dhopeshwarkar and Magar, 1952), probably due to the increased activity of ascorbic acid oxidase during the heating of the fruit tissue.

B. Processing and Storage of Canned Fruits

Many of the vitamins found in fruit, such as the B vitamins, Vitamin C and nicotinic acid, being water soluble, are distributed between the fruit and the syrup during heat processing and storage. This may lead to a loss of these vitamins if the syrup is not consumed with the fruit.

Another important factor affecting vitamin loss in canned fruit is the thermal stability of the vitamins. Vitamin A is fairly stable to heat (90–95% retention can generally be expected in canned fruits) and Vitamin B_1 is heat stable in acid media (90% retention in canned tomatoes). Vitamin B_2 and nicotinic acid are both stable to heat, but while the former is readily oxidized the latter is not affected by oxygen and is probably the least labile of all the vitamins. Vitamin B_6 is resistant to heat, oxygen and acids, whereas Vitamin E is heat stable but is inactivated in the presence of oxidizing agents. The rate of destruction of Vitamin B_1 is increased by the presence of copper and oxygen but the effect is less than in the destruction of Vitamin C (Huelin, 1958). The high retention of ascorbic acid in fruits canned under commercial conditions was demonstrated by Olliver (1936) using chemical methods and substantiated, in the case of blackcurrants, by animal feeding tests.

The effect of increasing headspace is, as might have been expected, to decrease the Vitamin C retention, but the temperature of the syrup at the time of closing the can is unimportant in this respect (Adam, 1941). Whilst oxygen is present in the can, the aerobic destruction of Vitamin C in the presence of metal catalysts, such as copper, will be the main reaction. Protection can be afforded if the oxygen is removed by combination with tin. Thus, the plain tin can preserves the Vitamin C to a greater extent than does a glass container or enamel lined can if appreciable amounts of oxygen are sealed in the can (Feaster *et al.*, 1949). Under conditions of low oxygen tension, the type of container has only a slight influence on ascorbic acid retention (Reister *et al.*, 1948).

When all the oxygen has been used up, anaerobic decomposition of Vitamin C may begin. This latter reaction is much slower than the aerobic destruction and is accelerated by fructose and its derivatives but not by copper. It appears that fructose is most effective in the furanose form (since fructose diphosphate has a much greater effect) and that the effect of sucrose is due to fructose liberated by hydrolysis (Huelin, 1958). Although copper does not appear to catalyse the anaerobic degradation of ascorbic acid, other metals do. Of these, lead and zinc, acting in their bivalent states, appear to be the most powerful catalysts (Finholt *et al.*, 1966). The total retention of Vitamin C in fruits during canning, taking into account the losses incurred at all stages of the canning process, ranges from 65% to 95%.

Losses of Vitamin C in canned fruit during storage are generally slight (0–10% in the first year) but this depends a great deal on storage time and temperature. Thus, it has been shown that ascorbic acid retention in canned apricots is exponentially related to storage time at 38 and 43°C (Brenner *et al.*, 1948). The results shown in Table I indicate that the longer the storage time and the higher the storage temperature the greater is the breakdown of Vitamin C in the canned fruit (Cameron *et al.*, 1955). The results suggest that storage temperatures of 10°C or below must be used for maximum retention of ascorbic acid (see Volume 1, Chapter 13).

TABLE I. Retention of ascorbic acid in canned fruits as a function of storage time and temperature

Product	Storage time (months)	Ascorbic acid (% retention) 10°C	18°C	26·6°C
Apricots	12	96	93	85
	24	94	90	56
Peaches	12	98	85	72
	24	98	80	53
Pineapple, sliced	12	100	95	74
	24	83	78	53
Tomatoes	12	95	94	82
	24	89	87	70

It has been demonstrated that carotene, nicotinic acid and riboflavin values may change to a small extent during storage of canned fruits but Cameron *et al.*, (1955) showed that retention of these vitamins is not markedly affected by storage temperature.

Other factors besides oxygen, fructose and storage temperature may well have a bearing on the breakdown of Vitamin C. As indicated in the section on colour changes in the processed fruit, ascorbic acid is one of the main factors involved in anthocyanin degradation. It follows, therefore, that Vitamin C

TABLE II. The values of the various vitamins which may be expected in some canned fruits (McCance and Widdowson, 1967)

Canned Fruit	Carotene (mg/100 g)	Vitamin D (I.U/100 g)	Thiamine (mg/100 g)	Riboflavine (mg/100 g)	Nicotinic acid (mg/100 g)	Vitamin C (mg/100 g)	Pantothenic acid (mg/100 g)	Vitamin B_6 (mg/100 g)	Biotin (μg/100 g)	Folic acid (μg/100g)	Vitamin B_{12} (μg/100g)
Apricots	1·0	0	0·02	0·01	0·3	5	0·10	0·05	—	Tr.	0
Blackberries	0·10	0	0·01	0·02	0·2	15	—	—	—	—	—
Cherries, red sour	0·50	0	0·02	0·02	0·2	4	—	—	—	—	—
Fruit salad	0·30[a]	0[a]	0·02[a]	0·01[a]	0·3[a]	3[a]	—	—	—	—	—
Loganberries	—	0	0·03	0·02	—	25	—	—	—	—	—
Olives	0·15	0	Tr.	Tr.	—	0	0·02	0·02	—	1	0
Peaches	0·25	0	0·01	0·02	0·6	4	0·05	0·02	0·2	Tr.	0
Pears	0·01	0	0·01	0·01	0·2	1	0·02	Tr.	—	Tr.	0
Pineapple	0·04	0	0·05	0·02	0·2	8	0·10	0·20	—	Tr.	0
Prunes	0·60	0	0·02	—	0·5	0	0·10	—	—	Tr.	0
Raspberries	—	0	0·01	—	—	10	—	—	—	—	—
Tomatoes	0·5	0	0·05	0·03	0·6	16	0·2	0·07	1·8	3	—

[a] Calculated assuming that canned fruit salad contains canned fruit in the following proportion: 35% apricots or peaches; 35% pears; 10% cherries; 10% grapes and 10% pineapple.

— signifies that no estimation has been made.

Tr. indicates that traces of the vitamin in question are known to be present.

will be less stable in canned fruits containing a high level of anthocyanin. Thus it has been shown that 40–60% of the original ascorbic acid is destroyed on processing strawberries (rich in anthocyanins), and after 15 weeks' storage at 37°C only about 40% of the quantity present at the beginning of storage remained (Kyzlink and Curdova, 1966). Dehydro-ascorbic acid is as active physiologically as Vitamin C but is much less stable being rapidly converted into 2,3-diketo-1-gulonic acid which no longer possesses anti-scorbutic value.

Table II shows the values of the various vitamins which may be expected in some canned fruits (McCance and Widdowson, 1967). It will be obvious from what has gone before that these results are only typical values and may be misleading in any specific instance due to the variation in losses during processing and storage and in the natural vitamin content of the fruit.

IV. INTERACTION BETWEEN FRUITS AND THEIR CONTAINERS

The effect of the various constituents of the fruit on the tinplate container can produce changes in the composition of the fruit both directly and indirectly. The acids are the main corrosive agents in fruit, and direct changes involve the reaction between these acids and the metals employed in the construction of the can, whilst the indirect changes may be brought about by reaction of the corrosion products with natural constituents of the fruit. The two major metals, tin and iron, present in the body of the can, act as electrodes in the corrosion reactions which may take place within the can after processing.

A. Factors which Influence the Corrosivity of the Pack

Although corrosion in canned fruit is caused primarily by the fruit acids, there are other substances present which, by acting as depolarizers, can accelerate the corrosion reactions. These various factors will now be considered in more detail.

1. *Acid constituents*

Fruits fall into a wide range of acidity and the pH can range from 2·7 (plum and loganberry) through 4·3 (cherry and tomato) to 5·5 (persimmon). The extent of corrosion cannot be related to pH alone since it has been demonstrated (Hirst and Adam, 1937) that cherries at pH 4 produce more hydrogen swells (cans in which the ends are distorted due to internal pressure of hydrogen) than do gooseberries at pH 3. The nature of the acid present in the fruit is probably more significant than the pH, and the peculiarly severe corrosion exhibited by rhubarb has been attributed to the high level of oxalic acid in the petioles. The influence of other non-acidic constituents of the fruit can accelerate the corrosion process, and it has been suggested that the corrosivity of rhubarb is caused by a combination of oxalate and nitrate, the latter acting as a depolarizer (Culpepper and Moon, 1933).

2. *Corrosion accelerators–depolarizers*

Probably one of the most important depolarizers present initially in canned fruit is oxygen. The amount of the gas included in the can is closely related to the efficiency of the exhausting process in removing air from the headspace. However, it has been shown that the gas present in the tissues of fruit is mainly nitrogen, and only a small amount of oxygen is trapped mechanically by the fruit (Horner, 1933). The time necessary for all the oxygen to be reduced by the corrosion process depends on the type of can employed, varying from 5 days in a plain can to 30 days in a lacquered can (Adam and Horner, 1937). Addition of ascorbic acid to canned apple halves was found to reduce headspace oxygen and the pack showed no corrosion after 6 months' storage at room temperature (Hope, 1961). The amount of tin dissolved during this stage of corrosion is important since it has a direct effect on the amount of iron exposed and this can lead to concentration of the attack on the iron resulting in pinholing of the can (Tutton and Coonen, 1950).

It has often been observed that the amount of hydrogen produced by the corrosion of a can is much less than that expected from the equivalent of metal dissolved (Dickinson, 1961), and many theories have been put forward to account for the absence of the gas in the headspace. It is thought that, possibly, the hydrogen produced by corrosion of the tin may be utilized in the reduction of the stannous oxide film which forms on the can surface (Cheftel *et al.*, 1955). The presence of cathodic depolarizers in canned prunes has been demonstrated by polarographic techniques (Frankenthal *et al.*, 1959). In this product no hydrogen is evolved until corrosion of the iron commences. This is in contrast with the situation found in tomato juice where hydrogen is evolved as the tin corrodes.

Among the many factors which have been shown to accelerate corrosion are caramelization in cherries (Cheftel *et al.*, 1955) and severe dehydration in dried prunes (Connell *et al.*, 1957). Although the level of hydroxymethylfurfural increased parallel with the corrosivity in the case of prunes, it was claimed that this was not the corrosive agent. Consistency has been shown to have a marked effect on corrosion in canned apple sauce, and the reduction in tin content of high consistency packs has been attributed to polarization of the tin and iron electrodes due to the reaction products being unable to diffuse away easily (Lopez, 1965).

Rapid detinning in tomato paste has often caused concern in the canning industry and various factors have been studied. A decrease in corrosivity in South African tomatoes has been observed to correspond with increased ripeness, and this was accompanied by a fall in nitrate level (Anon, 1966), which appears to be an important factor in the corrosion of the cans. It has been observed that dehydroascorbic acid and low methoxyl pectin have a corrosion accelerating effect when added to tomato paste and it is possible that these compounds are formed during production of tomato paste (Hernandez, 1961).

Of the natural constituents of the fruit, the anthocyanins and related pigments are amongst the most important potential cathodic depolarizers, since they are easily reduced. It has been shown that under anaerobic conditions, such as would prevail in a can once the headspace oxygen was used up, tin was dissolved more rapidly by a citric acid solution in the presence of cyanidin-3-glucoside than in a pure acid solution. It has been suggested that the accelerated corrosion exhibited by raspberries and plums may be due in part to the cyanidin-3-glucoside present in these fruits. Other related polyphenols may act in a similar way, e.g. the quercetin present in canned plums could act as a corrosion accelerator (Salt and Thomas, 1957). Studies on the rate of can perforation in canned apple have shown that this is increased by the addition of quercetin to the product (Kohman, 1925).

3. *Pigment–metal ion complexes*†

In addition to acting as cathodic depolarizers in connection with their ready reduction, the anthocyanin pigments can also act as anodic depolarizers through their ability to form complexes with cations, particularly those of iron and tin salts. Corrosion of tin is increased by combination of the metal in insoluble complexes, but soluble tin salts act as powerful inhibitors of the corrosion of iron (Morris and Bryan, 1931).

The effect of soluble metal salts on the natural colour of strawberries, raspberries and blackcurrants has been studied and it was observed that, while iron salts produced discoloration in strawberries, tin salts had little effect; amounts up to 100 p.p.m. did not produce objectionable colours (Morris and Bryan, 1936). Tin salts have a greater effect on the pigments of raspberries, and stannous chloride has been shown to give raspberry juice a blue tint (Law, 1933). The changes in the pigments of various fruits under actual processing and storage conditions results in a distinct purpling of the pigments of many of the fruits due to combination with metal salts brought into solution by the corrosion process. The accelerated corrosion in black cherries has been attributed to the high anthocyanin content of this fruit, the pigments acting as depolarizers (Culpepper and Caldwell, 1927).

Analysis of samples of canned fruit after a period of storage usually shows a greater amount of tin in the drained fruit than in the syrup (Adam and Horner, 1937), and this would indicate that at least part of the tin is combined

† See also Chapter 17.

in an insoluble form with some constituent within the fruit. It has been shown that the complexing of metal salts with the anthocyanin pigment of fruit is less at a lower pH (Culpepper and Caldwell, 1927) and the higher corrosion rate in blueberries compared to strawberries is attributed to this difference in pH. The reason that tin salts have a much less marked effect on the pigment of strawberries than on that of raspberries is not, however, an effect of pH but is dependent on the structure of their anthocyanin pigment. Raspberries contain cyanidin glycosides which have *ortho*-dihydroxy groups in their structures and it is these groups which are involved in the formation of complexes with metals such as tin (see Fig. 5). The major pigment of the strawberry is

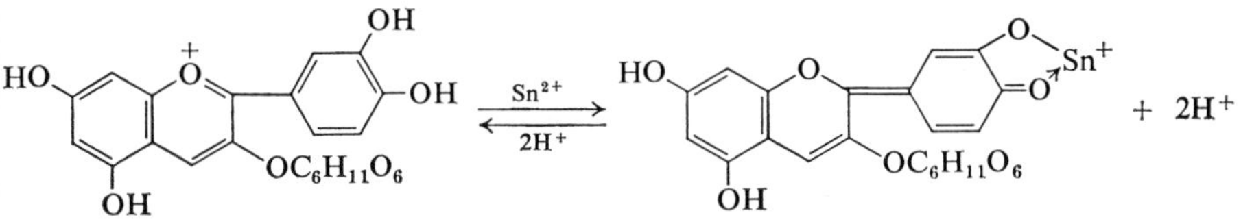

FIG. 5.

pelargonidin-3-glucoside (Geissman *et al.*, 1953) which does not possess the necessary *ortho*-dihydroxy groups for complex formation and, therefore, does not show the same shift to a blue colour on addition of tin as do glycosides of cyanidin (Fig. 5). The small effect of tin salts on strawberry pigments observed occasionally could be due to traces of other pigments present in the fruit which include glycosides of cyanidin (Blundstone and Crean, 1966), and the effect of iron salts has been attributed to the tannins present in the strawberry fruit (Morris and Bryan, 1936).

Combination of metal ions with tannins has been observed in other fruits, and discoloration in canned cranberry has been attributed to the formation of a complex between tannins present in the fruit and tin salts (Morse, 1927). It has also been observed that darkening in canned Maraschino cherries is more severe when fruit of a high tannin content is used. This could arise from underripe fruit, which tends to be higher in tannin, or be due to storage of fruit in wooden barrels before canning (Butland, 1952).

It is not necessarily the case that any reaction between corrosion products and the pigments of the fruit will produce undesirable colours, though this is usually so with the anthocyanin pigments. It has been claimed that the formation of yellow "lakes" between anthraquinone pigments and metal ions in rhubarb can improve the colour of those varieties which contain little or no anthocyanin (Gallop, 1965).

Any natural compound present in the fruit which can form a complex with metal ions could accelerate corrosion of that metal by removing these ions from solution.

4. *Enzyme systems and corrosion*

The sterilization process used for canned fruits is calculated on the basis of the minimum conditions required to kill any yeasts or mould spores present without having an adverse effect on the quality of the processed product, and is adequate to inactivate any enzyme systems present in the flesh of the fruit. However, it has been demonstrated (Dickinson, 1957) that, after such a process, enzyme activity is still present inside the kernel of stone fruits such as plums and cherries. As stated earlier, a β-glucosidase system has been isolated from plum kernels which will break down amygdalin, also present in the kernels, into benzaldehyde, glucose and hydrocyanic acid (Haisman and Knight, 1967b). It has also been shown that when canned plums are given a process insufficient to inactivate these enzyme systems they produce more hydrogen swells than cans given a higher process (Dickinson, 1961). Removal of the stones from plums before processing results in a reduction in the number of hydrogen swells, and suggests that the active enzyme systems within the kernel accelerates the corrosion reaction, the hydrocyanic acid acting as an anodic depolarizer.

B. Prevention of Undesirable Container-product Interactions

The problem of undesirable changes taking place as a result of action between constituents of the fruit and the container can be minimized in a number of ways. Amongst the more important of these are the selection of a suitable type of can from the wide choice available, and storage of the processed product at a suitable temperature.

Most of the reactions taking place within the can of fruit are normal chemical reactions to which the condition will apply that the rate will double for every 10°C (18°F) rise in temperature. It has been shown that the rate of formation of hydrogen swells increases many times as the storage temperature is raised from 0°C (32°F) to 36°C (97°F) (Kohman and Sanborn, 1929). Efficient cooling of the cans after processing is an important factor and, if the cans are cased and stacked whilst still hot, they will retain a high temperature in the centre of the stack for a considerable period (Kohman, 1931).

The use of a suitable type of container to protect the product as far as possible from undesirable effects will influence the choice of both can and lacquer system. It is a well-established fact that a corrosion problem in a plain can is not necessarily solved by the use of a lacquered can. An example of this type of behaviour is shown by canned cherries, which corrode plain cans fairly rapidly but lacquered cans even more quickly (Morris and Bryan, 1931). This could be associated with the reduction of the exposed tin area and consequent loss of anodic protection for the iron.

Special lacquers have been formulated for certain highly corrosive products, and the means of application of the lacquer can determine the degree of protection it gives. The tinplate may be lacquered before the cans are formed, but the subsequent can-making operations may damage the lacquer; alternatively, the lacquer can be applied by spraying the formed can. The latter is a more expensive method, but it has been shown to give low tin contents in canned rhubarb after 2 years' storage (Dickinson and Raven, 1962).

Two developments have occurred recently which affect the can itself. One, which is the use of pure tin to solder the side seam in a way that provides an excess of the metal along the seam of an otherwise lacquered can, is known as a "High Tin Fillet" can. This fillet acts as a sacrificial anode, protecting the can in those products in which rapid detinning is a problem (Hotchner and Kamm, 1967). The other development involves a move away from tin as the protective layer over the steel baseplate. The most satisfactory alternative layer so far produced is a chromium–chromic oxide layer, which has a very good lacquer affinity.

V. SELECTION OF VARIETIES OF FRUIT FOR CANNING

Generally speaking, most fruits have been bred for their field characteristics and for their suitability for sale on the fresh market rather than for their qualities when canned. This means that varieties of fruit are usually selected for canning from an organoleptic point of view, and little heed is paid to the vitamin content of the fruit or the interaction of the fruit with the can. This may be illustrated by quoting the case of the manufacture of canned tomato paste in South Africa. Here, the round varieties of tomatoes, such as Marglobe, have been replaced by pear-shaped varieties, such as Roma, in an attempt to increase the colour and consistency of the paste. As a result, an outbreak of hydrogen swells has been observed in South African tomato paste in recent years, attributable to the higher water-insoluble solids and nitrate levels in the pear-shaped varieties of tomatoes (Van der Merwe and Knock, 1968).

A. Selection for Organoleptic Qualities

The object is to produce a canned fruit which has organoleptic attributes similar to those of the fresh fruit and varieties must be selected with a view to the attainment of this object.

1. *Colour*

Varieties of fruit may be selected for their reduced tendency to discolour, either during preparation for canning or during canning and storing. An instance of the former type is found in the selection of the Sunbeam peach for its inability to brown enzymically during the peeling and slicing operations. This fruit is unique amongst peaches in not containing the polyphenolic substrate required for browning, although phenolase is present (Kertesz, 1933). Instances of the latter type are to be found in the selection of varieties of fruit for their minimal tendency to give pink and brown discolorations during canning and storing. Thus, the Bartlett pear rarely shows pink discoloration compared with other varieties (see p. 516). This is presumably due to the low leucoanthocyanin content of this variety compared with other varieties.

The Cambridge Favourite variety of strawberry is commonly canned in England because it is low in anthocyanin and shows little tendency to give non-enzymic brown discoloration. It can thus be dyed red with artificial colour and will retain this colour without the dulling caused by formation of brown pigments. Other varieties of strawberries, which are higher in anthocyanin, generally show an increased tendency to non-enzymic browning and are thus not suitable for canning either with or without added artificial colour.

In some cases it is impossible to select a variety which does not show discoloration on canning. For instance, all varieties of sweet black cherries tend to lose their colour on processing and develop a bluish hue which does not, however, appear to be due to the interaction of the anthocyanins present with tin salts (Adam, 1962).

2. *Texture*

Varieties of fruit may also be selected for their ability to withstand heat without losing firmness, or breaking down. Thus, the Cuthbert raspberry shows little tendency to "mush" on canning, whereas the Washington variety tends to "mush" badly (see p. 511). This difference does not, however, appear to be centred in the pectic fraction (Leinbach *et al.*, 1951). The firmer canned product given by clingstone peaches as compared with freestone varieties and by "pear" tomatoes as compared with "round" tomatoes has been attributed to differences in pectic substances (McColloch, *et al.*, 1950; Postlmayr *et al.*, 1956).

A serious defect, of unknown source, in some plum varieties, is the secretion of gum in the tissues surrounding the stone. This gum swells during canning and storing (eventually breaking through the plum flesh) and apparently absorbs colour from the fruit and syrup. The most susceptible variety, the Victoria, is also the most popular variety with consumers.

3. *Flavour*

Very little information is available on the selection of varieties of fruits from a point of view of flavour, and it is unfortunately true that flavour is apt to be regarded as a characteristic of secondary importance when breeding new varieties. Although the flavours of certain fruits, notably strawberries, alter on canning, it is usually the case that varieties which have a good flavour when eaten fresh are also good when canned. This is not the case, however, with gooseberries and apples, where the cooking varieties, with their higher acid content, have a much better flavour when canned than the milder dessert varieties.

It is not always possible to select a variety purely on the basis of a single organoleptic attribute since this may conflict with other properties of the fruit. Thus, although the clingstone varieties of peaches have a better texture than freestone peaches, for reasons described in an earlier section, they have also an inferior flavour compared with the freestone varieties. As mentioned earlier, "pear" tomatoes give a paste of improved colour and consistency but are more corrosive than "round" tomatoes, and may also impart a bitter flavour to the canned product.

There is obviously great scope for improvement in the area of variety selection, and we suggest that more attention be paid to the breeding of varieties with chemical characteristics which would make them particularly suitable for canning.

REFERENCES

Adam, W. B. (1941). *Ann. Rep. Fruit Veg. Cann. Quick Freez. Res. Ass., Campden*, p. 14.

Adam, W. B. (1949). *Ann. Rep. Fruit Veg. Cann. Quick Freez. Res. Ass., Campden*, p. 18.

Adam, W. B. (1962). *Recent Advan. Fd. Sci.* **2,** 83.

Adam, W. B. and Dickinson, D. (1959). *Fruit Veg. Cann. Quick Freez. Res. Ass., Campden, Sci. Bull.* No. 3.

Adam, W. B. and Horner, G. (1937). *J. Soc. chem. Ind. Lond.* **56,** 329T.

Adams, J. B. (1968). The Fruit and Vegetable Preservation Research Association, Campden. Unpublished results.

Al-Delaimy, K. A., Borgstrom, G. and Bedford, C. L. (1966). *Q. Bull. Mich. St. Univ.* **49** (2), 164.

Amerine, M. A., Pangborn, R. M. and Roessler, E. B. (1965). "Principles of the Sensory Evaluation of Food." Academic Press, New York.

Anderson, E, E., Esselen, W. B. and Fellers, C. R. (1949). *Fd Res.* **14**, 499.

Anon (1966). *Fd Preserv. Q.* **26,** 45.

Anthistle, M. J. and Dickinson, D. (1959). *Fruit Veg. Cann. Quick Freez. Res. Ass., Campden., Res. Leaflet* No. 4.

Blundstone, H. A. W. and Crean, D. E. C. (1966). *Fruit Veg. Cann. Quick Freez. Res. Ass., Campden, Tech. Bull.* No. 12.

Board, P. W. and Seale, P. E. (1954). *Fd Preserv. Q.* **14** (1), 2.

Borenstein, B. and Bunnell, R. H. (1966). *Adv. Fd Res.* **15**, 195.

Bowen, J. F., Strachan, C. C. and Moyls, A. W. (1954). *Fd Technol.* **8,** 471.

Brenner, S., Wodicka, V. O. and Dunlop, S. G. (1948). *Fd Technol.* **2,** 207.

Buch, M. L., Dryden, E. C., Hills, C. H. and Oyler, J. R. (1956). *Fd Technol.* **10,** 560.

Buch, M. L., Satori, K. G. and Hills, C. H. (1961). *Fd Technol.* **15,** 526.

Butland, P. (1952). *Fd Technol.* **6,** 208.

Cameron, E. J. and Esty, J. R. (1940). *Fd Res.* **5,** 549.
Cameron, E. J., Clifcorn, L. E., Esty, J. R., Feaster, J. F., Lamb, F. C., Monroe, K. H. and Royce, R. (1955). "Retention of Nutrients during Canning." Research Laboratories, National Canners Association, Washington.
Cheftel, H., Monovoisin, J. and Swirski, M. (1955). *J. Sci. Fd Agric.* **6,** 652.
Clark, E. J. and Dehr, A. (1947). *Fd Res.* **12,** 122.
Clegg, M. (1967, October). *Fd Technol., New Zealand,* p. 12.
Cohée, R. F. and Nelson, J. (1951). *Fd Ind.* **23** (3), 91.
Connell, J. C., McKirahan, R. D. and Willey, A. R. (1957). *Fd Technol.* **11**, 232.
Crang, A. and Sturdy, M. (1953). *J. Sci. Fd Agric.* **4,** 449.,
Cruess, W. V. (1952). *Fd Technol.* **6,** 110.
Cruess, W. V. (1958). "Commercial Fruit and Vegetable Products." McGraw-Hill Book Co. Inc., New York.
Culpepper, C. W. and Caldwell, J. S. (1927). *J. agric. Res.* **35,** 107.
Culpepper, C. W. and Moon, H. H. (1933). *J. agric. Res.* **46**, 387.
Dalal, K. B. and Salunkhe, D. K. (1964). *Fd Technol.* **18,** 1198.
Daravingas, G. and Cain, R. F. (1965). *J. Fd Sci.* **30**, 400.
Davis, E. G. and Kefford, J. F. (1955). *Fd Preserv. Q.* **15** (1), 15.
Dekazos, E. D. (1966a). *J. Fd Sci.* **31,** 956.
Dekazos, E. D. (1966b). *J. Fd Sci.* **31**, 226.
Dhopeshwarkar, G. A. and Magar, N. G. (1952). *J. sci. ind. Res., India* **11A,** 264.
Dickinson, D. (1957). *J. Sci. Fd Agric.* **8,** 721.
Dickinson, D. (1961). *Corros. Technol.* **8,** 20.
Dickinson, D. and Raven, T. W. (1962). *Fd Mf.* **37,** 480.
Doesburg, J. J. (1965). "Pectic Substances in Fresh and Preserved Fruit and Vegetables." I.B.V.T., Wageningen. Communication No. 25.
Fabian, F. W. and Henderson, R. H. (1950). *Fd Res.* **15,** 415.
Fagerson, I. S. (1954). *J. agric. Fd Chem.* **2,** 474.
Feaster, J. F., Tompkins, M. D. and Pearce, W. E. (1949). *Fd Res.* **14,** 25.
Finholt, P., Kristiansen, H., Krówczyński, L. and Higuchi, T. (1966). *J. Pharm. Sci.* **55,** 1435.
Frankenthal, R. P., Carter, P. R. and Laubscher, A. N. (1959). *J. agric. Fd Chem.* **7**, 441.
Furia, T. E. (1968). "The Handbook of Food Additives", p. 289. The Chemical Rubber Co., Ohio.
Gallop, R. A. (1965). *Fruit Veg. Cann. Quick Freez. Res. Ass., Campden, Sci. Bull.* No. 5.
Geisman, T. A., Jorgenson, E. C. and Harborne, J. B. (1953). *Chemy Indy* 1389.
Gillespy, T. G. and Thorpe, R. H. (1962). *Fruit Veg. Cann. Quick Freez. Res. Ass., Campden, Tech. Memo.* No. 44.
Gilpin, G. L., Dawson, E. H. and Siegler, E. H. (1954). *J. agric. Fd Chem.* **2,** 781.
Guerrant, N. B., Vavich, M. G., Fardig, O. B., Dutcher, R. A. and Stern, R. M. (1946). *J. Nutr.* **32,** 435.
Guyer, R. B. and Erickson, F. B. (1954). *Fd Technol.* **8**, 165.
Haisman, D. R. and Knight, D. J. (1967a). *J. Fd Technol.* **2,** 241.
Haisman, D. R. and Knight, D. J. (1967b). *Biochem. J.* **103,** 528.
Hanson, S. W. (1954). "Colour in Foods—a Symposium", p. 136. Nat. Acad. Sci., Nat. Res. Council, Washington.
Hargrave, P. D. and Hogg, N. J. (1946). *Sci. Agric.* **26,** 95.

Hernandez, H. H. (1961). *Fd Technol.* **15,** 543.
Hersom, A. C. and Hulland, E. D. (1969). "Canned Foods", 6th edn, p. 102. J. & A. Churchill Ltd., London.
Hewitt, E. J. (1963). *J. agric. Fd Chem.* **11,** 14.
Hirst, F. and Adam, W. B. (1937). *Fruit Veg. Pres. Res. Sta., Campden, Monograph* No. 1.
Hope, G. W. (1961). *Fd Technol.* **15,** 548.
Hope, G. W. (1964). *J. agric. Fd Chem.* **12,** 189.
Horner, G. (1933). *Ann. Rep, Fruit Veg. Cann. Quick Freez. Res. Sta., Campden,* p. 50.
Hotchner, S. J. and Kamm, G. G. (1967). *Fd Technol.* **21,** 901.
Hsu, C. P., Deshpande, S. N. and Desrosier, N. W. (1965). *J. Fd Sci.* **30,** 583.
Huelin, F. E. (1958). *Indian Fd Packer* **12** (12), 11.
Hultin, H. O. and Proctor, B. E. (1961). *Fd Technol.* **15,** 440.
Jennings, W. G., Leonard, S. and Pangborn, R. M. (1960). *Fd Technol.* **14,** 587.
Joslyn, M. A. and Goldstein, J. L. (1964). *J. agric. Fd Chem.* **12,** 511.
Joslyn, M. A. and Ponting, J. D. (1951). *Adv. Fd Res.* **3,** 1.
Joux, J. L. (1957). *Compt. Rend., Acad. Agric. Fr.* **43,** 506.
Kanujoso, B. W. T. and Luh, B. S. (1967). *Fd Technol.* **21,** 457.
Kertesz, Z. I. (1933). *N.Y. agric. exp. Sta. Tech. Bull.* No. 216.
Kohman, E. F. (1925). *Canning Age,* **6,** 191.
Kohman, E. F. (1931), *Fd Ind.* **3,** 77.
Kohman, E. F. and Sanborn, N. H. (1929). *National Canners Assoc. Bull.* No. 23L.
Kyzlink, V. and Curdova, M. (1966). *Sb. vys. Sk. chem.-technol. Potravinarska technol.* **9,** 41.
Kyzlink, V., Curdova, M. and Curda, D. (1962). *Sb. vys. chem.-technol. Šk. Praze, Potravinarska* **6,** 135.
Lamb, F. C., Pressley, A. and Zuch, T. (1947). *Fd Res.* **12,** 273.
Law, M. (1933), *Fd* **2,** 277.
Leinbach, L. R., Seegmiller, C. G. and Wilbur, J. S. (1951). *Fd Technol.* **5,** 51.
Li, K. C. (1966). *Proc. Con. Peach Processing Utilization, Georgia exp. Sta. Experiment (Griffin), Ga.* p. 31.
Loconti, J. D. and Kertesz, Z. I. (1941). *Fd Res.* **6,** 499.
Lopez, A. (1965). *Fd Technol.* **19,** 653.
Luh, B. S. (1960). *Fd Technol.* **14,** 173.
Luh, B. S., Leonard, S. J. and Patel, D. S. (1960). *Fd Technol.* **14,** 53.
Luh, B. S., Hsu, E. T. and Stachowicz, K. (1967). *J. Fd Sci.* **32,** 251.
McCance, R. A. and Widdowson, E. M. (1967). "The Composition of Foods", Medical Research Council, Special Report Series No. 297, 2nd imp., London.
McColloch, R. J., Nielsen, B. W. and Beavens, E. A. (1950). *Fd Technol.* **4,** 339.
McColloch, R. J., Keller, G. J. and Beavens, E. A. (1952). *Fd Technol.* **6,** 197.
MacKinney, G., Lukton, A. and Chichester, C. O. (1955). *Fd Technol.* **9,** 324.
Mahadeviah, M. (1966). *Indian Fd Packer* **20,** 5.
Mahdi, A. A., Rice, A. C. and Weckel, K. G. (1959). *J. agric. Fd Chem.* **7,** 712.
Mahdi, A. A., Rice, A. C. and Weckel, K. G. (1961). *J. agric. Fd Chem.* **9,** 143.
Makower, R. U. and Schwimmer, S. (1957). *J. agric. Fd Chem.* **5,** 768.
Markakis, P., Livingston, G. E. and Fellers, C. R. (1957). *Fd Res.* **22,** 117.
Mattick, L. R., Moyer, J. C. and Shallenberger, R. S. (1958). *Fd Technol.* **12,** 613.
Matz, S. A. (1962). "Food Texture", p. 177. The AVI Publishing Co. Inc. Connecticut.

Miers, J. C. (1966). *J. agric. Fd Chem.* **14,** 419.
Mitchell, J. H., Van Blaricom, L. O. and Roderick, D. B. (1948). *South Carolina agric. exp. Sta. Bull.* No. 372.
Mohammadzadeh-Khayat, A. A. and Luh, B. S. (1968). *J. Fd Sci.* **33,** 493.
Morris, T. N. and Bryan, J. M. (1931). *Fd Invest. Bd, Lond. Special Rep.* No. 40.
Morris, T. N. and Bryan, J. M. (1936). *Fd Invest. Bd, Lond. Special Rep.* No. 44.
Morse, F. W. (1927). *J. agric. Res.* **34,** 889.
Moyls, A. W. (1951). *Sci. agric.* **31,** 546.
Nelson, P. E. and Hoff, J. E. (1969). *J. Fd Sci.* **34,** 53.
Nortjé, B. K. (1966). *S. Afr. J. agric. Sci.* **9,** 681.
Olliver, M. (1936). *J. Soc. chem. Ind., Lond.* **55,** No. 24, 153T.
Olliver, M. and Rendle, T. (1934). *J. Soc. Chem. Ind, Lond.* **53,** No. 22, 166T.
Olliver, M. and Smith, G. (1933). *J. Bot., Lond.* **71,** 196.
Patron, A. (1956). *Fruits* **11,** 153.
Pederson, C. S. (1929). *N.Y. agric. exp. Sta. Tech. Bull.* No. 150.
Peterson, G. T. (1938). *Canner* **86** (No. 12, part 2), 72.
Ponting, J. D. (1960). *In* "Food Enzymes" (H. W. Schultz, ed.), p. 105. The AVI Publishing Co. Inc., Connecticut.
Ponting, J. D., Bean, R. S., Notter, G. K. and Makower, B. (1954). *Fd Technol.* **8,** 573.
Postlmayr, H. L., Luh, B. S. and Leonard, S. J. (1956). *Fd Technol.* **10,** 618.
Powers, M. J., Talburt, W. F., Jackson, R. and Lazar, M. E. (1958). *Fd Technol.*, **12,** 417.
Powers, J. J., Pratt, D. E., Downing, D. L. and Powers, I. T. (1961). *Fd Technol.* **15,** 67.
Put, H. M. C. (1964). *J. appl. Bact.* **27,** 59.
Ranganna, S. and Siddappa, G. S. (1961). *Fd Technol.* **15,** 204.
Ranganna, S., Setty, L. and Nagaraja, K. V. (1966). *Indian Fd Packer* **20,** 5.
Raven, T. W. (1962). *Fruit Veg. Cann. Quick Freez. Res. Ass., Campden, Res. Leaflet* No. 7.
Ross, E. (1955). *Fd Technol.* **9,** 18.
Reister, D. W., Wiles, G. D. and Coates, J. L. (1948). *Fd Ind.* **20,** 372, 494, 496.
Salt, F. W. and Thomas, J. G. N. (1957). *J. appl. Chem.* **7,** 231.
Savage, W. G. and Hunwicke, R. F. (1923). "Canned Fruit", *Fd Invest. Bd, Lond. Special Report* No. 16.
Shewfelt, A. L. (1965). *J. Fd Sci.* **30,** 573.
Siddappa, G. S. and Bhatia, B. S. (1956). *J. sci. ind. Res. India,* 15C, 118.
Singleton, V. L., Willis, A. G. and Young, H. Y. (1961). *J. Fd Sci.* **26,** 49.
Spencer, M. S. and Stanley, W. L. (1954). *J. agric. Fd Chem.* **2,** 1113.
Spiegelberg, C. H. (1940), *Fd Res.* **5,** 115.
Sterling, C. (1959). *Fd Technol.* **13,** 629.
Swain, T. (1962). *In* "The Chemistry of Flavonoids Compounds", (T. A. Geissman, ed.), p. 513. Macmillan, New York.
Tate, J. N., Luh, B. S. and York, G. K. (1964). *J. Fd Sci.* **29,** 829.
Thompson, P. (1950). *Fd Preserv. Q.* **10,** 17.
Tinsley, I. J. and Bockian, A. H. (1960). *Fd Res.* **25,** 161.
Townsend, C. T. (1939). *Fd Res.* **4,** 231.
Tutton, W. R. and Coonen, N. H. (1950). *Fd Can.* **10,** 42.
Van Buren, J. P. (1967). *J. Fd Sci.* **32,** 435.
Van Der Merwe, H. B., Wilmot, S. W. and Lombard, J. H. (1966). *Fd Ind. S. Afr.* **18,** 66.

Van Der Merwe, H. B. and Knock, G. G. (1968). *J. Fd. Technol.* **3,** 249.
Vaughn, R. H., Kreulevitch, I. H. and Mercer, W. A. (1952). *Fd Res.* **17,** 560.
Wade, P. (1964). *J. Sci. Fd Agric.* **15,** 51.
Wagenknecht, A. C., Scheiner, D. M. and Van Buren, J. P. (1960). *Fd Technol.* **14,** 47.
Wagner, J. R. and Miers, J. C. (1967). *Fd Technol.* **21,** 920.
Walker, J. R. L. and Reddish, C. E. S. (1964). *J. Sci. Fd Agric.* **15,** 902.
Whittenberger, R. T., Harrington, W. O. and Hills, C. H. (1968). *Canner Pckr,* **137,** (7), 33.
Whittenberger, R. T. and Hills, C. H. (1956). *Fd Engng* **28,** 53.
Wong, F. L. and Carson, J. F. (1966). *J. agric. Fd Chem.* **14,** 247.
Williams, C. C., Cameron, E. J. and Williams, O. B. (1941). *Fd Res.* **6,** 69.
Yang, H. Y., Steele, W. F. and Graham, D. J. (1960). *Fd Technol.* **14,** 644.

Chapter 16

Canned Citrus Products†

H. A. W. BLUNDSTONE, J. S. WOODMAN AND J. B. ADAMS

*Fruit and Vegetable Preservation Research Association,
Chipping Campden, England*

I	Introduction	543
II	Colour	544
	A. Browning Reactions Occurring During Storage	544
	B. Changes Occurring in the Carotenoid Pigments	546
III	Flavour	548
	A. Flavonoid Bitter Principles	548
	B. Lactone Bitter Principles	550
	C. Other Flavour Principles	552
	D. Flavour Retention	554
IV	Texture	555
	A. Firmness and Disintegration of Segments	555
	B. Factors Affecting the Drained Weights of Canned Segments	556
	C. Cloud	556
V	Vitamin Changes during Processing and Storage	560
	A. Changes in Vitamin C	561
	B. Changes in Other Vitamins	565
VI	Fruit Can Interactions	567
	References	569

I. INTRODUCTION

A considerable proportion of a typical year's citrus crop is not sold as fresh fruit but as processed products of various kinds. In this chapter, however, we will limit ourselves to those products which are heat processed in tinplate containers or bottles and sold to the public in that form. These are mainly juices and segmented fruits in syrup.

In the U.S.A., during 1966, 103 million gallons of single strength juice were heat processed and slightly less than this (83 million gallons) were sold in the frozen state. A total of 159 million gallons of concentrate were also produced. It is reported (Anon, 1969) that 60% of orange production and 30–40% of grapefruit and lemon production is processed into juices. Grapefruit and orange segments account for a considerable part of processed citrus products and during 1967 the U.K. imported 25·1 thousand tons of

† See also Chapter 3.

orange segments and 30·2 thousand tons of grapefruit segments. These figures clearly demonstrate the importance of canning to the citrus industry.

It is particularly difficult at the present time to deal with the biochemical changes taking place on processing and storage of citrus products as two of the major areas of interest are developing very rapidly. Recent publications (Scott *et al.*, 1965; Mizrahi and Berk, 1970) have changed the concepts of the nature of orange cloud and the factors affecting its stability; other publications (Clegg, 1964, 1966; Clegg and Morton, 1965) have demonstrated the differences and similarities between the browning reactions occurring in citrus products and the Maillard type reactions common in other less acid products. Rapid developments can be expected in both these areas over the next few years.

On the other hand, comparatively little work has been done over the past two decades on nutritional aspects of canned citrus products. More research would appear to be necessary on the breakdown of ascorbic acid and its relationship to browning. Little is known about the relative importance of volatile and non-volatile flavours or about their stability during storage, and, in this area, further work seems to be required on the non-enzymic hydrolysis of flavonoid bitter principles to less bitter compounds.

For convenience, this chapter has been divided into sections dealing with colour, flavour, texture, vitamins and fruit–can interactions but the division is quite arbitrary. The problems of cloud stability, for example, have been dealt with under texture because of the involvement of pectins and pectic enzymes, but it could equally well be classified under colour or flavour. The division between this chapter and the one entitled "Canned Fruits other than Citrus" (Chapter 15) is necessary because some of the problems experienced in canned citrus products, such as flavour changes and cloud stability, are specific to citrus fruits. Areas of overlap may be found particularly in the corresponding sections on vitamin changes and fruit–can interaction.

No attempt has been made in this chapter to discuss possible microbiological changes since research in this area is very limited. This is probably because spoilage organisms do not generally contribute to the organoleptic properties of canned citrus products.

The selection of varieties of citrus fruits for their biochemical attributes appears to be even more limited than for fruits other than citrus and, again, no separate discussion has been attempted on this topic.

II. COLOUR

A. Browning Reactions Occurring During Storage

The main colour problem facing the canned citrus industry is the non-enzymic browning occurring during storage and this is particularly troublesome in the more acid lighter-coloured products such as lemons and grapefruits. Factors having a major effect on the rate and type of browning are acidity, temperature, presence or absence of oxygen and nature of the container. Some authors have not specified these conditions adequately and frequently the changes in the composition of headspace gases in cans are unknown. However, it is well known that certain conditions can lead to very rapid loss of oxygen (Pulley and von Loesecke, 1939) as, for example, in plain cans, whereas in lacquered cans the rate of loss of oxygen is much lower.

For some products the discoloration on storage can be quite rapid, particularly at elevated temperatures. Using a Hunter Colour Difference Meter, Huggart and Wenzel (1955) showed that grapefruit sections darkened slightly in 12 months at 21°C and that considerable darkening occurred after as little as 6 months at higher temperatures (Huggart *et al.*, 1955). In single-strength orange juice, these changes occur more slowly and Pineapple orange and Valencia orange juices in plain cans have been reported to show no significant change in colour when stored at 32°C for 12 months (Huggart and Wenzel, 1955). In enamelled cans, however, darkening is noticeable in 6 months at 27°C (Curl, 1947). The effect of elevated temperatures on browning is very noticeable where juices (particularly concentrates) have been inadequately cooled after processing; the product may retain its heat for sufficiently long to form a central brown core which is clearly visible if the can is opened carefully.

A number of reports indicate that more than one type of reaction may be involved in this browning. The better known sugar–amino acid reactions of the Maillard type seem, however, to be of minor importance as the acidities of these products are generally too high. In glucose-glycine model systems heated at 100°C for 24 hours, Maillard browning has been reported (Ellis, 1959) to occur only above pH 4. Sugar-acid browning may, however, be significant. It has long been known that oxidation reactions can be involved (Joslyn *et al.*, 1934) and that under aerobic conditions the ascorbic acid disappears completely before the onset of browning (Tressler and Joslyn, 1954) whilst, under relatively anaerobic conditions, browning may be observed when as much as 85–90% of the ascorbic acid remains unchanged (Curl, 1947; Tressler and Joslyn, 1954). Natarajan and MacKinney (1949) reported that glass-packed orange juice reacts differently under aerobic and anaerobic conditions. In the presence of oxygen at 37°C, both furfural and hydroxymethylfurfural are formed in measurable quantities, whilst at lower temperatures, or in the absence of oxygen, neither of these aldehydes could be detected in the juice (see also Chapter 17).

A number of factors affect the rate of browning and Huggart *et al.* (1957) have investigated these extensively in the case of canned grapefruit segments. Although the chemical composition of the fruit changed with maturity, these changes were less important than chance variations from batch to batch. By grouping the samples irrespective of maturity, browning was shown to occur more frequently in samples with a high titratable acidity and low pH; this was confirmed by showing that cans with added citric acid discoloured more rapidly than controls with no added acid. On the other hand, raising the pH by addition of sodium hydroxide did not reduce browning measurably which could indicate that one reactant in the browning reaction is the citrate ion since the concentration of this ion would not be reduced by the addition of alkali. If citrate is a reactant, however, increasing the pH would be expected to give more browning; it is not clear whether this was observed or not.

These authors (Huggart *et al.*, 1957) further reported that badly packed cans of grapefruit segments with a small headspace, low vacuum and high acidity gave a poor colour. All the samples used in this work were analysed for ascorbic acid but it was not possible to correlate the ascorbic acid content with the degree of browning.

It has long been known that lemon juice is particularly susceptible to browning in the presence of air (Hamburger and Joslyn, 1941), and it now appears that the main browning under these conditions is due to reactions between ascorbic and citric acids with succinic and tartaric acids being slightly less active; at an advanced stage of the reaction amino acids increase browning considerably (Clegg, 1964).

Clegg and Morton (1965) have demonstrated a strong qualitative similarity between this reaction and the Maillard reaction. Both lemon juice and a series of model systems were analysed for carbonyl compounds by thin-layer chromatography of their 2,4-dinitrophenylhydrazones. In all cases about a dozen compounds were separated which were qualitatively similar to intermediates isolated from Maillard reactions. Using model systems, it was shown that the α,β-unsaturated carbonyl compounds crotonaldehyde and methyl vinyl ketone were particularly active in promoting the browning reaction. Amino acids were also shown to promote browning, apparently by reaction with carbonyls produced from the reaction between ascorbic and citric acids.

In a later paper Clegg (1966) suggested that, in model systems, citric acid is one of the reactants leading directly to the formation of brown polymers and does not have the role of catalyst or source of carbonyl compounds.

B. Changes Occurring in the Carotenoid Pigments†

The characteristic coloration of mature citrus fruits is mainly due to carotenoid pigments, principally xanthophyll esters together with small amounts of carotenes (Curl and Bailey, 1954, 1955), the actual colour

† See also Chapter 21.

depending on the ratio of the two types of carotenoid pigment present and on the amounts of these pigments finding their way into the final product. (Braverman, 1949) Grapefruit juice contains only traces of these pigments (Tressler and Joslyn, 1954).

The pigments are mainly located in the flavedo and endocarp but nearly all the tissues are pigmented to some extent. As colour is such an important parameter of quality—in many cases it is considered as important as flavour—the subject has received considerable attention; Curl and Bailey (1954, 1955, 1956) carried out an extensive survey of the pigments and their distribution in citrus fruits. This work has been well summarized by Kefford (1959) who gives a table of the percentages of thirty-four different pigments found in oranges, tangerines and pink-fleshed Ruby Red grapefruits, as well as in fresh orange juice and in a sample of canned orange juice which had been aged 3 years at room temperature.

Comparison of the analyses (Curl and Bailey, 1955, 1956) of these last two samples indicated that the xanthophyll epoxides isomerized to the corresponding furanoxides during storage. The results are given in Table I with tentative identifications of the various components.

TABLE I. Comparison of some carotenoid pigments of fresh frozen with aged canned orange juice

Component	% of total carotenoids	
	Fresh frozen	Canned (aged 3 years at room temp.)
Dihydroxy carotenes		
Lutein	6	5
Zeaxanthin	9	13
Furanoxide type		
Flavoxanthin-like	1	2
Trollein	15	14
Mutatoxanthin	7	19
Trollichromes-like	12	36
Auroxanthin	22	19
Epoxide type		
Trollixanthin-like	26	0

Although the isomerization of the xanthophyll epoxides was well advanced in this sample of canned juice, the colour was reported to be "good". Clearly the colour change brought about by this reaction is not commercially important.

A more difficult problem is caused by the changing pigmentation found in oranges (Curl and Bailey, 1956) and in mandarins (Ito *et al.*, 1968) at different stages of maturity. This has been solved by blending the juices so that a standard colour is obtained throughout the season (Huggart and Barron, 1966).

III. FLAVOUR

There would appear to be four main groups of chemical constituents contributing to the taste and flavour of citrus products. These are organic acids, of which citric acid is the major representative, sugars, bitter principles and volatile flavour constituents, mainly terpenes and carbonyls.

The sugar : acid ratio is an important consideration in selecting citrus fruit, particularly grapefruit, for processing. In Californian grapefruit the ratio of sugar (° Brix) to acid (% citric) is about 6·0 at maturity, whereas in Florida the sugar and acid both tend to be lower and a ratio of more than 7·0 is usual. The acidity of grapefruit tends to decrease towards the end of the season (Tressler and Joslyn, 1954) and proper selection of fruit for processing should take account of this.

Citrus products contain bitter principles which are important from the standpoint of flavour; these can be divided into two classes, the flavonoid type, e.g. naringin, and the lactone type, e.g. limonin.

A. Flavonoid Bitter Principles†

For many years the intense bitterness associated with naringin was not well understood especially as other closely related flavonoids did not have such intense bitter tastes. Two isomeric glycosides of naringenin have been isolated, both containing one mole each of the sugars rhamnose and glucose, and whilst one of these, naringin, is intensely bitter, the other, naringenin rutinoside, is not.

Recent work by Horowitz and Gentili (1969) has shown that it is the glycosidic linkage of the two sugars comprising the disaccharide unit which largely determines the bitterness. The neohesperidosyl group found in naringin consists of rhamnose and glucose linked $1 \rightarrow 2$, whereas in the rutinoside the linkage is the more usual $1 \rightarrow 6$ (Fig. 1). The neohesperidosyl group is always associated with, although not essential for, bitterness, as can be seen by the fact that naringenin-7-β-D-glucoside (prunin) which although lacking the disaccharide linkage still exhibits some bitterness.

The location of the bitter glycosides in the fruit is important in determining whether this bitterness is carried into products such as juice or segments prepared from them. Naringin is the main bitter principle of grapefruit where it has been found in the albedo layer and in the carpellary membranes (Griffiths and Lime, 1959). Thus the method of extraction of the juice or

† See also Volume 1, Chapter 11

preparation of segments profoundly influences the amount of bitterness found in the product. If the fruit is completely hand peeled, much of the naringin is removed with the albedo layer and outer membranes, but this method gives a lower yield of segments than if the segments are lye peeled. Here the fruit is scalded to soften the skin, and then the albedo layer is removed by an alkali dip and subsequent water spray to remove the loosened tissues. The efficiency of this process determines the bitterness of the product.

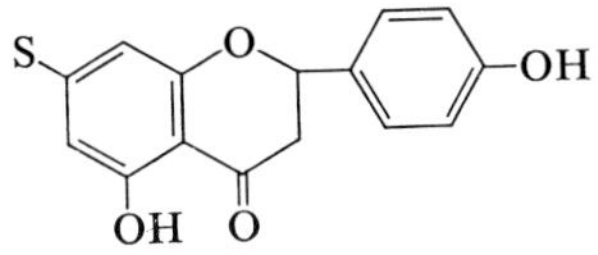

I

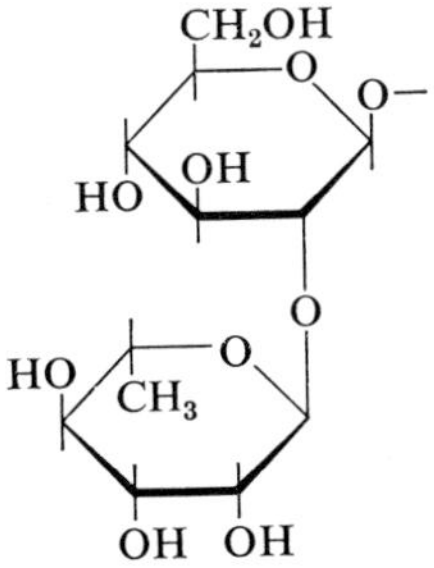

II Neohesperidosyl
(1→2 linkage)

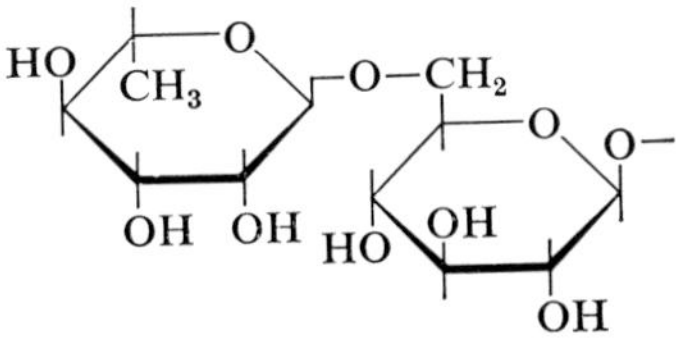

III Rutinosyl
(1→6 linkage)

S = OH Naringenin
(non-bitter)

S = II Naringin
(bitter)

S = III Naringenin-7 rutinoside
(non-bitter)

FIG. 1. Naringenin and its glycosides

Hagen *et al.* (1966) have studied the change in flavanone content with harvest date and have shown that the level of naringin is highest in immature fruit. Fruit sampled in November had only 38% of the level of bitter flavanone glycosides found in fruit harvested in July. However, it was shown that part of this decrease in bitter flavanones can be related to the increase in physical size of the fruit over this period. An early solution to the problem of bitterness in canned grapefruit was the selection of mature fruit for processing (Fellers 1929).

Naringin can occur in the processed product as a fine crystalline suspension which can impart a cloudy appearance to the syrup (Fellers, 1929), or it can give rise to larger crystals giving "glassy" lumps in the fruit segments.

Other bitter flavanone glycosides, all containing the characteristic neohesperidosyl group, have been isolated from citrus fruits. These include poncirin found in *Poncirus trifoliata* and neohesperidin isolated from Seville oranges (Horowitz and Gentili, 1969).

It is not desirable to exclude completely the pulp from processed fruit to prevent bitterness, as an inferior product may result. Griffiths and Lime (1959) have shown that in order to obtain a juice of satisfactory appearance, the pulp should be blended and added to the juice to give 22–26% suspended solids. However, this may result in undue bitterness due to excessive naringin content. These workers suggest the use of the enzyme naringinase which, at low temperatures (ca. 4°C), partially hydrolyses the naringin to give the less bitter prunin or, at higher temperatures (ca. 50°C), effects complete hydrolysis to the non-bitter aglycone, naringenin. Ting (1958) has shown that commercial pectinol enzyme is also capable of hydrolysing naringin and, when used to assist filtration, also has the effect of reducing bitterness in citrus juices. Thomas *et al.* (1958) have prepared naringinase from micro-organisms and shown that complete hydrolysis of naringin to the aglycone is unnecessary; partial hydrolysis to prunin sufficiently reduces bitterness.

Physical adsorption methods have also been proposed as means of controlling bitterness. Polyvinylpyrrolidone resins have a high affinity for flavonoids, and these resins have been proposed as a technique for debittering citrus juices (Griffiths *et al.*, 1963).

B. Lactone Bitter Principles

The main bitter principle of bitter orange is a dilactone limonin which has been isolated from seeds (Emerson, 1949) and peel (Maier and Dreyer, 1965). It has been observed that juice prepared from Navel oranges does not exhibit bitterness immediately after processing, but that an intense bitter flavour develops after standing for a few hours (Emerson, 1949). This can lead to serious problems when processing this fruit. Originally two theories were proposed to account for the delayed bitterness of citrus products. One theory ascribed the delay to a physical phenomenon dependent on the solubility of the bitter principle; Kefford (1959) suggested that it took a few hours for the relatively insoluble compound responsible for the bitterness to diffuse into the juice in sufficient amounts to give a detectable bitter flavour.

The second theory was that a chemical change occurred in which a non-bitter precursor was converted into a bitter compound by the process of extraction of the juice. It is this type of reaction which is now accepted as the explanation for the delayed bitterness; the pH has an important influence on the reaction. In the growing fruit the bitter precursor is separated from the acids of the sap and no reaction occurs, but when the juice is prepared this precursor comes into contact with the acidic juice and the reaction proceeds liberating the bitter compound. Emerson (1949) suggested that the precursor might be a glycoside, but more recently Maier and Margileth (1969) have isolated a non-bitter monolactone of limonoic acid from citrus fruits. They showed that this was rapidly converted to the dilactone limonin at pH 3·0, and that slow conversion occurred at pH 5·6 (Fig. 2). Maier *et al.* (1969) have isolated an enzyme from orange seeds (limonin D-ring lactone hydrolase) which catalyses conversion of the monolactone to the bitter limonin dilactone.

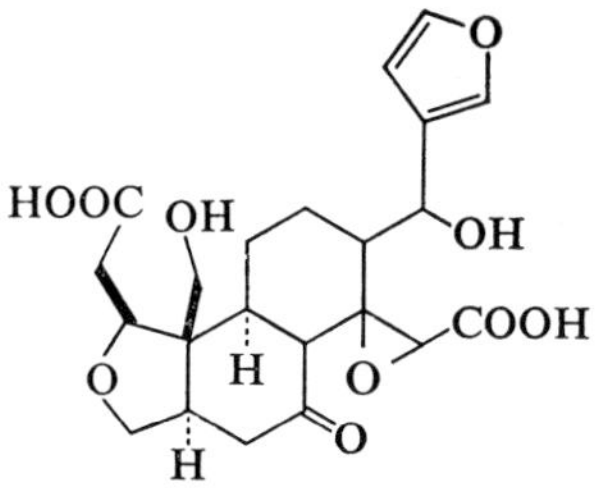

Limonoic acid

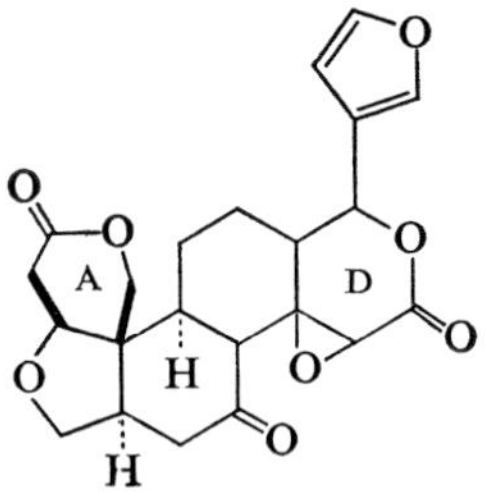

Limonin (dilactone)

FIG. 2.

There are a number of ways in which the incidence of bitterness in citrus products can be reduced. Precautions can be taken during the preparation of the product to prevent excessive extraction of the bitter principles from the rag and pulp; a special burr has been used for this purpose (Anon., 1962). This, in conjunction with raising the pH to about 4, prevents formation of the limonin dilactone. Siddappa and Bhatia (1958) have shown that slicing Coorg oranges before peeling gave a lower incidence of bitterness in the canned product and led to less damage to the segments. The use of pectic enzymes to reduce bitterness has been suggested (Anon., 1962). The dispersed colloids are coagulated by use of the enzymes, and in the subsequent precipitation they carry the bitter principles with them.

The use of polyamide adsorbants to remove the limonin from Navel orange juice has been investigated by Chandler *et al.* (1968). The method was found to be successful in that the bitterness was reduced by stirring juice with dry polyamide powders and subsequent centrifugation. Although the method is convenient for the removal of bitterness, a loss of up to 25% of ascorbic acid may also occur.

C. Other Flavour Principles

In addition to bitter principles there are several groups of compounds which contribute significantly to the flavour of citrus products, and certain of these can, by undesirable changes, lead to the development of off-flavours.

1. *Terpenes*†

A group of compounds which contributes to the volatile flavour of citrus fruits is the terpenes, of which the most common example is d-limonene. This terpene is present in oranges and is found in appreciable quantity in freshly prepared juice. Rymal *et al.* (1968) have shown that when juice is stored at 27°C the amount of d-limonene falls steadily, while at the same time there is an increase in the concentration of α-terpineol. Blair *et al.* (1952) considered that the hydration of d-limonene first to α-terpineol, and finally to terpin then cineole, gave rise to the off-flavours detected in canned orange juice on storage, which were described as "terebinthine-like".

Terpenes also occur widely in grapefruit juice and Kirchner and Miller (1953) have studied the change in concentration of such compounds on the storage of grapefruit juice. In this juice a decrease in d-limonene is also found on storage, and there is a corresponding increase in oxygenated compounds.

2. *Lipids*

Nagy and Nordby (1970) have examined the effect of storage on the lipid constituents of orange juice. They relate the increase in free fatty acids to hydrolysis of phospho-lipids in the juice. A comparison of the lipids of fresh and pasteurized orange juice (Huskins and Swift, 1953) showed little difference. These workers thought that the change in flavour brought about by pasteurization of juice was unlikely to be closely related to the lipid constituents.

3. *Carbonyl compounds*†

The total carbonyl content of orange juice has been studied by Senn (1963) who observed an initial rise in the amount of these compounds. On cool storage this was followed by a gradual decline which coincided with a flavour deterioration but Senn did not think the carbonyls were necessarily of prime importance in the juice flavour.

† See Volume 1, Chapter 10.

The water-soluble volatile constituents of grapefruit juice have been studied by Kirchner *et al.* (1953), who compared juice freshly prepared with that immediately after canning and also after a period of storage in cans. They observed an increase in the amount of furfural, methanol and acetic acid on storage (Table II).

TABLE II. Volatile constituents of fresh and canned grapefruit juice

	mg/kg of juice as described			Reference
	Fresh	Freshly canned	Stored canned	
Limonene	15·71	17·70	11·17	Kirchner and Miller (1953)
α-Caryophyllene	0·10	0·10	0·12	
β-Caryophyllene	1·40	1·40	0·87	
α-Terpineol	0·03	0·88	2·02	
Linalool monoxide	0·37	2·03	8·95	
Acetaldehyde	1·45	0·33	0·6	Kirchner *et al.* (1953)
Acetone	nil	nil	0·1	
Furfural	nil	tr.	8·20	
Ethanol	400	400	460	
Methanol	0·2	0·2	23	
Acetic acid	nil	1·9	23·3	
$C_6H_8O_2$ acid A	nil	4·8	2·9	
$C_6H_8O_2$ acid B	nil	1·9	1·6	

One carbonyl compound that has been extensively studied is nootkatone, a ketone found in the peel oil of grapefruit (Berry *et al.*, 1967). It has a very pungent aromatic odour, and the level in grapefruit juice is critical as regards flavour. If it is present at a level of 6–7 p.p.m., a good full-flavoured juice is obtained, but at higher levels an undesirable bitterness develops.

4. *Effect of oxygen*

Many of the changes occurring in the volatile constituents described in the preceding sections are oxidative in nature (Nolte and von-Loesecke, 1940; Ting and Newhall, 1965). The amount of oxygen in the headspace and product will be of great significance in the development of off-flavours due to these oxidative changes. Boyd and Peterson (1945) have shown that, although elimination of oxygen from both the headspace and can contents delayed the development of off-flavours, it did not prevent the eventual flavour deterioration often associated with citrus products.

Anti-oxidants. Various compounds have been investigated in attempts to prevent undesirable oxidative changes. The anti-oxidant activity of asparagine and cholesterol has been examined (Nolte *et al.*, 1942) and these compounds had a similar effect to the exclusion of oxygen by physical means in that they delayed the onset of undesirable flavour changes. Other anti-oxidants found to have a similar effect include tyrosine and methyl glucamine (Riester *et al.*, 1945).

Inclusion of peel oil in the juice reduces the incidence of off-flavours and has encouraged the study of various fruit tissues for natural anti-oxidants. If peel oil is excluded from the juice, a weak-flavoured product results, and up to 300 p.p.m. of peel oil is desirable to give a full-flavoured product (Blair *et al.*, 1952). The presence of natural anti-oxidants in the peel of the fruit has been inferred by Charley (1963), who observed that properly formulated comminuted juices containing 800 p.p.m. peel oil retain a good flavour and aroma. Curl and Veldhuis (1947) claimed that the presence of peel oil masked other changes taking place in the juice.

The distribution of natural anti-oxidant activity in various tissues of citrus fruits has been studied by Ting and Newhall (1965). The albedo layer, though high in flavonoids, had a low anti-oxidant activity, suggesting that the natural anti-oxidants were not of a flavonoid nature. The flavedo layer was found to be the most effective from an anti-oxidant point of view and, of the various fruits examined, extracts from grapefruit showed the highest anti-oxidant activity. The effectiveness of the commercial anti-oxidant, α-tocopherol was also examined and, at the 10 p.p.m. level in model systems, this completely suppressed the oxidation of d-limonene. α-Tocopherol has also been shown to be effective in preventing oxidation in citrus oils (Kenyon and Proctor, 1951).

D. Flavour Retention

Different methods have been proposed for the improvement or retention of flavour in canned citrus products. With concentrates, low temperature processes, particularly freeze concentration, have been found to make notable improvements. Where evaporative techniques are used, the volatile flavour constituents are liable to be lost with the distillate. Bomben *et al.* (1966) have described a technique by which these volatile constituents can be recovered and subsequently used for enhancement of the flavour of citrus extracts. This recovery technique employs a special "scrubbing" arrangement for the recovery of those volatiles which would normally be lost with the non-condensable gases. The storage of aroma solutions under various conditions has been investigated by Guadagni *et al.* (1970). Although the strength of the aroma remained fairly constant for up to one year at 7°C, a significant change in aroma character occurred under these conditions. At −18°C no significant change occurred in 88 weeks' storage.

Sugars have been used by Riester *et al.* (1945) who showed that addition of sucrose, fructose or glucose to canned juices helped to maintain a satisfactory flavour.

The storage temperature has a marked effect on the retention of a satisfactory flavour by canned products, and in this respect citrus fruits are no exception. Huggart *et al.* (1955) have examined the effect of storage at temperatures ranging from 0° to 32°C and found no change in flavour after 12 month's storage at temperatures below 15·5°C whereas at 32°C the flavour had deteriorated markedly after 9 months. It has been observed that the warehouses in Florida are, on average, 5–6°C warmer than those in California; this has led to a higher quality product from the cooler storage.

IV. TEXTURE

A. Firmness and Disintegration of Segments

In citrus fruits packed as segments or rings, textural characteristics are important, since excessive disintegration or lack of firmness of the fruit pieces makes the pack unattractive to the consumer.

Some segmented products have the carpellary membranes removed before canning and the methods chosen for achieving this have a marked effect on the amount of breakdown, hand peeling being the most advantageous in this respect (Suryaprakasa Rao *et al.*, 1969a). The separation of the segments is facilitated by immersing the fruit in boiling water 12–36 hours before peeling (Elstein, 1969). A number of different peeling methods have been compared and their effects on the amount of breakdown occurring in cans reported (Suryaprakasa Rao *et al.*, 1969b).

The action of hydrochloric acid and sodium hydroxide on the various types of tissue making up the segment membranes have been described and it was concluded that more rapid peeling and cleaner segments were obtained with better segment integrity if both reagents were used for peeling the segments (Suryaprakasa Rao *et al.*, 1969c).

Sawayama *et al.* (1964) have shown that the use of hard water makes the segment-peeling process more difficult; segments pick up calcium and magnesium ions from the water particularly under alkaline conditions. It was thought that the difficulty experienced in segment peeling could partly be explained by the activation of the pectinmethylesterase by these divalent ions. The enzyme would then liberate free carboxyl groups for binding more ions.

Two techniques have been developed and found useful by Mannheim and Bakal (1968) for studying the textural changes taking place during storage of canned grapefruit. The firmness of segments is assessed by their resistance to the passage of a shearing and compression device in a modified Kramer Shear Press. In a second apparatus, segments are subjected to a fairly rapid flow of water which tends to break the segments and to float off the smaller particles. In this technique the fruit is subjected to controlled forces which simulate mishandling in transport, and so measure the tendency to breakdown rather than the amount of breakdown which has taken place. Using these techniques, Bakal and Mannheim (1968) showed that the firmness of the segments could be increased by the addition of calcium chloride to the syrup and/or by vacuum syruping.

Ludin *et al.* (1969) and Levi *et al.* (1969) used the same techniques and confirmed the effect of calcium chloride noting that it decreased the tendency of the fruit to disintegrate. In contrast to this they also showed that although the firmness of grapefruit sections canned without the addition of calcium chloride increased on storage, the tendency to breakdown also increased and that these changes took place more rapidly at higher storage temperatures.

B. Factors Affecting the Drained Weights of Canned Segments

The legislation of a growing number of countries requires that canners guarantee to the consumer a certain minimum drained weight; consequently, factors affecting this have received some attention.

Immediately after processing, segments start to equilibrate with the syrup and Bakal and Mannheim (1968) report that during the first 4–6 days grapefruit segments lose water by osmosis, the drained weight falling to 60–70% of the filled weight. The segments then increase in weight by up to 10%, probably by absorption of sugar, reaching their equilibrium value 2–3 weeks after canning. The rate of this equilibration depends on temperature, being slower at low temperatures (Ludin *et al.*, 1969).

The loss of weight by osmosis is considerably reduced by using less concentrated sugar syrup (Bakal and Mannheim, 1968); vacuum syruping and the use of dry sugar also increases the drained weight.

Levi *et al.* (1969) pointed out that the use of calcium chloride to firm segments and to reduce the amount of breakdown had the serious disadvantage of reducing the drained weight of the product. This was counteracted to some extent by the addition of low methoxyl pectin, and it was reported that the use of these two products together had no effect on flavour or colour even after an extended period of storage. Drained weights and texture are areas of vital commercial significance but are not well understood. It would be useful to have more information on these topics which could be helpful in solving technological problems.

C. Cloud

The stability of cloud in citrus products is of great importance to the industry. Techniques for stabilizing the cloud in pulpy juices such as orange, for keeping the syrup clear in canned segmented products and for clarifying juices which are traditionally sold as clear pulp-free products have been developed. Joslyn and Pilnik (1961) have reviewed the work on the stability of orange cloud.

1. *The nature of orange cloud*

Much of the early work has been based on the implicit assumption that cloud is derived from various tissues (mainly rag and albedo) by disintegration during juicing; recent work by Scott *et al.* (1965) has, however, questioned the validity of this assumption.

These workers separated the cloud into three fractions by centrifugation. They considered that only the small amount of material which was most easily sedimented consisted of finely divided "rags". The remaining fractions were analysed for lipids and then, after extraction with further quantities of hexane, alcohol and acetone, analysed for nitrogen, pectins, phosphorus, hemicellulose and cellulose. The extraction procedure would remove any flavonoid compounds before the analyses were carried out. The results of these analyses are given in Table III together with the analyses of other juice components. It would appear from these figures that cloud can contain only very small proportions of albedo, rag or pulp.

TABLE III. Approximate composition of orange juice fractions (%)

	Cloud	Albedo	Rag	Pulp	Juice
Lipids (hexane soluble)	ca. 25	7	3	11	35
As per cent hexane, alcohol and acetone insolubles					
Nitrogen	ca. 7	0·7	0·7	2·8	6·0
Phosphorus (as P_2O_5)	ca. 1·5	trace	0·1	0·3	2·0
Pectins	ca. 80	44	51	57	80
Cellulose + hemicellulose	2–5	45	48	38	3

The lipid content was found to change rapidly during the first few hours after extraction as the peel oil is absorbed onto the surface of the cloud.

Mizrahi and Berk (1970) also separated the cloud into two fractions by centrifugation and showed that the finer fraction consisted largely of crystals of hesperidin having dimensions down to as small as 0·05 μm. Grapefruit cloud, as mentioned in the section on flavour, can contain fine crystals of naringin.

2. *Factors affecting the stability of cloud*

It has long been known that the pectic substances of citrus juices play a major part in the stability of cloud and many theories have been proposed as to their exact role. Particular attention has been paid to the increase in viscosity of the aqueous phase brought about by the pectins and to their precipitation as calcium salts (Stevens *et al.*, 1950). These pectins are strongly affected by the presence of pectic enzymes; polygalacturonase reduces the molecular weight and thus the effect of pectins on the viscosity, whilst pectinmethylesterase liberates free carboxyl groups thus aiding combination with calcium ions and precipitation.

Mizrahi and Berk (1970) have tested the effect of serum viscosity by comparing the stability of cloud (as judged by the time taken for clarification) in heated orange juice with its stability in the same juice after dilution. Although the viscosity of the juice was markedly decreased by dilution, the stability remained unchanged and thus, although heating increases the viscosity of the serum, the increase in stability of the cloud brought about by heating could not be due to the increased viscosity.

In their conclusions Mizrahi and Berk (1970) noted that although heating increased the viscosity of the serum, reduced the particle size of the cloud and affected its electrical charge, none of these effects was sufficient to account for the increased stability and that the principal factor controlling stability was the degree of hydration of the particles. It was claimed that the clarifying action of pectinmethylesterase could be explained in terms of incomplete hydration of the cloud particles.

It has been reported (Yufera *et al.*, 1965) that when orange cloud is resuspended in water containing normal levels of sugar and acid, and with the pH adjusted to that of orange juice, the suspension is quite stable. This has been confirmed and extended to show that addition of pectin, calcium ions or 0·4% potassium chloride initiates clarification (Baker and Bruemmer, 1970). The cloud could be further stabilized by the removal or destruction of serum-soluble pectins; the authors explained the action of potassium chloride as being due to its ability to promote the solubility of pectin. In the light of the more recent work of Mizrahi and Berk (1970), it seems more probable that the effect could be attributed to the action of these ions on the surface of the cloud particles.

Although these theories depart considerably from the earlier held position that cloud consists of suspended rag and albedo tissue, it is still clear that pectinmethylesterase plays an important role in determining the stability of cloud and, therefore, the rapid test for the activity of this enzyme proposed by Kew and Veldhuis (1950) is still of great practical use from this point of view and for estimating the degree of heat treatment that a sample of citrus juice may have received.

3. *Inactivation of pectinmethylesterase*

The inactivation of this enzyme is of such importance in regulating cloud stability and avoiding excessive detrimental effect on the flavour that many workers have published detailed data on the subject. Table IV presents some figures showing the relationship between temperature, pH and the time taken for complete inactivation of pectinmethylesterase (Joslyn and Sedky, 1940; Stevens *et al.*, 1950).

TABLE IV. Time (min) taken for complete inactivation of pectinmethylesterase in various varieties of orange

	Temperature (°C)			
pH	80	85	90	94
4·2	>10	8–10	0	1
		1	2	
3·8			2	0·25
3·3		2	0·25	
2·5	>10	10	1	1

Rouse and Atkins (1952) noted that juices containing larger amounts of pulp (obtained by mixing juices after screening through different mesh sizes) required slightly increased heat treatments in order completely to inactivate the enzymes. Although shorter heat treatments should theoretically be possible with concentrates, it is recommended (Atkins and Rouse, 1954) that no reduction is made in practice as it is difficult to ensure that a concentrate is treated evenly throughout.

4. *Gel formation*

It has frequently been observed that a canned juice which has clarified will sometimes form a gel and it is clear that these two phenomena are related. If a juice is inadequately pasteurized before canning, the remaining pectinmethylesterase activity will promote clarification giving rise by demethylation to a high content of pectic acid in the serum. Wenzel *et al.* (1951) showed that, given the right conditions of pH, a concentrate could form a gel with divalent ions if their concentration were high enough. These workers related gel formation to pectin concentration, pH, divalent ion content and enzyme activity and showed that these primary factors were dependent on variety, type of pulp, quantity of pulp, degree of concentration and storage temperature. The effect of high pulp concentration was also demonstrated by Olsen *et al.* (1951) and was found to be particularly important in seedy varieties such as Duncan grapefruits and Pineapple oranges.

It has been shown that the finishing process (Vandercook *et al.*, 1966) has a marked effect on the soluble pectin content of citrus juices, probably by altering the amount of albedo passing into the juice. This should, therefore, be critical in the control of gel formation.

5. *Flavonoid compounds in the cloud*

As already noted, a portion of the cloud in orange juice consists of hesperidin crystals (Mizrahi and Berk, 1970) and this component can cause particular difficulties in canned orange juice and in canned mandarin segments. Von Loesecke (1954) reported that an unattractive sludge, sometimes found in canned orange juices, contained fine hesperidin crystals. The syrups in canned mandarin segments (Ito *et al.*, 1969) and canned orange segments (Okado and Ono, 1969) have been observed to go cloudy from the same cause, and hesperidin has even been observed to crystallize as white flecks on the segments.

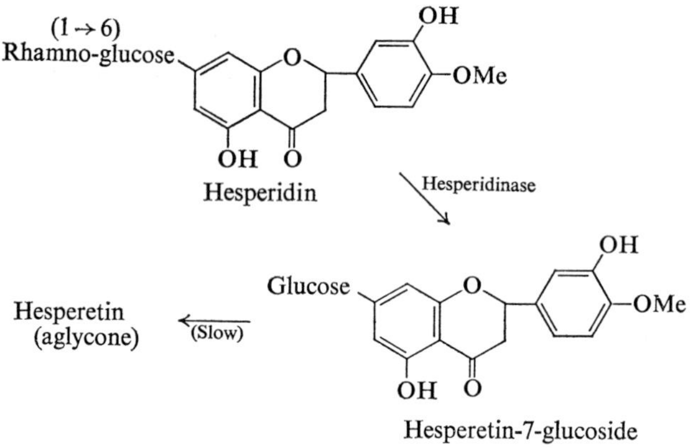

(1→6)
Rhamno-glucose
OH
OMe
OH O
Hesperidin
Hesperidinase
OH
Glucose
OMe
Hesperetin
(aglycone)
(Slow)
OH O
Hesperetin-7-glucoside

FIG. 4. Enzymic hydrolysis of hesperidin.

Hesperidinase added to the syrups solves the problem and a patent (Okado and Ono, 1969) has been taken out for the addition of a relatively heat-stable hesperidinase to the syrups before pasteurization. Shimoda *et al.* (1968) showed that hesperidinase was able to maintain the clarity of syrups for at least 2 years by converting the hesperidin to hesperetin-7-glucoside which was itself slowly hydrolysed to hesperetin (Fig. 4). The products of these reactions, although they would be expected to be even less soluble than hesperidin itself, do not crystallize under these conditions.

V. VITAMIN CHANGES DURING PROCESSING AND STORAGE

Since citrus fruits contain 85–92% water (Kefford, 1965) they cannot be important sources of major nutrients. Their importance in the diet depends upon the presence of minor nutrients, especially Vitamin C. Ascorbic acid content is, therefore, an important attribute of citrus fruits for processing and the major part of this section will be devoted to discussing the factors affecting the loss of this vitamin and the methods used to restrict its disappearance.

A. Changes in Vitamin C

A discussion of the reactions of ascorbic acid during canning can conveniently be divided into two parts, those occurring prior to heat treatment and the degradative changes which take place during processing and storage.

1. *Disappearance of Vitamin C prior to heat processing*†

(a) *Peeling losses.* With the exception of canned marmalades, canned citrus products contain very little of the flavedo or albedo portions of the fruit. This is unfortunate from the nutritional point of view as the concentration gradient of ascorbic acid declines from flavedo towards endocarp. In an analysis of several varieties of Italian oranges the following Vitamin C values have been reported (Rauen *et al.*, 1943):

Yellow peelings (flavedo):	175–292 mg/100 g
White peelings (albedo):	86–194 mg/100 g
Fruit flesh (endocarp)	44–74 mg/100 g

A similar distribution has been found in limes (Ribeiro *et al.*, 1942), oranges and grapefruits from Florida (Atkins *et al.*, 1945).

The effect of lye-peeling on ascorbic acid retention in citrus fruits does not appear to have received much attention, which is rather surprising in view of the high losses incurred at this stage during the canning of clingstone peaches (Lamb *et al.*, 1947). As Vitamin C is extremely unstable to heat under alkaline conditions, the lye treatment used should be only just sufficient to remove the outer carpellary membrane of the fruit.

(b) *Enzymic changes.* Ascorbic acid oxidase, cytochrome oxidase, peroxidase and phenolase can all lead to the oxidation of Vitamin C. The first three of these have been identified in oranges (Tressler and Joslyn, 1954). Ascorbic acid oxidase has also been more recently identified in other citrus fruits (Vines and Oberbacher, 1963; Dawson, 1966). These enzymes systems are presumably balanced, in intact fruits, by reductases, and it is the interplay of these oxidizing and reducing systems which controls the final level of ascorbic acid. However, when this enzyme balance is disturbed by cellular disruption, as occurs, for example, during the extraction of fruit for juices, it is the reductases which suffer the greatest impairment. The oxidases can then, theoretically, destroy all of the ascorbic acid, unless they are inhibited. This can be done by holding juices for short times at low temperatures during the

† See also Volume 1, Chapter 13.

blending stage, de-aerating the juice to remove the oxygen required by the oxidation reaction, and finally pasteurizing the juice at a sufficiently high temperature to inactivate the oxidizing enzymes.

Enzymic action has also been suggested as the cause of rapid losses of ascorbic acid in orange skins during the preparation of marmalade. It was found that boiling the grated peel with water for a short time substantially reduced the loss of the vitamin (Huelin and Stephens, 1944). Recently Baker and Bruemmer (1968) isolated an atypical enzyme system from mature oranges which degrades ascorbic acid without oxygen consumption and which is sensitive to ascorbic acid oxidase inhibitors. However, activity has only been observed in preparations from carefully neutralized juice and not from conventionally extracted juice so that the enzyme would not be expected to contribute to Vitamin C destruction in orange juice.

(c) *Non-enzymic changes.* Oxygen can react with ascorbic acid in the presence of trace amounts of copper to give dehydro-ascorbic acid and hydrogen peroxide. This catalytic effect of copper is enhanced by iron and is discussed in Chapter 17. The hydrogen peroxide formed during this reaction can lead to further oxidation of Vitamin C either directly or indirectly, by breakdown to water and oxygen, both reactions being catalysed by cupric ion (Weissberger and LuValle, 1944).

Dehydro-ascorbic acid retains full Vitamin C activity but is unfortunately more thermolabile than ascorbic acid and would therefore be easily destroyed during heat processing. It is thus essential for Vitamin C retention that citrus juices be kept in a de-aerated condition for as much of their preparation as possible and that they always be held in stainless steel or glass-lined containers.

It has been shown that flavonols with a 3-hydroxyl-4-carbonyl group in the pyrone ring or 3′,4′-dihydroxy group in the B ring show anti-oxidant properties by virtue of their ability to complex with metal ions (Clements and Anderson, 1966). However, the main flavonoids of citrus fruits, hesperidin and naringin, do not possess either of these groupings and thus cannot stabilize ascorbic acid by chelation of metal ions.

2. *Changes in Vitamin C during heat processing and storage*†

(a) *Aerobic destruction.* Oxidative destruction of Vitamin C can take place during heat processing of citrus products and during their storage in cans. This is in the main a non-enzymic process as the oxidative enzyme systems mentioned previously are destroyed by heat under acid conditions.

(i) Effect of headspace oxygen. Oxygen in the headspace of the can has been shown to have a detrimental effect on the ascorbic acid concentration in de-aerated sweetened orange juice (Boyd and Peterson, 1945). Generally

† See also Volume 1, Chapter 13.

speaking, a small headspace and a low oxygen tension will have a beneficial effect on Vitamin C retention in canned citrus products. De-aeration of juices is, of course, essential for ascorbic acid stabilization, as a juice which has not been de-aerated, even though filled hot, may contain more oxygen than is normally present in the headspace of properly filled cans.

(ii) Effect of the container. It has been shown in a number of investigations that for retention of ascorbic acid it is preferable to can citrus products in plain cans rather than completely lacquered cans (Curl, 1949; Moore *et al.*, 1951; Shiga and Kimura, 1956; Khan *et al.*, 1963). Although the slower cooling of bottled grapefruit and orange juices does not lower the ascorbic acid concentration immediately, the glass-packed products have been shown to be definitely inferior in ascorbic acid content to canned ones after 6 months' storage at 4·5°C and 27°C (Moore *et al.*, 1949). In the same research it was demonstrated that the dehydro-ascorbic acid values represented only 1–2% of the total ascorbic acid, indicating that the oxidized form of the vitamin was breaking down further, probably into diketogulonic acid.

The beneficial influence of the tinplate in plain cans undoubtedly lies in its ability to react preferentially with oxygen under acid conditions. The effect of this initial protection of ascorbic acid is found in the high tin contents, shortly after processing, in cans closed in air. Cans closed under an atmosphere of nitrogen contain much less tin at this stage (Riester *et al.*, 1945).

(iii) Effect of pH. Ascorbic acid is rapidly destroyed under neutral and alkaline conditions, but in citrus products where the highest pH is about 4·0 (observed in some canned orange juices), the vitamin is relatively stable. However, during the making of marmalade, raw peel is cooked prior to mixing with a juice in order to avoid excessive heating of the mixture. If this cooking procedure is carried out in 0·2% citric acid solution to reduce the breakdown of Vitamin C, the resultant marmalade may contain up to 25–30 mg/100 g. The effect of the citric acid here is, at least partly, due to its lowering the pH of the cooking medium; there is some evidence that it helps retain the ascorbic acid during storage of the marmalade (Lincoln and McCay, 1945).

(b) *Anaerobic destruction.* After a period of relatively rapid degradation of ascorbic acid in pasteurized orange juice in the presence of free oxygen, anaerobic destruction takes over (Kefford *et al.*, 1959). The rate of this reaction is virtually independent of pH in the range 1–11 with a slight increase in the pH range 3–4. The reaction is accelerated by fructose, fructose-6-phosphate and fructose-1,6-diphosphate. Sucrose also increases the rate of reaction, probably by hydrolysis to fructose (Huelin, 1953). Caramelized fructose has also been implicated in the aerobic and anaerobic degradation of ascorbic acid (Isaac, 1944). Furfural and carbon dioxide appear to be major products of decomposition, especially at high temperatures and acidities (Huelin, 1953). Concentrated products with a higher concentration of fructose and related substances would be expected to lose Vitamin C more rapidly and Table V indicates that this is so (Curl, 1947).

TABLE V. Percentage retention of ascorbic acid in canned orange juices at different concentrations

Storage Temperature (°C)	Storage Period (months)	Soluble solids (%) 13[a]	30	49	71
4·5	12	97	97	97	93
15·5	12	85	92	88	83
26·5	12	70	51	16	6
38·0	6	67	39	10	8
49·0	1	56	15	9	8

[a] Single strength juice contained 13% soluble solids.

(c) *Effect of temperature on Vitamin C stability.* Between 1940 and 1950 numerous investigations were made on the effect of temperature on the retention of ascorbic acid in canned foods, and a number of these were concerned with canned citrus products, for example, the one by Clifcorn (1948). More recently, the subject has been discussed in a U.S.D.A. handbook on the chemistry and technology of citrus products (Anon., 1962).

It would appear that immediately after heat processing more than 90% retention of Vitamin C can be expected in canned citrus products, i.e. very little breakdown takes place during preparation of the fruit or during the heat processing. However, subsequent retention is markedly dependent on storage temperature. This can be seen for canned orange juice (both single strength and concentrates) by inspection of Table V. To retain as much Vitamin C as possible it has been recommended that the product be stored between 4° and 10°C (Feaster *et al.*, 1950). The rates of aerobic and anaerobic destruction of ascorbic acid would undoubtedly be increased by increasing the storage temperature, but Evenden and Marsh (1948) showed that the rate of oxidation of ascorbic acid in orange juice in the temperature range 0–10°C is slow enough to permit satisfactory retail distribution over a period of 2–3 months.

(d) *Fortification of citrus products.* It will be appreciated from the preceding discussion that Vitamin C may be well retained in canned citrus products so long as care is taken with the processing and storage conditions. The basic stability of ascorbic acid in these products is probably due, at least in part, to the high concentration of polybasic or polyhydroxy acids, such as citric and malic, which can chelate metal ions of iron and copper. It has been shown

(Inagaki, 1944) that the presence of sugars such as sucrose, glucose or fructose increases the capacity of these acids to stabilize ascorbic acid.

However, the level of Vitamin C in citrus products may vary widely with variety, stage of maturity, climate, soil condition, geographic location and related factors, and under unfavourable conditions such variables may lead to low values of ascorbic acid in the processed product. Where standards of identity permit, the product may be enriched by addition of ascorbic acid so that one serving provides the minimum daily requirement for adults (30 mg) (Bunnell, 1968). There is some evidence, however, which indicates that Vitamin C added to citrus juices has a lower stability than the naturally present vitamin (Inagaki, 1943; Matsui and Ito, 1953).

B. Changes in Other Vitamins

Canned citrus products contain small amounts of provitamin A and of B vitamins but the amount of research carried out on the stability of these essential nutrients during the various canning processes is small compared with that carried out on the breakdown of Vitamin C.

1. *Provitamin A*

Vitamin A, as such, has not been shown to be present in any of the common fruits used for juice manufacture, but its carotenoid precursors α-, β- and γ-carotene, and cryptoxanthin are fairly widespread; of these, β-carotene is found in the highest concentration in citrus fruits. During filtration of fruit juices a large amount of this provitamin A can be lost since it is largely present in the suspended particles of the juice (Joslyn, 1937). Although Vitamin A may be destroyed by heating, the carotenes are much more heat stable (Tressler and Joslyn, 1954). It has, however, been suggested that large losses of carotenoids can be caused by various enzymic and non-enzymic oxidation reactions and that these can lead to decreases in provitamin A value (Diemair and Postel, 1964). However, the actual changes occurring in the carotenoids of citrus fruits during processing and storage, as already discussed in the section on colour, suggests that little diminution of provitamin A content does, in fact, take place.

Synthetically produced, water-dispersible forms of β-carotene are now available for enrichment of the colour and nutrient content of juices and drinks (Bauernfeind *et al.*, 1962). These appear to be stable in heat-processed juices packed in glass or tinplate containers, even in the presence of reducing agents, such as ascorbic acid (Bauernfeind, 1958).

2. *The B vitamins*

(a) *Vitamin* B_1. The factors affecting the thermal decomposition of Vitamin B_1 (thiamine) in pure buffer solutions have been summarized by

Huelin (1958). The rate of degradation increases with increasing pH and with increasing concentration of buffer salts, but this latter effect varies with the type of buffer used. The presence of free copper ions leads to an increase in the rate, as does oxygen at temperatures greater than 70°C, whereas complex copper-containing anions (occurring in tartrate, citrate or glycine buffers) may lead to a decrease in rate. Increasing the concentration of thiamine itself appears to have some stabilizing effect.

Vitamin B1 may also occur as thiamine pyrophosphate and in combination with protein. The pyrophosphate is less stable than free thiamine, whereas the protein complex would appear to have enhanced thermal stability (Huelin, 1958). The low pH of canned citrus products has a stabilizing effect on thiamine storage temperature appears to be the most critical parameter affecting the stability of this vitamin. Thus, canned grapefruit and orange juices can be stored for 2 years at 18·5°C or below with only 6% loss of thiamine, whereas 2 years' storage at 27·5°C leads to 17% loss (Sheft *et al.*, 1949).

It has also been suggested that some phenolic compounds, such as quercetin, have a strongly destructive effect on Vitamin B1 (Hasegawa, 1955), but the effect of the citrus flavonoids does not appear to have been investigated.

TABLE VI. Vitamin content of fresh and canned citrus juices

Fruit	Carotene (mg)	Vitamin D (mg)	Thiamine (mg)	Riboflavin (mg)	Nicotinic acid (mg)	Ascorbic acid (mg)
Grapefruit or grapefruit juice, fresh	tr.	0	0·05	0·02	0·2	40
Grapefruit or grapefruit juice, canned	tr.	0	0·04	0·01	0·2	Flesh 25 Juice 35
Lemons or lemon juice, fresh	0	0	0·02	tr.	0·1	50
Limes, fresh	0	0	tr.	tr.	0·2	25
Oranges, fresh	0·05	0	0·10	0·03	0·2	50
Orange juice, fresh	0·05	0	0·08	0·02	0·2	50
Orange juice, canned	0·05	0	0·07	0·02	0·2	40

(b) *Other B vitamins.* In buffer solutions Vitamin B2 (riboflavin) appears to be completely stable in the pH range 2–5 at temperatures below 120°C. Below pH 2, slow reversible reduction takes place (riboflavin is regenerated on aeration), whereas above pH 5 thermal hydrolysis leads to irreversible breakdown of the riboflavin molecule (Farrer and MacEwan, 1954). Vitamin

B2 would thus be expected to be stable in canned citrus products. Riboflavin is, however, decomposed by light and may break down to some extent prior to canning, or during storage of bottled products.

Biotin, folic acid, pyridoxin and inositol appear to be relatively stable during canning of orange and grapefruit juices (Krehl and Cowgill, 1950). Folic acid, however, is unstable to light.

Pantothenic acid shows appreciable decomposition on prolonged heating and both it and folic acid become less stable as the pH is decreased (Frost, 1943; Dick *et al.*, 1948). Nicotinic acid appears to be generally well retained in processed foods (Huelin, 1958).

Table VI gives some typical values of the more common vitamins to be found in fresh and canned citrus juices (McCance and Widdowson, 1967).

VI. FRUIT CAN INTERACTIONS†

The canning of citrus products, particularly grapefruit, has always been troublesome from the corrosion point of view. Lacquered cans tend to give rise to hydrogen swells more rapidly than plain cans. Stevenson (1934) quotes a failure rate in lacquered cans as high as 32% in 18 months from this cause, whilst in plain cans no swells were detected in this time. Kefford *et al.* (1959) have observed that hydrogen swells appeared in some lacquered cans after 5 months' storage at 30°C, whilst no swells were found in plain cans stored for up to 1 year at this temperature. The different behaviour of the plain and lacquered container is dependent on the relative areas of exposed tin. In a lacquered can the exposed tin is similar in area to the iron exposed at the sharp contours of the seams. In this situation the iron acts as an anode and is corroded, whereas in a plain can the much larger area of exposed tin takes the role of anode, so protecting the iron from attack.

Although the above workers have shown that it is the lacquered can which is the most likely to produce hydrogen swells, Frankenthal *et al.* (1959) have shown that hydrogen is produced by grapefruit juice packed in plain cans.

When oxygen is present in the can, headspace corrosion has been observed to proceed more rapidly initially (Kefford *et al.*, 1959) but after a period of storage the difference in corrosion rate becomes insignificant. Under these conditions, with oxygen present in the headspace, metal may be dissolved without the production of hydrogen (Sera and Perfetti, 1963). When the oxygen is exhausted hydrogen may be produced as corrosion proceeds with the dissolution of tin (Frankenthal *et al.*, 1959). The amount of hydrogen liberated is generally rather lower than the equivalent of metal dissolved. Sera and Perfetti (1963) have claimed that this deficiency is due, at least in

† See also Chapter 17.

part, to diffusion of hydrogen out of the can, but further evidence to support this theory has not been produced.

Bakal and Mannheim (1966) have studied the effects of processing variables, including de-aeration, on can corrosion by grapefruit juice. They showed that the rate of tin dissolution was far higher in cans having a normal headspace than in cans hot-filled to leave no headspace. De-aeration of the juice prior to filling gave a lower rate of corrosion than the controls. These workers attributed the observed effects to the varying amounts of oxygen in the packs, and concluded that removal of oxygen gave a prolonged shelf life to the product.

Unusual cases of accelerated corrosion with orange drinks in Japan have led Horio *et al.* (1966) to investigate the relationship between nitrate content and corrosion in acid packs. In experiments with model systems they found that addition of 10 p.p.m. of nitrate nitrogen gave tin contents of over 160 p.p.m. in 11 days at pH 2·5. In a study of the nitrate content of well water used in making orange drinks following an outbreak of food poisoning in Japan, Horio *et al.* (1967) found that some water contained 10 p.p.m. of nitrate. Samples of orange drink which caused poisoning were found to contain 300 p.p.m. of tin. These authors attributed the rapid detinning in these cases to the nitrate content of the well water used. Analysis of the can contents showed that those cans exhibiting the rapid detinning also contained increased amounts of ammonia. This led them to suggest that the nitrate acts as a cathodic depolarizer and is reduced firstly to nitrite and finally to ammonia.

Another constituent which has been implicated in the rapid detinning of canned orange drinks is monosodium fumarate. Kakizaki and Mori (1969) have found that the contents of cans containing this acidifying agent exceeded 150 p.p.m. of tin after a few months' storage. However, hydrogen evolution was less when monosodium fumarate rather than malic or citric acid was added to the product; these workers suggested that mono-sodium fumarate was acting as a depolarizer in the corrosion reaction.

Differences resulting in the use of plain or lacquered cans may affect flavour and colour of the product. In plain cans, bleaching of orange juice can occur due to the reducing action of tin, and a characteristic off-flavour may develop (Boyd and Peterson, 1945). In grapefruits, the reducing conditions prevailing inside a plain can have a bleaching effect which will counteract any yellowing of the fruit due to oxidation (Braverman, 1949).

Several workers have designed test cells to measure the electrical currents set up when a corrosion reaction occurs. Grapefruit juice has been selected by a number of these workers as a good electrolyte for their studies. Kamm *et al.* (1961) used grapefruit juice in a special cell for the alloy–tin–couple test (ATC) in which the current flowing between a tinplate sample detinned down to the alloy layer and a pure tin electrode was measured. The test was designed to give a rapid assessment of the performance of a tinplate specimen and so eliminate the time delay associated with long-term storage tests. Carter and Butler (1961) used the ATC test to study corrosion by grapefruit juice and found a good correlation between the ATC current and the shelf life of the pack.

REFERENCES

Anon (1962). "Chemistry and Technology of Citrus, Citrus Products and Byproducts." U.S.D.A., A.R.S., Washington, Agriculture Handbook No. 98.
Anon (1969). *Fd Inds Man.* (Woollen, A. H., ed.), 20th edition. Leonard Hill, London.
Atkins, C. D. and Rouse, A. H. (1954). *Fd Technol.* **8,** 498.
Atkins, C. D., Wiederhold, E. and Moore, E. L. (1945). *Fruit Prod. J.* **24,** (9) 260, 281.
Bakal, A. and Mannheim, H. C. (1966). *Israel J. Technol.* **4,** 262.
Bakal, A. and Mannheim, H. C. (1968). *Israel J. Technol.* **6,** 269.
Baker, R. A. and Bruemmer, J. H. (1968). *Proc. Fla St. hort. Soc.* **81,** 269.
Baker, R. A. and Bruemmer, J. H. (1970). *Citrus Ind.* **51** (1), 6.
Bauernfeind, J. C. (1958). *Symp. on Fruit Juice Concentrates* (*Bristol*), 265.
Bauernfeind, J. C., Osadca, M. and Bunnell, R. H. (1962). *Fd Technol.* **16** (8), 101.
Berry, R. E., Wagner, C. J. and Moshonas, M. G. (1967). *J. Fd Sci.* **32,** 75.
Blair, J. S., Godar, E. M., Masters, J. E. and Reister, D. W. (1952). *Fd Res.* **17,** 235.
Bomben, J. L., Kitson, J. A. and Morgan, A. I. (1966). *Fd Technol.* **20,** 1219.
Boyd, J. M. and Peterson, G. T. (1945). *Ind. Engng Chem.* **37,** 370.
Braverman, J. B. S. (1949). "Citrus Products." Interscience, New York.
Bunnell, R. H. (1968). *J. Agric. Fd Chem.* **16,** 177.
Carter, P. R. and Butler, T. J. (1961). *Corrosion* **17,** 94.
Chandler, B. V., Kefford, J. F. and Ziemelis, G. (1968). *J. Sci. Fd Agric.* **19,** 83.
Charley, V. L. S. (1963). *Fd Technol.* **17,** 987.
Clegg, K. M. (1964). *J. Sci. Fd Agric.* **15,** 878.
Clegg, K. M. (1966). *J. Sci. Fd Agric.* **17,** 546.
Clegg, K. M. and Morton, A. D. (1965). *J. Sci. Fd Agric.* **16,** 191.
Clements, C. A. B. and Anderson, L. (1966). *Ann. N.Y. Acad. Sci.* **136,** 339.
Clifcorn, L. E. (1948). *Adv. Fd Res.* **1,** 39.
Curl, A. L. (1947). *Canner* **105** (13), 14, 18.
Curl, A. L. (1949). *Fd Res.* **14,** 9.
Curl, A. L. and Bailey, G. F. (1954). *J. Agric. Fd Chem.* **2,** 685.
Curl, A. L. and Bailey, G. F. (1955). *Fd Res.* **20,** 371.
Curl, A. L. and Bailey, G. F. (1956). *J. Agric. Fd Chem.* **4,** 156.
Curl, A. L. and Veldhuis, M. K. (1947). *Fruit Prod. J.* **26,** 329, 342.
Dawson, C. R. (1966). *In* "The Biochemistry of Copper" (J. Persach, P. Aisen and W. E. Blundberg, eds). Academic Press, New York.
Dick, M. I. B., Harrison, I. T. and Farrer, K. T. H. (1948). *Aust. J. exp. Biol. med. Sci.* **26,** 239.
Diemair, W. and Postel, W. (1964). *Int. Fed. Fruit Juice Prod. Rept. No.* 5, *Vienna* , 135.

Ellis, G. P. (1959). *Adv. Carbohyd. Chem.* **14,** 84, 87.
Elstein (1969). Israeli Patent No. 30834.
Emerson, O. H. (1949). *Fd Technol.* **3,** 248.
Evenden, W. and Marsh, G. L. (1948). *Fd Res.* **13,** 244.
Farrer, K. T. H. and MacEwan, J. L. (1954). *Aust. J. Biol. Sci.* **7,** 73.
Feaster, J. F., Braun, O. G., Feister, D. W. and Alexander, P. E. (1950). *Fd Technol.* **4,** 190.
Fellers, C. R. (1929). *The Canner* **69** (18), 11.
Frankenthal, R. P., Carter, P. R. and Laubscher, A. N. (1959). *J. Agric. Fd Chem.* **7,** 441.
Frost, D. V. (1943). *Ind. Engng Chem. Analyt. edn,* **15,** 306.
Griffiths, F. P. and Lime, B. J. (1959). *Fd Technol.* **13,** 430.
Griffiths, F. P., Redman, G. H. and Lime, B. J. (1963). *Proc. of Citrus Processing Conference*, U.S.D.A., p. 14.
Guadagni, D. G., Bomben, J. L. and Mannheim, H. C. (1970). *J. Fd Sci.*, **35,** 279.
Hagen, R. E., Dunlap, W. J. and Wender, S. H. (1966). *J. Fd Sci.* **31,** 542.
Hamburger, J. J. and Joslyn, M. A. (1941). *Fd Res.* **6,** 599.
Hasegawa, E. (1955). *Vitamins* (*Japan*) **8,** 415.
Horio, T., Iwamoto, Y. and Oda, K. (1966). *Rep. Toyo Inst. Fd. Technol., Japan,* **7,** 19.
Horio, T., Iwamoto, Y. and Shiga, J. (1967). *C.I.P.C. 5th Int. Congress, Vienna,* Paper 11.
Horowitz, R. M. and Gentili, B. (1969). *J. Agric. Fd Chem.* **17,** 696.
Huelin, F. E. (1953). *Fd Res.* **18,** 633.
Huelin, F. E. (1958). *Indian Fd Packer* **12** (12), 11, 19.
Huelin, F. E. and Myee Stephens, I. (1944). *Fd Manuf. and Distrib.* (*Aust.*), Feb.
Huggart, R. L. and Barron, R. W. (1966). *Fd Technol.* **20,** 677.
Huggart, R. L. and Wenzel, F. W. (1955). *Fd Technol.* **9,** 27.
Huggart, R. L., Wenzel, F. W. and Moore, E. L. (1955). *Fd Technol.* **9,** 268.
Huggart, R. L., Wenzel, F. W. and Moore, E. L. (1957). *Fd Technol.* **11,** 638.
Huskins, C. W. and Swift, L. J. (1953). *Fd Res.* **18,** 360.
Inagaki, C. (1943). *J. agric. Chem. Soc., Japan* **19,** 451, 464.
Inagaki, C. (1944). *J. agric. Chem. Soc., Japan* **20,** 363.
Isaac, W. E. (1944). *Nature, Lond.* **154,** 269.
Ito, S., Izumi, Y. and Murata, N. (1969). *Canners' J., Japan* **48,** 225.
Ito, S., Murata, N. and Izumi, Y. (1968). *Bull. hort. Res. Stn, Okitsu, Japan* **B8,** 25.
Joslyn, M. A. (1937). *Fruit Prod. J.* **16,** 234.
Joslyn, M. A. and Sedky, A. (1940). *Fd Res.* **5**, 223.
Joslyn, M. A. and Pilnik, W. (1961). "The Orange, its Biochemistry and Physiology" (W. B. Sinclair, ed.) University of California, Riverside.
Joslyn, M. A., Marsh, G. L. and Morgan, A. F. (1934). *J. biol. Chem.* **105,** 17.
Kakizaki, K. and Mori, M. (1969). *Canners' J., Japan* **48,** 607.
Kamm, G. G., Wiley, A. R., Beese, R. E. and Krickl, J. L. (1961). *Corrosion* **17,** 106.
Kefford, J. F. (1959). *Adv. Fd Res.* **9,** 286.
Kefford, J. F. (1965). *Fd Pres. Quarterly* **25** (3), 41.
Kefford, J. F., McKenzie, H. A. and Thompson, P. C. O. (1959). *J. Sci. Fd Agric.* **10,** 51.
Kenyon, E. M. and Proctor, B. E. (1951). *Fd Res.* **16,** 365.
Kew, T. J. and Veldhuis, M. K. (1950). *Proc. Fla St. hort. Soc.*, **62,** 162,

Khan, S. A., Ahmad, M. and Khan, S. (1963). *West Pakistan J. agric. Res.* **1** (3), 27.
Kirchner, J. G. and Miller, J. M. (1953). *J. Agric. Fd Chem.* **1,** 512.
Kirchner, J. G., Miller, J. M., Rice, R. G., Keller, G. J. and Fox, M. M. (1953). *J. Agric. Fd Chem.* **1,** 510.
Krehl, W. A. and Cowgill, G. R. (1950). *Fd Res.* **15,** 179.
Lamb, F. C., Pressley, A. and Zuch, T. (1947). *Fd Res.* **12,** 273.
Levi, A., Samish, Z., Ludin, A. and Hershkowitz, E. (1969). *J. Fd Technol.* **4,** 179.
Lincoln, R. and McCay, C. M. (1945). *Fd Res.* **10,** 357.
Ludin, A., Samish, Z., Levi, A. and Hershkowitz, E. (1969). *J. Fd Technol.* **4,** 171.
McCance, R. A. and Widdowson, E. M. (1967). "Composition of Foods", 2nd imp. M.R.C. Published by H.M.S.O.
Maier, V. P. and Dreyer, D. L. (1965). *J. Fd Sci.* **30,** 874.
Maier, V. P. and Margileth, D. A. (1969). *Phytochem.* **8,** 243.
Maier, V. P., Hasegawa, S. and Hera, E. (1969). *Phytochem.* **8,** 405.
Mannheim, H. C. and Bakal, A. (1968). *Fd Technol.* **22,** 332.
Matsui, O. and Ito, S. (1953). *J. Utilization agric. Prod.* **1,** 4.
Mizrahi, S. and Berk, Z. (1970). *J. Sci. Fd Agric.* **21,** 251.
Moore, E. L., Atkins, C. D., Huggart, R. L. and MacDowell, L. G. (1951). *Citrus Ind.* **32,** (4), 5; **32,** (5), 8, 11, 14.
Moore, E. L., Wiederhold, E. and Atkins, C. D. (1949). *Fruit Prod. J.* **23,** 270, 285.
Nagy, S. and Nordby, H. E. (1970). *J. Agric. Fd Chem.* **18,** 593.
Natarajan, C. P. and MacKinney, G. (1949). *Fd Technol.* **3,** 373.
Nolte, A. J. and von Loesecke, H. W. (1940). *Fd Res.* **5,** 457.
Nolte, A. J., Pulley, G. N. and von Loesecke, H. W. (1942). *Fd Res.* **7,** 236.
Okada, S. and Ono, M. (1969). U.S. Patent No. 3,484,255.
Olsen, R. W., Huggart, R. L. and Asbell, D. M. (1951). *Fd Technol.* **5,** 530.
Pulley, G. N. and von Loesecke, H. W. (1939). *Ind. Engng Chem.* **31,** 1275.
Rauen, H. M., Devescovi, M. and Magnari, N. (1943). *Z. Untersuch. Lebensm* **85, 257.**
Reister, D. W., Braun, O. G. and Pearce, W. E. (1945). *Fd Inds.* **17,** 742.
Ribeiro, R. F., Bonoldi, V. and Ribeiro, O. F. (1942). *Rev. Faculdada Med. Vet. Univ. São Paulo (Brazil)* **2,** 23.
Rouse, A. H. and Atkins, C. D. (1952). *Fd Technol.* **6,** 291.
Rymal, K. S., Wolford, R. W., Ahmed, E. M. and Dennison, R. A. (1968). *Fd Technol.* **22,** 1592.
Sawayama, Z., Shimoda, Y. and Hayashi, H. (1964). *Rpt Toyo Inst. Fd Technol., Japan* **7,** 104.
Scott, W. C., Kew, T. J. and Veldhuis, M. K. (1965). *J. Fd Sci.* **30,** 833.
Senn, V. J. (1963). *J. Fd Sci.* **28,** 531.
Serra, G. and Perfetti, G. A. (1963). *Fd Technol.* **17,** 350.
Sheft, B. J., Griswold, R. M., Tarlowsky, E. and Halliday, E. G. (1949). *Ind. Engng Chem.* **41,** 144.
Shiga, I. and Kimura, K. (1956). *Rpt. Oriental Canning Instit., Japan* **4,** 55.
Shimoda, Y., Oku, M., Mori, D. and Sawayama, Z. (1968). *Rpt Toyo Inst. Fd Technol., Japan* **8,** 130.
Siddappa, G. S. and Bhatia, B. S. (1958). *Fd Sci. (Mysore)* **7,** 112.
Stevens, J. W., Pritchett, D. E. and Baier, W. E. (1950). *Fd Technol.* **4,** 469.
Stevenson, A. E. (1934). *Ind. Engng Chem.* **26,** 823.

Suryaprakasa Rao, P. V., Giridhar, N., Prasad, P. S. R. K. and Nageswara Rao, G. (1969a). *Indian Fd Packer* **23,** (1), 26.

Suryaprakasa Rao, P. V., Giridhar, N., Prasad, P. S. R. K. and Nageswara Rao, G. (1969b). *Indian Fd Packer* **23** (2), 42.

Suryaprakasa Rao, P. V., Giridhar, N., Prasad, P. S. R. K. and Nageswara Rao, G. (1969c). *Indian Fd Packer* **23** (3), 5.

Thomas, D. W., Smythe, C. V. and Labbee, M. D. (1958). *Fd Res.* **23,** 591.

Ting, S. V. (1958). *J. Agric. Fd Chem.* **6,** 546.

Ting, S. V. and Newhall, W. F. (1965). *J. Fd Sci.* **30,** 57.

Tressler, D. K. and Joslyn, M. A. (1954). *In* "The Chemistry and Technology of Fruit and Vegetable Juice Production", pp. 234, 388. AVI, New York.

Vandercook, C. E., Rolle, L. A., Postlmayr, H. L. and Atterberg, R. A. (1966). *J. Fd Sci.* **31,** 58.

Vines, H. M. and Oberbacer, M. F. (1963). *Pl. Physiol., Lancaster* **38,** 333.

von Loesecke, H. W. (1954). *In* "The Chemistry and Technology of Fruit and Vegetable Juice Production," (Tressler, D. K. and Joslyn, M. A., eds), p. 421. AVI, New York.

Weissberger, A. and LuValle, J. E. (1944). *J. Amer. chem. Soc.* **66,** 700.

Wenzel, F. W., Moore, E. L., Rouse, A. H. and Atkins, C. D. (1951). *Fd Technol.* **5,** 454.

Yufera, E. P., Mosse, J. K. and Royo-Iranzo, J. (1965). *Proc. 1st Internat. Congr. of Fd Sci. & Technol., London* **2,** 337.

Chapter 17

Fruit Juices

A. POLLARD AND C. F. TIMBERLAKE

Department of Agriculture and Horticulture, University of Bristol, England

I	Introduction	573
	A. Juice Production	573
	B. Types of Juice	574
II	Fruit and Juice Composition	575
	A. The Quality of the Fruit	575
	B. The Composition of the Juice	576
III	Processing Operations	579
	A. Juice Extraction	579
	B. Juice Treatments	586
	C. Storage of Juices	591
IV	Chemical Changes in Stored Juices	595
	A. Metals and Metallic Complexes in Fruit Juices	595
	B. Visual Appearance	605
	C. Stability of Ascorbic Acid	607
V	Conclusions	611
	References	612

I. INTRODUCTION

A. Juice Production

The conversion of fruits into juice was originally developed as a method for making use of supplies surplus to the fresh fruit market but, while it still fulfils this function, juice production is now firmly established in its own right. In the United States two-thirds of the citrus crop is processed, and elsewhere new plantings have been designed for juice production. The volume of juice made has continued to rise over the past 10 years and, as shown by the current review of the Commonwealth Secretariat (Anon., 1968), it was of the order of 1000 million gallons in 1966. An appreciable amount of this, possibly one-fifth, is converted into concentrates. The total does not include the volume of grape juice fermented to wine; this can exceed 6000 million gallons.

The United States is the chief world producer being, in some years, responsible for three-quarters of the total. Citrus juices account for over half the United States production, other main products being from pineapple, apple and grape. In Northern Europe and in Canada apple juice predominates, total world production being of the order of 200 million gallons. The manufacture of citrus and apple juices is now increasing in other fruit-growing areas in the Eastern Mediterranean and in the Southern Hemisphere; in the developing countries the use of tropical fruits is now receiving more attention.

The manufacture of juices and especially of their concentrates extends the season of fruit consumption and assists in equalizing supplies from one year to another: it has encouraged an increasing international trade in products for direct consumption and for manufacture. Juice concentrates and comminuted products from citrus are widely used as bases for the preparation of a wide range of fruit drinks, while apple and grape juice concentrates are being used increasingly for later fermentation. As a result of this expanding trade the question of product definition is now being considered by the FAO/Codex Alimentarius Commission, by the EEC and other international and national bodies.

The ten-fold rise in juice production in the past 30 years has run parallel with developments in technology that have drawn upon and stimulated advances in chemistry, microbiology and engineering. The methods of manufacture are now well established although undergoing continuous improvement. Developments and methods up to 1960 have been discussed in detail by Tressler and Joslyn (1961) and specific aspects have been described in later reviews and in numerous reports of Congresses and Symposia held by the International Federation of Fruit Juice Producers. The detailed biochemistry of individual fruits has been discussed in earlier chapters; the present account will be confined mainly to aspects of the subject that seem to the authors to be of significance at the present time.

B. Types of Juice

A fruit juice may be defined as the liquid expressed by pressure or other mechanical means from the edible portion of the fruit. It will frequently be turbid, containing cellular components in colloidal suspension with variable amounts of finely divided tissue. It may also contain oily or waxy material and carotenoid pigments derived from the skin or rind of the fruit. Some juices, for example orange juice, are consumed in a naturally cloudy state: clarification would impair the appearance and flavour of the juice. The juices of apples and of berries have traditionally been consumed in clear condition and the problem has been to clarify them effectively and to maintain them in a brilliant condition throughout their storage life. In recent years unclarified or "opalescent" juices have, however, also come into production.

The juices of some fruits lack much of the flavour and colour of the fruit itself, for these are associated with insoluble cellular material. Fruits rich in carotenoids, such as the apricot and the tomato, are immediate examples; they are more usually converted into pulpy products by disintegrating the whole fruit to give purees or nectars.

All types of juice are inherently unstable; they rapidly undergo microbiological attack by organisms already present on the fruit or gaining access to the product during processing; they are also subject to enzymic and non-enzymic changes. It is thus essential to destroy the micro-organisms at an early stage, or to prevent their development, and to restrict chemical change by heat treatments to inactivate enzymes or by refrigeration. The storage of juice in a stable condition involves the maintenance of large stocks of finished product in containers for distribution, or the installation of expensive refrigerated tankage. As a result, increasing volumes of juice are now converted into concentrates that are both more stable and require less storage space. The concentrates may be diluted for further processing as juice or as fermented products, or they may be distributed unchanged through the frozen-food chain.

The range of products is particularly wide in the citrus field, as described by Royo-Iranzo (1965) in a review of current methods of processing. They include not only the conventional pasteurized single-strength juices in cans for direct consumption, but also chilled juices with a short storage life and frozen concentrates for dilution. In addition, both single strength and concentrated juices are distributed in larger containers as bases for the manufacture of a variety of soft drinks. They may be used in conjunction with comminuted products containing albedo and flavedo components. Dehydrated juices in powder form are also on the market.

II. FRUIT AND JUICE COMPOSITION

A. The Quality of the Fruit

Fruit intended for juice production should be of quality comparable with that suitable for direct consumption. It should be adequately ripe, with a suitable balance of acid and sugar, and with fully developed aroma and flavour. In practice, large bulks of fruit are rarely uniform in quality; washing and sorting treatments are needed to remove specimens showing damage or microbiological attack. It may not always be possible to store fruit to optimum maturity for, as discussed by Lüthi (1968), in apples internal breakdown may give rise to the formation of excessive amounts of ethanol and other metabolic products. As shown by Guinot (1967), unsuitable storage conditions lead to rotting with the production of abnormal levels of methanol (from pectin), ethanol, acetoin and volatile acids. Certain *Acetomonas* species infecting apples produce 2,5-D-threo-diketohexose (5-ketofructose) and other carbonyl compounds that combine with sulphur dioxide and render it ineffective for microbiological control when it is added to juices intended for fermentation (Carr *et al.*, 1963; Burroughs, 1964).

The amounts of such metabolites, in particular of acetoin and diacetyl, in citrus juices have been used as criteria of fruit quality and process sanitation (Sharf, 1967; Murdock, 1968). Similarly in apple juices the amounts of alcohol and acetoin have been used as quality indices (Zubekis, 1965; Hill and Fields, 1966). The microbial flora of fruits, notably apples and grapes, have mainly been studied in relation to their use for fermented products. The general microbiology of fruit juices has been reviewed by Ingram and Lüthi (1961) and it will be only briefly touched on in this chapter.

B. The Composition of the Juice

1. *Fruit composition and juice quality*

The many factors influencing fruit composition have been discussed in earlier chapters; they include the effects of genetic make-up, fruit maturity, cultural and nutritional conditions, climate and weather. The composition of a particular type of juice may thus show wide variations according to the provenance of the fruit: this raises problems in defining norms for juice products in international trade. The variations encountered with citrus products have been discussed by Koch and Sajak (1964) and by Primo-Yufera and Royo-Iranzo (1967). Many studies of the composition of fruit juices have been directed to the detection of adulteration of commercial juice products, particularly from citrus fruits (Sawyer, 1963; Stanley, 1965; Primo-Yufera and Royo-Iranzo, 1967). Betaine content has been one criterion of composition (Lewis, 1966); some others remain unpublished as this would invite further sophistication.

The suitability of a particular species or cultivar for juice is dependent upon the balance of acid and sugar, the type and amount of phenolic components, the aroma constituents and the amounts of vitamins present, in particular, of ascorbic acid. Deficiencies or excesses of one or more components may be overcome by blending, but it is desirable that bulks of fruit for manufacture should be of suitable and uniform composition. The suitability of citrus cultivars, species and rootstocks is related, not only to sugar and acid content, but also to the amounts of bitter flavanones passing into the juice, notably naringin in grapefruit and limonin in oranges (Sections IV 3, V 3). Methods for their removal have been suggested, by adsorption on polyamides (Chandler *et al.*, 1968) or by enzymic means, the bitter naringin being converted to the non-bitter aglycone naringenin (Olsen and Hill, 1965).

It has been found recently that the bitterness developing in the juices of early season Navel orange cultivars is the result of secondary changes following juice extraction (Maier and Beverly, 1968). In the carpellary membranes and albedo the limonin is present as the mono-lactone which is not bitter; in the juice under the acid conditions present, this undergoes conversion into the bitter di-lactone of limonin. The transformation is accelerated by heat treatments. The problem can thus be overcome if excessive maceration of the tissues is prevented during juice extraction and if the coarser fractions are eliminated from the juice.

The citrus flavanones can be utilized as by-products of juice manufacture; for example, hesperidin which often separates out during juice concentration. Others may serve as intermediates for further synthesis. As described by Horowitz (1964), the dihydrochalcones of naringin and of some other flavanones are intensely sweet and Krbechek *et al.* (1968) have shown that naturally occurring flavanones can be used as a starting material for their preparation.

The suitability of apple cultivars for juice is largely dependent upon their content of acid and sugar, which is, in turn, influenced by post-harvest changes. The fall in acid during storage may sometimes be beneficial but, as described later, it is accompanied by pectin changes that give rise to problems in processing. The types of apple grown in France and England for fermentation to cider are usually unsuitable for other juice products because of their high content of phenolics which may be ten times that of dessert apples (Burroughs, 1962). These increase the tendency to enzymic browning during juice extraction, and the polymers of the catechins and leucoanthocyanins contribute excessive bitterness and astringency to the juice.

As described later, the astringent leucoanthocyanins of some perry pears also limit the uses of the fruit. With some other fruits the problem of excessive "tannin" character can be overcome by careful methods of juice extraction as in the pomegranate where these components are associated with structural tissues. The astringency of the persimmon can be reduced by modifying the metabolism of the fruit by interrupting the process of respiration (Nakayama and Chichester, 1963). The subject of astringent phenolics in fruits and their products has been reviewed by Joslyn and Goldstein (1964).

Some fruit cultivars are especially valuable for juice products since they contain desirable and characteristic aroma components such as the esters of *trans*:2-*cis*:4-decanoic acid present in the Bartlett (William) pear (Jennings *et al.*, 1964). It has often been the practice to blend such fruits with others lacking in character, but the increasing availability of aroma concentrates now offers the possibility of adding aroma to enhance flavour when legislation permits. Lüthi (1968) has suggested that the emanations of fruit in store could be collected and used in this way.

2. *Components of special interest*

Many juices are consumed primarily as beverages, others for nutritional reasons, especially as sources of ascorbic acid (see Volume 1, Chapter 13). Although different types of fruit fall within fairly well-defined categories as sources of ascorbic acid, the range of values can be wide. In blackcurrants, extensively used in Europe for vitamin products, the range extends from about 100 mg to over 250 mg ascorbic acid/100 g; as a result the low-level cultivars are being little propagated. The values for any one cultivar also vary according to maturity, climate and nutritional factors (Chapter 11). Other fruits may show a still wider range of ascorbic acid content: in rose hips, called into use during the last war, the variation is over five-fold, and in the guava the variation may be more than ten-fold.

Fruits contain a wide range of other vitamins but, with the exception of those containing carotenoids, their contribution to the dietary requirements is low in comparison with their function as sources of ascorbic acid. The amounts of some components may, however, represent several per cent of the daily requirement, for example, of thiamine in orange and pineapple, of riboflavin and nicotinic acid in apricot, peach and blackcurrant, and of pantothenic acid and biotin in many of the common fruits (McCance and Widdowson, 1960a). The amounts of these vitamins present in the juice will vary according to processing methods and heat treatments, but the values found for grape juice by Schneyder (1964) were similar to those found in fresh grapes. The values for wine were similar, suggesting that where a fermented juice beverage is consumed in some quantity the intake of vitamins may be of importance; a similar conclusion was drawn for cider by Jacquet and Le Breton (1966).

The presence of biologically active flavonoids in fruits and juice products has aroused interest for more than 30 years and, although the term Vitamin P is largely obsolete, interest in the nutritional value of this class of compound has persisted. A symposium on the pharmacology of plant phenolics was held in 1958 when the effects of flavonoids on capillary structure (Lockett, 1959) and on vascular resistance (Lavollay and Neumann, 1959) were discussed (see Volume 1, Chapter 11). More recent work has been summarized by Charley (1966) showing that, although a specific dietary requirement for such phenolics is yet to be demonstrated, they appear to have an effect in diminishing capillary fragility which may aid recovery from bruising, as in subjects such as baseball players. The function of fruit phenolics as inhibitors of oxidation in juice systems is discussed later in this chapter.

III. PROCESSING OPERATIONS

A. Juice Extraction

1. *Methods of juice extraction*

(a) *Berry fruits.* Juice has been extracted from grapes, and wine from their fermented pulps, by simple means since ancient times. The method has been essentially the enclosure of the pulp in a cloth, basket or other perforated container, and the application of mechanical pressure. Juice has similarly been pressed from hard fruits after they have been disintegrated by milling or grinding. Until the last decade, this general principle has been mainly followed, and improvements in press design have been directed to increasing the efficiency of the process. In recent years, however, labour difficulties have led to the development of a wider range of equipment capable of continuous operation or of some degree of automation.

In some of the newer types the pulp is pressed continuously between perforated rollers or in bands of permeable material. Others use a screw to press the pulp against an outer perforated casing, or use centrifugal force to separate juice and tissue. Automated cylindrical piston presses with internal drainage tubes are also widely used.

Detailed accounts of methods of juice extraction and of the equipment have been given in a number of reviews, for example, Celmer (1961); Schaller (1965); Lüthi (1965) and Swindells and Robbins (1966). The basic principles of the extraction of juice from fruit pulps have also been investigated, notably in a mathematical treatment by Körmendy (1964, 1965). The process of extraction by diffusion has similarly been studied in detail by Ott (1965).

The traditional method of extraction is the hydraulic pressing of a series of layers of pulp enclosed in cloth and separated by racks. Although this is laborious the pulp is not exposed to the air during pressing and the juice obtained is relatively clear, for the mass of fruit tissue itself acts as a filtering medium. By contrast, in the procedures where the pulp is mechanically agitated and forced through screens, as in some of the newer types of press, oxidation can be considerable; moreover, the juice often contains fine suspended tissue thus introducing oxidizing enzymes associated with the cellular debris.

In the United States there is a long-established manufacture of juice from the Concord grape owing much of its flavour to the presence of methyl anthranilate. In Europe production is mainly from *Vinifera* cultivars although use is also made of hybrids derived from crosses with American species; the latter are not considered suitable for wine.

Juice is readily extracted from grapes for the juice drains from the crushed fruit, and the skins and stems themselves act as a press-aid. The pulps of pigmented grapes such as Concord are usually heated before pressing to extract anthocyanins from the skin, the treatment being regulated to avoid excessive extraction of other components of the tannin fraction. The amounts of other extractives, including pectin, are also increased and heat-resistant pectin enzyme preparations are added to the pulp before pressing. Light-coloured grapes are pressed without heating; this gives a fresher flavour provided that enzymic oxidation can be prevented. Ascorbic acid may be added as an oxidation inhibitor, either making the addition to the fruit as pressed or to the juice.

Although crushed grapes and the pulps of apples and pears readily yield juice when subjected to pressure, many berry fruits such as currants, strawberries and raspberries give highly pectinous pulps, from which little juice drains away. Whereas apple juices usually contain less than 0·1% of soluble pectin, berry juices may contain up to five times this amount. If such fruits are heated, the viscosity is lowered and some juice can be extracted, but the customary, more effective method is to treat the milled pulp with pectin-degrading enzymes to render the mass fluid; juice is then readily extracted under pressure (Robbins, 1968). Blackcurrants are often depectinized at an elevated temperature (45°C) to give a more rapid breakdown of pectin and greater extraction of pigment and other solubles (Koch, 1955). A preliminary flash-heating to a higher temperature (85°C) has been found advantageous, giving a higher separation of ascorbic acid and anthocyanins in the juice (Wucherpfennig and Bretthauer, 1966).

The subject of pectin degradation is discussed in greater detail later.

(b) *Pome fruits*. Apples are still pressed mainly in the traditional rack and cloth hydraulic presses but newer types are now coming into use. One of the problems here is the extensive oxidation occurring in screw presses and centrifuges; the use of a nitrogen atmosphere within the machine has been suggested. The efficiency of juice extraction is influenced by the degree and type of disintegration of the apples and this is a subject calling for further investigation. Subjecting the pulp to high frequency vibration has been found to increase the proportion of ruptured cells and to increase juice yield (Flaumenbaum and Sejtpaeva, 1965), but the method raises problems in commercial operation.

Stored apples become increasingly difficult to press as the period following harvest increases. The detailed changes in the protopectin of the cell wall have been discussed by Joslyn (1962). From the point of view of juice extraction the main effect is that the soluble pectin rises and the pulp softens and becomes increasingly viscous and impossible to press without the use of fillers or filter aids. The loss of structural pectin from the cell wall material leads to disintegration and softening of the tissue structure so that the pulp loses the rigidity required in pressing. The weakening of the structure may follow the migration of calcium or other divalent cations from the cell wall rather than demethylation of the pectin by the pectinesterase of the fruit, for this would have the reverse effect by increasing the carboxyl linkages available (Deuel and Stutz, 1958; Neal, 1965). The use of calcium salts for firming apple tissues is well established in the food industries and the addition of lime appears to have been traditional in the pressing of over-ripe apples for cider intended for distillation.

Commercial processes for the extraction of juice from pears follow methods essentially similar to those used for apples. One difference, however, is that pears contain less intercellular gas and are denser than water; they cannot be washed by flotation. The rapid onset of ripening also causes problems, for dessert pears when ripe give a slimy pectinous pulp difficult to press. Such pears are not used extensively for juice, a main product being pulpy nectars, although the manufacture of liqueurs and aroma concentrates from the juices of highly flavoured cultivars such as Bartlett (Williams Bon Chretien) is becoming important.

The largest volume of pear juice and juice concentrates is now obtained from different types of pear originally grown in Europe for fermentation to perry or distilled spirits. The history and use of these pears have been the subject of a monograph (Luckwill and Pollard, 1963). Such pears have a granular texture due to the presence of lignified stone cells and are readily pressed. A point of special interest is the high content of leucoanthocyanin in the astringent cultivars; it can amount to 1% of the juice and appears to be a polymer of 5,7,3′,4′-tetrahydroxyflavan-3,4-diol (Kieser *et al.*, 1953). This substance passes into the juice where it undergoes further polymerization to give voluminous gelatinous precipitates, but the process is usually incomplete and gives rise to hazes and deposits in fermented perries; the astringency falls as the leucoanthocyanin is precipitated. In earlier days, the slow process of grinding the fruit in a stone mill rendered much of the leucoanthocyanin insoluble. In more modern methods of juice extraction the same effect can be produced by allowing the pulp to stand before pressing; both oxidation and precipitation with proteins appear to be concerned.

(c) *Citrus fruit*. The main citrus juices in production are from orange and grapefruit and the literature on the subject is extensive. Current production methods for orange juice have been described by Veldhuis (1961) and for grapefruit juice by Burdick (1961).

Juice extraction involves the collection of the juice contained within the interior sacs without at the same time incorporating excessive amounts of the albedo of the peel or the inner pith and vascular materials that contribute to bitterness and increase the pectinesterase in the juice. The juice sacs contain oil but the main source of oil in the juice is derived from the outer rind or flavedo; the amount passing into the juice must be restricted. Juice is extracted by reaming the halved fruits or by compressing the whole fruit on to a plunger in the form of a circular knife. The oil from the disrupted peel cells runs down the outside of the fruit and is collected as an emulsion. An alternative method is to rasp the peel for oil collection before the juice extraction. The composition of the juice may vary considerably according to the method of extraction used (Danziger and Mannheim, 1967).

Similar methods of juice extraction are used for grapefruit and lemons, but lemons may also be crushed and passed through a screw press, the juice and oil emulsion then being strained from the solid matter. Limes are more usually crushed and strained and the juice fraction allowed to clarify spontaneously, the final product being clear rather than turbid.

The past 15 years have seen the development of comminuted citrus products made by disintegrating the whole fruit or selected portions of it. These contain both juice and oil with a variable amount of tissue in fine suspension, the system being stabilized by passage through a homogenizer. The development and uses of such comminuted products for the manufacture of citrus drinks and squashes has been described by Charley (1963). Although these products contain relatively large amounts of peel oil, the off-flavours associated with the oxidation of terpenes in juice do not develop and it is assumed that the solid tissues contribute natural anti-oxidants. The phenolics may restrict non-enzymatic oxidation by chelating metal ions, but other components present in citrus oil, possibly related to the tocopherols, have also been implicated (Ting and Newhall, 1965).

(d) *Tropical fruits*. The methods used for extracting juice or pulps from tropical fruits vary widely according to their structure and composition. The most important juice fruit is the pineapple and the methods of production have been described by Mehrlich (1961). The extraction of juice is often combined with the manufacture of canned fruit packs and makes use of material that would otherwise be discarded. The juice is separated from the solid tissues by screw-extractors, filters or centrifugal machines.

The passion fruit is now taking on an increasing importance as a source of juice. The fruit is small and has a hard rind, so that juice extraction presents problems; centrifugation of the cut fruit is one method adopted. Further problems are the presence of starch and the heat-sensitivity of the aroma components (Pruthi, 1963; Charley, 1968). Unless the amount of starch can be reduced by centrifugal or other means following juice extraction, the formation of gels makes subsequent preservation by heat treatment difficult. Preservation of the juice by freezing overcomes both these problems.

The present and potential uses of tropical fruits in the developing countries have also been discussed by Charley (1968) and Seale (1967). Although many have attractive flavours, the aroma components tend to be heat sensitive and the fruits themselves are often of inconvenient shape or size; they also present problems in transport and storage. Use is already made of some of these fruits, as, for example, guava, mango, papaya, banana and acerola, but more usually for pulpy products than for juices. The more highly flavoured fruits can be used in blends with others and this can overcome the problem of sterilizing products made from low-acid fruits with a pH near 4·5. The acerola (*Malpighia glabra* L.) is of interest by reason of its high ascorbic acid content which varies from 1·4% to 3·5%; its chief use is as an additive in other juice products. The banana has as yet been little used for liquid products but the possibilities have been explored (Dupaigne, 1964). The products may either be pulpy suspensions or clear juices prepared by depectinization of the pulp.

2. *Changes occurring during juice extraction*

The process of milling fruit disorganizes the cellular structure and brings enzymes normally associated with structural components, or otherwise segregated, into contact with soluble substrates and with oxygen from the air or from the intercellular spaces (Smock and Neubert, 1950). Fruits rich in phenolases and their substrates, such as apples, rapidly brown when the tissues are disintegrated and this continues in the extracted juice, especially when it contains tissue particles in suspension. A clear juice extracted by rapid compression of a whole apple, or by a process involving rapid filtration (Gottauf and Duden, 1965), can retain its unoxidized pale colour for a long period.

Other changes that may occur are the oxidation of ascorbic acid, where suitable enzyme systems are present, and other oxidative changes that can affect quality and flavour. Pectin changes may also occur, although these are more usually associated with later stages of processing and will be discussed under the heading of "Juice Treatments".

Some changes in the appearance and flavour of fruit juices are inevitable during manufacture and some have come to be preferred by the consumer as, for example, intensification of colour and flavour. Such changes are, however, not readily halted at an optimum stage and are often excessive or lead to secondary changes that are definitely deleterious. The present trend is towards the restriction of enzymic and other changes during processing and to retain the original character of the fruit.

(a) *The oxidation of phenolics.*† The phenolase enzymes of apples have been the subject of much study, and although the following discussion will be devoted mainly to this fruit, many of the findings are of wider application. The members of the phenolase complex in apples may differ in their properties (Walker and Hulme, 1966; Constantinides and Bedford, 1967), but their main

† See also Volume 1, Chapter 11.

effect is the oxidative browning of (+)-catechin, (−)-epicatechin and chlorogenic acid, although other phenolics may also be involved (Van Buren *et al.*, 1965). The capacity for undergoing oxidation also depends upon the structure of the phenolic. Glycosidation may restrict the oxidation of flavonols by phenolases; for example, quercetrin is neither a substrate for apple phenolase (Shannon and Pratt, 1967) nor for potato phenolase (Baruah and Swain, 1959). The aglycone quercetin is, however, a substrate for the potato enzyme. According to Mayer (1962), the aglycones may only function as a substrate if capable of undergoing keto-enol tautomerism.

The oxidation of the phenolics leads in turn, by reaction with the intermediate quinones, to the oxidation of ascorbic acid; it is rare to find measurable amounts in a normally extracted apple juice although some varieties are relatively rich in ascorbic acid (Pollard *et al.*, 1946). The phenolics differ in their ability to react in this manner and the subject has been reviewed by Monties (1966).

The quinones formed by phenolase oxidation can react, not merely with ascorbic acid, but with a range of other juice components. As discussed by Swain (1965), these may include compounds with amino or sulphydryl groups to give coloured or insoluble products. The quinone produced from one particular phenol can also undergo coupled oxidation with another phenol of lower oxidation–reduction potential; this may be the mechanism leading to the decoloration of anthocyanins which are themselves poor substrates for phenolase (Peng and Markakis, 1963). However, according to other workers (Sakamura *et al.*, 1966) some phenolases may oxidize anthocyanins preferentially.

(b) *The inhibition of phenolase activity.* The oxidation of apple juice phenolics can be restricted by chilling the fruit, by extracting the juice in an inert atmosphere or in the presence of added ascorbic acid or sulphur dioxide. An effective method used in the manufacture of cloudy or "opalescent" apple juice has been to add ascorbic acid to the juice as soon as pressed at the rate of about 0·5 g/litre. This bleaches the juice by reduction of the quinones already formed and retards oxidation sufficiently to allow time for flash-pasteurization and permanent inactivation of the enzymes; the loss of ascorbic acid is relatively small (Kieser *et al.*, 1960). The current methods of production have been described by Moyls (1964) and by Ruck and Kitson (1965).

Although ascorbic acid has been found to inactivate the phenolase of potato (Baruah and Swain, 1953), apple (Täufel and Voigt, 1964) and tobacco leaf (Pierpont, 1966), no inhibition has been reported for grapes (Bayer *et al.*, 1957) or for apple or potato by other workers (Duden and Siddiqui, 1966). In the commercial production of apple juice the main restriction of oxidation is almost certainly the depletion of available oxygen if the juice remains unagitated. As shown by Burroughs (1968), the dissolved oxygen in a freshly pressed juice can fall from 4·1 p.p.m. to 0·3 p.p.m. in 15 min.

Daepp (1964) found that the de-aeration of apple juices offers little protection against oxidation since this occurs during the extraction process. In citrus juices the uptake of oxygen is much less rapid (Lüthi, 1955), and vacuum de-aeration or nitrogen sparging can be used. As pointed out by Kefford (1964), vacuum de-aeration, often combined with the removal of excess peel oil, effectively reduces oxidation in citrus juices and is of value where they are to be bottled or frozen; in cans, any oxygen present is rapidly taken up at the tin surface. The use of glucose oxidase has been suggested for removing oxygen in the later stages of juice manufacture, but, unfortunately, juices are usually filled hot into the final container or pasteurized as soon as filled. There is thus little opportunity for the enzyme to be effective before it is inactivated.

Although the use of sulphur dioxide in foodstuffs is undergoing reappraisal from the point of view of nutrition, it is still used extensively in foods to inhibit oxidation and browning. It has also been added as a preservative for the bulk transport of juices that are later desulphited or used as bases for manufacturing purposes. Its main use has been for the suppression of undesirable micro-organisms prior to the yeast fermentation of wines and ciders.

The anti-oxidant effect of sulphur dioxide in fruit juices lies mainly in its inactivation of phenolase (Embs and Markakis, 1965). It also reacts with the intermediate quinones, but to a lesser extent than does ascorbic acid.

Phenolase activity can also be inhibited by copper chelating agents such as 1-phenyl-2-thiourea or diethyl dithiocarbamates (Knapp, 1965) or by polyvinylpyrrolidone acting as a competitive inhibitor (Walker and Hulme, 1965). N-vinyl-2-pyrrolidone and 2,3-naphthalenediol have been suggested as inhibitors of browning for apple slices (Harel *et al.*, 1966). Cysteine at the rate of 16–24 mg/100 ml was found to inhibit browning in apple juices (Walker and Reddish, 1964) but neither cysteine nor the other inhibitors have as yet been commercially adopted in juice production.

Alternative methods for inhibiting the oxidation of phenolics have been put forward by Finkle and co-workers, involving the modification of the substrate itself. One is the enzymic methylation of the 3-hydroxyl group of 3,4-dihydroxy phenols; the other is the cleavage of the aromatic ring (see also Volume 1, Chapter 8). It was shown by Finkle and Nelson (1963a) that methylation of a phenolic by a m-*O*-methyltransferase of animal origin, in the presence of a methyl donor, gave a product no longer susceptible to oxidation by phenolase. A similar enzyme was demonstrated in plant tissue and by its action caffeic acid was converted to the 3-*O*-methyl derivative, ferulic acid (Finkle and Nelson, 1963b). Such treatments at pH 8 effectively prevented the oxidative browning of apple juice or apple tissues (see Chapter

18). A further method suggested for the prevention of oxidative browning in apple juice is the enzymatic cleavage of the aromatic ring of the phenolic substrate (see Finkle, 1967, and Chapter 18).

B. Juice Treatments

1. *Clarification*

A freshly pressed juice contains variable amounts of fine cellular debris with colloidal material, pectic substances, gums, proteins and other components. The system is initially stabilized by hydration of the particles and by their electrical charges or those due to adsorbed ions, and by the presence of soluble pectin. Spontaneous clarification usually ensues following the formation of protein–tannin complexes, insoluble pectates, or multiple changes giving precipitates containing suspended material and a range of juice components.

Although the colloidal systems in juices are inherently unstable, natural clarification is usually slow and it is necessary to accelerate the process if a clear juice is required. Direct centrifugation or filtration is sometimes possible but is often uneconomic; the presence of soluble pectin renders the juice viscous and filtration is slow. Moreover, unless the juice has been stabilized by heat, further changes leading to loss of clarity are probable. It is usual to subject the juice to fining procedures or to depectinization. The juice can then be filtered and stabilized by heat treatment.

(a) *Stabilization*. If a juice is to remain turbid it is essential that spontaneous clarification be prevented. Examples of such products are citrus and tomato juices; in these, as discussed later, the juice is given an effective heat treatment at an early stage of processing to inactivate the enzymes present. The alternative method of low temperature storage restricts enzymatic changes but requires that the product be kept at such temperatures throughout all stages of processing and distribution. It is thus of more limited application, and usually follows heat stabilization rather than replaces it.

(b) *Fining*. The filtration of cloudy juices can often be made possible if the suspended matter is removed by a fining process. In this a substance carrying a charge opposite to that of the stabilizing colloids or capable of reacting with one of the components is added to the juice. This provokes the formation of a precipitate that carries down with it other material in suspension. Gelatin is commonly used in this way to combine with juice phenolics to form insoluble complexes, the optimum amount being found by preliminary test since an excess stabilizes the system. Its effectiveness is dependent upon a number of factors including temperature, pH and the concentration of ferric and other ions. Synthetic polyamides could similarly be used as discussed by Wucherpfennig and Bretthauer (1962).

In some juices, the removal of protein by the addition of bentonite is a more effective stabilizing treatment. Its main use, however, is in the removal of proteins responsible for after-clouding of wines in bottle or following heat treatment. The nature of the grape proteins has been studied by Koch, Diemair and co-workers using electrophoresis and ultra-centrifugation and they have also reviewed the field (Diemair *et al.*, 1962).

2. *Enzymatic changes*

(a) *Pectin enzymes.*† The changes leading to the degradation of pectin in juices and other fruit products have stimulated the detailed study of pectin and of pectin enzymes. The great expansion of the citrus juice industry in the period 1940–50 raised the problem of the stabilization of the juice cloud that was necessary if the appearance and flavour of juices were to be preserved and if the gelling of concentrates was to be prevented. Similarly, developments in apple, grape and berry juices focused attention on the problems of juice extraction and clarification, with the use of commercial enzymes.

In orange juices the main problem was the efficient heat inactivation of pectinesterase at an early stage of processing, and time–temperature relations were worked out (Rouse and Atkins, 1952). As a result, heat treatment is now included in the processing sequence, even when the product is subsequently stabilized by low temperature storage. This prevents the demethylation of the pectin and its precipitation as pectate gels, and maintains the cloud in suspension, the latter consisting mainly of pectinous and lipid material derived from the juice sacs (Scott *et al.*, 1965). In contrast, spontaneous clarification of lime juice is allowed to proceed since clear juices are mainly required.

In tomatoes, pectin degradation can be particularly rapid, for in addition to pectinesterase, an active polygalacturonase is present (Hobson, 1962, 1964). The breakdown of pectin with losses of juice viscosity and consistency readily occurs in juice manufacture in the period between disintegration of the tissues and heat inactivation of the enzymes. This problem has been studied by Wagner *et al.* (1968) who have proposed the acidification of the tissue with hydrochloric acid to pH values below 2·5 followed by readjustment with sodium hydroxide when the product has been stabilized by heat. This inactivates the pectinesterase and partially inactivates the polygalacturonase which is also indirectly suppressed by the inhibition of pectin demethylation. The presence of sodium chloride is no disadvantage, for it is normally added to tomato juice. While polygalacturonase activity has been demonstrated in relatively few fruits, notably tomato, pear, pineapple and avocado (Hobson, 1962), pectinesterase is of more common occurrence; the

† See also Volume 1, Chapter 3.

literature has been reviewed by Vas *et al.* (1967) and by Leuprecht and Schaller (1968). In many apple cultivars its activity is low, 1% or less of the activity of tomato, but its activity is measurable (Pollard and Kieser, 1951), and although its optimum activity at pH 6·6 is far above that of juice, its effects in juices can still be demonstrated. In the types of apple used for the manufacture of unfermented juice, the pectinesterase activity is inadequate to produce an effective degradation of pectin. As discussed later, it is usual to add pectin enzyme preparations from mould sources.

In contrast, many of the apples grown extensively in Europe for fermentation to cider are much higher in pectinesterase and the enzyme plays a major part in the natural clarification of ciders made by the traditional methods (Jacquin, 1955). The juice enzyme, acting in conjunction with polygalacturonase produced by members of the yeast micro-flora, effectively degrade the pectin present (Pollard and Kieser, 1959). The properties of the pectinesterase of cider apples have been used empirically for many years, both in England and more especially in France, for the clarification of juices at an early stage of fermentation. By allowing the milled pulp to stand for a period before pressing (maceration) the amounts of pectin and of the enzyme passing into the juice increase. On standing for some days a calcium pectate gel separates from the juice, carrying with it yeasts and suspended material, the enzyme action being hastened by the addition of calcium and sodium ions. The main purpose of this "défécation" procedure is the removal of nitrogen as amino acids by the proliferating yeasts, making the liquor less fermentable and suitable for the production of naturally sweet ciders (Jacquin, 1955; Beech, 1958). The process is difficult to control precisely and is now going out of use.

The enzymic pectin changes in grape juices are similar to those in apple juices, pectinesterase hydrolysing pectin with the liberation of methanol (Marteau *et al.*, 1961). On fermentation to wine, polygalacturonase enzymes from yeast again continue the degradation of pectin. In a later study of natural clarification of apple and grape juices, Marteau (1967) observed that a marked fall in viscosity preceded the sedimentation of suspended matter in grape juice and he attributed this to the presence of polygalacturonase in the fruit. Although Hobson (1962) could not detect this enzyme in the grapes studied, its presence in other cultivars is not excluded. In the commercial production of grape juices the addition of fungal enzymes for rapid clarification is now usual.

The depectinization of fruit pulps to facilitate juice extraction, and of juices for clarification, is carried out routinely by adding commercial enzymes of fungal origin. These contain varying proportions of pectinesterase, polygalacturonase and pectin-*trans*-eliminase enzymes and the different preparations are designed for specific purposes. Some are intended for the clarification of juices low in pectin, others for depectinizing highly viscous fruit pulps at ambient temperatures or in the region of 45°C or higher (Koch, 1955); for the latter purpose enzymes showing suitable heat-stability are chosen. The classification of the pectin-degrading enzymes has been discussed by Nyiri (1968) in a review of methods for their production. Other recent work on the depectinization of pulps and juices has been reviewed by Reed (1966).

Although pectin can be efficiently removed in juice extraction and clarification treatments by the choice of suitable enzyme preparations, the process still remains to some extent empirical. The presence of pectinesterase in the fruit itself may assist degradation by the added enzymes, or it may interfere by rendering pectin insoluble as pectate gels; such gels may in turn adsorb enzymes and remove them from the sphere of operation (Pollard and Kieser, 1959). Reactions of the enzymes with juice phenolics may also lead to enzyme inactivation by largely non-specific interactions between oxidized phenolics and enzyme protein. The subject was investigated by a number of workers some 10 years ago in relation, not only to enzyme inactivation in juice systems, but to the mechanism of the resistance shown by some fruit to invasion by fungal pathogens that secrete pectolytic enzymes (Byrde *et al.*, 1960). In juice systems, a leucoanthocyanin was found to reduce markedly the activity of apple pectinesterase and of yeast polygalacturonase (Pollard *et al.*, 1958).

One of the recent developments in this field is the production of the so-called protopectinase enzymes that attack cell wall material rather than soluble pectin. They can be used to disintegrate fruit tissues in the manufacture of viscous pulpy products, an advantage being that the cells themselves suffer less disruption with a consequent reduction of enzymatic browning (Pilnik, 1969). A further development, possibly of wide application, is the production of insoluble enzymes, where the active enzyme is coupled to a support such as a cellulose derivative or a polyaminoacid. The subject is reviewed by Kay (1968), who lists some possible future uses of such preparations. These may eventually include the depectinization of fruit juice products, the removal of oxygen by glucose oxidase or the breakdown of undesirable protein.

(b) *Other enzymes.* A wide range of enzymes may pass into a juice from the fruit, especially where the cellular tissues are finely disintegrated. Some have an immediate effect upon the character of the juice when they are active during the process of extraction, as for example phenoloxidase. The activity of others is restricted by the low pH of the juice and, in many acid fruits, enzyme changes have not given rise to obvious problems. The increasing adoption of heat treatments at an early stage of processing has also helped to restrict deterioration arising from enzyme activity. The production of juices from the less acid fruits that have so far been little utilized may, however, present difficulties. The presence of enzymes in fruits may have significance other than in juice processing *per se*. Pineapples and papayas provide a valuable source of the proteolytic enzymes bromelin and papain which are used as meat tenderizers. These fruits do, however, give rise to problems in the preparation of their products since the enzymes attack the hands of the operatives. The action of papain in the handling of papayas has been overcome by the use of barrier creams containing zinc peroxide, for both oxygen and heavy metals act as inhibitors of the enzyme (Seale, 1967).

The extensive study of fruit aromas in recent years has again focused attention on the flavour changes that occur during the juice manufacture. In some juices the effects are not marked, or a particular aroma component with a pronounced character masks such changes that do take place. In others the changes may be sufficient to modify the character of the aroma. In some juices this occurs at an early stage of processing, during juice extraction, as in apples during the period when the phenoloxidase system is active. This enzyme has accordingly been suspected of being responsible for aroma change, in particular the formation of C_6 aldehydes in apple juice, but, as shown by Gottauf and Duden (1965), the two effects are not related. In apple juices extracted without oxidation the amounts of hexanal and of hexene-2-al-1 were very small, less than 0·1 mg/litre, whereas in oxidized juices several mg/litre were present. However, blocking the phenoloxidase system with oxalate inhibited browning but did not prevent the formation of the aldehydes. A similar production of hexene-2-al-1 in strawberry extracts prepared in the presence of oxygen had been noted by Winter *et al.* (1962). Drawert and co-workers have investigated this further and studied a range of fruits, including apple, pear and grape (Drawert *et al.*, 1965a, 1966). In extracts of these fruits where enzyme activity was inhibited by methanol they found small amounts of both aldehydes, the level varying according to the stage of fruit development and maturity. Where enzymes were not inactivated the amounts reached high levels. By additions of unsaturated fatty acids it was demonstrated that hexene-2-al-1 was derived from linolenic acid and hexanal-1 from linoleic acid. The presence of these aldehydes and of hexene-3-ol-1 accounts for the "grassy" odour of some apple juices, especially those from under-ripe fruit; their presence in grape juices is also of significance during wine fermentation where they are reduced to the corresponding alcohols (Drawert and Rapp, 1966).

Other enzymatic changes occurring during juice extraction are associated with the hydrolysis of esters to the corresponding alcohols, usually with a loss of aroma (Drawert *et al.*, 1965b). The level of esters in the aroma fraction is, however, also influenced by the maturity of the fruit (Osman and Höfner, 1966), hydrolysis of esters being evident on long storage.

As evident from the work of these various authors, the aroma components of juices may differ in marked degree from those of the intact fruit and some reassessment of the published gas chromatographic data may be necessary. The effects of processing methods on juice aromas have been discussed in recent reviews by a number of workers (Gierschner and Baumann, 1968; Prillinger *et al.*, 1968).

C. Storage of Juices

1. *Methods of storage*

The processing sequence from fruit to juice may continue without interruption until the product reaches the final container in which is to be distributed. This has been common practice for citrus juices and for many others in the United States and Canada; in Europe it has been more usual to hold large bulks of juice in store for later bottling. The second method makes it possible to press fruit of variable composition throughout the season and yet to attain a standard juice by blending at the time of bottling or canning.

In Europe, the method followed for bulk storage of apple or grape juice was for many years the Boehi process, where juice was impregnated with carbon dioxide at the rate of 8 g/litre, equivalent to about 8 atmospheres pressure, and held in large lined tanks at ambient or cellar temperature (Borgström, 1961). This inhibited the growth of yeasts and moulds but, as eventually became apparent, lactic acid bacteria were capable of growth under these conditions producing as main products lactic and acetic acids from sugars. The breakdown of citric acid to these acids and of malic acid to lactic acid, with other subsidiary changes, also brought about a lowering of quality. An alternative method also in use was the flash-pasteurization of juices into tanks where it was allowed to cool. This gave a sterile product but submitted the juice to severe heat treatment with the production of cooked flavours and of artifacts such as hydroxymethylfurfural (HMF).

The production of HMF in stored juice products has received considerable attention for, although it does not itself give any appreciable off-flavour, it is a useful indication of quality; a high level suggests that heat treatment has been excessive. The formation of HMF following reactions between amino acids and sugars was followed by Koch and Kleesat (1961) who noted the highest amounts in grape juices, with blackcurrant, apple and orange in decreasing order. A similar study by Erdelyi *et al.* (1967) indicated that quality deterioration was only apparent when the level of HMF exceeded 5 mg/litre.

In many factories the methods of bulk storage have now been modified: Boehi storage is carried out at lower temperatures, near 0°C, where the effective concentration of carbon dioxide is equivalent to 3 atmospheres pressure. Juice intended for storage without pressure is now cooled rapidly after flash-heating and filled under sterile conditions into sterilized tanks and stored at low temperature. A later modification developed in France has been the storage of apple and grape juices after flash-pasteurization but without refrigeration. The juice is held in tanks under nitrogen to prevent the growth of moulds arising from the germination of heat-resistant spores (Menoret, 1962; Pollard and Beech, 1966a). With this type of storage it is essential to keep the juice under close supervision and to avoid the entry of micro-organisms through air-locks following changes in atmospheric pressure. Oxygen must be absent and nitrogen alone is not as effective in inhibiting mould growth as carbon dioxide or mixtures of carbon dioxide and nitrogen (Beech *et al.*, 1964a).

Bulk storage has been in use in the United States for many years for grape juice which is held at temperatures between $-6°C$ and $-2°C$; tartrate removal takes place at the same time. The juice is flash-pasteurized and cooled and stored in closed tanks or in open tanks irradiated by ultra-violet lamps to prevent mould growth. The development of psychrophilic yeasts has, however, raised problems under these conditions.

2. *Storage of juices as concentrates*

(a) *The use of concentrates.* Concentrated fruit juices have been made on a large scale for many years; they provide a convenient means for the transport of juice in international trade. Concentrates with a soluble solids content above 65° Brix are relatively free from microbiological spoilage; those of lower density are preserved by low temperature storage or by the addition of chemical preservatives. They have been used for reconstitution to juice for fermentation, for the preparation of fruit drinks and for other manufacturing processes.

Following improvements in concentration methods and, more especially in methods for the recovery of juice aromas, concentrates are now widely used for reconstitution to juice and as providing an alternative method of storage. There are still differences of opinion as to the designation of juices prepared in this way but they are being increasingly accepted as equivalent to juices made and stored by the more traditional means. With many juices it is not possible to differentiate between those prepared directly and those made from concentrate and aroma (Beech *et al.*, 1964b). With others, differences are perceptible, but further improvements in techniques and in the recovery of aromas will no doubt reduce their number. For example, improvements in aroma recovery from orange juice are now possible with the use of lower temperatures and efficient condensation of the distillate (Schultz *et al.*, 1967).

The storage of juice as concentrate gives a great saving in space and in the cost of storage; it also removes some of the uncertainties of other methods. It does not, however, avoid the need for temperature control for, as described later, low temperature storage is required for high density concentrates if deterioration is to be minimized. Low density concentrates require the conditions used for juice, and in Switzerland apple and pear juices are now increasingly stored under Boehi conditions as such "half-concentrates". The aroma is stored separately and, when returned to the reconstituted juice, is stated to give a product superior to that made from high density concentrates; the current developments have been described by Pollard and Beech (1966a). Concentration is also used for the storage of blackcurrant juice (Charley, 1962) and for other berry juices used by the preserves industries, usually with low temperature storage conditions. Orange juice concentrates have long been stored and distributed in a frozen condition and are now a main product.

Vacuum distillation is the most usual technique used in juice concentration and the methods in current use have been reviewed by many authors (Heid and Caston, 1961; Pollard and Beech, 1966b; Armerding, 1966; Robbins, 1967). Freezing techniques are also used, but to a lesser extent, the ice being separated from the concentrate by centrifugation or other means (Joslyn, 1961a). Newer freezing methods involve the use of evaporative freezing of the water *in vacuo* or of added low-boiling solvents at higher pressures (Muller, 1967). The removal of water by the formation of gas hydrates with halogenated hydrocarbons has also been described (Huang *et al.*, 1966). A different principle is used in reverse osmosis where the water is removed through a semi-permeable membrane under pressure (Morgan *et al.*, 1965; Willits *et al.*, 1967).

(b) *Storage problems.* In 1958 a symposium devoted to fruit juice concentrates emphasized that current problems were largely associated with deteriorative changes occurring during manufacture and storage. They included non-enzymic browning with the production of HMF and other artifacts, other oxidative changes and problems of microbial infection. Since then, micro-biological problems have been reduced by greater emphasis on plant hygiene and the use of lower storage temperatures. Non-microbiological deterioration has also been reduced by the greater use of heat treatments to inactivate enzyme systems and the wider adoption of improved storage methods for products.

Many interactions are possible in fruit juice concentrates, notably between sugars, organic acids, nitrogenous components and phenolics, the changes being greatest at higher temperatures of storage. Reactions involving sugars were studied by Vasatko and Pribela (1965) who noted the production of dihydroxyacetone, glyceraldehyde, hydroxymethylfurfural and melanoidins, with a decrease in sucrose. Pribela and Betusova (1964), in a review of nitrogen changes, studied in detail the loss of amino acids in stored apple and pear juice concentrates. After 60 days at 37°C the losses for most amino acids present exceeded 50% and those for glutamic acid, phenylalanine and lysine were even higher. An investigation by Kern (1962) showed that oxidation could lead to darkening at temperatures even below 20°C when concentrates were pumped from the evaporator to storage tanks. Emphasis on rapid cooling and anaerobic storage has since brought about significant improvements in the treatment and storage of concentrates in European factories.

(c) *The retention of juice aroma.* If a diluted concentrate is to resemble the juice from which it was made it is essential that the aroma be recovered before or during the evaporation process and returned unchanged when the concentrate is diluted. As discussed later, aroma concentrates are stable and are protected from interactions with other juice components. In a study of aroma changes in apple juices stored by different methods Wucherpfennig and Bretthauer (1964) found that quality changes were least when the aroma fraction was stored separately and returned later. This conclusion is now generally accepted and has been noted in other such tests (Beech *et al.*, 1964b). It has been pointed out, however, by Guadagni and Harris (1967) that the presence of aroma components can mask the production of off-flavours in sensory quality assessments. It has also been noted by the authors that small quality differences in juice products deprived of aroma can be detected by a tasting panel, whereas they are no longer perceptible when the aroma has been returned. A similar observation was made by Jennings and Jacob (1968), who suggest that the cooked aroma of heat-treated fruit products may not necessarily be due to the formation of artifacts but to substances initially present but subsequently masked by volatile components lost during heating.

The subject of fruit aromas has been discussed in detail in Volume I, Chapter 10, and the processing aspects have been recently discussed in a number of reviews (Lüthi, 1968; Gierschner and Baumann, 1968; Prillinger *et al.*, 1968).

The amount of volatile flavour components present varies from one type of fruit to another and, as pointed out by Lüthi (1968), may vary from less than 10 p.p.m. for peaches or strawberries to over 100 p.p.m. for some pineapple and grape cultivars. To recover the essential aroma of some fruits distillation of only some 10% of the volume of the juice may be required, as for example with apple juices; for others up to 40% may need to be distilled if essential components are to be recovered. This is so with Concord grape juice where the characteristic aroma resides largely in the high-boiling methyl anthranilate and where special fractionation techniques are required (Moyer and Saravacos, 1968).

Pribela and Strimiska (1967) showed that losses in aroma are involved during the whole sequence of juice-processing treatments and the trend is now to recover aroma fractions as soon as possible after juice extraction to minimize losses and changes in composition.

It has been found that aroma concentrates show good stability at temperatures near 0°C and a study of the subject by Guadagni and Harris (1967) has shown that a main reason is that at this temperature the mixture is usually homogeneous. At lower temperatures, near −4°C to −7°C, the water freezes out leaving the organic constituents exposed to oxidation. At still lower temperatures (−34°C) stability increases with the fall in the rate of chemical reaction.

Advances in the preservation and recovery of fruit aromas, and consequently of the quality of juice products, have followed the development not only of analytical techniques, but also of methods for the sensory evaluation of flavour. A further reason for advance, as may be seen from recent publications, is that in the laboratory the nose is once more taking its rightful place with the mass spectrometer so that the significance of the aroma components can be better assessed.

IV. CHEMICAL CHANGES IN STORED JUICES

Juices undergo changes on storage due to chemical reactions between their constituents. Although the reactions are largely non-enzymic, since any enzymes present have been previously inactivated, the extent to which any initial enzymic change has been allowed to proceed, such as phenolic oxidation, is likely to be a factor in the appearance and stability of the final product. Other factors involved are the amount of residual oxygen in the product, the metallic content (natural and adventitious), the temperature of storage and the extent of exposure to light. Deterioration of juices on storage has been discussed by Curl and Talburt (1961). They considered changes in flavour and appearance of more significance than loss in nutritive value. Some juices such as pineapple and tomato, in contrast to orange and apple, undergo only slow changes on storage. Changes of flavour and of colour, involving anthocyanin and carotenoid pigments, are discussed in Chapter 21 and in Volume I, Chapters 11 and 12. Discussion of storage changes will, therefore, be confined to the role of trace metals, particularly copper and iron, in oxidative deterioration, to appearance, as affected by haze and sediment formation, and browning reactions, and to the stability of ascorbic acid in juices and juice systems.

A. Metals and Metallic Complexes in Fruit Juices

1. *Metals naturally present in fruits*

Traces of metals are normally present in fruits as intrinsic biological components. McCance and Widdowson (1960b) analysed twenty-eight fruits and found contents of 0·3–2·6 p.p.m. copper and 1·9–12·7 p.p.m. iron respectively per fr. wt; similar data have been reported by Deschreider (1953), Erkama *et al.* (1953) and Zook and Lehmann (1968). Two general trends are evident; firstly iron content is invariably (and can be considerably) greater than copper, and secondly, iron is usually found in largest amounts in berry fruits. Concentrations of metals are much higher in the skin or peel than in the flesh of apples and tomatoes (Horner, 1941; Timberlake, 1957a). As these findings would suggest, traces of metals in the fruit are only partly soluble in the juice. The copper and iron contents of fruits and juices prepared from them, with rigorous precautions to avoid adventitious metallic contamination, are given in Table I (Timberlake, 1959a and unpublished). Great care has to

TABLE I. Natural metal-content of fruit and juice

Fruit	No. of samples	Copper[a] Fruit	Copper[a] Juice	Iron[a] Fruit	Iron[a] Juice
Apple	4	0·40–0·55	0·28–0·30	0·91–2·64	0·13–0·53
Strawberry	1	0·23	0·20	3·9	1·6
Blackcurrant	8	0·67–1·12	0·24–0·44	4·1–7·5	2·5–3·4

[a] p.p.m. on fr. wt. basis.

taken to avoid contamination with metals during pressing of juice (Timberlake, 1952) and at all other stages of manufacture (Joslyn, 1961b). In practice, therefore, the metal content of commercial juices tends to be somewhat higher than the figures given in Table I and can vary considerably (Kern, 1961).

2. *Ionic state of metals in juices*

It has been recognized for a long time that many of the organic components of fruits are capable of combining with metals and modifying their catalytic activity. In particular, the ionic state of copper and iron in wines has been of considerable interest since the 1930's when it was investigated in connection with the various casses and deposits which can occur (Ribereau–Gayon and Peynaud, 1961). In 1950, in a review of oxidation in fruit juices, Lavollay and Patron (1950) stressed how little was known of even the chief complexes present and their mode of action. Today, largely as a result of investigations carried out in model or isolated systems, we are in a better position to predict which complexes are likely to be present and their role in oxidative changes. This present account is confined largely to information which has become available since the previous reviews on this subject (Timberlake, 1957a, 1958). In general, the formula of a metal complex may be represented by $M_yH_pL_m(OH)_x$ (Bjerrum *et al.*, 1957) where M and L represent the metal and the ligand respectively. Stability constant data published to the end of 1960 have been collected by Martell (1964).

(a) *Organic acid complexes.* Cupric ions form weak complexes in acid solutions with simple monobasic acids, but stronger complexes with acids containing α-hydroxyl groups. The major fruit acids, citric, malic and tartaric, are the main components in fruit juices which can combine with copper. The nature and extent of complex formation depend upon a number of factors including pH, the concentrations of both copper and the organic acid, and the ionization constants of the organic acid. Although dimeric complexes can be formed under certain conditions, the simple monomeric complexes CuL probably predominate at the low metal and high acid concentrations characteristic of fruit juices; the amount of copper remaining in the free ionic state must be only a small percentage of the total. Under similar conditions, ferric iron, which forms much stronger complexes than those of copper, must exist practically entirely in the combined form, with only negligible amounts of the free aquo- or hydrated ions. Neutral and anionic trimers may now be present in addition to monomers and dimers. Ferrous ions form weaker complexes than either ferric or cupric ions with the organic acids; consequently, larger amounts of free ferrous aquo ions than cupric ions will be present under equivalent conditions. For all these complexes the order of stability is the same; citrate forms the strongest complexes followed by malate and then tartrate as summarized in Table II. Despite the reduced stability of its

TABLE II. Log K_s (stability constant) of acid-metal complexes

	CuL	$Cu_2L_2(OH)_2$	$Fe^{II}L$	$Fe^{III}L$	$Fe^{III}{}_2L_2(OH)_2$
Citrate	5·2	13·2	4·4	11·40	21·17
Malate	3·4	8·5	2·6	7·13	12·85
Tartrate	3·2	8·2	2·2	6·49	11·87
References	(1)	(2)	(3)	(3)	(3)

(1) Lefebvre (1957).
(2) Rajan and Martell (1967).
(3) Timberlake (1964a, b).

complexes, tartrate can be a more effective chelating agent than malate under certain conditions because of its lower basicity. In wines, malic acid has been considered the most important complexing agent, although citrate complexed copper and iron to a greater extent than malate under equivalent conditions (Rankine, 1960).

Since the fruit acids are multidendate ligands capable of forming polynuclear chelates, it is not surprising that different metals may be combined within the same complex. Such mixed metal chelates have been described, e.g. ternary chelates of iron(III) and tin(II) or copper and tin(II) with citrate or tartrate (Smith, 1965), chromium-cerium-citrate (Schulz *et al.*,

1966), copper-chromium-citrate (Irving and Tomlinson, 1968) and iron-aluminium-tartrate (Pyatnitski and Glushchenko, 1966).

(b) *Complexes of sugars and polyhydroxy compounds.* Although polyhydroxy compounds have long been known to combine with metals in alkaline media (Bourne *et al.*, 1959), it was not until 1963 that complex formation under more acid conditions was described (Charley *et al.*, 1963). The most stable complexes are formed with fructose; in neutral solution cupric fructose (1 : 2) and ferric fructose (2 : 2) complexes are formed (Aasa *et al.*, 1964). Under more acid conditions, complex formation is very slight; the only quantitative data available concern the weak 1 : 1 ferric-fructose complex (log $K_s = 0{\cdot}33$ at pH 2·5; Sarkar *et al.*, 1964). Nevertheless, the effect of sugars may not be insignificant because of the slow reduction of ferric to ferrous ions which may occur (Charley *et al.*, 1963).

The remarkable chelating ability of fructose for iron (III) in contrast to aldo-hexoses and disaccharides has been attributed to the structural features of carbons 1–3 of the open-chain form and is shared by other keto-sugars such as sorbose and tagatose (Davis and Deller, 1966). The ability of fructose to exist almost entirely in the open-chain form in aqueous solution (Antikainen, 1959) is probably an additional factor involved.

(c) *Complexes with phenolic compounds.* (i) Phenols and phenolic acids. Phenols and phenolic acids, such as (+)-catechin, caffeic and chlorogenic acids (Timberlake, 1959b) which contain the o-dihydroxyl or catechol grouping, combine with cupric ions in the manner of catechol (H_2L) itself (Nasanen and Markkanen, 1956; Timberlake 1957b; Murakami *et al.*, 1963; Neher, 1964; Jameson and Neillie, 1965; Athavale *et al.*, 1966; Dubey and Mehrotra, 1966; L'Heureux and Martell, 1966; Oka and Harada, 1967). Complexes of type CuL occur in acid solutions and are converted into CuL_2 with increasing pH values. When the acid moiety of the phenolic acid also combines with copper, it can compete with the catechol grouping. Thus, the copper is combined mainly with the quinic acid component of chlorogenic acid at low pH ($<3{\cdot}5$), but is yielded to the more powerfully chelating phenolic groups as the pH rises. The competitive effect of the quinic acid component is less pronounced with iron (III) since the ferric o-diphenol complexes are much stronger than those of copper and can accordingly be formed at lower pH values. Thus the green ferric-chlorogenate complex similar to the 1 : 1 ferric–catechol complex (Weinland and Binder, 1912) is formed at pH values of less than 3. Increasing pH produces colour changes to blue, violet and red as successive 2 : 1 and 3 : 1 (chlorogenic : ferric) complexes are formed. The complexes are difficult to study because equilibration is slow. Ferrous iron forms only weak complexes with catechol, caffeic and chlorogenic acids which are rapidly oxidized to the coloured ferric complexes at pH values approaching neutrality even under reducing conditions.

The effect of ferric complex formation with o-diphenolic compounds on valency changes is described later. Complex formation by other phenols has been described qualitatively (Gore and Newman, 1964) and quantitatively (Ernst and Menashi, 1963; Ernst and Herring, 1965; Jabalpurwala and Milburn, 1966; Park, 1966).

(ii) Flavonoid compounds. Flavonoid compounds contain three potential chelating sites: (a) the 3′,4′-dihydroxyl group, already partly discussed; (b) the 3-hydroxyl-4-carbonyl group; and (c) the 5-hydroxyl-4-carbonyl group. Characteristic bathochromic wavelength shifts are produced when aluminium combines with certain of these groups, so facilitating their structural identification in flavonoids (Jurd, 1962). Although the extent of complex formation may vary according to the nature of the solvent (Jurd, 1969), observations made during the course of these investigations provides some information on the relative stabilities of the three types of complexes. Thus, those with the 3′,4′-dihydroxyl grouping are the weakest, since they are the least tolerant of acid (Markham and Mabry, 1968). Of the two remaining sites, the 3-hydroxyl-4-carbonyl grouping appears to form stronger aluminium complexes than the 5-hydroxyl-4-carbonyl group. Thus 3-hydroxyflavone forms a much stronger complex than 5-hydroxyflavone (Jurd and Geissman, 1956). The stability of the former complex was attributed to the formation of a flavylium salt structure, stabilized by its quasi-aromatic nature; contributions to the stability by other possible structures involving the 7- or 4′-hydroxyl groups of polyhydroxyflavonols (such as kaempferol) were not evident, since these compounds gave wavelength shifts no different from 3-hydroxyflavone itself (about 60 nm). In contrast, contributions by hydroxyl groups outside the pyrone ring were considered significant in stabilizing 5-hydroxyl-4-carbonyl complexes, since much larger shifts (40 nm) occur with polyhydroxyflavones (not 3-substituted), such as apigenin, than 5-hydroxyflavone itself.

Flavanones and 3-hydroxyflavanones would be expected to form weaker complexes than the flavones, since saturation of the 2·3 double bond precludes not only flavylium salt formation but also stabilization of the 5-hydroxyl-4-carbonyl group by the 4′-hydroxyl group. The 3-hydroxyl group of 3-hydroxyflavanones is probably not involved in complex formation, since the magnitude of the aluminium shift (of the short wavelength band) given by dihydrokaempherol (22 nm) is practically the same as that given by its 3-rhamnoside (24 nm—personal communication from A. H. Williams) and the corresponding flavanone naringenin (21 nm—Horowitz and Jurd (1961); compare also eriodictoyl and taxifolin).

Greater stability of the 3-hydroxyl-4-carbonyl chelates has also been shown for zirconium (Hörhammer *et al.*, 1954) and for aluminium, which can form three complexes with quercetin (1 : 1, 2 : 1 and 3 : 1 quercetin–aluminium—Bayer *et al.*, 1966). When the 3-hydroxyl group is replaced by a methoxy or a glycoside group, complex formation occurs at the 5-hydroxyl-4-carbonyl group and is not only weaker but is restricted to formation of the 1 : 1 complex, because of steric hindrance.

However, with hydrogen and copper, there is evidence that suggests the contrary, viz. that stronger complexes are formed by the 5-hydroxyl-4-carbonyl than the 3-hydroxyl-4-carbonyl group. Thus, the pyrone ring of flavones formed stronger hydrogen bonds with the 5-hydroxyl than the 3-hydroxyl group (Simpson and Garden, 1952a, b). Simultaneous binding of both groups, although involving an overall loss (Shaw and Simpson, 1955), was stabilized by the 4′-hydroxyl group; its stabilizing effect was greater on the 3-hydroxyl than on the 5-hydroxyl-4-carbonyl grouping, in contrast to that described previously for aluminium. Further, Saxena and Seshadri (1957) found that 5-hydroxy- and 5-hydroxy-7-methoxyflavone formed stronger copper complexes than 3-hydroxy- and 3-hydroxy-7-methoxyflavone respectively. 5-hydroxy-7-methoxyisoflavone gave a weaker copper complex than 5-hydroxy-7-methoxyflavone, but the pK values (hydrogen bonding) of both compounds were identical. These authors found that 1 : 1 complexes only were formed, whereas only a 2 : 1 (3-hydroxy flavone : copper) complex was found by Detty *et al.* (1955).

A further method of assessing the chelating ability of flavonoids has been by their effect on the copper-catalysed oxidation of ascorbic acid; the assumption being that the greater the chelating ability the greater the inhibition. Most investigations have been carried out at pH values near 7. The findings with the flavonoid aglycones largely agree with those already described for aluminium. Thus, inhibition of the reaction was greatest with the 3-hydroxyl-4-carbonyl group, weak with the 5-hydroxyl-4-carbonyl group and very slight with the 3′-4′-dihydroxyl group (Letan, 1966b). Clemetson and Andersen (1966) attributed no chelation capacity to the 5-hydroxyl-4-carbonyl group, since hesperidin gave no inhibition. However, they found a pronounced inhibition by rutin (greater than quercetrin), which is unlikely to be due only to chelation with the 3′,4′-dihydroxyl group which they thereby imply. Rutin and quercetrin (but not naringin) were effective inhibitors at pH 3·6–5·3 (Davidek, 1963). On the other hand, only weak inhibition by rutin, quercetrin and naringin compared with quercetrin and other flavonols has been reported (Heimann and Heinrich, 1959; Samoradova-Bianki, 1965). The reliability of the method has been questioned (Timberlake, 1960a) and such considerations may account for the inconsistent results obtained.

(iii) Anthocyanins. The formation of metal salts of the anthocyanins gives rise to colour changes in canned fruits (Culpepper and Caldwell, 1927; Heintze, 1960) and fruit juices (Széchényi, 1964). Complex formation is greater with trivalent metals such as ferric iron and aluminium than with bivalent metals such as ferrous iron. The blue iron and aluminium complexes of cyanin and cyanidin have been isolated and characterized (Bayer *et al.*, 1960); they are formed between the 3′,4′-dihydroxyl grouping (in the form of the anhydro base) and the metal. The corresponding copper complexes were brown, indicating oxidation by the cupric ions. Metal anthocyanin complexes can also form ternary complexes (which play an important role in flower petal colours) and which are of two different types—complexes with polysaccharides, and complexes with flavonoid materials. Thus, the pigment of the blue cornflower, isolated and named "protocyanin" by Bayer (1958), contained iron and aluminium (1 equivalent of each *per* mole of cyanin) in a macromolecular co-ordination complex with cyanin (19%) and carbohydrate (80%; mainly galacturonic acid, rhamnose and glucose). This "chromosaccharide" was considered to be the first naturally occurring metal chelate described in which the macromolecular carrier was polysaccharide instead of protein (Bayer *et al.*, 1966). The "protocyanin" examined by Japanese workers, however, contained magnesium, iron, potassium, peptide, carbohydrate and flavonoid material, but no aluminium (Saito *et al.*, 1961; Saito and Hayashi, 1965). Asen and Jurd (1967) described a crystalline deep blue pigment also from cornflowers which they named "cyanocentaurin", since it differed from Bayer's protocyanin in consisting of an iron complex of 4 moles of cyanin and 3 moles of a "bisflavone" glucoside, the aglycones of which were identified as genkwanin and (tentatively) 7-O-methyl vitexin. An essentially similar structure was found for "commelinin", another blue flower pigment consisting of a metal (magnesium)–anthocyanin (delphinidin 3,5-diglucoside acylated with p-coumaric acid)–flavonoid (6-C-glucopyranosylgenkwanin 4′-O-glucoside) complex (Mitsui *et al.*, 1959; Takeda *et al.*, 1966). Blue complexes have also been obtained in model systems containing cyanidin 3-glucoside, aluminium and chlorogenic acid or quercetrin (Jurd and Asen, 1966). Complex formation did not occur in the presence of citrate (pH 5·45), presumably because of chelation of the aluminium as a more stable citrate complex. It is interesting that Bayer *et al.* (1966) found that the blue colour of protocyanin was reduced on addition of quercetrin (pH 4·5); displacement of the metal from protocyanin to quercetrin was suggested. In view of the work of Jurd and Asen (1966), however, it is possible that quercetrin may have partly displaced galacturonic acid (probably a weaker ligand than cyanin) from the protocyanin with formation of a quaternary complex less coloured than protocyanin itself.

(iv) Leucoanthocyanins (proanthocyanidins). Nylon was found to remove both leucoanthocyanins and iron from model systems, ciders and perries of high phenolic (particularly leucoanthocyanin) content, possibly in a complexed form (Timberlake, 1961).

(v) Oxidized and condensed phenolics. The formation of soluble copper complexes of the o-quinones produced by tyrosinase oxidation of catechol, caffeic and chlorogenic acids has been described. Complex formation decreased on quinone polymerization due to reduction in the number of active sites (Arthur and McLemore, 1956, 1959). The formation of insoluble copper complexes of oxidized and condensed apple phenolics (mainly (−)-epicatechin and leucoanthocyanins) has been demonstrated (Kieser *et al.*, 1957).

(d) *Amino acids, peptides and proteins.* Amino acids and peptides form 1 : 1 copper complexes in acid solutions which are transformed into 2 : 1 (amino acid : copper) complexes with increasing pH. No stereoselective effects were found in the 2 : 1 complexes (Ritsma *et al.*, 1965; Gillard *et al.*, 1966). The 1 : 1 complexes become less stable with increasing peptide polymerization, but their structures are under discussion (Kim and Martell, 1966; Brunetti *et al.*, 1968). In acid solutions, amino acids form 1 : 1 ferric complexes which are stronger and 1 : 1 ferrous complexes which are much weaker than the corresponding copper complexes (Perrin, 1958a, b).

Protein–metal complexes are of interest with regard to the formation of casses, hazes and sediments in wines, beer and fruit juices. In proteins, complexing of metals with side-chains such as carboxyl, imidazole and sulphydryl groups is more likely than with terminal amino and carboxyl groups. Potential chelating groups include those containing nitrogen (ε-amino, the guanidino group of arginine, amide), oxygen (amide, phenolate of tyrosine, hydroxyl groups in serine and threonine) and sulphur (the thio-ether of methionine and the disulphide of cysteine). The sulphydryl groups of proteins may be oxidized by metal ions with sufficiently high oxidation–reduction potentials, such as Cu^{2+} and Fe^{3+}, following the formation of a reactive complex (Gurd and Wilcox, 1956).

(e) *Ascorbic acid.* Ascorbic acid forms complexes with many metals (Susic, 1955; Stolyarov and Amantova, 1964; Veselinovic and Susic, 1965), some of a very transient nature (e.g. with cupric and ferric ions) which are intermediates in the oxidation of the ascorbic acid and reduction of the metal.

(f) *Mixed ligand complexes.* Not only are mixed metal complexes with a single ligand possible, but also mixed ligand complexes with a single metal. Examples already cited are the aluminium complexes of both anthocyanins and other flavonoid compounds; others include copper complexed with both amino and phenolic acids (Perrin *et al.*, 1967). For the evaluation of multiple equilibria (multi-metal–multi-ligand), a computer programme has been described (Perrin, 1965).

(g) *Valency changes.* Cuprous ions are inherently unstable in aqueous solution, undergoing disproportion into metallic copper and cupric ions, and, in the presence of air, being rapidly oxidized to the cupric state. They will remain in the reduced state only if stabilized by the formation of complexes, usually insoluble, as, for example, in the form of cuprous sulphite haze which occurs in sulphited juices and wines (Rentschler and Tanner, 1960). Sulphydryl groups are likely ligands for cuprous ions (Hemmerich, 1966), and have been reported to combine with copper in freshly made beer, followed by slow dissociation of the complexes on ageing (Chapon, 1965). Mixed valence complexes are also possible (Klotz *et al.*, 1958). In the absence of cuprous stabilizing compounds, ascorbic acid does not reduce cupric ions except in catalytic amounts (Nord, 1955); thus, blue cupric–citrate remains unchanged on the addition of ascorbic acid. Therefore, unless juices contain large amounts of cuprous-combining material the bulk of the copper must be in the cupric state and complexed largely with organic acids, amino acids and related compounds and, to a lesser extent, phenolic compounds.

In contrast to cuprous copper, ferrous ions and ferrous complexes are stable in aqueous solution under certain conditions. Reduction of ferric iron which in juices is likely to be entirely in the combined form—mainly with organic acids, phenolic compounds and, to a lesser extent, amino acids and fructose—will yield appreciable amounts of ferrous ion in its free ionic state in equilibrium with its weak ferrous complexes. For example, a simple calculation indicates that 6 p.p.m. of ferric tartrate, entirely combined in a medium of 0·01 M tartrate, at pH 3·5, yields on reduction 4·5 p.p.m. ferrous ion and 1·5 p.p.m. ferrous tartrate. Ascorbic acid will reduce ferric ions and its complexes with organic acids and phenolic compounds to the ferrous state.

Factors affecting valency changes of copper and iron in juices include the extent of complex formation, particularly with organic acids and phenolic compounds (which form latent metal-reducing systems), oxygen, light and reducing agents such as ascorbic acid. Depending upon the conditions, traces of copper and iron in fruit juices can be in a continual state of flux between their valency states and can give rise to oxidative changes which can have a deleterious effect on quality.

3. *Catalytic effects of metals and their complexes*

(a) *Organic acids*. The catalytic activity of cupric ions is reduced by formation of relatively inert complexes with the major fruit acids (citric, tartaric and malic acids). However, cupric and other metal ions can catalyse the decarboxylation of β-oxo acids such as oxalacetic, dihydroxyfumaric, acetonedicarboxylic and dihydroxytartaric acids by formation of transient intermediate complexes (Prue, 1952; Hay, 1963). The reaction has been discussed by Williams (1959) in a general way relating to selective catalysis and inhibition by metal ions in biological systems.

In contrast to those of copper, the iron complexes of the major fruit acids are much more reactive. Since much stronger complexes are formed with ferric than ferrous ions, the ferric–ferrous oxidation–reduction potential is lowered (Clarke, 1960) and rapid oxidation of the ferrous complexes to ferric occurs.

Moreover, the ferric complexes are photochemically reducible. The valency changes of iron induced by oxygen and sunlight are accompanied by oxidation of the organic acid. Among the products of tartrate decomposition which have been identified are dihydroxymaleic acid, dioxosuccinic acid, tartronic acid, mesoxalic acid, mesoxalic semi-aldehyde, glyoxal, glycolaldehyde, glyoxylic acid and oxalic acid (Fenton, 1897; Benrath, 1917; Wieland and Franke, 1928; Baraud, 1954). The effect of these changes on potentials and flavours of wines has been discussed (Deibner, 1957). Photochemical decomposition of ferric–citrate gives acetonedicarboxylic- and acetoacetic acids and acetone (Frahn, 1958). Ferric–malate is less affected by light but acetaldehyde and formaldehyde were identified as decomposition products by Benrath (1917), More recent chromatographic examination shows that considerable quantities of unidentified dicarbonyls are produced in these photochemical reactions. The decomposition products can bind sulphur dioxide (Timberlake, 1964c).

(b) *Phenolic compounds.* Ferric and cupric complexes of catechol and phenolics containing the catechol grouping, e.g. iron monophenol complexes (Williams, 1959), have a latent tendency to undergo electron transfer within their complexes with formation of the reduced metal and a radical form of the oxidized phenolic (the semi-quinone anion). Thus, at the demand of reagents (e.g. 2·2′-dipyridyl, 2·2′-diquinolyl) which combine preferentially with the reduced form of the metal, the reactions can be induced to go to completion (Timberlake, 1959a; Resnik *et al.*, 1961), unless there is competition by other substances which chelate the higher valence state (Banga 1938; Timberlake, 1957c). In the presence of oxygen, the reduced form is reoxidized, so catalysing the oxidation of the phenol. It is interesting that Banga (1938) postulated that oxygen was bound to the initially formed complex in a way similar to that recently suggested for ascorbate oxidation (Khan and Martell, 1967a). Although oxidation in the iron–catechol system itself soon reaches equilibrium further reactions of the quinones (as already described under enzymic reactions) may result. Copper is particularly effective in oxidizing phenols (Kaeding, 1963). Secondary condensation of the oxidized phenolics, also, catalysed by cupric ions (Baruah and Swain, 1953), can cause sedimentation of condensed phenolic material (Kieser *et al.*, 1957).

(c) *Hydrolysis of esters and peptides.* Copper catalyses the hydrolysis of ethyl glycinate (Connor *et al.*, 1965) and diglycine (Grant and Hay, 1965) in acid solutions. The reactive species in each case was the 1 : 1 copper complex.

(d) *Aromatic hydroxylation.* Hydroxylation of aromatic compounds by hydrogen peroxide occurs in the presence of ferrous ion (Fenton's reagent—probably via the hydroxyl radical) and ferric iron and an ene-diol such as catechol (Hamilton's system; Hamilton *et al.*, 1966; Norman and Smith,

1965; Mason, 1965). Hydrogen peroxide can be formed by metal-catalysed oxidation of ascorbic acid or phenolic compounds, but it is not certain to what extent it can participate in these reactions in juices.

B. Visual Appearance

1. *Hazes and deposits*

The copper and iron casses of wines have been extensively studied (Joslyn and Lukton, 1953; Tanner and Vetsch, 1956; Amerine, 1958; Rentschler, 1963). Copper–leucoanthocyanin deposits occur in fruit juices but not in wines (Rentschler and Tanner, 1960). Deposits of degraded phenolic (mainly leucoanthocyanin and (−)-epicatechin) and protein material can occur on storage of clarified apple juices (Kieser *et al.*, 1957; Johnson *et al.*, 1968). Sedimentation is promoted by copper and tin(II). A reversible chill haze of apple juice has also been described (Monties and Barret, 1965).

2. *Browning reactions*

Sugar–amine interactions of the Maillard type (Reynolds 1963, 1965) and the resultant flavour changes (Hodge, 1967) are not as prominent in juices as in dried fruits, because of their aqueous nature and low pH. Organic acids and ascorbic acid are more involved in interactions with the sugars and amino acids. Discussion will be confined largely to information which has become available since the recent review on the subject (Reynolds, 1965).

5-Hydroxymethyl-2-furaldehyde (HMF) is produced in fruit juices from sugars, particularly ketoses, by heating during processing and can give rise to browning reactions with amino compounds (Diemair and Jury, 1965) and sugars, and undergo further polymerization and rearrangement both in the presence and absence of oxygen (Koch and Kleesat, 1961; Kern, 1964). The amounts formed increase with increasing heat treatment and increasing juice amino acid and sugar contents (least in orange and greatest in red grape juice), and normally should not exceed 5 mg/litre in juices and 10 mg/litre in concentrates. Only much larger quantities (100 mg/litre) can be detected organoleptically.

During 3-year storage a concentrated lemon juice of high sucrose (56° Brix) but low ascorbic acid content browned five- to ten-fold, suffered loss of amino nitrogen and acidity, and produced HMF and 3-deoxyglucosone presumably from the fructose formed from sucrose inversion (Kato *et al.*, 1963).

In lemon juice and comparable low sugar (3% glucose) model systems, ascorbic acid and citric acid have been implicated in oxidative browning (Clegg, 1964, 1966; Clegg and Morton, 1965). Clegg showed that browning was mainly due to reactions of reactive carbonyl compounds, produced by ascorbic acid breakdown, with amino acids and was increased by citric acid. The most reactive intermediates were α, β-unsaturated carbonyls, and in the early stages of browning, dicarbonyls of the glyoxal type; furfural, though formed, was not considered to be active. The formation of carbonyls of similar type from citric acid by operation of the ferric–ferrous cycle in the presence of light and air (described earlier), although not implicated in these particular studies, should, however, not be overlooked if juices are exposed to light. Malic acid (Livingstone, 1953) and tartaric acid (Kato and Sakurai, 1964) also give rise to browning. An alternative degradative path for ascorbic acid, besides the usual one via dehydro-ascorbic acid, was suggested by the finding of Kurata *et al.* (1967) that ascorbic acid browned more intensely than dehydroascorbic acid (pH 2·2–4·6). Browning was little affected by glucose but greatly increased by glycine addition. Under more acid conditions (5% sulphuric acid), furfural, ethylglyoxal, 2-oxo-3-deoxy-L-pentono-γ-lactone and L-xylosone were isolated (Kurata and Sakurai, 1967b).

Where anthocyanins are also present with ascorbic acid, as in strawberry, raspberry and blackcurrant juices and concentrates (Pollard *et al.*, 1955), browning can occur as a result of their degradation, to an extent increasing with the temperature of storage. Anthocyanins are not very stable and are affected by heat, light, oxygen, traces of metals, ascorbic acid (largely as a peroxide generating system) and sugar breakdown products (Starr and Francis, 1968). Their degradation is discussed in detail in Chapter 21. The anthocyanidins which are formed from the anthocyanins by mild acid hydrolysis (although to what extent hydrolysis does occur in fruit juice is uncertain) are very much less stable than the anthocyanins and are rapidly transformed into colourless enol forms of the carbinol bases (Timberlake and Bridle, 1966). Further slow transformations occur, presumably to the corresponding keto forms, with browning. Other phenolics such as catechins, leucoanthocyanins (Swain, 1965) and caffeic acid (Clegg and Morton, 1968) may also undergo oxidation to brown pigments concurrently with anthocyanins.

Bisulphite is commonly used to inhibit browning reactions in citrus juices such as lemon (see p. 585). It will combine with intermediate reactive carbonyls (Reynolds, 1965); more recent work, however, suggests that a free radical may be involved (Song and Chichester, 1967). Iron-catalysed reactions are also inhibited, presumably by maintenance of the iron in the reduced state. However, bisulphite cannot be employed satisfactorily in anthocyanin-containing juices, since colourless anthocyanin complexes, although weaker than those of other flavylium compounds (Timberlake and Bridle, 1967) may be formed to extents sufficient to bring about juice decolourization. Because of this factor and the general instability of anthocyanins, some commercial juice products are coloured with permitted azo-dyes. Many of these, however, are also not particularly stable in the presence of ascorbic acid (Lueck, 1965) or bisulphite. The possible use of other colouring agents having superior stability characteristics, e.g. carotenoids (Borenstein and Burnell, 1966; Kläui and Manz, 1967) and flavylium salts (Timberlake and Bridle, 1968) are, therefore, of interest.

C. Stability of Ascorbic Acid†

1. *General*

Since fruits are a primary source of ascorbic acid in the diet, the preservation of the vitamin content of juices, both natural and enriched, is a matter of some importance. The major loss of ascorbic acid during juice manufacture is due to the activity of phenolase. It is significant that fruits which discolour and are high in phenolase activity, e.g. apples, pears, plums, apricots, peaches, pineapples, sweet cherries, cranberries, etc., give juices or nectars which contain little or no ascorbic acid (Bauernfeind, 1953). Oxidation of ascorbic acid by the peroxidase system is possible (Mapson, 1945; Keilin and Hartree, 1955; Heimann and Heimann, 1965) but its extent would depend upon the availability of hydrogen peroxide. Ascorbic acid oxidase is present in fruits, e.g. citrus (Vines and Oberbacher, 1963; Dawson, 1966) and tomato (Joslyn, 1961b), but little of the enzyme appears in the juice (Huelin and Stephens, 1948; Atkinson and Strachan, 1950). The content and stability of ascorbic acid vary according to the nature of the juice (Pollard 1950; Diemair and Postel, 1964). It is more stable in blackcurrant than in lemon juice (Clegg and Morton, 1968), in orange than in apple juice (Noel and Robberstad, 1963) and in pineapple than in lime juice (Uprety and Revis, 1964), probably due to variations in the amounts present of natural inhibitors (such as flavonoids) of non-enzymic (or enzymic) oxidation (Herrmann, J., 1958; Jackson and Wood, 1959; Kuusi, 1961; Spanyar *et al.*, 1964b, 1966). Its stability may also be enhanced by the presence of dehydroascorbate reductase, e.g. in sweet orange and sweet lemon juices (Hassib and Ragab, 1962a, b) or catalase (Keilin and Hartree, 1955; Ross and Chang, 1958).

The question of fortification of juices with ascorbic acid has been discussed by Bunnell (1968). Although he states that addition of ascorbic acid does not upgrade colour or flavour, Beech *et al.* (1964b) found that these qualities can be improved with apple juice and diluted concentrates if ascorbic acid is added immediately prior to bottling. Addition of ascorbic acid (120 p.p.m.) to apple juice at the bottling stage gave a product of better flavour and more acceptable colour than either the untreated juice or juice which had received the same amount of ascorbic acid during enzyme treatment (Table III—unpublished work).

† See also Volume 1, Chapter 13.

Ascorbic acid prolongs the shelf life of a product by reacting with residual oxygen and retarding the development of off-flavours. However, on storage of juices it can give rise to secondary changes such as browning which have already been discussed.

TABLE III. Fortification of apple juice

Juice	Ascorbic acid (p.p.m.) Added	Remaining	Lost	% Colour (at 440 nm)
1	—	—	—	100
2	120 during enzyming	36	84	29
3	120 at bottling	95	25	55

2. *Model systems*

(a) *Aerobic oxidation.* It is well known that traces of copper catalyse the oxidation of ascorbic acid in acid solutions, and that the effect is enhanced by iron (Timberlake, 1960a; Spanyar *et al.*, 1963, 1964a). The catalytic species is generally considered to be the cupric aquo-ion since organic acid and synthetic chelating agents reduce catalysis according to their ability to combine with copper (Timberlake, 1960a; Onishi and Hara, 1964; Khan and Martell, 1967a). At higher pH values under conditions when inorganic copper is insoluble, oxidation by copper bound to amino acids, peptides and proteins can occur, the more firmly bound the copper, the more active the catalyst (Scaife, 1959). The mechanism of copper catalysis has been assumed for many years to involve formation of a transient cupric–ascorbate complex, followed by a rate-determining electron transfer within the complex from ascorbate to cupric copper (Weissberger and Luvalle, 1944). Khan and Martell (1967a) now consider that oxygen is bound to the initially formed complex, and that electron transfer occurs through the metal ion to oxygen. Both mechanisms may be operative (Ogata *et al.*, 1968). Ferric iron catalyses oxidation of the neutral species of ascorbic acid more than copper (Khan and Martell, 1967a), which may account in part for the hitherto unexplained catalytic effect of iron at very low pH values (Huelin and Stephens, 1946, 1947). Iron salts also play an important role in the oxidation of ascorbic acid by hydrogen peroxide (Ross and Chang, 1958). In contrast to those of copper, the iron complexes of synthetic chelating agents can enhance the oxidation of ascorbic acid (Timberlake, 1960a; Khan and Martell, 1967b).

The protective effect of some flavonoid compounds on the copper-catalysed oxidation of ascorbic acid has been attributed to their ability to chelate copper and has been described earlier in this chapter. However, other anti-oxidant characteristics of these compounds involving reaction with free radicals may be important factors in the stabilization of ascorbic acid in some aqueous citrate solutions comparable with blackcurrant juice (Clegg and Morton,

1968).† In the latter, flavonol aglycones such as quercetin exhibited a protective effect, both with and without added copper, whereas flavonol glycosides such as rutin were without appreciable effect. In apple juice fortified with ascorbic acid, however, rutin had a protective effect (Charlampowicz and Gajewski, 1965) It is well known that ascorbic acid has a destructive effect upon the anthocyanin pigments, but the nature of the converse effect, viz. the effect of anthocyanins on ascorbic acid stability, is not as definite. It has been suggested that anthocyanins protect ascorbic acid either by combining with hydrogen peroxide formed during its oxidation, which otherwise would oxidize the ascorbic acid itself (Spanyar *et al.*, 1964c; Zyzlink, 1965) or by chelating copper (Davidek, 1960, 1963). It should be stressed that these findings were based not on work with pure anthocyanins, but rather on extracts of fruit juices which may have contained other substances capable of inhibition, apart from the anthocyanins.

However, blackcurrant anthocyanins considerably increased ascorbic acid oxidation in the presence of copper, both with and without iron, the products formed by interaction of anthocyanin and peroxide thus appearing to catalyse the oxidation of ascorbic acid (Timberlake, 1960a). Clegg and Morton (1968) also found that blackcurrant anthocyanins accelerated ascorbic acid oxidation, but only without added copper; with its addition, a slight retardation occurred. Blueberry anthocyanins slightly accelerated non-enzymic oxidation, but had no effect on the enzymic oxidation of ascorbic acid (Schillinger 1966). Further work is evidently required on this point.

(b) *Anaerobic.* The rate of anaerobic degradation of ascorbic acid in contrast to the aerobic degradation hardly varies from pH 1–11. However, a small but definite maximum occurs at pH 4, coincident with the pK_a value, and suggests the existence of a complex between ascorbic acid and its monohydrogen anion (Finholt *et al.*, 1963). Such complex formation was also indicated in the autoxidative pathway (Levandoski *et al.*, 1964). During acidic anaerobic degradation, ascorbic acid yielded approximately equimolar amounts of carbon dioxide and furfural (Yamamoto and Yamamoto, 1964; Finholt *et al.*, 1965a, b), with intermediate formation of 3-deoxy-L-pentosone (Kurata and Sakurai, 1967a). Lead and aluminium are the most powerful inorganic catalysts of anaerobic degradation (Finholt *et al.*, 1966), but were effective only at concentrations far in excess of those likely to be found in fruit juices.

3. *Fruit juices*

Oxidation of ascorbic acid in blackcurrant juice appears to be entirely due to non-enzymic processes catalysed by copper and enhanced by iron and

† The mechanism by which flavonoids interfere with chain reactions of autoxidation in fruit juices has been further discussed recently by Harper *et al.* (1969).

anthocyanins present (Timberlake, 1960a, b). There is no evidence of any loss as a result of phenolase activity as occurs in apple juice. Thus, Herrmann, K. (1958) reported little phenolase (only 1–2% of the peroxidase activity) in blackcurrants, while phenolase activity was not demonstratable in fruit at its own or higher pH values (Kuusi, 1961; Timberlake 1960b and unpublished results; Heimann *et al*, 1958). Lack of activity might be attributed to inhibition by fruit components of any phenolases present similar to that in apple (Hooper and Ayres, 1950). Other phenolases may not be inhibited, however, since addition of potato phenolase to blackcurrant juice greatly increased oxygen uptake and ascorbic acid loss (Timberlake, unpublished results). The stability of ascorbic acid in blackcurrant juice has been attributed to lack of enzymic oxidation, the low natural copper content of the juice coupled with its high citrate content (Timberlake, 1960b) and the antioxidant effect of phenolics such as flavonol aglycones (Clegg and Morton, 1968)

Heimann and Wisser (1962) found that, whereas two atoms of oxygen were consumed per mole of ascorbic acid oxidized in freshly expressed blackcurrant juice (ten times diluted with acetate buffer), only one oxygen atom was consumed in the correspondingly diluted heated juice. The extra atom of oxygen consumed was attributed to oxidation of undetermined juice components by oxidative enzymes present in the diluted fresh juice, such oxidized components being unable to react with ascorbic acid in the manner described earlier for phenolase. Such enzymes would protect ascorbic acid, which would thus be more stable in unpasteurized than pasteurized juice. However, these authors did not consider the formation and subsequent reactions of hydrogen peroxide known to be formed during the non-enzymic oxidation of ascorbic acid. Depending upon the subsequent fate of this hydrogen peroxide (its decomposition or reactions with ascorbic and dehydroascorbic acids, their degradation products and other substances) so the number of atoms of oxygen consumed for each mole of ascorbic acid oxidized can vary between one and two (Scaife, 1959; Timberlake, 1960a). The finding of two atoms of oxygen consumed by the fresh juice could then equally well be interpreted as due to oxidation of other juice constituents by hydrogen peroxide either enzymically, i.e. by peroxidase, or non-enzymically. However, in some recent tests (Timberlake, unpublished) using freshly expressed blackcurrant juices and fruit extracts at pH values of 3–5 and a variety of experimental methods, there was little difference in the number of oxygen atoms consumed per mole of ascorbic acid oxidized between unpasteurized and pasteurized samples, but both oxygen uptakes and ascorbic acid losses were invariably greater in the heated than the unheated samples. The increased rate of oxidation depended upon the extent of the heat treatment, flash-pasteurization of juice producing only a slight increase. It is interesting that ascorbic acid is also more stable in fresh orange juice than after pasteurization (Shillinglaw and Levine, 1943), While heat inactivation of oxidation inhibitors might be possible, a more likely effect of heat would be the stimulation of non-enzymic breakdown of ascorbic acid and other components (such as sugars), with increased formation of reactive products such as hydrogen peroxide and hydroxymethylfurfural. It can be concluded therefore, in agreement with Heimann and Wisser but for different reasons, that ascorbic acid is likely to be more stable in unpasteurized than pasteurized blackcurrant juice. Heat stimulation of oxygen uptakes also occurred in blackcurrant juice after oxidation of all its ascorbic acid; in this system oxidation of other juice constituents was occurring (Timberlake, 1960b). Similar conclusions were reached by Wucherpfennig and Bretthauer (1963, 1966) who measured oxygen and ascorbic acid losses during pasteurization of blackcurrant juice. These authors examined filling techniques and emphasized the need to avoid undue aeration, high temperature, metallic contamination and exposure to light to preserve the nutritional status of the juice. Non-enzymic fixation of oxygen by orange juice components, other than ascorbic acid, which are not normally autoxidizable has also been described (Huet, 1965). Ascorbic acid oxidation in lemon and lime juices was attributed to non-enzymic reactions (Hassib and Ragab, 1962b). In orange juice, added Vitamin K_5 accelerated ascorbic acid loss on storage (Rushing and Senn, 1965). Acerola juice contains a high level of ascorbic acid (1·4–3·5%) and swelling of cans due to evolution of carbon dioxide can be a problem in its processing. The carbon dioxide originates solely from ascorbic acid (largely by mono-decarboxylation of carbon 1) under both aerobic and non-aerobic conditions (Chan *et al.*, 1966).

V. CONCLUSIONS

The problems involved in devising fruit juice standards for international usage have emphasized the difficulties of defining quality in terms of chemical composition. Such standards may act as safeguards against gross adulteration, but they do not at this stage adequately define organoleptic quality in terms of appearance, taste, aroma and the physical attributes loosely grouped as "body". This is now being generally appreciated, and the mere listing of juice components in ever-increasing numbers is now giving way to more sophisticated studies of their interactions and of their relation to quality as judged by the consumer. These studies should lead to further improvements in processing techniques; how far they will be adopted in practice will depend upon economic factors.

One of the topics of immediate concern is the need for further information on the physical behaviour and rheology of fruits and their pulps as they affect the efficiency and economics of juice extraction. The empirical approach has proved time-consuming and expensive. Much knowledge has been acquired on the occurrence and behaviour of phenolics, but certain practical aspects remain incompletely resolved. The property of astringency has been found associated with the pro-anthocyanidins or leucoanthocyanins, but the components responsible for bitterness are less clearly defined. Nor has the problem of natural colour retention in juice products yet been solved, although its urgency increases as the use of artificial colourants becomes further restricted.

REFERENCES

Aasa, R., Malmström, B., Saltman, P. and Vänngärd, T. (1964). *Biochim. Biophys. Acta* **88,** 430.

Amerine, M. A. (1958). *Adv. Fd Res.* **8,** 135.

Anon (1968). *In* "Fruit", pp. 168–186. Commonwealth Secretariat, London.

Antikainen, P. J. (1959). *Acta chem. scand.* **13,** 312.

Armerding, G. D. (1966). *Adv. Fd Res.* **15,** 305.

Arthur, J. C. and McLemore, T. A. (1956). *J. Am. chem. Soc.* **78,** 4153.

Arthur, J. C. and McLemore, T. A. (1959). *J. agric. Fd Chem.* **7,** 714.

Asen, S. and Jurd, L. (1967). *Phytochem.* **6,** 577.

Athavale, V. T., Prabhu, L. H. and Vartak, D. G. (1966). *J. inorg. nucl. Chem.* **28,** 1237.

Atkinson, F. E. and Strachan, C. C. (1950). *In* "Recent Advances in Fruit Juice Production" (V. L. S. Charley, ed.), pp. 114–124. Commonwealth Bureau of Horticulture and Plantation Crops, London.

Banga, I. (1938). *Z. physiol. Chem.* **254,** 165.

Baraud, J. (1954). *Annls Chim.* **9,** 535.

Baruah, P. and Swain, T. (1953). *Biochem. J.* **55,** 392.

Baruah, P. and Swain, T. (1959). *J. Sci. Fd Agric.* **10,** 125.

Bauernfeind, J. C. (1953). *Adv. Fd Res.* **4,** 359.

Bayer, E. (1958). *Chem. Ber.* **91,** 1115.

Bayer, E., Born, F. and Reuther, K. H. (1957). *Lebensmittelunters. u. -Forsch.* 105, 77.

Bayer, E., Nether, K. and Egeter, H. (1960). *Chem. Ber.* **93,** 2871.

Bayer, E., Egeter, H., Fink, A., Nether, K. and Wegman, K. (1966). *Angew. Chem. Int. Edn Engl.* **5,** 791.

Beech, F. W. (1958). *Soc. Chem. Ind.* Monograph No. 3, 37.

Beech, F. W., Kieser, M. E. and Pollard, A. (1964a). *Rep. agric. hort. Res. Stn Univ. Bristol* 1963, 147.

Beech, F. W., Kieser, M. E. and Pollard, A. (1964b). *Rep. agric. hort. Res. Stn Univ. Bristol* 1963, 139.

Benrath, A. (1917). *J. prakt. Chem.* **96,** 190.

Bjerrum, B., Schwarzenbach, G. and Sillen, L. G. (1957). *In* "Stability Constants. Part 1. Organic Ligands", Special Publication No. 6, p. vii. The Chemical Society, London.

Borenstein, B. and Burnell, R. H. (1966). *Adv. Fd Res.* **15,** 195.

Borgström, G. (1961). *In* "Fruit and Vegetable Juice Processing Technology" (D. K. Tressler and M. A. Joslyn, eds), pp. 179–183. AVI Publications, Connecticut.

Bourne, E. J., Nery, R. and Weigel, H. (1959). *Chemy Ind., Lond.* 998.

Brunetti, A. P., Lim, M. C. and Nancollas, G. H. (1968). *J. Am. chem. Soc.* **90,** 5120.
Bunnell, R. H. (1968). *J. agric. Fd Chem.* **16,** 177.
Burdick, E. M. (1961). *In* "Fruit and Vegetable Juice Processing Technology" (D. K. Tressler, and M. A. Joslyn, eds), pp. 874–902. AVI Publications, Connecticut.
Burroughs, L. F. (1962). *Rep. agric. hort. Res. Stn Univ. Bristol* 1961, 173.
Burroughs, L. F. (1964). *J. Sci. Fd Agric.* **15,** 176.
Burroughs, L. F. (1968). *Analyst* **93,** 618.
Byrde, R. J. W., Fielding, A. H. and Williams, A. H. (1960). *In* "Phenolics in Plants in Health and Disease" (J. B. Pridham, ed.), pp. 95–99. Pergamon Press, Oxford.
Carr, J. G., Coggins, R. A. and Whiting, G. C. (1963). *Chemy Ind., Lond.* 1279.
Celmer, R. F. (1961). *In* "Fruit and Vegetable Juice Processing Technology" (D. K. Tressler and M. A. Joslyn, eds), pp. 254–277. AVI Publications, Connecticut.
Chan, H. T., Yamamoto, H. Y. and Higaki, J. C. (1966). *J. agric. Fd Chem.* **14,** 483.
Chandler, B. V., Kefford, J. F. and Ziemelis, G. (1968). *J. Sci. Fd Agric.* **19,** 83.
Chapon, L. (1965). *J. Inst. Brewing* **71,** 299.
Charlampowicz, Z. and Gajewski, A. (1965). *Zeszyty Probl. Pospetow Nauk Rolniczych* **53,** 121; in *Chem. Abstr.* (1965) **63,** 12976g.
Charley, V. L. S. (1962). *Intern. Fruchtsaft-Union, Ber. Wiss-Tech. Komm.* **4,** 137.
Charley, V. L. S. (1963). *Fd Technol.* **17,** 33.
Charley, V. L. S. (1966). *Intern. Fruchtsaft-Union, Ber. Wiss-Tech. Komm.* **7,** 93.
Charley, P. J., Sarkar, B., Stitt, C. F. and Saltman, P. (1963). *Biochim. Biophys. Acta* **69,** 313,
Clarke, W. M. (1960). "Oxidation–reduction Potentials of Organic Systems", p. 463. Williams and Wilkins, Baltimore.
Clegg, K. M. (1964). J. Sci. *Fd Agric.* **15,** 878.
Clegg, K. M. (1966). *J. Sci. Fd Agric.* **17,** 546.
Clegg, K. M. and Morton, A. D. (1965). *J. Sci. Fd Agric.* **16,** 191.
Clegg, K. M. and Morton, A. D. (1968). *J. Fd Technol.* **3,** 277.
Clemetson, C. A. B. and Andersen, L. (1966). *Ann. N.Y. Acad. Sci.* **136,** 339.
Connor, W. A., Jones, M. M. and Tuleen, D. L. (1965). *Inorg. Chem.* **4,** 1129.
Constantinides, S. M. and Bedford, C. L. (1967). *J. Fd Sci.* **32,** 446.
Culpepper, C. W. and Caldwell, J. S. (1927). *J. agric. Res.* **35,** 107.
Curl, A. L. and Talburt, W. F. (1961). *In* "Fruit and Vegetable Juice Processing Technology" (D. K. Tressler and M. A. Joslyn, eds), pp. 410–446. AVI Publications, Connecticut.
Daepp, H. U. (1964). *Intern. Fruchtsaft-Union, Ber. Wiss-Tech. Komm.* **5,** 69.
Danziger, M. T. and Mannheim, H. C. (1967). *Fruchtsaft Ind.* **12,** 124.
Davidek, J. (1960). *Biokhimiya* **25,** 1105; in *Chem. Abstr.* (1961) **55,** 12773g.
Davidek, J. (1963). *Veda Vyzkum Prumyslu Potravinarskem* **12,** 179; in *Chem. Abstr.* (1965) **62,** 7027e.
Davis, P. S. and Deller, D. J. (1966). *Nature, Lond.* **212,** 404.
Dawson, C. R. (1966). *In* "The Biochemistry of Copper" (J. Peisach, P. Aisen and W. E. Blundberg, eds), pp. 305–337. Academic Press, New York.

Deibner, L. (1957). *Rev. Ferment. Ind. aliment.* **12,** 57, 231.
Deschreider, A. R. (1953). *Conserva* (*The Hague*), **1,** 291.
Detty, W. E., Heston, B. O. and Wender, S. H. (1955). *J. Am. chem. Soc.* **77,** 162.
Deuel, H. and Stutz, E. (1958). *Adv. Enzymol.* **20,** 341.
Diemair, W. and Jury, E. (1965). *Z. Lebensmittelunters. u. -Forsch.* **127,** 249.
Diemair, W. and Postel, W. (1964). *Wiss. Veroeffentl. Deut. Ges. Ernaehrung* **14,** 248.
Diemair, W., Koch, J. and Sajak, E. (1962). *Z. Lebensmittelunters. u. -Forsch.* **116,** 327.
Drawert, F. and Rapp, A. (1966). *Vitis* **5,** 351.
Drawert, F., Heimann, W., Emberger, R. and Tressl, R. (1965a). *Z. Naturf.* **20b,** 497.
Drawert, F., Heimann, W., Emberger, R. and Tressl, R. (1965b). *Naturwissenshaften* **52,** 304.
Drawert, F., Heimann, W., Emberger, R. and Tressl, R. (1966). *Ann. Chem.* **694,** 200.
Dubey, S. N. and Mehrotra, R. A. (1966). *J. Indian chem. Soc.* 43, 73.
Duden, R. and Siddiqui, R. (1966). *Z. Lebensmittelunters. u. -Forsch.* **132,** 1.
Dupaigne, P. (1964). *Intern. Fruchtsaft-Union, Ber. Wiss-Tech. Komm.* **5,** 27.
Embs, R. J. and Markakis, P. (1965). *J. Fd Sci.* **30,** 753.
Erdélyi, E., Dworschák, E., Vas, K., Lindner, L., Telegdy-Kovats, M. and Szóke-Szotyori, K. (1967). *Fruchtsaft Ind.* **12,** 54.
Erkama, J., Salminen, A. and Sinkonnen, I. (1953). *Suomen Kemistilehti* **26B,** 20.
Ernst, Z. L. and Herring, F. G. (1965). *Trans Faraday Soc.* **61,** 1.
Ernst, Z. L. and Menashi, J. (1963). *Trans. Faraday Soc.* **59,** 2838.
Fenton, H. J. H. (1897). *J. chem. Soc.* **71,** 375.
Finholt, P., Paulssen R. B. and Higuchi, T. (1963). *J. Pharm. Sci.* **52,** 948.
Finholt, P., Paulssen, R. B., Alsos, I. and Higuchi, T. (1965a). *J. Pharm. Sci.* **54,** 124.
Finholt, P., Alsos, I. and Higuchi, T. (1965b). *J. Pharm. Sci.* **54,** 181.
Finholt, P., Kristiansen, H., Krówczyński, L. and Higuchi, T. (1966). *J. Pharm. Sci.* **55,** 1435.
Finkle, B. J. (1967). Abstr. VII Int. Congr. Biochem. (Tokyo), J-290.
Finkle, B. J. and Nelson, R. F. (1963a). *Nature, Lond.* **197,** 902.
Finkle, B. J. and Nelson, R. F. (1963b). *Biochim. Biophys. Acta* **78,** 747.
Flaumenbaum, B. L. and Sejtpaeva, S. K. (1965). *Konserv. i Ovoshchesushil' Prom.* **20,** 4; in *Fruchtsaft Ind.* **10,** 149.
Frahn, J. L. (1958). *Aust. J. Chem.* 11, 399.
Gierschner, K. and Baumann, G. (1968). *In* "Aroma- und Geschmacks-Stoffe-in Lebensmitteln", pp. 49–89. Forster Verlag, Zurich.
Gillard, R. D., Irving, H., Parkins, R. M., Payne, N. C. and Pettit, L. D. (1966). *J. chem. Soc. A, Inorg. Phys. Theoret.* 1159.
Gore, P. H. and Newman, P. J. (1964). *Analytica chim. Acta* **31,** 111.
Gottauf, M. and Duden, R. (1965). *Z. Lebensmittelunters. u. -Forsch.* **128,** 257.
Grant, I. J. and Hay, R. W. (1965). *Aust. J. Chem.* **18,** 1189.
Guadagni, D. G. and Harris, J. (1967). *Fd Technol.* **21,** 454.
Guinot, G. Y. (1967). *Ind. Aliment. Agr.* (*Paris*) **84,** 1609.
Gurd, F. R. N. and Wilcox, P. E. (1956). *Adv. Protein Chem.* **11,** 311.
Hamilton, G. A., Hanifin, J. W. and Friedman, J. P. (1966). *J. Am. chem. Soc.* **88,** 5269.
Harel, E., Mayer, A. M. and Shain, Y. (1966). *J. Sci. Fd Agric.* **17,** 389.

Harper, K. A., Morton, A. B. and Rolfe, E. J. (1969). *J. Fd Technol.* **4,** 255.
Hassib, M. and Ragab, H. (1962a). *Z. Lebensmittelunters. u. -Forsch.* **116,** 397.
Hassib, M. and Ragab, H. (1962b). *Z. Lebensmittelunters. u. -Forsch.* **116,** 492.
Hay, R. W. (1963). *Rev. pure appl. Chem.* **13,** 157.
Heid, J. L. and Caston, J. W. (1961). *In* "Fruit and Vegetable Juice Processing Technology" (D. K. Tressler and M. A. Joslyn, eds), pp. 278–313. AVI Publications, Connecticut.
Heimann, W. and Heimann, A. (1965). *Wiss. Veroeffentl. Deut. Ges. Ernaehrung* **14,** 211.
Heimann, W. and Heinrich, B. (1959). *Fette, Seifen, Anstrichmittel* **61,** 1024.
Heimann, W. and Wisser, K. (1962). *Z. Lebensmittelunters. u. -Forsch.* **116,** 313.
Heimann, W., Wucherpfennig, K. and Reintjes, H. J. (1958). *Nahrung* **2,** 117.
Heintze, K. (1960). *Dt. Lebensmitt. Rdsch.* **56,** 194.
Hemmerich, P. (1966). *In* "The Biochemistry of Copper" (J. Peisach, P. Aisen and W. E. Blundberg, eds), pp. 15–34. Academic Press, New York.
Herrmann, J. (1958). *Flüssiges Obst* **25,** IX/21-IX/28.
Herrmann, K. (1958). *Z. Lebensmittelunters. u. -Forsch.* **108,** 152.
Hill, E. and Fields, N. L. (1966). *Fd Technol.* **20,** 77.
Hobson, G. E. (1962). *Nature, Lond.* **195,** 804.
Hobson, G. E. (1964). *Biochem. J.* **92,** 324.
Hodge, J. E. (1967). *In* "Chemistry and Physiology of Flavours" (H. W. Schultz, E. A. Day and L. M. Libbey, eds), pp. 465–491. AVI Publications, Connecticut.
Hooper, F. C. and Ayres, A. D. (1950). *J. Sci. Fd Agric.* **1,** 5.
Hörhammer, L., Hänsel, R. and Hieber, W. (1954). *Naturwissenschaften* **41,** 529.
Horner, G. (1941). *J. Soc. chem. Ind.* **60,** 62.
Horowitz, R. M. (1964). *In* "Biochemistry of Phenolic Compounds" (J. B. Harborne, ed.), pp. 545–571. Academic Press, London.
Horowitz, R. M. and Jurd, L. (1961). *J. org. Chem.* **26,** 2446.
Huang, C. P., Fennema, O. and Powrie, W. D. (1966). *Cryobiology,* **2,** 240, in *Chem. Abstr.* (1966) **65,** 7617g.
Huelin, F. E. and Stephens, I. M. (1946). *Nature, Lond.* **158,** 703.
Huelin, F. E. and Stephens, I. M. (1947). *Aust. J. Exp. Biol. Med. Sci.* **25,** 17.
Huelin, F. E. and Stephens, I. M. (1948). *Aust. J. Sci. Res.* **B1,** 58.
Huet, R. (1965). *Fruits (Paris)* **20,** 331.
Ingram, M. and Lüthi, H. (1961). *In* "Fruit and Vegetable Juice Processing Technology" (D. K. Tressler and M. A. Joslyn, eds), pp. 117–163. AVI Publications, Connecticut.
Irving, H. M. N. H. and Tomlinson, W. R. (1968). *Chem. Commun.* 497.
Jabalpurwala, K. E. and Milburn, R. N. (1966). *J. Am. chem. Soc.* **88,** 3224.
Jackson, G. A. D. and Wood, R. B. (1959). *Nature, Lond.* **184,** 902.
Jacquet, J. and Le Breton, J. (1966). *Compt. Rend.* **52,** 1054.
Jacquin, P. (1955). *Ann. Technol. Agric.* **4,** 67.
Jameson, R. F. and Neillie, W. F. S. (1965). *J. inorg. nucl. chem.* **27,** 2623.
Jennings, W. G. and Jacob, M. (1968). *Schweiz. Z. Obst-Weinbau* **104,** 19.
Jennings, W. G., Creveling, R. K. and Heinz, D. E. (1964). *J. Fd Sci.* **29,** 730.
Johnson, G., Donnelly, B. J. and Johnson, D. K. (1968). *J. Fd Sci.* **33,** 254.
Joslyn, M. A. (1961a). *In* "Fruit and Vegetable Juice Processing Technology" (D. K. Tressler and M. A. Joslyn, eds), pp. 314–333. AVI Publications. Connecticut.

Joslyn, M. A. (1961b). *In* "Fruit and Vegetable Juice Processing Technology" (D. K. Tressler and M. A. Joslyn, eds), pp. 64–116. AVI Publications, Connecticut.
Joslyn, M. A. (1962). *Adv. Fd Res.* **11,** 2.
Joslyn, M. A. and Goldstein, J. L. (1964). *Adv. Fd Res.* **13,** 179.
Joslyn, M. A. and Lukton, A. (1953). *Hilgardia* **22,** 451.
Jurd, L. (1962). *In* "The Chemistry of Flavonoid Compounds" (T. A. Geissman, ed.), pp. 107–155. Pergamon Press, Oxford.
Jurd, L. (1969). *Phytochem.* **8,** 445.
Jurd, L. and Asen, S. (1966). *Phytochem.* **5,** 1263.
Jurd, L. and Geissman, T. A. (1956). *J. Org. Chem.* **21,** 1395.
Kaeding, W. W. (1963). *J. Org. Chem.* **28,** 1063.
Kato, H. and Sakurai, Y. (1964). *Nippon Nogei Kagaku Kaishi* **38,** 536; in *Chem. Abstr.* (1965) **63,** 8648e.
Kato, H., Ichihata, H. and Fujimaki, M. (1963). *J. agric. Chem. Soc. Japan* **37,** 220 in *Chem. Abstr.* **62,** 16885d.
Kay, G. (1968). *Process Biochem.* **3** (8), 36.
Kefford, J. F. (1964). *Intern. Fruchtsaft-Union Ber. Wiss-Tech. Komm.* **5,** 53.
Keilin, D. and Hartree, E. F. (1955). *Biochem. J.* **60,** 310.
Kern, A. (1961). *Intern. Fruchtsaft-Union Ber. Wiss-Tech, Komm.* **3,** 33.
Kern, A. (1962). *Schweiz. Z. Obst-Weinbau* **71,** 190.
Kern, A. (1964). *Intern. Fruchtsaft-Union Ber. Wiss-Tech. Komm.* **5,** 203.
Khan, M. M. T. and Martell, A. E. (1967a). *J. Am. chem. Soc.* **89,** 4176.
Khan, M. M. T. and Martell, A. E. (1967b). *J. Am. chem. Soc.* **89,** 7104.
Kieser, M. E., Pollard, A. and Williams, A. H. (1953). *Chemy Ind., Lond.* 1260.
Kieser, M. E., Pollard, A. and Timberlake, C. F. (1957). *J. Sci. Fd Agric.* **8,** 151.
Kieser, M. E., Pollard, A. and Sissons, D. J. (1960). *Rep. agric. hort. Res. Stn Univ. Bristol* 1959, 145.
Kim, M. K. and Martell, A. E. (1966). *J. Am. chem. Soc.* **88,** 914.
Kläui, H. and Manz, U. (1967). *Beverages* **8,** 16.
Klotz, I. M., Czerlinski, G. H. and Fiess, H. A. (1958). *J. Am. chem. Soc.* **80,** 2920.
Knapp, F. W. (1965). *J. Fd Sci.* **30,** 930.
Koch, J. (1955). *Flüssiges Obst* **22,** 9.
Koch, J. and Kleesat, R. (1961). *Fruchtsaft Ind.* **6,** 107.
Koch, J. and Sajak, E. (1964). *Fruchtsaft Ind.* **9,** 26.
Körmendy, I. (1964). *J. Fd Sci.* **29,** 631.
Körmendy, I. (1965). *Fruchtsaft Ind.* **10,** 246.
Krbechek, L., Inglett, G., Holik, M., Dowling, B., Wagner, R. and Riter, R. (1968). *J. agric. Fd Chem.* **16,** 108.
Kurata, T. and Sakurai, Y. (1967a). *Agric. biol. Chem.* (*Tokyo*) **31,** 170.
Kurata, T. and Sakurai, Y. (1967b). *Agric. biol. Chem.* (*Tokyo*) **31,** 177.
Kurata, T., Wakabayashi, H. and Sakurai, Y. (1967). *Agric. biol. Chem., Tokyo* **31,** 101.
Kuusi, T. (1961). *Valtion Tek. Tutkimuslaitos Tiedotus*, Sarja IV **34,** 5.
Lavollay, J. and Neumann, J. (1959). *In* "The Pharmacology of Plant Phenolics" (J. W. Fairbairn, ed.), pp. 103–122. Academic Press, London.
Lavollay, J. and Patron, A. (1950). "Proceedings of the 2nd International Congress of Fruit Juice Producers, Zurich", pp. 47–76.
Lefebvre, J. (1957). *J. Chim. Phys.* **54,** 553.

L'Heureux, G. A. and Martell, A. E. (1966). *J. inorg. nucl. Chem.* **28,** 481.
Letan, A. (1966a). *J. Fd Sci.* **31,** 518.
Letan, A. (1966b). *J. Fd Sci.* **31,** 395.
Leuprecht, H. and Schaller, A. (1968). *Fruchtsaft Ind.* **12,** 1, 60.
Levandoski, N. G., Baker, E. M. and Canham, J. E. (1964). *Biochemistry* **3,** 1465.
Lewis, W. M. (1966). *J. Sci. Fd Agric.* **17,** 316.
Livingston, G. E. (1953). *J. Am. chem. Soc.* **75,** 1342.
Lockett, M. F. (1959). *In* "The Pharmacology of Plant Phenolics" (J. W. Fairbairn, ed.), pp. 81–89. Academic Press, London.
Luckwill, L. C. and Pollard, A. (editors) (1963). "Perry Pears". University of Bristol.
Lueck, H. (1965). *Z. Lebensmittelunters. u. -Forsch.* **126,** 193.
Lüthi, H. (1955). *Flüssiges Obst* **22,** 4.
Lüthi, H. R. (1965). *Intern. Fruchtsaft-Union, Ber. Wiss-Tech. Komm.* **6,** 87.
Lüthi, H. R. (1968). *Schweiz. Z. Obst-Weinbau* **104,** 34, 60.
Maier, V. P. and Beverly, G. D. (1968). *J. Fd Sci.* **33,** 488.
Mapson, L. W. (1945). *Biochem. J.* **39,** 228.
Markakis, P. and Embs, R. J. (1966). *J. Fd Sci.* **31,** 807.
Markham, K. R. and Mabry, T. J. (1968). *Phytochem.* **7,** 1197.
Marteau, G. (1967). *Ann. Nutr. Aliment.* **21,** B223.
Marteau, G., Scheur, J. and Oliveri, C. (1961). *Ann. Technol. Agric.* **10,** 161; in *Chem. Abstr.* (1962) **56,** 6473i.
Martell, A. E. (1964). "Stability Constants of Metal–Ion Complexes. Section II. Organic Ligands." Chemical Society Special Publication No. 17. London.
Mason, H. S. (1965). *A. Rev. Biochem.* **34,** 595.
Mayer, A. M. (1962). *Phytochem.* **1,** 237.
McCance, R. A. and Widdowson, E. M. (1960a). *In* "The Composition of Foods" (M.R.C. Spec. Rept. Series No. 297), pp. 170–186. H.M.S.O., London.
McCance, R. A. and Widdowson, E. M. (1960b). *In* "The Composition of Foods" (M.R.C. Spec. Rept. Series No. 297), pp. 70–81. H.M.S.O., London.
Mehrlich, F. P. (1961). *In* "Fruit and Vegetable Juice Processing Technology" (D. K. Tressler and M. A. Joslyn, eds), pp. 746–786. AVI Publications, Connecticut.
Menoret, Y. (1962). *Inds aliment. agric.* **79,** 419.
Mitsui, S., Hayashi, K. and Hattori, S. (1959). *Proc. Japan Acad.* **35,** 169.
Monties, B. (1966). *Annls Physiol. vég.* (*Paris*) **8,** 101.
Monties, B. and Barret, A. (1965). *Ann. Technol. Agric.* **14,** 167.
Morgan, A. I., Lowe, E., Merson, R. L. and Dunkee, E. L. (1965). *Fd Technol.* **19,** 1790.
Moyer, J. C. and Saravacos, G. D. (1968). *Intern. Fruchtsaft-Union VII Congr. Rep.*, 109.
Moyls, A. (1964). *Intern. Fruchtsaft-Union Ber. Wiss-Tech. Komm.* **5,** 239.
Muller, J. G. (1967). *Fd Technol.* **21,** 49.
Murakami, Y., Nakamura, K. and Tokunaga, M. (1963). *Bull. chem. Soc. Japan* **36,** 669.
Murdock, D. I. (1968). *Fd Technol.* **22,** 90.
Nakayama, T. O. M. and Chichester, C. O. (1963). *Nature, Lond.* **199,** 72.
Näsänen, R. and Markkanen, R. (1956). *Suomen Kemistilehti* **B29,** 119.
Neal, A. E. (1965). *J. Sci. Fd Agric.* **16,** 604.
Neher, E. (1964). *Vezyr. Vegyip. Eqy. Közl.* **8,** 316. English summary.

Nelson, R. E. and Finkle, B. J. (1964). *Phytochem.* **3,** 321.
Noel, G. L. and Robberstad, M. T. (1963). *Fd Technol.* **17,** 947.
Nord, H. (1955). *Acta Chem. Scand.* **9,** 442.
Norman, R. O. C. and Smith, J. R. L. (1965). *In* "Oxidases and Related Systems" (T. E. King, H. S. Mason and M. Morrison, eds), Vol. 1, pp. 131–156. John Wiley, London.
Nyiri, L. (1968). *Process Biochem.* **3** (8), 27.
Ogata, Y., Kosugi, Y. and Morimoto T. (1968). *Tetrahedron* **24,** 4057.
Oka, Y. and Harada, H. (1967). *Nippon Kagaku Zasshi* **88,** 441; in *Chem. Abstr.* (1967) **67,** 36838b.
Olsen, R. W. and Hill, E. G. (1965). *Citrus Ind.* **46,** 21; in *Fruchtsaft Ind.* (1967) **12,** 18.
Onishi, I. and Hara, T. (1964). *Bull. Chem. Soc. Japan* **37,** 1317.
Osman, A. E. and Höfner, W. (1966). *Z. Lebensmittelunters. u. -Forsch.* **129,** 139.
Ott, J. (1965). *Fruchtsaft Ind.* **10,** 79.
Park, M. V. (1966). *J. chem. Soc.* 816.
Peng, C. Y. and Markakis, P. (1963). *Nature, Lond.* **199,** 597.
Perrin, D. D. (1958a). *J. chem. Soc.* 3120.
Perrin, D. D. (1958b). *J. chem. Soc.* 3125.
Perrin, D. D. (1965). *Nature, Lond.* **206,** 170.
Perrin, D. D., Sayce, I. G. and Sharma, V. S. (1967). *J. chem. Soc. A. Inorg. Phys. Theoret.* 1755.
Pierpont, W. S. (1966). *Biochem. J.* **98,** 567.
Pilnik, W. (1969). *Flüssiges Obst* **36,** 39.
Pollard, A. (1950). *In* "Recent Advances in Fruit Juice Production" (V. L. S. Charley, ed.), pp. 125–144, Commonwealth Bureau of Horticulture and Plantation Crops, London.
Pollard, A. and Beech, F. W. (1966a). *Rep. agric. hort. Res. Stn Univ. Bristol,* 1965, 259.
Pollard, A. and Beech, F. W. (1966b). *Process Biochem.* **1** (4), 229, 238.
Pollard, A. and Kieser, M. E. (1951). *J. Sci. Fd Agric.* **2,** 30.
Pollard, A. and Kieser, M. E. (1959). *J. Sci. Fd Agric.* **10,** 253.
Pollard, A., Kieser, M. E. and Bryan, J. D. (1946). *Rep. agric. hort. Res. Stn Univ. Bristol* 1945, 200.
Pollard, A., Kieser, M. E. and Timberlake, C. F. (1955). *Fd Manuf.* **30,** 355.
Pollard, A., Kieser, M. E. and Sissons, D. J. (1958). *Chemy Ind., Lond.* 952.
Pribela, A. and Betusova, M. (1964). *Fruchtsaft Ind.* **9,** 15.
Pribela, A. and Strimiska, F. (1967). *Flüssiges Obst* **34,** 83, 136.
Prillinger, F., Madner, A. and Kovacs, J. (1968). *Mitt (Klosterneuburg) Ser. A (Rebe Wein)* **18,** 98.
Primo-Yufera, E. and Royo Iranzo, J. (1967). *Intern. Fruchtsaft.-Union, Ber. Wiss-Tech. Komm.* **8,** 199.
Prue, J. E. (1952). *J. chem. Soc.* 2331.
Pruthi, J. S. (1963). *Adv. Fd Res.* **12,** 203.
Pyatnitski, I. V. and Glushchenko, L. M. (1966). *Ukr. Khim. Zh.* **32,** 1220; in *Chem. Abstr.* (1967) **66,** 32402m.
Rajan, K. S. and Martell, A. E. (1967). *J. inorg. nucl. Chem.* **29,** 463.
Rankine, B. C. (1960). *Aust. J. Appl. Sci.* **11,** 305, 500.
Reed, G. (1966). *In* "Enzymes in Food Processing" (M. L. Anson, C. O. Chichester, E. M. Mrak and G. F. Stewart, eds), pp. 73–87; 301–322. Academic Press, New York.

Rentschler, H. (1963). *Annls. Technol. agric. Numero Hors Ser.* 1. **12,** 267.
Rentschler, H. and Tanner, H. (1960). *Schweiz. Z. Obst-Weinbau* **69,** 368.
Resnik, R., Cohen, T. and Fernando, Q. (1961). *J. Am. chem. Soc.* **83,** 3344.
Reynolds, T. M. (1963). *Adv. Fd Res.* **12,** 1.
Reynolds, T. M. (1965). *Adv. Fd Res.* **14,** 167.
Ribéreau-Gayon, J. and Peynaud, E. (1961). *In* "Traité d'Oenologie", Vol. 11, pp. 305–381. Librarie Polytechnique Ch. Béranger, Paris et Liège.
Ritsma, J. H., Wiegers, G. A. and Jellinek, F. (1965). *Rec. Trav. Chim.* **84,** 1577.
Robbins, R. H. (1967). *Process Biochem.* **2** (6), 47.
Robbins, R. H. (1968). *Process Biochem.* **3** (5), 38.
Ross, E. and Chang, A. T. (1958). *J. agric. Fd Chem.* **6,** 610.
Rouse, A. H. and Atkins, C. D. (1952). *Fd Technol.* **6,** 291.
Royo Iranzo, J. (1965). *Intern. Fruchtsaft-Union, Ber. Wiss-Tech. Komm.* **6,** 175.
Ruck, J. A. and Kitson, J. A. (1965). *Intern. Fruchtsaft-Union, Ber. Wiss-Tech. Komm.* **6,** 433.
Rushing, N. B. and Senn, V. J. (1965). *J. Fd Sci.* **30,** 178.
Saito, N. and Hayashi, K. (1965). *Sci. Rept. Tokyo Kyoiku Daigaku,* **12B,** 39.
Saito, N., Mitsui, S. and Hayashi, K. (1961). *Proc. Japan Acad.* **37,** 485.
Sakamura, S., Shibusa, S. and Obata, Y. (1966). *J. Fd Sci.* **31,** 317.
Samoradova-Bianki, G. B. (1965). *Biokhimiya* **30,** 248; in *Chem. Abstr.* (1965) **63,** 3545a.
Sarkar, B., Saltman. P., Benson, S. and Adamson, A. (1964). *J. inorg. nucl. Chem.* **26,** 1551.
Sawyer, R. (1963). *J. Sci. Fd Agric.* **14,** 302.
Saxena, G. M. and Seshadri, T. R. (1957). *Proc. Indian Acad. Sci. Sect. A.* **46,** 218.
Scaife, J. F. (1959). *Can. J. Biochem. Physiol.* **37,** 1049.
Schaller, A. (1965). *Fruchtsaft Ind.* **10,** 263.
Schillinger, A. (1966). *Z. Lebensmittelunters. u. -Forsch.* **129,** 65.
Schneyder, J. (1964). *Mitt. (Klosterneuburg) Ser. A (Rebe Wein)* **14,** 282.
Schulz, T. H., Black, D. R., Bomben, J. L., Mon, T. R. and Teranishi, R. (1967). *J. Fd Sci.* **32,** 698.
Schulz, W. W., Mendel, J. E. and Phillips, J. F. (1966). *J. inorg. nucl. Chem.* **28,** 2399.
Scott, W. C., Kew, T. J. and Veldhuis, M. K. (1965). *J. Fd Sci.* **30,** 833.
Seale, P. E. (1967). *Fd Technol., Australia* **19,** 233, 239.
Shannon, C. T. and Pratt, D. E. (1967). *J. Fd Sci.* **32,** 479.
Sharf, J. M. (1967). *Intern. Fruchtsaft-Union, Ber. Wiss-Tech. Komm.* **8,** 143.
Shaw, B. L. and Simpson, T. H. (1955). *J. chem. Soc.* 655.
Shillinglaw, C. A. and Levine, M. (1943). *Fd Res.* **8,** 453.
Simpson, T. H. and Garden, L. (1952a). *J. chem. Soc.* 4638.
Simpson, T. H. and Garden, L. (1952b). *J. chem. Soc.* 5027.
Smith, T. D. (1965). *J. chem. Soc.* 2145.
Smock, R. M. and Neubert, A. M. (1950). *In* "Apples and Apple Products" (Z. I. Kertesz, ed.), pp. 264–265. Interscience Publishers, New York.
Song, P. S. and Chichester, C. O. (1967). *J. Fd Sci.* **32,** 107.
Spanyar, P., Kevei, E. and Blazovich, M. (1963). *Z. Lebensmittelunters. u. -Forsch.* **120,** 1.
Spanyar, P., Kevei, E. and Blazovich, M. (1964a). *Z. Lebensmittelunters. u. -Forsch.* **123,** 418.
Spanyar, P., Kevei, E. and Blazovich, M. (1964b). *Z. Lebensmittelunters. u. -Forsch.* **124,** 405.

Spanyar, P., Kevei, E. and Blazovich, M. (1964c). *Z. Lebensmittelunters. u. -Forsch.* **126,** 10.

Spanyar, P., Kevei, E. and Blazovich, M. (1966). *Z. Lebensmittelunters. u. -Forsch.* **132**, 129.

Stanley, W. L. (1965). *Intern. Fruchtsaft-Union, Ber. Wiss-Tech. Komm.* **6,** 207.

Starr, M. S. and Francis, F. J. (1968). *Fd Technol.* **22,** 1293.

Stolyarov, K .P. and Amantova, I. A. (1964). *Vestn. Leningr. Univ.* 19, *Ser, Fiz. i. Khim No.* 2, 113; in *Chem. Abstr.* (1964) **61,** 10195h.

Susic, M. V. (1955). *Bull. Inst. Nucl. Sci. "Boris Kidrich" (Belgrade)* No. 81, 65.

Swain, T. (1965). *Intern. Fruchtsaft-Union, Ber. Wiss-Tech. Komm.* **6,** 221.

Swindells, R. and Robbins, R. H. (1966). *Process Biochem.* **1** (9), 457, 469.

Széchényi, L. (1964). *Ind. aliment. Agric. (Paris)* **81,** 309.

Täufel, K. and Voigt, J. (1964). *Z. Lebensmittelunters. u. -Forsch.* **126,** 19.

Takeda, K., Mitsui, S. and Hayashi, K. (1966). *Bot. Mag. Tokyo* **79,** 578.

Tanner, H. and Vetsch, V. (1956). *Schweiz. Z. Obst-Weinbau* **65,** 238, 261.

Timberlake, C. F. (1952). *Rep. agric. hort. Res. Stn Univ. Bristol* 1951, 160.

Timberlake, C. F. (1957a). *J. Sci. Fd Agric.* **8,** s66.

Timberlake, C. F. (1957b). *J. chem. Soc.* 4987.

Timberlake, C. F. (1957c). *J. Sci. Fd Agric.* **8,** 159.

Timberlake, C. F. (1958). "Metal Complexes in Fruit Juices", Proc. Symposium Fruit Juice Concentrates, Bristol, pp. 291–301. Juris-Verlag, Zürich.

Timberlake, C. F. (1959a). "Studies of Copper and its Complexes in Relation to Storage Changes in Fruit Juices". Ph.D. Thesis, University of Bristol.

Timberlake, C. F. (1959b). *J. chem. Soc.* 2795.

Timberlake, C. F. (1960a). *J. Sci. Fd Agric.* **11,** 258.

Timberlake, C. F. (1960b). *J. Sci. Fd Agric.* **11,** 268.

Timberlake, C. F. (1961). *Rep. agric. hort. Res. Stn Univ. Bristol* 1960, 147.

Timberlake, C. F. (1964a). *J. Chem. Soc.* 1229.

Timberlake, C. F. (1964b). *J. chem. Soc.* 5078.

Timberlake, C. F. (1964c). *Rep. agric. hort. Res. Stn Univ. Bristol* 1963, 39.

Timberlake, C. F. and Bridle, P. (1966). *Nature, Lond.* **212,** 158.

Timberlake, C. F. and Bridle, P. (1967). *J. Sci. Fd Agric.* **18,** 479.

Timberlake, C. F. and Bridle, P. (1968). *Chemy. Ind., Lond.* 1489.

Ting, S. V. and Newhall, W. F. (1965). *J. Fd Sci.* **30,** 57.

Tressler, D. K. and Joslyn, M. A. (1961). "Fruit and Vegetable Juice Processing Technology." AVI Publications, Connecticut.

Uprety, M. C. and Revis, B. (1964). *J. Pharm. Sci.* **53,** 1248.

Van Buren, J., Senn, G. and Neukom, H. (1965). *Intern. Fruchtsaft-Union, Ber. Wiss-Tech. Komm.* **6,** 245.

Vas, K., Nedbalek, M., Scheffer, H. and Kovaks-Proszt, G. (1967). *Fruchtsaft Ind.* **12,** 164.

Vasatko, J. and Pribela, A. (1965). *Isv. Vysshich. Uchebn. Zavedenii Pishchevaya Teknol* (6), 17; in *Chem. Abstr.* (1966) **64,** 9930h.

Veldhuis, M. K. (1961). *In* "Fruit and Vegetable Juice Processing Technology" (D. K. Tressler and M. A. Joslyn, eds), pp. 838–873. AVI Publications, Connecticut.

Veselinovic, D. S. and Susic, M. V. (1965). *Glas. Hem. Drus. Beograd* **30,** 63; in *Chem. Abstr.* (1967), **66,** 41148k.

Vines, H. M. and Oberbacher, M. F. (1963). *Plant Physiol., Lancaster* **38,** 333.

Wagner, J. R., Miers, J. C., Sanshuck, D. W. and Becker, R. (1968). *Fd Technol.* **22,** 1484.
Walker, J. R. L. and Hulme, A. C. (1965). *Phytochem.* **4,** 677.
Walker, J. R. L. and Hulme, A. C. (1966). *Phytochem.* **5,** 259.
Walker, J. R. L. and Reddish, C. E. S. (1964). *J. Sci. Fd Agric.* **15,** 902.
Weinland, R. F. and Binder, K. (1912). *Chem. Ber.* **45,** 148.
Weissberger, A. and Lu Valle, J. E. (1944). *J. Am. chem. Soc.* **66,** 700.
Wieland, H. and Franke, W. (1928). *Ann. Chem.* **464,** 101.
Williams, R. J. P. (1959). *In* "The Enzymes" (P. D. Boyer, H. Lardy and K. Myrbäck, eds), Vol. 1. 2nd edition, pp. 391–442. Academic Press, New York and London.
Willits, C. O., Underwood, J. C. and Merten, U. (1967). *Fd Technol.* **21,** 24.
Winter, M., Palloy, E., Hinder, M. and Willhalm, B. (1962). *Helv. chim. Acta* **45,** 2186.
Wucherpfennig, K. and Bretthauer, G. (1962). *Fruchtsaft Ind.* **7,** 40.
Wucherpfennig, K. and Bretthauer, G. (1963). *Flüssiges Obst.,* **30,** XI/6-XI/8; XII/2-XII/5.
Wucherpfennig, K. and Bretthauer, G. (1964). *Intern. Fruchtsaft-Union, Ber. Wiss-Tech. Komm.* **5,** 105.
Wucherpfennig, K. and Bretthauer, G. (1966). *Fruchtsaft Ind.* **11,** 11.
Yamamoto, R. and Yamamoto, E. (1964). *Yakuzaigaku* **24,** 309; in *Chem. Abstr.* (1965), **63,** 1663e.
Zook, E. G. and Lehmann, J. (1968). *J. Am. diet. Ass.* **52,** 225.
Zubekis, E. (1965). *Fruchtsaft Ind.* **10,** 198.
Zyzlink, V. (1965). *Nahrung* **9,** 417.

Chapter 18

Dehydrated Fruit

D. McG. McBEAN, M. A. JOSLYN AND F. S. NURY

Division of Food Research, C.S.I.R.O., New South Wales, Australia; Department of Nutritional Sciences, University of California, Berkeley, California, U.S.A.; Department of Food Science, Frenso State College, Frenso, California, U.S.A.

I	Introduction	623
II	Maturity and Variety	625
III	Colour Changes	626
	A. Enzymic	627
	B. Non-enzymic	629
IV	Taste and Flavour	629
V	Preparation for Drying	630
	A. Harvesting and Handling	631
	B. Pre-drying Treatments	634
	C. Sulphuring	637
VI	Drying	637
	A. Moisture Transfer	637
	B. Physical Barriers to Moisture Transfer	639
	C. Biochemical Factors Limiting Moisture Transfer	641
VII	Storage	642
VIII	Rehydration	647
IX	Conclusion	648
	References	648

I. INTRODUCTION

The water content of fruits during their development and maturation is maintained by the bearing tree or vine which supplies a steady flow of nutrients as well as water. This water balance is disturbed at about the time of the onset of senescence. Water is then lost from the fruits at a faster rate than it is supplied by the tree, causing them to shrivel and eventually fall to the ground.

Under arid conditions such fruits continue to lose water; if the relative humidity is in the region of 60%, equilibrium will be established at 10–15% water content. At these levels they will remain free from spoilage for some considerable time. This simple fact undoubtedly explains how inhabitants of the eastern Mediterranean area came to preserve some of their excess fruit production of grapes, dates and figs very many years ago. Subsequently, they improved on this "natural" process by harvesting the fruits before they shrivelled on the tree, laying them out in the sun to "dry". A high sugar content was essential to the success of this process. The sun's energy is still employed today in drying the majority of this type of fruit. The method of drying is called sun-drying.

Towards the end of the 19th century the type of fruits dried was expanded to include apricots, peaches, plums (prunes), apples and pears. For good results, however, it was found necessary to supply heat from an artificial source, since sun-drying is only efficient in arid areas. The term "dehydration" was applied to these "artificial" methods of drying which were introduced at the beginning of the present century. Later, additional procedures such as cutting, skin treatments to crack the skin and the exposure to fumes from burning sulphur were introduced to expedite drying, to improve the colour of the product and to reduce microbiological spoilage. Vacuum drying techniques, including programmed pressure–temperature cycles to promote puff drying, were introduced later. Since World War II many sophisticated drying methods have been developed such as foam-mat, fluidized bed, osmotic—and freeze-drying. These developments have led to the feasibility of dehydration of more fruits, including bananas, pineapples, papayas, citrus fruits and berries. Although modern methods of freeze-drying have greatly improved the quality of frozen meats, vegetables and fruits, vacuum and freeze-drying are expensive and, so far as the authors are aware, have not gone beyond the experimental stage as regards fruits. Reconstituted vacuum or freeze-dried fruits, have little, if any, advantage over air-dried fruits, all losing the turgor of the fresh material. Thus it is clear why the more sophisticated and expensive drying methods have not been widely applied commercially.

In spite of the early inception of drying and dehydration as a method of food preservation, our knowledge of the physical and biological factors involved in determining the adaptability of a particular variety or kind of fruit to preservation by drying, of the changes occurring during preparation for and actual dehydration, and of the subsequent storage deterioration is still largely based on empirically determined, rather than scientifically known, facts. As late as 1953, Richert emphasized that most of our knowledge of dried fruit is based on technological rather than scientific evidence. This is also quite apparent in the comprehensive historical review of the principles of fruit and vegetable dehydration by Phaff (1951). Since that time the technology of drying and dehydration has been developed largely by evaluation of the existing early practices and their improvement, rather than on the basis of fundamental investigations of the physical, chemical and biochemical phenomena that influence this process.

The older publications of Cruess and his collaborators in California are still valuable in depicting the development of the present practices. These are presented in the bulletins and circulars of the California Agricultural Experiment Station, such as Cruess *et al.* (1920); Cruess and Christie (1921); Christie and Barnard (1925); Christie (1926); Nichols and Christie (1930a, b) and Nichols (1933). More recent publications of this type are those of Cruess (1943); Mrak and Long (1941); Long *et al.* (1940); Perry *et al.* (1946) and Mrak and Phaff (1947). Specific details for drying and dehydration of fruits were summarized by Mrak and Phaff (1949); Mrak and Perry (1948a, b) and Phaff and Mrak (1948). Drying and dehydration also were investigated very early in Oregon by Wiegand and Bullis (1929, 1931), and in the U.S. Department of Agriculture by Caldwell (1923). One of the important advances was the early demonstration by Eidt (1938) in Canada that two-stage dehydration is applicable to, and preferable to, the old kiln drying of apples. The technological data relating to this type of dehydration is summarized in the texts of von Loesecke (1955), Morris (1947), Van Arsdel and Copley (1964), and in chapters in various reference works such as those of Mrak (1942); Mrak and Mackinney (1951); Cruess (1958); Joslyn (1963) and Brekke and Nury (1964).

II. MATURITY AND VARIETY

Best-quality dried fruits are invariably produced from raw material which is fully mature. In terms of structural and chemical definition, fruits for drying should have reached that stage when they are still firm enough to withstand harvesting and processing practices without bruising or breakage of the skin whereby enzyme systems are stimulated to produce browning. Once present this colour defect cannot be removed. By this stage, green colour due to chlorophyll should have disappeared and the characteristic natural colour should be near its maximum. In addition, starch should have been converted to sugar, and the sugar : acid ratio should be near its highest level. Flavour volatiles should also be near their peak concentration. While these general biochemical requirements are known, the optimum harvest time is usually judged subjectively and with the personal bias which is inherent in such a method. Objective methods have been devised and tested, but these usually lack the combination of reliability, speed and simplicity which seems essential for field usage. Probably the most useful field method is that in which the change in soluble solids is followed using a refractometer. Soluble solids generally increase as fruits mature, but it is not always reliable to compare the changes in one season with those of another; district-to-district and even orchard-to-orchard variations may be large.

Even though the chief requirements of fruits for drying are known, few attempts have been made to breed varieties which are more suitable for drying.

One prime prerequisite is high solids content so that the yield of dried product is high, and the amount of water to be removed is low. Types of fruit containing large amounts of sugar are most suitable for drying.

The limited knowledge concerning the role of constituents in determining the suitability of a particular variety of fruit for dehydration and of the changes occurring during drying and storage is well illustrated in investigations on the effect of maturity. As early as 1915 Bioletti reported that yield of raisins was influenced by the maturity of the grapes when harvested for drying. This was confirmed by Cruess *et al.* (1920); Cruess and Christie (1921); Christie and Barnard (1925); Nichols and Christie (1930a) and Jacob (1942). There was however, little basic difference between the early data of Bioletti and those of Jacob.

III. COLOUR CHANGES

A. Enzymic†

The enzyme-catalysed oxidative darkening which occurs in whole fruits such as dates, figs and prunes during drying and in apples, apricots, peaches and pears when cut, is due largely to polyphenolase action on the naturally occurring substrates such as chlorogenic acid and catechins in apple and pear fruits, caffeoyl-shikimic acid in dates, 3,4-dihydroxyphenyl ethylamine in bananas and catechins in grapes (Joslyn and Ponting, 1951; Pridham, 1963). In early studies, peroxidase was also believed to be involved in such enzymic browning. Thus, Hussein *et al.*, (1942) implicated grape peroxidase in the darkening of raisins, and Chari *et al.* (1948) considered it to be involved in the darkening of prunes. However, the work of Rahman and Joslyn (1954) and Deibner and Rifai (1963) refutes this suggestion.

Phenolase activity decreases rapidly with increase in soluble solids content, particularly sucrose content. Rahman and Joslyn (1954) reported inhibition of both prune phenolase and peroxidase by sucrose or invert sugar, and Ponting and Stanley (1968) found that sucrose inhibited the phenolase of apples. Dates, grapes, figs and prunes, which are generally dried without any enzyme inhibition, darken during water removal and are acceptable in this form. In these circumstances the initial enzyme activity is repressed and ultimately stopped by the increasing concentration of solutes in the plant tissue. With other fruits in which darkening is unacceptable, means of preventing it must be employed. For this purpose, heat in the form of hot water or steam (blanching) is not, at present, used extensively in the dried fruit industry. Sulphur dioxide applied principally to prevent non-enzymic browning during drying and storage is, however, widely used and is highly inhibitory to the polyphenolase. (See Section VC). Gincarevic and Hawker (1971) recently reported that browning of sultanas during drying under mild conditions is

† See also Volume 1, Chapter 11.

caused by the action of O-diphenol oxidase on substrates released from the skin tissues. Dipping or spraying with an oil emulsion prevents browning by increasing the sugar concentration.

It has long been known that sulphite inhibits polyphenol oxidase-catalysed browning, but the mechanism of this effect of sulphite has only recently been elucidated. Embs and Markakis (1965) reported that sulphite prevented oxidation of catechol by mushroom polyphenolase by combining with the enzymatically produced o-quinones and preventing their condensation to melanins. Markakis and Embs (1966) subsequently reported that ascorbic acid facilitates sulphite inhibition of phenolase.

Although enzyme activity is suppressed in those fruits to which sulphite is added such as apricots, peaches, apples and pears, the main value of the SO_2 treatment here is the suppression of non-enzymic browning; SO_2 treatment is dealt with in Section VC.

B. Non-enzymic

1. *Chlorophyll and carotenoids*†

It has long been known that dehydrated cut fruit, such as apricots, peaches and pears yield a dried product that is not as acceptable as when the same fruit is sun-dried. Part of the difference is due to the apparent higher retention of chlorophyll or to its conversion into pheophytin during dehydration. During ripening on the tree, particularly when exposed to sunlight, chlorophyll is decomposed to colourless, still unknown products. If half of a fruit is protected from light by leaves it still retains some chlorophyll but this also is destroyed during sun-drying. When such fruit is dehydrated the chlorophyll in the leaf-shaded half is either partly retained or decomposed to the brown-coloured pheophytin. The nature of the apparently photochemical decomposition of chlorophyll in greenish fruit during sun-drying is not known. Many attempts have been made to duplicate this process on dehydration; the use of artificial light during the process has failed to give a satisfactory breakdown of the chlorophyll. However, Mrak and Phaff (1943) showed that blanching of cut fruit prior to sulphuring (see later) resulted in the desired deep-orange colour in apricots and peaches.

Both the plastid pigments, chlorophyll and the carotenoids, are involved in the colour and colour retention of dried fruits. Mackinney (1937) reported that, although the chlorophyll content of green Sultanina grapes is much lower than that in the leaves, the ratio of chlorophyll "a" to chlorophyll "b" is higher in the fruit than in the leaves. Even after sulphuring and drying the grapes retain some chlorophyll, and Mackinney found that the level was not much lower in amber-coloured raisins than in greenish ones.

Data are available on the carotenoids in fresh fruits but data on changes in

† See Volume 1, Chapter 12.

these compounds during drying and subsequent storage are limited. Carotenoids are responsible for the yellow to reddish pigments which give dried apricots and peaches their pleasing colours. Apricots contain principally β-carotene, phytoene, phytofluene, lycopene, cryptoxanthin and lutein (Curl, 1960). The characteristic colour of sun-dried apricots is reddish while that of dehydrated fruit is yellow although their spectral absorption curves are identical (McBean, unpublished data).

2. *Browning (the Maillard reaction) and other colour changes*

The condensation of reducing sugars with amino acids, a process accelerated by heat, is responsible for much of the darkening which occurs in fruit tissues during storage; it is also very probably responsible for the major part of the darkening occurring during the drying of fruits.

Nearly all fruits have low protein contents but generally those species which are dried have levels which are less than 1% of the fresh weight. However, this small amount of material is of vast biochemical significance since enzymes consist of or contain proteins. In addition, the amino groups of protein or of free amino acids are involved in darkening of the tissue during and after drying. Free amino acids and related amines have also been found in fruits (see Volume 1, Chapter 5). There are many gaps in our knowledge of amino compounds in fruits and a comprehensive survey of those present in fruits used for drying would provide useful information relative to quality changes. Many of the latter are due to non-enzymatic, Maillard-type, browning reactions which often produce sufficient pigments to render the dried fruits inedible after a term of storage.

Possible courses of the browning reaction have been considered in a number of reviews, notably by Hodge (1953) and by Reynolds (1963, 1965) who suggested that the organic acids present in the fruit may participate in the reaction. Recent data are reported by Cole (1967).

The mechanism of the inhibition of the Maillard reaction by SO_2 was investigated by Song and Chichester (1966, 1967) and by Song *et al.* (1966). They reported that bisulphite inhibited the reaction prior to the one giving "steady-state" browning, and suggested that an active form of bisulphite combined with an intermediate compound involved in the overall browning reaction. They considered that free radicles formed from bisulphite were concerned in the process. Bisulphite reaction with sugar intermediates to form stable sulphonated products is another explanation of the inhibition process (Braverman, 1953; Ingles, 1959; McWeeney *et al.*, 1969).

Anthocyanin pigments in fruits occur principally in and just below the skin although in freestone peaches they may also be present in the pit cavity. The colour imparted by the anthocyanins is generally red and becomes intensified as the fruits mature. These somewhat unstable compounds change colour during drying. If sulphuring is not employed they become greyish. On the other hand in fruits where SO_2-treatment is needed as a preservative, while the anthocyanins may be bleached initially, the colour they impart may return in an enhanced form as much of the SO_2 becomes lost. The chemistry of the bleaching of anthocyanins by SO_2 was investigated by Jurd (1964) and by Timberlake and Bridle (1967) (see Chapter 16). Leucoanthocyanins (see Volume 1, Chapter 11) are present in most fruits used for drying but little is known concerning the changes they undergo during the drying process.

IV. TASTE AND FLAVOUR†

Fully ripe fruit is, for preference, used for drying. This is when its flavour is at its best and it is important to try to retain as much of this natural flavour as possible. The flavour of fruits is due mainly to volatile components spread throughout the tissue. While a large number of these compounds have been identified with the aid of modern methods of gas chromatography, it has not yet become entirely possible to express the characteristic flavour of a particular fruit in terms of a specific volatile compound or group of compounds.

As might be expected, there is a gradual reduction in volatile flavour compounds during drying, while, at the same time, other compounds impart a flavour to the product as result of chemical reactions taking place at normal drying temperatures (32–82°C). O'Neal (1950) investigated changes in volatile permanganate-reducing substances obtained by aeration of fresh and dehydrated prunes, and found that the volatile reducing substances (VRS) increase during ripening on the tree at first slowly and then more rapidly until they level off when the prunes are mature. The VRS content for French prunes at optimum maturity was 20–23, for Sugar and Imperial varieties 17–28 and for Robe de Sargent 13–17 micro-equivalents per 100 g. The VRS were largely carbonyl compounds, mainly acetaldehyde and glyoxal. After dehydration the VRS content increased to as high as 67 micro-equivalents per 100 g prunes, but, on prolonged storage, decreased to levels as low as 15. The carbonyl compound identified in dehydrated French prunes was crotonaldehyde, believed to be formed by the condensation of two molecules of acetaldehyde.

Most fruits which are dried contain appreciable amounts of organic acids (see Volume I, Chapter 4); these contribute to their flavour and taste characteristics, while the sugar : acid ratio is often used as a criterion in judging maturity.

Fruits contain significant quantities of Vitamin C and carotene (provitamin A) and smaller amounts of the B vitamins (see Volume I, Chapter 13). Some loss of vitamin content is inevitable during drying. The presence of SO_2 enhances the retention of Vitamin C to the extent that 50% of the original

† See also Volume 1, Chapter 10.

content may be retained in the dried product (Morgan and Field, 1929, 1930). On the other hand, thiamin is totally destroyed by SO_2 (Morgan *et al.*, 1935). Losses of carotene during drying may be as low as 10–15%.

V. PREPARATION FOR DRYING

A. Harvesting and Handling

It has already been mentioned that fruit should be mature with good colour and flavour development before being harvested for dehydration. The unwelcome presence of chlorophyll in such fruits as apricots, peaches and plums has also been discussed. It seems possible that a treatment with ethylene (see Volume I, Chapter 16) might overcome this problem and give a uniformly ripe starting material.

Mature fruits, suitable for drying, are in an active metabolic condition and, in addition, they are generally soft enough to be susceptible to damage caused by mishandling. Thus, they should be treated as gently as possible from the time they are taken from the tree or vine until drying is well advanced. Enzymic breakdown and deterioration are particularly likely at this stage because many fruits are harvested at temperatures in the 27–43°C range, at which temperatures polyphenol oxidase is highly active. Most fruits for drying are usually picked by hand, but with ever-increasing labour costs there is pressure for the development of more mechanical harvesting methods. At present such methods best apply to those fruits which form a pronounced abscission layer in the stem when they reach maturity. This is best demonstrated in the case of prunes which are generally shaken from the trees onto canvas collectors or are picked up from the ground after dropping naturally. Some bruising is, however, unavoidable by the use of such procedures and this may be accentuated when the fruits are transported in bulk to the drying factory in deep beds. Weights of 500–2000 lb of fruit are held in bins in which a layer of fruit may be up to 3 feet below the surface. Delay at any of these steps may well be critical. Hand picking does not entirely prevent damage because most fruits are picked by contract workers whose income depends on the speed with which they harvest and their practices often result in bruising of fruits and breakage of skins. Rates of respiration and transpiration are increased greatly by all such handling procedures but little effort has so far been made to define these changes in biochemical terms. Soft fruits are stored only when emergencies preclude immediate drying operations. Apples are sometimes stored at low temperatures for relatively short times (1–3 months) before drying in order to spread the load at the factory.

Fruits are usually rinsed in cold water at the drying plant to remove surface dust and spray residues, but the time of immersion is generally so short that little reduction in flesh temperature occurs and the biochemical significance of such a treatment is minor.

B. Pre-drying Treatments

Processing treatments applied to fruits before they are dried are usually necessary to ensure a reasonably short drying time and to limit heat-induced deteriorative changes to a minimum. For example, the time needed to dry a whole apple or pear would be so long that the end-product would be dark-coloured and not in a suitable form for subsequent usage. Thus the consumer has come to accept dried fruits in certain forms which have generally been dictated by technological necessity. Apples are peeled, cored and sliced into rings or segments; pears, apricots and peaches are halved while grapes, figs, prunes, berries and cherries are dried as whole fruits.

Peeling the apple fruit removes the heavy layer of wax which would inhibit moisture transfer. The wax on the exterior of banana skins is also completely removed when the fruit is peeled. Pears are halved and usually dried unpeeled. They are notoriously slow to dry because of this skin, their large size even as halves and, possibly, because of their cellular structure. Grapes may be dried whole or dipped into hot water, dilute lye or subjected to the so-called soda-dip process. These treatments "crack" the waxy layer on the skin and increase moisture transfer. Prunes may similarly be "checked" by dipping into hot water (Imperial) or hot dilute lye solution (French). There is a consumer demand for the development of products which are simpler to use. Such new products should reconstitute readily and, therefore, be porous.

There are a number of reasons why removal of water from fruit tissue becomes increasingly difficult as drying continues. A study of a typical drying curve shows that there is no constant rate drying phase; the falling rate phase starts immediately after evaporation commences. This is caused by the initial, uniform water distribution throughout fruit tissue. Water close to the drying surface diffuses readily to the outside where evaporation occurs. However, water deeper within the tissue has increasing difficulty in diffusing to the surface due to (i) the greater distance to travel; (ii) the collapse of cells which seals diffusion pathways; (iii) interruption of flow by entrapped air pockets, (iv) increasing viscosity as solids content, and particularly sugar concentration, rises and (v) the waxy layer at the epidermis if this is not removed or "checked" (see Section VIB).

The distance for water vapour to travel is reduced by sub-division of fruit into smaller pieces; at the same time this increases drying rate by raising the surface to weight ratio. However, the sub-division which includes cutting of cells induces enzymic darkening by bringing together the enzyme, substrate and oxygen. This is pronounced in apple tissue and is prevented by dipping in a solution containing sodium bisulphite or by exposure to the fumes from burning sulphur for a short time (see p. 634). The collapse of cells and inclusion of air will be discussed under drying.

Enzymic darkening is also prevented by heat treatments. This procedure is not widely used in fruit drying, but experimental work has shown that cooking in steam or water (blanching), results in expulsion of much of the interstitial and intracellular gases in the plasmolysis of tissue and in the softening of cell walls. From a practical viewpoint it also results in a reduction in drying times.

Mrak and Phaff (1943) reported that, while unblanched dehydrated fruit is opaque in appearance, the blanched fruit has a desirable translucent appearance. They found, however, that while steam blanching of cut fruits (except apples) was feasible, blanching of whole fruits such as prunes and grapes was not because of losses of syrup. However, Lazar *et al.* (1963) introduced blanching after a preliminary drying stage in a dry–blanch–dry process; this reduced the syrup loss so that blanching could be applied even to grapes.

In California, raisins are produced by cutting bunches of Thompson seedless grapes from the vine and placing them on paper trays in the sun. In the absence of any treatment to the skin, it needs at least 2 weeks to reduce the moisture content to 10–15%. The dried product is dark brown in colour due to both enzymic and non-enzymic darkening, but it has a distinctly visible waxy layer still present on the surface.

In Australia, the product from Thompson seedless grapes is known as the sultana and in that country a pre-drying procedure based on older Mediterrean practices is used. The bunches are harvested into perforated metal baskets which are immersed for 1–4 min in a solution containing about 2·5% potassium carbonate and 2·0% of an emulsion containing C_{14}–C_{18} fatty acid ethyl esters. The temperature of dipping is between 32° and 43°C, the ambient level. After draining, the grapes are spread on tiered racks which are usually roofed so that the grapes are in direct sunlight only in the early mornings and late afternoons. This treatment not only results in a near three-fold increase in the rate of water removal compared with undipped fruit, but the dried product is amber to light brown in colour. Chambers and Possingham (1963), in attempting to explain the reason for the faster drying rate of dipped fruit, postulated that the dipping solution, by flooding the minute spaces in the wax layer, makes it hydrophilic and provides a liquid pathway whereby on dehydration of the tissue water can escape by diffusion. In the drying of undipped fruit it is presumed that water on reaching the hydrophobic layer vaporizes before transferring across the barrier. The reason why dipped fruits are lighter in colour than undipped is obscure but this effect is probably due to reduced polyphenolase activity. This, in turn, may be due to increased pH or, as suggested by Radler (1964), to the more rapid rise in solid matter associated with the faster drying rate. Chambers and Possingham (1963) showed that little of the wax was actually removed from the cuticle by the cold dip and suggested it was merely a change in the nature of the layer which permitted faster water transferance. Radler (1964) showed that the faster drying rate did not occur in the presence of potassium carbonate alone. In order to produce a light-coloured dried fruit from Thompson seedless grapes, the Californian practice is to dip the fruit in dilute (0·2–0·3%) sodium hydroxide solution for 2–3 sec at about 94°C, rinse, expose to the fumes from burning sulphur for 2–4 hours and then dehydrate. Other dips using sodium hydroxide are also used in Australia, mainly on larger varieties of grapes such as Waltham Cross. The use of caustic soda removes some of the wax and may even cause breaks in the cuticle, particularly when temperatures near boiling are employed. The use of these elevated temperatures, even for a short time, would destroy some of the polyphenolase, particularly in the skin which is known to be a rich source of the enzyme.

It was previously general practice to dip prunes in a boiling 0·1–0·3% sodium hydroxide solution before drying. This process removed about half the total wax leaving the residue in a disorganized form; it also produced small cracks or "checks" in the cuticle. Using this treatment a reduction in drying time of 10% or 2–3 hours resulted (Bain and McBean, 1967) when prunes were dried in a conventional counter-flow dehydrator in which fruit enters the cold end of the drying tunnel (50–55°C) and progresses toward the hotter end (75°C) where they emerge. The temperature at the hot end is limited to 75°C to prevent excessive non-enzymic browning. The prune wax, as a whole, does not melt until about 65°C which occurs only after 10–12 hours in the 22–30 hour cycle. Thus, during the earlier stages of drying the cuticular wax would retain its ability to act as a water barrier. In recent years, the caustic soda treatment has generally been omitted on the grounds that the extra cost of the process was not justified by the slightly increased drying rate which resulted. Gentry *et al.* (1965) showed that prunes could be dried in 12–16 hours by introducing them into the hot end of a tunnel and moving them parallel with the air flow towards the cool end. Due to initial high evaporative cooling at the surface of the fruits, drying temperatures up to 90°C could be used without causing serious heat damage to the tissue. The faster drying rate in the latter system is due to increased rates of evaporation and diffusion as well as changes in the structure of the wax layer. The temperature at the surface of the prunes rises to more than 65°C soon after their entry into the dehydrator causing the wax to melt and lose its water-barrier properties. This finding was corroborated by McBean *et al.* (1966) who showed that, under parallel-flow conditions of dehydration, untreated prunes dried at the same rate as prunes which had been de-waxed by light petroleum treatment; even though the wax was still present on undipped fruit, it no longer possessed water-barrier properties at the higher temperatures. The subject of the surface lipid layers on plant organs is being intensively studied at the present time (see Volume 1, Chapter 9). Some of the more important questions relate to the site of generation of the wax, the method of transport to the surface of the wax is synthesized below the cuticle, and the influence on water transmission of the chemical and physical nature of these thin but complex layers. Bain and McBean (1969) have found that wax is present on the surface of prunes only 3 weeks after full blossom when the fruits are about 0·5 cm in length. At this stage, however, the layer is in the form of a smooth sheet with no projections. The latter start to develop in the 4th and 5th week after blossom and are believed to impart greater resistance to water transmission.

C. Sulphuring

Probably the most significant pre-treatment applied to many fruits before drying is that commonly known as "sulphuring", i.e. the application of sulphur dioxide or sulphite solutions to fruits such as apples, pears, apricots, peaches and nectarines which are referred to as cut fruits since they are sub-divided before drying. Figs and prunes are not normally sulphured, while a small proportion of Californian dried grapes (10,000–15,000 tons) are only treated in order to produce a light-coloured dried product. The main purpose of SO_2 is to reduce browning (see Section IIIA and B) in fruits while they are drying and during their subsequent storage. Sulphur dioxide cannot completely stop these browning reactions which adversely affect colour and texture, but it can retard them sufficiently to allow the dried products to remain acceptable for about a year at 21°C (Nury *et al.*, 1960e).

In addition to retarding non-enzymic browning (see p. 628), SO_2 has a marked influence on enzymic darkening† and particularly that due to polyphenolase. Using buffered catechol solutions, sodium bisulphite and a partially purified polyphenolase preparation, Ponting (1960) found that even 1 p.p.m. of SO_2 caused a 20% reduction in enzyme activity and that 10 p.p.m. inactivated the enzyme completely. Working with apple juice containing catechol, he found that some SO_2 reacted preferentially with enzymic oxidation products, probably quinone, and that higher levels of SO_2 were needed to inhibit browning. It is clear that a relatively low level of SO_2 will inactivate polyphenolase but it is particularly necessary with tissue to obtain penetration of the inhibitor for it to be fully effective. Some of the fruits which are dried do not contain enough tissue oxygen to cause internal browning; this is fortunate because even gaseous SO_2 does not penetrate these fruits very well. This fact is shown in Fig. 1, which indicates the steep concentration gradient in peach tissue following its exposure to SO_2 gas.

In practice, the most usual procedure for sulphuring is the exposure of fruits to the fumes from burning sulphur in an enclosure with just enough admission

† See Chapter 20.

of air to maintain combustion and to provide mixing by convection currents. Exposure times vary from as low as 2–4 hours for grapes and apricots in some areas of cultivation up to 72 hours for pears in areas where they dry slowly. The concentration of SO_2 attained in the enclosed atmosphere is usually between 1 and 3% (v/v) (McBean *et al.*, 1967). Various procedures are used with apples which are generally dipped in a sulphite-containing solution immediately after peeling and slicing. Bisulphite or metabisulphite solutions are used as their pH values are not too different from that of the fruit. In Australia, however, immediately after preparation, apples are exposed to the fumes from burning sulphur for about 30 min. No matter how the preservative is applied its main aim with apples is to prevent enzymic darkening at the start of processing. During drying, which is done slowly in deep beds on slatted floors, apple pieces are often exposed to a little gaseous SO_2 to restrict darkening. The main sulphuring treatment, which is aimed at preventing non-enzymic darkening and requires up to 3000 p.p.m. of SO_2, is given by spraying a concentrated sulphite-containing solution on the apple after drying and just before it is packed.

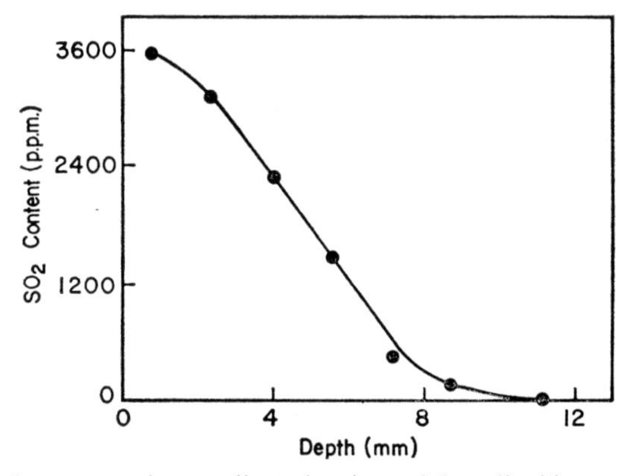

FIG. 1. A typical concentration gradient showing sulphur dioxide at various depths in clingstone peach tissue following its exposure for 1 hour at 1% SO_2 gas level.

A comprehensive report on the earlier work on the absorption and retention of SO_2 in fruits was made by Long *et al.* (1940); the subject was also discussed by Mrak *et al.* (1942). It was not until 1964 that experiments were described which aimed at defining the significance of the many factors which affect the uptake of gaseous SO_2 by fruit tissue. These experiments by McBean *et al.* (1964) showed that rate of absorption is influenced by many characteristics of the fruit, although time of exposure and concentration of SO_2 were the most important factors involved. Their results indicate that, (i) the rate of SO_2 penetration through the fruit cuticle is 3 to 8 times slower than that through the cut flesh; (ii) firm, immature tissue absorbs SO_2 more rapidly than softened mature tissue; (iii) the more open-textured fruits, apricots, absorbed SO_2 most rapidly followed in order by peaches and pears; (iv) when the absorbing surface was disorganized, such as by blanching, SO_2 uptake was slower. These observations suggest that penetration is partly due to gaseous diffusion. Sulphur dioxide is readily soluble in water (7% w/w at 25°C) so that some must dissolve in the aqueous phase of fruit tissue (82–88% water) and then diffuse into the tissue.

The limit of SO_2 concentration in dried fruits is usually in the range of 2000 and 3000 p.p.m. for cut fruits and 1500 p.p.m. for golden raisins. Such limits are invariably specified as total SO_2, but Joslyn and Braverman (1954) pointed out that the preservative must exist in both free and combined forms; the relative proportions of these are determined by such factors as the composition of the fruit, its pH and its temperature during drying. McBean (1967) showed that between 10 and 20% of the total SO_2 absorbed by apricot or peach tissue was in the combined form at the conclusion of sulphuring and that the rate of combination was linear.

Desorption of SO_2 commences immediately after fruit is removed from the sulphuring environment and continues until drying is considered complete. By this time only a fraction of absorbed SO_2 is retained. McBean *et al.* (1965) showed that as little as 5% might be retained under slow drying conditions and up to 40% if water removal was very rapid. Such observations imply that under rapid drying rates diffusion pathways for the escape of gaseous SO_2 are sealed rapidly. At the same time further chemical fixation is probably occurring. Subsequently, McBean (1967) found that free SO_2 contents diminished throughout the whole drying period, whereas levels of combined SO_2 remained relatively unchanged and reached 80–90% of the total in the dried fruit. Immediately following sulphuring, conditions which favour desorption exist; the concentration of SO_2, and particularly free SO_2, is high near the surface of the fruit which is now moving from an atmosphere with a high partial pressure of the gas to one containing negligible amounts of SO_2. It is interesting that Chace *et al.* (1930, 1933) reported that blanching of apricots before sulphuring did not affect either absorption or retention of SO_2, but blanching after sulphuring increased retention; McBean has found that a 10 sec post-sulphuring blanching doubles the retention of SO_2 in the same fruit.

The introduction of such an active chemical agent as SO_2 into living fruit tissue must cause many other biochemical changes but few of these have been investigated. During sulphuring, fruit tissue softens appreciably, the skin is loosened and plasmolysis occurs releasing cellular fluids. These observations imply degradation of cell walls and intercellular layers as well as alteration to the osmotic equilibrium within the tissue. There is a marked gaseous exchange, much of the normally occurring gases being replaced by intercellular exudate.

No data have been published on changes in respiration rates induced by the sulphur-containing atmosphere, even though concentrations of SO_2 up to 50 p.p.m. are used to inhibit microbial activity on cool-stored fresh grapes. Finally, research on the influence of SO_2 on plant colouring and flavouring compounds has received little attention (see Section IIIB2).

Nichols and Reed (1931) reported that in sulphured dried fruit stored in cans, even in absence of air, the sulphur dioxide content decreased but the total sulphur remained constant. Sorber (1944) confirmed this result but found that less than half of the total sulphur present could be accounted for as sulphur dioxide even immediately after drying. During storage the sulphur dioxide content decreased until after 10·5 years less than 10% was in this form. More than 50% of the total sulphur found at the end of 10·5 years was not recovered either as sulphur dioxide or as sulphate sulphur. Hägglund *et al.* (1929) found that bisulphites act as oxidants on sugars and organic acids, being reduced to elemental sulphur, and Ingles (1959) suggested that this reaction may take place in sulphited foods. However, Bloch (1963) was unable to detect any elemental sulphur in commercially dried apricots. He did find, however, that of 1000 p.p.m. of elemental sulphur added to a purée of dried apricots, only 200 p.p.m. remained after 9 months at 37°C. Thus it is possible that any sulphur arising from the reduction of SO_2 could be an intermediary and subsequently converted to another sulphur compound. Ephraim (1943) showed that elemental sulphur reacts with sulphurous acid to form polythionic acids ($H_2S_2O_6$) which are unstable and yield sulphuric acid, sulphur dioxide and sulphur. Ultimately much of the sulphur would be in the form of sulphate. In a series of experiments, Bloch noted that SO_2 was converted to sulphate in dried fruits during storage and that, even during the drying process, no significant loss of gaseous SO_2 occurred from saran-coated cellophane packages. He also confirmed Stadtman's observation (1948) that no sulphate was formed from SO_2 in the absence of oxygen. It seems, therefore, that while sulphate is an important end-product of SO_2 in dried fruits, the complete fate of this preservative is still obscure.

VI. DRYING

A. Moisture Transfer

1. *Physical*

Theoretical aspects of dehydration have been considered by Van Arsdel (1963). The most important operation in dehydration is the mass transfer of moisture. In living fruit tissues water loss largely occurs by diffusion from the cells to the intercellular spaces, followed by diffusion, probably as vapour, from the intercellular spaces through pores (lenticels or stomata) in their skin (see Fogg, 1965). During dehydration, evaporation commences from the outer surface of the tissue, water is gradually drawn, mainly as vapour, from the cell through the plasma membrane and cell wall and then partly via the intercellular spaces out through the skin when this has not been removed before drying. The overall process no doubt involves water both as vapour and as liquid with movement both by diffusion and by capillary action. As drying proceeds, cell walls collapse and some capillaries contract and ultimately seal off, resulting in a falling rate of water transfer. Solutes move to different sites and their concentrations rise. Components such as reducing sugars and amino acids are probably brought into intimate contact and the stage is set for the initial phases of the "browning" reaction. This reaction can proceed even when the residual water level is reduced to a few per cent. If water is removed too rapidly from the drying surface, capillaries may be closed so quickly that water movement from the interior is outstripped. This condition is called "case-hardening" and causes substantially reduced drying rates. Fundamental aspects of these transfer processes were discussed in 1958 at a Society of Chemical Industry Symposium but data were limited to model systems or vegetables. Crank (1958), for instance, discussed the mathematical diffusion studies on model systems relevant to dehydration and concluded that the so-called "case hardening" could not occur. In the dehydration of fruit it has long been believed that when evaporation at the surface occurs more rapidly than diffusion from the centre to the evaporating surface then a surface barrier to moisture transfer will develop ("case hardening"). This led to the restriction of fruit dehydration to counter current atmospheric drying for fruits such as grapes and prunes. Eidt (1938), however, very early reported that case hardening was not a limiting factor with apples, and Joslyn (unpublished data, 1942–44) agreed with this finding for apricots, prunes or pears. The extent to which case hardening is a serious factor in the dehydration of fruits will depend on the design and operating temperatures of the drier.

During the early stages of drying, evaporation rate is so high that the surface remains relatively cooler than air temperature. In the later stages, the temperature of the dried tissue gradually approaches that of the drying air. This increase in temperature accelerates certain biochemical changes many of which are not clearly understood but which generally result in deleterious changes in flavour, texture and colour. Thus, the temperature of the material in the terminal stages of drying is a most important factor in determining the quality of the product.

2. *Biochemical*

Apart from increased solids concentration due to water removal, some individual components also undergo changes. Partial inversion of sucrose occurs in those fruits which contain large amounts of sucrose particularly if the acid content is also high. These changes which may result in a more hygroscopic dried fruit with altered taste, texture or appearance are not always undesirable. For example, in dates during drying, sucrose inversion occurs and condensation of the reducing sugar with nitrogen compounds (see earlier) besides giving a brown colour also imparts a "caramel" flavour to the product.

The difference in the flavour between fresh and dried fruits is partly due to the loss of some volatile components during water removal. Nury and Salunkhe (1968), using gas–liquid chromatography, showed that there was a marked decrease in isobutanol, isovaleraldehyde, methyl-isobutyl ketone and n-hexanol in apples dried at 71°C. No other studies along these lines have yet been reported.

Even when no SO_2 treatment is given (e.g. with prunes and raisins), enzyme activity declines as drying proceeds presumably as a result of the increasing concentration of various fruit constituents as well as through direct heat denaturation. There exist, however, few reliable data on the rate of enzyme deactivation or on the state of the enzyme in its desiccated environment. With fruit dried in the sun, the temperatures within the tissue are unlikely to reach levels which would destroy the enzymes completely.

While dried fruits generally have a flavour different from that of their fresh starting material they may still possess their main characteristics and be acceptable in their own right, although one common off-flavour described as "hay-like" is present in some types of fruit; its origin is unknown.

B. Physical Barriers to Moisture Transfer

Most fruits, apples, grapes, pears and prunes have a pronounced waxy bloom. This consists essentially of waxes deposited at the outer surface or cuticle and lipids in the epidermal cells both of which restrict water transfer. The lipid constituents of fruits has been discussed in Volume 1, Chapter 9. The cuticular wax of grapes was investigated by Chambers and Possingham (1963), of apples by Meigh (1964) and of prunes by Bain and McBean (1967, 1969). This waxy layer, which acts as a barrier to movement of moisture repelling water externally and conserving it internally, has a diversity of physical form (Juniper and Bradley, 1958) and varies in thickness. The waxy layer on apples is relatively heavy and may be up to 3 μ thick (Skene, 1963). The cuticular wax on grapes and prunes is also quite heavy. The pronounced bloom on these fruits is due to light scattering at the surface of the lipid layers. The structural form of this layer as well as its thickness is important in determining water repellency of the layer. The bloom on grapes consists of multi-fingered platelets, while that on prunes consists of a basal "platy" medium from which fragile projections protrude.

The small surface hairs on the skin of fruit like peaches also impede water vapour loss by acting as a barrier against air movement around the lenticels or stomata present in the skin of many fruits. Maxie *et al.* (1967) report that peaches lose less moisture than nectarines under the same conditions of temperature and relative humidity. They state that removal of the "fuzz" will increase moisture loss in two ways: (i) by eliminating the barrier against air movement around the lenticels and stomata, and (ii) the broken hairs leave new openings in the skins of the fruit through which water vapour can be lost.

Loss of moisture by diffusion into the surrounding atmosphere depends on the existence of a vapour-pressure difference between the relative humidity in the intercellular spaces (approximately 100%) and that in the surrounding air. Air movement increases moisture transfer by blowing away the water vapour as it emerges from the lenticel and thus effectively reducing the relative humidity in the surrounding atmosphere.

The occurrence, structure and function of lenticels in the skin of fruits is discussed in general by Esau (1953, 1960), Hayward (1938) and Eames and MacDaniels (1947). Esau (1960) reported the occurrence of stomata and true lenticels in apples, cherries and nectarines. Hayward (1938) called attention to the absence of lenticels in grapes. Bennet-Clark (1959) discussed the water relations of plant cells in general and Heath (1959) and Zelitch (1967) discussed the water relations of stomatal cells and the mechanism of stomatal movement.

The older data on the structure of fruits are summarized in the reference text of Winton and Winton (1933) but more recent data are limited and scattered.

The tissue of those fruits which are dried is composed mainly of soft parenchymatous cells carried on weak supporting structures. Minor variations in this structural form are largely responsible for the texture of particular fruits and are possibly important in the degree of cellular collapse which occurs during drying and in the extent of imbibition of water during rehydration. These parenchymatous cells are usually many-sided with a fairly prominent cell wall composed of polysaccharides of which cellulose is the most abundant. They are partly filled by a large sap-filled vacuole. Between the cells there are air spaces which may constitute up to 30% of the total tissue volume. The cells are held together by pectin of the middle lamella which acts as an intercellular cement. These pectic materials change during maturation of the fruit and by the time fruits are ready for drying, the tissue has softened appreciably.

The barrier of moisture transfer due to case hardening has been discussed on p. 638.

The design of fruit dehydrators based on actual determination of mass and heat transfer factors was investigated in some detail for prunes by Guillou (1942). Perry (1944) reported data on heat and vapour transfer in the dehydration of prunes, including temperature and moisture distribution during dehydration. The older data, typified by the results of Cruess *et al.* (1920) and

Nichols and Christie (1930a) for grapes and Wiegand (1924) and Christie (1926) for prunes and concerned chiefly with such factors as tray loading, maturity and size and shape of fruit, and drying conditions, need further examination using modern techniques and embracing a wider variety of fruits.

C. Biochemical Factors Limiting Moisture Transfer

Water is by far the most abundant constituent of fruits, ranging from 90% in berries down to 70% in bananas (see Table I); prunes grown without irrigation contain only 65% water. Within fruit species, the water level varies widely depending on environmental and cultural factors. In general, water is distributed uniformly through the edible portion of the fruit with the exception of the skin which usually contains less than the tissue it encloses.

TABLE I. The average water content of various fruits (g/100 g) at harvest

Fruit	Water	Fruit	Water
Apple (flesh only)		Greengage (without stone)	78·2
dessert	84·3	Lemon (including skin)	85·2
cooking	85·3	Longanberry	85·0
Apricot (flesh and skin)	86·0	Melon (flesh only)	
Avocado (flesh only)	81·3	Cantaloupe	92·8
Banana (flesh only)	70·7	Honeydew	92·6
Blackberry	83·4	Orange (flesh only)	86·1
Blueberry	82·3	Passion fruit (flesh and seeds)	73·3
Cherry (without stone)	81·4	Peach (without stone)	86·2
Cranberry	87·2	Pear (flesh only)	
Currant		dessert	84·1
black	77·4	cooking	85·1
red	82·8	Pineapple (flesh only)	84·8
white	83·3	Plum (without stone)	
Damson (without stone)	77·5	dessert	84·1
Fig (green)	84·6	cooking	85·1
Gooseberry	86·3	Quince (flesh only)	84·2
Grape (without seeds)		Raspberry	83·2
black	80·7	Strawberry	89·4
white	79·3	Tangerine (flesh only)	86·7
Grapefruit (flesh only)	89·8	Tomato	94·1

(Compiled from McCance & Widdowson, 1960, and other sources.)

The diffusion of water in the tissue during drying is influenced by the state of the water present. In fruit tissues most of the water present may be "free", but appreciable amounts of water may be held as water of hydration by such constituents as dextrose or maltose, organic acids such as tartaric acid and tartrates and certain other salts. Much of the bound water may be held by hydrogen bonding to particular constituents. While the concept of "bound" water has been questioned by Kuprianoff (1958), there is much evidence in its favour. Ward (1963) discussed the nature of the forces between water and the macromolecular constituents of food. The naturally occurring high molecular weight polymers such as cellulose, pectin, hemicelluloses which markedly influence textural quality in fruit products and the changes which they undergo during dehydration largely influence texture. The irreversibility in water removal is related to conformational changes in these macromolecular components. Shimazu and Sterling (1961) investigated changes in cellulose and calcium pectate in model systems during dehydration and found an increase in crystallinity. Consequently it would be reasonable to consider that this increase might be responsible for the inability of dried fruits to rehydrate to their original condition. However, while supporting evidence for this effect was reported for dehydrated carrots by Sterling and Shimazu (1961), similar evidence for fruit is lacking.

VII. STORAGE

The deteriorative changes in flavour, texture and colour initiated during drying and possibly during pre-drying procedures continue when fruit is held or stored after drying. Such deterioration increases with increasing time and temperature of storage. Fruit may even be pre-conditioned for such changes while it is still on the tree, e.g. when it experiences heat or water stress due to high ambient temperatures just before harvest.

Dried fruits stored for a long time even at moderate temperatures develop off-flavours which render them unacceptable. However, few precise data have been published which indicate the compounds responsible for this defect.

Texture changes also occur during storage of dried fruits. These are indicated by an increasing reluctance of the tissue to absorb water but the exact causes for this phenomenon are not clearly understood.

The changes in colour which occur in dried fruits during storage have been the subject of extensive investigations (Mackinney, 1946). Some have aimed at elucidation of the complex series of reactions which ultimately end with the production of brown pigments, while others have determined the practical limitations which define the acceptability of the stored product. Notable among the former has been the work of Reynolds (1963, 1965) who has considered in detail possible reactions which occur before the appearance of brown pigmentation. The influence of such factors as time of storage, temperature, SO_2 level, moisture content and oxygen availability on the rate of browning has been extensively studied (Stadtman *et al.*, 1946a, b, d). In those fruits which require SO_2, studies have generally involved the rate of darkening and the loss of the preservative as these are intimately associated. No satisfactory substitute for SO_2 has been found for the preservation of colour and flavour in dried fruits although many possible alternatives have been tested. The SO_2 content of fruits continues to decrease gradually after drying; the rate of decline increases by a factor of 2–3 for each 10°C rise in storage temperature. The higher the storage temperature the greater will be the initial SO_2 required, but at 40°C and above, even 3000 p.p.m. will not maintain apricots in an edible condition for more than a few weeks.

Stadtman *et al.* (1946a) reported that, in absence of oxygen, the rate of darkening of dried apricots increased with decrease in moisture content over the range 40–10% and reached a maximum at a moisture content of between 5 and 10%. As might be expected, exposure to oxygen decreases the storage life of dried apricots (Stadman *et al.*, 1946b).

Stadman *et al.* (1946b) found an apparent activation energy of 26 kg cal for the darkening of apricots and Nury and Brekke (1963) reported 24–26 kg cal for raisins. Barger *et al.* (1948) suggested that the best temperature for the storage of dried fruits generally was 0°C at a relative humidity of 55%; Schrader and Thompson (1947) and Thompson and Schrader (1949) recommended 0·8°C for dried apples. In practice, the moisture content, availability of oxygen and exposure to sunlight are less important than the levels of SO_2 and the time–temperature régimes.

Data on the chemical, physical and organoleptic changes occurring in commercially dried fruit, packaged in Saran-treated cellophane bags and cartons with foil laminated paper overwrap, stored under a variety of atmospheric conditions were obtained for figs (Nury *et al.*, 1960a), raisins and sulphur-bleached raisins ("Golden raisins"), prunes (Nury *et al.*, 1960c) and apricots (Nury *et al.*, 1960d). Typical data for raisins are shown in Table II (starting material) and in Figs. 2 and 3 for colour changes taking place during storage at 10, 21 and 32°C. The relative humidity was maintained at 60%, representing an equilibrium moisture content of 16%, and colour was determined by measuring the absorbance at 440 mμ of a 50% ethanol extract of the dried fruit.

TABLE II. Analyses of raisins used in the stability study described in the text

	Raisins	Golden raisins
Moisture (%)	16·08	16·05
Crude fibre (%)	0·98	0·82
Nitrogen (%)	0·54	0·46
Ash (%)	1·89	1·77
Sugar content:		
% total	70·6	71·2
% reducing	70·2	70·7
% fructose	39·2	40·0
% glucose	31·0	30·7
% sucrose	0·4	0·5
SO_2 (p.p.m.)	—	1450

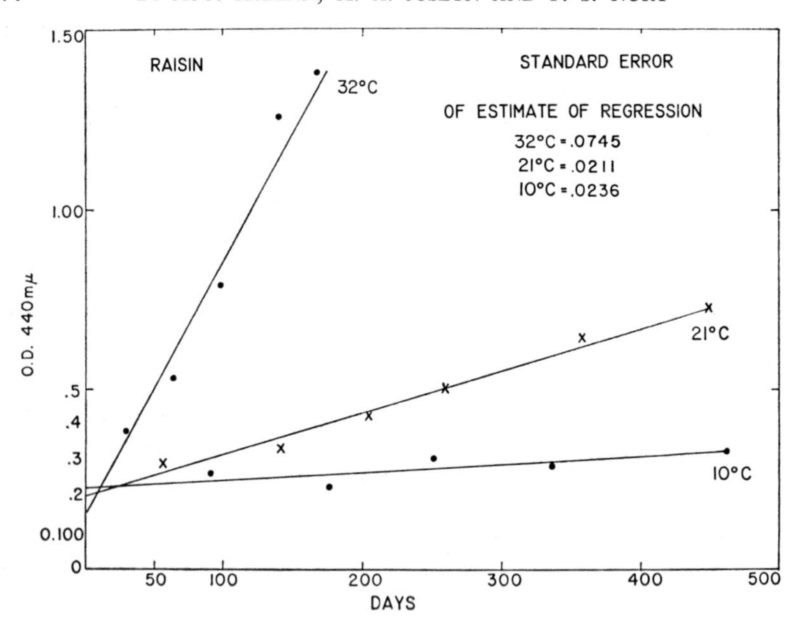

FIG. 2. Colour changes in unsulphured raisins during storage at various temperatures (10, 21 and 32°C).

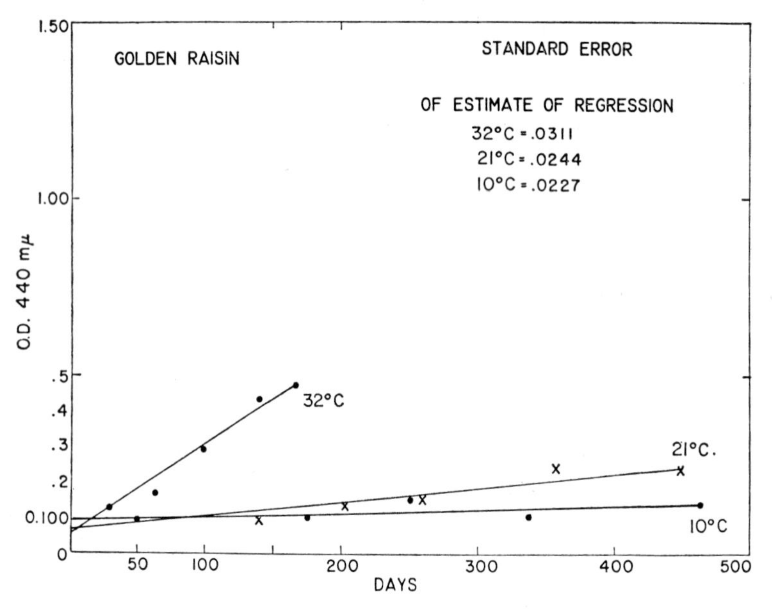

FIG. 3. Colour changes in sulphured or Golden raisins during storage at various temperatures (10, 21 and 32°C).

The rate of development of pigmentation was low at 10°C and rapid at 32°C indicating the marked influence of temperature on deteriorative changes. From these results it was considered that absorbance changes of 0·3 for Golden and 0·9 for unsulphured raisins represented objective values defining the limits of acceptability.

Apparent activation energies for the darkening of raisins were calculated using the Arrhenius equation

$$\Delta Ha = \frac{2{\cdot}3RT_1T_2}{T_2 - T_1}\log\frac{k_2}{k_1},$$

where k_2 and k_1 are the rates for increase in optical density at temperatures T_2 and T_1, respectively, and R is the gas constant. In the temperature range of 10–32°C, apparent activation energies, in kcal/mole, were 26 for regular raisins and 24 for Golden bleached raisins; the corresponding Q_{10} values are 4·2 and 3·9.

Reasonable estimates of the quality or shelf life of raisins at a given temperature, or at an integrated series of temperatures, can be made by measuring the alcohol-soluble colour and applying the indicated rate factors (see Figs 2 and 3 legends). Assuming the end-points noted in Figs 2 and 3, the shelf life of the unsulphured raisins in these experiments was estimated at about 95 days at 32°C and about 600 days at 21°C; for golden raisins the shelf life would be about 80 days at 32°C and about 500 days at 21°C. A panel of experienced judges from the dried fruits industry rated regular raisins stored for 90 days at 32°C equal to raisins stored at 21°C for 450 days thus confirming the laboratory studies.

Maier and Schiller (1960) tested the applicability of both reflectance spectroscopy and solvent extraction methods of measuring darkening of dates. A rotating sample reflectance method was found to be sensitive to initial darkening and to correlate closely with visual observed darkness. Extracting the dates with aqueous methanol and measurement of absorbance of the extract in the region of 450–600 mμ was not related to visually observed darkening but provided a better index of the later stages of darkening where the reflectance method was not sufficiently sensitive.

The decrease in sulphur dioxide level during storage has been used as another index of deterioration. Nury and co-workers found that apricots and sulphured raisins were no longer acceptable when their SO_2 levels had fallen to 700 and 500 p.p.m. respectively from initial values of 3000 and 1500 p.p.m.

Maier and Schiller (1961a) reported on some chemical changes occurring during storage of Deglet Noir dates at 49°C. Darkening was found to be the combined result of both oxidative (10–20%) and non-oxidative browning. The latter results in a decrease in pH. Enzymic changes which occurred included inversion of sucrose by invertase and oxygen absorption and carbon dioxide production due to phenolase. Maier and Schiller (1961b) reported that the oxidative darkening had an apparent energy of activation of 23·4 kcal/mole, while the non-oxidative had 34·5. Oxidative browning in ground date tissue was enzyme catalyzed.

Maier and Metzler (1965a, b) reported data on the changes occurring in date polyphenols during maturation and storage and their relation to browning. The flavans and caffeoyl-shikimic acid were found to undergo the greatest decrease during maturation and browning. These phenolics were most susceptible to enzymic browning. The soluble leucoanthocyanidins were found to be converted into insoluble leucoanthocyanidins during growth and the latter were involved in non-enzymic oxidative browning during storage.

In recent years there has been an increasing demand for dried fruits with higher than the usual levels (15–20%) of moisture. Such fruits are softer in texture and more readily prepared for eating. For example, dried prunes and figs having a moisture content of about 16% may be treated in boiling water or steam to raise their water content to 40%. At this moisture content the product will readily support the growth of yeasts and moulds. If, however, they are sprayed with sorbic acid or diethylpyrocarbonate before packaging or if ethylene oxide or propylene oxide (Phaff *et al.*, 1946) is included in the final flexible transparent packages, such microbial growth spoilage is prevented (Nury *et al.*, 1960b). In Australia, where neither of these preservatives is permitted, a process involving rapid sealing of hot reconstituted prunes in plastic pouches has been used for a number of years (McBean and Pitt, 1965). However, this method of protection is not as efficient as the use of sorbate and the product spoils soon after the package is opened and allowed to stand at room temperature. The most common moulds found growing on dried prunes are species of *Aspergillus*; *Xeromyces bisoris* is less common.

Dried fruits which have been stored for long periods often exhibit crystallization of sugar, either as granules within the tissue or as a white incrustation on the surface. Such defects occur more in raisins, figs, dates and prunes than in cut fruits. Mrak and Baker (1934) reported that 25–44% of the "sugar" on prunes and 6–21% on figs was composed of yeast cells. The balance was glucose. More recently Miller and Chichester (1960) reported that the crystalline deposit on prunes and figs contained glucose and fructose, with traces of citric and malic acids, lysine, asparagine and aspartic acid. Apricot and peach sugars contained, in addition, some sucrose and large amounts of asparagine and aspartic acid. Deposits from raisins were similar to those from figs with the exception that they also contained large amounts of tartaric acid. Dried fruits containing soluble solids contents of greater than 80%, most of which is sucrose, glucose and fructose, act as supersaturated solutions, but the precise conditions causing crystallization have not been well defined. Water level, time and temperature of storage all have some influence on this problem. Once crystallization has occurred it is difficult to reverse the process without damaging the dried fruit. Humectants such as glycerol and sorbitol are sometimes applied to the surface of dried fruits in efforts to prevent "sugaring" but they are of doubtful benefit.

The moisture level at which dried fruit would have the desired storage life depends on the reduction of available moisture below 25–30%, at which level microbial growth and activity and tissue enzyme activity become appreciable. Available data on the water relations of enzyme activity were summarized by Acker (1962) and those of food-spoilage micro-organisms by Scott (1957). As stated above, at higher moisture levels chemical preservatives must be used; alternatively, the micro-organisms capable of developing may be destroyed by heat (Bolin and Boyle, 1967).

Objective methods for the commercial standardization of raisins were reported early by Chace and Church (1927). Subsequently Wiegand and Bullis (1929, 1931) proposed the use of specific gravity as an index of quality for Oregon prunes, and Nichols and Reed (1932) used a similar procedure for California prunes. Gas–liquid chromatography has still not been widely applied in this field.

VIII. REHYDRATION

The conditions influencing absorption of moisture by dried fruits were investigated by Nichols (1935) for moisture increase during processing of dry prunes in boiling water. Nury *et al.* (1963) developed an improved procedure for rehydration using steam and boiling water followed by a short immersion in cold water, and applied it to figs and raisins as well as prunes. Further studies on chemical changes in the cell walls and means of improving rehydration of dried tissue are needed.

The textural characteristics of fruits are altered appreciably during drying (Hohl, 1948). The collapse of cells associated with water removal makes the dried tissue tough and resistant to pressure. On rehydration, dried fruits are flaccid and do not generally regain turgor. Crafts (1944) studied changes during drying of fruit tissue on a heated stage mounted on a microscope. Lee *et al.* (1966, 1967) and Nury and Salunkhe (1968) investigated the histological differences between air-dried and vacuum-dried fruits. They all reported expansion of intercellular gases, the escape of gas and the creation of spaces containing entrapped air. Crafts also observed a to-and-fro movement of small columns of liquid as more air entered the tissue. The cell walls became flaccid and finally rigid. They were often coated with syrupy cell contents. Nury and Salunkhe (1968) showed that cell walls shrank during drying and some breakage occurred. They also showed that if drying was done under vacuum, large air-free intercellular spaces were formed and that cells although shrunken were not so severely collapsed as those of air-dried fruits. The porous nature of vacuum-dried material permits rapid and more complete reconstitution than the collapsed nature of air-dried fruit. Sterling (1963) observed that dried cell-walls are not very permeable to water, suggesting a denaturation of the cell wall material.

IX. CONCLUSION

Dried fruits differ appreciably from their fresh counterparts. In spite of this they have become, and remain, well accepted due mainly to attributes such as long storage life without refrigeration and reduction in weight and volume achieved by the removal of water. Changing demands in preservation and distribution, calling for more readily prepared foods, are steadily increasing pressure on dried fruits and there is a slow but steady decline in world production of traditional dried fruits.

It seems likely that the age-old methods of drying which still account for the majority of dried fruit production today will have to be replaced by procedures which will give products more in keeping with modern demands. Freeze-drying may even become economically possible. Improvement in dried fruit products requires a greater knowledge of the many biochemical changes which occur in fruits growing on the tree, at harvest and during pre-processing treatments as well as during processing and subsequent storage. The recent growth of interest in fruit biochemistry, brought about as a result of the development of precise methods for the separation and identification of chemical constituents, should, therefore, greatly benefit the fruit-growing industry.

REFERENCES

Acker, L. (1962). *Adv. Fd Res.* **11,** 263.
Bain, J. M. and McBean, D. McG. (1967). *Aust. J. biol. Sci.* **20,** 895.
Bain, J. M. and McBean, D. McG. (1969). *Aust. J. Biol. Sci.* **22.**
Barger, W. R., Pentzer, W. T. and Fisher, C. K. (1948). *Fd Inds* **20** (3), 337.
Bennet-Clark, T. A. (1959). *In* "Plant Physiology" (F. C. Steward, 3d.), Vol. II, pp. 105–191. Academic Press, New York and London.
Bioletti, F. T. (1915). *Int. Congr. Vitic.* (S.F.), pp. 307–314.
Bloch, F. (1963). 23rd Annual Meeting, Northern California Section, I.F.T.
Bolin, H. R. and Boyle, F. P. (1967). *J. Sci. Fd Agric.* **18,** 289.
Braverman, J. B. S. (1953). *J. Sci. Fd Agric.* **4,** 540.
Brekke, J. E. and Nury, F. S. (1964). *In* "Food Dehydration" (W. B. Van Arsdel and M. J. Copley, eds), Vol. II, pp. 407–507. AVI Publishing Co., Westport, Connecticut.
Caldwell, J. S. (1923). *U.S. Dep. Agric. Bull.* **1141,** 62 pp.
Chace, E. M. and Church, C. G. (1927). *U.S. Dep. Agric. Tech. Bull.* **1,** 23 pp.

Chace, E. M., Church, C. G. and Sorber, D. G. (1930). *Ind. Engng. Chem. ind. Edn* **22,** 1317.

Chace, E. M., Church, C. G. and Sorber, D. G. (1933). *Ind. Engng. Chem. ind. Edn* **25,** 1366.

Chambers, T. C. and Possingham, J. V. (1963). *Aust. J. biol. Sci.* **16,** 818.

Chari, C. N., Natarajan, C. P., Mrak, E. M. and Phaff, H. J. (1948). *Fruit Prod. J. Am. Vin. Mfr* **27,** 206.

Christie, A. W. (1926). *Calif. Agric. Exp. Sta. Bull.* **404,** 1. Revised 1929 by P. F. Nichols. 32 pp.

Christie, A. W. and Barnard, L. C. (1925). *Calif. Agric. Exp. Sta. Bull.* **388,** 60 pp.

Cole, J. (1967). *J. Fd. Sci.* **32,** 245.

Crafts, A. S. (1944). *Fd Res.* **9,** 442.

Crank, J. (1958). *In* "Fundamental Aspects of the Dehydration of Foodstuffs", pp. 37–41. Society of Chemical Industries, London.

Cruess, W. V. (1943). *Ind. Engng. Chem. ind. Edn* **35,** 53.

Cruess, W. V. (1958). "Commercial Fruit and Vegetable Products", pp. 540–618, 648–680. McGraw-Hill, New York.

Cruess, W. V. and Christie, A. W. (1921). *Calif. Agric. Exp. Sta. Bull.* **330,** 50.

Cruess, W. V., Christie, A. W. and Flosfeder, F. C. H. (1920). *Calif. Agric. Exp. Sta. Bull.* **322,** 421.

Curl, A. L. (1960). *Fd Res.* **25,** 190.

Deibner, L. and Rifai, H. (1963). *Mitt. Klosterneuburg, Ser. A* **13,** 56, 113.

Eames, A. J. and MacDaniels, L. H. (1947). "An Introduction to Plant Anatomy", pp. 262–266. McGraw-Hill, New York.

Eidt, C. C. (1938). *Can. Dep. Agric. Pub.* **625,** 36 pp.

Embs, R. J. and Markakis, P. (1965). *J. Fd Sci.* **30,** 753.

Ephraim, F. (1943). *In* "Inorganic Chemistry", 4th edition.

Esau, K. (1953). "Plant Anatomy", p. 594. John Wiley & Sons, New York.

Esau, K. (1960). "Anatomy of Seed Plants", pp. 68–70, 323. John Wiley & Sons, New York.

Fogg, G. E. (editor) (1965). "The State and Movement of Water in Living Organisms." Soc. Expt. Biol. Symp. No. 19, 432 pp. Cambridge University Press, Cambridge, England.

Gentry, J. P., Miller, M. W. and Claypool, L. L. (1965). *Fd Technol.* **19,** 121.

Gincarevic, M. and Hawker, J. S. (1971). *J. Sci. Fd Agric.* **22,** 270.

Guillou, R. (1942). *Agric. Engng, St. Joseph, Mich.* **23,** 313.

Hägglund, E., Johnson, T. and Silander, S. (1929). *Ber. dt chem. Ges.* **62,** 84.

Hayward, H. E. (1938). "The Structure of Economic Plants." Macmillan, New York.

Heath, O. V. S. (1959). *In* "Plant Physiology" (F. C. Steward, ed.), Vol. II, pp. 193–250. Academic Press, New York and London.

Hodge, J. E. (1953). *J. agric. Fd Chem.* **1,** 928.

Hohl, L. A. (1948). *Fd Technol.* **2,** 158.

Hussein, A. A., Mrak, E. M. and Cruess, W. V. (1942). *Hilgardia* **14,** 349.

Ingles, D. L. (1959). *Aust. J. Chem.* **12,** 97.

Jacob, H. E. (1942). *Hilgardia* **14,** 321.

Joslyn, M. A. (1963). *In* "Food Processing Operations" (M. A. Joslyn and J. L. Heid, eds), Vol. II, pp. 545–584. AVI Publishing Co., Westport, Connecticut.

Joslyn, M. A. and Braverman, J. B. S. (1954). *Adv. Fd Res.* **5,** 97.

Joslyn, M. A. and Ponting, J. D. (1951). *Adv. Fd Res.* **3,** 1.

Juniper, B. E. and Bradldy, D. E. (1958). *J. Ultrastruct. Res.* 2, 16.

Jurd, L. (1964). *J. Fd Sci.* **29,** 16.
Kuprianoff, J. (1958). *In* Soc. Chemy Indy, "Fundamental Aspects of the Dehydration of Foods", pp. 14–23. Macmillan, New York.
Lazar, M. E., Barta, E. J. and Smith, G. S. (1963). *Fd Technol.* **17,** 120.
Lee, C. Y. and Salunkhe, D. K. (1967). *J. Sci. Fd Agric.* **18,** 566.
Lee, C. Y., Salunkhe, D. K. and Nury, F. S. (1966). *J. Sci. Fd Agric.* **17,** 393.
Lee, C. Y., Salunkhe, D. K. and Nury, F. S. (1957). *J. Sci. Fd Agric.* **18,** 89.
Long, J. D., Mrak, E. M. and Fisher, C. D. (1940). *Calif. Agric. Exp. Sta. Bull.* **636,** 56 pp.
Mackinney, G. (1937). *Pl. Physiol., Lancaster* **12,** 1001.
Mackinney, G. (1946). *Ind. Engng. Chem. ind. Edn* **38,** 324.
Maier, V. P. and Metzler, D. M. (1965a). *J. Fd Sci.* **30,** 80.
Maier, V. P. and Metzler, D. M. (1965b). *J. Fd Sci.* **30,** 747.
Maier, V. P. and Schiller, F. H. (1960). *Fd Technol.* **14,** 139.
Maier, V. P. and Schiller, F. H. (1961a). *J. Fd Sci.* **26,** 322.
Maier, V. P. and Schiller, F. H. (1961b). *J. Fd Sci.* **26,** 529.
Markakis, P. and Embs, R. J. (1966). *J. Fd Sci.* **31,** 807.
Maxie, E. C., Sommer, N. F. and Mitchell, F. G. (1967). *In* "Proceedings Fruit and Vegetable Perishable Handling Conference," pp. 5–9. University of California, Davis, California.
McBean, D. McG. (1967). *Fd Technol.* **21,** 1402.
McBean, D. McG. and Pitt, J. I. (1965). *Fd Preserv. Q.* **25,** 27.
McBean, D. McG., Johnson, A. A. and Pitt, J. I. (1964). *J. Fd Sci.* **29,** 257.
McBean, D. McG., Pitt, J. I. and Johnson, A. A. (1965). *Fd Technol.* **19,** 141.
McBean, D. McG., Miller, M. W., Pitt, J. I. and Johnson, A. A. (1966). *Fd Preserv. Q.* **26,** 2.
McBean, D. McG., Miller, M. W., Johnson, A. A. and Pitt, J. I. (1967). *Fd Preserv. Q.* **27,** 22.
McCance, R. A. and Widdowson, E. M. (1960). "The Composition of Foods." M.R.C. Special Report Series No. 297. H.M.S.O., London.
McWeeney, D. J., Biltcliffe, D. O., Powell, R. C. T. and Spark, A. A. (1969). *J. Fd Sci.* **34,** 641.
Meigh, D. F. (1964). *J. Sci. Fd Agric.* **15,** 436.
Miller, M. W. and Chichester, C. O. (1960). *Fd Res.* **25,** 424.
Morgan, A. F. and Field, A. (1929). *J. biol. Chem.* **82,** 579.
Morgan, A. F. and Field, A. (1930). *J. biol. Chem.* **88,** 9.
Morgan, A. F., Kimmell, L., Field, A. and Nichols, P. F. (1935). *J. Nutr.* **9,** 369.
Morris, T. N. (1947). "The Dehydration of Food." D. Van Nostrand Co., New York.
Mrak, E. M. (1942). *Fd Inds* **21,** 48.
Mrak, E. M. and Baker, E. E. (1934). "Proc. Third Int. Cong. Microbiol", pp. 707–708.
Mrak, E. M. and Long, J. D. (1941). *Calif. Agric. Exp. Sta. Circ.* **350,** 69 pp.
Mrak, E. M. and Mackinney, G. (1951). *In* "The Chemistry and Technology of Foods" (M. B. Jacobs, ed.), Vol. III, pp. 1773–1821. Interscience Publishers Inc., New York.
Mrak, E. M. and Perry, R. L. (1948a). *Calif. Agric. Exp. Sta. Circ.* **383,** 11 pp.
Mrak, E. M. and Perry, R. L. (1948b). *Calif. Agric. Exp. Sta. Circ.* **381,** 11 pp.
Mrak, E. M. and Phaff, H. J. (1943). Unpublished data. Cited by Phaff (1951).
Mrak, E. M. and Phaff, H. J. (1947). *Fd Technol.* **1,** 147.
Mrak, E. M. and Phaff, H. J. (1949). *Calif. Agric. Exp. Sta. Circ.* **392,** 19 pp.

Mrak, E. M., Fisher, C. D. and Bornstein, B. (1942). *Fruit Prod. J. Am. Fd Vin. Mfr* **21,** 175, 199, 217, 219, 237, 297.
Nichols, P. F. (1933). *Calif. Agric. Exp. Sta. Circ.* **75,** 37 pp.
Nichols, P. F. (1935). *Fruit Prod. J. Am. Vin. Mfr* **14,** 211, 240, 332, 370.
Nichols, P. F. and Christie, A. W. (1930a). *Calif. Agric. Exp. Sta. Bull.* **500,** 31 pp.
Nichols, P. F. and Christie, A. W. (1930b). *Calif. Agric. Exp. Sta. Bull.* **485,** 46 pp.
Nichols, P. F. and Reed, H. M. (1931). *Western Canner and Packer* **23,** 11.
Nichols, P. F. and Reed, H. M. (1932). *Hilgardia* **6,** 561.
Nury, F. S. and Brekke, J. E. (1963). *J. Fd Sci.* **28,** 95.
Nury, F. S. and Salunkhe, D. K. (1968). *U.S. Dep. Agric. ARS* **74–45,** 19 pp.
Nury, F. S., Miller, M. W. and Brekke, J. E. (1960e). *Fd Technol.* **14,** 113.
Nury, F. S., Taylor, D. H. and Brekke, J. E. (1960a). *U.S. Dep. Agric. ARS* **74–16,** 15 pp.
Nury, F. S., Taylor, D. H. and Brekke, J. E. (1960b). *U.S. Dep. Agric. ARS* **74-17,** 26 pp.
Nury, F. S., Taylor, D. H. and Brekke, J. E. (1960c). *U.S. Dep. Agric. ARS* **74-18,** 14 pp.
Nury, F. S., Taylor, D. H. and Brekke, J. E. (1960d). *U.S. Dep. Agric. ARS* **74-19,** 20 pp.
Nury, F. S., Bolin, H. R. and Brekke, J. E. (1963). *Fd Technol.* **17,** 98.
O'Neal, R. (1950). "The Volatile Organic Compounds of Prunes." M.Sc. Thesis in Food Science, University of California, Berkeley, California.
Perry, R. (1944). *Transactions of the Am. Soc. mech. Engrs* **66,** 447.
Perry, R. L., Mrak, E. M., Phaff, H. J., Marsh, G. L. and Fisher, C. D. (1946). *Calif. Agric. Exp. Sta. Bull.* **698,** 1.
Phaff, H. J. (1951). *Chronica Bot.* **12,** 306.
Phaff, H. J. and Mrak, E. M. (1948). *Calif. Agric. Exp. Sta. Circ.* **382,** 1.
Phaff, H. J., Mrak, E. M., Allemann, R. and Whelton, R. (1946). *Fruit Prod. J. Am. Fd Mfr* **25,** 140, 155.
Ponting, J. D. (1960). *In* "Food Enzymes" (H. S. Schultz, ed.). AVI Publishing Co., Westport, Connecticut.
Ponting, J. D. and Stanley, W. L. (1968). I.F.T. 28th Annual Meeting Abstract.
Pridham, J. B. (editor) (1963). "Enzyme Chemistry of Phenolic Compounds." Pergamon Press, New York.
Radler, F. (1964). *J. Sci. Fd Agric.* **12,** 864.
Rahman, B. M. and Joslyn, M. A. (1954). *Mitt. Klosterneuburg, Ser. B.* **4,** 49.
Reynolds, T. M. (1963). *Adv. Fd Res.* **12,** 1.
Reynolds, T. M. (1965). *Adv. Fd Res.* **14,** 167.
Richert, P. H. (1953). *J. agric. Fd Chem.* **1,** 610.
Schrader, A. L. and Thompson, A. H. (1947). *Proc. Am. Soc. hort. Sci.* **49,** 125.
Scott, W. J. (1957). *Adv. Fd Res.* **7,** 84.
Shimazu, F. and Sterling, C. (1961). *J. Food Sci.* **26,** 291.
Skene, D. S. (1963). *Ann. Bot.* (N.S.) **27,** 581.
Song, P.-S. and Chichester, C. O. (1966). *J. Fd Sci.* **31,** 914.
Song, P.-S. and Chichester, C. O. (1967). *J. Fd Sci.* **32,** 98.
Song, P.-S., Chichester, C. O. and Stadtman, F. H. (1966). *J. Fd Sci.* **31,** 906.
Sorber, D. G. (1944). *Fruit Prod. J. Am. Vin. Mfr*, 32 234, 251.
Stadtman, E. R. (1948). *Adv. Fd Res.* **1,** 325.
Stadtman, E. R., Barker, H. A., Haas, V. and Mrak, E. M. (1946a). *Ind. Engng Chem. ind. Edn* **38,** 541.

Stadtman, E. R., Barker, H. A., Mrak, E. M. and Mackinney, G. (1946b). *Ind. Engng. Chem. ind. Edn* **38,** 99.

Stadtman, E. R., Barker, H. A., Haas, V., Mrak, E. M. and Mackinney, G. (1946c). *Ind. Engng Chem. ind. Edn* **38,** 324.

Sterling, C. (1963). *In* "Recent Advances in Food Science" (J. M. Leitch and D. N. Rhodes, eds), Vol. 3, pp. 259–281. Butterworth, London.

Sterling, C. and Shimazu, F. (1961). *J. Fd Sci.* **26,** 479.

Thompson, A. H. and Schrader, A. L. (1949). *Proc. Am. Soc. hort. Sci.* **54,** 73.

Timberlake, C. F. and Bridle, P. (1967). *J. Sci. Fd Agric.* **18,** 473, 479.

Van Arsdel, W. B. (1963). "Food Dehydration", Vol. I. AVI Publishing Co., Westport, Connecticut.

Van Arsdel, W. B. and Copley, M. J. (editors) (1964). "Food Dehydration", Vol. II. AVI Publishing Co., Westport, Connecticut.

von Loesecke, H. W. (1955). "Drying and Dehydration of Foods", 2nd edition. Reinhold Publishing Co., New York.

Ward, A. G. (1963). *In* "Recent Advances in Food Science" (J. M. Leitch and D. N. Rhodes, eds), Vol. 3, pp. 207–214. Butterworth, London.

Wiegand, E. H. (1924). *Oregon Agric. Exp. Sta. Bull.* **205,** 26 pp.

Wiegand, E. H. and Bullis, D. E. (1929). *Oregon Agric. Exp. Sta. Bull.* **252,** 47 pp.

Wiegand, E. H. and Bullis, D. E. (1931). *Oregon Agric. Exp. Sta. Bull.* **291,** 35 pp.

Winton, A. L. and Winton, K. B. (1933). "The Structure and Composition of Foods", Vol. II: Vegetables, Legumes, Fruits. John Wiley & Son, New York.

Zelitch, I. (1967). *Am. Scient.* **55,** 472.

Chapter 19

Freezing Preservation

BERNARD J. FINKLE

Western Regional Research Laboratory, U.S.D.A., Albany, California, U.S.A.

I Introduction 653
II Physical and Chemical Aspects of Freezing Preservation 655
A. Preservation by Freezing 655
B. Factors in Freezing Damage 656
C. Cold Stability 663
D. Effects of Thawing 663
E. Effects of Freeze-protective Additives 664
F. Causes of Tissue Injury 665
III Enzymological Considerations in Freezing 667
A. Isolated Enzymes 667
B. Enzymes in Tissues 669
IV Factors in Freezing Preservation of Fruit 671
A. Texture Changes 671
B. Colour Changes 672
C. Flavour Changes 674
D. Effects of Conditions of Freezing 674
E. Effects of Additives 675
V Fruit Stability during Frozen Storage 677
A. Colour and General Appearance 677
B. Flavour 678
VI Conclusions 680
References 682

I. INTRODUCTION

Commercially frozen fruits and juices represent about 40% of the consumption of products coming from the U.S. frozen food industry (National Frozen Food Association, 1969). These items answer a significant consumer need in providing good quality products in practical storage form. There are, however, still important uncertainties of both a basic and an applied nature with respect to freezing effects on fruit tissues, the clarification of which could lead to the improvement of frozen products. Examples of practical requirements still not fully achieved in marketed frozen products are the avoidance, upon thawing, of a mushy texture in strawberries and the prevention, without adding an undesired amount of SO_2, of internal discoloration of apples. There are other fruits that are not available frozen because they cannot meet market quality requirements.

Extended studies have been made of the gross effects of freezing preservation on packaged fruits, e.g. time–temperature tolerance as described in Section V, but few research efforts have addressed themselves to related causative factors, particularly those of a biochemical or physiological nature. Information of potential value to an understanding of the problems involved has, however, been achieved from work on the resistance to freezing of other plant organs. There are, for example, studies on the "freeze-hardening" reaction of the many kinds of plant tissues that seasonally make a metabolic adjustment toward internal avoidance of freezing damage. Twig tissue is normally killed by exposure to only a few degrees below 0°C, as in early autumn, but the cells of naturally freeze-hardened fruit tree twigs, or pieces of similar tissue that have been artificially hardened by immersion in sucrose-containing solutions, can subsequently be exposed to severe freezing and still retain their viability (Tumanov and Krasavtsev, 1966), even down to liquid hydrogen temperature (−253°C) under some conditions (Krasavtsev and Khvalin, 1959). Findings on animal organ preservation, such as the successful use of particular freezing rates and of chemical additives, may also be pertinent. These potentially valuable comparative approaches with respect to the prevention of freezing damage to fruit tissue appear to be worth consideration by reason of the scarcity of research directly on fruits and because injury is "no doubt basically the same in all organisms" (Levitt, 1964) as in also its prevention (Samygin and Matveeva, 1967).

In the present chapter it is, therefore, proposed to discuss the limited amount of biochemical information pertaining to the freezing preservation of fruit against a background of related studies of other living tissues, particularly of mammalian tissue and the thin-walled parenchyma of cold-hardened woody species.

Fundamental differences between the structure and chemical composition of plant tissues compared with animal tissues must of necessity limit the value of such an approach. Major differences also appear within the plant kingdom itself and Reeve and Brown (1966), in their consideration of some structural and histochemical changes related to frozen fruits and vegetables, have emphasized the importance of differences in freezing effects between various plants and between their immature and mature, specialized tissues. In addition they have stressed the contrast between freeze damage and freeze killing of living plant tissue on the one hand and the problems of commercial freezing preservation of fruits and vegetables on the other. These reservations must be borne in mind in interpretating results on the comparative basis proposed in the present instance. The main emphasis here will be in trying to delineate the chemical, biochemical and physical factors which may be manipulated both in a tissue itself and in the freezing conditions in order to minimize damage (Sections II and III). Data on the freezing of fruit will be considered in Section IV and following sections.

An extensive literature review of fruit preservation by freezing by Joslyn (1966) supplements his earlier coverage (Joslyn and Diehl, 1952) and includes descriptions of the classic observations on fruit by J. G. Woodroof. Fennema and Powrie (1964) have discussed physical and general aspects related to the freezing of plant tissue, and in the same volume Burke and Decareau (1964) described fundamental principles, processes, and effects of freeze-drying treatments. Holdsworth (1968) gives concise coverage emphasizing freeze-processing literature, while volumes edited by Tressler *et al.* (1968) and by Van Arsdel *et al.* (1969) describe more fully process-related and commercial aspects of the freezing of fruit.

II. PHYSICAL AND CHEMICAL ASPECTS OF FREEZING PRESERVATION

A. Preservation by Freezing

The advantages of frozen fruits are the result of low storage temperature (often recommended as below −17·8°C for food (Clark, 1960)), low humidity (0·96 mm Hg at this temperature), and rigidity of structure. The latter not only hinders the circulation of reactants in the tissue but also largely eliminates the movement of gases such as oxygen into the tissue (Scholander *et al.*, 1953; Rakitina, 1965). Thus, after freezing to adequately low temperatures (see Section V), both reaction rates and bacterial growth are slowed down with consequent stabilization of chemical and structural quality. Along with this benefit of stabilization, however, deleterious changes during the freezing and thawing processes may take place which become most apparent after thawing. These changes generally manifest themselves as loss of characteristic texture, sometimes a cracking of the brittle frozen fruit, leakage of cell constituents and external and internal darkening reactions. Other changes which take place as judged by more sophisticated criteria (such as breakdown in selectivity of cellular membranes and loss of cell viability) may, superficially, appear to be of less importance for frozen fruit than for other types of frozen tissues. However, when the intensive activity and the progress taking place in other areas of frozen tissue research are considered (such as the cryogenic storage of blood or whole organs), it would appear that we stand at an early stage of development with respect to optimal conditions for storage of fruits by freezing.

B. Factors in Freezing Damage

The source of damage to frozen fruit and other plant tissues is far from clear although it is generally agreed that in the living tissues the stage of freezing where ice forms causes the most damage. The findings of Mazur and Miller (1967) with yeast suspensions illustrate this point: when cooled to −20°C, uncrystallized (supercooled) samples survived 100%, while samples that froze spontaneously at or above this temperature were 99% destroyed. Some plant tissues, however, are severely damaged even in the non-frozen stage at temperatures close to 0°C. General injuries during cold storage of various fruits (Weier and Stocking, 1949; Lutz and Hardenburg, 1968), including death of the subepidermal cells of members of the squash family and the internal breakdown of apples (see Volume 1, Chapter 18), are examples. An extreme case with a plant tissue was evident in early experiments with a myxomycete (*Physarium polycephalum*). The fungus showed protoplasmic gelation and death even after as little as 5 sec at 0°C in still-liquid aqueous solution (Gehenio and Luyet, 1939).

Temperature and rate of freezing and displacement of cellular water are believed to be major factors in the freezing damage to both plant and animal tissues. These have been considered in a series of studies in which viability of tissue after thawing is the chief criterion used. But while this criterion *per se* is of little concern to the general problem of preservation of frozen fruit, it may still be used to advantage as an ultimate standard. There may also be some usefulness in the viable preservation of frozen fruit, for example, for purposes related to maintaining fruit tissue cultures (Schroeder, 1961).

1. *Temperature of freezing*

A remarkable trimodal pattern of damage appears in many tissues. Reports on, for instance, the thin-walled parenchyma cells of various trees and on red blood cells agree that the tissues studied can survive freezing at temperatures just below the freezing point and also at very cold temperatures, but they are much less viable at *intermediate* sub-zero temperatures. Figure 1 illustrates this for thin sections of freeze-hardened mulberry twig cortical parenchyma and Fig. 2 illustrates a similar situation for human erythrocytes. In the mulberry parenchyma, maintenance of the tissue at temperatures of −20°C to −40°C is particularly damaging, the amount of damage increasing with time. To the red blood cells the most damaging region of temperature is near −10°C to −20°C (Fig. 2). A concurrent observation that has been widely made in many tissues, both plant and animal, is that in this intermediate frozen temperature region, large crystals of ice are formed during freezing or during warming from a lower temperature (Mazur, 1966; Sakai and Otsuka, 1967). At much colder temperatures only small crystals or no crystals are formed. Much has been made of the large crystals as a factor in tissue damage (see below), but the point to be stressed here is that they appear in a temperature region coincident with tissue damage, while very rapid freezing to below this temperature region avoids tissue damage. The boundaries of the intermediate frozen temperature region are variable, dependent on the type of tissue studied, degree of hydration, solute content and other factors, but the existence of such an intermediate cold temperature region of minimum survival (or

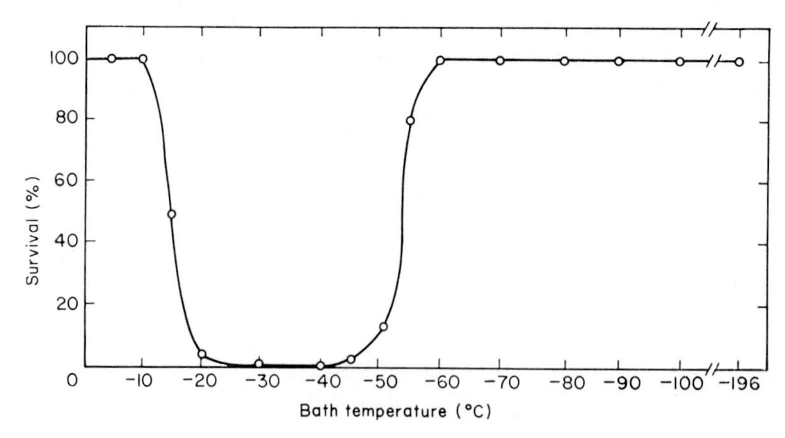

FIG. 1. Survival of mulberry cortical parenchyma cells cooled and rewarmed rapidly. An unmounted tissue section at room temperature held with forceps was rapidly immersed in isopentane baths at various temperatures and kept there for 20 sec before being rewarmed rapidly (from Sakai and Otsuka, 1967).

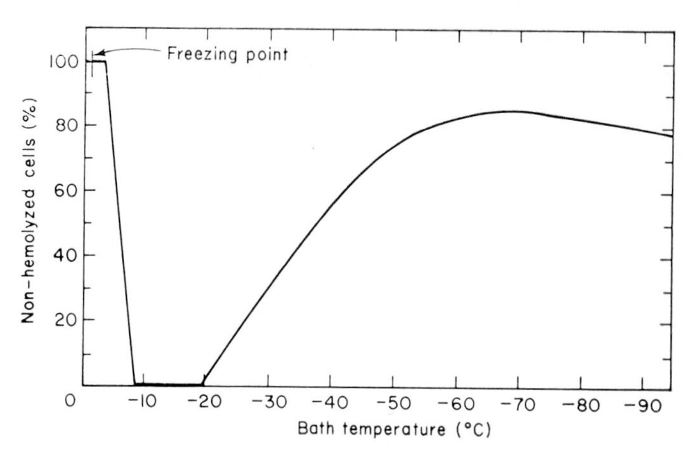

FIG. 2. This curve relates the condition of human erythrocytes following freezing in a glass capillary tube to the temperature of the freezing bath (redrawn from Luyet *et al.*, 1963).

maximum destruction) appears in various carefully controlled studies of tissue freezing.

2. *Rate of freezing*

Another important factor in such studies is the rate of temperature change during freezing. The range of freezing rates tested has been very great during some of the most critical studies of both plant and animal tissues. The curve for freeze-hardened mulberry parenchyma sections (Fig. 3) (Sakai *et al.*, 1968) is similar in very many respects to that for survival of yeast cells (Fig. 4) as well as for human erythrocytes and other cells and tissues. The division of the rate of freezing curve into three general regions is to be particularly noted, i.e. high survival at extremely fast rates of cooling (greater than 10^4 °C/min), low survival at intermediate rates, and again high survival at still lower cooling rates. Many experimenters have not been aware of this trimodal aspect of the curve so that descriptions such as "rapid" or "slow" freezing, etc. have led to difficulties in interpretation. Of particular interest and common to diverse tissues, is the fact that a freezing rate as rapid as 10^2–10^4 °C/min is often insufficient to give high survival rates. At still higher rates of cooling, the increased speed with which the temperature falls through the harmful intermediate frozen temperature region decreases injury (Meryman, 1966a). Freezing rates required for high survival, namely greater than 10^4 °C/min are, of course, attainable only under rigorous experimental conditions in which highly dispersed cells or very thin sections are immersed in the freezing medium. These high rates of freezing are probably attainable with very thin sections of fruit and it would be valuable to test the structural survival of fruit tissue under such conditions.

Also common to such freezing rate curves (Figs 3 and 4) is the maintenance of high survival and structural integrity at very *low* rates of cooling (e.g. at 1°–10°C/min). This is well documented for many types of tissues (Luyet, 1965). It is recognized that, while under conventional conditions the slow freezing of fruit and other multicellular, thick structures may often be harmful (Fennema and Powrie, 1964), the use of freeze-protective agents may prevent damage (see Sections III and IV).

3. *Water displacement*

An explanation frequently given for the lack of lethality during slow freezing of tissues is based on evidence that (i) under slow freezing, extracellular ice formation takes place first (Chandler, 1913; Mazur, 1965a); (ii) free water from the cell interior then diffuses outward where it freezes, partially dehydrating the cell; and then (iii) with free water removed, the partially dehydrated cells resist forming intracellular crystalline ice so that freezing takes place without damage (Ching and Slabaugh, 1966; Sakai *et al.*, 1968). That the rate of freezing affects the amount of decrease in volume (that is, volume of water) in the cell is shown for single yeast cells in Fig. 5. Analogous patterns of water-exit have been demonstrated in many tissues. A view of unfrozen and frozen fir needle parenchyma cells, in longitudinal section, is shown in Fig. 6 and illustrates the effect on thin-walled tissue (somewhat similar to fruit

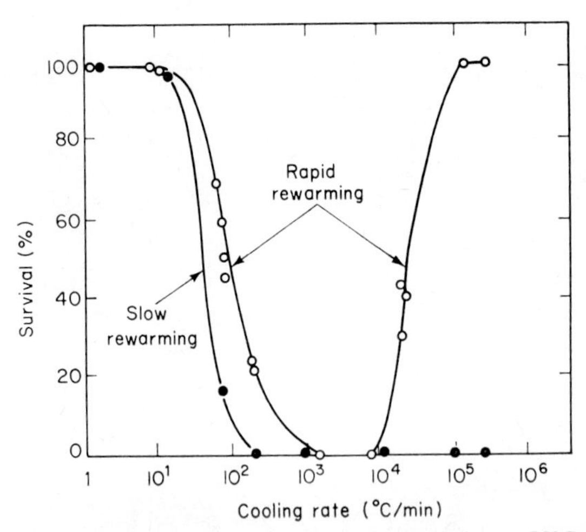

FIG. 3. Survival of mulberry cortical parenchyma cells cooled to −75°C at indicated rates on the abscissa and rewarmed either rapidly or slowly. At slow cooling rates, tissue sections mounted with water between cover glasses were cooled by various methods. At cooling rates higher than 100,000°C/min, an unmounted tissue section held with small forceps was rapidly immersed in liquid nitrogen or liquid isopentane. Rapid rewarming: in water at 30°C. Slow rewarming: in air at 0°C (from Sakai *et al.*, 1968).

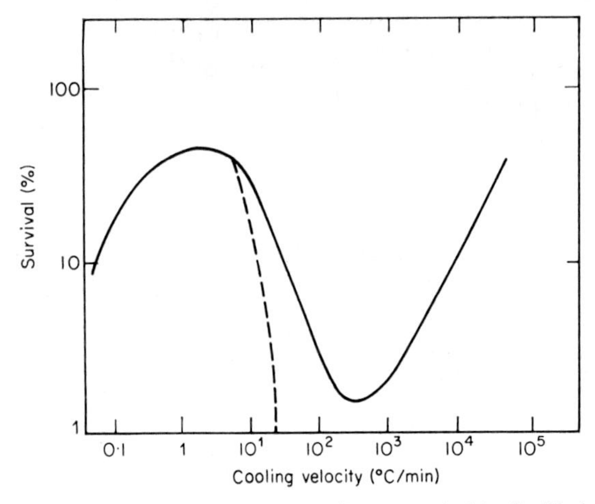

FIG. 4. Survival of *Saccharomyces cerevisiae* cells suspended in distilled water, cooled to −30°C or below and rewarmed either rapidly (——) or slowly (– – –) (after Mazur, 1965a).

parenchyma) of the large water loss that can occur during freezing. The freezing conditions used in the examples shown in both figures produced viable cells.

Analogies have been drawn between the stabilization to freezing, to heating (Levitt, 1958; Holm-Hansen, 1967), and also to drought (Santarius and Heber, 1967) in tissues that have been appropriately dehydrated, but the

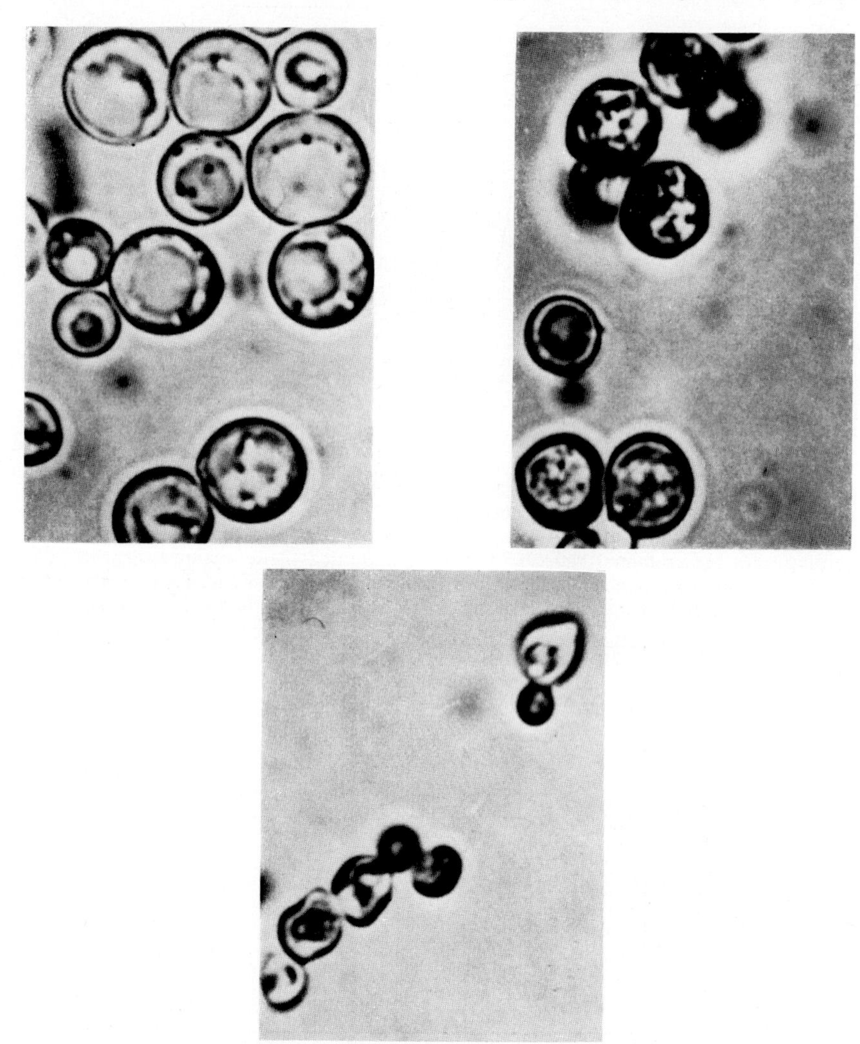

FIG. 5. Effect of rate of freezing on *Saccharomyces cerevisiae* cells. Centre and right micrographs are of cells substituted with cold ethanol after slow and rapid freezing. The light micrograph on the left is of untreated cells. (Ultra-rapidly cooled yeast appears normal morphologically when examined by the electron microscope) (from Mazur, 1965a).

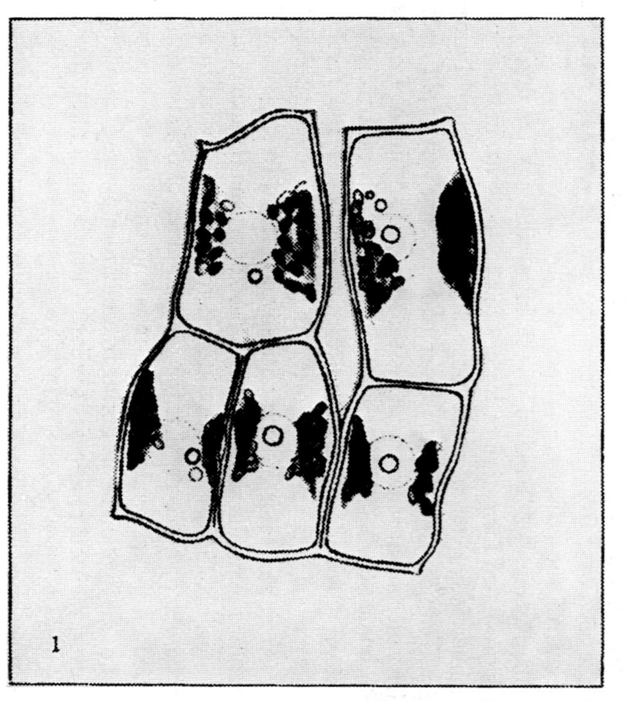

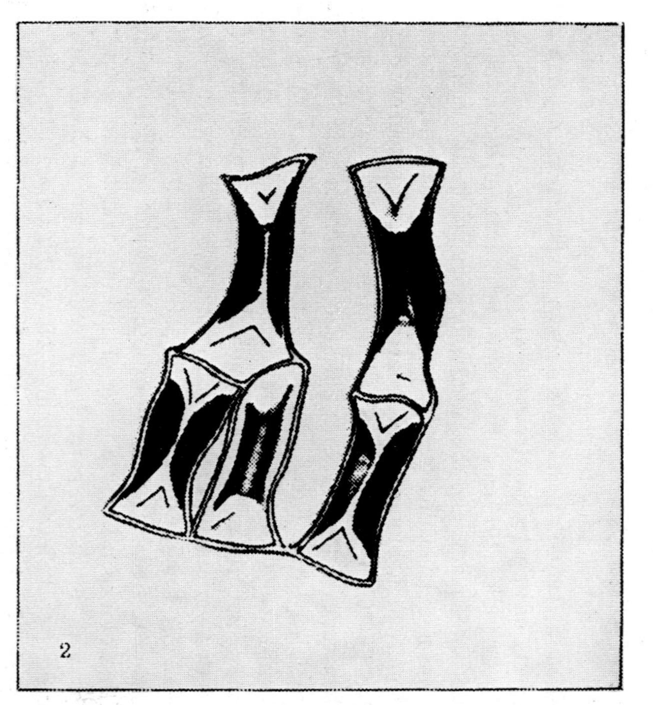

FIG. 6. Parenchyma cells in a longitudinal radial section of a fir needle. 1. Before freezing; 2. After freezing, at $-40°C$; cell volume has decreased, and the intercellular spaces have increased in size (magnif 680) (from Tumanov and Krasavtsev, 1959).

biochemical modifications that take place to counteract cell damage, in these several types of resistance, are not clear. In many such cases, the benefits of a variety of tissue-dehydrating treatments or of protective chemical additives are attributed, at least in part, to free water displacement in some form. Striking effects of apparent water removal in freeze protection have been observed by Chandler (1913) and by Ching and Slabaugh (1966), the former using wilted whole plants and the latter white pine pollen. It was found that the pollen was not viable after freezing at -25°C when the water content was greater than 40%, and at a reversibly controllable water content of 30–33% it still showed 26–39% loss of germination after freezing. When the latter frozen pollen was examined by X-ray diffraction, no crystals of ice were found drawing attention to the effect of lethal factors other than ice crystallization. When frozen at a still lower hydration, survival was 100%. A hydration threshold may, similarly, explain the results of Chandler and Hildreth (1935) who found that various types of pollen grains were killed when frozen to -15° to -17°C, if briefly exposed to water at 0°C before freezing, but showed good resistance to freezing when not exposed to water, or if they were first coated with a film of oil before being placed in contact with water. A set of findings with mulberry parenchyma cells appear to have the same interpretation (see "Slow Rewarming" curve, Fig. 3). These cells when frozen very slowly (1°–10°C/min), could be rewarmed and thawed either slowly or quickly with a high rate of survival. By contrast, the figure shows that when the cells were frozen at faster rates, even at very fast rates, there was no survival if the cells were rewarmed only slowly. Yeast cells, too, reacted similarly (Fig. 4). Removal of the intracellular water during slow freezing has prevented formation of ice crystals in the cells, in the normally destructive intermediate temperature range slowly passed through during the protracted warming period. On the other hand, during very rapid freezing, intracellular water is not removed and is thus trapped within the frozen cell. Then, during slow thawing as the temperature is gradually raised, crystallization, crystal growth and concurrent disruptive effects are the result. No studies appear to have been made to test the effect on fruit tissues of such contrasting freezing rates or of such critically controlled freezing conditions as were used for the results shown in Fig. 3. An approach to such conditions for apples (Sterling, 1968), but with different results, is discussed in Section IV. It is to be noted, however, that the plant findings of Sakai *et al.* (1968) were observed on tissues rendered freeze-resistant, and that the biochemical changes that occur in plant tissue during freeze-hardening are not understood. Often, however, their effect appears similar to that brought about by use of freeze-protective additives (Section IIE).

The movement and clustering of water and water vapour and the crystalline forms of water, complex temperature-related phenomena with great physical and chemical impact on the structures and reactions concerned, have been discussed by Scheraga (1965) and by Karow and Webb (1965). When frozen rapidly to below −120°C, ice freezes entirely in the non-crystalline "vitreous" form (Rapatz and Luyet, 1965).

C. Cold Stability

Various tissues, including fruit as discussed later, are relatively stable to storage once in the frozen form, with few or no further changes taking place at very cold temperatures. At temperatures down to about −70°C, frozen tissues may still contain water in the liquid state in decreasing amount, as determined by NMR analysis (Sussman and Chin, 1966). At below −120°C the amorphous vitreous form of ice is stable; above this temperature crystal formation and recrystallization, as well as other physical reactions, can occur even in the frozen state (Luyet and Gehenio, 1965), and at increasing speed with increasing temperature. The occurrence of chemical reactions may be considered negligible at very cold temperatures; but when frozen tissue is stored at higher temperatures or as it approaches the thawing point, this is far from the case, as is described below.

D. Effects of Thawing

During the rise of temperature associated with the thawing of tissues, physical and chemical changes take place at increasing rates. The rapid succession of these changes may partly account for the damage to tissue already described as taking place in the intermediate temperature range a few tens of degrees below 0°C. In addition, at close to the thawing temperature, intensive recrystallizations occur, with numerous solute–solvent interface changes comparable to multiple freezing and thawing, in what Luyet and Gehenio (1965) have described as a "quasi-fluid" state. The thawing stage of the total freezing process has long been recognized as critical (Clayton, 1932; Meryman, 1966b; Tumanov and Krasavtsev, 1966). Inability of the cell to redistribute appropriately the recrystallizing and inflowing water (Luyet and Gehenio, 1965; Yoshida and Sakai, 1967) and adequately to rehydrate its constituent materials suggest a chaotic situation. Cell damage can, however, sometimes be prevented by a lingering exposure at the thawing temperature, as, for example, Yoshida and Sakai (1967) have shown in leaves. Under other conditions as already noted, rapid thawing is favourable (Silver *et al.*, 1964; Sakai *et al.*, 1968). Rate of change appears as an all-important factor in maintaining structural and biochemical integrity of the frozen cell. Important determinants here are the rate at which the material had been brought to the frozen state and the pre-freezing condition of the cell, including its content of freeze-protective substances (Section IV).

Microwave diathermy for large samples, where rapid thawing becomes increasingly beneficial, has been recommended to avoid damage (Silver *et al.*, 1964). Also, the thawing of tissue under high pressure to avoid large thermal gradients in the tissue has been tested recently (Persidsky and Leef, 1969). Figures 3 and 4 illustrate an effect of rate of thawing on mulberry parenchyma and yeast cells, respectively; the outstanding importance of the intermediate frozen temperature range, already discussed as producing a harmful effect, is worth re-emphasizing. In the work of Sakai *et al.* (1968) cells of mulberry prefrozen at $-10°C$ were frozen quickly in liquid nitrogen and thawed quickly with 100% survival, but a pause during thawing of even 1 min in the neighbourhood of $-30°C$ was lethal. On examining such tissue by electron microscopy, after using the technique of freeze-substitution with alcohol (replacing ice deposits), evidence of cellular damage and of the formation of ice crystals was seen. It is notable that this physiologically destructive temperature range is not far from the range usually recommended for the storage of frozen food.

E. Effects of Freeze-protective Additives

Additives of various types were found many years ago to have a beneficial effect in tissue preservation. Maximow (1912) reported the use of an extract of red cabbage, and also pure non-polar and ionic solutions of various types, in protecting otherwise non-resistant plant tissues (including leaf sections from tropical *Tradescantia discolor*) against freezing injury at very cold temperatures. The work of Chandler (1913) and of Parker (1959) on the freeze-hardening of plant tissues produced many ideas about the freeze-protection of tissues, particularly of their cell membranes. These ideas have since been amplified and applied most incisively in the study of animal tissue problems. Hardening studies on plants are, however, now yielding pertinent information. Table I shows the damage (observed as leakage from the anthocyanin-containing cells) sustained by sections of collard stem when exposed to two rates of slow freezing while immersed in solutions of glycerol for 12–24 hours; the sections were thawed to 2°C slowly over a 24 hour period (Samygin and Matveeva, 1965). The effects of glycerol concentration, freezing temperature and freezing rate are illustrated. The benefits of various protective additives on the leaves and stems of several cabbage-like varieties of *Brassica*—succulent tissues not distant cytologically from fruits—have been studied in this and a subsequent paper (Samygin and Matveeva, 1967). The results confirm the observations made by Sakai *et al.* (see Fig. 3) that, when exposed to an extremely slow (intracellularly dehydrating) rate of freezing, tissue that is freeze-hardened, or freeze-protected, can then pass safely into the $-30°C$ region both on freezing and thawing, with little harm. Still other factors may enter the picture, however, when high concentrations of a freeze-protective agent are used, as demonstrated in Table I, where a reversal to increasing damage at the highest concentrations of glycerol is seen.

The effects on plant and animal tissues of many cryoprotective compounds, of which dimethylsulphoxide, sugars, ethylene glycols, glycerol, and polymers such as polyvinyl alcohol (Gahan *et al.*, 1967) and polyvinylpyrrolidone have been especially useful, have been described (Persidsky *et al.*, 1965; Mazur, 1966; Meryman, 1966b; Samygin and Matveeva, 1967). Effective action by these compounds is reported to be through the decrease of free water in the cells, as well as by specific physicochemical effects in orienting and clustering

TABLE I. The effect of glycerol solutions of varying concentration on the stem sections of collard (non-hardened) when frozen at different rates and to different temperatures

Solution concentration		Damage to sections, as a %, when frozen at varying rates to the temperatures indicated					
		1° in 10–15 min			5° in 12 hours		
g/cm³	Δt°	−30°	−40°	−50°	−30°	−40°	−50°
0·05	−0·9	100	100	100	< 100	< 100	100
0·095	−1·8	0	70	100	20	50	70
0·18	−3·7	0	30	100	0	0	50
0·22	−4·6	0	50	< 100	0	< 10	80
0·27	−5·6	< 10	< 100	< 100	< 10	50	90

From Samygin and Matveeva (1965).

water agglomerates, modifying the surface properties or otherwise preserving the biochemical integrity of, for instance, membranes (Karow and Webb, 1965; Doebbler, 1966; Olien, 1967). Fennema and Powrie (1964) emphasized the desirability of minimizing the volume of water that changes into ice, such as by the addition of glycerol, thereby minimizing volume changes and the concentrating effect on solutes. Rapatz and Luyet (1965) suggested that particular additives stabilize the reportedly less damaging amorphous state of ice at temperatures well above their stable temperature as a pure-water ice (−120°C); thus, by the use of additives, the formation of ice crystals at intermediate sub-zero temperatures can be avoided. The concepts described in these recent articles on the properties of freeze-protective additives are interesting counterparts to the early hypotheses reviewed by Chandler (1913).

F. Causes of Tissue Injury

The basic cause of tissue injury or death is little understood, although most experimental results indicate that the removal or reorientation of free water is involved. Two primary causes of freeze damage (which may be expected to be interrelated) have strong experimental support. Maximow (1912) produced experimental evidence to show that protection against membrane damage alone would preserve *Tradescantia* tissue against freezing damage. Much modern work is in accord with this emphasis (Rowe, 1966; Heber, 1968). At the same time, the results of Sakai *et al.* (1968) and of others (Joslyn and Diehl, 1952; Sterling, 1968) have focused attention on the disruption of cell structure through formation of ice. Other factors given experimental prominence as effecting freeze-damage, besides those already mentioned as causing damage during thawing, are, excessive concentration of solutes, particularly ionic material, with "salting-out" effects (Lovelock, 1953), large effects on tissue pH with its important secondary effects on metabolism (Van Den Berg, 1966), and effects on the charge, surface properties and associations of biological polymers (Lovelock, 1957; Melnick, 1964); water removal from protoplasmic colloids during freezing, beyond the reversible limit for rehydration (Scholander *et al.*, 1953; Santarius and Heber, 1967); protein denaturation and effects on sulphur bonding (Levitt, 1964); enzyme inactivation and loss of biosynthetic capabilities (discussed in Section III); and a combination of these factors, such as is, perhaps, illustrated by the concept of a minimum "critical volume" of the cell, below which there is irreversible damage (Williams and Meryman, 1969). Several of the factors are obviously somewhat interdependent, hence the emphasis given often depends on the conditions under which experimental freezing is performed, as well as on the type of tissue studied. A consequence of, and also a delicate method for measuring (Mazur, 1965b), cell injury is the leakage of cell fluid, including, in many plants, anthocyanin pigments (Samygin and Matveeva, 1965), out of the tissue. Evidence from yeast cells (Mazur, 1965b) and liver mitochondria (Lusena and Dass, 1966) indicates that the membrane destruction which allows loss of fluid is on an all or nothing basis for each cell.

An important apparent contradiction between two factors in cell injury discussed above requires experimental elucidation. During freezing, liquid water removal leading to a high concentration of solutes, etc., is considered a cause of injury (e.g. Lovelock, 1953); but partial desiccation (Ching and Slabaugh, 1966) or slow initial freezing to encourage extracellular crystallization of the water diffusing out of the cell, thus avoiding intracellular ice crystal formation, has frequently been cited to be highly protective (Sakai *et al.*, 1968). Biochemical interpretation, and the directions taken in experimentally selecting the desired properties of new freeze-protective additives, will depend on which result of water removal, the concentrating effect on solutes or the avoidance of intracellular ice crystallization, is closer to being a truly primary factor during the freezing of tissue. It is interesting that the role of ice crystallization, and related ideas about the importance of mechanical disruption by ice as being a cause of cell damage, have been challenged by

Manax *et al.* (1964) on the basis that tissues often withstand freezing when the duration of exposure is not excessive; disruption through the impact of ice crystals ought not to be reversible, nor time dependent after freezing.

III. ENZYMOLOGICAL CONSIDERATIONS IN FREEZING

A. Isolated Enzymes

Tappel (1966), using solutions of the purified enzymes, studied the reactions of the enzymes peroxidase (from turnip) and bovine alkaline phosphatase at temperatures down to −30°C, both frozen and unfrozen. These studies focused on the large effects of concentrated inorganic and organic solutes on reaction rates, a situation that obtains during the freezing process, when ice formation withdraws liquid water from solution. There may be an increase (when enzyme–substrate concentration is rate-limiting) or decrease in reaction rate as the components of a reaction solution are concentrated. In the presence of highly concentrated solutes (e.g. glycerol or NaCl) there occurred a greater decrease in activity than expected, indicative of increased hydrogen bonding and concomitant conformational changes in enzyme at low temperatures. Increased viscosity at low temperatures, especially in the presence of additives like glycerol, also contributed to a decrease in reaction rate. In animal tissue, the release of a complex mixture of mitochondrial and lysosomal enzymes through disruption of the organelle membranes (the phenomenon of "latency") also occurs during freezing and thawing, with resultant tissue damage. Another approach has involved examination of reaction rates at ultracold temperatures. Empirically, no particular cut-off point has been found below which enzymatic activity stops. Bielski and Freed (1965) used mixtures of alcohols and water to show that enzyme activities can proceed even at temperatures below −120°C. In biological systems, however, one is dealing with highly aqueous solvent systems that exist largely in the frozen state even at a few degrees below zero. Evidence from such systems, described in section IIE, points to the presence of liquid water only down to about −70°C (Sussman and Chin, 1966), probably as small pools embedded in ice.

Generally, enzyme activity displays temperature dependence as expressed in the Arrhenius equation, and reaction rate decreases as the temperature is lowered. It was, therefore, surprising to find that reactions sometimes have a higher rate of activity in frozen solutions than in unfrozen liquid solutions at even a higher temperature. Several papers of this nature both with non-enzymic (Brown and Dolev, 1963; Butler and Bruice, 1964; Fennema, 1966; Kiovsky and Pincock, 1966; Grant and Alburn, 1967) and enzymic (Grant and Alburn, 1966) reaction systems show such an effect. Under some conditions even a ten-fold or more increase in rate is found in the frozen condition compared with that in unfrozen solutions. This effect is illustrated in Fig. 7 (Brown and Dolev, 1963). Changes in rates can be attributed to the effects of more highly concentrated reactants (Butler and Bruice, 1964; Kiovsky and Pincock, 1966), including sometimes large changes in reaction pH as water is removed from the cell through extracellular ice crystallization. A change in pH may be caused by a change in concentration or even by precipitation of buffering salts as water is withdrawn (Chilson *et al.*, 1965), or by effects of cold temperatures *per se* on equilibrium constants governing the pH of the cellular mixture. Kinetic evidence, however, indicates specific catalytic action by ice structures as also being important (Grant and Alburn, 1967). Thus a complex of changes—in metabolite form (Kiovsky and Pincock, 1966), enzyme reactivity, pH, etc.—can occur in frozen material to affect markedly the course and rate of reaction.

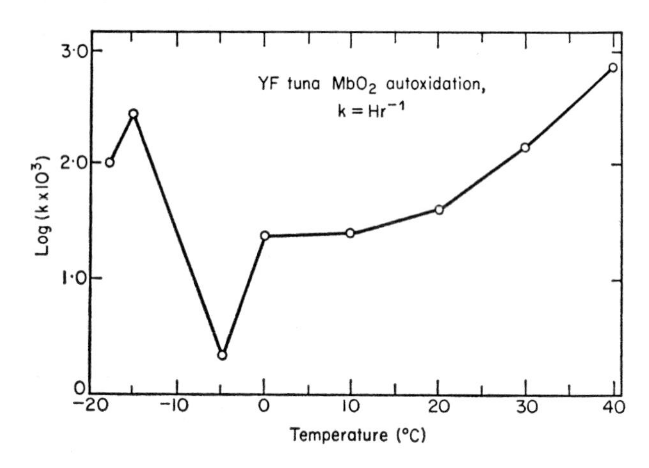

FIG. 7. Rates of oxidation of purified yellowfin tuna oxymyoglobin at various temperatures (after Brown and Dolev, 1963).

These reaction changes can occur in the frozen state both with or without modification of the chemical structures of enzymes. Chilson *et al.* (1965), Greiff and Kelly (1966), McCown *et al.* (1968) and Young (1968) described freezing studies on several dehydrogenases and other enzymes, the latter two papers being on plant enzymes. In the report of McCown *et al.* on several enzymes in *Dianthus* leaf and stem and in frozen extracts, the multiple forms, subunits and coenzyme associations of the enzymes demonstrated various stabilities toward freezing; some were quite unstable. Drying, too, can be harmful. A group of dehydrogenases frozen in liquid nitrogen and then freeze-dried was found to display far less activity than the enzymes when merely frozen in liquid nitrogen (Rickles *et al.*, 1966). Generalization or prediction about a given enzyme, more especially so in its complex natural environment, is not possible at this time. Even non-freezing cold temperatures may, by a change in aggregation or conformation, inactivate an enzyme (Pullman *et al.*, 1960). This may also be brought about by freezing temperatures in specific sub-zero ranges (Curti *et al.*, 1968). Cold inactivation of enzymes is sometimes reversible by restoration to a warm temperature (Curti *et al.*, 1968) or after addition of enzyme co-factors removed through freezing (Melnick, 1964; Roberts *et al.*, 1966). Of possible interest here is the similarity, in detail, of the temperature curve for killing of a tissue, as expressed in Fig. 1, to the temperature inactivation curve of a crystalline enzyme, presented by Curti *et al.* (1968): namely, the flavine adenine dinucleotide-containing L-amino acid oxidase found in a snake venom.

It is of interest to note that inactivation of enzymes may be avoided by the use of additives of the same types as protect whole cells and organs (Chilson *et al.*, 1965), possibly to some extent by similar mechanisms. It appears also that, as with tissues, the conditions of freezing and thawing are important factors in the stability of enzymes. High protein concentration, low rate of freezing (1°C/min) and fast thawing were enumerated by Greiff and Kelly (1966) as stability factors for purified beef heart lactic dehydrogenase. Melnick (1964) tested many enzymes, including lactic dehydrogenase, in several types of human tissue and often found an opposite effect under his conditions of freezing, namely greater stability at high freezing rates. It appears that the actual amounts of chemical species available as reactants might vary considerably, being dependent even upon the rate of freezing.

B. Enzymes in Tissues

Many of the effects described for isolated enzymes and their co-reactants might also be expected to obtain in organized intracellular structures and in tissue itself. A detailed demonstration of the changes in hydrogen ion concentration in several plant organs following freezing was presented by Van Den Berg (1966). Both the permeability of membranes (Araki, 1965; Heber, 1967; Young, 1968) and the functioning of their complex associations of enzymes (Heber, 1967) and associated proteins (Levitt, 1966) are drastically affected by low temperature. When the membranes of organelles are ruptured by freezing, the enzymes spill out, with serious effects. Tappel's (1966) emphasis on the adverse effects of lysosome rupture and the importance of avoiding this in frozen animal tissue may likewise apply to plants, where "spherosomes" appear to have some of the functions of animal lysosomes (Balz, 1966). Another enzyme association which seems susceptible to destruction by freezing is the phosphorylation complex involved in the generation of adenosine triphosphate by mitochondria and, photosynthetically, by chloroplasts. Heber and Santarius (1964) pointed out that freezing injury cannot be due merely to membrane breakage since photosynthetic phosphorylation is not harmed in osmotically ruptured isolated chloroplasts, whereas freezing the membranes destroys their phosphorylative activity (Heber, 1967, 1968).

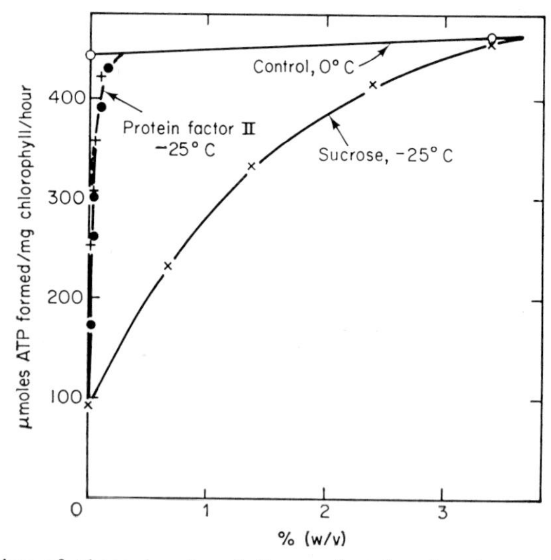

FIG. 8. Protection of photophosphorylation against freezing, by sucrose and protein factor II. Control at 0°C with sucrose. The protein factor was pre-incubated with the membranes for 20 min prior to freezing (from Heber, 1968).

Here again, the conventional additives also offer protection against the destructive and inactivating effects of freezing on both isolated enzymes in solution and on organized sub-cellular structures, including those in plants (Dickinson *et al.*, 1967; Heber and Ernst, 1967). Bekina and Krasnovskii (1968) demonstrated that chloroplasts retained even their most labile function, photosynthetic phosphorylation, for months unchanged when frozen at −79° or −196°C in 50% glycerol. One day's storage at −196°C without the additive caused complete loss of activity.

Heber (1968) isolated from freeze-hardy spinach and other plants specific heat-stable, low molecular weight proteins capable of protecting the vital phosphorylative process in chloroplast membranes with several thousand times the effectiveness of sucrose (Fig. 8). He also found (Heber, 1967) that dilute solutions of sucrose (0·1 M) protect preparations of washed chloroplast membranes against freezing. This protection extended to maintenance of membrane permeability, proton transfer, photophosphorylation, and, by inference, of the coupling of phosphorylation to electron transport. These parallel effects on such critically interrelated biochemical functions are interpreted as indicating that the primary effect of freezing injury is on membrane permeability, which in turn manifests itself biochemically as a defect in adenosine triphosphate formation. The direction of this approach, using the chloroplast and its membrane as model experimental material, promises to yield new insight into the basic factors involved in freeze-damage and its control.

IV. FACTORS IN FREEZING PRESERVATION OF FRUIT

Frozen fruit products have, in general, been favourably received although they are often considered only "second-best" to the fresh fruit in quality. Nutritionally, the frozen product is frequently as good as the starting materials (Derse and Teply, 1958). Perhaps on the questionable assumption that little further improvement in quality is to be expected, few basic biochemical or physiological studies have been carried out with the view to a better understanding of the primary problem of frozen fruits, the structure-destroying reactions. The off-colour and off-flavour producing reactions can probably be considered as secondary effects of structural disruptions and the consequent interactions of tissue components.

A major advantage of frozen fruit lies in the maintenance of flavour characteristics and, particularly, in the absence of "cooked" flavours. Thus the process of blanching, which inactivates enzymes in the treated product, is resorted to less often with fruits than with vegetables and other frozen products. As a result, enzymatic, especially oxidative reactions, are not generally controlled in the frozen-fruit product, except by the use of only partially satisfactory additives.

The structural and metabolic disorders occurring at just below freezing temperatures can be considerable. Freezing (ice formation) may be initiated in a general range from $-1°$ to $-7°C$ (Whiteman, 1957; Golovkin and Chernishov, 1965), depending on the fruit and freezing conditions. At the latter temperature, respiration has still been detected in plant tissues (Zeller, 1951).

A. Texture Changes

The most prominent change associated with frozen and thawed fruits is in their texture, most noticed in turgid, non-fibrous fruits like pears and strawberries. The change in texture on freezing is primarily attributed to a loss of turgidity through collapse of the individual cells owing to disruption of the cell membrane (Weier and Stocking, 1949; Matz, 1965). The intact cell membrane, besides being of primary importance to fruit texture, also acts as a barrier to the extension of ice crystallization within the cells (Lusena and Cook, 1953). As a result, water is osmotically withdrawn from the cell interior to feed the growing extracellular crystals. Surface tension forces have been described as yet another factor tending to exclude the transmission of ice crystals into the cells at moderate freezing temperatures. Mazur (1965b) calculated that the freezing point of external ice is lowered on entering the membrane pores, thus diminishing the entry of seed crystals into the cell and, again, encouraging extracellular accumulation of crystalline ice.

Effects on texture due to chemical changes in the cell wall or its constituent pectins during freezing appear generally to be small (Al-Delaimy *et al.*, 1966; Joslyn, 1966). Some firming of texture may, however, be accomplished when desired by introduction of calcium salts to increase the content of calcium-pectate in the cell wall (Guadagni, 1949). Weier and Stocking (1949) attributed important textural changes during freezing to the separation of cells from each other by dissolution or tearing of the middle lamella which they suggested it may be possible to strengthen. Little is known as to the cell-separating mechanism. It may be due simply to the pressure exerted during enlargement of pockets of extracellular ice (Sterling, 1968).

The small size of ice crystals formed, and other factors already mentioned as associated with ultra-rapid cryogenic freezing, may account for the superior texture and drip characteristics of various small fruits and fruit segments frozen in liquid nitrogen or dichlorodifluoromethane (Webster and Benson, 1966; Wolford *et al.*, 1967; Aref, 1968). Recently developed commercial methods for the quick freezing of food, including fruits, involve treatment of the material with a spray of liquid nitrogen.

B. Colour Changes

Enzymatic darkening caused by the action of phenolase is a serious colour problem with many frozen fruits and has received considerable attention (Joslyn, 1966). Off-colour formation through oxidation is often most intense at cut surfaces, but with open structured oxygen-rich tissues (air spaces in apple make up about one-quarter of the fruit volume according to Reeve, 1953) darkening can take place throughout the frozen sections when thawed. It is assumed that the phenolase reaction is initiated when freezing disrupts cellular, or vacuolar, membranes. A complex mixture of oxidative enzymes and their substrates are thereby intermixed and, in the presence of air, oxidation of the substrates takes place. These substrates include caffeic acid esters, such as the several chlorogenic acids and dactyliferic acid (in dates), or they may be catechins (in apple) or simple catechol derivatives, like dopamine (in banana). They are enzymatically oxidized to quinone-type polymers, giving a brown or black colour to the thawed fruit.

Several approaches to the treatment of fruits for prevention of darkening after freezing (on subsequent thawing) have been made. For example: inhibition or inactivation of the enzyme; introduction of reducing agents to counteract oxidative effects; and removal of air. Such treatments and their effects have been reviewed by Joslyn (1966). Except for SO_2 when used at relatively high concentrations, most of the permissible treatments employing additives are only able temporarily to avoid oxidative darkening. Soon after thawing, phenolase activity proceeds beyond the capacity of the customary amounts of additives used to cope with, or mask, its continuing activity. Removal of air prior to freezing is a good method of preventing oxidative darkening, particularly since ice is sparingly permeable to oxygen (Rakitina, 1965); protection is, however, terminated once the package is opened and thawed. SO_2 treatment is more permanent in effect, but may, except at low concentrations, soften the fruit, add an unpleasant odour and flavour and, by recent reports, act as a specific mutagen (Shapiro *et al.*, 1970; Hayatsu and Miura, 1970).

More recently another approach has been made, namely, to modify the structures of the substrates which darken, so that darkening is permanently prevented. This type of treatment holds promise but so far has been but little applied. As described in Volume 1, Chapter 8, two enzymatic reactions, and also a treatment of fruit surfaces involving merely a pH adjustment via a dip into slightly alkaline buffer such as K_2HPO_2 or $NaHCO_3$ (Bolin *et al.*, 1964; Nelson and Finkle, 1964), have been successfully used to modify permanently the phenolase substrates so that no enzymatic darkening reaction can occur. The enzymatic reactions involve (i) O-methylation, by a catechol O-methyltransferase, of the naturally occurring catechol substrates present in the fruit, such that the modified catechols cannot then react with phenolase (Nelson and Finkle, 1964), and (ii) oxidative ring-opening of catechol substrates through the addition of protocatechuate-3,4-dioxygenase, resulting in non-benzenoid, non-reacting compounds (Kelly and Finkle, 1969; Finkle *et al.*, 1971). Only air and small amounts of ascorbic acid or other suitable reducing agent are required in the latter case for the ring-opening reaction to proceed enzymatically to completion. These enzymatic methods have been successfully applied to the permanent prevention of the darkening of fresh apple juice (Nelson and Finkle, 1964; Kelly and Finkle, 1969) and might also be useful in retaining the light colour of frozen juices. The use of these enzymes is somewhat limited since they only operate satisfactorily in the region of pH 8. It is not easy to see how these methods could be used in frozen apples since internal oxidative darkening readily occurs on thawing. Cryogenically frozen fruits of a less open structure and lower air content, such as peaches, might be more suitable for the application of these processes. (F. P. Boyle and R. Jackson, unpublished results).

The other colour changes which may occur in frozen plant tissue—the chemical breakdown of chlorophyll, carotenoids and anthocyanins—are not so important in freezing fruit as in freezing other plant material (see Section

V). A post-thawing blue coloration of strawberries (Holdsworth, 1968) may well be due to a metal–phenolic complexation or a shift in metal complexing pattern of the anthocyanin pigments (see Chapter 17) following freezing damage to the cells. The avoidance of blue discoloration by the addition of citric acid lends support to this explanation (Jurd and Asen, 1966).

C. Flavour Changes

Changes in both the flavour and aroma constituents of frozen fruits are for the most part generated during prolonged storage (Section V). No marked changes in flavour arising simply from freezing have been reported in fruits, except after an extended freezing process such as often occurs during commercial case-freezing (Guadagni, 1960). However, a specialized kind of flavour change brought about by freezing was described by Joslyn and Goldstein (1964), namely, the loss of astringency and general phenolic content from unripe persimmons when puréed and then frozen.

D. Effects of Conditions of Freezing

The effect of freezing rate has received less detailed attention in fruit tissues than in other organs, but the effects are possibly similar to those occurring in other plant tissues, with some modification as a result of specific structural features. Thus a contrast in the response of different structural elements toward freezing was described by Single (1964) for various organs of wheat; rates of ice penetration varied between 120 cm/min (leaves) and 0·05 cm/min (1 cm ears). The volume of thin-walled parenchyma tissue and the amount of intercellular space may be particularly great in some fruits (e.g. strawberries—see Szczniak and Smith, 1969). These can affect both the rate of removal of heat and the amount of structural distortion during freezing.

Borgstrom (1968) states that with strawberries, raspberries and sliced peaches there is little difference in quality whether frozen slowly, rapidly or at intermediate rates. The rates for freezing were not defined, however, and they may vary widely with different effects as discussed in Section II. It has been pointed out by Morris and Barker (1939), for instance, that freezing of packaged strawberries in brine at $-20°C$ takes place more slowly than freezing by direct immersion of the fruit itself into a $-20°C$ syrup, with better results on storage of the directly immersed product. Recently, claims have been made of improved quality using the cryogenic freezing and individual quick freezing (I.Q.F.) of fruits, including types of fruits that are of unacceptable quality when frozen by conventional methods (Holdsworth, 1967; Wolford *et al.*, 1967; Aref, 1968).

A recent study of apple sections frozen under carefully defined conditions was carried out by Sterling (1968). He froze relatively small cylinders of fruit tissue (1 cm diameter by 1 cm long) and measured the internal changes of temperature under several sets of cooling conditions. The frozen cylinders were thawed in distilled water at room temperature. He found that (see Section II) cell damage was minimized at high freezing rates, such as on plunging the tissue into a bath at −196° or −78°C. From his data, however, the cooling rate at the latter temperature appears to be in the region of 20°C/min, a moderately slow rate in relation to conditions adopted for experiments recorded in Fig. 3. Slower freezing still, e.g. in air at −17°C, produced extensive damage and tissue softening. Histologically, the pattern of cell tearing and crushing was interpreted as being due to the pressures exerted on the cells during ice formation. Of studies made on the freezing of fruits, this work comes closest in experimental treatment to that of Sakai *et al.* (1968), discussed in Section II, in which, again, very high freezing rates gave minimum damage. Sakai and his co-workers reported, too, that exceedingly slow changes in temperature also gave little damage. As already mentioned in Section II, the overall effect of extremely slow freezing—a dehydrative process—is probably akin to the natural freeze-protective properties developed during the well-known cold-hardening treatments carried out with many plants. It seems not unlikely that this hardening could be imitated, beneficially, with fruit tissue. Attainment of Sakai's ultra-slow, controlled freezing conditions could probably be achieved experimentally only by using very thin sections whereby water could be removed from within the cells before intracellular freezing can take place. Such experiments with fruit, with and without the addition of cryoprotective agents, would be of interest to the critical evaluation of the cellular changes which result in damage.

Partial dehydration of fruit cells and consequent preservation of their structural integrity may be the explanation of the beneficial effects that concentrated (40–60%) sucrose syrups have on the texture, drip loss and flavour of frozen apple slices (Guadagni, 1949) and various other fruits. Wiegand (1931) claimed that the best results in terms of quality were obtained when strawberries were held in syrup at −1°C for several days before being frozen. It was later shown by Joslyn (1949) that losses of up to 40% in drained weight occurred with several types of fruit treated in this manner, thus indicating the extent of their dehydration.

Tissue cracking and distortion due to non-uniform pressures of ice is a problem of ultra-rapid cryogenic freezing. It has been avoided in some instances, for example in tomatoes (Webster and Benson, 1966), by incompletely freezing the tomatoes in liquid nitrogen, then removing them and finishing off the freezing process at −12°C.

E. Effects of Additives

Various physico-chemical causes of freezing damage and the ameliorative effects of additives have been discussed in Section II. As mentioned in Section

III, Bekina and Krasnovskii (1968) found that they could preserve the extremely delicate and complex photosynthetic phosphorylation system functioning in chloroplasts by deep freezing in 50% glycerol. It appears possible that with less extensive treatment than is required for preserving the phosphorylating complex, most of the tissue characteristics required for retaining gross cell structure and good eating quality of frozen fruit could be preserved.

Sterling (1968) (see p. 674) tested the effects of various types of additives on the texture of apples which were frozen at various temperatures down to −78°C and then thawed. Of the additives tested (e.g. at freezing temperatures between −5° and −17°C), 2–3 M concentrations of glycerol, monomeric and and polymeric ethylene glycol and NaCl proved the most effective in maintaining fruit firmness, but their effectiveness was very much dependent on the rate of freezing. Much of the damage was caused during freezing rather than during thawing. The added solutes reduced the ice-caused damage. Solutions of ethanol, dimethylsulphoxide and fluorocarbonic fluid "L-1644" were also effective but to a lesser extent than NaCl which, at a concentration of 2 M, resulted in a degree of firmness on thawing close to that of unfrozen apple. To what extent the additives penetrated the 1 cm cylinders of tissue was not stated. Armbruster (1967) stated that the texture of fresh and frozen strawberries is determined by the structural integrity of the epidermis, parenchyma and supporting tissue. Using pre-freezing treatments of a 40% sucrose syrup or a 5 : 1 ratio of fruit to dry sucrose (both treatments gave similar results), she found that, on freezing, the epidermal and xylem cells were not ruptured but the parenchyma cells were changed by rupture without cell separation. The degree of parenchyma cell wall rupture during freezing determined the extent of textural damage. Little other work appears to have been done on the use of cryo-additives in the preservation of the structure of frozen fruits.

Added substances may not only protect the cells from disruptive disorganization but they also permit freezing of tissue without serious damage to take place at slower freezing rates and at higher freezing temperatures than would be possible without the additive. This is particularly important in relation to the temperatures usually employed in the commercial freezing of fruits. The freezing rates described in studies by Samygin and Matveeva (1965, 1967) (see p. 664) and by Luyet (1965), using additives, were not dissimilar to those that are often used in ordinary food-freezing practice. It can be anticipated, therefore, that through increased information concerning protection by added substances and more attention to damage-controlling conditions during freezing, a higher quality of undamaged frozen fruit might be introduced without requiring a major overhaul in modern freezing-plant facilities.

The implications of a little-studied aspect of protection against freeze-damage warrants attention, namely, protective treatments in the field. Applications of growth-retarding chemicals such as decenylsuccinic acid

(Kuiper, 1964) and other compounds (Toman and Mitchell, 1968) in low concentrations (e.g. 10^{-3} M) have shown a freeze-protective (and also drought-protective) effect. Damage-prevention was shown to be transmitted even to organs remote from the site of application.

V. FRUIT STABILITY DURING FROZEN STORAGE

A basic reason for the success of freezing-preservation of fruits is that physical and physiological processes and biochemical reactions which normally take place in fruit preserved by other methods are generally stopped, or slowed to extremely low rates in the frozen material. This greatly reduced biochemical activity during frozen storage is, however, one of the reasons for the difficulties experienced in identifying biochemical pathways in frozen fruits. Indeed, the major problems in frozen fruit technology occur during freezing and thawing and post-thawing handling rather than during storage at appropriate temperatures. Studies of many factors have shown no chemical, physical or sensory change in frozen fruits that could be ascribed solely to frozen storage, provided the temperature was sufficiently low. During a long-term comprehensive study of the time–temperature tolerance (TTT) of frozen fruits conducted at our laboratory, it was found that temperatures of −29°C, or lower, prevented measurable chemical or sensory changes for periods of up to 5 years (Guadagni, 1969). However, commercial frozen fruits are seldom, if ever, held at such a low temperature, and measurable changes in quality may be found in commercially stored and distributed products. While the frozen-food industry in the U.S.A. normally strives to maintain frozen foods at −18°C or lower, temperatures well above this are often encountered even today.

A. Colour and General Appearance

Colour and appearance are extremely important quality attributes in unblanched, frozen, cut tree fruits, because they are subject when thawed to enzymatic browning or blackening (Section IV) which, if severe enough, can readily cause consumer rejection of a commercial product. It should be emphasized that browning does not result in a loss of the naturally occurring carotenoid pigments in such fruits as peaches and apricots; the polymerized quinones effectively mask the bright yellow-orange colour of the carotenoids. Extraction of the pigments from these fruits before and after severe enzymatic browning has shown essentially the same values for total carotenoid pigments (D. G. Guadagni, unpublished results). In sweet or sour red pitted cherries the colour changes can be due both to enzymatic oxidation of initially colourless phenolic compounds, such as chlorogenic acid and "catechol tannins", and to enzymatic degradation of the anthocyanin pigments themselves (Guadagni *et al.*, 1958), for which an anthocyanase present in fruit tissue has been described (Mori *et al.*, 1966).

The anthocyanin pigment content of strawberries and caneberries generally remains stable during extended frozen storage, although a small amount of degradation may occur, dependent on the conditions of freezing, storage and thawing (Guadagni *et al.*, 1961). The most serious defect of these fruits is in structural appearance rather than in colour. Strawberries and raspberries, particularly, present a collapsed, mushy appearance on thawing. Studies on thin slices, as suggested above (Section IV), may determine whether structural integrity can be maintained at extreme freezing rates. A factor which detracts from the appearance of syrup-packed, frozen coloured berry fruits during storage at higher (but still freezing) temperatures is a transfer of anthocyanin pigments from the fruit to the surrounding syrup. The degree of this pigment-transfer in raspberries is exponentially related to temperature (Guadagni *et al.*, 1960) and is all but eliminated below −18°C. In addition to colour, other natural constituents, such as acids and flavouring matter, are also transferred to the syrup. The net effect is to reduce colour, tartness and flavour in the fruit and to intensify these in the syrup.

In frozen citrus concentrate, the most important factor affecting acceptability of the reconstituted beverage is cloud stability. If the juice is not pasteurized, pectin esterase demethylates the pectin and the methyl groups are replaced by cations, such as calcium and magnesium. This causes the cloud to settle or, in severe cases of demethylation, the concentrate actually forms a low methoxyl pectin gel. Cloud loss or gelation does not occur at temperatures of −18°C or lower but increases with increasing temperature (Guadagni, 1969). Similarly, demethylation of pectin by pectin esterase is associated with increases in the firmness of sour red pitted cherries stored above −18°C (Guadagni *et al.*, 1958). Some varieties of sliced, sugar-packed strawberries showed gelation during storage at −6·7°C (D. G. Guadagni, unpublished).

B. Flavour

Next to bright natural colour and normal textural appearance, a near-fresh taste and aroma are usually the most prized characteristics of frozen fruits. The chemistry of the slowly developing off-flavours in some stored frozen fruits has not been elucidated, largely because of the difficulty in measuring the exceedingly slow enzymatic and autoxidative reactions taking place in the fruit at very low temperatures. However, sensory methods using well trained taste panels can yield statistically significant results in the field (Guadagni, 1957). By these methods, changes in the taste and aroma of many frozen fruits and fruit products can be reliably detected long before even the most sensitive of instrumental methods give any indication of change.

The first significant sensory flavour change noticed in stored frozen fruits is a loss of characteristic fresh aroma; this is usually followed by the development of some off-flavour, the nature of which depends on the type of fruit. Some of these off-flavours have been described by Joslyn (1966). The most important factor governing the rate of these changes is the storage temperature. Table II shows the storage time required to produce essentially the same degree of flavour change in a number of frozen fruits and fruit products stored at different temperatures. From these data it appears that the chemical and biochemical systems in raspberries are not as active as those of the other fruits held at −18°C, raspberries being the most stable of the fruits listed. The degradative systems in blueberries are apparently much more active in producing reaction products since here the flavour is only about one-fourth as stable as in raspberries. A different situation may occur in citrus fruit juices, namely an adverse effect of very cold storage. An unpleasant odour and "tallowy" flavour in the juice disappears only very slowly when stored at −34°C, but much more quickly at −18°C (Blair *et al.*, 1957).

TABLE II. Average stability of frozen fruits and fruit products held at various temperatures [a]

Product	Type of pack	Stability (days)		
		−18°C	−12°C	−7°C
Apples	Pie filling	360	250	60
Boysenberries	Pie filling	375	210	45
	Bulk, no sugar	405	125	45
	Retail, syrup	650	160	35
Blueberries	Pie filling	175	77	18
Cherries	Pie filling	490	260	60
Peaches	Pie filling	490	280	56
	Retail, syrup	360	45	6
Blackberries	Bulk, no sugar	630	280	50
Raspberries	Bulk, no sugar	720	315	70
	Retail, syrup	720	110	18
Strawberries	Bulk, dry sugar	630	90	18
	Retail	360	60	10

[a] Stability based on flavour changes, except for retail-pack peaches, for which it is based on colour changes.
From Guadagni (1969).

Other than temperature, the exclusion of oxygen from the packaged products is one of the most important factors affecting both colour and flavour stability of most frozen fruits. The exclusion of oxygen may be accomplished with a good type of package (hermetic container), through packaging under vacuum or inert atmosphere, or by submerging the fruit in syrups with or without antioxidants such as ascorbic acid. All of these procedures greatly extend the colour stability of light coloured frozen fruits, such as apples, peaches, apricots and pears (Guadagni, 1969). While sugar is an important adjunct for the acceptability of frozen fruit products, it does not appear to improve flavour stability nor does it reduce the rate of development of off-flavour under adverse storage conditions (Guadagni, 1969). With freeze-dried fruit, because of its high sugar content, dehydration to a very low moisture level is required for product stability (Burke and Decareau, 1964).

VI. CONCLUSIONS

Although considerable progress has taken place in the past few years we are, nevertheless, far from being able to claim that all fruits (or perhaps any) can be frozen and returned to a thawed condition comparable with the unfrozen commodity in all respects, most particularly in texture, "mouth feel", flavour and colour. Of apparent irrelevance to fruit preservation, yet appropriate to the goal and technology that it represents, is the reported achievement of freezing new-born ground squirrels to a solid ice (oesophageal temperature below $-10°C$), followed by their complete revival (Popovic and Popovic, 1963).

At the present time even high quality frozen fruit products, with a few exceptions (e.g. figs), are usually considerably modified in texture unless processed with large amounts of additives, with resultant effects on flavour, etc.

Further advancements in the freezing and storage of fruits will probably come about both through empirical testing and through increased knowledge of the particular factors, physical, chemical and metabolic, that contribute to fruit cell damage. In this respect information from the investigation of tissues other than fruit, both technical manipulations and detailed findings, is of relevance. However, biochemical knowledge has only just begun to enter the field of plant tissue freezing injury and there are still few aspects where the level of biochemical knowledge is adequate to offer useful guidelines.

Some biochemical changes taking place during freezing and correlations of these changes with physical occurrences that are taking place in the medium have already been described. Particularly noteworthy is the suggestion that many effects on tissue (including loss of cell viability and some instances of unexpectedly high enzyme activities) occur in the temperature region close to $-20°C$, a temperature at which frozen fruit is often stored. Since the conditions of freezing and the temperature of storage both have biochemical effects that are important in minimizing tissue damage, the work of Ching and Slaubaugh (1966) on fir pollen is challenging (see p. 662). It promises a method of isolating other factors in freeze damage from the process of ice crystallization itself as a factor, through the reversible control of water content in living cells.

The patterns of tissue freezing itself and of freezing damage undoubtedly involve changes in the concentrations, states of aggregation, etc. of essential components. Fruits have some unique characteristics to contribute in this field, for example, in their high sugar content as it affects freezing patterns.

It is not certain that the results of the freezing of thin sections of parenchyma tissue from woody skins, and other cells of yeast cells, erythrocytes etc. will necessarily apply also to the parenchyma fruits. Nevertheless, similar work carried out on fruit tissues would undoubtedly provide useful data for the better understanding of the adverse effects of freezing, on the physical and chemical characteristics of these tissues, and their control. Thin-tissue sections or clusters of cells offer efficient heat exchange and hence control of experimental freezing rate; this in turn, controls the degree of intracellular dehydration, the biochemical reactions that operate under specified conditions, and the kind and quantity of damage. On thin slices, the effects of partial dehydration could also be ascertained by means of special techniques, such as the use of high-osmotic bathing solutions, infiltration of solutions or vacuum evaporation. Frozen storage and revival of plant tissues in liquid suspension culture was recently reported by Quatrano (1968). When the key factors of damage control are learned by such experiments, control measures may be initiated which could, perhaps, also be applied to thick sections or even to large segments or whole fruits. It has been found by the author and his co-workers that 5 mm diameter cylinders of red beet (*Beta vulgaris* L.) and cantaloupe (*Cucumis melo* L.), as well as several kinds of whole fruits and vegetables, frozen at an ultra-slow rate of 0·1–0·2°/hour, are greatly preserved in structure after thawing (Finkle *et al.*, 1971; Sayre *et al.*, 1971).

An important aspect of freezing damage control lies in the use of additives to prevent cellular disruption and its consequent features, such as tissue flabbiness and discoloration, but a great deal still remains to be learned about their specific effects. New, consumable freeze-protective compounds of specific desired biochemical and organoleptic characteristics may be developed as the characteristics required for particular protective effects become better known. The additives used must be both palatable and non-toxic, and, preferably, effective in small amounts. Additives may perhaps be found that would, at low concentrations, produce promoter or hormone-like secondary effects such as appear to occur during the development of freeze-hardiness in metabolizing tissues (Steponkus, 1968) and growing plants. The unusual presence of stachyose and rhamnose (Parker, 1959; Sakai, 1966) or of low molecular weight heat-stable proteins (Heber, 1968) or soluble polysaccharides (Olien, 1967) with freeze-protective properties in freeze-hardy tissues merits special attention; likewise, numerous reports of changes in the contents of non-specific protein, carbohydrate, RNA and DNA fractions—see, for instance, Jung *et al.* (1967).

Some compounds work at the surface of cut tissue and membranes (e.g. enzymes, polyvinylpyrrolidones, polyalcohols and, perhaps, sucrose); others like glycerol and dimethylsulphoxide penetrate the tissue or organelle. Deep-tissue penetration, such as into large fruit sections, offers a particularly difficult challenge which could perhaps be controlled by means of hormone-like additives or freeze-protective gaseous compounds. A novel mode of enhancing tissue penetration makes use of the application, by infiltration into the plant organ, of monomeric substances (acrylamides) and their polymerization *in situ*. This has been reported for vegetables by Schwimmer (1969) as a model for sustaining the texture of dehydrated products. The use, in a similar manner, of compounds such as vinylpyrrolidone and their polymerization within the tissue might be a method of introducing freeze-protective polymers into fruits prior to their freezing, thus diminishing freeze-damage.

Also of interest here are recent investigation of freezing or thawing under high pressure, which has been reported to avoid destructive thermal gradients (Persidsky and Leef, 1969). The combined use of careful freezing technology with the addition of additives promises greater control of damaging conditions in fruit even perhaps to the point of tissue viability.

In terms of this goal, Weier and Stocking (1949) have reminded us that "dead cells cannot be expected to have the firmness of living cells".

ACKNOWLEDGEMENT

Appreciation is expressed to D. G. Guadagni and to M. S. Brown for helpful discussions and information.

REFERENCES

Al-Delaimy, K. A., Borgstrom, G. and Bedford, C. L. (1966). *Quart. Bull. Mich. agric. exp. Sta.* **49**, 164.
Araki, T. (1965). *Low Temp. Sci.* **B23**, 97–109; Abst. in *Cryobiol.* **3**, 472, 1967.
Aref, M. M. (1968). *J. Can. Inst. Fd Technol.* **1**, 11.
Armbruster, G.(1967). *Proc. Am. Soc. hort. Sci.* **91**, 876.
Balz, H. P. (1966). *Planta* **70**, 207–236.
Bekina, R. M. and Krasnovskii, A. A. (1968). *Biokhimiya* **33**, 178; *Biochemistry* (Consultants Bureau) (1968), **33**, 152.
Bielski, B. H. J. and Freed, S. (1965). *Cryobiol.* **2**, 24.
Blair, J. S., Godar, E. M., Reinke, H. G. and Marshall, J. R. (1957). *Fd Technol., Champaign* **11**, (2), 61.
Bolin, H. R., Nury, F. S. and Finkle, B. J. (1964). *Bakers Digest* **38** (3), 46.
Borgstrom, G. (1968). "Principles of Food Science", Vol. 1. MacMillan, New York.
Brown, W. D. and Dolev, A. (1963). *J. Fd Sci.* **28**, 211.

Burke, R. F. and Decareau, R. V. (1964). *Adv. Fd Res.* **13,** 1.
Butler, A. R. and Bruice, T. C. (1964). *J. Am. chem. Soc.* **86,** 313.
Chandler, W. H. (1913). *Mo. agric. exp. Sta. Res. Bull.* No. 8, 141.
Chandler, W. H. and Hildreth, A. C. (1935). *Proc. Am. Soc. hort. Sci.* **33,** 27.
Chilson, O. P., Costello, L. A. and Kaplan, N. O. (1965). *Fedn Proc.* **24** (2, Part 3), S55.
Ching, T. M. and Slabaugh, W. H. (1966). *Cryobiol.* **2,** 321.
Clark, H. (1960). *In* "Conference on Frozen Food Quality", pp. 81–82. ARS 74–21, U.S. Agric. Res. Service.
Clayton, W. (1932). "Colloid Aspects of Food Chemistry and Technology", Chapter 12. Blakiston's Son & Co., Inc., Philadelphia.
Curti, B., Massey, V. and Zmudka, M. (1968). *J. biol. Chem.* **243,** 2306.
Derse, P. H. and Teply, L. J. (1958). *J. agric. Fd Chem.* **6,** 309.
Dickinson, D. B., Misch, M. J. and Drury, R. E. (1967). *Science, N.Y.* **156,** 1738.
Doebbler, G. F. (1966). *Cryobiol.* **3,** 2.
Fennema, O. (1966). *Cryobiol.* **3,** 197.
Fennema, O. and Powrie, W. D. (1964). *Adv. Fd Res.* **13,** 219.
Finkle, B. J., Nuzaki, M. and Fujisawa, H. (1971). *Phytochem.* **10,** 235.
Finkle, B. J., Pereira, E. S. B. and Brown, M. S. (1971). *Int. Inst. Refrig. Bull. Annexe*. In press.
Gahan, P. B., McLean, J., Kalina, M. and Sharma, W. (1967). *J. exp. Bot.* **18,** 151.
Gehenio, P. M. and Luyet, B. J. (1939). *Biodynamica* **2** (55), 1.
Golovkin, N. A. and Chernishov, V. M. (1965). Refrig. Sci. Technol., Supplement, 1965–4, 649; Abst. in *Cryobiol.* (1967). **4,** 41.
Grant, N. H. and Alburn, H. E. (1966). *Nature, Lond.* **212,** 194.
Grant, N. H. and Alburn, H. E. (1967). *Archs Biochem. Biophys.* **118,** 292.
Greiff, D. and Kelly, R. T. (1966). *Cryobiol.* **2,** 335.
Guadagni, D. G. (1949). *Fd Technol.* **3,** 404.
Guadagni, D. G. (1957). *Fd Technol.* **11,** 471.
Guadagni, D. G. (1960). *In* "Conference on Frozen Food Quality", pp. 24–29; ARS 74-21, U.S. Agric. Res. Service.
Guadagni, D. G. (1969). *In* "Quality and Stability in Frozen Food" (W. B. Van Arsdel, M. J. Copley and R. L. Olson, eds). Interscience Publications, John Wiley and Sons, New York.
Guadagni, D. G., Nimmo, C. C. and Jansen, E. F. (1958). *Fd Technol.* **12,** 36.
Guadagni, D. G., Kelly, S. H. and Ingraham, L. L. (1960). *Fd Res.* **25,** 464.
Guadagni, D. G., Downes, N. J., Sanshuck, D. W. and Shinoda, S. (1961). *Fd Technol.* **15,** 207.
Hayatsu, H. and Miura, A. (1970). *Biochem. Biophys. Res. Commun.* **39,** 156.
Heber, U. (1967). *Pl. Physiol., Lancaster* **42,** 1343.
Heber, U. (1968). *Cryobiol.* **5,** 188.
Heber, U. and Ernst, R. (1967). *In* "Cellular Injury and Resistance in Freezing", (E. Asahina, ed.) Vol. 2, pp. 63–77. Proc. Internatl Conf. Low Temperature Sci., Institute Low Temperature Science, Hokkaido Univ., Sapporo, Japan.
Heber, U. and Santarius, K. A. (1964). *Pl Physiol., Lancaster* **39,** 712.
Holdsworth, S. D. (1967). *Fd Manuf.* **42** (7), 42.
Holdsworth, S. D. (1968). *Tech. Bull. No. 13*, Fruit and Vegetable Preservation Res. Assn, Chipping Campden, England.
Holm-Hansen, O. (1967). *Cryobiol.* **4,** 17.
Joslyn, M. A. (1949). *Fd Technol.* **3,** 8.

Joslyn, M. A. (1966). *In* "Cryobiology" (H. T. Meryman, ed.), p. 565. Academic Press, New York.
Joslyn, M. A. and Diehl, H. C. (1952). *A. Rev. Pl. Physiol.* **3,** 149.
Joslyn, M. A. and Goldstein, J. L. (1964). *J. agric. Fd Chem.* **12,** 511.
Jung, G. A., Shih, S. C. and Shelton, D. C. (1967). *Cryobiol.* **4,** 11.
Jurd, L. and Asen, S. (1966). *Phytochem.* **5,** 1263.
Karow, Jr A. M. and Webb, W. R. (1965). *Cryobiol.* **2,** 99.
Kelly, S. H. and Finkle, B. J. (1969). *J. Sci. Fd Agric.* **20**, 629.
Kiovsky, T. E. and Pincock, R. E. (1966). *J. Am. chem. Soc.* **88,** 4704.
Krasavtsev, O. A. and Khvalin, N. N. (1959). *Doklady Akad. Nauk SSSR, Bot. Sci. Sect.* **127,** 235. *Biol. Abst.* (1961). **36,** No. 2534.
Kuiper, P. J. C. (1964). *Science, N.Y.* **146,** 544.
Levitt, J. (1958). *Protoplasmologia* **VIII** (6), 1.
Levitt, J. (1964). *Cryobiol.* **1,** 11.
Levitt, J. (1966). *Cryobiol.* **3,** 243.
Lovelock, J. E. (1953). *Biochim. biophys. Acta* **10,** 414.
Lovelock, J. E. (1957). *Proc. Roy. Soc.* **B147,** 427.
Lusena, C. V. and Cook, W. H. (1953). *Archs Biochem. Biophys.* **46,** 232.
Lusena, C. V. and Dass, C. M. S. (1966). *Can. J. Biochem.* **44,** 775.
Lutz, J. M. and Hardenburg, R. E. (1968). Agric. Handbook No. 66, U.S. Dept. Agric.
Luyet, B. J. (1965). *Fedn Proc.* **24** (2, Part 3), S315.
Luyet, B. J. and Gehenio, P. M. (1965). *Cryobiol.* **2,** 21.
Luyet, B. J., Rapatz, G. L. and Gehenio, P. M. (1963). *Biodynamica* **9,** 95.
McCown, B. H., Hall, T. C. and Beck, G. E. (1968). Abst. Soc. Cryobiol., Aug 1968, in *Cryobiol.* **4,** 275.
Manax, W. G., Bloch, J. H., Longerbeam, J. K., Eyal, Z. and Lillehei, R. C. (1964). *Cryobiol.* **1,** 157.
Matz, S. A. (1965). "Water in Foods". Avi Publishing, Westport, Connecticut.
Maximow, N. A. (1912). *Ber. dt. bot. Ges.* **30,** 52, 293, 504.
Mazur, P. (1965a). *Fedn Proc.* **24** (2, Part 3), S175.
Mazur, P. (1965b). *Ann. N.Y. Acad. Sci.* **125,** 658.
Mazur, P. (1966). *In* "Cryobiology" (H. T. Meryman, ed.), p. 213. Academic Press, New York.
Mazur, P. and Miller, R. H. (1967). *Cryobiol.* **3,** 365.
Melnick, P. J. (1964). *Cryobiol.* **1,** 140.
Meryman, H. T. (1966a). *Cryobiol.* **2** , 165.
Meryman, H. T. (1966b). *In* "Cryobiology" (H. T. Meryman, ed.), pp. 1–114. Academic Press, New York.
Mori, Y., Awaya, S. and Okura, K. (1966). *Kaseigaku Zasshi* **17,** 142; *Chem. Abst.* (1966). **65,** No. 9625.
Morris, T. N. and Barker, J. (1939). *Rept Fd Invest. Bd., Lond. for 1938*, 190.
Natl Frozen Fd Assn (U.S.). (1969). *In* "Frozen Food Fact Book and Directory", p. 48. New York.
Nelson, R. F. and Finkle, B. J. (1964). *Phytochem.* **3,** 321.
Olien, C. R. (1967). *A. Rev. Pl. Physiol.* **18,** 387.
Parker, J. (1959). *Botan. Gaz.* **121,** 46.
Persidsky, M. D. and Leef, J. (1969). *Biophys. J.* **9,** FAM-J4.
Persidsky, M. D., Richards, V. and Leef, J. (1965). *Cryobiol.* **2,** 74.
Popovic, P. and Popovic, V. (1963). *Am. J. Physiol.* **204,** 949.

Pullman, M. E., Penefsky, H. S., Datta, A. and Racker, E. (1960). *J. biol. Chem.* **235,** 3322.
Quatrano, R. S. (1968). *Pl. Physiol., Lancaster* **43,** 2057.
Rakitina, Z. G. (1965). *Fiziol. Rast.* **12,** 909; *Soviet Plant Physiol.* (1965). (Consultants Bureau) **12,** 795.
Rapatz, G. L. and Luyet, B. J. (1965). *Cryobiol.* **2,** 20.
Reeve, R. M. (1953). *Fd Res.* **18,** 604.
Reeve, R. M. and Brown, M. S. (1966). *Cryobiol.* **3,** 214.
Rickles, N. H., Bennett, J. S. and Hillier, M. A. (1966). *Cryobiol.* **2,** 219.
Roberts, L. W., Baba, S. and Urban, K. (1966). *Pl. Cell Physiol., Tokyo* **7,** 177.
Rowe, A. W. (1966). *Cryobiol.* **3,** 12.
Sakai, A. (1966). *J. hort. Sci.* **41,** 207.
Sakai, A. and Otsuka, K. (1967). *Pl. Physiol., Lancaster* **42,** 1680.
Sakai, A., Otsuka, K. and Yoshida, S. (1968). *Cryobiol.* **4,** 165.
Samygin, G. A. and Matveeva, N. M. (1965). *Fiziol. Rast.* **12,** 516; *Soviet Plant Physiol.* (Consultants Bureau) (1965), **12,** 446.
Samygin, G. A. and Matveeva, N. M. (1967). *Fiziol. Rast.* **14,** 1048; *Soviet Plant Physiol.* (Consultants Bureau) (1967), **14,** 879.
Santarius, K. A. and Heber, U. (1967). *Planta* **73,** 109.
Sayre, R. N., Brown, M. S., Finkle, B. J. and Pereira, E. S. B. (1971). *Int. Inst. Refrig. Bull. Annexe.* In press.
Scheraga, H. A. (1965). *Ann. N.Y. Acad. Sci.* **125,** 253.
Scholander, P. F., Flagg, W., Hock, R. J. and Irving, L. (1953). *J. Cell. Comp. Physiol.* **42,** Suppl. 1, p. 1.
Schroeder, C. A. (1961). *Bot. Gaz.* **122,** 198.
Schwimmer, S. (1969). *Fd Technol* **23,** 975.
Shapiro, R., Servis, R. E. and Welcher, M. (1970). *J. Amer. chem. Soc.* **92,** 424.
Silver, R. K., Lehr, H. B., Summers, A., Greene, A. E. and Coriell, L. L. (1964). *Proc. Soc. exp. Biol. Med.* **115,** 453.
Single, W. V. (1964). *Aust. J. Agric. Res.* **15,** 869.
Steponkus, P. L. (1968). *Cryobiol.* **4,** 276.
Sterling, C. (1968). *J. Fd Sci.* **33,** 577.
Sussman, M. V. and Chin, L. (1966). *Science, N.Y.* **151,** 324.
Szczniak, A. S. and Smith, B. J. (1969). *J. Texture Studies,* **1,** 65.
Tappel, A. L. (1966). *In* "Cryobiology" (H. T. Meryman, ed.), pp. 163–177. Academic Press, New York.
Toman, F. R. and Mitchell, H. L. (1968). *J. agric. Fd. Chem.* **16,** 771.
Tressler, D. K., Van Arsdel, W. B. and Copley, M. J. (1968). "The Frozen Preservation of Foods", 4th edition. Avi Publishing, Westport, Connecticut.
Tumanov, I. I. and Krasavtsev, O. A. (1959). *Fiziol. Rast.* **6,** 654; *Soviet Plant Physiol.* (Consultants Bureau) (1960). **6,** 663.
Tumanov, I. I. and Krasavtsev, O. A. (1966). *Fiziol. Rast.* **13,** 144; *Soviet Plant Physiol.* (Consultants Bureau) (1966). **13,** 127.
Van Arsdel, W. B., Copley, M. J. and Olson, R. L. (1969). "Quality and Stability of Frozen Food." Interscience, John Wiley and Sons, New York.
Van Den Berg, L. (1966). *Cryobiol.* **3,** 236.
Webster, R. C. and Benson, E. J. (1966). U.S. Patent No. 3,250,630.
Weier, T. E. and Stocking, C. R. (1949). *Adv. Fd Res.* **2,** 297.
Whiteman, T. M. (1957). Marketing Res. Report No. 196, U.S. Dept. Agric.
Wiegand, E. H. (1931). Oregon Agric. Exp. Station, Station Bulletin 278.

Williams, R. J. and Meryman, H. T. (1969). Abst., 11th Intern. Bot. Cong., Aug.–Sept. 1969, p. 240.

Wolford, E. R., Ingelsbe, D. W. and Boyle, F. P. (1967). Proc. XII Intern. Cong. Refrigeration (Madrid), Paper 4.22, Vol. 3, p. 459.

Yoshida, S. and Sakai, A. (1967). *Low Temp. Sci.* **B25,** 71, Abst. in *Bull. Intern. Inst. Refrig.* (1968), **48,** 1192.

Young, Roger (1968). *In* "International Citrus Symposium." Riverside, California. In press.

Zeller, O. (1951). *Planta* **39,** 500.

Chapter 20

Effects of Radiation Treatments

I. D. CLARKE

Joint FAO/IAEA Division of Atomic Energy in Food and Agriculture, Vienna, Austria

I	Introduction	687
	A. Food Irradiation	687
	B. Properties of Ionizing Radiation	688
II	Extension of Storage Life of Fresh Fruits	690
	A. Texture	690
	B. Respiration	692
	C. Ethylene Production	694
	D. Pigment Formation	695
	E. Physiological Breakdown	696
	F. Enzyme Studies	696
	G. Ripening Processes	697
III	Microbiological Aspects	698
IV	The Radurization of Fruit Juices	699
V	Public Health Clearance	701
VI	Prospect for Commercial Application	701
	References	702

I. INTRODUCTION

A. Food Irradiation

The preservation of food by ionizing radiation has been a subject of interest for more than thirty years, but it was not until the development of large isotopic and machine sources that extensive research could begin on the feasibility of preserving food by irradiation. The initial stimulus for the application of radiation for this purpose stemmed from the fact that micro-organisms could be killed with only a negligible increase in temperature. It was soon realized that the process offered no solution for residual enzyme activity, which ultimately would have to be controlled by low temperature storage or by a short-term heat treatment to obtain optimum storage life.

The sensitivity of cell division to radiation is the basis of a number of other applications, independent of the bactericidal action of radiation. The inhibition of sprouting in potatoes, onions, carrots and the sterilization of insects for the preservation of stored grains or for the disinfestation of fruits are examples of such applications. A dose in the region of 10 krad will control sprouting in most cases while a dose of 25 krad will control most insect pests. Control of mould growth on the other hand requires a dose in excess of 100 krad. Usually, the quality of the fruit is markedly affected by doses higher than 200 krad and, as will be described later, it is sometimes possible to induce a form of damage in fruits that is only expressed later during storage. Doses between 500 and 1000 krad are recommended for the elimination of pathogenic organisms, such as *Salmonella* spp., from animal feed. Sterilizing doses for meats are between 4 and 5 Mrad, for this is the dose required to reduce the surviving population of spores of *Clostridium botulinum* by a factor of 10^{12}. This is based upon the well-known 12 D concept, which originated in the canning industry.

It is possible to anticipate now that the possibilities of greatest commercial interest include the extension of shelf-life of fresh fish (Ronsivalli *et al.*, 1965), chicken (Rhodes, 1965), strawberries, mushrooms (Clarke, 1959) and the disinfestation of cereal grains (Cornwell, 1966). Fruit pests, such as the Hawaiian fruit fly on papayas and the mango seed weevil (Maxie and Abdel-Kader, 1966; Shea, 1965), can be controlled by radiation doses which can also delay the ripening processes in certain fruits. The use of the radiation treatment provides a means of overcoming quarantine barriers, which are based upon the refusal by the importing government to permit entry of fruit products contaminated by pests.

Of particular concern to many governments which import foodstuffs is the contamination of food by harmful pathogenic organisms but, because of the low pH of fruits, problems of this nature are unlikely to occur in fruit products.

This review is not intended to cover all aspects of fruit irradiation, but should be read as a supplement to other reviews (Maxie and Abdel-Kadar, 1966; Romani, 1966; Sommer and Fortlage, 1966).

B. Properties of Ionizing Radiation

It is intended to describe only those properties necessary for readers who have no previous knowledge of the radiations used for processing food, to understand the text of this review.

1. *Units*

The most useful unit of radiation for food technologists is the rad, which is defined as follows:

$$1 \text{ rad} = 100 \text{ ergs/g of material irradiated,}$$
$$1 \text{ krad} = 1000 \text{ rad,}$$
$$1\ m\text{rad} = 1{,}000{,}000 \text{ rad.}$$

Other units have been used:

$$1 \text{ rep} = 84\text{–}94 \text{ ergs/g water or tissue,}$$
$$1 \text{ roentgen} = 84 \text{ ergs/g air.}$$

The great advantage of using the rad is that it refers specifically to the amount of energy absorbed by the material.

The unit of activity is the curie which is defined as:

$$1 \text{ curie} = 3{\cdot}67 \times 10^{10} \text{ disintegrations/sec}$$

or, alternatively, 1 curie will give a dose of approximately 1 rad 1 metre distance in material of unit density in 1 hour.

$$1 \text{ MeV} = 1{\cdot}6 \times 10^{-6} \text{ ergs.}$$

Descriptions of the methods used in measuring the absorbed dose are available (Brownell and Hine, 1956; Attix *et al.*, 1966).

2. *Radiation sources*

The radiation energies of main interest for the preservation of food are the gamma rays from certain isotopic sources, electrons produced by machine sources, X-rays from machine sources and the conversion of electrons to X-rays by bombarding a metal target, such as copper, by electrons. This latter type is unlikely to be of commercial significance because of the low efficiency in the conversion to X-rays. The choice of source is determined in part by the depth of penetration required. γ-rays and X-rays penetrate easily, whereas electrons penetrate to only a limited extent. For example, electrons of 10 MeV can only penetrate to a depth of 4 cm of material of unit density. Energy levels above 10 MeV may induce a small amount of radioactivity in foods and the presence of this activity may be sufficient to prevent governmental approval of these foods for public consumption.

Both electron and γ-sources have advantages and disadvantages. Machine sources have very high dose rates, much greater than is available with present day ^{60}Co sources of even the highest specific activity.

In the case of fruits, however, where the mycological problems are frequently associated with the contamination of the surface, Benn (1967) has shown that it is possible to exploit the weak penetrating ability of electrons. Few fruits can withstand the necessary rolling and handling procedures for uniform dosage to the surface by electrons. Many mould spores can survive in the depression at the button end of the fruit and it may prove difficult if not impossible to ensure that the base of the depression receives the dose required for the inactivation of the spores.

3. *Chemical changes induced by radiation*

Electron, γ and X-ray radiation bring about both physical and chemical changes in matter. It is generally accepted that free radicals of a very short lifetime, approximately 10^{-11}–10^{-8} sec mediate most of the chemical changes. For example, the radiolysis of water can be summarized as follows:

$$H_2O \rightsquigarrow H_2,\ H_2O_2,\ H \text{ and } OH\cdot.$$

The reducing and oxidative properties of the radicals, H and OH·, probably account for the major chemical changes occurring in irradiated foods. Secondary reactions may occur with the products of water radiolysis, such as the oxidation of lipids by hydrogen peroxide. The final products of radiolysis depend upon the chemical nature of the system being irradiated and many reactions may occur in pure systems which would not occur in cells and, of course, the reverse is true.

II. EXTENSION OF STORAGE LIFE OF FRESH FRUITS

The organoleptic acceptability or desirability of fruits depends upon many processes which occur during growth and maturation. Radiation can bring about changes in metabolism either immediately after treatment or at some later time during storage. In a number of cases, the modification of metabolism induced by radiation may cause a delay in the ripening processes which could prove useful commercially.

In this review, the effects of radiation on the biochemistry of the fruits and the effects on the viability of mould spores are treated separately.

A. Texture

Plant material has a marked tendency to soften after irradiation. Only in the case of lettuce has an increase in firmness been observed immediately after irradiation (Massey *et al.*, 1961). According to Glegg *et al.* (1956) the threshold dose is 34·7 krad for softening in Gravenstein apples, 160 krad in Chantenay carrots and 316 krad in Detroit red beets. It has been reported that the threshold doses for several varieties of apples varied between 4·2 and 107 krad (Boyle *et al.*, 1957). These varietal differences may, however, be due to the variation in the degree of ripeness at the time of the experiments. The effect of radiation on texture would probably be less in soft fruit than in firm fruit. It is probable that nearly all fruits soften at doses below 200 Krad. The well-known correlation between the degree of polymerization of the cell wall pectic substances and the texture of fruits led to an investigation of the changes in the pectic substances in irradiated fruits.

Evidence has been obtained since the work of Kertesz *et al.* (1964) that the softening effect of radiation can be associated with an increase in the water-soluble pectins, which supports the concept of the degradation of protopectin (Clarke, 1962; Massey *et al.*, 1965; Dennison and Ahmed, 1966). However, Clarke (1968) reported that it has not proved easy to provide direct evidence that the increase in the water-soluble pectins is due to the degradation of the higher molecular pectins to low molecular weight pectins. It is not clear if the degradation of cellulose and other cell wall components also plays a part. The change in texture after irradiation occurs so quickly that the mediation by enzymes seems unlikely although, as will be seen on p. 697, radiation may increase the activity of pectin esterase.

Recently Dzamic and Jankovic (1966) showed that dry preparations of pectin are depolymerized by γ-rays whereas aqueous solutions are depolymerized and also de-esterified. Wahba and Massey (1966) found that moisture content had a great influence on degradation by γ-radiation. Kertesz *et al.* (1964) γ-irradiated cylinders cut from apples. From zero to 2060 krad firmness of the tissue decreased strongly. Total uronides remained more or less constant at about 0·35% but the cold water-soluble fraction increased from 25 to 59% of the total at the expense of protopectin. The viscosity of all fractions decreased. Rouse and Dennison (1968) exposed Valencia oranges and Duncan grapefruit to 150 and 300 krad γ-radiation. Pectin determinations were then made in component parts (peel, membranes, juice sacs). In both fruits, water- and oxalate-soluble pectin increased at the expense of protopectin. Acid extracts of these component parts were also compared. Yields were not affected by irradiation; in Valencia oranges they were approximately 25% of alcohol-insoluble solids (AIS) of peel, 32% of membranes and 20% of juice sacs. Gel forming strength generally decreased which, since the degree of esterification was affected in the peel pectin only to the extent of approximately 10% decrease of methoxyl, suggested that depolymerization was occurring. The results obtained on grapefruits were very similar. It is interesting to note that the same experiments performed on fruit which was harvested on later dates showed less influence of irradiation. The explanation offered by the authors is that the increased sugar content acts as protecting agent as previously described for pectin solutions by Kertesz *et al.* (1956).

On storage after irradiation, some fruits maintain their texture better than controls and, in spite of the initial deterioration, may, therefore, at the end of the storage period be firmer than the controls (Romani and Bowers, 1963). Very little information is available concerning the nature of the changes in texture during post-radiation storage. Excessive variation between fruit samples often obscures changes due to treatment and it is possible that considerable chemical change could occur without being detected by the current chemical methods. Alternatively, relatively small chemical changes may bring about considerable changes in texture, thereby reducing the possibility of detecting a correlation between texture and chemical constitution in irradiated fruits during storage.

Because of the probable involvement of calcium in maintaining the structural integrity of the cell wall (Markakis and Nicholas, 1967), attempts have been made to increase the firmness of irradiated fruits by dipping the fruits in solutions of calcium chloride. Evidence has been obtained that the induced softening of blueberries can be reversed by washing the fruits in 0·5% calcium chloride for 1 hour and some benefit is claimed for increasing the firmness of irradiated strawberries (Al-Jasim *et al.*, 1968), but the present role of calcium in reversing the softening has not been established.

B. Respiration

An increase in the rate of carbon dioxide production is a marked feature of the maturation process in climacteric fruits. Radiation induces an immediate increase in the respiration rate of most fruits (Romani and Bowers, 1963; Romani, 1964; Maxie *et al.*, 1965; Bussel and Maxie, 1966; Maxie *et al.*, 1966). This might suggest a premature onset of the climacteric phase in irradiated fruits, such as peaches and nectarines, but sometimes the climacteric processes are quite clearly not involved, as in pears. Even in those fruits which exhibit a premature onset of the climacteric phase, the total amount of carbon dioxide produced may be in excess of the amount normally produced during the climacteric. Radiation appears to induce some form of modification in the respiratory apparatus or mechanisms which result in an enchanced rate of gaseous exchange, because oxygen uptake may also be stimulated as well as carbon dioxide production. (Massey and Bourke, 1967). It is unfortunate that most of the work reported so far only deals with the rate of carbon dioxide production without reference to the corresponding oxygen uptake. Reports of the effects of γ-radiation on tomato fruit have been confined mainly to studies of the respiration rate, ethylene production and ripening behaviour; little attention has been paid to biochemical changes. Abdel Kader *et al.* (1968) showed that up to 500 krad irradiation stimulated respiration in proportion to the severity of the dose; at 600 krad and above, depression of respiration occurred. Both this group of workers and Lee *et al.* (1968) noted that the burst of activity declined within a week. They also showed that immature fruit subjected to low doses of radiation subsequently showed a normal climacteric pattern as did Young (1965). In some fruits the stimulation of respiration following irradiation may be reflected in the oxidative capacity of the mitochondria (Romani, 1966), but this has not been demonstrated for the tomato.

The rise in carbon dioxide production following radiation is immediate and may persist for some days. The magnitude of the rise is also dependent upon the stage of maturity of the fruit at the time of irradiation. In general, for climacteric fruits including bananas (Young, 1965) and tomatoes (Van Kooy, 1968), the earlier the treatment is given in the pre-climacteric stage, the greater is the stimulated rise in carbon dioxide production. As the fruit passes through the climacteric, then very little response may be detected following radiation. Few determinations are available of the oxygen uptake and respiratory quotient of irradiated fruits, so that it is possible that some of the adverse effects that have been reported may prove to be the result of a stimulated anaerobic metabolism. The correspondence of the stimulated carbon dioxide production following irradiation with the processes involved in the climacteric phase must await further elucidation. It would be expected that an acceleration of the climacteric would be correlated with the acceleration of the ripening process which occurs, for example, in peaches and nectarines. As the rate of carbon dioxide production is increased in pears, bananas, mangoes and avocados, where the ripening processes can be inhibited, it must not be assumed that the stimulated respiratory activity is due to processes which control the climacteric phase. Radiation may influence other ripening processes, because radiation can also stimulate the production of ethylene which is another phenomenon associated with the climacteric stage.

It is generally accepted that enzyme activity is relatively resistant to radiation at the doses likely to be used in the processing of fruits. Also, it is well authenticated that the electron transport chain can be relatively resistant to radiation although oxidative phosphorylation may be quite sensitive (Clarke and Hayes, 1965; Metlitsky and Petrash, 1968). This sensitivity may not be due entirely to a direct effect upon the mitochondria for it may be mediated by other factors as in mammalian systems (Ord and Stocken, 1962; Van Bekkum *et al.*, 1962). A hypothesis was proposed that could link the stimulated carbon dioxide production with the onset of senescence and accelerated ripening by analogy with the effect of DNP which induces loss of phosphorylative control by the electron chain (Clarke, 1962). The disturbance of ATP synthesis by DNP bears some resemblance to the effect of radiation as both result in an increase of carbon dioxide production. As already noted, the phosphorylative enzymes in mitochondria are much more sensitive than the electron chain and it seemed possible that the radiation treatment could disturb the respiratory control of the mitochondria with the consequent burst in respiratory activity. It was thought that this could account for the accelerated loss of malic acid in apples (Maxie *et al.*, 1965). In this particular case, the acid loss occurred over a considerable storage period and therefore does not correspond to the pattern of the carbon dioxide production. However, a disturbed metabolic control could bring about an apparent accelerated ripening. Evidence against this hypothesis has been obtained with pear mitochondria which quite clearly demonstrated the resistance of the phosphorylative and oxidative processes in fruit mitochondria to radiation (Miller *et al.*, 1967; Romani *et al.*, 1967, 1968; Van Kooy, 1968). To account for this resistance, a second hypothesis postulated that the stimulated respiratory exchange was related to the repair of damage induced by the treatment. In support of this hypothesis, an increase of protein synthesis was observed in isolated organelles from irradiated pre-climacteric pear tissues (Romani and Fisher, 1966). Earlier work had shown that the protein content of irradiated pears was increased during storage (Clarke and Fernandez, 1961). There is no direct evidence that this excess protein may be involved in repair processes. In senescing (i.e. post-climacteric) pears the ability to repair radiation damaged mitochondria is lost (Romani *et al.*, 1968).

Neither of these two hypotheses explains the inhibition at the succinic dehydrogenase step of the tricarboxylic acid cycle. Evidence for the sensitivity of the particular step to radiation has been obtained for carrot and apple tissue (Massey, 1968). Some earlier work with apples by Hulme (unpublished) indicated the accumulation of succinate in apples after radiation, which would support the observation made on the sensitivity of this step. Unfortunately, it was not possible to confirm this result in a later season (Fernandez and Clarke, 1962). The first hypothesis does not explain how in fruits, such as pears, it is possible to stimulate the respiratory exchange and yet still induce a delay in the rate of ripening,because this would be precluded by the loss of respiratory control. Yet if the effect of radiation is to induce a non-specific damage to the respiratory system, which may or may not require ATP for the repair of the damage, it is difficult to understand a specific block at the succinic dehydrogenase level. The possible transfer of the principle respiratory pathway from the Emden–Meyerhof–Parnas scheme to the pentose phosphate shunt system (Faust *et al.*, 1967) seems to warrant further investigation because it would result in a stimulated respiratory exchange without involving the tricarboxylic acid cycle. If only one of these respiratory mechanisms was stimulated, then it should be possible to detect which one by the accumulation of specific intermediates.

C. Ethylene Production

It has been clearly shown that ethylene can induce the ripening process (see Volume 1, Chapter 16) and recent progress has been made in identifying the enzymes and co-factor requirements involved in the production of ethylene (Mapson and Wardale, 1968). As in the case of carbon dioxide production, ethylene production is stimulated by radiation. Maxie *et al.* (1966) showed that doses between 100–400 krad produce an immediate increase in ethylene production in many fruits. In this particular series of experiments, pears which had received 400 krad produced ethylene before the normal climacteric. After reaching a maximum in about 3 days, the rate of ethylene production declined. Pears that had received doses which did not affect the rate of ripening still produced more ethylene than the control fruits at the climacteric peak. Little or no ethylene was produced by fruits which received 300–400 krad. The production of ethylene is an immediate response to the radiation attack and has been compared to mechanical damage (Maxie and Sommer, 1968) which is also known to stimulate the production of ethylene. For example, Maxie *et al.*, (1965) found that ethylene production as well as respiration rate is stimulated in lemons some time after treatment with 50–100 krad giving the effect of a pseudo-climacteric (see Volume 1, Chapter 17).

The sensitivity of fruits to ethylene may vary after the radiation treatment. Pears are rendered less sensitive to ethylene than controls (Maxie *et al.*, 1966), while bananas retain their sensitivity, but require a longer exposure time to ethylene than the control fruit for the equivalent concentration (Sommer and Maxie, 1966). This observation is likely to be important if irradiation is used under commercial conditions where large quantities of fruit are being stored together.

The amounts of ethylene produced may vary according to the time of irradiation during the climacteric. Accelerated ripening is sometimes associated with ethylene production, but, in contrast, the ripening of pears and tomatoes may be delayed by irradiation doses which stimulate ethylene production. Lee *et al.* (1968) showed that irradiation of tomatoes with between 200 and 400 krad caused an evolution of ethylene half that observed during the ripening of normal fruit. Within 3–6 days, the evolution of ethylene fell to trace levels and, later on, when the fruit began to colour, ethylene production was higher than in ripening control fruits. In contrast, immature fruit exposed to 700 krad, or fruit at later stages of maturity subjected to moderate doses, subsequently evolved ethylene and at a slower rate than the controls.

From the foregoing it is clear that the stimulation of ethylene production in irradiated fruits could be satisfactorily explained on the basis of physical damage to the cell membranes. The origin of the ethylene remains obscure but the consequences of its production may be considerable in terms of the shelf life of the fruit.

D. Pigment Formation

It has been known for some time that radiation can destroy pigments in fruits and fruit juices (Hannan, 1955), with the extent of destruction dependent on the dose and conditions during radiation. Most of the work describing this destruction is concerned either with the effects on fruit juices, where the chemical environment is very different from that occurring in fresh fruits, or else with the effects of very high doses, which would not be suitable for the extension of shelf-life of fresh fruit (Horubala, 1968). Doses of 200–300 krad have no apparent effect upon the colour of fruits, although bleaching has been detected at doses above this. The formation of a deep-red pigment has been observed by Clarke (1959) to occur in irradiated pears during storage, while Maxie and Abdel-Kader (1966) observed an increase in red pigmentation, in excess of normal pigment formation, in irradiated peaches and nectarines. No explanation has been given for this phenomenon. It has been known for some time that lycopene formation can be inhibited in tomatoes, but since this is also associated with an inhibition of the ripening process, it is not possible to know if the synthetic pathway has been directly affected by radiation. Lukton and Mackinney (1956) found that β-carotene and lycopene are remarkably resistant to irradiation (see also Chapter 21, p. 711).

E. Physiological Breakdown

The physiological damage induced in fruits by radiation may be expressed in a variety of ways. In apples, a dose of 1·0 Mrad will cause considerable browning in the cortex, presumably due to an increase in membrane permeability and the activity of the phenoloxidases. The physiological changes induced by radiation may bring about the death of the cells at a considerable time after the treatment.

Clarke (1968) showed that apples grown in the United Kingdom always have a tendency after irradiation to a type of breakdown which is initiated in the middle of the apple between the carpels. Eventually the brown lesions coalesce and a ring of necrotic tissue surrounds the core. The symptoms appear very similar to those of core flush. This type of radiation-induced breakdown has only been observed in United Kingdom apples and the results contrast markedly with those obtained in North America, where no breakdown of this nature has been observed and, in fact, the data indicate that doses below 100 krad reduce scald and core flush. The time of picking of the fruit affects the incidence of this type of breakdown in the apple varieties, Cox's Orange Pippin and Tydeman's Late Orange (United Kingdom).

Citrus fruits have also been shown to exhibit a physiological breakdown of the flesh resulting in pitting with black areas. This kind of breakdown has been observed mainly in citrus from Israel by Monselise and Kahan (1968) and in South Africa by the present author. Lemons exhibit two types of injury after irradiation. Cavities appear between the walls of the flesh after 40 days at 15°C and necrosis of the button end also develops (Maxie *et al.*, 1964). Peel injury can be modified by dose, storage temperature and duration of treatment. Also, peel injury of grapefruit, pineapple oranges and Valencia oranges was reduced by combining the radiation treatment with a hot-water dip at 50–55°C according to Dennison and Ahmed (1966).

Skin damage has also been observed in bananas by Teas *et al.* (1962) using doses above 25 krad.

F. Enzyme Studies

Very few studies have been reported on the effects of radiation on specific enzyme systems in fruits. The activities of peroxidase and catalase have been studied by Monselise and Kahan (1968) in stored oranges and grapefruits. No significant changes were detected immediately after irradiation, but after 8 days at 18–20°C an increase in peroxidase activity was detected in the albedo of oranges and grapefruit. In irradiated oranges, catalase activity decreased in comparison with control fruit, whereas in grapefruit the reverse was the case. It was not possible to correlate the changes in these enzymes with the incidence of the breakdown of the peel, but it is probable that such a correlation would exist in view of the browning reactions associated with cell death.

A number of studies have been reported on the effect of radiation on pectin esterase activity† in mangoes (Dennison and Ahmed, 1967), cherries (Somogyi and Romani, 1964), peaches (Shewfelt, 1965) and in the membranes of oranges and grapefruit (Rouse and Dennison, 1968). These studies showed that the activity of PE increased in these fruits. In spite of this, textural changes (softening) may take place more slowly during storage (as opposed to the quick softening immediately after irradiation—see p. 691) in treated fruits than in control fruits, so that in some cases irradiated fruit becomes the firmer of the two. A possible explanation of this apparent paradox is that ionic bonding causes firming following the high degree of demethylation induced by the accelerated action of PE.

It has been suggested that PE activity may be associated with the concentration of indole-acetic acid in the cell wall and this may be connected in some way with the activation of the enzyme (Romani, 1966). Although nothing is known about the effect of radiation on growth substances in fruits, the synthesis of indole-acetic acid has been shown to be sensitive to radiation in other plant tissues (Brownell and Hine, 1956).

G. Ripening Processes

Many of the effects of the radiation treatment described in the previous section have been concerned directly with delayed or accelerated ripening processes. It is not possible to give here an overall generalization which will describe the effect of radiation on fruits. In some climacteric fruits, it is quite clear that the delay of ripening induced by radiation is dependent on the time of irradiation within the climacteric phase. If pears or bananas are treated before the climacteric then it will be possible to detect an effect; however, no response is elicited when the treatment is given after the climacteric. There is general agreement that radiation retards the ripening of tomato fruit (Maxie and Abdel-Kader, 1966; Abdel-Kader *et al.*, 1968), but in many points of detail the reports are conflicting, particularly with regard to the response to given levels of radiation. It would appear that the pattern of ripening after irradiation is a function of the physiological age of the fruit at the time of

† See Volume 1, Chapter 3.

exposure. A marked extension of the storage life of tomatoes has been achieved by Mathur (1967) by irradiation with 75 krad at the pre-climacteric respiration minimum with storage at 9–10°C and prompt post-irradiation treatment with indole-3-acetic acid to combat decay. On the other hand, the ripening processes may be accelerated as is the case for peaches and nectarines, and, here again, the time of irradiation will determine, to a large extent, the magnitude of the response to radiation. Berry fruits, such as strawberries, which are probably the most promising fruits for irradiation from the commercial point of view, are picked when mature but not yet fully ripe and little or no disturbance of the ripening process has been observed with these fruits. This again is consistent with the supposition that the disturbance of the ripening process is greater the earlier the time of irradiation. Some irradiated fruits may never ripen normally under normal storage conditions. In experiments carried out by the author (Clarke, 1959) apples and pears never developed their full flavour after storage at 3°C, whereas some pears did develop their flavour after much higher doses if they were kept at 20°C (Clarke, 1959).

III. MICROBIOLOGICAL ASPECTS

The general aspects of this topic have been adequately covered elsewhere (Sommer and Fortlage, 1966). As pointed out in the introduction, the potential usefulness of radiation in food processing is due to the fact that radiation can kill micro-organisms without raising the temperature of the food to any significant extent. Fungal spores are not particularly resistant to radiation, whereas bacterial spores may be very resistant. Spores are normally more resistant than vegetative cells. D_{10} values, which represent the doses required to reduce the viable population by a factor of ten, are referred to in Table I.

It should be realized that the radiation sensitivity of any cell depends upon many factors and it is not possible to give a single value which would be constant for the species under all possible conditions of irradiation. Also, not all radiation survival kinetics are truly exponential, some, such as those of *A. citri*, are definitely sigmoidal (Sommer and Maxie, 1966). A sigmoidal dose–response curve means that the cells can absorb a certain amount of radiation without suffering any loss in viability. This is an important factor to be taken into account when estimates are calculated for the decimal reduction factor for any one dose.

The possibility that radiation may induce an increase in the susceptibility to mould attack in fruit tissue has not been studied extensively. Tomatoes were first reported by Burns (1956) as having an increased susceptibility to mould attack and further evidence for this has been obtained for cauliflower and asparagus (Truelson, 1963). These latter results, like those for carrot tissue

(Skou, 1963), were interpreted as being due primarily to an increase in the permeability of the cell membranes producing local leaks which would eventually provide a good medium for the growth of micro-organisms. Strawberries do not exhibit an increased susceptibility to mould attack after irradiation (Chaluty *et al.*, 1966). The lethal effects of the treatment on the mould spores tend to obscure the possible enhancement of growth of the surviving spores by the treatment. A stimulation of mould growth has been observed on sweet cherries by Cooper and Salunkhe (1963) and by Massey *et al.* (1965), and evidence is available that *Penicillium* attack on lemons may also be promoted by low doses (Kuprianoff, personal communication). In both cases, the promotion of growth may be due either to a metabolic change in the spores themselves or to the provision of a good medium for growth as already postulated or even to some reduction in the concentration of inhibitory compounds present ın fruit tissue, such as phenolic substances.

TABLE 1. Resistance of fungal spores to radiation

Species	D_{10}	Reference
Pullularia pullulans	135	(1)
Rhinocladiella app.	53	(1)
Phialophora mustea	24–48	(1)
Mucor fruticolor	35	(2)[a]
Penicillium terrestre	35–72	(1)
Cladospora sphaerosperma	52	(1)
Cladospora herbarum	100	(2)[a]
Alternaria citri	40	(2)[a]
Monilinia fructicola	40	(2)[a]
Geotrichum candidum	50	(2)[a]
Phomopsis citri	20	(2)[a]
Botrytis cinerea	40	(2)[a]
Diplodia natalensis	40	(2)[a]
Rhizopus stolonifer	60	(2)[a]

[a] Estimated from published survival curves.
(1) Hammerschmid (personal communication).
(2) Sommer and Maxie (1966).

Some success has been reported for increasing the effectiveness of the radiation treatment on killing mould spores by combining the radiation treatment with a hot-water dip at about 50°C (Sommer and Maxie. 1966).

IV. THE RADURIZATION OF FRUIT JUICES

"Pasteurization" is a term used to describe the microbiological objective of a heat process aimed at extending the storage life of a product without achieving commercial sterility. Objections were made to the use of this term for a process involving a radiation treatment and the term "radurization" was proposed as an alternative to pasteurization.

The possibility of replacing the conventional heat process by a radiation process was worth investigating since the flavour of fruit juices is relatively sensitive to heat and any alternative method which would result in a stable product of equivalent quality would be valuable. Though some fruit juices can withstand considerable doses of radiation, up to 0·5 Mrad, without a marked loss of quality, it was soon realized that doses greater than 1·0 Mrad would be needed to produce a juice with satisfactory microbiological stability (Clarke, 1959; Fernandez *et al.*, 1966). At these high doses, considerable loss in quality was observed in juices of normal concentration because of adverse changes occurring in the flavour. When concentrated juices were irradiated (Kiss, 1967), it was observed that some could withstand a dose as high as 1·3 Mrad without any undue loss of flavour. These results have been confirmed for apple and grape juice by Levi (personal communication).

Stehlik and Kaindl (1966) considered the effectiveness of combining a radiation treatment of 0·3 Mrad, with a 50°C heat treatment. This combined treatment offered the possibility of reducing the heat processing to a minimum, while at the same time achieving the necessary inactivation of the micro-organisms, primarily yeasts. Zehnder *et al.* (1968) showed that the heat treatment increased the effectiveness of the radiation treatment. The converse is also true since Balkay *et al.* (1968) found that radiation treatment will increase the heat sensitivity of yeasts. A heat treatment of 50°C for 20 min and a dose of 0·3 Mrad was found to give a reduction in the number of viable yeast cells of approximately 10^7, which is the normal degree of inactivation required for a pasteurizing or radurizing treatment.

Kovacz (1968) investigated the production of off-flavour in apple juice by radiation and found that the off-flavour was less in the irradiated *filtered* juice than in the irradiated normal juice still containing suspended protein. The amount of off-flavour for a specific dose could be increased by adding grape proteins to the juice before irradiating. Further investigation with specific proteins established that sulphur-containing proteins produced more off-flavour than a protein which did not contain sulphur, such as protamine. Supplementing the juice with sulphur-containing amino acids before irradiating also increased the production of off-flavour (Levi and Radola, 1968).

Changes in the volatiles of apple and grape juices have also been investigated in irradiated juices. No differences could be detected in some volatile fractions of five Spanish apple varieties even after a dose of 2·5 Mrad, although this was probably due to the limitations of the techniques used. In others, a linear increase of acetaldehyde with dose was observed for both apple and grape juice. As the acetaldehyde increased it was observed that the concentration of 2-hexanol decreased (Barrera *et al.*, 1967). So far there has been no positive identification of substances responsible for the off-flavours in irradiated juices.

V. PUBLIC HEALTH CLEARANCE

Few detailed studies are available concerning the wholesomeness or safety for human consumption of irradiated fruits. No irradiated fruit has yet been cleared by a Public Health Authority for sale to the general public. The fruits that have been tested in U.S. programmes and the animal species used are described briefly elsewhere (Raica and Howie, 1966). A complete 2-year study with three animal species is being conducted by Ramey (1968) with irradiated strawberries and a similar study with irradiated papayas is due to begin. Acute toxicity tests with fruits which have received 2·79 and 5·58 Mrad have been conducted on rats for 8 weeks. These fruits include dried apricots, sour cherries, melon peaches, pears, raisins and strawberries. Long-term toxicity studies have been conducted on rats, monkeys and dogs with fruit compote, a mixture of dried fruits, oranges and peaches (Raica and Howie, 1966). Some adverse effects have been observed, but these effects have nearly always been observed in experiments with serious flaws in their design.

A complete 2-year study with two animal species, rats and dogs, has been conducted with peaches, plums and nectarines and no evidence was obtained of any loss of nutritional adequacy or of any toxicity (Gabriel and Edmonds, 1968). A long-term study with bananas has been carried out by Ramey (1968) and no adverse effects have been reported after 10-months of feeding the irradiated diet to rats and dogs.

VI. PROSPECT FOR COMMERCIAL APPLICATION

The short-term nature of the annual production of fruits necessarily involves a processing plant lying idle for long periods of the year. For the irradiation plant to be fully utilized it will be necessary to irradiate alternative products or fruits which ripen at different times of the year.

It is possible to increase the shelf life of most berry fruits, such as strawberries, raspberries and blackcurrants, by a factor of two at ambient and refrigerated temperatures with a dose between 0·2 and 0·25 Mrad. The cost of treating strawberries with a dose of 0.25 Mrad has been estimated to be of the order of 2·6 cents/lb (Dietz, 1967). As considerable losses occur in the transporting and storing of such fruits, it is feasible that these fruits may be processed commercially, when large-scale radiation trials have been completed and when the fruits involved are cleared for public consumption. A prerequisite, of course, would be a suitable commercial organization which would allow the processing of sufficient quantities of fruit at one central place. These particular fruits also have a high economic value at certain times of the year and it seems likely that the cost of radiation processing could easily be absorbed under normal conditions.

The possibility of delaying the ripening of bananas by a low dose is of interest to a number of countries. However, this is a fruit where the time of giving the treatment is of paramount importance and it remains to be investigated how successfully the radiation process can be incorporated into present-day handling procedures. Provided that certain economic conditions are satisfied, then the cost of irradiating bananas may be as low as 0·07 cent/lb, which should not be a deterrent to this particular radiation application (Dietz, 1967).

The disinfestation of papayas is believed to offer the promise of opening up a new market in the U.S.A. for papayas from Hawaii. A dose of 75 krad is sufficient to control the Hawaiian fruit fly and this treatment has the additional advantage of extending the storage life of the fruit by delaying the ripening processes. A similar situation exists for mangoes, where quarantine regulations of the U.S.A. forbid the importing of this fruit infested with the mango seed weevil. As in the case of papayas, the dose needed to control the insect is sufficient to delay the ripening process and thereby prolong the storage life of mangoes. The United States Atomic Energy Commission has estimated that the extension of storage life of 6–8 days, which can be achieved by radiation, may reduce the transport costs by 8·5 cents/lb by permitting a change in the mode of transport from air to surface Ramey (1968). As the treatment dispenses with the need for chemical disinfestation, the total savings may amount to $0.9 million in one year.

The application of radiation to food preservation will progress slowly until the first irradiated food achieves commercial success, and, if confidence is established that toxicity risks have been eliminated, many more commercial applications may develop.

REFERENCES

Abdel-Kader, A. S., Morris, L. L. and Maxie, E. C. (1968). *Proc. Am. Soc. hort. Sci.* **92,** 553.

Al-Jasim, H., Markakis, P. and Nicholas, R. C. (1968), Preservation of Fruit and Vegetables by Radiation, STI/PUB/149, 125. IAEA, Vienna.

Attix, F. H., Roesch, W. C. and Tochilin, E. (1966). *In* "Radiation Dosimetry", (F. H. Attix, ed.) Vols 1, 2, 3, 2nd Edition. Academic Press, New York and London.

Balkay, A., Zehnder, J. and Clarke, I. D. (1968), in preparation.

Barrera, R., de la Cruz, F., Gasco, L. and Manero, J. (1967). Seibersdorf Project Report No. 20. ENEA, Paris.

Benn, T. R. (1967). *In* "Radiation Preservation of Foods" (E. S. Josephson and J. H. Frankfort, eds), p. 126. Advances in Chemistry Series 65. American Chemical Society, Washington, D.C.

Boyle, F. P., Kertesz, Z. I. and Connor, M. A. (1957). *Fd Res.* **22,** 89.

Brownell, G. L. and Hine, G. J. (1956). *In* "Radiation Dosimetry". (F. H. Attix, ed.) Academic Press, New York and London.

Burns, E. E. (1956). "Maturation Changes in Tomato Fruits Induced by Ionizing Radiation". Ph.D. Thesis., Purdue University, U.S.A.
Bussel, J. and Maxie, E. C. (1966). *Amer. Soc. hort. Sci.* **88,** 151.
Chaluty, E., Maxie, E. C. and Sommer, N. F. (1966). *Amer. Soc. hort. Sci.* **88,** 365.
Clarke, I. D. (1959). *Int. J. appl. Radiat. Isotopes* **6,** 175.
Clarke, I. D. (1962). *Proc. 1st Int. Congr. Fd Sci. and Technol.* **3,** 551.
Clarke, I. D. (1968). Preservation of Fruit and Vegetables by Radiation, STI/PUB/149, 65. IAEA, Vienna.
Clarke, I. D. and Fernandez, S. J. G. (1961), *Int. J. appl. Radiat. Isotopes* **11,** 186.
Clarke, I. D. and Hayes, J. (1965). *Radiat. Res.* **24,** 152.
Cooper, G. M. and Salunkhe, D. K. (1963). *Fd Technol.* **17,** 123.
Cornwell, P. B. (1966). Fd Irrad. Proc. FAO/IAEA Symp. STI/PUB/127, 455. IAEA, Vienna.
Dennison, R. A. and Ahmed, E. A. (1966). Fd Irrad. STI/PUB/127, 619, IAEA, Vienna.
Dennison, R. A. and Ahmed, E. M. (1967). *J. Fd Sci.* **32,** 702.
Dietz, G. R. (1967). *In* "Radiation Preservation of Foods" (E. R. Josephson and J. H. Frankfort, eds), p. 118. Advances in Chemistry Series 65. American Chemical Society, Washington, D.C.
Dzamić, M. D. and Janković, B. R. (1966). *Int. J. appl. Radiation Isotopes* **17,** 561–566.
Faust, M., Chase, B. R. and Massey Jr., L. M. (1967). *Amer. Soc. hort. Sci.* **90,** 25.
Fernandez, H. Stehlik, G. and Kaindl, K. (1966). Seibersdorf Project Report No. 4. ENEA, Paris.
Fernandez, S. J. G. and Clarke, I. D. (1962). *J. Sci. Fd Agric.* **13,** 23.
Gabriel, K. L. and Edmonds, R. S. (1968). NYO 3422, Vol. 1-3.
Glegg, R. E., Boyle, F. P., Tuttle, L. W., Wilson, D. E. and Kertsz, Z. L. (1956). *Radiat. Res.* **5,** 127.
Hannan, R. S. (1955). "Scientific and Technological Problems Involved in Using Ionizing Radiations for the Preservation of Food." Department of Scientific and Industrial Research, Special Report No. 61. H.M.S.O., London.
Horubala, A. (1968). Preservation of Fruit and Vegetables by Radiation, STI/PUB/149, 57. IAEA, Vienna.
Kertesz, Z. I., Morgan, B. H., Tuttle, L. W. and Lavin, M. (1956). *Radiation Res.* **5,** 372–381.
Kertesz, Z. I., Glegg, R. E., Boyle, F. P., Parsons, F. G. and Massey Jr, L. M. (1964). *J. Fd Sci.* **29,** 40.
Kiss, I. (1967). Annual Report, Aug. 1966–July 1967, Res. Contr. 442/RB. IAEA, Vienna.
Kovacs, J. (1968). Personal communication.
Lee, T. H., McGlasson, W. B. and Edward, R. A. (1968). *Radiat. Bot.* **8,** 259.
Levi, A. and Radola, B. (1968). Personal communication.
Lukton, A. and Mackinney, G. (1956). *Fd Technol. Champaign* **10,** 630.
Mapson, L. W. and Wardale, D. A. (1968). *Biochem. J.* **107,** 433.
Markakis, P. and Nicholas, R. C. (1967). 7th Annual Joint AIBS Committee, USAEC Contractors' Meeting, Division of Isotopes Development and Division of Medicine and Biology. USAEC, Washington, D.C.
Massey Jr, L. M. (1968). Radiation Preservation of Fruit and Vegetables by Radiation, STI/PUB/149, 105. IAEA, Vienna.

Massey Jr., L. M. and Bourke, R. J. (1967). *In* "Radiation Preservation of Foods" (E. S. Josephson and J. H. Frankfort, eds) Advances in Chemistry Series 65. American Chemical Society, Washington, D.C.
Massey, L. M., Tallman, D. F. and Kertesz, Z. I. (1961). *J. Fd Sci.* **26,** 389.
Massey Jr, L. M., Robinson, W. B., Spaid, J. F., Splittstoesser, D. F., Van Buren, J. P. and Kertesz, Z. I. (1965). *J. Fd Sci,* **30,** 759.
Mathur, P. B. (1967). *Int. J. appl. Radiat. Isotopes* **18,** 663.
Maxie, E. C. and Abdel-Kader, A. (1966). *Adv. Fd Res.* **15,** 105.
Maxie, E. C. and Sommer, N. F. (1968). Radiation Preservation of Fruit and Vegetables by Radiation, STI/PUB/149, 39. IAEA, Vienna.
Maxie, E. C., Eaks, L. L. and Sommer, N. F. (1964). *Radiat. Bot.* **4,** 405.
Maxie, E. C., Eaks, L. L., Sommer, N. F., Rae, H. L. and El-Batel, S. (1965). *Pl. Physiol., Lancaster* **40,** 407.
Maxie, E. C., Sommer, N. F., Muller, C. J. and Rae, H. L. (1966). *Pl. Physiol., Lancaster* **41,** 437.
Maxie, E. C., Johnson, C. F., Boyd, C., Rae, H. L. and Sommer, N. F. (1966). *Amer. Soc. hort. Sci* **89,** 91.
Metlitsky, L. V. and Petrash, L. P, (1968). *Prikladnaja Biohimya i Mikrobiologiya* **4,** 3.
Miller, L. A., Prasad, R. and Romani, R. J. (1967). *Radiat. Bot.* **7,** 47.
Monselise, S. P. and Kahan, R. S. (1968). Preservation of Fruit and Vegetables by Radiation, STI/PUB/149, 93. IAEA, Vienna.
Ord, M. G. and Stocken, L. A. (1962). *Biochem. J.* **84,** 600.
Raica Jr, N. and Howie, D. L. (1966). Food Irradiation, STI/PUB/127, 119. IAEA, Vienna.
Ramey, J. T. (1968). *In* "Status of the Food Irradiation Programme. Hearings before the Sub-committee on Research Development and Radiation of the Joint Committee on Research Development and Radiation of the Joint Committee on Atomic Energy, Congress of the United States, 2nd Session." U.S. Government, Printing Office, Washington, D.C.
Rhodes, D. N. (1965). *Br. Poult. Sci.* **6,** 265.
Romani, R. J. (1964). *Radiat. Bot.* **4,** 299.
Romani, R. J. (1966). *Adv. Fd Res.* **15,** 57.
Romani, R. J., and Bowers, J. B. (1963). *Nature, Lond.* **197,** 509.
Romani, R. J. and Fisher, L. K. (1966). *Archs Biochem. Biophys.* **117,** 645.
Romani, R. J. and Fisher, L. K., Miller, L. A. and Breidenbach, R. W. (1967). *Radiat. Bot.* **7,** 41.
Romani, R. J. and Yu, I. K., Ku, L. L., Fisher, L. K., Dehgan, N. (1968). *Pl. Physiol., Lancaster* **43,** 1089.
Ronsivalli, L. J., Steinberg, M. A. and Seagran, L. H. (1965). *In* "Radiation Preservation of Food", p. 69. Publication 1273. NAS–NRC, Washington, D.C.
Rouse, A. H. and Dennison, R. A. (1968). *J. Fd Sci.* **33,** 258.
Shea, K. G. (1965). *In* "Radiation Preservation of Foods", p. 283. Publication 1273. NAS–NRC, Washington, D.C.
Shewfelt, A. L. (1965). *J. Fd Sci.* **30,** 573.
Skou, J. P. (1963). *Physiologia Pl.* **16,** 423.
Sommer, N. F. and Fortlage, R. J. (1966). *Adv. Fd Res.* **15,** 147.
Sommer, N. F. and Maxie, E. C. (1966). Food Irradiation, STI/PUB/127, 571. IAEA, Vienna.

Somogyi, L. P. and Romani, R. J. (1964). *J. Fd Sci.* **29,** 366.
Stehlik, G. and Kaindl, K. (1966). Food Irradiation, STI/PUB/127, 299. IAEA, Vienna.
Teas, H. J., Qunitana, D. C. and Oliver, J. C. (1962). *In* "Abstract 2nd Int. Congr. Rad. Res", p. 179. Oliver & Boyd, Edinburgh.
Truelson, R. A. (1963). *Fd Technol., Champaign* **17** (3), 100.
Van Bekkum, D. W., de Vries, M. J. and Klouwen, H. M. (1962). *Int. J. Radiat. Biol.* **9,** 449.
Van Kooy, J. G. (1968). Preservation of Fruits and Vegetables by Radiation, STI/PUB/149, 129. IAEA, Vienna.
Wahba, I. J. and Massey Jr, L. M. (1966). *J. Polymer Sci.* **A4,** 1751.
Young, R. E. (1965), *Nature, Lond.* **205,** 1113.
Zehnder, J., Balkay, A. and Clarke, I. D. (1968). In preparation.

Chapter 21

Pigment Degeneration During Processing and Storage

C. O. CHICHESTER* AND ROGER McFEETERS†

Department of Food Science and Technology,
University of California, Davis, California, U.S.A.

I	Introduction ..	707
II	Carotenoids ..	708
III	Flavonoids ..	711
IV	Chlorophylls ..	714
	References ..	717

I. INTRODUCTION

In the early development of food processing, pomologists and food technologists were concerned to a major extent with flavour and gross changes in the appearance of fruits. However, although they were particularly aware of changes which might occur in the colour of the fruits being processed, they seldom investigated the chemical or biochemical changes taking place, for, within wide limits, the colour changes were often relatively minute compared with other factors. As an example, the losses in the pigments of products such as raspberries or boysenberries when processed into preserves may approach 70–80% without an objectionable change in visual coloration. In many cases changes in coloration (and thus the pigments) which were induced seemed inevitable, and the product was accepted on the basis of the colour of the processed product. For instance, sun-dried apricots and pears are translucent in appearance and are substantially different from those processed on a large scale in dehydrators. Indeed, the appearance of the former tends to be more appealing to the average eye than those which are processed under less extreme conditions when considerably more of the pigments characterizing the fresh fruit are retained. Typical also is the appearance of a strawberry preserve processed under conditions prevailing in industries of many years

* Present address: Department of Food and Resource Chemistry, University of Rhode Island, Kingston, Rhode Island 02881, U.S.A.

† Present address: Department of Food Science, Michigan State University, East Lansing, Michigan, 48823, U.S.A.

ago. Virtually all of the anthocyanin pigments are degraded and replaced by the characteristic brownish-red pigments of the Maillard browning reaction.

The ultimate fate of the pigmented substances or the reactions which they undergo has been studied primarily in senescence or storage of the fresh material prior to processing. The majority of the investigations which have been reported relative to processing have been confined to measuring the losses which ensue in utilizing different processing or storage techniques. Most fruits and vegetables are processed in the presence of added carbohydrates (which are in relatively high concentration), and, since the fruits themselves contain high concentrations of sugars, conditions for browning are optimized. The pigmented substances exist in relatively small concentration compared with many other constituents of the fruit and of added ingredients, and consequently the investigation of their chemical changes during processing may be difficult. This is particularly so since the majority of changes that occur during processing are non-biological reactions and may occur between a large number of different constituents and the pigmented material.

The availability of newer techniques in chemical and biochemical investigations, coupled with an increased interest in chemical changes of pigmented substances in fruits, should represent a field in which significant advances can take place. Our knowledge of the chemical phenomena occurring in pigments during the processing of fruits is *in toto* rather meagre, but can be added to with much less difficulty than a few years ago.

Three major classes of pigments occur in fruits: the carotenoid, the anthocyanin and the chlorophyll pigments. These classes themselves will be considered rather than the processing techniques which might affect the various classes of pigments in different manners; individual processing operations have been reviewed recently and include information on pigment degradation (Goldblith, 1966; Romani, 1966; Joslyn, 1966; Singleton and Esau, 1969; see also Chapter 18). In many cases analogous plant products must be referred to, since often there is a dearth of information directly relating to fruits, and it is probably a valid assumption that the compounds will react in a similar manner whether they be in fruits or other plant materials.

The chemistry of carotenoid compounds is dealt with in detail in Volume 1, Chapter 12, and the anthocyanins in Volume 1, Chapter 11.

II. CAROTENOIDS

The stability of carotenoids during processing of fruit tissue varies widely, depending upon the type of carotenoid. For example, in tomatoes the major carotenoids are the hydrocarbon compounds such as β-carotene, lycopene and their various isomers, whereas in apricots, oranges, apples, pears and strawberries the oxygenated compounds appear to be the principal carotenoids

(Khan and Mackinney, 1953; Curl and Bailey, 1956a, b; Curl, 1959; Curl, 1960; Galler and Mackinney, 1965; Mackinney, 1966). Much of the work recorded in the literature on the changes in carotenoids during processing or prior to processing has involved the measurement of apparent decreases in carotenoid content in terms of decrease in optical absorption at 450 nm, no analysis being made of the changes which may be occurring in individual carotenoids. Not only do the results not consider the losses of individual carotenoids, but the simple measurement of optical absorption does not reflect changes which may occur in appearance due to partial destruction of different types of carotenoids.

Under conditions of low pH, carotenoids undergo isomerization. In fruits which contain poly-*cis*-carotenoids this will cause a deepening in colour (Joyce, 1954; Curl and Bailey, 1956b; Weckel *et al.*, 1962). Gortner and Singleton (1961) and Singleton (1963) demonstrated that the degree of injury in pineapple, such as would be expected in processing, released sufficient fruit acids to cause an increase in the *cis* isomers of the carotenoids present. No other changes took place, but this resulted in the lightening of the fruit colour. In orange juice, isomerization results in the formation of the furanoid pigments from the epoxy carotenoids, which in a similar manner would change the observable colour of the fruit juice. In acid fruits which have been heated during processing, it is probable that equilibrium mixtures of any set of isomers would result.

If fruit is processed in the presence of oxygen, it can be assumed that the carotenoids will be oxidized. Under conditions where oxidations catalysed by enzymes are excluded, oxidation occurs in a relatively non-specific manner. In the absence of oxygen the carotenoids are very stable substances. Cole and Kapur (1957a, b) found that the degradation of lycopene is related to the temperature and the availability of oxygen, and that traces of copper led to a marked increase in the rate of oxidation. They characterized these changes by isolating the 2,4-dinitrophenylhydrazones of acetone, methyl-heptenone, levulinic acid, levulinic aldehyde and what was probably glyoxal. Falconer *et al.* (1964) found that in carrots an off-flavour was developed in the presence of oxygen which was correlated with the loss of β-carotene. Land (1962) oxidized β-carotene and identified such products as acetaldehyde, 2-methylpropanol, butanol, diacetyl, and pentanol, while Fishwick (1962) identified colourless solids and pale-yellow oils which had their major absorptions in the ultraviolet band. Hunter and Krakenberger (1947) oxidized β-carotene in benzene and reported the production of 5,6-monoepoxides, 5,6,5′,6′-diepoxides and semi-carotenones. Other investigators have suggested the production of epoxides and apo-carotenals. In many cases the initial attack of the oxygen appears to occur on the rings.

From a recent study, it appears that the 4,4′ positions, if these are free, are the site of the initial reaction and that the kinetics of the oxidation of the hydrocarbon carotenoids are compatible with a free radical mechanism (El Tinay, 1969). The complexity of the products which have been isolated as either volatile or non-volatile products would strongly favour such a mechanism. The initial attack on the end rings of the carotenoids also argues against the presence of ionone in an oxidizing carotenoid system. The reported presence of methyl-heptenone, however, does pose a problem, unless it arises from another portion of the molecule. By postulating a free radical mechanism, each of the portions which initially produced oxidized products could be reoxidized to yield smaller fragments of great diversity. The diversity of the products from a small amount of pigment has made it extremely difficult to identify them in fruits, since in many cases they resemble other components of the fruit material. In processing in the presence of oxygen, such as in dehydration or in thermal processing in the presence of oxygen, one would expect the formation of many products which might significantly alter or modify the flavour of the processed fruit as well as its appearance.

When carotenoid-containing materials are preserved by drying or freeze-drying, moisture plays a rather critical role in the stability of the carotenoid pigments (Goldblith *et al.*, 1963; Kimura and Shioda, 1963). Decreasing moisture tends to stabilize the carotenoids until a point is reached where stability rapidly decreases. At very low levels of moisture, as are encountered in freeze-drying, the carotenoids are particularly susceptible to degradation (Mackinney, 1966). It has been postulated that the stability is increased by having a film of water present in the dehydrated materials. This in a sense may act as a barrier to the free movement of oxygen to the pigments. The individual stability of the pigments varies rather widely, although this has not been studied extensively in fruit. In alfalfa, Livingston *et al.* (1968) suggest that xanthophyll losses are several-fold higher than hydrocarbon losses; lutein appeared to be the most stable of the xanthophylls. An interesting observation by Purcell and Walter (1968) suggests that in dehydrated foods the oxidation of the carotenes is more complicated than that encountered during their oxidation *in vitro*.

Although it has been reported that in the processing of plant materials (fruits included) light has an effect on the stability of carotenoids, there has been comparatively little work done in this area (Feller and Buck, 1941; Ulrich and Mackinney, 1968). Tsukida *et al.* (1966) reported that the light irradiation of all-*trans* β-carotene in various solvents produced retrodehydro-carotenes and mutatochromes. Hasegawa *et al.* (1969) suggested that intense light in the presence of a photosensitizer produced isomeric carotenoids, as well as some epoxy carotenoids. Recent work indicates that a major product of the light degradation of carotenoids is semi-β-carotenone. Visser (1969) suggested that such a photo-oxidation would again prefer the 4,4′ positions.

The carotenoids in fruits tend not to be as sensitive to enzymatic degradation as they are in other plant or animal systems. However, although the enzymatic oxidation of these pigments may be of less consequence during processing of fruit tissue, it can be significant in processes wherein the material is unblanched. The most common catalytic agents for carotenoid destruction are the unsaturated fat oxidases and peroxidases. The lipoxidases, operating on a fat-oxidizing system, appear to attack in a coupled oxidation (Blain, 1963) and the products are generally long-chain aldehydes. The role of the carotenoids in such a system appears to be that of an antioxidant of the fats, although these effects are not normally large in fruit products. The peroxidases appear to be able directly to attack carotenoids, although they too may act through a coupled oxidation of a fatty acid. Friend and Mayer (1960) obtained an enzyme from sugar beet which was capable of the destruction of both crocin and β-carotene. Later reports by Friend and Dicks (1967; Dicks and Friend,1968) indicated that the crocin-destroying system did not require the presence of lipid material for the destruction of carotenoids. It is somewhat debatable whether this is a true oxidation system, since they could not detect the presence of hydrogen peroxide. Relatively speaking, however, the participation of these enzymes in fruits is of minor importance.

Studies on the destruction of carotenoids by irradiation processing illustrate the effect of solvents upon the stability of carotenoids. Lukton and Mackinney (1956) irradiated carotenoids in a large number of solvents and found that the stability of the pigments in the solvents was considerably lower than that in either the solid state or in a biological matrix. Some protection was afforded by irradiation of the fruit (tomato) under nitrogen, but even in the absence of nitrogen little destruction occurred, even at high dosage rates. In a sampling of whole tomatoes and tomato purées containing β-carotene the effect of irradiation was slight. The stability of the carotenoids in the solid state was roughly equivalent to those in the fruit, and these in turn were very much more stable than the carotenoids dissolved in various solvents. High doses of radiation do not induce any pigment destruction in thermally processed peaches or apricots (Salunkhe *et al.*, 1959).

III. FLAVONOIDS

The flavonoids, particularly the anthocyanins, differ significantly in stability from the other two major classes of pigments, the lipid-soluble chlorophylls and the carotenoids. Their behaviour is such that they are considerably more stable in fruit tissues than either the carotenoids or chlorophylls, although this is not an invariable rule (Goodwin, 1958).

The stability of anthocyanins during processing is particularly dependent upon the pH of the media. The aglucone of the anthocyanins is lower in stability than that of the glucose. Huang (1956) suggested that the anthocyanins can exist in ketonic form, which form is particularly susceptible to breakdown. Earlier, Sondheimer (1953) suggested an equilibrium between the glucoside and its pseudo-base. Recent hypotheses concerning the effect of pH on colour and structural form have been made by Jurd and Geissman (1963), Timberlake and Bridle (1966, 1967a) and Harper and Chandler (1967a, b). The flavylium salts exist in equilibrium between the ionic form, the anhydro base, the carbinol base, or the open-ringed chalcones. The ionic form is by far the most stable, and hence the stability of fruit colour under low pH conditions. The pseudo-base, or carbinol, is quite unstable, and the oxidation of anthocyanins in food during processing or storage is narrowly proportioned to the percentage of the pigment in the pseudo-base form (Lukton *et al.*, 1956; Jurd, 1964a, b). Substitution of a hydroxyl at position 3 of the oxygenated ring facilitates the conversion of these pigments. Under processing conditions, particularly those that involve heat, the equilibrium is rapidly shifted to the pseudo-base, which is depleted through oxidation mechanisms. The substitution of the hydroxyl in position 3 by glucosidation increases the stability of the pigment, thus explaining the difference in stability between the aglucones of the anthocyanins and their glucosides.

In many processes, such as freezing or the production of wine, sulphur dioxide is utilized in moderate concentration to inhibit microbiological spoilage or to prevent the accumulation of browning reaction products. The sulphur dioxide will at least partially reversibly decolourize the anthocyanins. The negative bisulphite ion adds directly to the anthocyanin at either position 2 or 4, producing a colourless sulphonic acid. This acidification partially restores the coloration by shifting the equilibrium to the coloured form by reducing the bisulphite ion concentration. Removal of the sulphite will under most conditions restore a portion of the colour. Selective oxidation of the bisulphite will also restore coloration in partially bleached fruit materials (Jurd, 1964a; Timberlake and Bridle, 1967b).

As indicated above, the aglucones are less stable than the glucosides of the flavonoids, and thus, during processing, the release of glucosidases (first established as an enzyme present in *Aspergillus* by Huang, 1955, 1956) rapidly decolourizes the pigment by splitting the sugar and rendering the pigment far more susceptible to oxidation. In recent years it has been found that many fruit and plant tissues contain enzymes capable of splitting the glycosidic linkage in anthocyanins (Forsyth and Quesnel, 1957; Van Buren *et al.*, 1960; Peng and Markakis, 1963; Sakamura and Obata, 1963).

Under processing conditions where the anthocyanin is subjected to high temperatures in the presence of other fruit constituents, the compounds can undergo a coupled degradation. In the presence of furfural and hydroxymethyl furfural, as well as other products of the browning reaction, the anthocyanins are degraded, but may enter the polymerized products of the browning reaction as a co-polymer. Maximum stability is achieved at low pH. For instance, the stability for cyanidin-3-glucoside is highest at about pH 1·8 (Daravingas and Cain, 1968).

Dalal and Salunkhe (1964) and Daravingas and Cain (1965) showed in canned apricots, cherries and raspberries that there were considerable losses of anthocyanins due to interactions with other products of the fruit. The browning reaction products with which the anthocyanins may react have spectra which visually resemble that of the original pigments; consequently, the loss of the pigment and its replacement by browning reaction polymers does not change the appearance of the fruit product as rapidly as one would expect if the anthocyanin were merely being degraded to less coloured products. In a reaction which has a similar visual effect it has been suggested that anthocyanins will link C–4 to C–6 or C–8 of catechin in dilute acetic acid solution, indicating the formation of a dimer (Jurd, 1967; Jurd and Lundin, 1968). This could presumably be carried further, and co-polymerization with other materials would give rise to products possessing a coloration similar either to the browning reaction products or to the anthocyanin itself, although without the characteristic fine structure in the spectrum. The tannins and anthocyanins could also link under processing or storage conditions to give other oxidation reaction products, which would account for the general retention of reddish colours in heat-processed products where presumably little or no anthocyanin is retained. In a similar manner the preferential oxidation of the anthocyanins is reported to spare ascorbic acid, although in some cases ascorbic acid may contribute to the polymerization of the anthocyanin with other products (Saburov and Ulyanova, 1967).

As early as 1927 it was reported that anthocyanin pigments would react under processing conditions to yield pink or purple discoloration (Culpepper and Caldwell, 1927). The ability of these pigments to chelate metallic compounds is well known, but this reaction during processing has only recently been a problem with peaches. Luh *et al.* (1962) reported that canned peaches with a high anthocyanin content reacted under processing conditions with stannous ions to form blue or purple discolorations. The removal of the tin from solution by the high anthocyanin-containing peaches induced rapid detinning, and subsequent corrosion of the can. This general reaction had been known earlier in berry fruits which contained a high concentration of anthocyanin, but had not been noted in fruits in which the anthocyanin is normally a very minor component (see Chapter 15).

In processing by irradiation, Markakis *et al.* (1959) showed that approximately 0·5 Mrad of cathode rays destroyed 55% of the anthocyanins in strawberry juice. In the whole fruit, Trueleson (1960) did not detect a decrease in red colour of strawberries at doses of 200 krad, although at doses exceeding

350 krad the fruit lightened. In many cases when the fruits are irradiated prior to ripening, it was found by Maxie *et al.* (1964) that the anthocyanins increased in peaches and nectarines, suggesting an increased synthesis of these pigments due to irradiation. The effect of irradiation on whole fruit versus the isolated pigment solutions has been noted also in the destruction of the carotenoids of tomatoes.

IV. CHLOROPHYLLS

Considerable work has been done on the measurement of chlorophyll changes during processing of fruits and vegetables, but an understanding of the mechanisms of these changes is still very limited. Fruit products are generally processed after ripening when chlorophyll has, in most cases, already disappeared. However, there are a few minor cases in which chlorophyll may be present during processing. These include gooseberries, pickled peppers, cucumbers and pickled green tomatoes. Therefore, degradation of chlorophylls during processing and ripening will be discussed briefly with reference to both fruits and vegetables since the reactions involved are similar.

The changes in the chlorophyll pigments which occur during and after processing of plant products have been recently reviewed (Chichester and Nakayama, 1965). Basically, three reactions of chlorophyll occur in processed fruits and vegetables. The first is removal of magnesium by hydrogen ions to give pheophytins. In most fruits, because of the low pH, this is the principal reaction. In the presence of copper or zinc ions, it is possible to replace the magnesium with these ions to give a very stable green-coloured complex (Schanderl *et al.*, 1965). Secondly, there is hydrolysis of the phytol group catalysed by chlorophyllase to give chlorophyllides. Both of these reactions can occur on the same molecule to give pheophorbides. Finally, there is destruction of the porphyrin ring. There is essentially nothing known concerning the products or mechanisms of this destruction except that it is an oxidative decomposition.

Loss of chlorophyll from frozen peas has been related to lipoxidase action by Wagenknecht and Lee (1956, 1958) and Lee and Wagenknecht (1958). Walker (1964) found chlorophyll degradation in frozen beans was related to fat peroxidation in a manner analogous to the carotenoid degradation previously studied by several workers (Sumner, 1942; Blain *et al.*, 1953; Blain and Styles, 1959; Blain and Barr, 1961; Tappel, 1954). Holden (1965) made an extensive study of chlorophyll bleaching in a system containing a fatty acid, lipoxidase and a legume seed extract. Chlorophyll was rapidly bleached in this mixture and no pheophytins, chlorophyllides or pheophorbides were detected. Oxygen was necessary for the reaction; antioxidants inhibited the reaction but cyanide did not. No reports of the products of chlorophyll degradation in this reaction have been reported.

Cho (1966) carried out an investigation of the kinetics of pheophytin formation from chlorophyll in an acetone–aqueous buffer medium. He showed the reaction is a general acid catalysis which is second order with respect to acid and first order with respect to chlorophyll. The greater stability of chlorophyll *b* to pheophytinization compared with chlorophyll *a* was explained by the following mechanism:

$$\mathrm{Chl} + 2\mathrm{H}^{+} \underset{}{\overset{K_{eq}}{\rightleftharpoons}} [(\mathrm{ChlH_2})^{++}] \xrightarrow{K_2} \text{pheophytin} + \mathrm{Mg}^{++}$$

in which the rate is determined by K_{eq}. The K_{eq} of chlorophyll *b* is about ten times smaller than for chlorophyll *a* because of the contribution of resonance structures. These structures place a greater positive charge on the nitrogen atoms of chlorophyll *b* that are attacked by protons than on the equivalent nitrogen atoms of chlorophyll *a*.

Strain and Manning (1942) were the first to report the presence of minor components of a chlorophyll mixture when chromatographed on powdered sugar columns and showed that these are slightly less absorbed than the respective chlorophylls *a* and *b*. These bands had visible spectra identical with chlorophylls *a* and *b* and were designated chlorophylls *a′* and *b′*. Since then there has been considerable speculation concerning the structure of these compounds (Chichester and Nakayama, 1965). Katz *et al.* (1968) have, however, obtained evidence from proton magnetic resonance spectra of the chlorophylls *a′* and *b′* showing that these compounds are C_{10} epimers of the respective chlorophyll. Evidence for a similar transformation in bacterio-chlorophyll was also obtained.

Degradation of chlorophyll during the storage of plant tissues is still a completely unsolved problem despite the fact that chlorophyll loss is commonly used as a measure of senescence, and, when present, is important in the preparation for processing. There have, however, been a few observations in recent years which may provide some starting points. Hoyt (1966) subjected chopped ryegrass (*Lolium perenne*) to a number of different treatments and measured chlorophyll loss. Conditions of high humidity and normal temperatures resulted in rapid chlorophyll degradation. Freezing, boiling, desiccation and waterlogging all stopped chlorophyll loss. Cooling slowed the degradation, but it returned to normal after rewarming. These data are consistent with an enzymatic mechanism of degradation. Wicliff and Aronoff (1963) showed that there is no turnover of chlorophyll in mature bean leaves. If this is true in other plants, it may indicate that the chlorophyll degradative system is inducible at the beginning of senescence. However, Perkins and Roberts (1963) reported that turnover occurred in dicotyledons but not in monocotyledons. Unfortunately their conclusion is based only upon incorporation of radioactivity into mature leaves and does not include data on the kinetics of the loss of radioactivity from the leaves. Pennington *et al.* (1967) have recently characterized C_{10}-hydroxychlorophyll *a* and *b*, formed, apparently enzymatically, in high yield in ground dandelion leaves in 50% methanol. However, there is no evidence for formation of these compounds during periods of chlorophyll degradation in normal tissues.

It has recently been shown (McFeeters and Schanderl, 1968) that a variety of bell peppers (*Capsicum frutescens*) will degrade chlorophyll *a* injected as a water suspension into the fruit. The distribution of radioactivity in various fractions when ^{14}C–chlorophyll *a* was injected was determined but no products of degradation were isolated and identified. The data are consistent with the possibility that this degradation is physiological, but further work is needed for confirmation or otherwise.

Another topic which is pertinent to the mechanism of chlorophyll degradation is the question of whether chlorophyllase is an enzyme in the pathway of chlorophyll biosynthesis or degradation. Most studies have shown that increase of chlorophyllase activity parallels or precedes chlorophyll biosynthesis and loss of chlorophyllase activity parallels chlorophyll degradation (Holden, 1961; Shimizu and Tamaki, 1962; Sudinya, 1963; Ramirez and Tomes, 1964; Böger, 1965; Chiba *et al.*, 1967; Stobart and Thomas, 1968). There have been two reports (Looney and Patterson, 1967; Rhodes and Wooltorton, 1967) that a rise of chlorophyllase activity during the climacteric in apples and bananas is correlated with a loss of chlorophyll from the tissue. These results are subject to some question since apple and banana peel are very high in phenolic compounds which can inactivate a variety of enzymes (Loomis and Battaile, 1966); these phenolics could decrease during the climacteric in which case enzyme inactivation during extraction would be reduced. However, Rhodes and Wooltorton took precautions to prevent enzyme inactivation by phenolics during their extraction procedures. The subsequent decline in activity could be a result of chlorophyllase degradation usually seen in ageing tissues. *In vitro*, chlorophyllase is observed to catalyse the hydrolysis of chlorophyll to chlorophyllide, but Willstätter and Stoll (1928) and Shimizu and Tamaki (1963) reported the *in vitro* synthesis of chlorophyll from chlorophyllide and phytol catalysed by chlorophyllase. However, this was done under conditions which were not physiological and other workers were not able to observe synthesis under different conditions (Holden, 1961; Klein and Vishniac, 1961) or to repeat these observations (Chiba *et al.*, 1967). Therefore, proof of enzymatic chlorophyll synthesis *in vitro* must still be considered an open question.

Recently Chiba *et al.* (1967) have observed enzymatic chlorophyll synthesis from phytol and ethyl chlorophyllide. Ethyl chlorophyllide is not known to occur in plants, but this experiment lends credence to Granick's (1967) idea that one of the reactants needs to be activated for chlorophyll synthesis to be favoured.

There is some preliminary evidence that chlorophyllase does not catalyse the initial step of chlorophyll degradation in cut ryegrass (McFeeters and Chichester, 1968). However, it is not possible to say that chlorophyllase is not involved in chlorophyll degradation because it may catalyse removal of phytol at a later step in the degradative pathway. The first step of the degradation requires the presence of oxygen, but it is not known whether oxygen is directly involved in the reaction or whether it is needed indirectly for maintenance of the degradative system.

The conclusion is that at the present time the evidence favours a biosynthetic rather than a degradative role for chlorophyllase, but the biosynthetic function has not been proved nor has a degradative mechanism been disproved. Chlorophyllase is an enzyme which has received little attention even though its existence was first reported in 1913. There is a need for considerably more information about the structure and properties of the enzyme to aid in the design of experiments to elucidate its role in plant tissue.

REFERENCES

Blain, J. A. (1963). *In* "Carotine und Carotinoide". Dr. Dietrich Steinkoff Verlag, Darmstadt.

Blain, J. A. and Barr, T. (1961). *Nature, Lond.* **190,** 538.

Blain, J. A. and Styles, E. C. C. (1959). *Nature, Lond.* **184,** 141.

Blain, J. A., Hawthorn, J. and Todd, J. P. (1953). *J. Sci. Fd Agric.* **4,** 580.

Böger, P. (1965). *Phytochem.* **4,** 435.

Chiba, Y., Aiga, I., Idemori, M., Satoh, Y., Matsushita, K., and Sasa, T. (1967). *Plant Cell Physiol.* **8,** 623.

Chichester, C. O. and Nakayama, T. O. M. (1965). *In* "Chemistry and Biochemistry of Plants Pigments" (T. W. Goodwin, ed.), p. 438. Academic Press, New York.

Cho, D. H. (1966). Ph.D. Thesis, University of California, Davis.

Cole, E. R. and Kapur, N. S. (1957a). *J. Sci. Fd Agric.* **8,** 360.

Cole, E. R. and Kapur, N. S. (1957b). *J. Sci. Fd Agric.* **8,** 366.

Culpepper, C. W. and Caldwell, J. S. (1927). *J. agric. Res.* **35,** 107.

Curl, A. L. (1959). *Fd Res.* **24,** 413.

Curl, A. L. (1960). *Fd Res.* **25,** 190.

Curl, A. L. and Bailey, G. F. (1956a). *J. agric. Fd Chem.* **4,** 156.

Curl, A. L. and Bailey, G. F. (1956b). *J. agric. Fd Chem.* **4,** 159.

Dalal, K. B. and Salunkhe, D. K. (1964). *Fd Technol.* **18,** 1198.

Daravingas, G. and Cain, R. F. (1965). *J. Fd Sci.* **30,** 400.

Daravingas, G. and Cain, R. F. (1968). *J. Fd Sci.* **33,** 138.

Dicks, J. W. and Friend, J. (1968). *Phytochem.* **7,** 1933.

El Tinay, A. H. (1969). Ph.D. Thesis, University of California, Davis.

Falconer, M. E., Fishwick, M. J., Land, D. G. and Sayer, E. R. (1964). *J. Sci. Fd Agric.* **15,** 897.

Fellers, C. R. and Buck, R. E. (1941). *Fd Res.* **6,** 135.

Fishwick, M. T. (1962). *Abs., First Int. Cong. Food Sci. Technol.* London.

Forsyth, W. G. C. and Quesnel, V. C. (1957). *Biochem. J.* **65,** 177.

Friend, J. and Dicks, J. W. (1967). *Phytochem.* **6,** 1193.

Friend, J. and Mayer, A. M. (1960). *Biochim. Biophys. Acta* **41,** 422.

Galler, M. and Mackinney, G. (1965). *J. Fd Sci.* **30,** 393.
Goldblith, S. A. (1966). *Adv. Fd Res.* **15,** 277.
Goldblith, S. A., Karel, M. and Lusk, G. (1963). *Food Technol.* **17** (2), 139.
Goodwin, T. W. (1958). *Biochem. J.* **68,** 503.
Gortner, W. A. and Singleton, V. L. (1961). *Fd Sci.* **26,** 53.
Granick, S. (1967). *In* "Biochemistry of Chloroplasts", Vol. II (T. W. Goodwin, ed.), p. 373. Academic Press, New York.
Harper, K. A. and Chandler, B. V. (1967a). *Aust. J. Chem.* **20,** 731.
Harper, K. A. and Chandler, B. V. (1967b). *Aust. J. Chem.* **20,** 745.
Hasegawa, K., Macmillan, J. D., Maxwell, W. A., and Chichester, C. O. (1969). *Photochem. Photobiol.* **9,** 165.
Holden, M. (1961). *Biochem. J.* **78,** 359.
Holden, M. (1965). *J. Sci. Fd Agric.* **16,** 312.
Hoyt, P. B. (1966). *Pl. Soil* **25,** 167.
Huang, H. T. (1955). *J. agric. Fd Chem.* **3,** 141.
Huang, H. T. (1956). *J. Am. Chem. Soc.* **78,** 2390.
Hunter, R. F. and Krakenberger, R. M. (1947). *J. Chem. Soc.* **1,** 1.
Joslyn, M. A. (1966). *In* "Cryobiology" (H. T. Meryman, ed.), p. 565. Academic Press, London and New York.
Joyce, A. E. (1954). *Nature, Lond.* **173,** 311.
Jurd, I. (1963). *J. org. Chem.* **28,** 987.
Jurd, I. (1964a). *J. Fd Sci.* **29,** 16.
Jurd, I. (1964b). *Hornblower* **16,** 1344.
Jurd, I. (1967). *Tetrahedron* **23,** 1057.
Jurd, I. and Geissman, T. A. (1963). *J. org. Chem.* **28,** 2394.
Jurd, I. and Lundin, R. (1968). *Tetrahedron* **24,** 2653.
Katz, J. J., Norman, G. D., Suec, W. A., and Strain, H. H. (1968). *J. Am. Chem. Soc.* **90,** 6841.
Khan, M. and Mackinney, G. (1953). *Pl. Physiol.* **28,** 550.
Kimura, S. and Shioda, K. (1963). *J. Fd Sci. Technol.* **10** (5), 169.
Klein, A. O. and Vishniac, W. (1961). *J. Biol. Chem.* **236,** 2544.
Land, D. G. (1962). *Abs., First Int. Cong. Food Sci. Technol.* London.
Lee, F. A. and Wagenknecht, A. C. (1958). *Fd Res.* **23,** 584.
Livingston, A. L., Knowles, R. E., Nelson, J. W. and Kohler, G. O. (1968). *J. agric. Fd Chem.* **16** (1), 84.
Loomis, W. D. and Battaile, J. (1966). *Phytochem.* **5,** 423.
Looney, W. E. and Patterson, M. E. (1967). *Nature, Lond.* **214,** 1245.
Luh, B. S., Chichester, C. O. and Leonard, S. J. (1962). *Abs., First Int. Cong. Food Sci. Technol.* London.
Lukton, A. and Mackinney, G. (1956). *Fd Technol.* **10,** 630.
Lukton, A., Chichester, C. O. and Mackinney, G. (1956). *Fd Technol.* **10,** 472.
Mackinney, G. (1966). *Qualitas Pl. Mater. veg.* **13,** 228.
Markakis, P., Livingston, G. E. and Fagerson, I. S. (1959). *Fd Res.* **24,** 520.
Maxie, E. C., Sommer, M. F. and Brown, D. S. (1964). *U.S. Atomic Energy Comm. Res. & Develop. Rept. No. UCD-34P80-2.*
McFeeters, R. F. and Chichester, C. O. (1968). *Abs., Pacific Slope Biochem. Conf.* p. 58.
McFeeters, R. F. and Schanderl, S. H. (1968). *J. Fd Sci.* **33,** 547.
Peng, C. Y. and Markakis, P. (1963). *Nature, Lond.* **199,** 597.
Pennington, F. C., Strain, H. H. and Katz, J. J. (1967). *J. Am. Chem. Soc.* **89,** 3875.

Perkins, H. J. and Roberts, D. W. A. (1963). *Can. J. Bot.* **41,** 221.
Purcell, A. E. and Walter Jr, W. M. (1968). *J. agric. Fd Chem.* **16** (14), 650.
Ramirez, D. A. and Tomes, M. L. (1964). *Bot. Gaz.* **125,** 221.
Rhodes, M. J. C. and Wooltorton, L. S. C. (1967). *Phytochem.* **6,** 1.
Romani, R. J. (1966). *Adv. Fd Res.* **15,** 57.
Saburov, N. V. and Ulyanova, D. A. (1967). *Dakl. TSKHA* **132,** 259.
Salunkhe, D. K., Gerber, R. K. and Pollard, L. H. (1959). *Proc. Am. Soc. Hort. Sci.* **74,** 423.
Sakamura, S. and Obata, Y. (1963). *Agric. Biol. Chem.* **27** (2), 121.
Schanderl, S., Marsh, G. L. and Chichester, C. O. (1965). *J. Fd Sci.* **30,** 312.
Shimizu, S. and Tamaki, E. (1962). *Bot. Mag., Tokyo* **75,** 462.
Shimizu, S., and Tamaki, E. (1963). *Arch. Biochem. Biophys.* **102,** 152.
Singleton, V. L. and Esau, P. (1969). *Adv. Fd Res.* **17,** 1.
Singleton, V. L. (1963). *Food Technol.* **17** (6), 112.
Sondheimer, E. (1953). *J. Am. Chem. Soc.* **75,** 1507.
Stobart, A. K. and Thomas, D. R. (1968). *Phytochem.* **7,** 1963.
Strain, H. H. and Manning, W. M. (1942). *J. biol. Chem.* **146,** 275.
Sudyina, E. G. (1963). *Photochem. Photobiol.* **2,** 181.
Sumner, R. J. (1942). *J. biol. Chem.* **146,** 215.
Tappel, A. L. (1954). *Arch. Biochem. Biophys.* **50,** 473.
Timberlake, C. F. and Bridle, P. (1966). *Nature, Lond.* **212,** 158.
Timberlake, C. F. and Bridle, P. (1967a). *J. Sci. Fd Agric.* **18,** 473.
Timberlake, C. F. and Bridle, P. (1967b). *J. Sci. Fd Agric.* **18,** 478.
Trueleson, T. A. (1960). *Rept Danish Atomic Energy Comm. Risö* **16,** 73.
Tsukida, K., Yokota, M. and Ikeuchi, K. (1966). *Vitamin* **33,** 174.
Ulrich, J. M. and Mackinney, G. (1968). *Photochem. Photobiol.* **7,** 315.
Van Buren, J. B., Scheiner, D. M. and Wagenknecht, A. C. (1960). *Nature, Lond.* **185,** 165.
Visser, A. (1969). Ph.D. Thesis, University of California, Davis, California.
Wagenknecht, A. C. and Lee, F. A. (1956). *Fd Res.* **21,** 605.
Wagenknecht, A. C. and Lee, F. A. (1958). *Fd Res.* **23,** 25.
Walker, G. C. (1964). *J. Fd Sci.* **29,** 383.
Weckel, K. G., Santos, B., Herman, E., Laferrierel, L., and Gabelman, W. H. (1962). *Fd Technol.* **16** (8), 91.
Wicliff, J. L. and Aronoff, S. (1963). *In* "Studies on Microalgae and Photosynthetic Bacteria" (Japanese Society of Plant Physiologists, eds), p. 441. University of Tokyo Press.
Willstäter, R. and Stoll, A. (1928). "Investigations on Chlorophyll" (translated by F. M. Schertz and A. R. Merz). Science Press Printing Co., Lancaster, Pa.

Chapter 22

Quality

GERALD G. DULL AND A. C. HULME

U.S.D.A., Southeastern Marketing and Nutrition Research Division, Athens, Georgia, U.S.A.; A.R.C. Food Research Institute, Norwich, England

From the point of view of the producer and consumer of fruit, all the biochemical and physiological research programmes have as their justification the accumulation of knowledge of the means by which maximum yield and quality of the end-product—the ripe fruit—can be attained; the two aims are not necessarily compatible, but quality has received far too little attention. Indeed, in many cases the meaning of quality in biochemical terms is still quite vague.

To the average consumer the difference in subjective terms between a high quality and a low quality, but still unblemished, ripe fruit may be considerable, but biochemistry is not yet able fully to explain the difference. The consumer may readily accept the inferior sample but how wide is the difference between "accept" and "prefer"? Often, with fruits such as the apple, the difference is mainly a question of colour, and the impetus to an intensive investigation of the total biochemical factors involved in overall quality is slight. This situation needs to be changed.

Several factors are involved in overall quality. The most important of these are maturity, cultural factors and post-harvest treatment. With a given set of trees in a given season the best that can be achieved at present—apart, perhaps, through treatment of the trees during the development of the fruit with "metabolic regulators", a procedure still in its infancy—is to pick the fruit at optimal ripeness. Nothing further can be done to improve the quality (melons may be an exception, see Chapter 5) and post-harvest treatments do well if they can maintain this quality until the fruit reaches the consumer. Where the fruit is of a type for which long-term storage is a possibility, the fruit must be picked slightly unripe and storage and post-storage ripening conditions chosen for maximum development of quality (flavour); even then it is usually impossible to attain the same quality as obtained in the freshly picked, ripe fruit. Avocado pears are an exception here since they will not ripen on the tree (see Chapter 1). Particularly with fruits such as citrus and bananas it is impossible, however carefully storage and transport conditions are controlled, to achieve the same quality as that of fruit picked ripe and consumed immediately. Within a variety, cultural conditions (soil type, manurial treatment and climate) can modify flavour considerably. Again, genetic potential will determine optimal quality but the genetic potential can be manipulated in the selection of varieties. This has, indeed, been done for many, many years in, for example, the development of modern apple and banana varieties from the crab apple and wild types of *Musa acuminata* and *Musa balbsiana*. Until, however, thorough and well-planned investigations of the biochemical factors involved in quality have been made there is not even a yard-stick by which the effects on quality of all the factors involved can be measured.

It cannot be over-emphasized that, since the object of such investigations would be to relate biochemical patterns to quality in organoleptic terms, it is essential that fully trained taste panels are employed to determine this quality. At a later stage it may be valuable to bring in groups of "lay" consumers.

Clearly, although volatile constituents of fruits play a large part in flavour and aroma (see Volume 1, Chapter 10), acidity (organic acids), sweetness (sugars), astringency (phenolics and terpenoids) and texture (polysaccharides) will also be important. Other factors, some of which are no doubt as yet unknown, such as those present in miracle fruits and the serendipity berry (see Chapter 14), will come into the picture as quality research progresses.

Let us consider the present position of quality factors where state of ripeness is the main consideration. Quality judgements are assessments of the culmination of compositional trends during development. The trends are primarily dependent on the inherent genetic potential of the variety within a species. Whether or not these quality judgements are based on post-harvest trends (e.g. banana, avocado) or trends occurring during growth on the plant (e.g. pineapple, orange, grape), they form the basis for determining an optimum quality range. The optimum quality range can be defined as the compositional range in which some 75% (a completely arbitrary figure) of the people queried would accept a fruit as being ripe. Here again, the word "accept" is not to be confused with the word "prefer".

When the quality of fresh fruit in a produce market is measured against an optimum quality range, the results are all too frequently disappointing in terms of satisfying the customer, i.e. it is outside the optimum quality range. To illustrate more specifically the concept of optimum quality range, consider the selected data in Table I which represent an idealized picture of compositional trends in developing peaches.

In the case of Sample 1 in Table I, no one would accept these fruit as being ripe. A few people might consider Sample 3 as ripe. However, the great majority of the consumers would recognise and accept Sample 5 as being ripe.

It is also recognized that fruit pass through their prime and enter senescence. The acceptance level as this occurs would be expected to decrease. This is characterized by Sample 7.

TABLE I. Tabulation of idealized chemical and physical changes during the development of the peach

Sample No.	Days after flower	Weight (g)	Total sugars (%)	Dry matter (%)	Pressure test (lb)	Colour	Acceptance (%)
1	85	30	3·0	13·5	18·0	Green	0
2	96	58	4·7	13·1	11·5	Greenish–yellow	0
3	106	80	5·3	13·5	9·5	Yellow–green	20
4	113	115	6·1	13·0	7·5	Yellow	50
5	117	152	7·2	12·4	5·9	Yellow–orange	95
6	120	155	7·9	12·9	2·5	Orange	90
7	124	157	8·0	13·0	0·5	Orange	60

It is concluded, *a priori*, that each fruit has a compositional range within which a majority of people would consider the fruit as ripe and acceptable. For the purpose of illustration, allow that a 75% acceptance level would be a desirable goal. When a fruit is consumed, the quality parameters should have specific values within the range of 75–100% acceptance. This would be defined as the optimum quality range. A critical aspect of this concept is the need to consider several quality parameters simultaneously. The use of a single parameter such as skin colour, total sugar content or a period in the pattern of respiration has not been a generally successful means of characterizing a specific stage of fruit development with all types of fruit.

The more subtle aspects of quality during the period of ripening are even more difficult to assess in biochemical terms at the present time. Apart from colour, sugar : acid ratios, starch content and hardness, attempts are being made to relate changes in the nature and content of a whole range of volatile, odoriferous compounds to ripening but many of these are still known merely as peaks on a GLC chromatogram. In certain specialized cases—limonoids in citrus generally, the bitter narigin in grapefruits (see Chapter 3), oleuropein the bitter principle of olives (see Chapter 7) etc.—organoleptically active elements can be traced to specific compounds. However, most of these attributes of quality are the result of random observations or are *ad hoc* criteria of consumer acceptability in general terms. Almost all existing grading schemes for apples are based on size and degree of skin coloration.

The one-well-documented and consistent criterion of approaching ripeness with certain types of fruits is a sudden increase in respiration rate—the respiration climacteric (see Volume 1, Chapter 17). However, even fruits in where this phenomenon occurs it is merely the result of a whole range of biochemical changes, many of which have been described earlier in this volume. Respiration rate is not indicative of the more subtle changes involved in quality. For example, we may have two sets of Cox's Orange Pippin apples with identical respiration climacterics and yet one set may develop excellent ripe quality and the other only a mediocre quality. Respiration is a measure of metabolic activity. Since phenomena other than ripening (ethylene, radiation, temperature, bruising) might certainly be expected to increase the metabolic rate and, therefore, the respiration rate, a change in the respiration rate of a fruit need not necessarily be indicative of ripening processes. By the same token, the absence of a major respiration peak (non-climacteric fruits) carries questionable significance. It is possible with some types of fruit that metabolic rates throughout development are essentially steady and ripening processes take place over a sufficiently long period that no burst of respiratory activity exists and, therefore, cannot be detected. At times, it appears that more effort is spent in trying to establish whether or not a fruit is a climacteric type than is spent in understanding those changes in the physical and chemical character of a fruit which go into making that fruit reach a peak of perfection for human consumption.

Consider the situation in which fruit such as the peach, tomato and pineapple are frequently held under post-harvest conditions designed to ripen the fruit. Post-harvest colour changes in these fruits are frequently equated to ripening. In fact, colour change is but one of several changes which characterize the ripening processes. Furthermore, ripening processes which proceed in the fruit while it is attached to the plant need not be, and undoubtedly are frequently not, the same as those changes which occur in the fruit ripening off the plant. The confusion can be compounded when the term "climacteric" is applied to a post-harvest ripening situation which is not clearly defined in terms of the quality of the final product. The respiratory climacteric as an indicator of physiological activity has its uses. When applied to indicate *optimal* ripening of fruit it is much too vague.

CONCLUSION

The purpose of this short concluding chapter is to create an awareness of some very practical aspects of fruit biochemistry, to stimulate research in defining quality and to set the stage for clearer communication between scientists and the fruit trade. The ultimate goal of providing nutritious and aesthetically pleasing fruit to the consumer should not be minimized. There is clearly a need for including objective data on fruit composition and sensory evaluation when dealing with post-harvest ripening or storage of fruit. The development of mechanical harvesting as an integral part of a fruit production programme gives rise to an even greater need for methods of separating fruit into maturity categories. The challenges in the field of quality are many and they must be met with renewed vigour.

It will be apparent from the chapters in Part II of this volume that the fruit processing industry is already well aware of the importance of quality in its products. While little insight has been obtained into the fundamental biochemical changes occurring during processing and during storage of the processed product, procedures have been continually modified to lead to products as close as possible to the original fruit or to a product acceptable in its own right (juice, purée, etc.). Uniformity of ripeness at the time of processing is still a major problem with some types of fruit. Optimal quality in a processed fruit may require a different stage of maturity or, indeed, a different variety of fruit from that giving optimal quality when consumed fresh. New varieties are being developed for processing (e.g. the Cambridge Favourite strawberry) and the establishment of standards in parameters such as sugar : acid ratio, colour, astringency, texture (firmness) etc. of the presently available starting material has, over the years, resulted in much improved products; new methods of processing (e.g. quick freezing, vacuum dehydration) have also been developed. Nevertheless, as for fruit with the "fresh" market, a more subtle biochemical evaluation, than is at present possible, of the starting material, allied with sensory evaluation of the processed product would, undoubtedly, lead ultimately to a considerable improvement in quality. Economics are involved here: will the cost of research be recouped in a readier sale of the commodity? We ourselves believe that the answer will be in the affirmative. All aspects of quality in fruit products must improve and not stagnate or regress.

Author Index

Numbers in *italics* refer to pages on which the full reference is given

A

Aasa, R., 598, *612*
Abbot, O. D., 13, 14, *61*, 151, *163*, 214, *230*
Abbott, A. J., 380, *407*, 496, *505*
Abdel-Gawad, H. A., 416, 424, *431*
Abdel-Kadar, A., 688, 695, 697, *704*
Abdel-Kader, A., 438, *479*
Abdel-Kader, A. S., 473, *475*, 692, 697, *702*
Abdul-Baki, A. A., 465, 467, *475*
Abeles, F. B., 148, *167*, 342, 345, *369*, *371*
Aberg, B., 441, *475*
Abernethy, J. L., 418, *433*
Acker, L., 647, *648*
Adam, W. B., 516, 524, 527, 530, 531, 532, 536, *537*, *539*
Adams, J. B., 514, 515, 518, *537*
Adamson, A., 598, *619*
Addicott, F. T., 148, *161*
Addoms, R. M., 413, 417, *431*
Agarwal, J. D., 418, *431*
Agnihotri, B. N., 251, *253*, *254*
Aharoni, Y., 152, *161*
Ahmad, M., 563, *571*
Ahmada, F., 303, 322, *324*
Ahmed, E., 382, *407*
Ahmed, E. A., 690, 696, *703*
Ahmed, E. M., 552, *571*, 697, *703*
Aiga, I., 716, *717*
Airan, J. W., 440, *475*
Ajon, G., 160, *161*
Akamine, E. K., 322, *323*
Akkerman, A. M., 457, *475*
Akuta, S., 400, *407*
Albach, R. F., 130, 131, 132, *166*
Alban, E. K., 440, 443, 444, 445, 447, 448, 449, *476*, *477*
Alberala, J., 123, *161*
Alberding, G. E., 136, 137, *162*, *169*
Alberg, C. L., 270, *277*
Albrecht, J. J., 504, *506*
Albrigo, L. G., 417, *431*, *432*, 492, *505*
Alburn, H. E., 667, 668, *683*
Al-Delaimy, K. A., 418, 430, *431*, 519, *537*, 672, *682*
Aldini, R., 444, *475*
Aldrich, W. W., 147, *166*
Alexander, P. E., 564, *570*
Alexandrowa, R. S., 217, *230*
Al-Jasim, H., 692, *702*
Allemann, R., 646, *651*
Allen, F. W., 418, 420, 421, 422, 423, *431*, *432*, *434*
Allinger, H. W., 215, 217, *229*, *231*
Almendinger, V. V., 160, *161*
Alquier-Bouffard, A., 184, *204*
Alsos, I., 609, *614*
Amantova, I. A., 602, *620*
Amerine, M. A., 521, *537*, 605, *612*
Andersen, L., 600, *613*
Anderson, D. G., 27, *60*, 222, *229*
Anderson, E. E., 81, 82, *103*, 524, *537*
Anderson, J. A., 91, *101*
Anderson, L., 562, *569*
Anderson, R. E., 420, *431*, 443, 444, 445, *475*
Andre, P., 419, *435*
Andreotti, R., 460, *475*
Anet, E. F. L. J., 389, *407*
Angelini, P., 362, *369*
Anker, A., 247, *253*
Anon, 532, *537*, 543, 551, 564, *569*, 573, *612*
Ansiaux, J. R., 460, *475*
Anthistle, M. J., 516, 517, *537*
Antikainen, P. J., 598, *612*
Appleman, D., 16, *60*, 119, *162*
Appleman, W. E., 46, *60*
Aprees, T., 187, *204*

Araki, T., 669, *682*
Arakji, O. A., 404, *407*
Arasimovich, V. V., 217, 221, 222, *229*, *231*
Aref, M. M., 672, 674, *682*
Arigoni, D., 133, *161*
Armbruster, G., 493, *505*, 676, *682*
Armerding, G. D., 593, *612*
Arneson, P. A., 453, *475*
Arnon, D. I., 460, *475*
Aronoff, S., 715, *719*
Arthur, J. C., 602, *612*
Asai, K., 282, *301*
Asbell, D. M., 559, *571*
Ascham, L., 442, *477*
Asen, S., 469, 471, *479*, 601, *612*, *616*, 674, *684*
Asenjo, C. F., 13, *60*
Ashby, D. L., 342, *369*
Aslanyan, G. Sh., 222, *229*
Asmaeva, A. P., 222, *229*
Athavale, V. T., 598, *612*
Atkins, C. D., 127, 136, 139, 159, 160, *161*, *167*, *169*, 559, 561, 563, *569*, *571*, *572*, 587, 619
Atkinson, F. E., 607, *612*
Attaway, J. A., 136, 137, 138, 146, *162*, *169*
Atterberg, R. A., 559, *572*
Attix, F. H., 689, *702*
Aumann, H., 401, *408*
Austin, W. W., 341, *370*
Avakyan, A. G., 211, *229*
Avakyan, S. O., 222, *229*
Avron, M., 46, 47, 48, 53, 55, *60*
Awaya, S., 678, *684*
Axelrod, B., 161, *162*, 383, *409*
Ayres, A. D., 393, *407*, 610, *615*
Azatyan, S. A., 222, *229*

B

Baba, S., 669, *685*
Baccharach, A. L., 401, *407*
Badenhop, A. F., 451, 452, *478*
Badran, A. M., 76, 85, *101*
Baier, W. E., 558, 559, *571*
Bailey, F. G., 709, *717*
Bailey, G. F., 116, 117, *163*, 546, 547, *569*
Bain, J. M., 94, *101*, 113, 114, 119, 152, *162*, 337, 338, 363, *369*, 633, 639, *648*
Bakal, A., 555, 556, 568, *569*, *571*
Baker, E. E., 646, *650*
Baker, E. M., 609, *617*
Baker, G. A., 211, 212, 217, *229*
Baker, J. E., 44, 45, *60*
Baker, L. R., 456, *475*
Baker, R. A., 558, 562, *569*
Baksay, L., 210, *229*
Balatsouras, G. D., 260, 262, 263, 270, 271, 273, 274, *277*, *279*
Balatsouras, V. D., 271, *279*
Baldini, E., 260, *277*
Baldrati, G., 451, *477*
Baldwin, J. T., 282, *301*
Balestrieri, G., 267, *279*
Balkay, A., 700, *702*, *705*
Ball, A. K., 134, 161, *162*
Ballinger, W. E., 381, *408*
Balls, A. K., 161, *165*
Balz, H. P., 669, *682*
Balzer, I., 442, *477*
Banerjee, H. K., 248, *253*
Banga, I., 604, *612*
Barabas, L. J., 137, *162*
Baraud, J., 604, *612*
Barber, G. A., 25, *62*
Barcus, D. E., 42, *60*
Barger, W. R., 156, *164*, 643, *648*
Barke, R. E., 461, *475*
Barker, A. G., 391, 397, *407*
Barker, H. A., 642, 643, *651*, *652*
Barker, J., 87, *101*, 674, *684*
Barker, W. G., 67, *102*
Barnabas, J., 440, *475*
Barnard, L. C., 625, 626, *649*
Barnell, E., 69, 84, *100*
Barnell, H. R., 69, 79, 84, *102*
Barnett, R. C., 222, *229*
Baronowski, P. E., 428, *435*
Barr, R. A., 69, 70, 86, 87, 89, 90, 91, 97, 98, *104*
Barr, T., 714, *717*
Barrera, R., 700, *702*
Barret, A., 605, *617*
Barron, E. J., 25, *60*
Barron, R. W., 547, *570*
Barta, E. J., 632, *650*
Bartan, H. R., 133, *161*
Bartel, E. E., 417, *434*

Bartholomew, E. T., 123, 132, *162*
Baruah, P., 584, 604, *612*
Basham, C. W., 470, 471, 474, *475*
Bassett, I. P., *166*
Bassler, G. C., 326, *331*
Batchelor, L. D., 126, *168*
Bates, F. L., 79, *102*
Bate-Smith, E. C., 325, *330*, 362, *371*, 400, *407*
Batjer, L. P., 338, 341, 342, *369*, *372*
Battaile, J., 716, *718*
Baudisch, W., 442, *475*
Bauernfeind, J. C., 565, *569*, 607, *612*
Baumann, G., 591, 594, *614*
Bauer, J. R., 94, *102*
Bayer, E., 584, 599, 601, *612*
Beadle, N. C. W., 441, *475*
Bean, R. C., 8, *60*, 157, *162*
Bean, R. S., 508, *540*
Beattie, H. G., 388, *409*
Beaudreau, C. A., 308, *323*
Beavens, E. A., 134, 151, *166*, 511, 536, *539*
Beck, G. E., 668, *684*
Becker, R., 587, *621*
Bedford, C. L., 418, 423, 426, 430, *431*, *432*, 519, *537*, 583, *613*, 672, *682*
Beech, F. W., 588, 592, 593, 594, 607, *612*, *618*
Beecham Products, 382, *407*
Beese, R. E., 568, *570*
Beeson, K. C., 460, 461, *475*
Beevers, H., 354, *372*
Beidler, L. M., 486, *506*
Beisel, C. G., 160, *161*
Bekina, R. M., 670, 676, *682*
Bell, R. A., 329, *331*
Belluci, G., 444, *475*
Benaro, J. R., 245, *254*
Benn, T. R., 689, *702*
Bennet-Clark, T. A., 640, *648*
Bennett, J. S., 668, *685*
Bennett, R. D., 456, *475*
Benrath, A., 604, *612*
Benson, E. J., 672, 675, *685*
Benson, S., 598, *619*
Ben-Yehoshua, S., 30, 31, *60*
Berk, Z., 544, 557, 558, 560, *571*
Berkowitz-Hundert, R., 221, 222, *229*
Beroza, M., 326, *330*
Berry, R. E., 553, *569*
Bertrand, A., 192, *204*
Betusova, M., 593, *618*
Bevenue, A., 22, *60*, 440, *481*
Beverly, G. D., 133, *165*, 577, *617*
Bewley, W. F., 471, *475*
Bhalla, P. R., 212, *230*
Bhat, S. S. 234, *253*
Bhatia, B. S., 240, 242, 245, 251, *254*, 515, *540*, 551, *571*
Biale, J. B., 15, 21, 28, 29, 31, 32, 33, 34, 35, 36, 37, 38, 41, 42, 44, 45, 46, 47, 48, 49, 50, 51, 52, 53, 54, 55, 56, 57, 58, *60*, *61*, *62*, *63*, 74, 75, 76, 77, *102*, *104*, *105*, 152, *162*, *169*, 197, *204*, 244, *253*, 318, 319, 320, 321, *323*, 348, 349, *373*, 431, *434*, 438, 449, 469, *475*, *480*
Bialogowski, J., 119, *162*
Bianco, V. V., 215, 216, 217, *229*
Bicknell, F., 441, 442, *475*
Bidmead, D. S., 450, *475*
Bieleski, R. L., 55, *60*
Bielski, B. H. J., 667, *682*
Bierl, B. A., 326, *330*
Biggs, R. H., 149, *169*
Billet, D., 247, *253*
Billington, A. E., 391, 397, *407*
Binder, K., 598, *621*
Bini, G., 212, *229*
Bioletti, F. T., 626, *648*
Bisson, C. S., 215, 217, *229*, *231*
Biswas, T. D., 455, *475*
Bjerrum, B., 596, *612*
Black, D. R., 136, *167*, 592, *619*
Black, O. R., 361, *370*
Blackman, F. F., 94, *102*
Blaim, K., 452, *475*
Blain, J. A., 711, 714, *717*
Blair, J. S., 552, 554, *569*, 679, *682*
Blake, J. R., 74, 76, 77, *102*, *104*
Blake, M. A., 413, 417, *431*
Blazovich, M., 426, *435*, 607, 608, 609, *619*, *620*
Blesa, A. C., 222, *229*
Bloch, F., 637, *648*
Bloch, J. H., 667, *684*
Block, K., 428, *435*
Blomquist, V., 418, *431*
Blonquist, H. V., 393, 397, *407*

Blundstone, H. A. W., 533, *537*
Board, P. W., 517, *537*
Bochain, A. H., 396, *409*, 513, *540*
Boe, A. A., 426, *435*, 442, 444, 452, 453, 455, 456, 473, *475*, *476*
Böger, P., 716, *717*
Bogin, E., 157, 158, 159, 160, 161, *162*
Boharti, G. S., 443, 461, *475*
Bohn, G. W., 208, 213, *229*
Boidron, J. N., 192, *204*
Boland, F. E., 393, 397, *407*, 418, *431*
Bolin, H. R., 647, *648*, *651*, 673, *682*
Bollard, E. G., 415, *433*
Bomben, J. L., 554, *569*, *570*, 592, *619*
Bomber, J., 426, *435*
Bonastre, J., 195, *204*
Bonner, J., 31, 55, *62*, 456, 474, *475*, *481*
Bonner, W. D., 348, *372*
Bonnie, H-Sun., 364, *369*
Bonoldi, V., 561, *571*
Borbolla y Alcala, J. M. R., de la, 261, 262, 265, 266, 267, 270, 271, 272, 274, *277*
Borchers, E. A., 452, *475*
Borenstein, B., 515, *537*, 607, *612*
Borgstrom, G., 519, *537*, 591, *612*, 672, 674, *682*
Born, F., 584, *612*
Born, R., 133, *162*
Bornstein, B., 635, *650*
Borstrom, G., 418, 430, *431*
Bose, A. N., 319, 322, *323*
Bouard, J., 179, *204*
Bould, C., 405, *407*
Bourke, R. J., 692, *704*
Bourne, E. J., 598, *612*
Boutonnet, C. E., 467, *480*
Bowen, J. F., 524, *537*
Bowers, J. B., 691, 692, *704*
Bowler, E., 5, *62*
Boyd, C., 692, 694, 695, *704*
Boyd, J. M., 553, 562, 568, *569*
Boyer, J. S., 314, *323*
Boyle, F. P., 647, *648*, 672, 674, *685*, 690, 691, *702*, *703*
Boyle, J. S., 472, *481*
Bradley, D. B., 443, 444, 445, 447, 461, *475*
Bradley, D. E., 639, *649*
Bradley, M., 414, *431*
Bradley, M. V., 414, 415, 417, *431*, *434*
Brady, C. J., 75, 76, 77, 80, 89, 90, 94, 95, 96, 97, 98, 99, *102*, *106*
Brannaman, B. L., 417, *432*
Bratley, C. O., 151, 154, *162*, *166*, 406, *407*
Braun, O. G., 554, 563, 564, *570*, *571*
Braverman, J. B. S., 128, *162*, 547, 568, *569*, 628, 636, *648*, *649*
Breidenbach, R. W., 430, *434*, *704*
Brekke, J. E., 625, 634, 643, 646, 647, *648*, *651*
Brenner, M. W., 69, 74, 79, *103*, *104*
Brenner, S., 528, *537*
Bretthauer, G., 580, 586, 594, 611, *621*
Bridle, P., 606, 607, *620*, 629, *652*, 712, *719*
Brieskorn, C. H., 457, 458, 460, *475*, *476*
Broadbent, L., 472, *476*
Broderick, J. J., 427, *431*
Brogden, W. B., Jr., 136, *164*
Brogdon, J. L., 440, 443, 444, 450, *481*
Brooks, C., 154, 156, *162*, *166*, 406, *407*
Brossard, J., 292, 293, *301*
Brouwer, J. N., 486, *505*
Brown, A. P., 442, *476*
Brown, D. S., 362, *369*, 713, *718*
Brown, H. D., 400, *408*, 461, 462, *477*
Brown, L. C., 212, 217, *229*
Brown, M. S., 654, 681, *683*, *685*
Brown, R. M., 211, *229*
Brown, W. D., 667, 668, *682*
Brownell, G. L., 689, 697, *702*
Brownlee, R. G., 326, *330*
Brubacher, L., 469, *481*
Bruce, D. W., 307, *323*
Bruemmer, J. H., 146, *162*, 558, 562, *569*
Bruice, T. C., 667, 668, *682*
Brun, W. A., 78, *103*
Brunetti, A. P., 602, *613*
Bryan, J. D., 442, *480*, 584, *618*
Bryan, J. M., 532, 533, 534, *540*
Bruyn, J. W., de, 440, *476*
Bruzeau, F., 196, 198, *205*
Bryan, J. D., 388, *407*
Buch, M. L., 420, *432*, 511, 512, *537*

Buchanan, J. R., 362, *369*
Büchi, W., 193, *204*
Buchmann, E., 276, *277*
Buck, R. E., 442, *479*, 710, *717*
Buckley, E. H., 70, 83, 85, 90, 91, 92, *102*, *105*
Budowski, P., 118, *163*
Buford, W. R., 223, *232*
Buhler, D. R., 38, *60*, 465, *476*
Buigues, N. M., 136, *165*
Bukovac, M. J., 415, *432*
Bulen, W. A., 443, *476*
Bulger, J., 80, *103*
Bullis, D. E., 625, 647, *652*
Bullock, R. M., 421, *432*
Bulstrode, P. C., 362, *370*
Burdick, E. M., 581, *613*
Buren, J. P., van, 690, 699, *704*
Burg, E., 243, *253*
Burg, E. A., 31, 36, 37, 38, 40, *60*, *61*, 70, 74, 75, 76, 95, *102*, 146, 148, 153, *162*, 319, *323*, 469, 470, 471, *476*
Burg, S. F., 243, *253*
Burg, S. P., 31, 36, 37, 38, 40, *60*, *61*, 70, 74, 75, 76, 95, 96, *102*, 146, 148, 153, *162*, 319, *323*, 344, *369*, 469, 470, 471, *476*
Burke, J. H., 108, 109, 110, *162*
Burke, R. F., 655, 680, *682*
Burnell, R. H., 515, *537*, 565, *569*, 607, *612*, *613*
Burns, E. E., 213, *231*, 698, *702*
Burrell, R. C., 443, 444, 447, 448, 449, 457, 461, 462, *476*, *477*, *482*
Burroughs, L. F., 340, *369*, 396, 398, *407*, 448, *476*, 576, 577, 585, *613*
Buslie, B. S., 138, 146, *162*
Bussel, J., 692, *702*
Butkus, V., 389, 391, *407*
Butland, P., 533, *537*
Butler, A. R., 667, 668, *682*
Butler, T. J., 569, *569*
Buttery, R. G., 451, *476*
Buttram, J., 149, *162*
Byrde, R. J. W., 589, *613*
Bystrom, B. G., 5, *62*

C

Cadahia Cicuendez, P., 260, *279*
Cain, J. C., 156, *167*
Cain, R. F., 400, *408*, 513, *538*, 713, *717*
Cairncross, S. E., 81, *102*
Caldwell, E., 219, *232*
Caldwell, J. S., 378, 381, 386, *408*, 532, 533, *538*, 600, *613*, 625, *648*, 713, *717*
Cama, A. R., 246, *254*
Cameron, E. J., 523, 525, 526, 528, *538*, *541*
Cameron, S. H., 119, *162*
Camp, A. F., 123, 144, 150, *167*
Campbell, C. W., 32, 34, *61*
Campbell, R. C., 414, *432*
Campo Sanchez, E., del 260, *279*
Canham, J. E., 609, *617*
Cantarelli, C., 262, 274, *277*
Canzonery, F., 264, *277*
Cappellini, P., 426, 427, *433*
Caprio, J. M., 144, *162*
Carangal, A. R., Jr., 443, 444, 447, 448, 449, *476*
Cardwell, A. B., 442, *477*
Cargill, B. F., 489, *505*
Carles, J., 179, 184, *204*
Carlone, R., 415, *432*
Carns, H. R., 148, *161*
Carr, J. G., 576, *613*
Carré, M. H., 385, *407*
Carson, J. F., 448, 452, *482*, 521, *541*
Carter, P. R., 531, *538*, 567, 569, *569*, *570*
Casas, A., 123, *161*
Casoli, K., 399, 400, *407*
Castillo, M. T., 15, *61*
Caston, J. W., 593, *615*
Catlin, P. B., 264, *279*, 415, 417, 425, 430, *432*, *434*
Celmer, R. F., 579, *613*
Chace, W. G., 156, *164*
Chachin, K., 299, *301*
Chaigneau, G., 247, *253*
Chaluty, E., 699, *702*
Chambers, T. C., 632, 639, *649*
Chan, H. T., 611, *613*
Chance, B., 42, 49, *61*, 462, *476*
Chandler, B. V., 118, 133, *162*, 400, *407*, 552, *569*, 577, *613*, 712, *718*
Chandler, W. H., 412, *432*, 658, 662, 664, 665, *683*
Chang, A. T., 607, 608, *619*

Chang, S. S., 362, *373*
Chaplin, M. H., 416, *432*
Chapman, H. D., 141, 144, *162*
Chapon, L., 603, *613*
Chari, C. N., 626, *649*
Charlampowicz, Z., 609, *613*
Charley, P. J., 598, *613*
Charley, V. L. S., 388, 391, 393, 397, *407*, 554, *569*, 578, 582, 593, *613*
Charlson, A. J., 12, *61*
Chase, E. M., 27, *61*, 152, *162*, 215, *229*, 636, 647, *648*, *649*
Chase, B. R., 358, *370*, 694, *703*
Cheema, G. S., 234, 239, 243, *253*
Cheftel, H., 531, *538*
Cheldelin, V. H., 224, *229*
Chen, S. D., 221, *229*
Chen, Y. T., 390, *408*
Chernishov, V. M., 671, *683*
Chesnut, V. K., 426, 427, *434*
Chiba, Y., 716, *717*
Chibnall, A. C., 359, *369*
Chichester, C. O., 218, *232*, 261, 273, *279*, 289, *301*, 363, *369*, 400, *409*, 418, *433*, 453, *476*, 499, *506*, 514, *539*, 577, 606, *617*, *619*, 628, 646, *650*, *651*, 710, 712, 713, 714, 715, 717, *717*, *718*, *719*
Child, R. D., 342, *370*, *372*
Childer's, N. F., 362, *373*, 417, *433*
Chilson, O. P., 668, 669, *683*
Chima, T. S., 196, *204*
Chin, L., 663, 667, *685*
Ching, T. M., 658, 662, 666, 680, *683*
Cho, D. H., 715, *717*
Chow, C. T., 40, *61*
Christensen, B., 465, *476*
Christensen, B. E., 224, *229*, 465, *481*
Christian, W. A., 127, *166*, 380, 382, 386, 388, *409*
Christie, A. W., 625, 626, 640, *649*, *651*
Christopher, E. P., 213, *231*
Chun, H. H. Q., 314, *324*
Church, C. G., 6, 16, *61*, 152, *162*, 215, *229*, 636, 647, *648*, *649*
Clagett, C. O., 344, *369*
Clark, E. J., 524, *538*
Clark, H., 655, *683*
Clark, R. B., 157, *162*
Clarke, I. D., 688, 690, 693, 694, 695, 696, 698, 700, *702*, *703*, *705*
Clarke, W. M., 603, *613*
Claypool, L. L., 417, 418, 420, 421, *431*, *432*, *433*, 492, *505*, 633, *649*
Clayton, W., 663, *683*
Clegg, K. M., 544, 546, *569*, 605, 606, 607, 608, 609, 610, *613*
Clegg, M., 516, *538*
Clements, C. A. B., 562, *569*
Clements, R. L., 23, *61*, 121, 123, 128, 129, *162*, 350, *370*
Clemetson, C. A. B., 600, *613*
Clendenning, K. A., 464, *476*
Clifcorn, L. E., 526, 528, *538*, 564, *569*
Clijsters, H., 338, 355, *370*
Clutter, M. E., 442, *476*
Co, H., 400, 402, 403, *408*
Coates, J. L., 527, *540*
Coates, M. E., 401, *407*
Cocking, E. C., 441, 450, 459, 464, *477*, 503, *506*
Coggins, C. W., Jr., 144, 147, *163*, *164*, *165*
Coggins, R. A., 576, *613*
Coheé, R. F., 518, *538*
Cohen, A., 144, *163*
Cohen, M., 147, *165*
Cohen, T., 604, *619*
Colagrande, O., 182, *204*
Cole, E. R., 709, *717*
Collazo de Rivera, A. L., 245, *254*
Collins, F. L., 305, *323*
Collins, J. L., 303, *323*
Collins, R. P., 451, *478*
Conn, E. E., 27, *60*, 222, *229*
Connell, D. W., 327, 328, *330*, *331*
Connell, J. C., 531, *538*
Conners, C. H., 413, *432*
Connor, M. A., 690, *702*
Connor, W. A., 604, *613*
Conover, R. A., 472, *476*
Conrad, C. M., 382, *408*
Constantinides, S. M., 418, *432*, 583, *613*
Cook, A. H., 401, *409*
Cook, W. H., 671, *684*
Cooley, J. S., 406, *407*
Coombe, B. G., 413, 415, *433*
Coonen, N. H., 531, *540*
Cooper, A. J., 467, *476*
Cooper, G. M., 426, *435*, 699, *703*

Cooper, H., 149, *163*
Cooper, W. C., 148, *163*, 341, *370*
Copley, M. J., 625, *652*, 655, *685*
Coppock, G. E., 148, 149, *169*
Cordon Casanueva, J. L., 266, *278*
Cordonnier, R., 193, *204*
Corey, E. J., 133, *161*
Coriell, L. L. 663, 664, *685*
Cornwell, P. B., 688, *703*
Costello, L. A., 668, 669, *683*
Couey, M., 420, *432*
Coulson, D. M., 329, *330*, *331*
Cowgill, G. R., 567, *571*
Craft, C. C., 154, *166*, 419, *432*, 470, *479*
Crafts, A. S., 647, *649*
Crance, C., 334, *370*
Crane, J. C., 147, *163*, 413, 414, 415, 416, 417, 424, *431*, *432*, *433*, *434*
Crane, M. B., 442, *476*
Crang, A., 386, 388, *408*, 517, *538*
Crank, J., 638, *649*
Crean, D. E. C., 533, *537*
Creveling, C. R., 85, 91, *104*
Creveling, R. K., 329, *330*, 426, 429, *432*, 577, *615*
Criddle, R. G., 26, 28, *62*
Croteau, R. J., 404, *408*
Cruess, W. V., 261, 262, 270, 273, *277*, 441, 445, *480*, 509, 512, 516, *538*, 625, 626, 640, *649*
Crutchfield, C. A., 133, *169*
Cruz, F., de la, 700, *702*
Cruz Valero, A., 265, *277*
Culp, R., 282, *301*
Culpepper, C. W., 378, 381, 386, *408*, 531, 532, 533, *538*, 600, *613*, 713, *717*
Cultrera, R., 399, 400, *407*
Cummings, G. A., 417, *432*
Cummings, K., 4, 5, *61*
Cummins, S. E., 503, *506*
Curda, D., 514, *539*
Curdova, M., 514, 530, *539*
Curl, A. L., 116, 117, 123, *162*, 218, *229*, 292, *301*, 392, *408*, 419, 426, *432*, *434*, 455, *476*, 545, 546, 547, 554, 563, 564, *569*, 595, *613*, 628, *649*, 709, *717*
Currence, T. M., 211, 215, *229*, 442, *476*
Curti, B., 669, *683*
Curtis, R. C., 388, *407*
Curwen, D., 417, *432*
Czerlinski, G. H., 603, *616*
Czerny, J., *372*
Czygan, F. C., 455, 456, *476*

D

Dabbah, R., 495, *506*
Daepp, H. U., 585, *613*
Daghetta, A., 426, *432*
Daji, J. A., 237, 244, *254*
Dalal, K. B., 442, 444, 451, 452, 453, 455, *476*, 514, 515, *538*, 713, *717*
Dall'Agro, G., 399, 400, *401*
Dalzel, A., 485, *506*
Dani, P. G., 239, 243, *253*
Daniell, W. F., 486, *506*
Danielson, L. L., 208, 213, *229*
Danziger, M. T., 582, *613*
Daravingas, G., 400, *408*, 513, *538*, 713, *717*
Darkanabaev, T. B., 462, *476*
Das, N. B., 455, *475*
Das, S. K., 418, *432*
Daskaloff, C., 442, *476*
Dass, C. M. S., 666, *684*
Date, W. B., 251, *253*, 418, *431*
Datta, A., 669, *684*
Davenport, J. B., 12, 16, 17, 21, *61*, 359, *370*
David, J. J., 419, *432*
Davidek, J., 600, 609, *613*
Davidson, O. W., 413, *431*
Davies, D. D., 187, *204*
Davies, J. N., 440, 441, 443, 444, 445, 446, 447, 448, 449, 457, 458, 460, 461, 462, 467, 468, *476*, *477*, *482*
Davies, J. W., 441, 450, 459, 464, *477*
Davies, R., 317, 318, *323*, 421, *434*
Davis, D. W., 440, 443, 444, 450, *481*
Davis, E. G., 518, *538*
Davis, G. N., 207, 208, 209, 212, 213, 217, *229*, *232*
Davis, L., 419, *435*
Davis, P. S., 598, *613*
Davis, R. M., Jr., 211, 212, 217, *229*
Davis, W. B., 112, 132, 136, 160, *163*
Davison, P. K., 459, *482*
Davison, R. M., 415, *432*

Dawson, C. R., 561, *569*, 607, *613*
Dawson, E. H., 523, *538*
De, H. N., 249, *253*
De, S., 319, 322, *323*
Deasy, C. L., 326, *330*
Debnath, J. C., 249, *253*
De Candolle, A., 412, *432*
De Candolle, A. P., 486, *506*
Decareau, R. V., 655, 680, *682*
Dedolph, R. R., 76, *102*, 213, 215, *230*
Dehean, N., 421, 430 *435*, 693, 694, *704*
Dehr, A., 524, *538*
Deibner, L., 604, *614*, 626, *649*
Deidda, P., 263, *278*
Dekazos, E. D., 420, *432*, 510, *538*
Deller, D. J., 598, *613*
De Moura, J., 420, *434*
Denisen, E. L., 455, 456, *477*
Denne, M. P., 337, *370*
Dennison, R. A., 455, *480*, 552, *571*, 690, 691, 696, 697, *703*, *704*
Denny, F. E., 36, *61*, 154, *163*, 215, 224, *229*
Dent, K. P., 495, *506*
Derse, P. H., 671, *683*
Deschreider, A. R., 596, *614*
Deshpande, P. B., 419, 426, *432*, *435*
Deshpande, S. N., 512, *539*
Desrosier, N. W., 512, *539*
De Swardt, G. H., 80, *102*
Deszyck, E. J., 118, 124, 125, 128, 145, 146, *163*, *167*, *168*
Detty, W. E., 600, *614*
Devel, H., 193, *204*, 581, *614*
Deuel, H., 193, *204*, 581, *614*
Develter, E., 261, 262, 273, *277*
Devescovi, M., 561, *571*
Dewey, D. H., 383, 386, 388, 390, 405, *410*
Dhaliwal, A. S., 426, *435*
Dhopeshwarkar, G. A., 527, *538*
Dick, M. I. B., 567, *569*
Dickinson, D., 516, 517, 524, 531, 534, 535, *537*, *538*
Dickinson, D. B., 462, 466, 467, *477*, 670, *683*
Dicks, J. W., 711, *717*
Diehl, H. C., 655, 666, *683*
Diemair, W., 565, *569*, 587, 605, 607, *614*
Dietz, G. R., 701, 702, *703*
Dilley, D. R., 94, *102*, 341, 347, 349, 350, 353, *370*, 471, *477*
Dillman, C. A., 160, *161*
Dittmar, H. F. K., 15, *61*, 419, *433*
Do, J. Y., 419, 427, *432*, *435*, 451, 452, 473, *475*, *476*
Doebbler, G. F., 665, *683*
Doeden, D., 91, *101*
Doesburg, J. J., 512, *538*
Dolendo, A. L., 16, 20, 24, 30, *61*
Dolev, A., 667, 668, *682*
Donnelly, B. J., 605, *615*
Doolittle, S. P., 208, 213, *229*
Dostal, H. C., 420, *434*, 471, 473, *477*
Douglas, H. C., 271, 272, *279*
Dourmichidze, S. V., 202, *204*
Dowling, B., 500, 504, *506*, 577, *616*
Downes, N. J., 678, *683*
Downing, D. L., 519, *540*
Doyle, W. P., 465, *477*, *481*
Drawert, F., 84, *102*, 185, *204*, 361, *370*, 590, *614*
Drews, H., 399, *409*
Dreyer, D. L., 133, 134, *163*, *165*, 550, *571*
Drury, R. E., 462, 464, 466, *477*, 670, *683*
Dryden, E. C., 512, *537*
Dubey, S. N., 598, *614*
Duden, R., 583, 584, 590, *614*
Dugger, W. M., Jr., 147, *165*
Dull, G. G., 309, 310, 311, 318, 319, 320, 321, *323*, 438, *477*
Dunkee, E. L., 593, *617*
Dunlap, W. J., 130, 131, 132, *163*, *164*, *166*, 549, *570*
Dunlop, S. G., 528, *537*
Dupaigne, P., 583, *614*
Dupuy, P., 185, *204*
Durbin, R. D., 453, *475*
Dutcher, R. A., 245, *254*, 526, *538*
Dutcher, R. A., 245, *254*, 526, *538*
Duvekot, N. S., 406, *408*
Dworschak, E., 591, *614*
Dzamić, M. D., 691, *703*

E

Eaks, I. L., 151, 155, 156, *163*
Eaks, L. L., 692, 693, 695, 696, *704*
Eames, A. J., 640, *649*
Eaton, F. M., 461, *477*

Eckey, E. W., 11, *61*
Edmonds, R. S., 701, *703*
Edney, K. L., 367, *371*
Edward, R. A., 692, 695, *703*
Edwards, G. J., 123, 136, 137, 153, 156, *162*, *164*, *167*, *168*
Edwards, R. A., 453, 454, 455, *477*
Edwards, V. M., 495, *506*
Egerton, L. J., 341, *370*
Egeter, H., 599, 601, *612*
Eidt, C. C., 625, 638, *649*
Eilati, S. K., 118, *163*
Eisen, J., 461, *478*
Eisner, J., 263, 275, 276, *278*
El Ansary, M. A. I., 247, *253*
El-Batel, S., 692, 693, 695, *704*
Elbe, J. H., von, 418, *435*
Elfvin, L. G., 44, 45, *60*
Ellis, G. P., 545, *569*
Ellis, R. J., 347, *372*
Ellis, S. C., 12, 16, 17, 21, *61*
Elmer, O. H., 153, *163*
El Saifi, A., 261, 262, 273, *277*
El-Sayed, A. R. S., 418, *432*
El Sherbeiny, A. E. A., 247, *253*
El Sissi, H. I., 247, *253*
Elstein, 555, *570*
El Tinay, A. H., 710, *717*
Emberger, R., 84, *102*, 361, *370*, 429, *432*, 590, *614*
Embleton, T. W., 141, *163*, *165*
Embs, R. J., 430, *432*, 585, *614*, *617*, 627, *649*, *650*
Emerson, O. H., 133, 134, *163*, 550, 551, *570*
English, H., 421, *433*
Enslin, P. R., 221, 222, *229*, *231*
Eny, D. M., 128, 141, 157, *167*
Ephraim, F., 637, *649*
Erdélyi, E., 591, *614*
Erickson, F. B., 516, 517, *538*
Erickson, L. C., 24, *61*, 128, 143, 157, 158, *162*, *163*, *169*, 417, *432*
Erkama, J., 596, *614*
Ernst, R., 670, *683*
Ernst, Z. L., 599, *614*
Esau, K., 640, *649*
Esau, P., 418, *433*, 708, *719*
Esselen, W. B., 388, 389. *408*, 524, *537*
Estrin, B., 393, 397, *407*
Esty, J. R., 523, 526, 528, *538*
Eugster, C., 329, *331*
Evans, H. J., 460, *477*
Evenden, W., 564, *570*
Eyal, Z., 667, *684*
Ezell, B. D., 222, *229*

F

Fabian, F. W., 525, *538*
Fagerson, I. S., 522, *538*, 713, *718*
Fahn, A., 67, *102*
Falconer, M. E., 709, *717*
Falk, K. B., 141, *166*
Fang, T. T., 244, *253*
F.A.O., 438, *477*
Fardig, O. B., 526, *538*
Farias, L. V., de, 222, *229*
Farrankop, H., 219, *232*
Farrer, K. T. H., 566, 567, *569*, *570*
Faust, M., 358, 365, *370*, 694, *703*
Feaster, J. F., 526, 527, 528, *538*, 564, *570*
Feister, D. W., 564, *570*
Feller, C. R., 710, *717*
Fellers, C. R., 388, 389, *408*, 441, 442, *479*, 514, 524, *537*, *539*, 549, *570*
Felsher, R. Z., 222, *232*
Fennema, O., 593, *615*, 655, 658, 665, 667, *683*
Fenton, H. J. H., 604, *614*
Ferenczy, L., 212, *229*
Ferguson, J. H. A., 440, *476*
Ferguson, W. E., 76, *102*
Fernandez, H., 700, *703*
Fernandez, S. J. G., 694, *703*
Fernandez Diez, M. J., 261, 262, 263, 265, 266, 267, 270, 271, 272, 273, 274, *277*, *278*
Fernandez Villasante, J., 265, *278*
Fernando, Q., 604, *619*
Ferrao, J. E. M., 262, *278*
Fideghelli, C., 426, 427, *433*
Fidler, J. C., 355, 362, *370*
Field, A., 629, 630, *650*
Fielding, A. H., 589, *613*
Fields, M. L., 440, 441, 443, 444, *479*
Fields, N. L., 576, *615*
Fildes, R., 212, *229*
Fiess, H. A., 603, *616*

Filipic, V. J., 329, *331*
Finch, A. H., 144, *165*
Finholt, P., 528, *538*, 609, *614*
Fink, A., 599, 601, *615*
Finkle, B. J., 368, *371*, *372*, 390, *408*, *409*, 585, 586, *614*, *618*, 673, 681, *682*, *683*, *684*, *685*
Firestone, D., 263, 275, 276, *278*
First, T., 67, *102*
Fisher, B. E., 329, *330*
Fisher, C. D., 625, 635, *650*, *651*
Fisher, C. K., 643, *648*
Fisher, D. F., 115, 116, 123, 140, 141, 144, 153, *164*, *166*
Fisher, J. F., 139, *163*
Fisher, L. K., 347, *373*, 421, 430, *435*, 693, 694, *704*
Fishwick, M. J., 709, *717*
Fishwick, M. T., 709, *717*
Flagg, W., 655, 666, *685*
Flath, R. A., 361, *370*
Flaumenbaum, B. L., 580, *614*
Fletcher, J. T., 472, *478*
Flood, A. E., 239, *254*, 335, 351, *370*
Florida Department of Agriculture, 110, 111, *163*
Flosfeder, F. C. H., 625, 626, 640, *649*
Fogg, G. E., 637, *649*
Fogleman, M. E., 472, *477*
Ford, E. S., 112, *163*
Fore, H., 391, *408*
Forshey, C. G., 440, 445, *477*
Forsyth, W. G. C., 712, *717*
Forti, G., 426, *432*
Fortlage, R. J., 688, 698, *705*
Fouassin, A., 400, *408*
Fox, J. E., 416, *433*
Fox, M. M., 553, *571*
Foy, J. M., 307, *323*
Foytik, J., 260, *278*
Frahn, J. L., 604, *614*
Francis, F. J., 362, 364, *369*, *370*, 400, 402, *408*, *409*, 606, *620*
Francke, A., 486, *505*
Franke, W., 604, *621*
Frankenthal, R. P., 531, *538*, 567, *570*
Frazier, J. C., 442, *477*
Freed, S., 667, *682*
Freeman, J. A., 440, 447, 448, *477*
Freiberg, S. R., 66, 69, 70, 71, 73, 76, 86, 87, 89, 90, 91, 97, 98, *102*, *104*
French, D., 79, *102*
French, R. B., 13, 14, *61*, 151, *163*, 214, *230*
Frenkel, C., 94, *102*, 347, 349, 350, *370*, 471, *477*
Friedman, J. P., 604, *614*
Friend, J., 711, *717*
Fripp, P. J., 90, *103*
Frost, D. V., 567, *570*
Fruits, 333, 335, *370*
Fryer, H. C., 442, *477*
Fugerson, I. S., 404, *408*
Fujimaki, M., 605, *616*
Fujisawa, H., 673, 681, *683*
Fujita, Y., 138, *165*
Fuleki, T., 400, *407*, *408*
Fukushima, Y., 223, *231*
Fuller, G. W., 451, *478*
Funaro, A., 264, *278*
Furia, T. E., 509, 516, *538*
Furuhashi, S., 294, *301*
Furz, J. R., 114, *163*

G

Gabelman, W. H., 709, *719*
Gabriel, K. L., 701, *703*
Gaddum, L. W., 126, *163*
Gahan, P. B., 665, *683*
Gajewski, A., 609, *613*
Galler, M., 392, *408*, 418, *433*, 709, *718*
Galliard, T., 28, *61*, 344, 350, 354, 356, 357, 359, 360, 361, 363, *370*, *371*, *373*
Gallo, P., 15, *61*
Gallop, R. A., 359, *371*, 533, *538*
Gane, R., 74, 77, *102*, 344, *370*
Ganz, D., 133, 134, *167*
Garber, M. J., 147, *164*
Garden, L., 600, *619*
Gardner, H. L., 307, *323*
Garcia-Blanco, J., 220, *230*
Gardner, F. E., 145, *167*
Garoglio, P. J., 255, 260, *278*
Garrie, J. B., 401, *408*
Garrison, S. A., 462, 464, 466, 468, *477*
Gasco, L., 700, *702*
Gates, J. E., 469, *480*
Gatet, L., 196, 199, *204*
Gawler, J. H., 327, *330*

Gaylord, F. C., 215, *230*
Gee, M., 312, 313, *323*
Gehenio, P. M., 656, 657, 663, *683*, *684*
Geisler, G., 196, 197, *204*
Geissman, T. A., 15, *61*, 533, *538*, 599, *616*, 712, *718*
Gentili, B., 130, 131, 132, *163*, *164*, 500, *506*, 548, 550, *570*
Gentry, J. P., 633, *649*
Georgiev, H. P., 442, *477*
Geraldson, C. M., 460, 461, *477*
Gerber, C., 195, 196, 198, 199, *204*, 222, *230*, 264, *278*
Gerber, R. K., 711, *719*
Gerhardt, F., 222, *229*, 341, *369*, 421, *433*
Ghosh, S., 245, *254*
Giannone, L., 451, *477*
Gierschner, K., 591, 594, *614*
Gigante, R., 471, *477*
Gilbart, D. A., 213, 215, *230*
Gililland, J. R., 271, 272, *278*, *279*
Gillard, R. D., 602, *614*
Gillespy, T. G., 525, *538*
Gilpin, G. L., 523, *538*
Gimeno, F., 15, *63*
Gindhari, Lal, 244, *254*
Ginsburg, L., 322, *323*
Giovanelli, J., 187, *204*
Giral, J., 15, *61*
Giridhar, N., 555, *572*
Gizis, E. J., 404, *408*
Glaziou, K. T., 313, *323*
Glegg, R. E., 690, 691, *703*
Gleisberg, W., 401, *408*
Glushchenko, L. M., 598, *618*
Godar, E. M., 552, 554, *569*, 679, *682*
Goddum, L. W., 141, *166*
Goeseels, Jr., P., 424, *435*
Goeschl, J. D., 36, *62*, 224, 225, 226, 227, *231*
Goldblith, S. A., 708, 710, *718*
Goldfine, H., 428, *435*
Goldstein, J. L., 84, 89, *102*, 289, *301*, 513, *539*, 577, *616*, 674, *684*
Goldsworthy, L. J., 132, *163*
Golovkin, N. A., 671, *683*
Gomez Herrera, C., 261, 262, 265, 266, 270, 271, 272, 274, *277*
Gonzalez Cancho, F., 261, 262, 265, 266, 270, 271, 272, 273, 274, *277*, *278*
Gonzalez Pellisso, F., 261, 262, 263, 265, 266, 267, 270, 271, 272, *277*
Good, L. J., 139, *169*
Goodall, H., 133, *163*
Goodes, J. E., 405, *408*
Goodwin, T. W., 139, *169*, 363, 364, *369*, *370*, 453, 455, 456, *476*, *477*, 711, *718*
Goor, A., 260, *278*
Gore, H. C., 196, *204*, 318, *323*
Gore, P. H., 599, *614*
Goren, R., 114, *163*
Gortner, W. A., 308, 309, 310, 311, 313, 314, 315, 316, 322, *323*, *324*, 329, *330*, *331*, 438, *477*, 709, *718*
Goto, S., 76, *102*
Gottauf, M., 583, 590, *614*
Gould, W. A., 443, 449, 451, 452, *478*
Govindan, P. R., 462, *477*
Gracian Tous, J., 276, *278*
Grado Cerezo A., de, 265, *277*, *278*
Graham, D. J., 430, *436*, 511, 518, *541*
Graham, M., 442, *479*
Granick, S., 363, *370*, 716, *718*
Grant, I. J., 604, *614*
Grant, N. H., 667, 668, *683*
Grant, W. C., 16, *61*
Gray, G. F., 471, *481*
Gray, G. P., 145, *163*
Greene, A. E., 663, 664, *685*
Greenham, D. W. P., 405, *408*
Gregory, D. W., 503, *506*
Greiff, D., 668, 669, *683*
Grierson, W., 153, 154, 156, *164*, *168*
Griffith, F. P., 118, 130, 131, 132, *165*
Griffiths, D. G., 350, *372*
Griffiths, F. P., 548, 550, *570*
Griffiths, L. A., 85, *102*
Griswold, R. M., 566, *571*
Grobois, M., 88, *102*
Grosch, W., 451, *480*, *481*
Gross, H., 6, 7, 8, 14, *61*
Groves, K., 385, *409*
Guadagni, D. G., 325, *330*, 361, 362, *370*, 420, 429, *433*, 452, *477*, 554, *570*, 594, 595, *614*, 672, 674, 675, 677, 678, 679, 680, *683*
Guerrant, N. B., 245, *254*, 526, *538*
Guichard, C., 179, *204*

Guillou, R., 640, *649*
Guimberteau, G., 181, 202, *205*
Guinot, G. Y., 576, *614*
Guilliard, T., 94, *103*
Gum, O. B., 461, 462, *477*
Gupta, S. S., 238, 239, *254*
Gurd, F. R. N., 602, *614*
Gurgen, K. H., 417, *434*
Gustafson, F. G., 210, *230*, 414, *433*
Gutierrez, G., 261, 262, 265, 266, 270, 271, 272, 276, *277*, *278*
Gutierrez y Fernandez Salguero, A., 265, *277*
Guyer, R. B., 516, 517, *538*
Guymon, J. F., 83, *102*
Guzman Garcia, R., 261, 262, 265, 266, 270, 271, 272, *277*
Gyr, J., 182, *204*

H

Haagen-Smit, A. J., 326, *330*
Haard, N. F., 78, 85, 94, 100, *103*
Haas, V., 642, 643, *651*, *652*
Haber, E. S., 469, *477*
Hackenbrock, C. R., 44, *61*
Hackney, F. M. V., 31, *63*
Hadwiger, L. A., 220, *230*
Haenseler, C. M., 472, *477*
Haeseler, G., 418, *433*
Hagen, R. E., 130, 131, 132, *164*, *166*, 549, *570*
Hägglund, E., 637, *649*
Haisman, D. R., 522, 523, 534, *538*
Hale, C. R., 182, *204*
Halevy, A. H., 118, *166*, 212, *230*
Hall, A. D., 334, *370*
Hall, A. P., 6, 7, 13, 14, *61*
Hall, C. B., 455, 456, 469, 472, *477*, *478*, *480*
Hall, C. V., 220, *230*
Hall, D. H., 132, *164*
Hall, E., 399, 400, *408*
Hall, E. G., 31, *63*, 73, *103*, 406, 407, *408*
Hall, G. D., 322, *323*
Hall, H. F., 472, *478*
Hall, T. C., 668, *684*
Hall, W. C., 153, *164*
Hallaway, M., 416, *434*
Haller, M. H., 153, *164*, 423, *433*
Halliday, E. G., 566, *571*
Halpin, J. E., 456, *481*
Hamburger, J. J., 546, *570*
Hamdy, M. M., 443, 449, *478*
Hamilton, G. A., 604, *614*
Hamner, C. L., 443, *478*
Hamner, K. C., 440, 441, 443, 447, *478*
Hane, M., 389, *408*
Hanifin, J. W., 604, *614*
Hannan, R. S., 695, *703*
Hänsel, R., 599, *615*
Hansen, E., 38, *60*, 224, *229*, 344, 350, 351, 363, *370*, *371*, 465, 469, 470, *476*, *478*, *481*
Hanson, J. B., 462, 466, 467, *477*
Hanson, S. W., 515, *538*
Hapka, S., 217, 219, *230*
Hara, T., 608, *618*
Harada, H., 598, *618*
Harborne, J. B., 399, 400, *408*, 533, *538*
Hardenburg, R. E., 208, *230*, 656, *684*
Hardin, G. J., 15, *62*
Harding, P. L., 123, 140, 141, 144, 151, 153, 156, *164*, 241, 242, 247, *254*
Hardy, F., 118, *164*
Hardy, P. J., 180, 185, *204*
Harel, E., 367, *371*, *372*, 585, *614*
Hargrave, P. D., 526, *538*
Harkett, P. J., 355, *373*
Harper, K. A., 400, *407*, 609, *615*, 712, *718*
Harper, R., 325, *330*, 362, *371*
Harrington, W. O., 510, *541*
Harris, A. T., 400, *408*
Harris, J., 594, 595, *614*
Harris, P. L., 69, 79, 87, *103*, *104*
Harris, R. S., 391, *409*, 418, *434*, 441, *480*
Harris, W. M., 462, *479*
Harrison, I. T., 567, *569*
Hartman, J. D., 215, *230*
Hartman, R. T., 473, *478*
Hartmann, C., 347, 349, 350, 353, 358, *371*, *373*, 424, *433*
Hartmann, H. T., 260, 264, 273, *278*, *279*, 425, *434*
Hartree, E. F., 607, *616*
Harvey, E. M., 150, 154, 155, *164*
Harvey, P. M., 362, *370*
Harvey, R. B., 74, *103*
Hasegawa, E., 566, *570*
Hasegawa, K., 710, *718*

Hasegawa, S., 551, *571*
Haskell, G., 401, *408*
Hassan, H. H., 442, *478*
Hasselstrom, T., 81, *103*
Hassib, M., 607, 611, *615*
Hatch, M. D., 345, *371*
Hatton, T. T., 238, 242, 252, *254*
Hatton, T. T., Jr., 32, 34, *61*
Hattori, S., 601, *617*
Hawke, J. C., 26, *61*
Hawker, J. S., 179, 184, *204*
Hawkins, L. A., 156, *164*
Hawthorn, J., 714, *717*
Hay, R. W., 603, 604, *614*, *615*
Hayashi, H., 555, *571*
Hayashi, K., 210, 211, *231*, 601, *617*, *619*, *620*
Hayes, J., 693, *703*
Haynes, D., 385, *407*
Hayward, H. E., 640, *649*
Hazatsu, H., *683*
Heath, O. V. S., 640, *649*
Heber, U., 660, 666, 669, 670, 681, *683*, *685*
Heftmann, E., 456, *475*
Hegarty, M. P., 69, 70, 86, 87, 89, 90, 91, 97, 98, *104*
Heid, J. L., 593, *615*
Heilman, A. S., 316, 322, *323*
Heimann, A., 607, *615*
Heimann, W., 84, *102*, 361, *370*, 590, 600, 607, 610, *614*, *615*
Hein, R. E., 451, *478*
Heinicke, R. M., 308, *323*
Heinrich, B., 600, *615*
Heintze, K., 600, *615*
Heinz, D. E., 429, *433*, 577, *615*
Heinze, P. H., 421, *434*, 457, *486*
Hemmerich, P., 603, *615*
Hendershott, C. H., 147, 148, *164*, *169*
Henderson, R. H., 525, *538*
Henderson, R. W., 79, *103*
Hendricks, L. C., Jr., 212, 217, *229*
Hendricks, S. B., 364, 365, *373*
Hendrickson, R., 113, 129, 130, 131, 134, 135, 136, *164*, *165*
Henning, G. J., 486, *505*
Henning, G. L., 144, *163*
Henry, D. W., 329, *330*
Henry, W. H., 148, 149, *163*, 341, *370*
Hepler, R. W., 444, 445, *481*
Hepton, J., 313, *323*
Hera, E., 551, *571*
Herman, E., 709, *719*
Hernandez, H. H., 532, *539*
Herregods, M., 469, *478*
Herring, F. G., 599, *614*
Herrman, K., von, 402, *408*, 457, *478*, 610, *615*
Herrmann, J., 607, *615*
Herschkowitz, E., 556, *571*
Hersom, A. C., 525, *539*
Hess, M., 219, *232*
Hester, J. B., 442, 472, *478*
Heston, B. O., 600, *614*
Hewitt, E. J., 81, *103*, 461, *478*, 522, *539*
Hicks, J. R., 362, *369*, 414, *432*
Hieber, W., 599, *615*
Hield, H. Z., 147, *163*, *164*, *167*
Higaki, J. C., 611, *613*
Higby, R. H., 133, 134, *164*
Higuchi, M., 282, *301*
Higuchi, T., 528, *538*, 609, *614*
Hilditch, T. P., 8, *61*
Hildreth, A. C., 662, *683*
Hilgeman, R. H., 144, 151, *164*, *165*
Hill, E., 576, *615*
Hill, E. G., 577, *618*
Hillier, M. A., 668, *685*
Hillis, W. E., 84, *104*
Hillis, W. F., 458, *481*
Hills, C. H., 420, 425, *432*, *434*, 510, 511, 512, 514, *537*, *541*
Hills, H. G., 451, *478*
Hilts, R. W., 267, *278*
Himel, C. M., 329, *331*
Hinder, M., 590, *621*
Hine, G. J., 689, 697, *702*
Hirst, F., 530, *539*
Ho, H., 79, *105*
Hoagland, D. R., 460, *475*
Hoban, N., 425, *434*
Hobson, G. E., 24, 46, 49, 50, 52, 53, 54, 55, *61*, *62*, 222, *230*, 308, *323*, 441, 450, 457, 459, 462, 464, 465, 466, 467, 468, 472, *478*, 587, 588, *615*
Hock, R. J., 655, 666, *685*
Hodge, J. E., 329, *330*, 605, *615*, 628, *649*
Hodgson, R. W., 110, *164*

Hoff, J. E., 452, *480*, 521, *540*
Hoffman, J. C., 211, *230*
Hoffman, M. B., 341, *370*
Hofman, A., 328, 329, *331*
Höfner, W., 590, *618*
Hogg, N. J., 526, *538*
Hoglan, F. A., 504, *506*
Hohl, L. A., 647, *649*
Holden, M., 363, *371*, 714, 716, *718*
Holdsworth, S. D., 655, 674, *683*
Holik, M., 500, *506*, 577, *616*
Hollingshead, R. S., 267, *278*
Holm, R. E., 345, *371*
Holm-Hansen, O., 660, *683*
Honda, S. I., 44, 45, *60*
Hooper, F. C., 610, *615*
Hoover, M. W., 213, *230*
Hope, G. W., 509, 523, 531, *539*
Hopkins, H., 461, *478*
Hörhammer, L., 402, *408*, 599, *615*
Hori, F., 297, *301*
Horio, T., 568, *570*
Horner, G., 531, 532, *537*, *539*, 596, *615*
Horowitz, R. M., 130, 131, 132, *163*, *164*, 500, *506*, 548, 550, *570*, 577, 599, *615*
Horspool, R. P., 119, 161, *168*
Horubala, A., 695, *703*
Hoshino, M., 294, *301*
Hotchner, S. J., 535, *539*
Howard, F. D., 214, *230*, 441, *482*
Howard, G. E., 328, *331*
Howie, D. L., 701, *704*
Hoyt, P. B., 715, *718*
Hsia, C., 418, *433*
Hsia, C. L., 399, *409*, 418, *433*
Hsu, C. P., 512, *539*
Hsu, E. T., 419, *433*, 517, *539*
Huang, C. P., 593, *615*
Huang, H. T., 400, *408*, 711, 712, *718*
Hudson, C. G., 12, *62*
Huecker, D. E., 440, 441, 443, 444, *479*
Huelin, F. E., 150, 152, 160, *164*, *168*, 359, 362, *371*, 527, 528, *539*, 562, 563, 564, 566, 567, *570*, 607, 608, *615*
Huet, R., 611, *615*
Huff, R., 465, *477*
Huffaker, R. C., 157, *164*
Huggart, R. L., 545, 546, 548, 555, 559, 563, *570*, *571*
Hughes, E. B., 380, *408*
Hulland, E. D., 525, *539*
Hulme, A. C., 21, 48, *61*, 69, 70, 86, 87, 89, 90, 91, 94, 97, 98, *103*, *104*, 129, 156, *164*, 239, *254*, 335, 336, 337, 338, 339, 340, 341, 342, 344, 345, 346, 347, 348, 349, 350, 351, 352, 353, 354, 355, 356, 357, 358, 359, 361, 363, 365, 366, 367, *370*, *371*, *372*, *373*, 388, 389, *408*, 462, 466, *480*, 583, 585, *621*, 694, *703*
Hultin, H. O., 80, 81, 94, 100, *103*, 521, *539*
Hume, H. H., 108, 110, *164*, 283, *301*
Hunter, G. L. K., 136, 139, *164*, *165*
Hunter, R. F., 709, *718*
Hunwicke, R. F., 525, *540*
Hurst, H., 469, *478*
Huskins, C. W., 134, *165*, 552, *570*
Hussein, A. A., 152, 160, *165*, 626, *649*
Hyde, R. B., 445, *481*

I

Ice, B. L., 401, *410*
Ichihata, H., 605, *616*
Idemori, M., 716, *717*
Igolen, G., 138, *165*
Iguina de George, A. M., 245, *254*
Iikubo, S., 284, 294, *301*
Ikeda, 138
Ikeda, R. M., 136, 137, *164*, *167*
Ikegami, T., 281, *301*
Ikeuchi, K., 710, *719*
Imparato-Gargano, E., 455, *478*
Inagaki, C., 565, *570*
Inatomi, H., 220, *230*
Ingelsbe, D. W., 672, 674, *685*
Ingles, D. L., 628, 637, *649*
Inglett, G., 500, *506*, 577, *616*
Inglett, G. E., 494, 504, *506*
Ingraham, L. L., 678, *683*
Ingram, M., 576, *615*
International Oil Council, 260, *278*
Inukai, F., 220, *230*
Iranzo, J. R., 141, 142, *169*
Irving, G. W., 453, *478*
Irving, H., 602, *614*
Irving, H. M. N. H., 598, *615*
Irving, L., 655, 666, *685*

Isaac, W. E., 563, *570*
Isawa, I. T., 220, *230*
Isherwood, F. A., 88, *103*, 390, *408*
Issenberg, P., 76, 80, 81, 84, *103*, *104*, *105*
Ito, N., 222, *230*
Ito, S., 219, *230*, 287, 289, 295, 299, *301*, 547, 560, 565, *570*, *571*
Ivanoff, S. S., 213, *230*
Iverson, J. L., 275, 276, *278*
Iwahori, S., 177, *204*
Iwamoto, Y., 568, *570*
Iwanov, N. N., 217, *230*
Iwata, M., 299, *301*
Izquierdo Tamayo, A., 261, 262, 265, 266, 270, 271, 272, *277*
Izumi, Y., 547, 560, *570*

J

Jabalpurwala, K. E., 599, *615*
Jackson, D. I., 413, 415, 416, *433*
Jackson, G. A. D., 607, *615*
Jackson, R., 527, *540*
Jacob, F. C., 416, 424, *433*, *434*, *435*
Jacob, H. E., 626, *649*
Jacob, M., 594, *615*
Jacquet, J., 578, *615*
Jacquin, P., 451, *478*, 588, *615*
Jaffe, M. E., 6, 8, 14, *61*
Jain, N. L., 239, 240, 242, 244, 245, 246, 247, 250, *254*
James, W. E., 393, *408*
Jameson, R. F., 598, *615*
Jang, R., 123, 161, *162*, *165*, 383, 390, *409*
Janković, B. R., 691, *703*
Jansen, E. F., 23, 39, *61*, *62*, 123, 159, 160, 161, *165*, 678, *683*
Jaquin, P., 418, *433*
Jeger, O., 133, *161*
Jelenic, D., 462, *478*
Jellinok, F., 602, *619*
Jenkins, J. E. E., 472, *478*
Jennings, W. G., 329, *330*, 426, 427, 428, 429, *432*, 433, *435*, 512, *539*, 577, 594, *615*
Johnson, A. A., 633, 635, 636, *650*
Johnson, B. E., 78, *103*
Johnson, C. F., 692, 694, 695, *704*
Johnson, D. K., 605, *615*
Johnson, G., 605, *615*
Johnson, H., 156, *164*
Johnson, H. B., 223, *232*, 406, *407*
Johnson, J. H., 451, 452, *478*
Johnson, K. W., 219, *232*
Johnson, R. M., 245, 247, *254*
Johnson, R. M., Jr., 451, 452, *478*
Johnson, S. A., 16, *62*
Johnson, T., 637, *649*
Joliffe, V. H., 126, *167*
Jones, D. E., 85, *101*
Jones, H. A., 209, *230*
Jones, I. D., 417, *435*
Jones, J. D., 338, 344, 345, 346, 347, 349, 353, 359, *371*, *372*, *373*, 462, 466, *480*
Jones, J. K. N., 12, *61*
Jones, M. M., 604, *613*
Jones, W. W., 141, 143, 144, 145, *163*, *165*
Jorgenson, E. C., 533, *538*
Joshi, M. C., 314, *323*
Joslyn, M. A., 12, 13, 14, *61*, 120, *168*, 289, *301*, 365, *371*, 419, *433*, 508, 513, *539*, 545, 546, 547, 548, 557, 559, 561, 565, *570*, *572*, 574, 577, 580, 593, 596, 605, 607, *615*, *616*, *620*, 625, 626, 636, *649*, *651*, 655, 666, 672, 673, 674, 675, 679, *683*, *684*, 708, *718*
Josysch, D., 388, *408*
Joubert, F. J., 221, 222, *229*
Jouret, C., 424, *433*
Joux, J. L., 519, *539*
Joyce, A. E., 709, *718*
Jucker, E., 117, *165*
Jung, G. A., 681, *684*
Jungalwala, F. B., 246, *254*
Juniper, B. E., 639, *649*
Jurd, I., 712, 713, *718*
Jurd, L., 599, 601, *612*, *615*, *616*, 629, *649*, 674, *684*
Jurics, E. W., 219, 220, 221, *230*, 458, *478*
Juritz, C. J., 145, *165*
Jury, E., 605, *614*

K

Kaeding, W. W., 604, *616*
Kahan, R. S., 696, 697, *704*
Kaindl, K., 700, *703*, *705*
Kajanne, P., 401, *408*
Kajderowicz-Jarosinska, D., 456, *478*
Kajiura, M., 288, *301*
Kakiuchi, N., 287, *301*
Kakizaki, K., 568, *570*
Kalb, A. J., 419, *435*
Kalina, M., 665, *683*
Kalogueria, S. A., 255, *278*
Kamali, A. R., 417, *433*
Kamm, G. G., 535, *539*, 568, *570*
Kampelmacher, E. H., *704*
Kanapaux, M. S., 442, *480*
Kanujoso, B. W. T., 519, *539*
Kaplan, N. O., 668, 669, *683*
Kapoor, K. L., 251, *253*
Kapp, P. P., 459, *478*
Kapur, N. A., 709, *717*
Kapur, N. S., 243, 250, 251, *254*
Kar, B. K., 248, *253*
Karel, M., 710, *718*
Kargl, T. E., 456, *478*
Karlson, P., 181, *204*
Karow, A. M., Jr., 663, 665, *684*
Karrer, P., 117, *165*
Kaski, I. J., 442, *478*
Kasmire, R. F., 208, 211, 212, 213, 217, 224, *229*, *230*
Katakura, K., 262, 266, *279*
Katayama, O., 451, 452, *478*
Kato, H., 605, 606, *616*
Katz, J. J., 715, *718*
Kawatovari, T., 272, *278*
Kay, G., 589, *616*
Kaya, T., 220, *232*
Kefford, J. F., 133, 134, 151, *162*, *165*, 518, *538*, 547, 550, 552, 560, 563, 567, *569*, *570*, 577, 585, *613*, *616*
Keilin, D., 607, *616*
Keller, G. J., 511, *539*, 553, *571*
Kelley, J. N., 74, *103*
Kelly, R. T., 668, 669, *683*
Kelly, S. H., 368, *371*, 390, *408*, 673, *678*, *683*, *684*
Kemmerer, A. R., 218, *232*
Kendall, A., 386, *408*
Kennard, W. C., 236, *254*
Kenny, I. J., 210, *230*
Kent, M. J., 308, 313, 316, *323*
Kenworthy, A. L., 416, 417, *432*, *433*, 678, *683*, *435*
Kenyon, E. M., 554, *570*
Kepner, R. E., 193, *205*
Keränen, A. J. A., 400, 401, *409*
Kern, A., 594, 596, 605, *616*
Kerns, K. R., 305, *323*
Kertesz, Z. I., 365, *371*, 383, 404, *408*, *409*, 429, *433*, 512, 536, *539*, 690, 691, 699, *702*, *703*, *704*
Kesterson, J. W., 113, 130, 131, 134, 135, 136, *164*, *165*
Keulen, H. A., van, 440, *476*
Kevei, E., 426, *435*, 607, 608, 609, *619*, *620*
Kew, T. J., 544, 557, 558, *570*, *571*, 587, *619*
Khalifah, R. A., 76, *103*, 151, 156, *165*
Khan, M., 709, *718*
Khan, M. M. T., 604, 608, *616*
Khan, M. U. D., 117, *165*
Khan, S., 563, *571*
Khan, S. A., 563, *571*
Khavalin, N. N., 654, *684*
Kidd, F., 28, *61*, 152, *165*, 344, 345, 348, 350, 352, *371*, *372*, 424, *433*
Kidson, E. B., 461, 468, 471, 472, *478*
Kieser, M. E., 383, 403, 404, 405, *408*, 442, *480*, 581, 584, 588, 589, 592, 594, 602, 604, 605, 606, 607, *612*, *616*, *618*
Kikuta, Y., 9, 10, 16, 19, 20, 21, 24, *61*
Kim, M. K., 602, *616*
Kimble, K. A., 212, 217, *229*
Kimbrough, W. D., 381, *408*
Kimmell, L., 630, *650*
Kimura, K., 563, *571*
Kimura, S., 710, *718*
King, G. S., 139, *165*
King, J. R., 263, *279*
Kiovsky, T. E., 667, 668, *684*
Kirch, E. R., 442, *478*
Kirchner, J. G., 326, *330*, 552, 553, *571*
Kirkpatrick, J. D., 141, *163*
Kirshner, N., 91, *104*
Kiss, I., 700, *703*
Kitagawa, H., 299, *301*
Kitahara, M., 294, *301*
Kitson, J. A., 554, *569*, 584, *619*

Kläui, H., 607, *616*
Kleesat, R., 591, 605, *616*
Klein, A. O., 716, *718*
Klein, E., 426, *433*
Klein, I., 94, *102*, 347, 349, 350, *370*, 471, *477*
Kliewer, W. M., 179, 182, *204*
Klotz, I. M., 603, *616*
Klotz, L. J., 111, 147, *167*, *168*
Klouwen, H. M., 693, *705*
Knapp, F. W., 27, *61*, 585, *616*
Knee, M., 366, *372*
Knight, D. J., 522, 523, 534, *538*
Knight, L. D. M., 386, *408*
Knock, G. G., 535, *541*
Knorr, L. C., 127, 160, *167*
Knowles, R. E., 710, *718*
Kobayahsi, A., 80, *105*
Koch, B., 444, *479*
Koch, J., 576, 580, 587, 588, 591, 605, *614*, *616*
Koda, R., 400, *407*
Kodera, M., 210, *231*
Koenders, E. B., 329, *331*
Koeppen, B. H., 367, *372*
Kohler, G. O., 710, *718*
Kohman, E. F., 514, 532, 534, *539*
Kolankiewicz, J., 456, *479*
Komatsu, S., 138, *165*
Komazawa, T., 299, *301*
Konigsbacher, K. S., 81, *103*
Koo, R. C. J., 119, 141, *165*
Körmendy, I., 579, *616*
Kosugi, Y., 608, *618*
Kott, V., 424, *433*
Kovacs, J., 591, 594, *618*, 700, *703*
Kovaks-Proszt, G., 588, *620*
Krakenberger, R. M., 709, *718*
Kramer, M., 217, 219, *230*
Kramer, P. J., 314, *323*
Krasavtsev, O. A., 654, 661, 663, *684*, *685*
Krasnovskii, A. A., 670, 676, *682*
Krauss, B., 309, 310, 311, *323*
Krauss, B. H., 438, *477*
Krbechek, L., 500, *506*, 577, *616*
Krehl, W. A., 567, *571*
Kreulevitch, I. H., 524, *541*
Krezdorn, A. H., 147, *165*
Krickl, J. L., 568, *570*
Krishnamurthy, G. V., 240, 242, 244, *254*
Kristiansen, H., 528, *538*, 609, *614*
Krowczynski, L., 528, *538*, 609, *614*
Ku, H., 93, 94, 101, *103*
Ku, H. S., 462, 470, 474, *479*
Ku, L. L., 421, 430, 431, *433*, *435*, 693, 694, *704*
Kubo, R., 138, *165*
Kudrjawzewa, M. A., 217, *230*
Kuiper, P. J. C., 677, *684*
Kun, E., 53, *62*
Kunishi, A., 344, *372*
Kuprianoff, J., 641, *650*
Kurata, T., 606, 609, *616*
Kurihara, K., 486, *506*
Kushman, L. J., 381, *408*
Kuusi, T., 404, *408*, 607, 610, *616*
Kuykendall, R., 151, 156, *165*
Kyzlink, V., 514, 530, *539*

L

Labanauskas, C. K., 141, 147, *165*
Labat, le R. Pere, 485, *506*
Labee, M. D., 132, *168*, 550, *572*
La Belle, R. L., 424, *433*
Laferrierel, L., 709, *719*
Lafon-Lafourcade, S., 186, 187, *204*
La Forge, F. B., 12, *62*
Lafourcade, S., 194, *205*
Lamb, F. C., 526, 528, *538*, *539*, 561, *570*
Lambeth, V. N., 440, 441, 443, 444, *479*
Lamport, C., 400, *408*
Lance, C., 46, 49, 50, 52, 53, 54, 55, *61*, *62*, 348, *372*
Land, D. G., 325, *330*, 362, *371*, 709, *717*, *718*
Langer, T., 388, 399, *410*
Larson, R., 215, *229*
Laubscher, A. N., 531, *538*, 567, *570*
Laurencot, H. J., Jr., 129, *168*
Lavee, S., 415, *433*
Lavin, M., 691, *703*
Lavollay, J., 578, 596, *616*
Law, M., 532, *539*
Lawrence, J. M., 161, *162*, 385, *409*
Lawson, R. E., 211, *229*
Lazar, M. E., 527, *540*, 632, *650*

Le Breton, J., 578, *615*
Lecrenier, A., 424, *435*
Lee, C. Y., 657, *650*
Lee, F. A., 445, *479*, 714, *718*, *719*
Lee, T. H., 218, *232*, 692, 695, *703*
Leef, J., 664, 665, 682, *684*
Leeper, P. W., 210, *230*
Leeper, R. W., 329, *331*
Lefèbvre, A., 185, 195, *204*, *205*, 597, *616*
Lehmann, J., 596, *621*
Lehninger, A. A., 50, *62*
Lehninger, A. L., 347, 348, *372*
Lehr, H. B., 663, 664, *685*
Leinbach, L. R., 376, 382, 388, 394, 404, *409*, 511, 536, *539*
Leland, H. V., 121, 123, *162*
Leley, V. K., 237, 244, *254*
Léon, H., 191, *205*
Leonard, E. R., 28, *63*, 78, *103*, 223, *232*, 235, 237, 239, *254*
Leonard, S., 512, *539*
Leonard, S. J., 419, *434*, 441, 443, 444, 449, *480*, *481*, *482*, 516, 519, 536, *539*, *540*, 713, *718*
Leoncini, G., 264, *279*
Leopold, A. C., 146, 147, *165*, 471, 473, *477*
Letan, A., 600, *617*
Letham, D. S., 413, 415, *433*
Leuprecht, H., 588, *617*
Levandoski, N. G., 609, *617*
Levi, A., 556, *571*, 700, *703*
Levin, R. E., 272, *279*
Levine, A. S., 80, *103*
Levine, M., 610, *619*
Levine, W. G., 16, *62*
Levitt, J., 654, 660, 666, 669, *684*
Lewis, A. H., 460, *479*
Lewis, L. N., 147, *163*, *165*
Lewis, T. L., 338, 349, *372*
Lewis, W. M., 576, *617*
Ley, F. J., *703*
L'Heureux, G. A., 598, *617*
Li, K. C., 418, *433*, 521, *539*
Li, M. C., 217, 227, 228, *230*
Liang, H. K., 210, 217, 221, 222, 223, 227, 228, *230*
Lieberman, M., 344, 345, *372*, 469, 470, 471, *479*
Lillehei, R. C., 667, *684*
Lilleland, O., 413, 414, *433*
Lim, L., 424, 426, *433*
Lim, M. C., 602, *613*
Lime, B. J., 118, 130, 131, 132, *165*, 548, 550, *570*
Lincoln, R., 563, *571*
Lincoln, R. E., 117, *166*
Lindner, K., 217, 219, *230*
Lindner, L., 591, *614*
Lineweaver, H., 159, 160, *165*
Ling, L. C., 451, *476*
Lipe, W. N., 415, *433*
Lips, S. H., 41, *62*, 354, *372*
Littari, F. S., 152, *161*
Livingston, A. L., 710, *718*
Livingston, G. E., 514, *539*, 606, *617*, 713, *718*
Lockett, M. F., 578, *617*
Loconti, J. D., 512, *539*
Lodh, S. B., 319, 322, *323*
Loesecke, H. W., von, 66, 69, 75, 76, 77, 78, 79, 80, 86, 88, 89, 93, 97, *103*, *104*, 545, 553, 560, *571*, *512*, 625, *652*
Loewus, F. A., 182, 185, *205*, 390, *408*, *409*
Lombard, J. H., 519, *540*
Long, J. D., 625, 635, *650*
Long, M. I. E., 447, 461, 467, *482*
Longerbeam, J. K., 667, *684*
Loomis, W. D., 716, *718*
Looney, N. E., 80, *103*, 342, 349, 350, 363, *369*, *372*, 716, *718*
Lopez, A., 531, *539*
Losekoot, J. A., 329, *331*
Lovelock, J. E., 666, *684*
Lovenberg, W., 91, 92, *104*, 123, *168*, 447, *481*
Lovett-Janison, P. L., 141, *165*
Lowe, E., 593, *617*
Lower, R. L., 444, 445, *479*, *481*
Lozano, J. A., 430, *435*
Lü, C. S., 217, 227, 228, *230*
Luckwill, L. C., 338, 342, *372*, 437, 456, *479*, 581, *617*
Ludin, A., 556, *571*
Ludin, R. E., 136, *167*
Lueck, H., 607, *617*
Luh, B. G., 399, *409*

Luh, B. S., 16, 20, 24, 30, *61*, 379, 380, 383, 386, 388, 389, 391, 402, 405, *409*, 418, 419, 430, *432*, *433*, *434*, 441, 443, 444, 449, 457, 458, *480*, *481*, *482*, 508, 516, 517, 519, 536, *539*, *540*, 713, *718*
Lukton, A., 400, *409*, 514, *539*, 605, *616*, 696, *703*, 711, 712, *718*
Lulla, B. S., 86, *103*
Lundin, R., 713, *718*
Lusena, C. V., 666, 671, *684*
Lusk, G., 710, *718*
Lüthi, H., 576, 585, *615*, *617*
Lüthi, H. R., 575, 578, 579, 594, *617*
Luthra, J. C., 196, *204*
Lutz, J. M., 153, *164*, 208, *230*, 656, *684*
Lu Valle, J. E., 562, *572*, 608, *621*
Luyet, B. J., 656, 657, 658, 663, 665, 676, *683*, *684*, *685*
Lynch, S. J., 245, *254*
Lyon, C. B., 443, *478*
Lyon, J. L., 148, *161*
Lyons, J. M., 223, 224, 227, *230*, 466, 470, *479*
Lysenko, M. K., 462, *476*

M

Mabry, T. J., 599, *617*
Macara, T., 378, 380, *409*
MacDonald, I. R., 347
MacDonnell, L. R., 159, 160, 161, *165*
MacDowell, L. G., 563, *571*
MacEwan, J. L., 566, *570*
MacGillivray, J. H., 208, 214, 215, 216, *229*, *230*, *231*, *232*
MacKay, A. M., 81, *103*
MacKinney, G., 117, *165*, 292, 293, *301*, 392, 400, *408*, *409*, 418, *433*, 456, *482*, 514, *539*, 545, *571*, 625, 627, 642, 643, *650*, *651*, *652*, 696, *703*, 709, 710, 711, 712, *718*, *719*
Maclay, W. D., 123, *165*
MacLeod, W. D., Jr., 136, *165*
MacLinn, W. A., 441, 442, *479*
Macmillan, J. D., 710, *718*
MacQueen, K. F., 76, *102*
McArdle, F. J., 417, *432*
McBean, D. McG. 633, 635, 636, 639, 646, *648*, *650*
McCance, R. A., 92, *103*, 126, 143, *165*, *169*, 380, 386, 391, 392, 393, 394, 395, *409*, 529, 530, *539*, 567, *571*, 578, 595, *617*, 641, *650*
McCarthy, A. I., 76, 81, 82, 83, *103*, *104*, *105*
McCay, C. M., 563, *571*
McClendon, J. H., 443, 450, 459, *479*, *482*
McClung, A. C., 417, *435*
McColloch, L. P., 156, *162*, 406, *407*
McColloch, R. J., 511, 536, *539*
McCollum, J. P., 440, 442, 455, 456, 462, 464, 466, *477*, *478*, *479*
McComb, E., 312, 313, *323*
McComb, E. A., 23, *62*
McCornack, A. A., 155, *165*
McCown, B. H., 668, *684*
McCready, R., 312, 313, *323*
McCready, R. M., 23, *62*, 123, *165*, 383, *409*
McDaniel's, L. H., 640, *649*
McDaniels, L. W., 335, *372*
McElroy, W. D., 460, 461, *480*
McFadden, W. H., 136, *167*, 361, *370*, 426, *435*
McFeeters, R. F., 716, 717, *718*
McGlasson, W. B., 32, 56, *62*, 76, 77, 92, 97, 101, *104*, 208, 209, 210, 211, 212, 220, 223, 224, 227, 228, *230*, *231*, *232*, 692, 695, *703*
McHenry, E. W., 442, *479*
McKee, H. S., 340, *372*
McKenzie, H. A., 563, 567, *570*
McKirahan, R. D., 531, *538*
McLean, J., 665, *683*
McLemore, T. A., 602, *612*
McMeans, J. L., 148, *161*
Madner, A., 591, 594, *618*
Maeda, S., 298, *301*
Maezawa, T., 468, *480*
Magalhäes, Neto, B., 222, *229*
Magar, N. G., 527, *538*
Maggiora, L., 193, *203*
Magnari, N., 561, *571*
Mahadeviah, M., 516, *539*
Maharg, L., 441, 442, *480*
Mahdi, A. A., 443, *479*, 521, *539*
Mahdi, M., 266, 267, *279*
Mahecha, G., 262, *279*

Maheshwari, S. C., 212, *230*
Maier, V. P., 131, 133, *165*, 550, 551, *571*, 577, *617*, 645, 646, *650*
Makower, B., 508, *540*
Makower, R. U., 27, *62*, 510, *539*
Malmström, B., 598, *612*
Manax, W. G., 667, *684*
Mancha Perello, M., 267, *279*
Manchester, T. C., 161, *165*
Manek, P. V., 90, *103*
Manero, J., 700, *702*
Manion, J. T., 69, 79, *104*
Mann, L. K., 210, 224, *231*
Mann, W., 383, *409*
Mannheim, H. C., 554, 555, 556, 568, *569*, *570*, *571*, 582, *613*
Manning, W. M., 715, *719*
Manz, U., 607, *616*
Mapson, L. W., 74, 75, *103*, 344, 345, 366, *372*, 390, *408*, 441, 469, 470, 471, 474, *479*, 607, *617*, 694, *703*
Maquenne, M., 11, *62*
March, J. F., 345, *372*
Margileth, D. A., 551, *571*
Markakis, P., 400, 402, 403, *408*, 418, 420, 430, *432*, *434*, 514, *539*, 584, 585, *614*, *617*, *618*, 627, *649*, *650*, 691, 692, *702*, *703*, 712, 713, *718*
Markham, K. R., 599, *617*
Markkanen, R., 598, *617*
Markley, K. S., 139, *165*, 359, *372*
Markov, V. M., 213, *231*
Marks, R., 95, *102*
Marloth, R. H., 114, *165*
Marmoy, F. B., 460, *479*
Marsh, G. L., 145, *165*, 419, *432*, 545, 564, *570*, 625, *651*, 714, *719*
Marshall, J. R., 679, *682*
Marshall, P. B., 91, *103*
Marshall, R. E., 417, 418, 424, *432*, *433*, *434*
Marsteller, R. L., 322, *323*
Marteau, G., 588, *617*
Martell, A. E., 596, 597, 598, 602, 604, 608, *616*, *617*, *618*
Marth, P. C., 76, *103*
Martin, D., 338, 349, *372*
Martin, G. C., 338, 342, *372*
Martin, W. E., 151, *165*
Martinez Suarez, J. M., 276, *279*
Mason, H. S., 368, *372*, 605, *617*
Mason, J. L., 420, *434*
Massey, D. M., 440, 443, 445, 446, 458, 462, 467, *479*, *482*
Massey, L. M., Jr., 358, *370*, 690, 691, 692, 694, 699, *703*, *704*, *705*
Massey, V., 669, *683*
Massicot, G., 241, *253*
Masters, J. E., 552, 554, *569*
Masuda, T., 210, 211, *231*
Masui, M., 223, *231*
Matalack, M. B., 118, 134, *165*, *166*
Mathur, P. B., 251, *253*, 473, *479*, 698, *704*
Matienko, B. T., 210, 219, *231*
Matshenko, A., 241, *253*
Matsuda, Z., 138, *165*
Matsui, M., 294, *301*
Matsui, O., 565, *571*
Matsuoka, C., 299, *301*
Matsushita, A., 294, *301*
Matsushita, K., 716, *717*
Matthews, R. F., 451, *479*
Mattick, L. R., 521, *539*
Mattoo, A. K., 238, 248, 249, 252, *254*
Matui, H., 222, *231*
Matveeva, N. M., 654, 664, 665, 666, 676, *685*
Matz, S. A., 518, *539*, 671, *684*
Matzik, B., 388, 389, 399, 400, 401, *409*
Maugenet, J., 424, *433*
Maunsell, J., 380, *408*
Maurié, A., 182, 186, *205*
Maurya, V. N., 251, *254*
Maxie, E. C., 76, 80, *102*, *103*, 264, 273, *278*, *279*, 350, 358, *372*, 415, 417, 421, 424, 425, *432*, *433*, *434*, 438, 473, *475*, *479*, 639, *650*, 688, 692, 693, 694, 695, 696, 697, 698, 699, *702*, *704*, *705*, 714, *718*
Maximow, N. A., 664, 665, *684*
Maxwell, W. A., 710, *718*
May, D., 224, *230*
May, D. M., 212, 217, *229*
May, G. E., 212, 217, *229*
May, J. F., 494, *506*
May, S., 66, *103*
Mayer, A. M., 367, *371*, *372*, 584, 585, *614*, *617*, 711, *717*
Mayer, F., 401, *409*

Maynard, L. A., 440, 441, 447, *478*
Mazelis, M., 222, *231*
Mazliak, P., 8, 9, 10, 11, 16, 20, *62*, 88, *102*, 359, *372*
Mazur, P., 656, 658, 659, 660, 665, 666, 672, *684*
Mazur, V., *168*
Mazur, Y., 139, *166*, *168*
Mead, A., 366, *372*
Medina, M. G., 249, *254*
Meeuse, A. D. J., 221, *231*
Meheriuk, M., 470, *479*
Mehlitz, A., 388, 389, 399, 400, 401, *409*
Mehrlich, F. P., 582, *617*
Mehrotra, R. A., 598, *614*
Meigh, D. F., 344, 359, *372*, 470, *479*, 639, *650*
Meinert, U., 212, 217, *229*
Meisels, A., 139, *168*
Melnick, P. J., 666, 669, *684*
Mel'nik, A. V., 217, 221, *231*
Menashi, J., 599, *614*
Mendel, J. E., 597, *619*
Menoret, Y., 592, *617*
Mentzer, C., 247, *253*
Mercer, F. V., 94, *101*, 363, *369*
Mercer, W. A., 524, *541*
Mercier, D., 247, *253*
Meredith, F. I., 455, *479*
Merrill, A. L., 143, *168*, 391, 392, 393, 394, 395, *410*
Merson, R. L., 593, *617*
Merten, U., 593, *621*
Meryman, H. T., 658, 663, 665, 666, *684*, *685*
Messing, J. H. L., 461, *479*, *482*
Metcalf, J. F., 146, 159, *166*, *168*
Metlitsky, L. V., 693, *704*
Metzler, D. M., 131, *165*, 646, *650*
Meynhardt, J. T., 350, 358, *372*
Micke, W. C., 421, *434*
Miers, J. C., 452, *477*, *479*, 511, 521, *540*, *541*, 587, *621*
Milburn, R. N., 599, *615*
Miller, C. L., 86, *103*
Miller, E. V., 114, 115, 116, 117, 118, 143, 150, 151, 153, 154, 155, 156, *166* 316, 322, *323*, 406, *407*, 442, *476*
Miller, I., 141, *166*
Miller, J. M., 123, *166*, 552, 553, *571*
Miller, L. A., 693, *704*
Miller, L. P., 36, *61*, 224, *229*
Miller, M. W., 633, 634, 635, 646, *649*, *650*, *651*
Miller, R. H., 656, *684*
Miller, R. L., *166*
Millerd, A., 31, 55, *62*, 345, *371*
Mills, H. H., 417, *434*
Milner, H. W., 15, *62*
Minges, P. A., 467, 471, 472, *479*, *480*
Mineo, M., de, 263, *279*
Misch, M. J., 670, *683*
Misselhorn, K., 418, *433*
Mitchell, A. E., 417, 422, 424, *433*, *435*
Mitchell, F. G., 416, 421, 424, *434*, *435*, 639, *650*
Mitchell, H. L., 677, *685*
Mitchell, J. H., 514, 526, *540*
Mitchell, J. W., 76, *103*
Mitsui, S., 601, *617*, *619*, *620*
Miura, A., *683*
Miyabayashi, T., 289, *301*
Mizelle, J. W., 130, 131, 132, *166*
Mizrahi, S., 544, 557, 558, 560, *571*
Mlodecki, H., 456, *479*
Moats, W. A., 495, *506*
Mobley, E. O., 440, 443, 444, 450, *481*
Modi, V. V., 238, 246, 248, 249, 252, *254*
Mohammadzadeh-Khayat, A. A., 519, *540*
Molinari, O. Ch., 255, 260, *279*
Mollard, J., 444, *479*
Mon, T. R., 592, *619*
Monastra, F., 426, 427, *433*
Money, R. W., 127, *166*, 380, 382, 386, 388, *409*
Monovoisin, J., 531, *538*
Monroe, K. H., 526, 528, *538*
Monselise, S. P., 114, 118, 152, *161*, *163*, *166*, 696, 697, *704*
Monties, B., 584, 605, *617*
Monzini, A., 426, *432*
Moon, H. H., 378, 381, 386, *408*, 531, *538*
Moor, K., 406, *407*
Moore, E. L., 127, 139, 159, 160, *161*, *167*, *169*, 545, 546, 555, 559, 561, 563, *569*, *570*, *571*, *572*
Moore, J. G., 6, 7, 13, 14, *61*

Moore, M. D., 424, *434*
Moore, S., 308, *323*
Moore, T. C., 342, *372*
Moreau, L., 179, *204*
Morgan, A. F., 6, 7, 13, 14, *61*, 545, *570*, 629, 630, *650*
Morgan, A. I., 554, *569*, 593, *617*
Morgan, B. H., 691, *703*
Morgan, R. C., 219, *231*
Mori, D., 560, *571*
Mori, H., 283, *301*
Mori, M., 568, *570*
Mori, R., 327, 328, *331*
Mori, T., 219, *231*
Mori, Y., 678, *684*
Morimoto, T., 608, *618*
Morris, L. L., 224, *231*, 473, *475*, 692, 697, *702*
Morris, T. N., 532, 533, 534, *540*, 625, *650*, 674, *684*
Morse, F. W., 533, *540*
Morton, A. B., 605, 609, *615*
Morton, A. D., 401, *409*, 544, 546, *569*, 606, 607, 608, 609, 610, *613*
Morton, R. K., 79, *103*
Moser, F., 442, *476*
Moshonas, M.G., 137, *164*, *165*, 553, *569*
Mosse, Barbara, 334, *372*
Mosse, J. K., 558, *572*
Mossel, D. A., *704*
Motawi, K., El-Din H., 429, *434*
Moyer, J. C., 521, *539*, 594, *617*
Moyls, A., 584, *617*
Moyls, A. W., 514, 524, *537*, *540*
Mozingo, A. K., 263, *278*
Mrak, E. M., 625, 626, 627, 632, 635, 642, 643, 646, *649*, *650*, *651*, *652*
Mudd, J. B., 26, *62*
Mukerjee, P. K., 236, 238, 239, *254*
Muller, C. H., 437, *479*
Muller, C. J., 692, 694, 695, *704*
Muller, J. G., 593, *617*
Muller, K. H., 402, *408*
Muniz, A., 13, *60*
Murakami, Y., 598, *617*
Murakishi, H. H., 472, *479*
Muraoka, N., 219, *231*
Murata, N., 547, 560, *570*
Murata, T., 93, 94, 101, *103*
Murdock, D. I., 576, *617*
Murneek, A. E., 441, 442, *480*
Murphy, E., 13, *62*
Murray, E., 76, 81, *104*
Murray, K. E., 362, *371*
Mustard, M. J., 245, *254*
Myee Stephens, I., 562, *570*
Myers, M., 76, 81, *104*
Myers, M. J., 84, *103*

N

Nagaraja, K. V., 516, *540*
Nageswara Rao, G., 555, *572*
Nagy, S., 552, *571*
Naik, K. C., 234, *253*
Nakabayashi, T., 292, *301*
Nakamura, K., 598, *617*
Nakamura, M., 298, *301*
Nakamura, R., 294, 299, *301*
Nakayama, M., 456, *481*
Nakayama, T. O. M., 289, *301*, 363, *369*, 453, *476*, 577, *617*, 714, 715, *717*
Namikawa, J., 282, *301*
Nancollas, G. H., 602, *613*
Narasaki, T., 262, 266, *279*
Narayana, N., 237, 244, *254*
Narsimhau, P., 243, 250, *254*
Näsänen, R., 598, *617*
Nason, A., 460, 461, *480*
Natarajan, C. P., 545, *571*, 626, *649*
National Academy of Sciences, 490, *506*
Natl Frozen Fd Assn., 653, *684*
Naves, Y. R., 138, *166*
Neal, G. E., 349, 352, 353, 355, *372*, 383, *409*, 493, *506*, 581, *617*
Nedbalek, M., 588, *620*
Neher, E., 598, *617*
Neillie, W. F. S., 598, *615*
Nelson, E. C., 329, *330*
Nelson, E. K., 132, 139, *165*, *166*, 426, *434*
Nelson, E. N., 388, *409*
Nelson, J., 518, *538*
Nelson, J. M., 141, *165*
Nelson, J. W., 710, *718*
Nelson, P. E., 452, *480*, 521, *540*
Nelson, R. F., 368, *372*, 585, *614*, *618*, 673, *684*
Nery, R., 598, *612*

Nether, K., 599, 601, *612*
Nettles, V. F., 455, *480*
Neubert, A. M., 334, 335, *373*, 583, *619*
Neubert, P., 441, 449, *480*
Neukom, H., 584, *620*
Neukom, J., 365, *372*
Neumann, J., 578, *616*
Nevin, C. S., 452, *475*
Newhall, W. F., 123, 154, *164*, *167*, 553, 554, *572*, 582, *620*
Newman, B., 141, *166*
Newman, P. J., 599, *614*
Niavis, C. A., 88, *103*
Nicholas, R. C., 691, 692, *702*, *703*
Nichols, P. F., 264, *279*, 625, 626, 630, 637, 640, 647, *650*, *651*
Nicolea, H. G., 255, 260, *279*
Niederl, J. B., 74, *103*
Nielsen, B. W., 511, 536, *539*
Nightingale, G. T., 413, 417, *431*
Nimmo, C. C., 420, *433*, 678, *683*
Nip, W. K., 213, *231*
Niretina, N. V., 462, *476*
Nishida, T., 284, *301*
Nitsch, J. P., 414, *434*
Noda, L., 16, *60*
Noel, G. L., 607, *618*
Nolte, A. J., 553, *571*
Nomura, D., 138, *166*
Nord, H., 603, *618*
Nordby, H. E., 139, *163*, 552, *571*
Norman, G. D., 715, *718*
Norman, R. O. C., 604, *618*
Norman, S., 154, *166*
Norris, K. H., 424, *434*, 470, *479*
Nortje, B. K., 367, *372*, 516, *540*
Nosti Vega, M., 276, *278*
Nott, P. E., 247, *254*
Notter, G. K., 508, *540*
Nuccorini, R., 264, *279*
Nursten, H. E., 76, 81, *104*, 325, *331*
Nury, F. S., 625, 634, 639, 643, 646, 647, *648*, *650*, *651*, 673, *682*
Nuzaki, M., 673, 681, *683*
Nybom, N., 399, 400, *409*
Nyiri, L., 589, *618*

O

Obata, Y., 584, *619*, 712, *719*
Oberbacher, M. F., 146, 149, 156, 158, 160, *164*, *166*, *168*, 561, *572*, 607, *620*
Ochoa, S., 157, *166*
O'Connell, P. B. H., 75, 77, 80, 89, 90, 94, 95, 96, 97, 98, 99, *102*
Oda, K., 568, *570*
Ogata, K., 93, 101, *103*, 299, *301*
Ogata, Y., 608, *618*
Ogle, W. L., 213, *231*
Ognjanowa, A., 442, *476*
Oka, Y., 598, *618*
Okada, S., 560, *571*
Okamoto, T., 353, *372*
Okimoto, M. C., 304, *323*
Oku, M., 560, *571*
Okubo, M., 468, *480*
Okura, K., 678, *684*
Olien, C. R., 665, 681, *684*
Oliver, J. C., 696, *705*
Oliveri, C., 588, *617*
Olliver, M., 140, *166*, 379, 391, *409*, 418, *434*, 441, *480*, 487, 495, *506*, 525, 527, *540*
Olmstead, A. J., 36, 37, *60*, 74, 75, *102*, 152, *162*
Olsen, R. W., 559, *571*, 577, *618*
Olson, L. E., 427, *432*, 441, 442, 444, 447, 448, 449, 450, 451, 452, 453, 455, *476*, *482*
Olson, R. L., 655, *685*
O'Neal, R., 629, *651*
Onishi, I., 608, *618*
Ono, M., 560, *571*
Ono, Y., 138, *165*
Onslow, M. W., 429, *434*
Ord, M. G., 693, *704*
Organ, J. G., 442, *482*
Orr, M. L., 220, *231*
Ortega Nieto, J. M., 260, *279*
Osadca, M., 565, *569*
Osborne, D., 416, *434*
Oshima, Y., 289, *301*
Osman, A. E., 590, *618*
Ota, S., 308, *323*
Otsuka, K., 656, 657, 658, 659, 662, 663, 664, 666, 675, *685*
Ott, J., 579, *618*
Oury, B., 303, 322, *324*

Overath, P., 26, *62*
Oyler, J. R., 512, *537*
Ozawa, S., 138, *165*
Ozbun, J. L., 467, *480*

P

Pacheo, H., 247, *253*
Paillard, N., 361, *372*
Palloy, E., 590, *621*
Palmer, J. K., 32, *62*, 77, 78, 81, 82, 83, 85, 86, 87, 88, 89, 90, 92, 94, 96, 97, 98, 99, 101, *102*, *103*, *104*, *105*, 185, *205*
Palmieri, F., 449, *480*
Pangborn, R. M., 512, 521, *537*, *539*
Parija, P., 94, *102*
Park, M. V., 599, *618*
Parker, E. R., 143, 145, *165*
Parker, J., 664, 681, *684*
Parkins, R. M., 602, *614*
Parratt, J. R., 307, *323*
Parsons, C. S., 420, *431*, 469, *480*
Parsons, F. G., 690, 691, *703*
Patac de las Traviesas, L., 260, *279*
Patel, D. S., 516, *539*
Paterson, D. R., 213, *231*
Patron, A., 519, *540*, 596, *616*
Pattabhiraman, T. R., 247, *254*
Patterson, M. E., 80, *103*, 349, 350, 363, *372*, 716, *718*
Patwa, D. K., 246, *254*
Patwardhan, M. V., 248, *254*
Paulssen, R. B., 609, *614*
Payne, N. C., 602, *614*
Peacock, B. C., 74, 77, *104*
Pearce, W. E., 527, *538*, 554, 563, *571*
Pearson, J. A., 345, 348, 349, *371*, *372*
Pederson, C. S., 388, *409*, 443, *480*, 524, *540*
Pelle, C., 266, 267, *279*
Penefsky, H. S., 669, *684*
Peng, C. V., 420, *434*, 584, *618*, 712, *718*
Pennington, F. C., 715, *718*
Pennock, W., 245, *254*
Pentzer, W. T., 208, *231*, 421, *434*, 468, *482*, 643, *648*
Pereira, E. S. B., 681, *683*, *685*
Perfetti, G. E., 567, *571*
Peri, C., 192, *204*
Perkins, H. J., 715, *719*
Perret, A., 419, *435*
Perrin, D. D., 602, *618*
Perring, M. A., 340, *372*
Perry, R., 640, *651*
Perry, R. L., 625, *650*, *651*
Persidsky, M. D., 664, 665, 682, *684*
Peterson, G. T., 512, *540*, 553, 562, 568, *569*
Petrash, L. P., 693, *704*
Pettit, L. D., 602, *614*
Peynaud, E., 177, 181, 182, 186, 193, 194, 202, *205*, 596, *619*
Peyron, L., 139, *166*
Pflug, T. J., 362, *369*
Phaff, H. J., 23, 24, *62*, 383, *409*, 624, 625, 626, 627, 632, 646, *649*, *650*, *651*
Phillips, J. F., 597, *619*
Phinney, B. O., 415, *434*
Pienaar, W. J., 417, *434*
Pieringer, A. P., 137, *162*
Pierpoint, W. S., 584, *618*
Pilnik, W., 551, *510*, 589, *618*
Pincock, R. E., 667, 668, *684*
Piper, S. H., 359, *369*
Piringer, A. A., 457, *480*
Pitman, G., 265, 267, *279*
Pitt, J. I., 633, 635, 636, 646, *650*
Plancken, M. J., 329, *331*
Plastourgos, S., 272, *279*
Plaza, G., 66, *103*
Plicher, R. W., 141, *166*
Pojnar, E., 503, *506*
Poland, G. L., 69, 79, 87, *103*, *104*
Polansky, M. H., 13, *62*
Polgár, A., 219, *232*
Pollack, R. L., 420, 425, *434*
Pollard, A., 359, *369*, 383, 397, 401, 403, 404, 405, *408*, *409*, 442, *480*, 581, 584, 588, 589, 592, 593, 594, 602, 604, 605, 606, 607, *612*, *616*, *617*, *618*
Pollard, J. K., 86, 87, *104*
Pollard, L. H., 711, *719*
Polymenakos, N. G., 271, 274, *277*, *279*
Pomeroy, C. S., 147, *166*

Ponting, J. D., 508, 509, *539*, *540*, 626, 634, *649*, *651*
Pool, R. M., 177, *204*
Popenhoe, J., 242, *254*
Popovic, P., 680, *684*
Popovic, V., 680, *684*
Popper, C. S., 46, *60*
Porritt, S. W., 420, *434*
Porter, C. A., 129, *168*
Porter, D. R., 210, 215, 217, *230*, *231*
Porter, J. W., 117, *166*, 455, *480*, *481*
Possingham, J. V., 632, 639, *649*
Postel, W., 565, *569*, 607, *614*
Postlmayr, H. L., 419, *434*, 519, 536, *540*, 559, *572*
Potter, N. A., 350, *372*
Powell, L. E., 413, 415, *434*
Power, F. B., 264, *279*, 426, 427, *434*
Power, J. B., 503, *506*
Powers, I. T., 519, *540*
Powers, J. J., 160, *166*, 519, *540*
Powers, M. J., 527, *540*
Powrie, W. D., 593, *615*, 655, 658, 665, *683*
Prabhu, L. H., 598, *612*
Prakash, R., 212, *230*
Prasad, P. S. R. K., 555, *572*
Prasad, R., 693, *704*
Prater, A. D., 326, *330*
Pratt, C., 413, 415, *434*
Pratt, D. E., 160, *166*, 519, *540*, 584, *619*
Pratt, H. K., 16, 20, 24, 30, 32, 33, 36, 37, 56, *61*, *62*, *63*, *69*, 208, 209, 210, 211, 212, 218, 220, 223, 224, 225, 226, 227, 228, *230*, *231*, *232*, 449, 462, 466, 467, 470, 474, *479*, *480*, *482*
Prescott, F., 441, 442, *475*
Pressley, A., 526, *539*, 561, *571*
Preston, R. D., 313, *323*
Pretel, A., 222, *229*
Pribela, A., 593, 594, *618*, *620*
Pridham, J. B., 346, 347, 367, *371*, *373*, 626, *651*
Prillinger, F., 591, 594, *618*
Primer, P. E., 414, *432*
Primo, E., 123, *161*
Primo-Yufera, E., 576, *618*
Pritchett, D. E., 558, 559, *571*
Proctor, B. E., 81, *103*, 521, *539*, 554, *570*
Proebsting, E. L., Jr., 417, *434*
Prue, J. E., 603, *618*
Pruthi, J. S., 582, *618*
Puissant, A., 191, *205*
Pulley, G. N., 545, 553, *571*
Pullman, M. E., 669, *684*
Punsri, P., 414, 415, *432*
Purcell, A. E., 455, 456, *475*, *479*, 710, *719*
Put, H. M. C., 525, *540*
Pyne, A. W., 451, 452, *480*
Puski, G., 402, *409*
Py, C., 303, 322, *324*
Pyatnitski, I. V., 598, *618*

Q

Quackenbush, F. W., 218, *231*, 456, *478*
Quatrano, R. S., 681, *684*
Quesnel, V. C., 712, *717*
Quijano, R., 261, 262, 265, 266, 270, 271, 272, 276, *277*, *278*
Quinones, V. L., 245, *254*
Qunitana, D. C., 696, *705*

R

Rabourn, W. J., 218, *231*
Rabson, R., 69, 70, 86, 87, 89, 90, 91, 97, 98, *104*
Racker, E., 12, *62*, 669, *684*
Raddi, P., 212, *229*
Radler, F., 176, 195, 196, 197, *204*, *205*, 632, *651*
Radola, B., 700, *703*
Rae, H. L., 692, 693, 694, 695, *704*
Ragab, H., 607, 611, *615*
Rahman, B. M., 626, *651*
Raica, N., Jr., 701, *704*
Rajan, K. S., 597, *618*
Rakieten, M. L., 141, *166*
Rakitin, Y. V., 451, *480*
Rakitin, Yu. V., 220, *231*
Rakitina, Z. G., 655, 673, *685*
Ram, H. Y. M., 67, 68, 101, *104*
Ram, M., 67, 68, *104*
Ramakrishnan, C. V., 129, *168*

Ram Chandra, G., 470, *480*
Ramey, J. T., 701, 702, *704*
Ramirez, D. A., 453, 454, 456, *480*, 716, *719*
Ramsay, G. B., 208, *231*
Ramsey, J. C., 465, *480*, *481*
Ramsey, R. C., 128, *167*
Randall, J. M., 500, *506*
Ranganna, S., 515, 516, *540*
Rankine, B. C., 597, *618*
Rao, P., 247, *254*
Rapatz, G. L., 657, 663, 665, *684*, *685*
Rapp, A., 590, *614*
Rappaport, L., 224, *230*
Rasic, J., 462, *478*
Rasmussen, G. K., 128, 148, *163*, *166*, 341, *370*
Rattray, J. M., 208, *231*
Rauen, H. M., 561, *571*
Raven, T. W., 518, 522, 535, *538*, *540*
Rawlinson, W. A., 79, *103*
Raymond, D., 23, 24, *62*
Raymond, W. D., 245, 247, *254*
Rayner, D. S., 144, *162*
Reddish, C. E. S., 509, *541*, 585, *621*
Reddy, V. V. R., 246, 248, 249, *254*
Redman, G. H., 550, *570*
Reece, P. C., 148, *163*, 341, *370*
Reed, D. J., 342, *373*
Reed, G., 589, *618*
Reed, H., 114, *166*
Reed, H. M., 637, 647, *651*
Reed, L. B., 208, 213, *229*
Reeder, W. F., 252, *254*
Reeder, W. S., 32, 34, *61*
Reeve, R. M., 419, *434*, 493, *506*, 654, 672, *685*
Rehm, S., 221, 222, *229*, *231*
Reibeiz, C., 414, *434*
Reid, M. S., 218, *232*
Reinartz, H., 457, 458, 460, *475*, *476*
Reinke, H. G., 679, *682*
Reintjes, H. J., 610, *615*
Reister, D. W., 527, *540*, 552, 554, 563, *569*, *571*
Reitz, H. J., 140, *167*
Rendle, T., 525, *540*
Rentschler, H., 603, 605, *619*
Resnik, R., 604, *618*
Reuter, F. H., 453, 454, 455, *477*
Reuther, W., 145, *166*, *167*
Reuther, K. H., 584, *612*
Revis, B., 607, *620*
Reyes, P., 430, *434*
Reynard, G. B., 442, *480*
Reynolds, T. M., 389, *407*, 605, 606, *619*, 628, 642, *651*
Rhodes, D. N., 688, *704*
Rhodes, M. J. C., 48, *61*, 94, *103*, 222, *232*, 342, 344, 345, 347, 348, 349, 350, 354, 355, 356, 367, 361, 363, *370*, *371*, *372*, *373*, 716, *719*
Ribeiro, O. F., 561, *571*
Ribeiro, R. F., 561, *571*
Ribéreau-Gayon, G., 179, 180, 181, 182, 184, 185, 189, 200, *205*
Ribéreau-Gayon, J., 177, *205* 596, *619*
Ribéreau-Gayon, P., 182, 188, 189, 190, 191, 192, 200, *204*, *205*
Ricciuti, C., 420, *434*
Rice, A. C., 443, *479*, *480*, 521, *539*
Rice, R. G., 553, *571*
Richards, V., 665, *684*
Richert, P. H., 624, *651*
Richmond, A., 56, 57, 58, *62*, 349, *373*, 431, *434*, 449, *480*
Richtmyer, N., 12, *61*
Richtmyer, N. K., 12, *62*
Rickles, N. H., 668, *685*
Rifai, H., 626, *649*
Riter, R., 500, *506*, 577, *616*
Ritsma, J. H., 602, *619*
Ritter, C. M., 417, *432*
Rivas, N., 457, 458, *380*
Rivers, A. L., 426, *435*
Rivett, D. E. A., 221, *229*
Robb, J. A., 76, *102*
Robberstad, M. T., 607, *618*
Robbins, R. H., 579, 580, 593, *619*, *620*
Roberts, D. W. A., 715, *719*
Roberts, E. A., 76, 92, *104*
Roberts, J. A., 141, *166*
Roberts, J. B., 85, 86, *104*
Roberts, J. C., 247, *254*
Roberts, L. W., 669, *685*
Robertson, R. N., 21, 30, 31, 56, *60*, *63*, 337, 338, 345, 348, 349, *369*, *371*, *372*, 449, 467, *480*
Robertson, W. F., 423, 426, *431*
Robinson, B. J., 417, 425, *434*

Robinson, G. M., 85, *104*
Robinson, J., 210, *231*
Robinson, J. E., 74, 75, *103*
Robinson, R., 132, *163*, *165*
Robinson, W. B., 690, 699, *704*
Rockland, L. B., 118, 119, 123, 134, 151, *166*, *168*
Roderick, D. B., 514, 526, *540*
Rodin, J. E., 329, *331*
Rodin, J. O., 329, *331*
Roe, B., 146, *162*
Roesch, W. C., 689, *702*
Roessler, E. B., 215, *229*, 521, *537*
Rogai, F., 264, *279*
Rogers, B. J., 148, *163*, 341, *370*
Rohrer, D. E., 379, 380, 383, 386, 388, 389, 391, 402, 405, *409*
Rojahn, W., 426, *433*
Rolfe, E. J., 609, *615*
Rolle, L. A., 136, 137, *165*, *167*, 559, *572*
Romani, R. J., 35, 55, *62*, *63*, 75, *105*, 347, 350, 358, *372*, *373*, 416, 421, 424, 425, 426, 430, 431, *431*, *433*, *434*, *435*, 688, 691, 692, 693, 694, 697, *704*, *705*, 708, *719*
Romero, J. M., 263, *279*
Ronsivalli, L. J., 688, *704*
Rood, P., 423, *435*
Rosa, J. T., 209, 215, 217, *230*, *232*, 441, 443, 449, *480*
Rose, D. H., 153, 154, *164*, *165*, 468, *482*
Ross, E., 86, *103*, 520, *540*, 607, 608, *619*
Rossmiller, G. E., 489, *505*
Rottini, C. T., 267, *279*
Rouse, A. H., 127, 159, 160, *167*, *169*, 559, *569*, *571*, *572*, 587, *619*, 691, 697, *704*
Rousille, A., 264, *279*
Roux, E. R., 424, *435*
Rowan, K. S., 21, 56, *62*, 79, *104*, 212, 220, 227, *232*, 449, 467, *480*
Rowe, A. W., 666, *685*
Royce, R., 526, 528, *538*
Royo-Iranzo, J., 558, *572*, 575, 576, *618*, *619*
Rubatzky, V. E., 457, *480*
Rubinstein, B., 148, *167*
Ruck, J. A., 584, *619*
Rudich, Y., 212, *230*
Ruhland, W., 438, *475*
Rundle, R. E., 79, *102*
Rushing, N. B., 611, *619*
Ruszkowska, M., 462, *480*
Ryan, H. J., 145, *163*
Rygg, G. L., 150, 154, 155, 156, *164*
Rymal, K. S., 552, *571*
Ryugo, K., 414, 419, *435*

S

Sabater Garcia, F., 430, *435*
Saburov, N. V., 713, *719*
Sacher, J. A., 56, *62*, 77, 89, 94, 95, 96, 97, *104*, 470, *480*
Sadik, S., 467, 471, 472, *479*, *480*
Sagasegawa, H., 219, *230*
Sahai, P. N., 359, *369*
Saito, N., 601, *619*
Sajak, E., 576, 587, *614*, *616*
Sakai, A., 656, 657, 658, 659, 662, 663, 664, 666, 675, 681, *685*, *686*
Sakamura, S., 584, *619*, 712, *719*
Sakiyama, R., 443, 444, 445, 447, 461, *480*
Sakurai, Y., 606, 609, *616*
Saleh, N. A. M., 247, *253*
Salimen, A., 596, *614*
Salt, F. W., 532, *540*
Saltman, P., 598, *612*, *613*, *619*
Salunkhe, D. K., 419, 426, 427, *432*, *435*, 441, 442, 444, 447, 448, 449, 450, 451, 452, 453, 455, 456, 473, *475*, *476*, *482*, 514, 515, *538*, 639, 647, *650*, *651*, 699, *703*, 711, 713, *717*, *719*
Samish, Z., 133, 134, *167*, 556, *571*
Samoradova-Bianki, G. B., 600, *619*
Samygin, G. A., 654, 664, 665, 666, 676, *685*
Sanborn, N. H., 534, *539*
Sander, H., 453, 456, *480*
Sando, C. E., 359, 364, *372*, *373*, 441, 449, *480*
Sandret, F. G., 262, 266, *279*
Sankar, K. P., 244, *254*
Sanshuck, D. W., 587, *621*, 678, *683*
Santarius, K. A., 660, 666, 669, *683*, *685*
Santos, B., 709, *719*
Sanwal, B. D., 354, *373*

Saravacos, G., 449, *480*
Saravacos, G. D., 594, *617*
Sarkar, B., 598, *613*, *619*
Sarmiento, R., 326, *330*
Sarveshivara, R., 251, *254*
Sasa, T., 716, *717*
Sastry, L. V. L., 247, *254*
Sato, I., 220, *230*
Sato, T., 284, 299, *301*
Satoh, Y., 716, *717*
Satori, K. G., 420, *432*, 511, *537*
Saulnier-Blache, P., 196, 198, *205*
Savage, W. G., 525, *540*
Savastano, G., 270, *279*
Savchenko, A. P., 217, *229*
Sawayama, Z., 555, 560, *571*
Sawyer, R., 576, *619*
Sayce, I. G., 602, *618*
Sayer, E. R., 709, *717*
Sayre, C. B., 445, *479*
Sayre, R. N., 681, *685*
Saywell, L. G., 441, 445, *480*
Saxena, G. M., 600, *619*
Scaife, J. F., 608, 610, *619*
Scaramuzzi, F., 260, *277*
Schaller, A., 579, 588, *617*, *619*
Schaller, D. R., 418, *435*
Schanderl, S. H., 714, 716, *718*, *719*
Scheffer, H., 588, *620*
Scheiner, D. M., 510, *541*, 712, *719*
Scheraga, H. A., 663, *685*
Scheuerbrandt, G., 428, *435*
Scheuerman, R. W., 212, 217, *229*
Scheur, J., 588, *617*
Schiller, F. H., 645, 646, *650*
Schillinger, A., 609, *619*
Schmidt, C., 25, *62*
Schneider, G. W., 417, *435*
Schneyder, J., 578, *619*
Scholander, P. F., 655, 666, *685*
Scholz, E. W., 223, *232*
Schomer, H. A., 114, 115, 151, 156, *166*
Schormüller, J., 451, *480*, *481*
Schothorst, M. van, *704*
Schrader, A. L., 643, *651*, *652*
Schroeder, C. A., 4, 5, *61*, *62*, 211, *232*, 656, *685*
Schroeder, E., 12, *62*
Schuck, C., 221, *229*
Schultz, T. H., 136, *167*, 361, *370*, 592, *619*
Schultz, W. G., 500, *506*
Schulz, W. W., 597, *619*
Schutte, C. E., 417, *434*
Schwarzenbach, G., 596, *612*
Schweers, V. H., 212, 217, *229*
Schwimmer, S., 27, *62*, 510, *539*, 682, *685*
Schwob, R., 13, *62*
Scott, F. M., 5, *62*
Scott, G. W., 215, *232*
Scott, K. J., 76, 92, *104*
Scott, L. E., 382, *407*, 424, *434*
Scott, W. C., 544, 557, *571*, 587, *619*
Scott, W. J., 647, *651*
Scurti, F., 263, 264, *279*
Seagran, L. H., 688, *704*
Seale, P. E., 517, *537*, 582, 590, *619*
Seaton, H. L., 471, *481*
Sebrell, W. H., 391, *409*, 418, *434*, 441, *480*
Sedky, A., 559, *570*
Seegers, W. H., 16, *62*
Seegmiller, C. G., 376, 382, 383, 388, 390, 394, 404, *409*, 511, 536, *539*
Seifert, R. M., 451, *476*
Seifter, E., 160, *167*
Sejtpaeva, S. K., 580, *614*
Sekin, J., 158, 161, *167*
Sell, H. M., 383, 386, 388, 390, 405, *410*
Senn, G., 584, *620*
Senn, T. L., 418, *436*
Senn, V. J., 552, *571*, 611, *619*
Sephton, H. H., 12, *61*, *62*
Serini, G., 220, *232*
Serra, G., 567, *571*
Servis, R. E., *685*
Seshadri, T. R., 600, *619*
Seshagiri, P. F. V., 238, 239, *254*
Setty, L., 516, *540*
Sevenants, M. R., 427, *433*, *435*
Shafshak, S. A., 444, *481*
Shah, R., 246, *254*
Shain, Y., 367, *371*, *372*, 585, *614*
Shallenberger, R. S., 521, *539*
Shannon, C. T., 584, *619*
Shantha, H. S., 245, 248, *254*
Shao, Z. P., 217, 227, 228, *230*
Shapiro, R., *685*
Sharf, J. M., 576, *619*
Sharma, V. S., 602, *618*

Sharma, W., 665, *683*
Shaw, B. L., 600, *619*
Shaw, C. P., 81, 82, *103*
Shea, K. G., 688, *704*
Sheft, B. J., 566, *571*
Shelton, D. C., 681, *684*
Shepherd, A. D., 152, *162*
Sherman, M. S., 139, *165*
Shewfelt, A. L., 419, 430, *435*, 456, *481*, 519, *540*, 697, *704*
Shibusa, S., 584, *619*
Shiga, I., 563, *571*
Shiga, J., 568, *570*
Shih, S. C., 681, *684*
Shillinglaw, C. A., 610, *619*
Shimazu, F., 642, *651*, *652*
Shimizu, S., 716, *719*
Shimoda, Y., 555, 560, *571*
Shimokoriyama, M., 132, *167*
Shimura, I., 299, *301*
Shinano, S., 220, *232*
Shinoda, S., 678, *683*
Shioda, K., 710, *718*
Shitomi, H., 219, *231*
Showalter, R. K., 211, 213, *232*
Siddappa, G. S., 245, 251, *254*, 515, *540*, 551, *571*
Siddiqui, R., 584, *619*
Sideris, C. P., 314, *324*
Siegelman, H. W., 364, 365, *373*
Siegler, E. H., 523, *538*
Siiria, A., 404, *408*
Silander, S., 637, *649*
Sillen, L. G., 596, *612*
Sills, V. E., 388, *407*
Silver, R. K., 663, 664, *685*
Silverstein, M. S., 426, *432*
Silverstein, R. M., 326, 329, *330*, *331*
Simandle, P. A., 440, 443, 444, 450, *481*
Simmonds, N. W., 66, 67, 71, 72, 73, 76, 77, 78, 85, 92, *104*
Simone, M., 273, *278*
Simoni, R. D., 26, 28, *62*
Simpson, K. L., 261, 273, *279*
Simpson, T. H., 600, *619*
Sinclair, W. B., 116, 119, 123, 126, 128, 132, 141, 157, *162*, *167*, *168*
Singh, B. N., 238, 239, *254*
Singh, L. B., 233, 234, 235, 238, 239, 244, 245, 247, 251, *254*
Single, W. V., 674, *685*
Singleton, V. L., 307, 308, 311, 313, 314, 315, 316, 317, *323*, *324*, 329, *330*, 515, *540*, 708, 709, *718*, *719*
Sinkonnen, I., 596, *614*
Sisa, M., 282, *301*
Sissons, D. J., 383, 404, *408*, 584, 589, *616*, *618*
Sites, J. W., 119, 140, 141, 144, 145, 147, *165*, *167*
Sjoerdsma, A., 85, 91, 92, *104*, 123, *168*, 447, *481*
Sjöström, L. B., 81, *102*
Skene, D. S., 639, *651*
Skou, J. P., 698, *704*
Slabaugh, W. H., 658, 662, 666, 680, *683*
Smando, R., 354, *373*
Smillie, R. M., 75, 80, 89, 90, 94, 95, 96, 97, 98, 99, *102*
Smit, C. J. B., 14, *61*, 120, *168*
Smith, A. H., 119, *167*
Smith, A. J. M., 338, *373*
Smith, B. J., 674, *685*
Smith, F., 421, *433*
Smith, G., 487, *506*, 525, *540*
Smith, G. S., 632, *650*
Smith, J. A. B., 359, *369*
Smith, J. G., 151, *165*
Smith, J. H. C., 15, *62*
Smith, J. R. L., 604, *618*
Smith, M. A., 208, *231*
Smith, M. C., 219, *232*
Smith, O. E., 148, *161*
Smith, P. F., 145, *166*, *167*
Smith, P. G., 207, 209, *232*
Smith, P. R., 131, *167*, 472, *481*
Smith, T. D., 597, *619*
Smith, W. H., 156, *164*, 350, *371*, 406, *409*, 421, *435*
Smith, W. J., 91, *104*
Smith, W. L., 420, *431*
Smock, R. M., 334, 335, 364, *373*, 583, *619*
Smoot, J. J., 156, *164*
Smydzuk, J., 75, 77, 80, 95, 97, *102*
Smythe, C. V., 132, *168*, 550, *572*
Soler Martinez, A., 430, *435*
Solomos, T., 87, *101*
Somers, G. F., 443, 450, 459, *479*, *482*
Sommer, M. F., 713, *718*

Sommer, N. F., 76, *103*, 639, *650*, 688, 692, 693, 694, 695, 696, 698, 699, *702*, *704*, *705*
Somogyi, L., 430, *435*
Somogyi, L. P., 362, *373*, 697, *705*
Sondheimer, E., 400, *409*, 712, *719*
Sondheimer, F., 139, *166*
Song, P. S., 606, *619*, 628, *651*
Sontag, D., 138, *165*
Sorber, D. G., 429, *433*, 636, 637, *649*, *651*
Sorger, G. J., 460, *477*
Soriano, S., 272, *279*
Soule, M. J., 238, 241, 242, 247, *254*
Soumalainen, H., 400, 401, *409*
Souty, M., 419, *435*
Sowinska, H., 442, *481*
Spaid, J. F., 690, 699, *704*
Spalding, D. H., 469, *480*
Spanyar, P., 426, *435*, 607, 608, 609, *619*, *620*
Spencer, M. S., 450, 451, 470, 474, *479*, *480*, *481*, 521, *540*
Spiegelberg, C. H., 524, 525, *540*
Spitzer, K., 13, *62*
Splittstoesser, D. F., 690, 699, *704*
Spoehr, H. H., 15, *62*
Sprock, C. M., 416, 424, *433*, *434*, *435*
Spurr, A. R., 213, *232*, 438, 462, 472, *479*, *481*
Squires, C., 25, *60*
Squires, C. L., 25, *62*
Srecnivasan, A., 243, 250, *254*
Srere, P. A., 158, 161, *167*
Srivastara, H. C., 243, 250, 251, *254*
Srivastara, J. C., 251, *253*
Stachowicz, K., 399, *409*, 419, *433*, 517, *539*
Stadtman, E. R., 637, 642, 643, *651*, *652*
Stadtman, F. H., 628, *651*
Stafford, H. A., 27, *60*, 182, 185, *205*, 222, *229*
Stahl, A. L., 6, 7, *63*, 123, 150, 156, *167*, 244, 245, *254*
Stahly, E. A., 415, *435*
Stall, R. E., 472, *478*
Stanley, W. L., 136, 137, 139, *165*, *167*, *168*, 450, 451, *481*, 521, *540*, 576, *620*, 626, *651*
Stanton, D. J., 471, 472, *478*
Starr, M. S., 606, *620*
Stearn, C. R., Jr., 143, *167*
Steele, W. F., 430, *436*, 511, 518, *541*
Stehlik, G., 700, *703*, *705*
Stein, W. H., 308, *323*
Steinberg, M. A., 688, *704*
Stembridge, G. E., 366, *373*
Sten, M., 401, *408*
Stephan, H., 185, *204*
Stephens, I. M., 160, *164*, 607, 608, *615*
Stephens, T. S., 118, *165*
Stepka, W., 12, 13, *61*
Steponkus, P. L., 681, *684*, *685*
Sterling, C., 419, *435*, 499, *506*, 520, *540*, 642, 648, *651*, 652, 662, 666, 672, 674, 676, *685*
Stern, J. R., 157, *166*
Stern, R. M., 526, *538*
Stevens, J. W., 141, *167*, 558, 559, *571*
Stevens, K. L., 426, *435*
Stevenson, A. E., 567, *571*
Stevenson, C. D., 76, *102*
Steward, F. C., 67, 68, 69, 70, 76, 86, 87, 89, 90, 91, 97, 98, 101, *100*, *104* 460, 461, *478*, *480*
Stewart, I., 122, 123, *167*, *169*
Stewart, W. S., 147, *167*
Stitt, C. F., 598, *613*
Stobart, A. K., 716, *719*
Stocken, L. A., 693, *704*
Stocking, C. R., 656, 671, 672, 682, *685*
Stoler, S., 67, *102*
Stoll, A., 716, *719*
Stoll, M., 329, *331*
Stolyarov, K. P., 602, *620*
Stoner, W. N., 472, *481*
Storer, F. H., 419, *435*
Strachan, C. C., 524, *537*, 607, *612*
Strain, H. H., 15, *62*, 715, *718*, *719*
Straten, E. F., 444, *479*
Stratton, F. C., 78, 89, 97, *104*
Strimiska, F., 394, 395, *409*, 594, *618*
Stubbs, L. L., 472, *481*
Stumpf, P. K., 25, 26, 28, *60*, *61*, *62*, *63*
Sturdy, M., 386, 388, *408*, 517, *538*
Stutz, E., 581, *614*
Styles, E. C. C., 714, *717*
Subrahmanyan, V., 243, 250, *254*
Sudyina, E. G., 716, *719*

Suec, W. A., 715, *718*
Sugihara, J., 261, *277*
Sullivan, W. M., 90, 91, *102*
Summers, A., 663, 664, *685*
Sumner, R. J., 714, *719*
Sun, B., 80, *103*
Sunday, M. B., 123, *164*
Suryaprakasa Rao, P. V., 555, *572*
Susic, M. V., 602, *620*
Sussman, M. V., 663, 667, *685*
Sutherland, G. K., 308, 313, 316, *323*, *324*
Sutherland, J. L., 472, *481*
Suyama, Y., 220, *230*
Suzuki, M., 223, *231*
Swain, T., 84, *102*, *104*, 458, *481*, 523, *540*, 584, 604, 606, *612*, *620*
Sweeney, J. P., 440, 443, 444, 450, *481*
Swift, L. J., 123, 132, 133, 134, 139, *165*, *167*, 552, *570*
Swindells, R., 383, 393, *407*, *409*, 579, *620*
Swirski, M., 531, *538*
Sykes, S. M., 31, *63*
Szczniak, A. S., 674, *685*
Szechenyi, L., 600, *620*
Szoke, K., 217, 219, *230*
Szoke-Szotyori, K., 591, *614*
Szymczyk, F., 456, *479*

T

Tada, T., 289, *301*
Tager, J. M., 77, 78, 85, 98, *104*
Takahashi, T., 456, *481*
Takahashi, Y., 220, *230*
Takeda, K., 601, *620*
Takeuchi, Y., 294, *301*
Talburt, W. F., 527, *540*, 595, *613*
Tallman, D. F., 690, *704*
Tamaki, E., 716, *719*
Tanaka, S., 138, *165*
Tang, C. S., 426, 428, 429, *435*
Tanner, H., 603, 605, *619*, *620*
Tappel, A. L., 667, 669, *685*, 714, *719*
Tarlowsky, E., 566, *571*
Tarnovska, K., 455, 456, *481*
Tarutani, T., 298, *301*
Tate, J. N., 508, *540*
Taufel, K., 584, *620*
Taverna, G., 449, *481*
Tavernier, J., 451, *478*
Taylor, A. L., 115, *167*, 208, 213, *229*
Taylor, C. A., 114, *163*
Taylor, D. H., 643, 646, *651*
Taylor, O. C., 422, 424, *435*
Tazaki, K., 299, *301*
Teaotia, S. S., 251, *254*
Teas, H. J., 696, *705*
Telegdy-Kovats, M., 591, *614*
Teply, L. J., 671, *683*
Teranishi, R., 136, *167*, 592, *619*
Tettamanti, A. K., 131, *169*
Thomas, A. F., 329, *331*
Thomas, D. R., 716, *719*
Thomas, D. W., 132, *168*, 550, *572*
Thomas, J. G. N., 532, *540*
Thompson, A. E., 444, 445, *479*, *481*
Thompson, A. H., 341, *369*, 415, *435*, 643, *651*, *652*
Thompson, P., 517, *540*
Thompson, P. C. O., 563, 567, *570*
Thornton, N. C., 92, *104*, 406, *409*
Thorpe, R. H., 525, *538*
Timberlake, C. F., 402, 404, 405, *408*, *409*, 596, 597, 598, 600, 601, 602, 604, 605, 606, 607, 608, 609, 610, 611, *616*, *618*, *620*, 629, *652*, 712, *719*
Tindale, G. B., 150, 152, *168*
Ting, S. V., 112, 118, 119, 124, 125, 128, 129, 132, 139, 140, 145, 146, 156, *163*, *164*, *168*, 550, 553, 554, *572*, 582, *620*
Tinsley, I. J., 513, *540*
Tinsley, I. S., 396, *409*
Tisseau, M. A., 303, 322, *324*
Tityapova, I. G., 217, *232*
Tochilin, E., 689, *702*
Todd, G. W., 157, *162*
Todd, J. P., 714, *717*
Togo, S., 220, *230*
Tokunaga, M., 598, *617*
Tolkowsky, S., 108, 110, *168*
Toman, F. R., 677, *685*
Tomes, M. L., 219, *232*, 453, 454, 455, 456, *475*, *478*, *480*, *481*, 716, *719*
Tomkins, M. D., 527, *538*
Tomkins, R. G., 468, *481*

Tomlinson, W. R., 598, *615*
Tommasi, G., 263, 264, *279*
Tonsbeek, C. H. T., 329, *331*
Townsend, C. T., 524, *540*
Townsend, R. O., 214, *230*
Townsley, P. M., 120, *168*
Toy, L. R., 6, 7, *63*
Tressl, R., 84, *102*, 361, *370*, 429, 435, 590, *614*
Tressler, D. K., 545, 547, 548, 561 565, *572*, 574, *620*, 655, *685*
Trombly, H. H., 455, *481*
Tropical Products Institute, 504, *506*
Trout, S. A., 31, *63*, 150, 152, 153, *168*
Truelson, R. A., 698, *705*
Truelson, T. A., 713, *719*
Truscott, J. H. L., 469, *481*
Tseng, Kwang-Fang, 132, *166*
Tsubata, K., 451, 452, *478*
Tsukamoto, Y., 300, *301*
Tsukida, K., 710, *719*
Tsumaki, T., 297, *301*
Tukalo, E. A., 453, *481*
Tukey, H. B., 413, 414, *435*
Tuleen, D. L., 604, *613*
Tumanov, I. I., 654, 661, 663, *685*
Turrell, F. M., 111, *168*
Tutin, T., 264, *279*
Tuttle, L. W., 690, 691, *703*
Tutton, W. R., 531, *540*
Tuzson, P., 117, *169*, 219, *232*
Tyler, K. B., 212, *229*

U

Uchida I., 299, *301*
Udenfriend, S., 85, 91, 92, *104*, 123, *168*, 447, *481*
Ulrich, J. M., 710, *719*
Ulrich, R., 350, 358, *373*, 425, *435*
Ulyanova, D. A., 713, *719*
Underwood, J. C., 119, 134, 151, *166*, *168*, 329, *331*, 593, *621*
United Fruit Co., 73, *104*
Unrath, C. R., 416, *435*
Uota, M., 420, *435*
Uprety, M. C., 607, *620*
Urbach, G. E., 340, *372*
Urban, K., 669, *685*
Uriu, K., 417, *431*, *432*, 492, *505*
U.S. Dep. Agr, 109, 111, 140, *168*
Uzzan, A., 266, 267, *279*

V

Vajnberger, A., 462, *478*
Valdener, G., 247, *253*
Valenzuela, S., 80, *105*
Valeri, H., 15, *61*, *63*
Valmayor, R. V., 4, 6, *63*
Van Arsdel, W. B., 625, 637, *652*, 655, *685*
Van Bekkum, D. W., 693, *705*
Van Blaricom, L. O., 418, *436*, 514, 526, *540*
Van Buren, J., 584, *620*
Van Buren, J. B., 712, *719*
Van Buren, J. P., 418, *436*, 510, 511, *540*, *541*
Van Den Berg, L., 666, 669, *685*
Van Der Plank, J. E., 421, *434*
Vandercook, C. E., 117, 118, *169*, 559, *572*
Van Der Merwe, H. B., 519, 535, *540*, *541*
Van Goor, B. J., 473, *481*
Van Horn, C. W., 144, *165*
Van Kooy, J. G., 425, 430, *434*, 692, 693, *705*
Vänngärd, T., 598, *612*
Vannier, S. H., 136, 137, 139, *165*, *167*, *168*
van Overbeek, J., 146, 147, *168*
Varma, T. N. S., 129, *168*
Varner, J. E., 443, 444, 447, 448, 449, 474, *476*, *481*
Vartak, D. G., 598, *612*
Vas, K., 588, 591, *614*, *620*
Vasatko, J., 593, *620*
Vasquez Roncero, A., 262, 263, 267, 276, *279*
Vaughn, R. H., 261, 270, 271, 272, 273, *278*, *279*, 524, *541*
Vavich, M. G., 218, *232*, 526, *538*
Vazquez Ladron, R., 261, 262, 265, 266, 270, 271, 272, *277*
Veldhuis, M., 123, *167*

Veldhuis, M. K., 123, *163*, 544, 554, 557, 558, *569*, *570*, *571*, 581, 587, *619*, *620*
Veldstra, H., 457, *475*
Vendilo, G. G., 449, *481*
Vendrell, M., 76, *104*
Vennesland, B., 27, *60*, 222, *229*, *231*, *232*
Venstrom, D., 452, *477*
Vento, V., 220, *230*
Veselinovic, D. S., 602, *620*
Vetsch, V., 605, *620*
Vickery, H. B., 185, *205*
Vidal Sigler, A., 266, *278*
Villarreal, F., 443, 444, *481*
Vines, H. M., 112, 129, 146, 153, 156, 158, 159, 160, 161, *164*, *166*, *168*, 561, *572*, 607, *620*
Vinet, E., 179, *204*
Violante, P., 455, *478*
Vioque Pizarro, E., 267, *279*
Vishniac, W., 716, *718*
Visser, A., 710, *719*
Vogele, A. C., 454, 455, *481*
Voigt, J., 584, *620*
Von Abrams, G. J., 228, *232*
Vries, M. J., de, 693, *705*

W

Waalkes, T. P., 85, 91, *104*
Wada, M., 220, *232*
Wade, D., 383, *409*
Wade, N. W., 75, 76, 77, 80, 95, 97, *102*, *104*
Wade, P., 493, 495, *506*, 520, *541*
Wagenknecht, A. C., 418, *433*, 510, *541*, 712, 714, *718*, *719*
Wagner, C. J., 553, *569*
Wagner, J. R., 511, *541*, 587, *621*
Wagner, R., 500, *506*, 577, *616*
Wahba, I., J., 691, *705*
Wakabayashi, H., 606, *616*
Walker, E. D., 123, *165*
Walker, G. C., 714, *719*
Walker, J. R. L., 367, *373*, 457, 458, *481*, 509, *541*, 583, 585, *621*
Walkof, C., 445, *481*
Wallace, A., 157, 158, 159, 160, 161, *162*, *164*
Walter, T. E., 341, 364, *373*
Walter, W. M., Jr., 710, *719*
Wang, C. H., 38, *60*, 465, *476*, *477*, *480*, *481*
Wang, P. H., 217, *230*
Wang, P. M., 217, 227, 228, *230*
Wang, P. S., 217, 227, 228, *230*
Ward, A. G., 641, *652*
Ward, G. M., 442, *481*
Wardale, D. A., 344, *372*, 474, *479*, 694, *703*
Wardlaw, C. W., 6, 28, *63*, 67, 71, 72, 73, 76, 92, 93, *104*, 223, *232*, 235, 237, 239, *254*
Warner, J., 469, *481*
Warnerford, F. H. S., 118, *164*
Warth, A. H., 139, *168*
Watt, B. K., 143, *168*, 220, *231*, 391, 392, 393, 394, 395, *410*
Waynick, 114, *168*
Weatherby, L. G., 15, *63*
Weaver, R. J., 177, *204*
Webb, A. D., 193, *205*
Webb, R. A., 380, *407*, 496, *505*
Webb, W. R., 663, 665, *684*
Webber, H. J., 126, *168*
Webster, G. L., 442, *478*
Webster, R. C., 672, 675, *685*
Weckel, K. G., 443, *479*, 521, *539*, 709, *719*
Wedding, R. T., 119, 161, *168*
Weerdhof, T. v. d., 329, *331*
Wegman, K., 599, 601, *612*
Wehmer, C., 456, *481*
Weier, T. E., 656, 671, 672, 682, *685*
Weigel, H., 598, *612*
Weinland, R. F., 598, *621*
Weinstein, L. H., 129, *168*
Weissbach, H., 85, 91, *104*
Weissberger, A., 562, *572*, 608, *621*
Weizmann, A., 139, *166*, *168*
Wel, H., van der, 486, *505*
Welch, J. E., 207, 209, *232*
Welcher, M., *685*
Welti, D., 450, *475*
Wender, S. H., 130, 131, 132, *163*, *164*, *166*, 401, 402, *410*, 549, *570*, 600, *614*
Wenzel, F. W., 159, *169*, 545, 546, 555, 559, *570*, *572*
Wessels, J. H., 221, *231*

West, C., 28, *61*, 152, *165*, 344, 345, 348, 350, 352, *371*, *372*, 415, 424, *433*, *434*
West, G. B., 91, *104*, 447, 448, *481*
Westling, R., 487, *506*
Weurman, C., 366, *373*, 404, *410*
Wharton, D. C., 472, *481*
Wheaton, T. A., 122, 123, *167*, *169*, 466, *479*
Whelton, R., 646, *651*
Whitaker, T. W., 207, 208, 209, 213, *229*, *232*
White, H. L., 471, *475*
White, M. J., 117, *169*
Whiteman, T. M., 468, *482*, 671, *685*
Whiting, G. C., 388, 389, 390, *410*, 576, *613*
Whittenberger, R. T., 420, *434*, 510, 514, *541*
Wiant, J. S., 208, *231*, *232*
Wick, E. L., 76, 80, 81, 84, 89, *102*, *103*, *104*, *105*, 451, 452, *480*
Wicliff, J. L., 715, *719*
Widdowson, E. M., 92, *103*, 126, 143, *165*, *169*, 380, 386, 391, 392, 393, 394, 395, *409*, 529, 530, *539*, 567, *571*, 578, 595, *617*, 641, *650*
Wiederhold, E., 139, *161*, 561, 563, *569*, *571*
Wiegand, E. H., 625, 640, 647, *652*, 675, *685*
Wiegers, G. A., 602, *619*
Wieland, H., 604, *621*
Wiersum, L. K., 460, 461, 473, *481*
Wiggell, D., 472, *478*
Wilbur, J. S., 376, 382, 388, 394, 404, *409*, 429, *433*, 511, 536, *539*
Wilcox, P. E., 602, *614*
Wiles, G. D., 527, *540*
Wiley, A. R., 568, *570*
Wiley, R. C., 366, *373*
Wilhalm, B., 329, *331*, 590, *621*
Wilkinson, A. E., 377, *410*
Wilkinson, B., 461, *481*
Willey, A. R., 531, *538*
Williams, A. A., 325, *331*
Williams, A. H., 367, *373*, 461, *481*, 581, 589, *613*, *616*
Williams, B. L., 139, *169*, 401, 402, *410*
Williams, C. C., 525, *541*
Williams, E. F., 359, *369*
Williams, G. R., 42, 49, *61*, 462, *476*
Williams, K. T., 22, *60*, 440, *481*
Williams, M. W., 338, 342, *372*
Williams, O. B., 525, *541*
Williams, P. N., 8, *61*
Williams, R. J., 666, *685*
Williams, R. J. P., 603, 604, *621*
Willimott, S. G., 160, *169*
Willis, A. G., 515, *540*
Willis, W. W., 442, *477*
Willison, J. H. M., 503, *506*
Willits, C. O., 593, *621*
Willstäter, R., 716, *719*
Willühn, G., 455, 456, *476*
Wilmot, S. W., 519, *540*
Wilson, D. E., 690, *703*
Wilson, K. W., 133, *169*
Wilson, W. C., 147, 148, 149, *169*
Winsor, G. W., 440, 441, 443, 444, 445, 446, 447, 450, 458, 460, 461, 462, 467, 471, 472, *476*, *477*, *479*, *481*, *482*
Winston, J. R., 114, 115, 116, 117, 118, 140, 141, 143, 144, 150, 153, 154, *164*, *165*, *166*, *169*
Winter, M., 590, *621*
Winters, A. F., 236, *254*
Winton, A. L., 210, *232*, 417, *436*, 640, *652*
Winton, K. B., 210, *232*, 417, *436*, 640, *652*
Wiskich, J. T., 49, 50, 53, 54, 55, *63*, 347, *373*
Wisser, K., 610, *615*
Witte, P. J., 115, *167*
Wittwer, S. H., 441, 442, *480*
Wodicka, V. O., 528, *537*
Woidich, H., 388, 399, *410*
Wokes, F., 160, *169*, 443, *482*
Wolf, J., 86, *105*, 418, *436*
Wolfe, H. S., 6, 7, *63*
Wolford, E. R., 672, 674, *685*
Wolford, R. W., 136, 137, *162*, *169*, 552, *571*
Wong, C. Y., 212, *232*
Wong, F. L., 448, 452, *482*, 521, *541*
Wood, M., 219, *232*
Wood, R. B., 607, *615*
Woodbridge, C. G., 447, 448, *477*
Woodmansee, C. W., 443, 450, 459, *479*, *482*

Woodruff, R. E., 383, 386, 388, 390, 405, *410*
Woods, M. J., 467, 472, *482*
Woodward, C. F., 420, *434*
Woolhouse, H. W., 470, *480*
Wooltorton, L. S. C., 48, *61*, 94, *103*, 156, *164*, 222, *232*, 239, *254*, 335, 338, 342, 344, 345, 346, 347, 348, 349, 350, 351, 353, 354, 355, 356, 357, 361, 363, *370*, *371*, *372*, *373*, 388, 389, *408*, 462, 466, *480*, 716, *719*
Workman, M., 94, *102*, 364, *373*, 470, *480*, *482*
Worth, H. G. J., 383, *410*
Wright, D. N., 212, 217, *229*
Wright, J. A., 354, *373*
Wright, R. C., 468, *482*
Wu, M-A., 457, *482*
Wucherpfennig, K., 580, 586, 594, 610, 611, *615*, *621*
Wyman, H., 78, 83, 87, 88, *105*

Y

Yamaguchi, M., 214, *230*, 297, *301*, 441, *482*
Yamamoto, E., 609, *621*
Yamamoto, H. Y., 611, *613*
Yamamoto, M., 294, *301*, 456, *482*
Yamamoto, R., 609, *621*
Yamanaka, H., 299, *301*
Yamane, H., 299, *301*
Yamanishi, T., 80, *105*
Yamato, I., 451, 452, *478*
Yang, H. Y., 404, *407*, 430, *436*, 511, 518, *541*
Yang, S., 79, *105*
Yang, S. F., 26, *63*
Yasui, K., 282, *301*
Yates, A. R., 76, *102*
Yokota, M., 710, *719*
Yokoyama, H., 117, 118, 144, *163*, *169*
York, G. K., 508, *540*
Yoshida, S., 658, 659, 662, 663, 664, 666, 675, *685*, *686*
Yother, W. W., *166*
Young, G. T., 143, *167*
Young, H. Y., 314, *324*, 515, *540*
Young, J. O., 413, *435*
Young, L. B., 143, *169*
Young, R., 459, *482*
Young, R. E., 15, 21, 28, 32, 35, 36, 37, 39, 40, 46, 49, 50, 52, 53, 54, 55, *60*, *61*, *62*, *63*, 74, 75, 77, 85, *102*, *104*, 152, *162*, *169*, 318, 319, 320, 321, *323*, 348, *373*, 668, 669, *686*, 692, *705*
Yu, I. K., 347, *373*, 421, 430, *435*, 693, 694, *704*
Yu, M-H., 441, 447, 448, 449, 450, 451, 452, *476*, *482*
Yufera, E. P., 141, 142, *169*, 558, *572*

Z

Zahara, M., 212, *229*
Zechmeister, L., 117, *169*, 219, *232*
Zehnder, J., 700, *702*, *705*
Zelitch, I., 640, *652*
Zeller, O., 671, *686*
Zemplen, G., 131, *169*
Ziegler, M. R., 91, *101*
Ziemelis, G., 552, *569*, 577, *613*
Zilva, S. S., 442, *476*
Zinkiewicz, J., 462, *480*
Zmudka, M., 669, *683*
Zoller, H. F., 131, *169*
Zook, E. G., 596, *621*
Zscheile, F. P., 455, *480*
Zubekis, E., 576, *621*
Zuch, T., 526, *539*, 561, *571*
Zyzlink, V., 609, *621*

Subject Index

A

Abscisic acid, 148
  in peach and cherry, 415
Abscission,
  biochemistry of, 148–149
  of melon, 210
Acerola, high ascorbic acid content, 583
Acetyl esterase, 161
Achenes, 375, 378
Acids, *see* Organic acids
ACP, 28
Adenosine triphosphate (ATP) as inhibitor of enzymic browning, 510
Ageing,
  of apple tissue, 355
  of banana tissue, 96
  of citrus fruits, 155
Agronomy of fruit for processing, 489–490
Albedo, 112
Albedo browning, 155
Alcohol dehydrogenase, 355
Alcoholic components of citrus flavour materials, 137
Aldehydes in citrus fruits, 136
Amines of bananas, 91
Amino acids
  in apples, 339
  in avocado pears, 12
  in bananas, 70, 90
  in blackcurrant juice, 399
  in citrus fruits, 120, 121
  concentration in grapes, 186, 187
  copper complexes, 602
  ferrous and ferric iron complexes, 602
  in mangoes, 245
  mechanism of formation in grapes, 187, 188
  in melons, 220
  in olives, 262, 263
  in oranges, 119, 120, 121
  in persimmons, 293–294
  in pineapples, 307, 314, 315
  in soft fruits, 396, 398
  in strawberry juice, 399
  sulphur-containing, in citrus fruits, 123
  in tomatoes, 447–449
  in various citrus fruits, 121
Ammonium cations in grapes, 186, 187
Ammonium pyrrolidonecarboxylate, 521, 522
Amylase in mangoes, 249, 252
Anthocyanins, 516, 628, 708, 711–713
  in apples, 364–365
  biosynthesis, 364
  in blood oranges, 118
  characterization of wines by, 190
  chelation of metallic compounds by, 713
  complexes with metals, 600–601
  complexing with metals, 532, 533
  decolorization by sulphur dioxide, 712
  degradation, 606
    in canned fruits, 513–514
    factors affecting, 513
    reactions, 513
  destruction by cathode rays, 713
  formation, 191
  in grapes, 188
    changes during ripening, 191
  losses due to interactions with other fruit products, 713
  in olives, 262
  in peaches, 418
  pseudo-base, 712
  reactions to produce discoloration, 713
  in soft fruits, 397, 399, 400
  stability during frozen storage, 678
  stability in processing, dependent on pH, 711
  in stone fruits, 418
Anti-oxidants, 553–554

Apples (*see also* Pome fruits), 333
  ageing of peel discs, 355, 357
  amino acids, 339
  anthocyanins, 364–365
  carotenoids, 364
  cellulose, 365
  changes in chemical constituents during growth, 338
  changes in malic acid during growth, 352
  changes in peel anthocyanidins during development, 340
  chlorophyll, 362–363
  colour changes in, 363–365
  core flush, 337
  countries where grown, 335
  disintegration, 580
  effect of nitrogenous fertilizers on volatile compound production, 362
  ethylene production, 362
  flavour, 343
  flower and fruit, 336
  fruit morphology, 335, 336
  hemicelluloses of, 365
  hexosans of, 365
  history and distribution, 334–335
  juice, 574
    "grassy" odour in, 590
    opalescent, 584
    storage as half-concentrate, 593
  length of growth period, 338
  mitochondria, 346
  organic acids present in, 351
  original home, 334
  pectin, 365
  peel discs as model system for biochemical studies, 355
  peel and pulp lipids, 359, 360
  pentosans of, 365
  phenolic constituents, 367
  phenolic oxidizing enzymes, 367
  post-harvest light treatments, 364
  preparing for drying, 631
  respiration rate on and off tree, 346
  respiratory quotient of peel discs, 354
  rootstocks, 334
  suitability of cultivars for juice production, 577
  surface coating of skin, 359
  texture, 365–366
  varieties, 334
  volatile products, 361–362
Apricots, 417–418
  citric acid in, 417
  effect of growth regulators on maturity, 416
  lactones in, 428
  origin, 412
  parthenocarpy in, 414
  pre-drying treatment, 631
  polyphenolic compounds of, 418
  rate of darkening of dried, 643
  terpenes in, 427
  volatile constituents, 426
Arabinose in ripe grapes, 179
Aroma
  concentrates, 593
  of heat-treated fruit products, 594
  retention in juice concentrates, 594
Aromatic hydroxylation, 604
Ascorbic acid
  aerobic destruction, 562
  aerobic oxidation, 608
  anaerobic decomposition, 527, 563, 609
    inorganic catalysts of, 609
    substances which accelerate, 528, 563
  antioxidant activity synergized by sequestrants, 509
  in blackcurrants, 389, 390, 391
  changes during heat processing and storage, 562–565
  in cherries, 418
  in citrus fruit, 139, 140
  content of fruits, 578
  control of enzymic browning by, 509
  copper-catalysed oxidation, 562, 608
  degradative path, 606
  disappearance from citrus fruits before processing, 561–562
  effect of addition on can corrosion, 531
  effect of containers on, 563
  effect of pH in canned products, 563
  effect of temperature on stability, 564
  effect on shelf life, 608
  enzymic oxidation, 561–562
  fortification of fruit juices, 607
  in grapes, 194

loss during juice manufacture, 607
loss during marmalade preparation, 562
loss from canned fruit during storage, 528
loss from citrus fruits on peeling, 561
in mangoes, 238, 245, 251
in melons, 221
metal complexes, 602
model systems of degradation, 608
more stable in unpasteurized blackcurrant juice, 611
oxidation during juice extraction, 583, 584
oxidation in fruit juices, 609–611
as oxidation inhibitor, 580
partial extraction during canning or stewing, 496
percentage retention in orange juice, 564
in persimmons, 293, 294
a phenolase inactivator, 584
in pineapples, 314
in pome fruits, 368
protective effect of rutin, 609
reactions during canning, 561–564
in soft fruits, 391
in stone fruits, 418
stability, 607–611
synthetic, 499
in tomatoes, 441–443
Ascorbic acid oxidases, 561
inhibition of, 561–562
Aspergillus, heat-resistant species, 525
Astringency, 84, 289, 722
Auxins, affect on abscission, 146
Avocado gum, 15
Avocado oil
chemical changes in at maturity, 18
properties of, 11
Avocado pears, 1–63
amino acid incorporation in tissue discs, 56, 57
anatomy, 4–5
carbohydrates, 11–12
carbon dioxide evolution in, 29
chilling injury, 33
choice of storage conditions, 33
climacteric, 28–30
in tissue discs, 31–32
composition, 6–16
changes on development and in storage, 16–24
differences in RNA synthesis with ripening, 58
effects of changes in CO_2 concentration, 35
effects of irradiation, 39
effects of temperature:
on fruit set and development, 4
on respiration, 32
on ripening, 33
enzymes of, 26–28
ethylene in, 36–40
ethylene incorporation in, 38
fat content of, 3, 6
fat synthesis in, 24–26
fatty acid composition, 25
flower, 3
fruit growth, 5–6
fruit/seed ratio, 16
gas diffusion in, 30, 31
horticulture, 2–6
internal atmosphere, 30–31
lipids of, 8–11
changes in, 16–21
major components, 6–8
metabolic events in ripening, 50–58
metabolic pathways in, 40–49
mineral composition of, 14
mitochondria
activity of, 44–55
effect of cyanide on, 42
isolation of, 43
nitrogen content of, 22
nucleic acid synthesis in, 58
oxygen uptake of, 29, 34–35
pectin changes, 23–24, 30
pharmacologically active components, 15
phases of ripening process, 51
phosphate metabolism in tissue slices, 55–56
preparation of cytoplasm fractions, 42–44
protein metabolism, 56–58
proteins, 21–23
respiration, 39
gas exchange in, 28–32
regulation, 32–36

Avocado pears—*cont.*
  ripening, 3
  seasonal changes, 16
    in fat, 17
    in fatty acids, 18–20
    in lipid properties, 17, 19
    in moisture, 17
    in sugars, 18
  susceptibility to sub-zero temperatures, 3
  tannins, 14
  tricarboxylic acid cycle in, 42–49
  varieties, 2–3
  vitamin content of, 12–14
Avocado seed, 5
  polyphenolic compounds, 15

B

Bacillus coagulans, 524
Bacillus macerans, 524
Bacteria
  non-sporing, in canned fruit, 524–525
  spore-forming, in canned fruit, 524
Bacterial wilt in banana, 72
Bananas, 65–105
  amines, 91
  amino acids, 90
  carbohydrates, 78, 79
  cellulose, 79
  changes in total nitrogen, 70
  chilling injury, 93
  compositional changes during growth and development, 69–71
  development of parthenocarpic fruit, 68
  diseases, 72, 92
  effect of cultural practices on yield, 71
  effect of nitrogen fertilization on flavour, 71
  enzymatic browning, 85
  export, 66
  "green life", 74
  harvesting, 73
  hemicellulose, 79
  history, 65
  insect pests, 71
  irradiation of, 76
  isolation of mitochondria, 98, 100
  lipids, 88, 89
  major sugars, 79
  mitochondria, 100
  morphology, 67–69
  organic acids, 86, 87
  oxalic acid, 87
  oxo acids, 87, 88
  pectic substances in, 79
  peel structure, 68
  phenolic substances, 84–86
  pigments, 80
  post-harvest losses, 74
  principal growing areas, 65
  propagation, 67
  protection from fungal rots, 92
  proteins, 89
  pulp structure, 69
  reproductive phase, 67
  respiratory quotients, 77
  "ripes and turnings", 74, 75
  transpiration in mature green fruit, 78
  transport, 73
  total world production, 66
  varieties, 66–67
  vitamins, 92
  volatile constituents, 80–84
    chromatograms of, 82
    sensory impressions of, 81
  water relations, 78
  yields, 71
Banana ripening, 73
  biochemical mechanisms, 94–100
  delaying, 75, 76
  a differentiation process, 100
  experimental approaches, 101
  growth-regulating substances in, 76
  influence of high relative humidity, 94
  initiation, 74
  permeability changes in, 94–95
  protein synthesis in, 96–98, 99
    effect of inhibitors, 97
  respiration in, 76–87
  role of ethylene, 74, 75, 95, 96
  sugar accumulation in, 78
Bentonite as protein-removing agent, 587
Benzoic acid in cranberries, 389
Berry fruits, 375, 376–377
  contribution of seeds to insoluble solids, 382
  extracting juice from, 579–580
  pectinous pulps from, 580
  tannins, 402

Biochemical studies of fruits
developments in, 500–505
sample selection, 500
utilization of data, 503–504
Bioflavonoids of soft fruit, 401
Biotin, 13, 140, 194, 417, 567
Bisulphite treatment, 635
Bitter pit, 341
Blackberries, 376
flavonoids, 401
Blackcurrants, 377
ascorbic acid, 389, 390, 391, 578
ash content, 393, 396
changes in acidity during ripening, 390
changes in pectin during ripening, 384, 385
changes in solids content during maturation, 381
changes in weight with maturity, 379
depectinization at elevated temperature, 580
flavonoid content, 401
oxidation of ascorbic acid in juice, 609
polyphenol content, 393, 396
storage of juice concentrates, 593
sugar content, 387
Black heart of pineapples, 323
Blanching, minimizing vitamin loss in, 527
Bloom, 639
Blueberry, 378
Boehi process, 591, 593
Botanical structures, diversity in fruit, 500
Bovine serum albumin (BSA), 50
Boysenberry, 377
Bromelin, 308, 590
Browning
in canned fruits, 517
caused by irradiation, 696
of citrus products in storage, 544–546
conditions for, optimized in processing, 708
of dried fruit, 628–629
enzymic, *see* Enzymic browning
factors affecting, 545
Maillard, 545
non-enzymic, 510–511
oxidative, 584
Browning reaction, 638
bisulphite as inhibitor in citrus juice, 606
in juice, 605–607
Bunchy-top, 72
Byssochlamys fulva, 525
Byssochlamys nivea, 525

C

Calcium, role in maintaining structural integrity of cell wall, 691
Calcium salts for firming apple tissues, 581
Cambridge Favourite strawberry, 536
Canned fruits, 507–541
artificial dyes in, 518
astringency changes in, 523
bacteria in, 524
carotenoid changes in, 514–516
colour changes in, 513–518
discoloration of, *see* Discoloration in canned fruit
effect of sugar on texture, 520
flavour deterioration in, 520
fungi in, 525
microbial spoilage of, 523–525
pesticide residues in, 523
pigment degradation, 513–516
processing and storage, 513–525, 527–530
production, 507
reactions with container, 530–535
storage temperature, 518
texture changes in, 518–520
vitamin values, 529
Canning, 487
corrosion of cans, *see* Corrosion
effect of headspace oxygen on ascorbic acid, 562
effect of increasing headspace, 527
"high tin fillet" can, 535
hydrogen swells, 567
lacquers, 534
organoleptic changes during, 508–525
partial extraction of ascorbic acid in, 496
of peeled and unpeeled fruits, 526
preparation of fruit for, 508–513, 526–527
selection of fruit varieties for, 535–537
vitamin changes during, 526–530
Carbinol, 712

Carbohydrates
  in avocados, 11–12
  in bananas, 78
  catabolism, 181
  in citrus fruits, 123–127
  in grapes, 178–181
  in mangoes, 244
  in melons, 214
  photosynthetic production in grape vines, 179
  in pineapples, 306, 311
  in pome fruits, 358
  in soft fruits, 386
  in stone fruits, 418
  synthesis from malic acid in grapes, 180
Carbon dioxide, increased production upon irradiation, 692
Carbonic maceration in wine production, 202
Carbonyl compounds, 552–553
  in citrus oils, 137
β-Carotene
  in musk melon, 218
  in tomato, 453, 455
Carotenoids, 708–711
  in apples, 364
  in avocados, 13
  in bitter oranges, 116
  changes in canned fruits, 514–516
  changes on dehydration, 627–628
  changes in stored citrus products, 546–548
  in cherries, 418
  destruction by irradiation processes, 711
  distribution from solvents, 392
  effect of light on stability, 710
  effect of storage, in canned fruits, 515
  enzymatic degradation, 710
  in fresh frozen and aged canned orange juice, 547
  in grapefruits, 117
  isomerization, 515
    at low pH, 709
  in lemons, 118
  in limes, 118
  in mangoes, 245, 246
  in melons, 218, 219
  in orange juice, 115
  in orange peel, 115
  oxidization, 709
  oxygenated, 455
  in persimmons, 292–293, 297
  in soft fruits, 391
  stability:
    during processing, 708
    role of moisture, 710
  in stone fruits, 419
  in tomatoes, 465
  in Valencia oranges, 116
  vitamin A precursors, 565
Case hardening, 638
Casses, 605
Catalase, 26, 160, 368
  effect of irradiation on, 697
  in mangoes, 248, 249
Catecholamine, 84
Cell membranes, destruction of semi-permeability in canning, 518
Cell wall, importance in relation to fruit texture, 493
Cellulose
  in apples, 365
  in bananas, 79
  in tomatoes, 459
Chemical assay, 501
Chemical preservatives, 487
Cherries
  acids, 418
  "cracking" of sweet cherries, 421
  enzymes, 430
  origin, 412
  parthenocarpy in, 414
  pectins, 418
  polyphenols of, 418
  respiration rate and mitochondrial yield, 425
  scald, 420, 510
  sugars, 418
  tart, 412
  vitamin C content, 418
  volatile components, 426
Chilling injury,
  avocado pears, 33
  bananas, 93
  mangoes, 252
  melons, 208
  pineapples, 322
Chlorogenic acid, 339, 403, 419, 457, 598, 626

Chlorophyll, 708, 714–717
  in apples, 362–363
  breakdown, 363
  changes during processing, 714
  colour change in dehydrated fruit, 627
  degradation during storage, 715
  effect of cooking on, 515
  pheophytin formation from, 715
  reaction in processing, 714
Chlorophyllase, 349, 363, 494, 716
  biosynthetic rather than degradative role favoured, 717
Chromatographic fractionation, 326
Chromatography, gas–liquid, 361, 494
Chromatophores in citrus fruits, 112
Citrate-cleaving enzyme in mangoes, 249
Citrate synthetase, 158
Citric acid, 127, 129
  in apples, 351
  in apricots, 417
  in bananas, 86
  in grapes, 182
  lack of accumulation of in sweet lemons, 128
  in mangoes, 244
  in melons, 219
  in soft fruits, 389
  in stone fruits, 47
  synthesis in citrus fruits, 156–158
  in tomatoes, 443, 445
Citrulline, 220
Citrus cloud, stability after freezing, 678
Citrus fruits, 107–168
  acidity, 127
    effect of arsenicals on, 145
  ascorbic acid:
  loss in storage, 151
  loss on peeling, 561
  attempts to loosen fruit by chemical treatment, 148
  carbohydrates, 123–127
  carotenoids, 114
  changes during development and maturation, 113–149
  chemical changes during storage, 149–151
  chemical constituents, 114–143
  colour of, 114–118
    effect of nutrients on, 144
    effect of temperature on, 143
  commercial varieties, 109–111
  coumarins, 139
  drained weight of canned segments, 556
  edible pulp, 112
  effect of environment and culture on chemical composition, 143–146
  enzymes, 159–161
  essential oils, 113
  ethylene production in, 152, 153–154
  firmness of canned segments, 555–556
  flavonoids, 130–133, 548–550
  general composition of, 113
  general morphology, 111–113
  growth regulators, 146
  history, 108
  influence of rootstocks, 144
  inorganic constituents, 141–143
  juice extraction, 581–582
  lactone bitter principles, 550–552
  limonoids, 133–134
  lipids, 134–136
  lipid-solvent soluble compounds, 138–139
  low-temperature injury, 154
  metabolism, 156–161
  mineral composition, 143
  minimizing shrivelling, 150
  mitochondria, 158–159
  nitrogenous compounds, 119
  nomenclature, 108
  oil spotting, *see* Oleocellosis
  organic acids, 127–130
  "peel oil", 136, 554
  pH of juices, 127
  physiological disorders in storage, 154–156
  post harvest physiology, 149–156
  premature drop of, 146
    control, 147
  presently cultivated, 108
  quality
    effect of trace elements on, 145
    effect of nutrients on, 145
    influence of weather conditions on, 144
  respiratory activity in storage, 151–153
  seasonal changes in flavour components, 137
  steroids and triterpenoids, 139

Citrus fruits—*cont.*
storage temperatures, 150, 156
structure, 111
texture of canned fruits, 555–560
vitamins, 139–141
volatile compounds, 136–138
water balance with tree, 150
watery breakdown, 154
waxes, 139
world production, 108–109
Citrus juice, 574
concentrate, 543
frozen, 543
gel formation in canned, 559
heat processed, 543
main source of oil in, 582
pectic substances of, 557
vitamin content of canned, 566
Citrus products
ascorbic acid fortification, 564
browning reactions during storage, 544–546
canned, 543–572
corrosion problems, 567
nutritional aspects, 544
cloud in, 556–560
stability, 557–558
comminuted, 582
delayed bitterness, 550, 551
flavour
constituents contributing to, 548
effect of oxygen on, 553–554
flavour retention, 554
effect of storage temperature, 553
fruit can interactions, 567–569
Citrus seed oils, 134
chemical and physical characteristics, 135
fatty acid composition, 136
Clarification, 159
Clostridium botulinum, 524
Clostridium butyricum, 524
Clostridium pasteurianum, 524
Co-enzyme A, 160
Colour
of citrus products, 544–548
of frozen fruits, 677–678
of fruit for canning, 535–536
Colour changes
in canned fruits, 513–518
in dried fruits, 626–629
in fruit prepared for canning, 508–511
in fruit processing, 707
Commelinin, 601
Controlled atmosphere storage
of bananas, 76
of mangoes, 251, 252
of persimmons, 298
of pineapples, 320
of pome fruits, 336
of soft fruits, 406
of stone fruits, 420
of tomatoes, 469
Copper
in fruit and juice, 596
organic acid complexes, 597
Copper-catalysed oxidation of carotenoids, 709
Copper chelating agents as phenolase inhibitors, 585
Corrosion
accelerators, 531–532
amount of hydrogen produced by, 531
from citrus products, 567
effect of caramelization, 531
effects of processing variables, 568
and enzyme systems, 534
factors affecting corrosivity, 530–534
by oxygen, 531
relationship with nitrogen content, 568
Coumarins in citrus fruit, 139
Cowberries, 378
flavonoids, 401
Cranberries, 378
benzoic acid, 389
flavonoids, 402
pectic enzymes, 404
Crop yield, 495–496
Cryoprotective compounds, 665
Cryptoxanthin, 565
Cucumis melo, 208
Cucurbitaceae, 207
Currants, 375, 377–378
phenolic compounds, 402
Cyanocentaurin, 601
Cysteine as inhibitor of enzymic browning, 509
Cytokinins, 147
in peaches, 415
Cytological studies, 502

D

D_{10} values, 698, 699
Dates, measuring darkening of, 645
Degreening of citrus fruits, 154
Dehydrated fruits, 623–652, *see also* Dried fruits
  cut fruits, 634
    not as acceptable as sun-dried, 627
  decrease in sulphur dioxide content, 642, 643, 645
  deteriorative changes, 642
  enzymic colour changes, 626–627
  flavour, 629, 639
  non-enzymic colour changes in, 627–629
  porous nature of vacuum-dried, 648
  rehydration, 647–648
  storage, 642–647
    chemical and physical changes during, 642–647
Dehydration, 487, 624, *see also* Drying of fruits
  fully mature fruit required for, 625, 629
  harvesting and handling fruit for, 630–637
  high solids content a prerequisite, 626
  loss of vitamin content, 629
  loss of volatiles during, 639
  reduction in volatile flavour compounds, 629
  sugar–acid ratio for, 625
  theoretical aspects, 637
  two-stage, 625
Dehydrators, design of, 640
Dehydroascorbic acid, 562
Delayed light emission from stone fruits, 424
Depectinization, 580
  of fruit pulps, 588
Dewberries, 376
Dichogamy in avocado pears, 3
Dihydrochalcones, 500
2,5-Dimethyl-4-hydroxy-3(2H)-furanone, 329
Diosmin, 132
Discoloration in canned fruit
  enzymic browning, 517
  non-enzymic browning, 517
  pink, 516, 536
  purple, 516
Diseases of bananas, 92
  of citrus fruits, 154
  of tomatoes, 471
Disinfestation by radiation, 688
Dopamine
  in bananas, 84
  biosynthesis of, 85, 86
Dormin, *see* Abscisic acid
Dried fruits, *see also* Dehydrated fruits
  changes in storage, 643
  crystallization of sugar, 646
  darkening, 645
  decline on world production of traditional, 648
  denaturation of cell-wall, 648
  moisture absorption by, 647
Drying of fruits, 637–642, *see also* Dehydration
  biochemical aspects, 638–639
  changes during, 638
  physical aspects, 637–638
Drying methods, 624
Dupanol, *see* Sodium dodecylsulphate

E

Elderberry, rutin content of, 401
Electron transport chain, 40–42
  resistance to radiation, 693
Embden–Meyerhof glycolytic pathway, 465
Emulsin, 27
Environmental control of growth, 490
Enzymes, 497
  action during processing of fruit, 497–498
  activity resistant to radiation, 693
  in avocado fruit, 26–28
  avoidance of inactivation on freezing, 669
  and can corrosion, 534
  catalytic action by ice structures, 668
  in citrus fruits, 157–161
  effects of low temperatures in tissues, 669–671
  flavour modifying, 522–523
  inactivation by reaction with juice phenolics, 589
  "latency" phenomenon, 667

Enzymes—*cont.*
in mangoes, 248–250
in melons, 221–222
in olives, 263
pectin-degrading, 580
phenolic oxidizing, in pome fruits, 367–368
in pineapples, 307, 308
in pome fruits, 349
proteolytic, 590
reactions of isolated, at low temperatures, 667–669
reaction rates at ultracold temperatures, 667
in soft fruits, 404–405
in stone fruits, 429
temperature dependence of activity, 676
in tomatoes, 466
Enzymic browning, 508–511
in canned fruits, 517
control by ascorbic acid, 508, 509
of frozen fruits, 672
inhibitors, 509
of thawed frozen fruits, 677
prevention, 631, 632
Epidermis of citrus fruits, 111
Eriodicitrin, 132
D-Erythro-L-galacto-nonulose, 12, 15
D-Erythro-L-gluco-nonulose, 12, 15
Essential oils of citrus fruits, 113
Esters
in citrus oils, 136
metal-catalysed hydrolysis, 604
Ethanol in muskmelons, 220
"Ethrel", 491
Ethylene, 148, 343
antagonism with ethylene oxide in effects on green tomatoes, 471
in avocados, 36–40
effects on climacteric pattern, 37, 38
identification, 36
incorporation, 38
effect on banana ripening, 95, 96
effect on colour of citrus fruits, 154
effect of radiation on fruit sensitivity to, 695
function in tomato ripening, 469–471
production in avocado pears, 36
production in citrus fruits, 153–154
production in mangoes, 243
production by melons, 224, 226
production by pome fruits, 344–345, 362
production stimulated by radiation, 693, 694
response of olives to, 425
as ripening hormone, 37, 344
synthesis by tomato fruits, 474
Ethylene glycol as freeze-protective additive, 676

F

False fruits, 375, 378
Fat synthesis in avocado, 24–26
factors affecting, 25
Fatty acid synthetase, 26
Fermentation
alcoholic, by yeasts, 202
malolactic, 202
Flavedo, 111
Flavones, methylated, 132
Flavonoids, 548–550, 711–714
assessing chelating ability, 600
biologically active, 578
of citrus fruits, 130–133
decrease with fruit size increase, 131
complexes with metals, 599–600
in orange cloud, 560
of tomatoes, 457
Flavonols, methylated, 132
Flavonones
as by-products of juice manufacture, 577
change in content in citrus fruits with harvest date, 549
Flavour
changes in canned fruits, 520–523
changes in non-volatile compounds in canned fruits, 521
of citrus products, 548–555
cooked peaches, 521
of dried fruits, 629
effect of oxygen on citrus products, 553–554
enhancing in canned fruits, 522
of fruit for canning, 512, 536–537
retention in canned citrus products, 554–555
of stored frozen fruit, 678–680
Folic acid, 12, 194, 245, 567

Food preservation, 486–488
  general principles, 486–487
Freezing, 487, 653–686, *see also* Thawing of frozen products
  advancements in, 680
  deleterious changes during, 655
  effect of conditions, 674–675
  effect on enzymes, 667–671
  factors in freezing preservation of fruits, 671–677
  gross effect on packaged fruits, 654
  physical and chemical aspects of preservation, 655–667
  of thin sections, 681
  treatments to prevent darkening of fruit during, 672–673
  water displacement during, 658–663
Freezing damage, 656–663, 681
  causes of tissue injury, 665–667
  effect of freezing rate, 658, 660
  effects of freezing temperature, 656–658
  large ice crystals as factor in, 656, 657
  "salting-out" effects, 666
  through ice formation, 666
  trimodal pattern, 656, 657
Freezing protection
  additives, 670, 681
    effects of, 664–665
  effects of apparent water removal, 662
  by glycerin, 665
  of photophosphorylation, 670
Freezing techniques in juice concentration, 593
Frozen fruits
  absence of "cooked" flavours, 671
  colour, 677–678
  colour changes in, 672–674
  development of off-flavours in storage, 678, 679
  effects of additives, 675–677
  enzymatic browning, 677
  enzymatic darkening, 672
  flavour, 678–680
  flavour changes in, 674
  importance of exclusion of oxygen, 679
  loss of fresh aroma, 679
  stability during storage, 677–680
  stability at various temperatures, 663, 679
  texture changes, 671–672
  time–temperature tolerance, 677
Frozen fruit products,
  advantages of, 655
  citrus concentrate, 678
  rigidity of structure, 655
Fructose (*see under* Sugars)
  chelating ability for iron, 598
Fruit can interactions of citrus products, 567–569
Fruit quality, biochemistry of, 493–495
Fungicides, systemic, 490

G

Galacturonic acid, 160, 244, 389, 443
Gallic acid, in persimmons, 292
Gamma ray sources, 689
Gas chromatography
  of banana volatiles, 82
  of grape juice, 192, 193
  of pineapple volatiles, 325, 326
"Gas" storage, *see* Controlled atmosphere storage
Gelatin, use in juice fining, 586
Gelation of citrus juices, 159, 559
Genkwanin, 601
Geranial, 136
Geraniol, presence in ripe mangoes, 247
Gibberellic acid
  induction of parthenocarpy by, 414
  in stone fruits, 415
Gibberellin, 338
  postulated role in fruit set, 147
Gibberellin-like substances in apricot, 415
Glucose (*see under* Sugars)
  conversion to malate in grapes, 184
Glucose-6-phosphate dehydrogenase, 27
Glucosidases, 712
Glutamic acid decarboxylase, 161
D-Glycero-D-galacto-heptose, 12, 15
D-Glycero-D-manno-octulose, 12, 15
D-Glycero-L-galacto-octulose, 12, 15
Glycerol as freeze-protective additive, 676
Glycosides as bitter principle,
  of citrus fruits, 131
  of olives, 261
  of watermelons, 221
Gooseberries, 377

Grapes, 171–205, 639
ability to synthesize di-glucosides, 190
ammonium cations in, 186, 187
anatomy of the berry, 176–177
Cabernet, 193, 199
carbohydrates, 178–181
changes in photosynthetic CO_2 assimilation during development, 197
colour change at veraison, 175
components of skin, 176
composition:
influence of climate, 200, 201
influence of cultivation, 200
of pulp and skin, 178
constitution of, 174–178
dessert varieties, 174
development of, 174–178
effect of humidity on acid content, 200
effect of soil on quality, 199
extracting juice from, 579, 580
factors affecting maturation and quality, 199–200
"foxe" taste in, 193
French wine-making varieties of, 173
fungal attack on, 201
general composition of the raceme, 177, 178
gum, 193
hybrid, 174
maturation period, 175
Merlot, 200
mineral components, 194–195
Muscat, aroma of, 193
nitrogen compounds, 186–188
odoriferous compounds of, 192
organic acids, 181–186
distribution, 178
over-ripening, 175
Panse Muscat, 199
pectic enzymes, 193
pectins, 193
phenolic compounds, 188–192
Pinot, 193
pulp, 177
relation between oxygen uptake and malic acid utilization, 199
relations between stone numbers, size, and composition, 177
respiration:
during development, 195–199
influence of temperature, 196
seeds, 177
storage of dessert varieties, 203
vitamins, 194
volatile compounds, 192, 193
Grape juice, 203, 574
acidity of, 182
clarification by fungal enzymes, 588
gas chromatography of, 192, 193
Grape technology, 201–203
Grape vine
cultivation of, 172, 175
origin of, 172
soil for, 172
translocation of compounds from leaves to berries, 180
vegetative cycle, 174–175
Grapefruit juice,
effects of processing variables on corrosion by, 568
volatile constituents, 553
Grapefruits
common, 110
organic acids, 129
peel pigments, 117
pigmented, 111
pitting, 154
Growth regulators, 414–415, 431, 490
attempts to increase fruit set with, 147
in citrus fruit, 146–148
effects on stone fruit maturity, 416
use on pome fruits, 336, 341, 342
Growth retardants, freeze-protective effect, 677
Guava, ascorbic acid in, 578

H

Haze in fruit juices, 605
Hemicellulose
in apples, 365
in bananas, 79
in citrus fruits, 125
in pineapples, 313
in tomatoes, 459
Hesperidin, 130, 132, 600
as by-product of juice manufacture, 577
in orange cloud, 560
synthesis by oranges in storage, 150

Hexosans of apples, 365
Hexose monophosphate pathway, 465
Histological studies, 502
Hydroxymethylfurfural (HMF), 605
in stored juice products, 591, 593

I

Idioblasts in avocado, 5
Individual quick freezing, 674
Indolylacetic acid, effect on ethylene production, 344
Inorganic constituents of fruits, 14, 141–143, 194, 308, 393, 460
Inositol in canned citrus products, 567
Insects as banana pests, 71
Invertase in mangoes, 249, 252
Ionizing radiation, *see* Radiation
Irradiated fruit
Public Health clearance, 701
toxicity studies, 701
Irradiation, 487
of food, 687–688
increasing firmness of fruit subjected to, 692
pigment destruction by, 711
softening of plant material after, 690
Iron
content of fruits, 596
ferrous and ferric complexes with organic acids, 597

J

Juice, 499, 573–621
bitterness in Navel orange juice, 577
browning reactions, 605–607
centrifugal separation, 579
clarification, 586
colloidal systems in, 586
composition, 576–579
de-aeration, 585
effects of fruit variety and composition on quality, 576–578
enzymatic changes in, 587–591
fining, 586
fortification with ascorbic acid, 607, 608
fruit quality for juice production, 575–576
hazes and deposits in, 605
heat treatment to restrict enzyme activity, 589
inherent instability, 575
loss of quality at high radiation dose, 700
metal content, 596
oxidation in extraction machines, 580
protein removal, 586
radurization of, 699–700
spontaneous clarification, 586
stabilization, 586
storage, 575, 591–595
chemical changes in, 595–611
under nitrogen, 592
of pasteurized, 592
types of, 574–575
visual appearance of, 605–607
Juice concentrates, 573, 574, 575, 592–595
able to withstand high radiation dose, 700
concentration methods, 593
deteriorative changes, 593
low temperature storage of, 593
reconstitution to juice, 592
retention of aroma, 594
Juice extraction, 579–586
from berry fruits, 579–580
changes occurring during, 583–586
from citrus fruits, 581
by compressing whole fruit, 582
from grapes, 579, 580
by hydraulic pressing, 579, 580
methods, 579–583
from pears, 581
by perforated rollers, 579
from pineapples, 582
from pome fruits, 580–581
by reaming, 582
from tropical fruits, 582
Juice production, 573–574
by-products, 577
flavour changes during, 590
processing operations in, 579–595

K

Kaki tannin, 292
Ketones
in citrus fruits, 136
in pineapples, 326

Kinins, 147
  in peach, 415
Krebs cycle, *see* Tricarboxylic acid cycle

L

Lactones, 427–429, 550–552
Lactobacillus brevis, 524
Lemon, 111
  cold storage breakdown, 155
  curing, 149
  organic acids, 129
  sweet, 129
  yellow pigments, 118
Lenticels in fruit skin, 640
Leucoanthocyanins (proanthocyanins), 15, 189, 289, 367, 403, 419
  of perry pears, 577
Leucomostoc mesentenoides, 525
Leucomostoc pleofructi, 525
Lime, 111
  carotenoids, 118
  cold storage breakdown, 155
  organic acids, 129
  pitting, 155
d-Limonene, 552
(+)-Limonene, 136
Limonin, 133, 550, 551
  structure, 133
Limonoic acid, 551
Limonoids of citrus fruit, 133–134
Lipids,
  in avocados, 8–11
  in bananas, 88–89
  in citrus fruits, 134–136
  in orange juice, 552
  in pineapples, 307
  in pome fruits, 359–361
  in tomatoes, 459–460
Lipoxidase, 349, 466
Locular cavity, 438, 439
Loganberries, 377
Lutein in tomatoes, 453
Lycopene
  temperature dependence of synthesis, 455
  in tomatoes, 453, 456, 465
  in watermelons, 219
Lycoxanthin in tomatoes, 453
Lye peeling, vitamin loss in, 526

M

Maceration, 588
Maillard type browning reactions, 497, 545
  in dried fruits, 628
  inhibition by sulphur dioxide, 628
Malate decarboxylating system, 356
Malate effect, 356
  inhibition of, 356
Malate oxidation, 51, 53, 54
Maleic acid in cherries, 418
Malic acid, 46, 128, 129, 219, 313, 420
  in apples, 352
  in bananas, 86
  in citrus fruits, 129
  in grapes, 181, 182
    as basis of carbohydrate synthesis, 180, 181
    decrease at veraison, 184
    metabolism, 182–185
  synthesis, 183
  in soft fruits, 388
  in tomatoes, 443, 445
Malic enzyme, 27, 247, 349
Malonic acid, 128, 129
Mandarin orange, 109
Mangifera indica, see Mangoes
Mangoes, 233–254
  acid content, 237
  amino acids, 245
  ascorbic acid content, 238, 245
  average annual production in India, 234
  biochemical changes in storage, 250–253
  biochemical indices of maturity, 239–243
  botanical aspects, 234–235
  carbohydrates, 244
  carotene, 245, 246
  changes in physico-chemical characteristics during ripening, 240
  chemical composition, 244–250
  chilling injury in storage, 252
  comparative values of sugar/solids ratios for Indian and Florida, 242
  effect of post-harvest treatment with growth regulators, 251
  enzymes, 248–250
  geraniol in, 247

growth and development, 235–239
Indian and Florida, compared, 243
lack of biochemical knowledge of, 235
normal respiration of, 238
odoriferous compounds, 247
organic acids, 244
origin, 233
phenolic compounds, 247
propagation, 234
protein content, 245
stages of development, 235
stages of maturity, 239
thiamine content, 245
vitamins, 245–247
weight changes during development, 236
D-Manno-heptulose, 12, 15
Maturity indices, 73, 128, 213, 239, 267, 309–311, 421–423
Maturity of fruit for processing, 491–492
Melons (muskmelon), 207–232, *see also* Watermelons
acetaldehyde, 220
acetoin content, 220
amino acids, 220
anatomy, 210
ATP and ADP concentrations, 227
botanical aspects of, 209
butylene glycol, 220
cantaloupe, 208
carbohydrates, 214–217
changes in pigments in relation to fruit age, 218
characteristics as function of fruit age, 215
chilling injury to, 208
composition, 213–221
effect of honeydew on human mucous membranes, 221
enzymes, 221–222
ethanol, 220
ethylene production, 224, 226
flesh pigments, 217–219
gas exchange patterns, 222
growth and development, 210–212
growth regulation, 212
honeydew, 208
maturity and quality of, 213
nucleotides, 220
organic acids of, 219–220
origins, 209
pectic substances, 217
pollination, 210
proteinase in, 221
proteins, 220
respiration, 222, 223, 226
sugars, 213–217
tissue investigations, 228
vitamin content, 221
"Membranos stain", 155
Metabolic regulators, 721
Metal chelates, 597
Metals,
amino acid complexes, 602
catalytic effects, 603–605
complexes with ascorbic acid, 602
complexes with flavonoid compounds, 599–601
complexes with oxidized and condensed phenolics, 601
complexes with phenols and phenolic acids, 598
complexes with proteins, 602
complexes with sugars and polyhydroxy compounds, 598
ionic state in juices, 596
mixed ligand complexes, 602
naturally present in fruits, 595
organic acid complexes, 597
stability of complexes, 602–603
valency changes, 602
Miracle fruits, 485, 722
active principle of, 486
Miraculin, 486
Mitochondria
of apples, 346
of avocados, 45
isolation, 43
oxidation of Krebs cycle acids by, 46
of bananas, 98, 100
of citrus fruits, 158–159
of tomatoes, 462, 463
Mitochondrial oxidations, 44, 50–55
co-factor requirements, 46
detection of reaction products, 47
rate measurement, 462
terminal oxidase inhibition, 41
Mitotic figures in mature avocado fruit, 5

Modified air storage, 36, *see also* Controlled atmosphere storage
Moisture transfer, 637–639
  biochemical factors limiting, 641–642
  physical barriers to, 639–641
Monosodium fumarate, 568
Moulds
  in canned fruit, 525
  control of growth by radiation, 688
  on dried prunes, 646
  on grapes, susceptibility of fruit to attack following irradiation, 698
Mulberries, 377

N

"Nacconal", 511
Naringenin, 131, 548, 549, 577, 600
Naringin, 131, 132, 548, 549, 557
  Davis test for, 132
Nectars, 575
Neohesperidin, 550
Neral, 136
Niacin, 13, 140, 194, 214, 309, 417
Nicotinic acid, 578
  in processed foods, 567
Nitrogen,
  in citrus fruits, 119–123
  effects of fertilization on banana flavour, 71
  extraction from fruit, 22
  influence of fertilization on citrus fruits, 145
Nootkatone, 136, 553
Nouaison, 174
Nucleic acids, metabolism in avocados, 58
  in pome fruits, 344, 348
Nucleotides of melons, 220
Nutritional value, 495

O

Oil cells in avocados, 15
Oil glands in citrus fruits, 112
Oleanol, 264
Oleocellosis, 154
  of limes, 155
Oleuropein, 261
  constitution of, 262
Olives, 255–279
  amino acids, 262, 263
  anthocyanins, 262
  ash content, 266
  biochemical changes during processing, 270–274
  black, 260, 268
  California style, 268, 269, 273
  changes in flesh components during growth, 265
  composition, 260–263
    of seed, 263
    of woody part, 263
  enzymes, 263
  fermentation of Greek-style, 273
  fresh, 260–267
  fruit development and maturation, 263–267
  Greek naturally ripe, 269, 270, 273–274
  green, 260, 268
  history, 255, 260
  for oil extraction, 260, 274–276
    harvesting, 274
  pickling, 267, 268
  post-harvest problems, 274
  processed, 267–274
  processing methods, 267–270
  processing for oil, 274–276
  protein fraction of pulp, 263, 266
  respiratory quotient during growth, 264
  Spanish-style green, 267, 268, 270–273
  spoilage of Greek-style, 274
  storage, 274
  sugars, 261, 271
  total nitrogen content, 266
  varieties, 256–259, 260
Olive oil, 260, 500
  composition, 275, 276
  organoleptic spoilage, 276
  refining, 276
  virgin, 276
Oranges (*see also* Citrus fruits)
  acidless, 110
  amino acids, 119, 120, 121
  bitter, 109, 550
    oil constituents, 138
    rootstocks, 109
  blood orange pigments, 118
  common varieties, 110
  cultural practices and chemical constitution, 143

effect of ethylene on colour, 154
effects of low temperatures on, 156
flavonoids, 130
limonoids, 133
lipids, 134, 139
mandarin, 109, 110
maturation period, 114
Navel, 110
organic acids, 127–129
physiological disorders, 154
pigmented, 110
pitting, 154
regreening of, 144
respiration in stored, 152
seed oils, 135
separation layers of, 148
stages of development of Valencia, 113
steroids and triterpenoids, 139
sweet, 110
temperature coefficient for respiration, 152
vitamins, 139–140
volatile compounds, 136–139
Orange cloud
flavonoid components, 560
nature of, 544, 557
Orange juice
ash content of, 141, 142
bleaching by tin, 568
concentrates, 593
lipids, 134
percentage composition of fractions, 557
Organic acids, 722
in bananas, 56
in canned fruit, 530
complexes with metals, 597
in citrus fruits, 127–130
in grapes, 181–186
in mangoes, 244
in melons, 219
metal-catalysed reactions, 603–604
in pineapples, 313–314
in pome fruits, 350–358
in soft fruits, 389
in stone fruits, 417, 422
in tomatoes, 445
volatile, in orange juice, 137
Organoleptic qualities, 490, 493–495
for canning, 535–537
Orujo oil, 276
Oxalacetate, effects of pyruvate and succinate oxidation, 47, 48
Oxalacetic acid
as inhibitor of malate oxidation, 53
mode of action, 55
Oxalic acid, 87, 128, 129, 219, 309, 351, 389
Oxidases (*see also* Phenolase), 160, 307, 404, 429, 561
Oxidative phosphorylation
in avocado mesocarp, 48
by avocado mitochondria, 49
sensitivity to radiation, 693
α-oxoglutarate oxidation, 52
Oxygen
effects in fruit processing, 498
part played in corrosion of cans, 531
Oxygen tension,
effect on avocado pear respiration, 34, 35
effect on citrus fruit respiration, 152, 153
effect on malate oxidation in apple tissue, 356

P

Panama disease, 67, 72
Pantothenic acid, 13, 140, 194, 309, 417, 567, 578
Papain, 590
Papayas, disinfestation by irradiation, 702
Parenchyma tissues in citrus fruits, 112
Parthenocarpy, 414
Passerillage of grapes, 175
Passion fruit as source of juice, 582
Patrimoine of a vineyard, 199
Peach,
acceleration of ripening by radiation, 698
anthocyanin pigments, 418
browning, 429
enzymes, 430
flavour, 426
"fuzz", 640
histochemical changes in ripening, 419
lactones, 419, 427, 428
major aroma constituents, 427
organic acids, 419
origin, 412
parthenocarpy in, 414

Peach—*cont.*
  pectins, 419
  polyphenols, 419
  pre-drying treatment, 631
  substances present in pits, 419
  volatile constituents, 427
Pear (*see also* Pome fruits) 333
  dessert varieties, 335
  extracting juice from, 581
  flavour, 343
  juice storage as half-concentrate, 593
  organic acids present in, 351
  phenolic constituents, 367, 577
  phenolic oxidizing enzymes, 367
  pre-drying treatment, 631
  softening of, 366
Pectic enzymes, 23, 80, 159, 193, 222, 249, 365, 366, 384, 404, 430, 459, 580
  effect of chemical inhibitors on, 511
  in fruit juices, 587
Pectic substances
  in citrus fruits, 127
  in citrus juices, 126, 557
  enzymic changes in canning, 518
  non-enzymic changes in canning, 519
  in pineapples, 312–313
  seasonal changes in content of oranges, 127
  in tomatoes, 459
Pectin,
  in apples, 365
  in avocados, 23–24, 30
  in bananas, 79
  changes during ripening of black-currant, 384, 385
  in cherries, 418
  degree of methoxylation of raspberry, 383
  depolymerization by radiation, 691
  in grapes, 193
  in peaches, 419
  in persimmons, 288, 289
  in soft fruits, 382–386
    molecular weight, 383
  in tomatoes, 459
  use in therapeutics, 495
Pectinesterase (pectinmethylesterase) (*see also* Pectic enzymes), 159
  in "blotchy" tomatoes, 467
  effect of radiation on activity, 697
  inactivation of, 558, 559, 587
  role in cider clarification, 588
  role in cloud stability, 558
  texture of canned citrus fruits, 555
Peeling with lye, *see* Lye peeling
Penicillium, heat-resistant species, 525
Pentosans of apples, 365
Pentose phosphate pathway, 358
Peptides
  in grapes, 186
  metal-catalysed hydrolysis, 604
Peroxidase, 27, 160, 222, 248, 249, 307, 368, 429, 561
  effect of irradiation on, 697
  reactions at low temperatures, 667
Perseitol, 11, 15, 17
Persimmons, 281–301
  ascorbic acid content, 293, 294
  astringency removal:
    alcohol treatment for, 298
    carbon dioxide treatment for, 298, 300
    chemical coating for, 299
    by freezing, 299
    by ionizing radiation, 299
    mechanism, 299
    methods, 298–300
    physiological problems of, 299–300
    warm water treatment for, 298
  astringent varieties, 283, 284, 286
  average temperature for cultivation, 284
  carotenoids, 292, 293, 297
  changes during growth and maturation, 294
  commercially significant varieties, 283–288
  effect of removal of calyx lobes, 297
  Hachiya variety, 284
  history, 281
  Kaki (Japanese) variety, 281, 283
  lycopene content of, 293
  major constituents of fruit, 287, 288–294
  morphology, 282
  non-astringent varieties, 283, 284, 285
  pectins, 288, 289
  pollination constant varieties, 283
  pollination variant varieties, 283
  post-harvest treatments, 298–300

species used for fruit production, 281, 282
sugars, 287, 288, 294, 296
tannin cells, 289, 290, 291
tannins, 289, 292, 294, 295, 296
varieties, 284–288
Pesticide residues in canned fruit, 523
Peteca (of lemons and limes) 155
Phenolase (polyphenolase, polyphenoloxidase), 27, 85, 222, 367, 402, 430, 457, 472, 561, 589, 626, 672–673, 696
and destruction of ascorbic acid, 607
inhibition, 584
by sucrose, 626
substrate specificity in avocado, 27
Phenolic amines in citrus juices, 122
Phenolic acids as germination inhibitors, 457
Phenolic compounds
in avocado pears, 15
in apples, 367
in bananas, 84–86
complexes with metals, 598–602
effect on red wine, 202
enzyme reactions in fruit juices, 589
in grapes, 188–192
metabolism during maturation, 190–192
of mangoes, 241, 247
metal-catalysed reactions, 604
oxidation during juice extraction, 583
in pears, 367
in pineapple, 309, 390, 316
of pome fruits, 339
role in determination of wine character, 188
in soft fruits, 399
in stone fruits, 419
in tomato, 457–458
Pheophytin, formation from chlorophyll, 715
Phlobatannin, 118
Phosphatases, 161, 249, 350, 464
reactions at low temperatures, 667
Phosphate metabolism in avocado slices, 55
Physiological disorders of fruits, 93, 154, 253, 322, 337, 341, 350, 420, 471–472
Phytoene in muskmelon, 218
Phytofluene in muskmelon, 218
Pigments, 494
carotenoid, 546–548
degeneration during processing and storage, 707–719
degradation in canned fruit, 513–516
destruction by radiation, 695
formation in irradiated pears, 695
investigation of chemical changes during processing, 708
major classes, 708
Pigment–metal ion complexes, 532–534
Pineapples, 303–331
amino acids, 307
amino acid patterns during fruit development, 314, 315
ascorbic acid, 314
carbohydrates, 306, 311
carotenoids, 314
cell wall characteristics, 313
cell wall composition, 312
centres of production, 304
chemical changes in storage, 316, 317
chilling injury, 322
commercial varieties, 304
composition, 305–309
controlled atmosphere storage, 320
cultural practices, 304
definition of ripeness, 306, 309
developmental biochemistry, 309–311
effect of storage temperature, 317
enzymes, 307–308
ethylene response, 319
flavour, 325–331
of stored fruit, 317
flavour concentrate from, 328
fruit development and anatomy, 304
gas chromatography of, 325
inorganic constituents, 308
juice, 574
juice extraction, 582
lipids, 307
maturation, 310, 311
nitrogen balance sheet, 307
nitrogenous constituents, 314
organic acids, 309, 313, 314
origin, 303
pectic substances in, 312–313
phenolics, 309, 316

Pineapples—*cont.*
physiological disorders, 322
pigments, 314
post-harvest physiology, 316–323
pre-maturation, 310, 311
proteolytic enzymes in, 590
refrigeration, 321, 322
respiration, 317–319
respiration patterns, 317–321
ripening, 310, 311
senescence, 310, 311
storage, 321–322
titrable acid change in stored fruit, 317
vitamins, 309
volatile compounds, 325, 326, 327, 328, 330
in canned juice, 326
chromatographic isolation of, 326
spectrometry of, 326
"yellowing-up" of shell, 316
Plantains, 66, 79
Plums
damson, 412
low-temperature breakdown, 421
malic acid, 420
origins, 412
salicina, 412
Polygalacturonase, 23, 24, 160, 193, 222, 405, 430, 467, 587–589
absence from apple, 366
presence in pears, 366
Polyphenol oxidase (*see* Phenolase)
Polyphenolic compounds of avocado seeds, 15
Polysaccharides of citrus fruits, 125
Pome fruits
anthocyanins, 364
ascorbic acid, 368
biochemistry of climacteric, 355
carbohydrate metabolism, 358
carotenoids, 364
changes during ripening, 342–368
changes occurring during growth, 337–343
colour changes, 362
controlled atmosphere storage, 336
effect of manurial treatment, 340
ethylene production, 344–345
gibberellin spraying, 338
growth curve, 335
"growth regulators":
effect of, 341
effect on fruit drop, 342
treatment with, 336
history and distribution, 334–335
juice extraction, 580–581
lipid metabolism, 359–361
nitrogen content, 339
organic acid metabolism, 350–358
oxidizing enzymes, 367
phase of rapid cell division, 337
phenolic compounds, 339, 366–367
protein metabolism, 348–350
protopectinase, 365
respiration climacteric, 345–348
RNA metabolism, 348–350
"taste", 344
texture, 365, 366
vitamins, 368
world production, 333
Poncirin, 550
Potassium in citrus fruits, 141
effect on juice quality, 145
"Pourriture noble", 175
effect on grape composition, 201
Pre-drying treatments, 631–634
Preservation of foods, 486
Preservation of fruits, 487–488
processes, 496–500
Processing
effect of metal ions in, 498
effect of oxygen in, 498
enzyme action in, 497–498
handling of fruit for, 492
losses, 708
low-temperature, 497
maturity of fruit for, 491–492
projects aimed at improvement in, 504, 505
quality requirements for, 489
selection of fruit for, 488, 491
temperature, 496–497
Proteinase, 161, 308, 590
Proteins
in avocados, 8
metabolism, 56–58
seasonal changes, 21
in bananas, 89
synthesis in ripening, 96
in fresh olives, 263, 266

in grapes, 186
in melons, 220
metabolism in pome fruits, 348–350
metal complexes, 602
removal from juice, 587
separation, 23
in tomatoes, 449–450, 459
Protein synthesis, effect of inhibitors, 97, 347
Protocyanin, 601
Protopectin, 519
Protopectinase, 589
Provitamin A in canned citrus products, 565
Prunes, 633
acids of, 420
"checking", 631
heat injury to, 421
Purées, 575
Pyridoxin in canned citrus products, 567
Pyrrolidone carboxylic acid, 443
Pyruvate decarboxylase, 349, 353
Pyruvate oxidation, 53

Q

Quality
factors involved in, 721
influence of genetic potential, 722
meaning in biochemical terms still vague, 721
optimum range defined, 722, 723
"Quasi-fluid" state, 663
Quince, 333
Quinic acid, 86, 128, 129, 182, 351, 389

R

Radiation
chemical changes induced by, 689
disinfestation by, 688
effects on micro-organisms, 698–699
effects on respiration, 692
effects on ripening, 697–698
effects on specific enzyme systems, 696
and ethylene production, 693, 694-695
and ethylene sensitivity, 695
and pectinesterase activity, 697
and pigment formation, 695–696
preservation of food by, 687
properties of, 688–690
prospect for commercial application, 701–702
sources, 689
threshold dose for softening, 690
units of, 688–689
Radiation damage, 688
in apples, 696
in bananas, 696
in citrus fruits, 696
Radiation treatments, 687–705
Radiolysis, 690
Radurization of fruit juices, 699–700
Raffinose in grape vine, 179
Raisins
activation energies for darkening, 645
analyses of, 643
colour changes in, 644
production, 632
shelf life, 645
Raspberries, 376
enzymes of, 404
flavonoids, 401
"mush" in canning, 536
"mushiness" after freezing, 678
Reaction inactivation of enzymes, 509
Red blotch of citrus fruits, 155
Redcurrants, 377
Refractive index as measure of maturity of avocado, 20
Refrigerated transport, 492
Refrigeration of pineapples, 321, 322
Regreening of oranges, 144
Research projects most likely to be useful to processing industry, 504–505
Respiration
of apples, 346
of avocados, 28–35
of bananas, 76–78
of citrus fruits, 151, 152
effects of radiation on, 692–694
of grape pericarp, 195
of grapes, 196, 197, 199
of mangoes, 238
of melons, 222–223, 226
of olives, 264
of pineapples, 317–319
rate not indicative of quality, 724
of tomatoes, 462

Respiration climacteric, 28, 77, 223, 239, 345–348, 424–425, 462, 491, 723, 724
Respiratory Control Index, 347
Respiratory quotient
of apple peel discs, 354
in ripening bananas, 77
Reverse osmosis in juice concentration, 593
Riboflavin, 13, 140, 194, 245, 566, 578
Ripeness
physical methods of assessing, 491
Ripening
effect of radiation, 695, 697–698
phases in avocado, 51
uneven, 491
Ripening-prevention hormone, 38
RNA metabolism
in avocados, 58
in pome fruits, 348–350
Rose hips, ascorbic acid in, 578
Rutin in elderberry, 401

S

Sample selection, 500
Sauternes, 201
Senescence, 723
disturbance of water balance with onset, 623
Serendipity berry, 494, 722
Serotonin, 307
Shikimic acid, 86, 182, 351, 389
Sigatoka disease, 72
Sodium chloride
as freeze-protective additive, 676
as polyphenol oxidase inhibitor, 509
Sodium dodecylsulphate, nitrogen extraction with, 22
Soft fruits, 375–410
amino acids, 396, 398
anthocyanins, 397, 399, 400
ascorbic acid content, 391
carbon dioxide treatment, 406
carotene content, 391, 392
characteristics, 375–405
effects of cultural practice, 405
enzymes, 404–405
flavonoids, 397–402
growth changes in, 379
holding, 405–406
importance of picking at optimum maturity, 406
insoluble solids of, 381–382
meaning of term, 375
mineral constituents, 393, 394–395
nitrogenous constituents, 393, 396–397
organic acids, 388, 389
pectins, 382–386
phenolic compounds, 402–403
pigments, 397, 399–402
post-harvest changes, 405–407
seeds in, 378
short-term storage, 406
solids and moisture content, 380
sugar content, 386
tannins, 402–403
transportation, 406–407
vitamins, 390–393
volatile constituents, 403–404
Soil, effect on grape quality, 199
Solanine, 456
"Sparker" acid, 47
Spherosomes, 669
Split pits in peaches, 412
Squalene in olives, 267
Stachyose in grape vine, 179
Starch in citrus fruits, 126
Steroids
in citrus fruits, 139
in tomatoes, 452–453
Stomata, densities on mature banana peel, 78
Stone fruits, 411–436
acid content as maturity index, 422
botanical aspects of, 412
characteristic growth curve, 413
characteristics, 412
composition and biochemical changes during maturation, 417–420
controlled atmosphere storage, 420
delayed light emission as index of quality, 424
effect of cultural practices on maturation and quality, 417
endogenous hormones of, 415–416
enzymes, 429–430
firmness correlated with organoleptic acceptability, 423
firmness as maturity index, 422

flavour and quality, 425–429
growth and development, 413–416
growth patterns, 413
histochemical changes in, 419
lactones and flavour, 427
light transmission as index of quality, 424
maturation, 416–421
maturity indices, 421–424
metabolic response to hormone applications, 415
organelles, 430
organic acids, 417, 419
origins, 412
phases of fruit development, 413
post-harvest abnormalities in, 420–421
respiratory climacteric of, 424
responses to growth regulators, 414–415
ripening, 416–421
delay due to high nitrogen conditions, 417
senescence, 416–421
soluble solids content as maturity index, 422
soluble solids correlated with organoleptic acceptability, 423
sugars, 418
volatiles, 426, 427
Storage (*see also* Controlled atmosphere storage)
of avocado pear, 33
effects of unsuitable conditions, 576
extension of storage life by irradiation, 690–698
Strawberries, 378
flavonoids, 402
formula for berry weight, 380
"mushiness" after freezing, 678
tannins, 403
Sucrose
in grapes, 179
hydrolysis, 179
partial inversion on drying, 638
as phenolase inhibitor, 626
Sugars, 722
accumulation in ripening bananas, 78
as additive in processing, 499
in avocados, 8
biosynthesis of unusual, 12
in cherries, 418
in citrus fruits, 127
in citrus peel, 124, 125
complexes with metals, 598
crystallization in dried fruits, 646
effect on canned fruit texture, 520
in grapes, 178, 179
in lime and lemon fruits, 126
in mangoes, 244
in melons, 214–217
in olives, 261
in orange juice, 123
pathways of metabolism, 358
in persimmons, 288, 294, 296
in pineapples, 311
in pome fruits, 358
seasonal changes in composition of citrus fruit juices, 124
in soft fruits, 386–387
in tomatoes, 440–441
Sugar–acid ratio, 128, 213, 241, 386, 390
in canned fruits, 522
of citrus fruits for processing, 548
Sulphur dioxide
biochemical changes caused by, 636
decolorization of anthocyanins, 712
desorption, 636
fate of, in dried fruits, 637
influence on enzymic and non-enzymic darkening, 634
limit of concentration in dried fruits, 636
as preservative, 499
as polyphenolase inhibitor, 626
rate of decrease as index of deterioration of dried fruits, 645
use in foodstuffs, 585
Sulphur olive oil, 276
Sulphuring of fruit before drying, 634–637
procedure, 634
Sultanas, pre-drying procedure, 632
Sun-drying, 624
high sugar content essential, 624
Süssmost, 375
Synsepalum dulcificum, 486

T

D-Talo-heptulose, 12, 15
Tangelo, 110

Tangerine
  Dancy, 110
  organic acids, 129
Tangeritin, 132
Tangor, 110
Tannins (*see also* Phenolic compounds)
  astringent properties of, 191
  in avocados, 14
  in bananas, 84, 85
  in berry fruits, 402
  combination with metal ions, 533
  in grapes, 188, 191
  in mangoes, 247
  in persimmon, 289, 292, 294, 295, 296
  in strawberries, 403
Tartaric acid, 86, 173, 178
  formation from glucose, 186
  in grapes, 181, 182
    metabolism, 185–186
  synthesis, 185
Tartrate decomposition products, 604
Terpenes, 427, 552
Texture, 722
  in canned fruit,
  changes, 518, 520
  changes in prepared fruit, 511–512
  control of breakdown, 511
  effect of calcium salts on, 512, 672
  effect of heat, 518
  effect of sugar, 520
  effects of irradiation, 690–692
  of frozen fruit, 671–672
  of fruit for canning, 536
  maintenance in canned peaches, 519
Thawing of frozen products,
  deleterious changes during, 655
  effect on enzymes, 669
  effects of, 663–664
  by microwave diathermy, 663
  under high pressure, 664
Thiamine, 13, 140, 194, 309, 565
  in mangoes, 245
  in oranges and pineapples, 578
  protein complex, 566
  pyrophosphate, 566
Thickinase, 28
Tissue culture methods, 502
Tissue oxidation, electron chain inhibitors, 41
Tomatine, 453
  fall in content of tomato fruit on ripening, 456
Tomatoes, 437–482
  acid content of various varieties, 445
  amino acids, 447–449
  ascorbic acid, 441–443
    effect of light intensity on content, 442
    effect of potassium and manganese soil treatment on content, 442
  average composition of various varieties, 444
  blossom-end rot, 457, 472
  "blotchy" fruit, 467–468
  blotchy ripening, 471
  boron deficiency, 461
  calcium, 460
  carbohydrates, 440–443
  carotenoids, 455, 456
  cell wall,
    constituents, 459
    importance of, 493
  cellulose, 459
  changes during growth and maturation, 440–462
  changes in pigment concentration during maturation, 454
  chlorophyll in fruits, 453, 454, 456
    genetic control of content, 456
  climacteric fruit, 465
  composition of juices, 446
  controlled atmosphere storage, 469
  development, 438
  dominant role of potassium in, 460
  effect of minor elements in, 461
  effect of nitrogen treatment on lycopene content, 456
  effects of temperature during storage, 468–469
  ethylene action, mechanism of, 470
  ethylene synthesis by fruits, 474
  flavonoids of, 457
  graywall disease, 472
  hemicelluloses, 459
  increase of respiratory activity accompanying incipient ripening, 465
  internal browning, 472
  lipids, 459–460
  major metabolic pathways in fruits, 465

maturation, 438
metabolism during maturation and senescence, 462–468
mineral constituents, 460–462
minimum storage temperature, 469
mitochondria, 462, 463, 465
negative correlation between fruit weight and titrable acidity, 444
non-flavonoid pigments of, 453–457
organic acids, 443–447
organic acids in blotchy fruit, 468
origin, 438
pectic substance, 459
phenolic compounds, 457–458
pre-climacteric fruit, 464–465
proteins, 449–450, 459
relationships between potassium and acidity, 445
relationships between respiration, ripening and ethylene production, 470
respiratory activity, 462
ripening, 438
control of, 473–475
disorders, 471–473
effect of temperature on, 474
factors affecting, 468–473
function of ethylene in, 469–471
retardation by radiation, 697
sprays and dips affecting, 473
senescence, 438
senescent fruit, 466–467
starch, 441
steroids, 452–453
storage in nitrogen, 469
storage of ripe fruit in terms of colour, 456
sugars, 440–441
effect of defoliation on content, 440
effect of shading on content, 440
increase of content on ripening, 440
varieties, 438
vascular browning, 472
virus infections, 472
volatiles, 450–452
increase on ripening, 452
waxy patch, 417
Transpiration in mature green bananas, 78
Tricarboxylic acid cycle, 181
in apples, 345
in avocado, 42–49
in bananas, 78
inhibition by radiation, 694
Triterpenoids in citrus fruits, 139
Tyrosinase (*see also* Phenolase), 601

V

Vaccinium family, 378
Vacuum distillation to produce juice concentrates, 593
Veraison, 174, 175
decrease of malic acid concentration at, 184
glucose/fructose ratio at, 178
increase of respiratory rate in, 198
increase of sugar content at, 179
Vinification, 201–203
biochemical processes involved in, 202
Vitamins,
in avocado pears, 12–14
in bananas, 92
in canned fruits, 529
carotenoid precursors, 565
in citrus fruits, 139–141
changes in storage and processing, 560–567
heat stability of, 527
in soft fruits, 390–393
Vitamin B group,
in canned citrus products, 565–567
in grapes, 194
Vitamin C, *see* Ascorbic acid
Vitamin P, 401, 578
Vitis vinifera, 173
varieties of, 173
Volatile compounds, 722, 723
of bananas, 82
changes in canned fruits, 520–521
of citrus fruits, 136–138
formed in storage of processed fruits, 521
Volatile reducing substances of prunes, 629

W

Water,
average content of fruits, 641
"bound", 641

Water balance in fruits disturbed with onset of senescence, 623
Watermelons, 207, 208
  bitter principles, 221
  eating quality, 213
  gibberellin-like factor in, 212
  glycosides, 221
  kinin-like activity in, 212
  origin, 209
  pectic substances, 217
Waxes (*see also* Lipids)
  in citrus fruits, 139
  on outer surface of fruits, 639
White currant, 377
Whortleberry, 378
Wine, 172
  casses, 605
  characteristic of white, 202
  characterization by anthocyanins, 190
  manufacture, *see* Vinification
  production of red, 202
  production of white, 202
  sparkling, 202
Wood–Werkman reaction, 183

X

Xanthophylls in tomatoes, 453, 455
X-rays, sources, 689
Xylose in ripe grapes, 179

Y

Yeasts
  in canned fruits, 525
  effect of radiation on heat sensitivity, 700